# AN INTRODUCTION TO
# NUMERICAL ANALYSIS

# AN INTRODUCTION TO
# NUMERICAL ANALYSIS
## Second Edition

**Kendall E. Atkinson**
*University of Iowa*

**John Wiley & Sons**

*New York    Chichester    Brisbane    Toronto    Singapore*

*Library of Congress Cataloging in Publication Data:*

Atkinson, Kendall E.
    An introduction to numerical analysis/Kendall E. Atkinson.—
    2nd ed.
        p.    cm.
    Bibliography: p.
    Includes index.
    ISBN 0-471-62489-6
    1. Numerical analysis.    I. Title.
QA297.A84   1988
519.4—dc19

Printed in the United States of America

10 9 8 7 6 5 4 3

*To my mother, Helen Fleming Hart, and*
*to the memory of my father, Harold Eugene Atkinson*

# PREFACE

The main organization of this second edition is essentially the same as the previous one, with the addition of several new sections. All sections have been rewritten, sometimes quite extensively, and in some instances topics have been deleted. In addition, introductions to a number of new topics have been included. For example, trigonometric interpolation and the FFT (Section 3.8), numerical differentiation (Section 5.7), the method of lines (Section 6.9), boundary value problems (Section 6.11), the conjugate gradient method (Section 8.9), and the least squares solutions of systems of linear equations (Section 9.7). An appendix on mathematical software is also included which describes some of the better known software packages. The bibliography for each chapter has been updated to reflect the current textbooks and research literature.

With regard to mathematical software, I strongly recommend that the student make use, as much as possible, of programs in one of the high-quality commercial numerical-analysis program libraries. The most widely used ones are those of IMSL and NAG, both of which are discussed further in the appendix on mathematical software. In this text, a number of special-purpose numerical analysis packages are described. These are valuable when the source code is needed. Usually, however, the programs in the commercial libraries will be sufficient for most programming problems that students will encounter; in addition, the commercial libraries are generally easier to use. Also, most of the special-purpose packages that are described in the text have been absorbed into the major commercial libraries.

As before, I thank my many students. It has been very enjoyable to teach them, and to then learn from them as they go into their own areas of research. Some of the additions to this second edition were based on their research needs. I also thank my colleagues Alex Bogomolny, Florian Potra, and Ezio Venturino for their many helpful suggestions. The book was typed by Lois Friday, who has done an excellent job, for which I am very grateful. Finally, I thank my wife Alice and my daughters for their patience, forbearance, and support. It is much needed for a project such as this one.

*Iowa City, June 1987*                                                **Kendall E. Atkinson**

# PREFACE
## to the first edition

This introduction to numerical analysis was written for students in mathematics, the physical sciences, and engineering, at the upper undergraduate to beginning graduate level. Prerequisites for using the text are elementary calculus, linear algebra, and an introduction to differential equations. The student's level of mathematical maturity or experience with mathematics should be somewhat higher; I have found that most students do not attain the necessary level until their senior year. Finally, the student should have a knowledge of computer programming. The preferred language for most scientific programming is Fortran.

A truly effective use of numerical analysis in applications requires both a theoretical knowledge of the subject and computational experience with it. The theoretical knowledge should include an understanding of both the original problem being solved and of the numerical methods for its solution, including their derivation, error analysis, and an idea of when they will perform well or poorly. This kind of knowledge is necessary even if you are only considering using a package program from your computer center. You must still understand the program's purpose and limitations to know whether it applies to your particular situation or not. More importantly, a majority of problems cannot be solved by the simple application of a standard program. For such problems you must devise new numerical methods, and this is usually done by adapting standard numerical methods to the new situation. This requires a good theoretical foundation in numerical analysis, both to devise the new methods and to avoid certain numerical pitfalls that occur easily in a number of problem areas.

Computational experience is also very important. It gives a sense of reality to most theoretical discussions; and it brings out the important difference between the exact arithmetic implicit in most theoretical discussions and the finite-length arithmetic computation, whether on a computer or a hand calculator. The use of a computer also imposes constraints on the structure of numerical methods, constraints that are not evident and that seem unnecessary from a strictly mathematical viewpoint. For example, iterative procedures are often preferred over direct procedures because of simpler programming requirements or computer memory size limitations, even though the direct procedure may seem simpler to explain and to use. Many numerical examples are given in this text to illustrate these points, and there are a number of exercises that will give the student a variety of computational experience.

The book is organized in a fairly standard manner. Topics that are simpler, both theoretically and computationally, come first; for example, rootfinding for a single nonlinear equation is covered in Chapter 2. The more sophisticated topics within numerical linear algebra are left until the last three chapters. If an instructor prefers, however, Chapters 7 through 9 on numerical linear algebra can be inserted at any point following Chapter 1. Chapter 1 contains a number of introductory topics, some of which the instructor may wish to postpone until later in the course. It is important, however, to cover the mathematical and notational preliminaries of Section 1.1 and the introduction to computer floating-point arithmetic given in Section 1.2 and in part of Section 1.3.

The text contains more than enough material for a one-year course. In addition, introductions are given to some topics that instructors may wish to expand on from their own notes. For example, a brief introduction is given to stiff differential equations in the last part of Section 6.8 in Chapter 6; and some theoretical foundation for the least squares data-fitting problem is given in Theorem 7.5 and Problem 15 of Chapter 7. These can easily be expanded by using the references given in the respective chapters.

Each chapter contains a discussion of the research literature and a bibliography of some of the important books and papers on the material of the chapter. The chapters all conclude with a set of exercises. Some of these exercises are illustrations or applications of the text material, and others involve the development of new material. As an aid to the student, answers and hints to selected exercises are given at the end of the book. It is important, however, for students to solve some problems in which there is no given answer against which they can check their results. This forces them to develop a variety of other means for checking their own work; and it will force them to develop some common sense or judgment as an aid in knowing whether or not their results are reasonable.

I teach a one-year course covering much of the material of this book. Chapters 1 through 5 form the first semester, and Chapters 6 through 9 form the second semester. In most chapters, a number of topics can be deleted without any difficulty arising in later chapters. Exceptions to this are Section 2.5 on linear iteration methods, Sections 3.1 to 3.3, 3.6 on interpolation theory, Section 4.4 on orthogonal polynomials, and Section 5.1 on the trapezoidal and Simpson integration rules.

I thank Professor Herb Hethcote of the University of Iowa for his helpful advice and for having taught from an earlier rough draft of the book. I am also grateful for the advice of Professors Robert Barnhill, University of Utah, Herman Burchard, Oklahoma State University, and Robert J. Flynn, Polytechnic Institute of New York. I am very grateful to Ada Burns and Lois Friday, who did an excellent job of typing this and earlier versions of the book. I thank the many students who, over the past twelve years, enrolled in my course and used my notes and rough drafts rather than a regular text. They pointed out numerous errors, and their difficulties with certain topics helped me in preparing better presentations of them. The staff of John Wiley have been very helpful, and the text is much better as a result of their efforts. Finally, I thank my wife Alice for her patient and encouraging support, without which the book would probably have not been completed.

*Iowa City, August, 1978*                                                  **Kendall E. Atkinson**

# ABOUT THE AUTHOR

Professor Atkinson was born and raised in the state of Iowa; and his degrees are from Iowa State University and the University of Wisconsin. The author is a Professor of Mathematics at the University of Iowa, having been appointed in 1972. Previous to this, he was a faculty member at Indiana University, and he has also held visiting positions at Colorado State University, the Australian National University, the University of New South Wales, and the University of Queensland.

Professor Atkinson's research has been mainly on the numerical solution of integral equations, and he has published a book *A Survey of Numerical Methods for the Solution of Fredholm Integral Equations of the Second Kind* on this topic. More recently, he has been studying the use of integral equations in the numerical solution of boundary value problems for partial differential equations. He is very interested in education issues in numerical analysis, and is the author of another text *Elementary Numerical Analysis*, also published by John Wiley.

Professor Atkinson is an associate editor for the *SIAM Journal on Numerical Analysis* and the *Journal of Integral Equations and Applications*. He also has been an elected member of the SIAM Council and an appointed member of the SIAM Education Committee.

# CONTENTS

# FOUR
# APPROXIMATION OF FUNCTIONS  197

# FIVE
# NUMERICAL INTEGRATION  249

# SIX
# NUMERICAL METHODS FOR ORDINARY
# DIFFERENTIAL EQUATIONS  333

# ONE

# ERROR: ITS SOURCES, PROPAGATION, AND ANALYSIS

The subject of numerical analysis provides computational methods for the study and solution of mathematical problems. In this text we study numerical methods for the solution of the most common mathematical problems and we analyze the errors present in these methods. Because almost all computation is now done on digital computers, we also discuss the implications of this in the implementation of numerical methods.

The study of error is a central concern of numerical analysis. Most numerical methods give answers that are only approximations to the desired true solution, and it is important to understand and to be able, if possible, to estimate or bound the resulting error. This chapter examines the various kinds of errors that may occur in a problem. The representation of numbers in computers is examined, along with the error in computer arithmetic. General results on the propagation of errors in calculations are given, with a detailed look at error in summation procedures. Finally, the concepts of stability and conditioning of problems and numerical methods are introduced and illustrated. The first section contains mathematical preliminaries needed for the work of later chapters.

## 1.1 Mathematical Preliminaries

This section contains a review of results from calculus, which will be used in this text. We first give some mean value theorems, and then we present and discuss Taylor's theorem, for functions of one and two variables. The section concludes with some notation that will be used in later chapters.

**Theorem 1.1** (Intermediate Value)    Let $f(x)$ be continuous on the finite interval $a \leq x \leq b$, and define

$$m = \text{Infimum} f(x), \qquad M = \text{Supremum} f(x)$$
$$\phantom{m = }{\scriptstyle a \leq x \leq b} \qquad\qquad \phantom{M = }{\scriptstyle a \leq x \leq b}$$

Then for any number $\zeta$ in the interval $[m, M]$, there is at least one point $\xi$ in $[a, b]$ for which

$$f(\xi) = \zeta$$

In particular, there are points $\underline{x}$ and $\bar{x}$ in $[a, b]$ for which

$$m = f(\underline{x}), \qquad M = f(\bar{x})$$

**Theorem 1.2** (Mean Value)   Let $f(x)$ be continuous for $a \leq x \leq b$, and let it be differentiable for $a < x < b$. Then there is at least one point $\xi$ in $(a, b)$ for which

$$f(b) - f(a) = f'(\xi)(b - a)$$

**Theorem 1.3** (Integral Mean Value)   Let $w(x)$ be nonnegative and integrable on $[a, b]$, and let $f(x)$ be continuous on $[a, b]$. Then

$$\int_a^b w(x)f(x)\, dx = f(\xi) \int_a^b w(x)\, dx$$

for some $\xi \in [a, b]$.

These theorems are discussed in most elementary calculus textbooks, and thus we omit their proofs. Some implications of these theorems are examined in the problems at the end of the chapter.

One of the most important tools of numerical analysis is Taylor's theorem and the associated Taylor series. It is used throughout this text. The theorem gives a relatively simple method for approximating functions $f(x)$ by polynomials, and thereby gives a method for computing $f(x)$.

**Theorem 1.4** (Taylor's Theorem)   Let $f(x)$ have $n + 1$ continuous derivatives on $[a, b]$ for some $n \geq 0$, and let $x, x_0 \in [a, b]$. Then

$$f(x) = p_n(x) + R_{n+1}(x) \tag{1.1.1}$$

$$p_n(x) = f(x_0) + \frac{(x - x_0)}{1!} f'(x_0)$$

$$+ \cdots + \frac{(x - x_0)^n}{n!} f^{(n)}(x_0) \tag{1.1.2}$$

$$R_{n+1}(x) = \frac{1}{n!} \int_{x_0}^x (x - t)^n f^{(n+1)}(t)\, dt$$

$$= \frac{(x - x_0)^{n+1}}{(n + 1)!} f^{(n+1)}(\xi) \tag{1.1.3}$$

for some $\xi$ between $x_0$ and $x$.

**Proof**   The derivation of (1.1.1) is given in most calculus texts. It uses carefully chosen integration by parts in the identity

$$f(x) = f(x_0) + \int_{x_0}^x f'(t)\, dt$$

repeating it $n$ times to obtain (1.1.1)–(1.1.3), with the integral form of the remainder $R_{n+1}(x)$. The second form of $R_{n+1}(x)$ is obtained by using the integral mean value theorem with $w(t) = (x - t)^n$.  ∎

Using Taylor's theorem, we obtain the following standard formulas:

$$e^x = 1 + x + \frac{x^2}{2} + \cdots + \frac{x^n}{n!} + \frac{x^{n+1}}{(n+1)!} e^{\xi_x} \tag{1.1.4}$$

$$\cos(x) = 1 - \frac{x^2}{2!} + \frac{x^4}{4!} - \cdots + (-1)^n \frac{x^{2n}}{(2n)!}$$

$$+ (-1)^{n+1} \frac{x^{2n+2}}{(2n+2)!} \cos(\xi_x) \tag{1.1.5}$$

$$\sin(x) = x - \frac{x^3}{3!} + \frac{x^5}{5!} - \cdots + (-1)^{n-1} \frac{x^{2n-1}}{(2n-1)!}$$

$$+ (-1)^n \frac{x^{2n+1}}{(2n+1)!} \cos(\xi_x) \tag{1.1.6}$$

$$(1 + x)^\alpha = 1 + \binom{\alpha}{1} x + \binom{\alpha}{2} x^2 + \cdots + \binom{\alpha}{n} x^n$$

$$+ \binom{a}{n+1} \frac{x^{n+1}}{(1 + \xi_x)^{n+1-\alpha}} \tag{1.1.7}$$

with

$$\binom{\alpha}{k} = \frac{\alpha(\alpha - 1) \cdots (\alpha - k + 1)}{k!} \qquad k = 1, 2, 3, \ldots$$

for any real number $\alpha$. For all cases, the unknown point $\xi_x$ is located between $x$ and 0.

An important special case of (1.1.7) is

$$\frac{1}{1 - x} = 1 + x + x^2 + \cdots + x^n + \frac{x^{n+1}}{1 - x} \qquad x \neq 1 \tag{1.1.8}$$

This is the case $\alpha = -1$, with $x$ replaced by $-x$. The remainder has a simpler form than in (1.1.7); it is easily proved by multiplying both sides of (1.1.8) by $1 - x$ and then simplifying. Rearranging (1.1.8), we obtain the familiar formula for a finite geometric series:

$$1 + x + x^2 + \cdots + x^n = \frac{1 - x^{n+1}}{1 - x} \qquad x \neq 1 \tag{1.1.9}$$

Infinite series representations for the functions on the left side of (1.1.4) to (1.1.8) can be obtained by letting $n \to \infty$. The infinite series for (1.1.4) to (1.1.6) converge for all $x$, and those for (1.1.7) and (1.1.8) converge for $|x| < 1$. Formula (1.1.8) leads to the well-known infinite geometric series

$$\frac{1}{1-x} = \sum_{k=0}^{\infty} x^k \qquad |x| < 1 \tag{1.1.10}$$

The Taylor series of any sufficiently differentiable function $f(x)$ can be calculated directly from the definition (1.1.2), with as many terms included as desired. But because of the complexity of the differentiation of many functions $f(x)$, it is often better to obtain indirectly their Taylor polynomial approximations $p_n(x)$ or their Taylor series, by using one of the preceding formulas (1.1.4) through (1.1.8). We give three examples, all of which have simpler error terms than if (1.1.3) were used directly.

*Example 1.*   Let $f(x) = e^{-x^2}$. Replace $x$ by $-x^2$ in (1.1.4) to obtain

$$e^{-x^2} = 1 - x^2 + \frac{x^4}{2!} - \cdots + (-1)^n \frac{x^{2n}}{n!} + (-1)^{n+1} \frac{x^{2n+2}}{(n+1)!} e^{\xi_x}$$

with $-x^2 \le \xi_x \le 0$.

2.   Let $f(x) = \tan^{-1}(x)$. Begin by setting $x = -u^2$ in (1.1.8)

$$\frac{1}{1+u^2} = 1 - u^2 + u^4 - \cdots + (-1)^n u^{2n} + (-1)^{n+1} \frac{u^{2n+2}}{1+u^2}$$

Integrate over $[0, x]$ to get

$$\tan^{-1}(x) = x - \frac{x^3}{3} + \frac{x^5}{5} - \cdots + (-1)^n \frac{x^{2n+1}}{2n+1}$$

$$+ (-1)^{n+1} \int_0^x \frac{u^{2n+2}}{1+u^2} du \tag{1.1.11}$$

Applying the integral mean value theorem

$$\int_0^x \frac{u^{2n+2}}{1+u^2} du = \frac{x^{2n+3}}{2n+3} \cdot \frac{1}{1+\xi_x^2}$$

with $\xi_x$ between 0 and $x$.

3.   Let $f(x) = \int_0^1 \sin(xt)\,dt$. Using (1.1.6)

$$f(x) = \int_0^1 \left[ xt - \frac{x^3 t^3}{3!} + \cdots + (-1)^{n-1} \frac{(xt)^{2n-1}}{(2n-1)!} \right.$$

$$\left. + (-1)^n \frac{(xt)^{2n+1}}{(2n+1)!} \cos(\xi_{xt}) \right] dt$$

$$= \sum_{j=1}^n (-1)^{j-1} \frac{x^{2j-1}}{(2j)!} + (-1)^n \frac{x^{2n+1}}{(2n+1)!} \int_0^1 t^{2n+1} \cos(\xi_{xt})\,dt$$

with $\xi_{xt}$ between 0 and $xt$. The integral in the remainder is easily bounded by $1/(2n+2)$; but we can also convert it to a simpler form. Although it wasn't proved, it can be shown that $\cos(\xi_{xt})$ is a continuous function of $t$. Then applying the integral mean value theorem

$$\int_0^1 \sin(xt)\,dt = \sum_{j=1}^n (-1)^{j-1} \frac{x^{2j-1}}{(2j)!} + (-1)^n \frac{x^{2n+1}}{(2n+2)!} \cos(\zeta_x)$$

for some $\zeta_x$ between 0 and $x$.

**Taylor's theorem in two dimensions**    Let $f(x, y)$ be a given function of the two independent variables $x$ and $y$. We will show how the earlier Taylor's theorem can be extended to the expansion of $f(x, y)$ about a given point $(x_0, y_0)$. The results will easily extend to functions of more than two variables. As notation, let $L(x_0, y_0; x_1, y_1)$ denote the set of all points $(x, y)$ on the straight line segment joining $(x_0, y_0)$ and $(x_1, y_1)$.

**Theorem 1.5**    Let $(x_0, y_0)$ and $(x_0 + \xi, y_0 + \eta)$ be given points, and assume $f(x, y)$ is $n + 1$ times continuously differentiable for all $(x, y)$ in some neighborhood of $L(x_0, y_0; x_0 + \xi, y_0 + \eta)$. Then

$$f(x_0 + \xi, y_0 + \eta)$$

$$= f(x_0, y_0) + \sum_{j=1}^n \frac{1}{j!} \left[ \xi \frac{\partial}{\partial x} + \eta \frac{\partial}{\partial y} \right]^j f(x, y) \Bigg|_{\substack{x=x_0 \\ y=y_0}}$$

$$+ \frac{1}{(n+1)!} \left[ \xi \frac{\partial}{\partial x} + \eta \frac{\partial}{\partial y} \right]^{n+1} f(x, y) \Bigg|_{\substack{x=x_0+\theta\xi \\ y=y_0+\theta\eta}} \quad (1.1.12)$$

for some $0 \le \theta \le 1$. The point $(x_0 + \theta\xi, y_0 + \theta\eta)$ is an unknown point on the line $L(x_0, y_0; x_0 + \xi, y_0 + \eta)$.

**Proof**   First note the meaning of the derivative notation in (1.1.12). As an example

$$\left[\xi\frac{\partial}{\partial x} + \eta\frac{\partial}{\partial y}\right]^2 f(x, y)$$

$$= \xi^2 \cdot \frac{\partial^2 f(x, y)}{\partial x^2} + 2\xi\eta \cdot \frac{\partial^2 f(x, y)}{\partial x\, \partial y} + \eta^2 \cdot \frac{\partial^2 f(x, y)}{\partial y^2}$$

The subscript notation, $x = x_0$, $y = y_0$, means the various derivatives are to be evaluated at $(x_0, y_0)$.

The proof of (1.1.12) is based on applying the earlier Taylor's theorem to

$$F(t) = f(x_0 + t\xi, y_0 + t\eta) \qquad 0 \le t \le 1$$

Using Theorem 1.4,

$$F(1) = F(0) + \frac{F'(0)}{1!} + \cdots + \frac{F^{(n)}(0)}{n!} + \frac{F^{(n+1)}(\theta)}{(n+1)!}$$

for some $0 \le \theta \le 1$. Clearly, $F(0) = f(x_0, y_0)$, $F(1) = f(x_0 + \xi, y_0 + \eta)$. For the first derivative,

$$F'(t) = \xi\frac{\partial f(x_0 + t\xi, y_0 + t\eta)}{\partial x} + \eta\frac{\partial f(x_0 + t\xi, y_0 + t\eta)}{\partial y}$$

$$= \left[\xi\frac{\partial}{\partial x} + \eta\frac{\partial}{\partial y}\right] f(x, y)\Bigg|_{\substack{x = x_0 + t\xi \\ y = y_0 + t\eta}}$$

The higher order derivatives are calculated similarly.   ∎

**Example**   As a simple example, consider expanding $f(x, y) = x/y$ about $(x_0, y_0) = (6, 2)$. Let $n = 1$. Then

$$\frac{x}{y} = f(6, 2) + (x - 6)\frac{\partial f(6, 2)}{\partial x} + (y - 2)\frac{\partial f(6, 2)}{\partial y}$$

$$+ \frac{1}{2}\left[(x - 6)^2 \cdot \frac{\partial^2 f(6, 2)}{\partial x^2} + 2(x - 6)(y - 2)\frac{\partial^2 f(6, 2)}{\partial x\, \partial y}\right.$$

$$\left. + (y - 2)^2 \cdot \frac{\partial^2 f(6, 2)}{\partial y^2}\right]_{\substack{x = \delta \\ y = \gamma}}$$

$$= 3 + \frac{1}{2}(x - 6) - \frac{3}{2}(y - 2)$$

$$+ \frac{1}{2}\left[(x - 6)^2 \cdot 0 - 2(x - 6)(y - 2)\frac{1}{\gamma^2} + (y - 2)^2 \cdot \frac{2\delta}{\gamma^3}\right]$$

with $(\delta, \gamma)$ a point on $L(6, 2; x, y)$. For $(x, y)$ close to $(6, 2)$,

$$\frac{x}{y} \doteq 3 + \frac{1}{2}x - \frac{3}{2}y$$

The graph of $z = 3 + \frac{1}{2}x - \frac{3}{2}y$ is the plane tangent to the graph of $z = x/y$ at $(x, y, z) = (6, 2, 3)$.

**Some mathematical notation** There are several concepts that are taken up in this text that are needed in a simpler form in the earlier chapters. These include results on divided differences of functions, vector spaces, and vector and matrix norms. The minimum necessary notation is introduced at this point, and a more complete development is left to other more natural places in the text.

For a given function $f(x)$, define

$$f[x_0, x_1] = \frac{f(x_1) - f(x_0)}{x_1 - x_0} \qquad f[x_0, x_1, x_2] = \frac{f[x_1, x_2] - f[x_0, x_1]}{x_2 - x_0}$$

$$(1.1.13)$$

assuming $x_0, x_1, x_2$ are distinct. These are called the first- and second-order divided differences of $f(x)$, respectively. They are related to derivatives of $f(x)$:

$$f[x_0, x_1] = f'(\xi) \qquad f[x_0, x_1, x_2] = \tfrac{1}{2}f''(\zeta) \qquad (1.1.14)$$

with $\xi$ between $x_0$ and $x_1$, and $\zeta$ between the minimum and maximum of $x_0$, $x_1$, and $x_2$. The divided differences are independent of the order of their arguments, contrary to what might be expected from (1.1.13). More precisely,

$$f[x_0, x_1] = f[x_1, x_0]$$

$$f[x_0, x_1, x_2] = f[x_i, x_j, x_k] \qquad (1.1.15)$$

for any permutation $(i, j, k)$ of $(0, 1, 2)$. The proofs of these and other properties are left as problems. A complete development of divided differences is given in Section 3.2 of Chapter 3.

The subjects of vector spaces, matrices, and vector and matrix norms are covered in Chapter 7, immediately preceding the chapters on numerical linear algebra. We introduce some of this material here, while leaving the proofs till Chapter 7. Two vector spaces are used in a great many applications. They are

$$\mathbf{R}^n = \left\{ x = \begin{bmatrix} x_1 \\ \vdots \\ x_n \end{bmatrix} \middle| x_1, \ldots, x_n \text{ real numbers} \right\}$$

$$C[a, b] = \{ f(t) \mid f(t) \text{ continuous and real valued}, \ a \le t \le b \}$$

For $x, y \in \mathbf{R}^n$ and $\alpha$ a real number, define $x + y$ and $\alpha x$ by

$$x + y = \begin{bmatrix} x_1 + y_1 \\ \vdots \\ x_n + y_n \end{bmatrix} \qquad \alpha x = \begin{bmatrix} \alpha x_1 \\ \vdots \\ \alpha \dot{x}_n \end{bmatrix}$$

For $f, g \in C[a, b]$ and $\alpha$ a real number, define $f + g$ and $\alpha f$ by

$$(f + g)(t) = f(t) + g(t) \qquad (\alpha f)(t) = \alpha f(t) \qquad a \le t \le b$$

Vector norms are used to measure the magnitude of a vector. For $\mathbf{R}^n$, we define initially two different norms:

$$\|x\|_\infty = \max_{1 \le i \le n} |x_i| \qquad x \in \mathbf{R}^n \tag{1.1.16}$$

$$\|x\|_2 = \sqrt{x_1^2 + \cdots + x_n^2} \qquad x \in \mathbf{R}^n \tag{1.1.17}$$

For $C[a, b]$, define

$$\|f\|_\infty = \max_{a \le t \le b} |f(t)| \qquad f \in C[a, b] \tag{1.1.18}$$

These definitions can be shown to satisfy the following three characteristic properties of all norms.

1.  $\|v\| = 0$ if and only if $v = 0$, the zero vector
2.  $\|\alpha v\| = |\alpha| \cdot \|v\|$, for all vectors $v$ and real numbers $\alpha$
3.  $\|v + w\| \le \|v\| + \|w\|$, for all vectors $v$ and $w$

Property (3) is usually referred to as the *triangle inequality*. An explanation of this name and a further development of properties and norms for $C[a, b]$ is given in Chapter 4.

Norms can also be introduced for matrices. For an $n \times n$ matrix

$$A = \begin{bmatrix} a_{11} & a_{12} & \cdots & a_{1n} \\ a_{21} & a_{22} & \cdots & a_{2n} \\ \vdots & & & \\ a_{n1} & a_{n2} & \cdots & a_{nn} \end{bmatrix}$$

define

$$\|A\|_\infty = \max_{1 \le i \le n} \sum_{j=1}^{n} |a_{ij}| \tag{1.1.19}$$

With this definition, the properties of a vector norm will be satisfied. In addition, it can be shown that

$$\|AB\|_\infty \le \|A\|_\infty \|B\|_\infty \tag{1.1.20}$$

$$\|Ax\|_\infty \le \|A\|_\infty \|x\|_\infty \qquad \text{all } x \in \mathbf{R}^n \tag{1.1.21}$$

where $A$ and $B$ are arbitrary $n \times n$ matrices. The proofs of these results are left to the problems. They are also taken up in greater generality in Chapter 7.

*Example*   Consider the vector space $\mathbf{R}^2$ and matrices of order $2 \times 2$. In particular, let

$$A = \begin{bmatrix} 1 & -1 \\ 3 & 2 \end{bmatrix} \qquad x = \begin{bmatrix} 1 \\ 2 \end{bmatrix} \qquad y = Ax = \begin{bmatrix} -1 \\ 7 \end{bmatrix}$$

Then

$$\|A\|_\infty = 5 \qquad \|x\|_\infty = 2 \qquad \|y\|_\infty = 7$$

and (1.1.21) is easily satisfied. To show that (1.1.21) cannot be improved upon, take

$$x = \begin{bmatrix} 1 \\ 1 \end{bmatrix} \qquad y = Ax = \begin{bmatrix} 0 \\ 5 \end{bmatrix}$$

Then

$$\|y\|_\infty = 5 = \|A\|_\infty \|x\|_\infty$$

## 1.2   Computer Representation of Numbers

Digital computers are the principal means of calculation in numerical analysis, and consequently it is very important to understand how they operate. In this section we consider how numbers are represented in computers, and in the remaining sections we consider some consequences of the computer representation of numbers and of computer arithmetic.

Most computers have an *integer mode* and a *floating-point mode* for representing numbers. The integer mode is used only to represent integers, and it will not concern us any further. The floating-point form is used to represent real numbers. The numbers allowed can be of greatly varying size, but there are limitations on both the magnitude of the number and on the number of digits. The floating-point representation is closely related to what is called *scientific notation* in many high school mathematics texts.

The number base used in computers is seldom decimal. Most digital computers use the base 2 (binary) number system or some variant of it such as base 8 (octal) or base 16 (hexadecimal).

***Example*** **(a)**   In base 2, the digits are 0 and 1. As an example of the conversion of a base 2 number to decimal, we have

$$(11011.01)_2 = 1 \cdot 2^4 + 1 \cdot 2^3 + 0 \cdot 2^2 + 1 \cdot 2^1 + 1 \cdot 2^0 + 0 \cdot 2^{-1} + 1 \cdot 2^{-2}$$

$$= 27.25$$

When using numbers in some other base, we will often use $(x)_\beta$ to indicate that the number $x$ is to be interpreted in the base $\beta$ number system.

**(b)**   In base 16, the digits are $0, 1, \ldots, 9, A, \ldots, F$. As an example of the conversion of such a number to decimal, we have

$$(56C.F)_{16} = 5 \cdot 16^2 + 6 \cdot 16^1 + 12 \cdot 16^0 + 15 \cdot 16^{-1} = 1388.9375$$

The conversion of decimal numbers to binary is examined in the problems.

Let $\beta$ denote the number base being used in the computer. Then a nonzero number $x$ in the computer is stored essentially in the form

$$x = \sigma \cdot (.a_1 a_2 \cdots a_t)_\beta \cdot \beta^e \qquad (1.2.1)$$

with $\sigma = +1$ or $-1$, $0 \le a_i \le \beta - 1$, $e$ an integer, and

$$(.a_1 a_2 \cdots a_t)_\beta = \frac{a_1}{\beta^1} + \frac{a_2}{\beta^2} + \cdots + \frac{a_t}{\beta^t}$$

The term $\sigma$ is called the *sign*, $e$ is called the *exponent*, and $(.a_1 \cdots a_t)_\beta$ is called the *mantissa* of the floating-point number $x$. The number $\beta$ is also called the *radix*, and the point preceding $a_1$ in (1.2.1) is called the *radix point*, for example, decimal point ($\beta = 10$), binary point ($\beta = 2$). The integer $t$ gives the number of base $\beta$ digits in the representation. We will always assume

$$a_1 \ne 0$$

giving what is called the normalized floating-point representation. We will also assume that

$$L \le e \le U \qquad (1.2.2)$$

which limits the possible size of $x$. The number $x = 0$ is always allowed, requiring a special representation. Table 1.1 contains the values of $\beta$, $t$, $L$, and $U$ for a number of common computers. The use of $\beta$, $t$, $L$, and $U$ to specify the arithmetic characteristics is based on that in [Forsythe et al. (1977), p. 11]. Some computers use a different placement of the radix point (e.g., CDC CYBER). We have modified their exponent bounds so that the limits on the size of a floating-point number will be correct when using the theory of this section. We also include results for double precision representations on some computers that include it in their hardware. In Table 1.1 there are additional columns that will be explained later.

**Table 1.1    Floating-point representations on various computers**

| Machine | S/D | R/C | $\beta$ | $t$ | $L$ | $U$ | $\delta$ | $M$ |
|---------|-----|-----|---------|-----|-----|-----|----------|-----|
| CDC CYBER 170 | S | R | 2 | 48 | $-976$ | 1071 | 3.55E $-$ 15 | 2.81E14 |
| CDC CYBER 205 | S | C | 2 | 47 | $-28{,}626$ | 28,718 | 1.42E $-$ 14 | 1.41E14 |
| CRAY-1 | S | C | 2 | 48 | $-8192$ | 8191 | 7.11E $-$ 15 | 2.81E14 |
| DEC VAX | S | R | 2 | 24 | $-127$ | 127 | 5.96E $-$ 8 | 1.68E7 |
| DEC VAX | D | R | 2 | 53 | $-1023$ | 1023 | 1.11E $-$ 16 | 9.01E15 |
| HP-11C, 15C | S | R | 10 | 10 | $-99$ | 99 | 5.00E $-$ 10 | 1.00E10 |
| IBM 3033 | S | C | 16 | 6 | $-64$ | 63 | 9.54E $-$ 7 | 1.68E7 |
| IBM 3033 | D | C | 16 | 14 | $-64$ | 63 | 2.22E $-$ 16 | 7.21E16 |
| Intel 8087 | S | R | 2 | 24 | $-126$ | 127 | 5.96E $-$ 8 | 1.68E7 |
| Intel 8087 | D | R | 2 | 53 | $-1022$ | 1023 | 1.11E $-$ 16 | 9.01E15 |
| PRIME 850 | S | R | 2 | 23 | $-128$ | 127 | 1.19E $-$ 7 | 8.39E6 |
| PRIME 850 | S | C | 2 | 23 | $-128$ | 127 | 2.38E $-$ 7 | 8.39E6 |
| PRIME 850 | D | C | 2 | 47 | $-32{,}896$ | 32,639 | 1.42E $-$ 14 | 1.41E14 |

Note: S/D: single or double precision;
   R/C: rounding or chopping;
   $\beta$: number base (radix);
   $t$: digits in mantissa [see (1.2.1)];
   $L, U$: exponent limits [see (1.2.2)];
   $\delta$: unit round [see (1.2.12)];
   $M$: exact integers bound [see (1.2.16)].

**Chopping and rounding**    Most real numbers $x$ cannot be represented exactly by the floating-point representation previously given, and thus they must be approximated by a nearby number representable in the machine, if possible. Given an arbitrary number $x$, we let fl$(x)$ denote its machine approximation, if it exists. There are two principal ways of producing fl$(x)$ from $x$: chopping and rounding.
   Let a real number $x$ be written in the form

$$x = \sigma \cdot (.a_1 a_2 \cdots a_t a_{t+1} \cdots )_\beta \cdot \beta^e \qquad (1.2.3)$$

with $a_1 \neq 0$, and assume $e$ satisfies (1.2.2). The chopped machine representation of $x$ is given by

$$\text{fl}(x) = \sigma \cdot (.a_1 \cdots a_t)_\beta \cdot \beta^e \qquad (1.2.4)$$

The reason for introducing chopping is that many computers use chopping rather than rounding after each arithmetic operation.
   The rounded representation of $x$ is given by

$$\text{fl}(x) = \begin{cases} \sigma \cdot (.a_1 \cdots a_t)_\beta \cdot \beta^e & 0 \le a_{t+1} < \dfrac{\beta}{2} \quad (1.2.5) \\[2ex] \sigma \cdot [(.a_1 \cdots a_t)_\beta + (.0 \cdots 01)_\beta] \cdot \beta^e & \dfrac{\beta}{2} \le a_{t+1} < \beta \quad (1.2.6) \end{cases}$$

In the last formula, $(.0 \cdots 01)_\beta$ denotes $\beta^{-t}$. Although this definition of $\mathrm{fl}(x)$ is somewhat formal, it yields the standard definition of rounding that most people have learned for decimal numbers. A variation of this definition is sometimes used in order to have unbiased rounding. In such a case, if

$$(1) \quad a_{t+1} = \frac{\beta}{2} \quad \text{and} \quad (2) \quad a_j = 0 \quad \text{for} \quad j \geq t + 2$$

then whether to round up or down is based on whether $a_t$ is odd or even, respectively. This leads to the unbiased rounding rule that most people learn for rounding decimal numbers, but we will henceforth assume the simpler definition (1.2.5)–(1.2.6).

With most real numbers $x$, we have $\mathrm{fl}(x) \neq x$. Looking at the relative (or percentage) error, it can be shown that

$$\frac{x - \mathrm{fl}(x)}{x} = -\epsilon \tag{1.2.7}$$

with

$$-\beta^{-t+1} \leq \epsilon \leq 0 \qquad \text{chopped } \mathrm{fl}(x) \tag{1.2.8}$$

$$-\tfrac{1}{2}\beta^{-t+1} \leq \epsilon \leq \tfrac{1}{2}\beta^{-t+1} \qquad \text{rounded } \mathrm{fl}(x) \tag{1.2.9}$$

We will show the result (1.2.8) for chopping; the result (1.2.9) for rounding is left as a problem.

Assume $\sigma = +1$, since the case $\sigma = -1$ will not change the sign of $\epsilon$. From (1.2.3) and (1.2.4), we have

$$x - \mathrm{fl}(x) = (.00 \cdots 0 a_{t+1} \cdots)_\beta \cdot \beta^e$$

Letting $\gamma = \beta - 1$,

$$0 \leq x - \mathrm{fl}(x) \leq (.00 \cdots 0\gamma\gamma \cdots)_\beta \cdot \beta^e$$

$$= \gamma \left[ \beta^{-t-1} + \beta^{-t-2} + \cdots \right] \cdot \beta^e$$

$$= \gamma \left[ \frac{\beta^{-t-1}}{1 - \beta^{-1}} \right] \cdot \beta^e = \beta^{-t+e}$$

$$0 \leq \frac{x - \mathrm{fl}(x)}{x} \leq \frac{\beta^{-t+e}}{(.a_1 a_2 \cdots)_\beta \cdot \beta^e}$$

$$\leq \frac{\beta^{-t}}{(.100 \cdots)_\beta} = \beta^{-t+1}$$

This proves (1.2.8), and the proof of (1.2.9) is similar.

The formula (1.2.7) is usually written in the equivalent form

$$fl(x) = (1 + \epsilon)x \qquad (1.2.10)$$

with $\epsilon$ given by (1.2.8) or (1.2.9). Thus $fl(x)$ can be considered to be a small relative perturbation of $x$. This formula for $fl(x)$ also allows us to deal precisely with the effects of rounding/chopping errors in computer arithmetic operations. Examples of this are given in later sections. The definition of $fl(x)$ and the use of (1.2.10) is due to [Wilkinson (1963)], and it is widely used in analyzing the effects of rounding errors.

**Accuracy of floating-point representation**     We now introduce two measures that give a fairly precise idea of the possible accuracy in a floating-point representation. The first of these is closely related to the preceding error result (1.2.7)–(1.2.9) for $fl(x)$.

The *unit round* of a computer is the number $\delta$ satisfying: (1) it is a positive floating-point number, and (2) it is the smallest such number for which

$$fl(1 + \delta) > 1 \qquad (1.2.11)$$

Thus for any floating-point number $\hat{\delta} < \delta$, we have $fl(1 + \hat{\delta}) = 1$, and $1 + \hat{\delta}$ and 1 are identical within the computer's arithmetic. This gives a precise measure of how many digits of accuracy are possible in the representation of a number. Most high-quality portable computer programs use the unit round in order to note the maximal accuracy that is possible on the computer being used.

The unit round $\delta$ is easily calculated, and it is given by

$$\delta = \begin{cases} \beta^{-t+1} & \text{chopped definition of } fl(x) \\ \frac{1}{2}\beta^{-t+1} & \text{rounded definition of } fl(x) \end{cases} \qquad (1.2.12)$$

We show this for rounded arithmetic on a binary machine. First we must show

$$fl(1 + 2^{-t}) > 1 \qquad (1.2.13)$$

Write

$$1 + 2^{-t} = [(.10 \cdots)_2 + (.00 \cdots 010 \cdots)_2]2^1$$

$$\uparrow$$

$$\text{position } t + 1$$

$$= (.10 \cdots 010 \cdots)_2 \cdot 2^1 \qquad (1.2.14)$$

Form $fl(1 + 2^{-t})$, and note that there is a 1 in position $t + 1$ of the mantissa. Then from (1.2.6)

$$fl(1 + 2^{-t}) = (.10 \cdots 01)_2 \cdot 2^1 = 1 + 2^{-t+1}$$

$$\uparrow$$

$$\text{position } t$$

Thus (1.2.13) is satisfied, although

$$\text{fl}\,(1 + 2^{-t}) \neq 1 + 2^{-t}$$

The fact that $\delta$ cannot be smaller than $2^{-t}$ follows easily by reexamining (1.2.14). If $\hat{\delta} < \delta$, then $1 + \hat{\delta}$ has a 0 in position $t + 1$ of the mantissa in (1.2.14): and the definition (1.2.5) of rounding would then imply $\text{fl}\,(1 + \hat{\delta}) = 1$.

A second measure of the maximal accuracy possible in a floating-point representation is to find the largest integer $M$ for which

$$m \text{ an integer and } 0 \leq m \leq M \Rightarrow \text{fl}\,(m) = m \qquad (1.2.15)$$

This also implies $\text{fl}\,(M + 1) \neq M + 1$. It is left as a problem to show that

$$M = \beta^t \qquad (1.2.16)$$

The numbers $M$ and $\delta$ for various computers are given in Table 1.1, along with whether the computers round (R) or chop (C).

*Example*  For PRIME computers in single precision,

$$M = 2^{23} = 8388608.$$

Thus all integers with six decimal digits and most with seven decimal digits can be stored exactly in the single precision floating-point representation. For the unit round,

$$\delta = 2^{-22} \doteq 2.38 \times 10^{-7} \qquad \text{chopped arithmetic}$$

$$\delta = 2^{-23} \doteq 1.19 \times 10^{-7} \qquad \text{rounded arithmetic}$$

Users of the PRIME have both chopped and rounded arithmetic available to them, for single precision arithmetic.

In almost all cases, rounded arithmetic is greatly preferable to chopped arithmetic. This is examined in more detail in later sections, but the main reason lies in the biased sign of $\epsilon$ in (1.2.8) as compared to the lack of such bias in (1.2.9).

**Underflow and overflow**  When the exponent bounds of (1.2.2) are violated, then the associated number $x$ of (1.2.1) cannot be represented in the computer. We now look at what this says about the possible range in magnitude of $x$.

The smallest positive floating-point number is

$$x_L = (.10 \cdots 0)_\beta \cdot \beta^L = \beta^{L-1}$$

Using $\gamma = \beta - 1$, the largest positive floating-point number is

$$x_U = (.\gamma \cdots \gamma)_\beta \cdot \beta^U = (1 - \beta^{-t}) \cdot \beta^U$$

Thus all floating-point numbers $x$ must satisfy

$$x_L \leq |x| \leq x_U \qquad (1.2.17)$$

Within the Fortran language of most computers, if an arithmetic operation leads to a result $x$ for which $|x| > x_U$, then this will be a fatal error and the program will terminate. This is called an *overflow error*. In contrast, if

$$0 < |x| < x_L$$

then usually fl$(x)$ will be set to zero and computation will continue. This is called an *underflow error*.

***Example*** Consider evaluating $x = s^{10}$ on an IBM mainframe computer. Then there will be an underflow error if

$$s^{10} < 16^{-65}$$

Thus $x = s^{10}$ will be set to zero if $|s| < 1.49 \times 10^{-8}$. Also, there will be an overflow error if

$$s^{10} > 16^{63}$$

or equivalently,

$$|s| > 3.9 \times 10^7$$

## 1.3  Definitions and Sources of Error

We now give a rough classification of the major ways in which error is introduced into the solution of a problem, including some that fall outside the usual scope of mathematics. We begin with a few simple definitions about error.

In solving a problem, we seek an exact or true solution, which we denote by $x_T$. Approximations are usually made in solving the problem, resulting in an approximate solution $x_A$. We define the error in $x_A$ by

$$\text{Error}(x_A) \equiv \text{Error in } x_A = x_T - x_A$$

For many purposes, we prefer to study the percentage or *relative error* in $x_A$,

$$\text{Rel}(x_A) \equiv \text{relative error in } x_A = \frac{x_T - x_A}{x_T}$$

provided $x_T \neq 0$. This has already been referred to in (1.2.7), in measuring the error in fl$(x)$.

***Example***   $x_T = e = 2.7182818\ldots \qquad x_A = \frac{19}{7} = 2.7142857\ldots$

$$\text{Error}(x_A) = .003996\ldots \qquad \text{Rel}(x_A) = .00147\ldots$$

In place of relative error, we often use the concept of *significant digits*. We say $x_A$ has $m$ significant (decimal) digits with respect to $x_T$ if the error $x_T - x_A$ has magnitude less than or equal to 5 in the $(m + 1)$st digit of $x_T$, counting to the right from the first nonzero digit in $x_T$.

**Example (a)**   $x_T = \frac{1}{3}$     $x_A = .333$     $|x_T - x_A| \doteq .00033$

Since the error is less than 5 in the fourth digit to the right of the first nonzero digit in $x_T$, we say that $x_A$ has three significant digits with respect to $x_T$.

**(b)**   $x_T = 23.496$     $x_A = 23.494$     $|x_T - x_A| = .002$

The term $x_A$ has four significant digits with respect to $x_T$, since the error is less than 5 in the fifth place to the right of the first nonzero digit in $x_T$. Note that if $x_A$ is rounded to four places, an additional error is introduced and $x_A$ will no longer have four significant digits.

**(c)**   $x_T = .02138$     $x_A = .02144$     $|x_T - x_A| = .00006$

The number $x_A$ has two significant digits, but not three, with respect to $x_T$.

The following is sometimes used in measuring significant digits. If

$$\left| \frac{x_T - x_A}{x_T} \right| \leq 5 \times 10^{-m-1} \tag{1.3.1}$$

then $x_A$ has $m$ significant digits with respect to $x_T$. To show this, consider the case $.1 \leq |x_T| < 1$. Then (1.3.1) implies

$$|x_T - x_A| \leq 5 \times 10^{-m-1}|x_T| < .5 \times 10^{-m}$$

Since $.1 \leq x_T < 1$, this implies $x_A$ has $m$ significant digits. The proof for a general $x_T$ is essentially the same, using $x_T = \hat{x}_T \cdot 10^e$, with $.1 \leq |x_T| < 1$, $e$ an integer. Note that (1.3.1) is a sufficient condition, but not a necessary condition, in order that $x_A$ have $m$ significant digits. Examples (a) and (b) just given have one more significant digit than that indicated by the test (1.3.1).

**Sources of error**   We now give a rough classification of the major sources of error.

**(S1)   Mathematical modeling of a physical problem**   A mathematical model for a physical situation is an attempt to give mathematical relationships between certain quantities of physical interest. Because of the complexity of physical reality, a variety of simplifying assumptions are used to construct a more tractable mathematical model. The resulting model has limitations on its accuracy as a consequence of these assumptions, and these limitations may or may not be troublesome, depending on the uses of the model. In the case that the model is not sufficiently accurate, the numerical solution of the model cannot improve upon this basic lack of accuracy.

***Example*** Consider a projectile of mass $m$ to have been fired into the air, its flight path always remaining close to the earth's surface. Let an $xyz$ coordinate system be introduced with origin on the earth's surface and with the positive $z$-axis perpendicular to the earth and directed upward. Let the position of the projectile at time $t$ be denoted by $\mathbf{r}(t) = x(t)\mathbf{i} + y(t)\mathbf{j} + z(t)\mathbf{k}$, using the standard vector field theory notation. One model for the flight of the projectile is given by Newton's second law as

$$m\frac{d^2\mathbf{r}(t)}{dt^2} = -mg\mathbf{k} - b\frac{d\mathbf{r}(t)}{dt} \tag{1.3.2}$$

where $b > 0$ is a constant and $g$ is the acceleration due to gravity. This equation says that the only forces acting on the projectile are (1) the gravitational force of the earth, and (2) a frictional force that is directly proportional to the speed $|\mathbf{v}(t)| = |d\mathbf{r}(t)/dt|$ and directed opposite to the path of flight.

In some situations this is an excellent model, and it may not be necessary to include even the frictional term. But the model doesn't include forces of resistance acting perpendicular to the plane of flight, for example, a cross-wind, and it doesn't allow for the Coriolis effect. Also, the frictional force in (1.29) may be proportional to $|\mathbf{v}(t)|^\alpha$ with $\alpha \neq 1$.

If a model is adequate for physical purposes, then we wish to use a numerical scheme that will preserve this accuracy. But if the model is inadequate, then the numerical analysis cannot improve the accuracy except by chance. On the other hand, it is not a good idea to create a model that is more complicated than needed, introducing terms that are relatively insignificant with respect to the phenomenon being studied. A more complicated model can often introduce additional numerical analysis difficulties, without yielding any significantly greater accuracy. For books concerned explicitly with mathematical modeling in the sciences, see Bender (1978), Lin and Segal (1974), Maki and Thompson (1973), Rubinow (1975).

**(S2) Blunders** In precomputer times, chance arithmetic errors were always a serious problem. Check schemes, some quite elaborate, were devised to detect if such errors had occurred and to correct for them before the calculations had proceeded very far past the error. For an example, see Fadeeva (1959) for check schemes used when solving systems of linear equations.

With the introduction of digital computers, the type of *blunder* has changed. Chance arithmetic errors (e.g., computer malfunctioning) are now relatively rare, and programming errors are currently the main difficulty. Often a program error will be repeated many times in the course of executing the program, and its existence becomes obvious because of absurd numerical output (although the source of the error may still be difficult to find). But as computer programs become more complex and lengthy, the existence of a small program error may be hard to detect and correct, even though the error may make a subtle, but crucial difference in the numerical results. This makes good program debugging very important, even though it may not seem very rewarding immediately.

To detect programming errors, it is important to have some way of checking the accuracy of the program output. When first running the program, you should use cases for which you know the correct answer, if possible. With a complex program, break it into smaller subprograms, each of which can be tested separately. When the entire program has been checked out and you believe it to be correct, maintain a watchful eye as to whether the output is reasonable or not.

**(S3)**   Uncertainty in physical data   Most data from a physical experiment will contain error or uncertainty within it. This must affect the accuracy of any calculations based on the data, limiting the accuracy of the answers. The techniques for analyzing the effects in other calculations of this error are much the same as those used in analyzing the effects of rounding error, although the error in data is usually much larger than rounding errors. The material of the next sections discusses this further.

**(S4)**   Machine errors   By machine errors we mean the errors inherent in using the floating-point representation of numbers. Specifically, we mean the rounding/chopping errors and the underflow/overflow errors. The rounding/chopping errors are due to the finite length of the floating-point mantissa; and these errors occur with all of the computer arithmetic operations. All of these forms of machine error were discussed in Section 1.2; in the following sections, we consider some of their consequences. Also, for notational simplicity, we henceforth let the term *rounding error* include chopping where applicable.

**(S5)**   Mathematical truncation error   This name refers to the error of approximation in numerically solving a mathematical problem, and it is the error generally associated with the subject of numerical analysis. It involves the approximation of infinite processes by finite ones, replacing noncomputable problems with computable ones. We use some examples to make the idea more precise.

*Example* (a)   Using the first two terms of the Taylor series from (1.1.7),

$$\sqrt{1 + x} \doteq 1 + \tfrac{1}{2}x \qquad (1.3.3)$$

which is a good approximation when $x$ is small. See Chapter 4 for the general area of approximation of functions.

**(b)**   For evaluating an integral on $[0, 1]$, use

$$\int_0^1 f(x)\, dx \doteq \frac{1}{n} \sum_{j=1}^{n} f\left(\frac{2j - 1}{2n}\right) \qquad n = 1, 2, 3, \ldots \qquad (1.3.4)$$

This is called the *midpoint numerical integration rule*: see the last part of Section 5.2 for more detail. The general topic of numerical integration is examined in Chapter 5.

**(c)**    For the differential equation problem

$$Y'(t) = f(t, Y(t)) \qquad Y(t_0) = Y_0 \tag{1.3.5}$$

use the approximation of the derivative

$$Y'(t) \doteq \frac{Y(t + h) - Y(t)}{h}$$

for some small $h$. Let $t_j = t_0 + jh$ for $j \geq 0$, and define an approximate solution function $y(t_j)$ by

$$\frac{y(t_{j+1}) - y(t_j)}{h} = f(t_j, y(t_j))$$

Thus we have

$$y(t_{j+1}) = y(t_j) + hf(t_j, y(t_j)) \qquad j \geq 0 \qquad y(x_0) = Y_0$$

This is *Euler's method* for solving an initial value problem for an ordinary differential equation. An extensive discussion and analysis of it is given in Section 6.2. Chapter 6 gives a complete development of numerical methods for solving the initial value problem (1.3.5).

Most numerical analysis problems in the following chapters involve mainly mathematical truncation errors. The major exception is the solution of systems of linear equations in which rounding errors are often the major source of error.

**Noise in function evaluation**    One of the immediate consequences of rounding errors is that the evaluation of a function $f(x)$ using a computer will lead to an approximate function $\hat{f}(x)$ that is not continuous, although it is apparent only when the graph of $\hat{f}(x)$ is looked at on a sufficiently small scale. After each arithmetic operation that is used in evaluating $f(x)$, there will usually be a rounding error. When the effect of these rounding errors is considered, we obtain a computed value $\hat{f}(x)$ whose error $f(x) - \hat{f}(x)$ appears to be a small random number as $x$ varies. This error in $\hat{f}(x)$ is called *noise*. When the graph of $\hat{f}(x)$ is looked at on a small enough scale, it appears as a fuzzy band of dots, where the $x$ values range over all acceptable floating-point numbers on the machine. This has consequences for many other programs that make use of $\hat{f}(x)$. For example, calculating the root of $f(x)$ by using $\hat{f}(x)$ will lead to uncertainty in the location of the root, because it will likely be located in the intersection of the $x$-axis and the fuzzy banded graph of $\hat{f}(x)$. The following example shows that this can result in considerable uncertainty in the location of the root.

*Example*    Let

$$f(x) = x^3 - 3x^2 + 3x - 1 \tag{1.3.6}$$

which is just $(x - 1)^3$. We evaluated (1.3.6) in the single precision BASIC of a

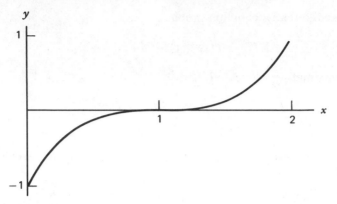

**Figure 1.1**    Graph of (1.3.6).

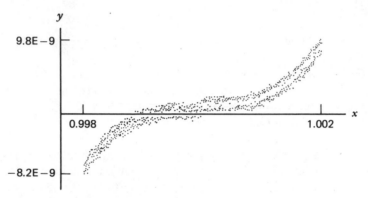

**Figure 1.2**    Detailed graph of (1.3.6).

popular microcomputer, using rounded binary arithmetic with a unit round of $\delta \doteq 2^{-24} = 5.96 \times 10^{-8}$. The graph of $\hat{f}(x)$ on $[0, 2]$, shown in Figure 1.1, is continuous and smooth to the eye, as would be expected. But the graph on the smaller interval $[.998, 1.002]$ shows the discontinuous nature of $f(x)$, as is apparent from the graph in Figure 1.2. In this latter case, $\hat{f}(x)$ was evaluated at 640 evenly spaced values of $x$ in $[.998, 1.002]$, resulting in the fuzzy band that is the graph of $f(x)$. From the latter graph, it can be seen that there is a large interval of uncertainty as to where $\hat{f}(x)$ crosses the $x$-axis. We return to this topic in Section 2.7 of Chapter 2.

**Underflow / overflow errors in calculations**    We consider another consequence of machine errors. The upper and lower limits for floating-point numbers, given in (1.2.17), can lead to errors in calculations. Sometimes these are unavoidable, but often they are an artifact of the way the calculation is arranged.

To illustrate this, consider evaluating the magnitude of a complex number,

$$|x + iy| = \sqrt{x^2 + y^2} \tag{1.3.7}$$

It is possible this may underflow or overflow, even though the magnitude $|x + iy|$

is within machine limits. For example, if $x_U = 1.7 \times 10^{38}$ from (1.2.17), then (1.3.7) will overflow for $x = y = 10^{20}$, even though $|x + iy| \doteq 1.4 \times 10^{20}$. To avoid this, determine the larger of $x$ and $y$, say $x$. Then rewrite (1.3.7) as

$$|x + iy| = |x| \cdot \sqrt{1 + \alpha^2} \qquad \alpha = \frac{y}{x} \tag{1.3.8}$$

We must calculate $\sqrt{1 + \alpha^2}$, with $0 \le \alpha \le 1$. This avoids the problems of (1.3.7), both for underflow and overflow.

## 1.4  Propagation of Errors

In this and the next sections, we consider the effect of calculations with numbers that are in error. To begin, consider the basic arithmetic operations. Let $\omega$ denote one of the arithmetic operations $+, -, \times, /$; and let $\hat{\omega}$ be the computer version of the same operation, which will usually include rounding. Let $x_A$ and $y_A$ be the numbers being used for calculations, and suppose they are in error, with true values

$$x_T = x_A + \epsilon \qquad y_T = y_A + \eta$$

Then $x_A \hat{\omega} y_A$ is the number actually computed, and for its error,

$$x_T \omega y_T - x_A \hat{\omega} y_A = [x_T \omega y_T - x_A \omega y_A] + [x_A \omega y_A - x_A \hat{\omega} y_A] \tag{1.4.1}$$

The first quantity in brackets is called the *propagated error*, and the second quantity is normally rounding or chopping error. For this second quantity, we usually have

$$x_A \hat{\omega} y_A = \text{fl}(x_A \omega y_A) \tag{1.4.2}$$

which means that $x_A \omega y_A$ is computed exactly and then rounded. Combining (1.2.9) and (1.4.2),

$$|x_A \omega y_A - x_A \hat{\omega} y_A| \le \frac{\beta}{2} |x_A \omega y_A| \beta^{-t} \tag{1.4.3}$$

provided true rounding is used.

For the propagated error, we examine particular cases.

*Case* (a)    Multiplication. For the error in $x_A y_A$,

$$x_T y_T - x_A y_A = x_T y_T - (x_T - \epsilon)(y_T - \eta)$$

$$= x_T \eta + y_T \epsilon - \epsilon \eta$$

$$\text{Rel}(x_A y_A) \equiv \frac{x_T y_T - x_A y_A}{x_T y_T} = \frac{\eta}{y_T} + \frac{\epsilon}{x_T} - \frac{\epsilon}{x_T} \cdot \frac{\eta}{y_T}$$

$$= \text{Rel}(x_A) + \text{Rel}(y_A) - \text{Rel}(x_A) \cdot \text{Rel}(y_A)$$

For $|\text{Rel}(x_A)|, |\text{Rel}(y_A)| \ll 1$,

$$\text{Rel}(x_A y_A) \doteq \text{Rel}(x_A) + \text{Rel}(y_A) \tag{1.4.4}$$

The symbol " $\ll$ " means "much less than."

*Case* **(b)**    Division. By a similar argument,

$$\text{Rel}\frac{x_A}{y_A} = \frac{\text{Rel}(x_A) - \text{Rel}(y_A)}{1 - \text{Rel}(y_A)} \tag{1.4.5}$$

For $|\text{Rel}(y_A)| \ll 1$,

$$\text{Rel}\frac{x_A}{y_A} \doteq \text{Rel}(x_A) - \text{Rel}(y_A) \tag{1.4.6}$$

For both multiplication and division, relative errors do not propagate rapidly.

*Case* **(c)**    Addition and subtraction.

$$(x_T \pm y_T) - (x_A \pm y_A) = (x_T - x_A) \pm (y_T - y_A) = \epsilon \pm \eta$$

$$\text{Error}(x_A \pm y_A) = \text{Error}(x_A) \pm \text{Error}(y_A) \tag{1.4.7}$$

This appears quite good and reasonable, but it can be misleading. The relative error in $x_A \pm y_A$ can be quite poor when compared with $\text{Rel}(x_A)$ and $\text{Rel}(y_A)$.

*Example*    Let $x_T = \pi$, $x_A = 3.1416$, $y_T = \frac{22}{7}$, $y_A = 3.1429$. Then

$$x_T - x_A \doteq -7.35 \times 10^{-6} \qquad \text{Rel}(x_A) \doteq -2.34 \times 10^{-6}$$

$$y_T - y_A \doteq -4.29 \times 10^{-5} \qquad \text{Rel}(y_A) \doteq -1.36 \times 10^{-5}$$

$$(x_T - y_T) - (x_A - y_A) \doteq -.0012645 - (-.0013) \doteq 3.55 \times 10^{-5}$$

$$\text{Rel}(x_A - y_A) \doteq -.028$$

Although the error in $x_A - y_A$ is quite small, the relative error in $x_A - y_A$ is much larger than that in $x_A$ or $y_A$ alone.

**Loss of significance errors**    This last example shows that it is possible to have a large decrease in accuracy, in a relative error sense, when subtracting nearly equal quantities. This can be a very important way in which accuracy is lost when error is propagated in a calculation. We now give some examples of this phenomenon, along with suggestions on how it may be avoided in some cases.

**Example** Consider solving $ax^2 + bx + c = 0$ when $4ac$ is small compared to $b^2$; use the standard quadratic formula for the roots

$$r_T^{(1)} = \frac{-b + \sqrt{b^2 - 4ac}}{2a} \qquad r_T^{(2)} = \frac{-b - \sqrt{b^2 - 4ac}}{2a} \qquad (1.4.8)$$

For definiteness, consider $x^2 - 26x + 1 = 0$. The formulas (1.4.8) yield

$$r_T^{(1)} = 13 + \sqrt{168} \qquad r_T^{(2)} = 13 - \sqrt{168} \qquad (1.4.9)$$

Now imagine using a five-digit decimal machine. On it, $\sqrt{168} \doteq 12.961$. Then define

$$r_A^{(1)} = 13 + 12.961 = 25.961 \qquad r_A^{(2)} = 13 - 12.961 = .039 \qquad (1.4.10)$$

Using the exact answers,

$$\text{Rel}\left(r_A^{(1)}\right) \doteq 1.85 \times 10^{-5} \qquad \text{Rel}\left(r_A^{(2)}\right) \doteq 1.25 \times 10^{-2} \qquad (1.4.11)$$

For the data entering into the calculation (1.4.10), using the notation of (1.4.7),

$$x_T = x_A = 13 \qquad y_T = \sqrt{168} \qquad y_A = 12.961$$

$$\text{Rel}(x_A) = 0 \qquad \text{Rel}(y_A) \doteq 3.71 \times 10^{-5}$$

The accuracy in $r_A^{(2)}$ is much less than that of the data $x_A$ and $y_A$ entering into the calculation. We say that significant digits have been lost in the subtraction $r_A^{(2)} = x_A - y_A$, or that we have had a *loss-of-significance error* in calculating $r_A^{(2)}$. In $r_A^{(1)}$, we have five significant digits of accuracy, whereas we have only two significant digits in $r_A^{(2)}$.

To cure this particular problem, of accurately calculating $r_A^{(2)}$, convert (1.4.9) to

$$r_T^{(2)} = \frac{13 - \sqrt{168}}{1} \cdot \frac{13 + \sqrt{168}}{13 + \sqrt{168}} = \frac{1}{13 + \sqrt{168}}$$

Then use

$$\frac{1}{13 + \sqrt{168}} \doteq \frac{1}{25.961} \doteq .038519 \equiv r_A^{(2)} \qquad (1.4.12)$$

There are two errors here, that of $\sqrt{168} \doteq 12.961$ and that of the final division. But each of these will have small relative errors [see (1.4.6)], and the new value of $r_A^{(2)}$ will be more accurate than the preceding one. By exact calculations, we now have

$$\text{Rel}\left(r_A^{(2)}\right) \doteq -1.03 \times 10^{-5},$$

much better than in (1.4.11).

This new computation of $r_A^{(2)}$ demonstrates then the loss-of-significance error is due to the form of the calculation, not to errors in the data of the computation. In this example it was easy to find an alternative computation that eliminated the loss-of-significance error, but this is not always possible. For a complete discussion of the practical computation of roots of a quadratic polynomial, see Forsythe (1969).

**Example**   With many loss-of-significance calculations, Taylor polynomial approximations can be used to eliminate the difficulty. We illustrate this with the evaluation of

$$f(x) = \int_0^1 e^{xt}\, dt = \frac{e^x - 1}{x} \qquad x \neq 0 \tag{1.4.13}$$

For $x = 0$, $f(0) = 1$; and easily, $f(x)$ is continuous at $x = 0$.

To see that there is a loss-of-significance problem when $x$ is small, we evaluate $f(x)$ at $x = 1.4 \times 10^{-9}$, using a popular and well-designed ten-digit hand calculator. The results are

$$e^x \doteq 1.000000001$$

$$\frac{e^x - 1}{x} \doteq \frac{10^{-9}}{1.4 \times 10^{-9}} \doteq .714 \tag{1.4.14}$$

The right-hand sides give the calculator results, and the true answer, rounded to 10 places, is

$$f(x) = 1.000000001$$

The calculation (1.4.14) has had a cancellation of the leading nine digits of accuracy in the operands in the numerator.

To avoid the loss-of-significance error, use a quadratic Taylor approximation to $e^x$ and then simplify $f(x)$:

$$e^x = 1 + x + \frac{x^2}{2} + \frac{x^3}{6}e^\xi \qquad 0 \leq \xi \leq x \leq 1$$

$$f(x) = 1 + \frac{x}{2} + \frac{x^2}{6}e^\xi \tag{1.4.15}$$

With the preceding $x = 1.4 \times 10^{-9}$,

$$f(x) \doteq 1 + 7 \times 10^{-10}$$

with an error of less than $10^{-18}$.

In general, use (1.4.15) on some interval $[0, \delta]$, picking $\delta$ to ensure the error in

$$f(x) \doteq 1 + \frac{x}{2}$$

is sufficiently small. Of course, a higher degree approximation to $e^x$ could be used, allowing a yet larger value of $\delta$.

In general, Taylor approximations are often useful in avoiding loss-of-significance calculations. But in some cases, the loss-of-significance error is more subtle.

*Example*    Consider calculating a sum

$$S = \sum_{j=1}^{n} x_j \qquad (1.4.16)$$

with positive and negative terms $x_j$, each of which is an approximate value. Furthermore, assume the sum $S$ is much smaller than the maximum magnitude of the $x_j$. In calculating such a sum on a computer, it is likely that a loss-of-significance error will occur. We give an illustration of this.

Consider using the Taylor formula (1.1.4) for $e^x$ to evaluate $e^{-5}$:

$$e^{-5} \doteq 1 + \frac{(-5)}{1!} + \frac{(-5)^2}{2!} + \frac{(-5)^3}{3!} + \cdots + \frac{(-5)^n}{n!} \qquad (1.4.17)$$

Imagine using a computer with four-digit decimal rounded floating-point arithmetic, so that each of the terms in this series must be rounded to four significant digits. In Table 1.2, we give these rounded terms $x_j$, along with the exact sum of the terms through the given degree. The true value of $e^{-5}$ is .006738, to four significant digits, and this is quite different from the final sum in the table. Also, if (1.4.17) is calculated exactly for $n = 25$, then the correct value of $e^{-5}$ is obtained to four digits.

In this example, the terms $x_j$ become relatively large, but they are then added to form the much smaller number $e^{-5}$. This means there are loss-of-significance

**Table 1.2    Calculation of (1.4.17) using four-digit decimal arithmetic**

| Degree | Term | Sum | Degree | Term | Sum |
|--------|------|-----|--------|------|-----|
| 0 | 1.000 | 1.000 | 13 | −.1960 | −.04230 |
| 1 | −5.000 | −4.000 | 14 | .7001E − 1 | .02771 |
| 2 | 12.50 | 8.500 | 15 | −2334E − 1 | .004370 |
| 3 | −20.83 | −12.33 | 16 | .7293E − 2 | .01166 |
| 4 | 26.04 | 13.71 | 17 | −.2145E − 2 | .009518 |
| 5 | −26.04 | −12.33 | 18 | .5958E − 3 | .01011 |
| 6 | 21.70 | 9.370 | 19 | −.1568E − 3 | .009957 |
| 7 | −15.50 | −6.130 | 20 | .3920E − 4 | .009996 |
| 8 | 9.688 | 3.558 | 21 | −.9333E − 5 | .009987 |
| 9 | −5.382 | −1.824 | 22 | .2121E − 5 | .009989 |
| 10 | 2.691 | .8670 | 23 | −.4611E − 6 | .009989 |
| 11 | −1.223 | −.3560 | 24 | .9607E − 7 | .009989 |
| 12 | .5097 | .1537 | 25 | −.1921E − 7 | .009989 |

errors in the calculation of the sum. To avoid this problem in this case is quite easy. Either use

$$e^{-5} = \frac{1}{e^5}$$

and form $e^5$ with a series not involving cancellation of positive and negative terms; or simply form $e^{-1} = 1/e$, perhaps using a series, and multiply it times itself to form $e^{-5}$. With other series, it is likely that there will not be such a simple solution.

**Propagated error in function evaluations**    Let $f(x)$ be a given function, and let $\hat{f}(x)$ denote the result of evaluating $f(x)$ on a computer. Then $f(x_T)$ denotes the desired function value and $\hat{f}(x_A)$ is the value actually computed. For the error, we write

$$f(x_T) - \hat{f}(x_A) = [f(x_T) - f(x_A)] + [f(x_A) - \hat{f}(x_A)] \quad (1.4.18)$$

The first quantity in brackets is called the *propagated error*, and the second is the error due to evaluating $f(x_A)$ on a computer. This second error is generally a small random number, based on an assortment of rounding errors that occurred in carrying out the arithmetic operations defining $f(x)$. We referred to it earlier in Section 1.3 as the noise in evaluating $f(x)$.

For the propagated error, the mean value theorem gives

$$f(x_T) - f(x_A) \doteq f'(x_T)(x_T - x_A) \quad (1.4.19)$$

This assumes that $x_A$ and $x_T$ are relatively close and that $f'(x)$ does not vary greatly for $x$ between $x_A$ and $x_T$.

*Example*    $\sin(\pi/5) - \sin(.628) \doteq \cos(\pi/5)[(\pi/5) - .628] \doteq .00026$, which is an excellent estimation of the error.

Using Taylor's theorem (1.1.12) for functions of two variables, we can generalize the preceding to propagation of error in functions of two variables:

$$f(x_T, y_T) - f(x_A, y_A) \doteq f_x(x_T, y_T)(x_T - x_A) + f_y(x_T, y_T)(y_T - y_A)$$

$$(1.4.20)$$

with $f_x \equiv \partial f/\partial x$. We are assuming that $f_x(x, y)$ and $f_y(x, y)$ do not vary greatly for $(x, y)$ between $(x_T, y_T)$ and $(x_A, y_A)$.

*Example*    For $f(x, y) = x^y$, we have $f_x = yx^{y-1}$, $f_y = x^y \log(x)$. Then (1.4.20) yields

$$x_T^{y_T} - x_A^{y_A} \doteq \epsilon \cdot y_T x_T^{y_T-1} + \eta[\log(x_T)] x_T^{y_T}$$

$$\text{Rel}(x_A^{y_A}) \doteq y_T[\text{Rel}(x_A) + \text{Rel}(y_A)\log(x_T)] \quad (1.4.21)$$

The relative error in $x_A^{y_A}$ may be large, even though $\text{Rel}(x_A)$ and $\text{Rel}(y_A)$ are small. As a further illustration, take $y_T = y_A = 500$, $x_T = 1.2$, $x_A = 1.2001$. Then $x_T^{y_T} = 3.89604 \times 10^{39}$, $x_A^{y_A} = 4.06179 \times 10^{39}$, $\text{Rel}(x_A^{y_A}) = -.0425$. Compare this with $\text{Rel}(x_A) = 8.3 \times 10^{-5}$.

**Error in data**  If the input data to an algorithm contain only $r$ digits of accuracy, then it is sometimes suggested that only $r$-digit arithmetic should be used in any calculations involving these data. This is nonsense. It is certainly true that the limited accuracy of the data will affect the eventual results of the algorithmic calculations, giving answers that are in error. Nonetheless, there is no reason to make matters worse by using $r$-digit arithmetic with correspondingly sized rounding errors. Instead one should use a higher precision arithmetic, to avoid any further degradation in the accuracy of results of the algorithm. This will lead to arithmetic rounding errors that are less significant than the error in the data, helping to preserve the accuracy associated with the data.

## 1.5   Errors in Summation

Many numerical methods, especially in linear algebra, involve summations. In this section, we look at various aspects of summation, particularly as carried out in floating-point arithmetic.

Consider the computation of the sum

$$S = \sum_{j=1}^{m} x_j \tag{1.5.1}$$

with $x_1, \ldots, x_m$ floating-point numbers. Define

$$S_2 = \text{fl}(x_1 + x_2) = (x_1 + x_2)(1 + \epsilon_2) \tag{1.5.2}$$

where we have made use of (1.4.2) and (1.2.10). Define recursively

$$S_{r+1} = \text{fl}(S_r + x_{r+1}) \qquad r = 2, \ldots, m-1$$

Then

$$S_{r+1} = (S_r + x_{r+1})(1 + \epsilon_{r+1}) \tag{1.5.3}$$

The quantities $\epsilon_2, \ldots, \epsilon_m$ satisfy (1.2.8) or (1.2.9), depending on whether chopping or rounding is used.

Expanding the first few sums, we obtain the following:

$$S_2 - (x_1 + x_2) = \epsilon_2(x_1 + x_2)$$

$$S_3 - (x_1 + x_2 + x_3) = (x_1 + x_2)\epsilon_2 + (x_1 + x_2)(1 + \epsilon_2)\epsilon_3 + x_3\epsilon_3$$

$$\doteq (x_1 + x_2)\epsilon_2 + (x_1 + x_2 + x_3)\epsilon_3$$

$$S_4 - (x_1 + x_2 + x_3 + x_4) \doteq (x_1 + x_2)\epsilon_2 + (x_1 + x_2 + x_3)\epsilon_3$$

$$+ (x_1 + x_2 + x_3 + x_4)\epsilon_4$$

**Table 1.3   Calculating $S$ on a machine using chopping**

| $n$ | True | SL | Error | LS | Error |
|-----|------|------|-------|------|-------|
| 10 | 2.929 | 2.928 | .001 | 2.927 | .002 |
| 25 | 3.816 | 3.813 | .003 | 3.806 | .010 |
| 50 | 4.499 | 4.491 | .008 | 4.479 | .020 |
| 100 | 5.187 | 5.170 | .017 | 5.142 | .045 |
| 200 | 5.878 | 5.841 | .037 | 5.786 | .092 |
| 500 | 6.793 | 6.692 | .101 | 6.569 | .224 |
| 1000 | 7.486 | 7.284 | .202 | 7.069 | .417 |

**Table 1.4   Calculating $S$ on a machine using rounding**

| $n$ | True | SL | Error | LS | Error |
|-----|------|------|-------|------|-------|
| 10 | 2.929 | 2.929 | 0.0 | 2.929 | 0.0 |
| 25 | 3.816 | 3.816 | 0.0 | 3.817 | − .001 |
| 50 | 4.499 | 4.500 | − .001 | 4.498 | .001 |
| 100 | 5.187 | 5.187 | 0.0 | 5.187 | 0.0 |
| 200 | 5.878 | 5.878 | 0.0 | 5.876 | .002 |
| 500 | 6.793 | 6.794 | − .001 | 6.783 | .010 |
| 1000 | 7.486 | 7.486 | 0.0 | 7.449 | .037 |

We have neglected cross-product terms $\epsilon_i \epsilon_j$, since they will be of much smaller magnitude. By induction, we obtain

$$S_m - \sum_1^m x_i \doteq (x_1 + x_2)\epsilon_2 + \cdots + (x_1 + x_2 + \cdots + x_m)\epsilon_m$$

$$= x_1(\epsilon_2 + \epsilon_3 + \cdots + \epsilon_m) + x_2(\epsilon_2 + \epsilon_3 + \cdots + \epsilon_m)$$

$$+ x_3(\epsilon_3 + \cdots + \epsilon_m) + \cdots + x_m \epsilon_m \qquad (1.5.4)$$

From this formula we deduce that the best strategy for addition is to add from the smallest to the largest. Of course, counterexamples can be produced, but over a large number of summations, the preceding rule should be best. This is especially true if the numbers $x_i$ are all of one sign so that no cancellation occurs in the calculation of the intermediate sums $x_1 + \cdots + x_m$, $m = 1, \ldots, n$. In this case, if chopping is used, rather than rounding, and if all $x_i > 0$, then there is no cancellation in the sums of the $\epsilon_i$. With the strategy of adding from smallest to largest, we minimize the effect of these chopping errors.

*Example*   Define the terms $x_j$ of the sum $S$ as follows: convert the fraction $1/j$ to a decimal fraction, round it to four significant digits, and let this be $x_j$. To make the errors in the calculation of $S$ more clear, we use four-digit decimal floating-point arithmetic. Tables 1.3 and 1.4 contain the results of four different ways of computing $S$. Adding $S$ from largest to smallest is denoted by LS, and adding from smallest to largest is denoted by $SL$. Table 1.3 uses chopped arithmetic, with

$$-.001 \leq \epsilon_j \leq 0 \tag{1.5.5}$$

and Table 1.4 uses rounded arithmetic, with

$$-.0005 \leq \epsilon_j \leq .0005 \tag{1.5.6}$$

The numbers $\epsilon_j$ refer to (1.5.4), and their bounds come from (1.2.8) and (1.2.9).

In both tables, it is clear that the strategy of adding $S$ from the smallest term to the largest is superior to the summation from the largest term to the smallest. Of much more significance, however, is the far smaller error with rounding as compared to chopping. The difference is much more than the factor of 2 that would come from the relative size of the bounds in (1.5.5) and (1.5.6). We next give an analysis of this.

**A statistical analysis of error propagation**   Consider a general error sum

$$E = \sum_{j=1}^{n} \epsilon_j \tag{1.5.7}$$

of the type that occurs in the summation error (1.5.4). A simple bound is

$$|E| \leq n\delta \tag{1.5.8}$$

where $\delta$ is a bound on $\epsilon_1, \ldots, \epsilon_n$. Then $\delta = .001$ or $.0005$ in the preceding example, depending on whether chopping or rounding is used. This bound (1.5.8) is for the worst possible case in which all the errors $\epsilon_j$ are as large as possible and of the same sign.

When using rounding, the symmetry in sign behavior of the $\epsilon_j$, as shown in (1.2.9), makes a major difference in the size of $E$. In this case, a better model is to assume that the errors $\epsilon_j$ are uniformly distributed random variables in the interval $[-\delta, \delta]$ and that they are independent. Then

$$E = n\left(\frac{1}{n}\sum_{1}^{n}\epsilon_j\right) = n\bar{\epsilon}$$

The sample mean $\bar{\epsilon}$ is a new random variable, having a probability distribution with mean 0 and variance $\delta^2/3n$. To calculate probabilities for statements involving $\bar{\epsilon}$, it is important to note that the probability distribution for $\bar{\epsilon}$ is well-approximated by the normal distribution with the same mean and variance, even for small values such as $n \geq 10$. This follows from the *Central Limit Theorem* of probability theory [e.g., see Hogg and Craig (1978, chap. 5)]. Using the approximating normal distribution, the probability is $\frac{1}{2}$ that

$$|\bar{\epsilon}| \leq .39\delta/\sqrt{n} \qquad |E| \leq .39\delta\sqrt{n}$$

and the probability is .99 that

$$|\bar{\epsilon}| \leq 1.49\delta/\sqrt{n} \qquad |E| \leq 1.49\delta\sqrt{n} \qquad (1.5.9)$$

The result (1.5.9) is a considerable improvement upon (1.5.8) if $n$ is at all large.

This analysis can also be applied to the case of chopping error. But in that case, $-\delta \leq \epsilon_j \leq 0$. The sample mean $\bar{\epsilon}$ now has a mean of $\delta/2$, while retaining the same variance of $\delta^2/3n$. Thus there is a probability of .99 that

$$\left(\frac{n}{2} - 1.49\sqrt{n}\right)\delta \leq E \leq \left(\frac{n}{2} + 1.49\sqrt{n}\right)\delta \qquad (1.5.10)$$

For large $n$, this ensures that $E$ will approximately equal $n\delta/2$, which is much larger than (1.5.9) for the case of rounding errors.

When these results, (1.5.9) and (1.5.10), are applied to the general summation error (1.5.4), we see the likely reason for the significantly different error behavior of chopping and rounding in Tables 1.3 and 1.4. In general, rounded arithmetic is almost always to be preferred to chopped arithmetic.

Although statistical analyses give more realistic bounds, they are usually much more difficult to compute. As a more sophisticated example, see Henrici (1962, pp. 41–59) for a statistical analysis of the error in the numerical solution of differential equations. An example is given in Table 6.3 of Chapter 6 of the present textbook.

**Inner products**    Given two vectors $x$, $y \in \mathbf{R}^m$, we call

$$x^Ty = \sum_{j=1}^{m} x_j y_j \qquad (1.5.11)$$

the *inner product* of $x$ and $y$. (The notation $x^T$ denotes the matrix transpose of $x$.) Properties of the inner product are examined in Chapter 7, but we note here that

$$\|x\|_2 = \sqrt{x^Tx} = \sqrt{\sum_{j=1}^{m} x_j^2} \qquad (1.5.12)$$

$$|x^Ty| \leq \|x\|_2\|y\|_2 \qquad (1.5.13)$$

The latter inequality is called the *Cauchy–Schwarz inequality*, and it is proved in a more general setting in Chapter 4. Sums of the form (1.5.11) occur commonly in linear algebra problems (for example, matrix multiplication). We now consider the numerical computation of such sums.

Assume $x_i$ and $y_i$, $i = 1,\ldots, m$, are floating-point numbers. Define

$$S_1 = \text{fl}(x_1 y_1)$$

$$S_{k+1} = \text{fl}(S_k + \text{fl}(x_k y_k)) \qquad k = 1,2,\ldots, m - 1 \qquad (1.5.14)$$

Then as before, using (1.2.10),

$$S_1 = x_1 y_1 (1 + \epsilon_1)$$

$$S_2 = [S_1 + x_2 y_2 (1 + \epsilon_2)](1 + \eta_2)$$

$$\vdots$$

$$S_m = [S_{m-1} + x_m y_m (1 + \epsilon_m)](1 + \eta_m)$$

with the terms $\epsilon_j$, $\eta_j$ satisfying (1.2.8) or (1.2.9), depending on whether chopping or rounding, respectively, is used. Combining and rearranging the preceding formulas, we obtain

$$S_m = \sum_{j=1}^{m} x_j y_j (1 + \delta_j) \tag{1.5.15}$$

with

$$1 + \gamma_j = (1 + \epsilon_j)(1 + \eta_j)(1 + \eta_{j+1}) \cdots (1 + \eta_m) \qquad \eta_1 = 0$$

$$\doteq 1 + \epsilon_j + \eta_j + \eta_{j+1} + \cdots + \eta_m \tag{1.5.16}$$

The last approximation is based on ignoring the products of the small terms $\eta_i \eta_k, \epsilon_i \eta_k$. This brings us back to the same kind of analysis as was done earlier for the sum (1.5.1). The statistical error analysis following (1.5.7) is also valid. For a rigorous bound, it can be shown that if $m\delta < .01$, then

$$|\gamma_j| \le 1.01(m + 1 - j)\delta \qquad j = 1, \ldots, m, \tag{1.5.17}$$

where $\delta$ is the unit round given in (1.2.12) [see Forsythe and Moler (1967, p. 92)]. Applying this to (1.5.15) and using (1.5.13),

$$|S - S_m| \le \sum_{j=1}^{m} |x_j y_j \gamma_j|$$

$$\le 1.01m \cdot \delta \|x\|_2 \|y\|_2 \tag{1.5.18}$$

This says nothing about the relative error, since $x^T y$ can be zero even though all $x_i$ and $y_i$ are nonzero.

These results say that the absolute error in $S_m \doteq x^T y$ does not increase very rapidly, especially if true rounding is used and we consider the earlier statistical analysis of (1.5.7). Nonetheless, it is often possible to easily and inexpensively reduce this error a great deal further, and this is usually very important in linear algebra problems.

Calculate each product $x_j y_j$ in a higher precision arithmetic, and carry out the summation in this higher precision arithmetic. When the complete sum has been computed, then round or chop the result back to the original arithmetic precision.

For example, when $x_i$ and $y_i$ are in single precision, then compute the products and sums in double precision. [On most computers, single and double precision are fairly close in running time; although some computers do not implement double precision in their hardware, but only in their software, which is slower.] The resulting sum $S_m$ will satisfy

$$S - S_m \doteq \delta S \qquad (1.5.19)$$

a considerable improvement on (1.5.18) or (1.5.15). This can be used in parts of a single precision calculation, significantly improving the accuracy without having to do the entire calculation in double precision. For linear algebra problems, this may halve the storage requirements as compared to that needed for an entirely double precision computation.

## 1.6  Stability in Numerical Analysis

A number of mathematical problems have solutions that are quite sensitive to small computational errors, for example rounding errors. To deal with this phenomenon, we introduce the concepts of *stability* and *condition number*. The condition number of a problem will be closely related to the maximum accuracy that can be attained in the solution when using finite-length numbers and computer arithmetic. These concepts will then be extended to the numerical methods that are used to calculate the solution. Generally we will want to use numerical methods that have no greater sensitivity to small errors than was true of the original mathematical problem.

To simplify the presentation, the discussion is limited to problems that have the form of an equation

$$F(x, y) = 0 \qquad (1.6.1)$$

The variable $x$ is the unknown being sought, and the variable $y$ is data on which the solution depends. This equation may represent many different kinds of problems. For example, (1) $F$ may be a real valued function of the real variable $x$, and $y$ may be a vector of coefficients present in the definition of $F$; or (2) the equation may be an integral or differential equation, with $x$ an unknown function and $y$ a given function or given boundary values.

We say that the problem (1.6.1) is *stable* if the solution $x$ depends in a continuous way on the variable $y$. This means that if $\{y_n\}$ is a sequence of values approaching $y$ in some sense, then the associated solution values $\{x_n\}$ must also approach $x$ in some way. Equivalently, if we make ever smaller changes in $y$, these must lead to correspondingly smaller changes in $x$. The sense in which the changes are small will depend on the norm being used to measure the sizes of the vectors $x$ and $y$; there are many possible choices, varying with the problem. Stable problems are also called *well-posed problems*, and we will use the two terms interchangeably. If a problem is not stable, it is called *unstable* or *ill-posed*.

***Example* (a)**    Consider the solution of

$$ax^2 + bx + c = 0 \qquad a \neq 0$$

Any solution $x$ is a complex number. For the data in this case, we use $y = (a, b, c)$, the vector of coefficients. It should be clear from the quadratic formula

$$x = \frac{-b \pm \sqrt{b^2 - 4ac}}{2a}$$

that the two solutions for $x$ will vary in a continuous way with the data $y = (a, b, c)$.

**(b)**    Consider the integral equation problem

$$\int_0^1 \frac{.75x(t)\, dt}{1.25 - \cos(2\pi(s + t))} = y(s) \qquad 0 \leq s \leq 1 \qquad (1.6.2)$$

This is an unstable problem. There are perturbations $\delta_n(s) = y_n(s) - y(s)$ for which

$$\underset{0 \leq s \leq 1}{\text{Max}} |\delta_n(s)| \to 0 \qquad \text{as} \quad n \to \infty \qquad (1.6.3)$$

and the corresponding solutions $x_n(s)$ satisfy

$$\underset{0 \leq s \leq 1}{\text{Max}} |x_n(s) - x(s)| = 1 \qquad \text{all} \quad n \geq 1 \qquad (1.6.4)$$

Specifically, define $y_n(s) = y(s) + \delta_n(s)$

$$\delta_n(s) = \frac{1}{2^n} \cos(2n\pi s) \qquad 0 \leq s \leq 1 \quad n \geq 1$$

Then it can be shown that

$$x_n(s) - x(s) = \cos(2n\pi s)$$

thus proving (1.6.4).

If a problem (1.6.1) is unstable, then there are serious difficulties in attempting to solve it. It is usually not possible to solve such problems without first attempting to understand more about the properties of the solution, usually by returning to the context in which the mathematical problem was formulated. This is currently a very active area of research in applied mathematics and numerical analysis [see, for example, Tikhonov and Arsenin (1977) and Wahba (1980)].

For practical purposes there are many problems that are stable in the previously given sense, but that are still very troublesome as far as numerical computations are concerned. To deal with this difficulty, we introduce a measure of stability called a *condition number*. It shows that practical stability represents a continuum of problems, some better behaved than others.

The condition number attempts to measure the worst possible effect on the solution $x$ of (1.6.1) when the variable $y$ is perturbed by a small amount. Let $\delta y$

be a perturbation of $y$, and let $x + \delta x$ be the solution of the perturbed equation

$$F(x + \delta x, y + \delta y) = 0 \tag{1.6.5}$$

Define

$$K(x) = \underset{\delta y}{\text{Supremum}} \frac{\|\delta x\| / \|x\|}{\|\delta y\| / \|y\|} \tag{1.6.6}$$

We have used the notation $\| \cdot \|$ to denote a measure of size. Recall the definitions (1.1.16)–(1.1.18) for vectors from $\mathbf{R}^n$ and $C[a, b]$. The example (1.6.2) used the norm (1.1.18) for measuring the perturbations in both $x$ and $y$. Commonly $x$ and $y$ may be different kinds of variables, and then different norms are appropriate. The supremum in (1.6.6) is taken over all small perturbations $\delta y$ for which the perturbed problem (1.6.5) will still make sense. Problems that are unstable lead to $K(x) = \infty$.

The number $K(x)$ is called the condition number for (1.6.1). It is a measure of the sensitivity of the solution $x$ to small changes in the data $y$. If $K(x)$ is quite large, then there exists small relative changes $\delta y$ in $y$ that lead to large relative changes $\delta x$ in $x$. But if $K(x)$ is small, say $K(x) \leq 10$, then small relative changes in $y$ always lead to correspondingly small relative changes in $x$. Since numerical calculations almost always involve a variety of small computational errors, we do not want problems with a large condition number. Such problems are called *ill-conditioned*, and they are generally very hard to solve accurately.

*Example* Consider solving

$$x - a^y = 0 \qquad a > 0 \tag{1.6.7}$$

Perturbing $y$ by $\delta y$, we have

$$\frac{\delta x}{x} = \frac{a^{y + \delta y} - a^y}{a^y} = a^{\delta y} - 1$$

For the condition number for (1.6.7),

$$K(x = \underset{\delta y}{\text{Supremum}} \left| \frac{\delta x / x}{\delta y / y} \right| = \underset{\delta y}{\text{Supremum}} \left| y \left( \frac{a^{\delta y} - 1}{\delta y} \right) \right|.$$

Restricting $\delta y$ to be small, we have

$$K(x) \doteq |y \cdot \ln(a)| \tag{1.6.8}$$

Regardless of how we compute $x$ in (1.6.7), if $K(x)$ is large, then small relative changes in $y$ will lead to much larger relative changes in $x$. If $K(x) = 10^4$ and if the value of $y$ being used has relative error $10^{-7}$ due to using finite-length computer arithmetic and rounding, then it is likely that the resulting value of $x$ will have relative error of about $10^{-3}$. This is a large drop in accuracy, and there is little way to avoid it except perhaps by doing all computations in longer precision computer arithmetic, provided $y$ can then be obtained with greater accuracy.

***Example***    Consider the $n \times n$ nonsingular matrix

$$Y = \begin{bmatrix} 1 & \dfrac{1}{2} & \dfrac{1}{3} & \cdots & \dfrac{1}{n} \\[2mm] \dfrac{1}{2} & \dfrac{1}{3} & \dfrac{1}{4} & \cdots & \dfrac{1}{n+1} \\[2mm] \vdots & & & & \vdots \\[2mm] \dfrac{1}{n} & \dfrac{1}{n+1} & & \cdots & \dfrac{1}{2n-1} \end{bmatrix} \qquad (1.6.9)$$

which is called the *Hilbert matrix*. The problem of calculating the inverse of $Y$, or equivalently of solving $YX = I$ with $I$ the identity matrix, is a well-posed problem. The solution $X$ can be obtained in a finite number of steps using only simple arithmetic operations. But the problem of calculating $X$ is increasingly ill-conditioned as $n$ increases.

The ill-conditioning of the numerical inversion of $Y$ will be shown in a practical setting. Let $\hat{Y}$ denote the result of entering the matrix $Y$ into an IBM 370 computer and storing the matrix entries using single precision floating-point format. The fractional elements of $Y$ will be expanded in the hexadecimal (base 16) number system and then chopped after six hexadecimal digits (about seven decimal digits). Since most of the entries in $Y$ do not have finite hexadecimal expansions, there will be a relative error of about $10^{-6}$ in each such element of $\hat{Y}$.

Using higher precision arithmetic, we can calculate the exact value of $\hat{Y}^{-1}$. The inverse $Y^{-1}$ is known analytically, and we can compare it with $\hat{Y}^{-1}$. For $n = 6$, some of the elements of $\hat{Y}^{-1}$ differ from the corresponding elements in $Y^{-1}$ in the first nonzero digit. For example, the entries in row 6, column 2 are

$$(Y^{-1})_{6,2} = 83160.00 \qquad (\hat{Y}^{-1})_{6,2} = 73866.34$$

This makes the calculation of $Y^{-1}$ an ill-conditioned problem, and it becomes increasingly so as $n$ increases. The condition number in (1.62) will be at least $10^6$ as a reflection of the poor accuracy in $\hat{Y}^{-1}$ compared with $Y^{-1}$. Lest this be thought of as an odd pathological example that could not occur in practice, this particular example occurs when doing least squares approximation theory (e.g., see Section 4.3). The general area of ill-conditioned problems for linear systems and matrix inverses is considered in greater detail in Chapter 8.

**Stability of numerical algorithms**    A numerical method for solving a mathematical problem is considered stable if the sensitivity of the numerical answer to the data is no greater than in the original mathematical problem. We will make this more precise, again using (1.6.1) as a model for the problem. A numerical method for solving (1.6.1) will generally result in a sequence of approximate problems

$$F_n(x_n, y_n) = 0 \tag{1.6.10}$$

depending on some parameter, say $n$. The data $y_n$ are to approach $y$ as $n \to \infty$; the function values $F_n(z, w)$ are to approach $F(z, w)$ as $n \to \infty$, for all $(z, w)$ near $(x, y)$; and hopefully the resulting approximate solutions $x_n$ will approach $x$ as $n \to \infty$. For example, (1.6.1) may represent a differential equation initial value problem, and (1.6.10) may present a sequence of finite-difference approximations depending on $h = 1/n$, as in and following (1.3.5). Another case would be where $n$ represents the number of digits being used in the calculations, and we may be solving $F(x, y) = 0$ as exactly as possible within this finite precision arithmetic.

For each of the problems (1.6.10) we can define a condition number $K_n(x_n)$, just as in (1.6.6). Using these condition numbers, define

$$\hat{K}(x) = \underset{n \to \infty}{\text{Limit}} \, \underset{k \geq n}{\text{Supremum}} \, K_k(x_k) \tag{1.6.11}$$

We say *the numerical method is stable* if $\hat{K}(x)$ is of about the same magnitude as $K(x)$ from (1.6.6), for example, if

$$\hat{K}(x) \leq 2K(x)$$

If this is true, then the sensitivity of (1.6.10) to changes in the data is about the same as that of the original problem (1.6.1).

Some problems and numerical methods may not fit easily within the framework of (1.6.1), (1.6.6), (1.6.10), and (1.6.11), but there is a general idea of stable problems and condition numbers that can be introduced and given similar meaning. The main use of these concepts in this text is in (1) rootfinding for polynomial equations, (2) solving differential equations, and (3) problems in numerical linear algebra. Generally there is little problem with unstable numerical methods in this text. The main difficulty will be the solution of ill-conditioned problems.

*Example*    Consider the evaluation of a Bessel function,

$$x = J_m(y) = \left(\frac{1}{2}y\right)^m \sum_{k=0}^{\infty} \frac{\left(-\frac{1}{4}y^2\right)^k}{k!(m+k)!} \qquad m \geq 0 \tag{1.6.12}$$

This series converges very rapidly, and the evaluation of $x$ is easily shown to be a well-conditioned problem in its dependence on $y$.

Now consider the evaluation of $J_m(y)$ using the triple recursion relation

$$J_{m+1}(y) = \frac{2m}{y}J_m(y) - J_{m-1}(y) \qquad m \geq 1 \tag{1.6.13}$$

assuming $J_0(y)$ and $J_1(y)$ are known. We now demonstrate numerically that this

**Table 1.5**  **Computed values of $J_m(1)$**

| $m$ | Computed $J_m(1)$ | True $J_m(1)$ |
|---|---|---|
| 0 | .7651976866 | .7651976866 |
| 1 | .4400505857 | .4400505857 |
| 2 | .1149034848 | .1149034849 |
| 3 | .195633535E $-$ 1 | .1956335398E $-$ 1 |
| 4 | .24766362E $-$ 2 | .2476638964E $-$ 2 |
| 5 | .2497361E $-$ 3 | .2497577302E $-$ 3 |
| 6 | .207248E $-$ 4 | .2093833800E $-$ 4 |
| 7 | $-$.10385E $-$ 5 | .1502325817E $-$ 5 |

is an *unstable numerical method* for evaluating $J_m(y)$, for even moderately large $m$. We take $y = 1$, so that (1.6.13) becomes

$$J_{m+1}(1) = 2mJ_m(1) - J_{m-1}(1) \qquad m \geq 1 \qquad (1.6.14)$$

We use values for $J_0(1)$ and $J_1(1)$ that are accurate to 10 significant digits. The subsequent values $J_m(1)$ are calculated from (1.6.4) using exact arithmetic, and the results are given in Table 1.5. The true values are given for comparison, and they show the rapid divergence of the approximate values from the true values. The only errors introduced were the rounding errors in $J_0(1)$ and $J_1(1)$, and they cause an increasingly large perturbation in $J_m(1)$ as $m$ increases.

The use of three-term recursion relations

$$f_{m+1}(x) = a_m(x)f_m(x) - b_m(x)f_{m-1}(x) \qquad m \geq 1$$

is a common tool in applied mathematics and numerical analysis. But as previously shown, they can lead to unstable numerical methods. For a general analysis of triple recursion relations, see Gautschi (1967). In the case of (1.6.13) and (1.6.14), large loss of significance errors are occurring.

## Discussion of the Literature

A knowledge of computer arithmetic is important for programmers who are concerned with numerical accuracy, particularly when writing programs that are to be widely used. Also, when writing programs to be run on various computers, their different floating-point characteristics must be taken into account. Classic treatments of floating-point arithmetic are given in Knuth (1981, chap. 4) and Sterbenz (1974).

The topic of error propagation, especially that due to rounding/chopping error, has been difficult to treat in a precise, but useful manner. There are some important early papers, but the current approaches to the subject are due in large part to the late J. H. Wilkinson. Much of his work was in numerical linear algebra, but he made important contributions to many areas of numerical analysis. For a general introduction to his techniques for analyzing the propagation of errors, with applications to several important problems, see Wilkinson (1963), (1965), (1984).

Another approach to the control of error is called *interval analysis*. With it, we carry along an interval $[x_l, x_u]$ in our calculations, rather than a single number $x_A$, and the numbers $x_l$ and $x_u$ are guaranteed to bound the true value $x_T$. The difficulty with this approach is that the size of $x_u - x_l$ is generally much larger than $|x_T - x_A|$, mainly because the possible cancellation of errors of opposite sign is often not considered when computing $x_l$ and $x_u$. For an introduction to this area, showing how to improve on these conservative bounds in particular cases, see Moore (1966). More recently, this area and that of computer arithmetic have been combined to give a general theoretical framework allowing the development of algorithms with rigorous error bounds. As examples of this area, see the texts of Alefeld and Herzberger (1983), and Kulisch and Miranker (1981), the symposium proceedings of Alefeld and Grigorieff (1980), and the survey in Moore (1979).

The topic of ill-posed problems was just touched on in Section 1.6, but it has been of increasing interest in recent years. There are many problems of indirect physical measurement that lead to ill-posed problems, and in this form they are called *inverse problems*. The book by Lavrentiev (1967) gives a general introduction, although it discusses mainly (1) analytic continuation of analytic functions of a complex variable, and (2) inverse problems for differential equations. One of the major numerical tools used in dealing with ill-posed problems is called *regularization*, and an extensive development of it is given in Tikhonov and Arsenin (1977). As important examples of the more current literature on numerical methods for ill-posed problems, see Groetsch (1984) and Wahba (1980).

Two new types of computers have appeared in the last 10 to 15 years, and they are now having an increasingly important impact on numerical analysis. These are *microcomputers* and *supercomputers*. Everyone is aware of microcomputers; scientists and engineers are using them for an increasing amount of their numerical calculations. Initially the arithmetic design of microcomputers was quite poor, with some having errors in their basic arithmetic operations. Recently, an excellent new standard has been produced for arithmetic on microcomputers, and with it one can write high-quality and efficient numerical programs. This standard, the *IEEE Standard for Binary Floating-Point Arithmetic*, is described in IEEE (1985). Implementation on the major families of microprocessors are becoming available; for example, see Palmer and Morse (1984).

The name supercomputer refers to a variety of machine designs, all having in common the ability to do very high-speed numerical computations, say greater than 20 million floating-point operations per second. This area is developing and changing very rapidly, and so we can only give a few references to hint at the effect of these machines on the design of numerical algorithms. Hockney and Jesshope (1981), and Quinn (1987) are general texts on the architecture of supercomputers and the design of numerical algorithms on them; Parter (1984) is a symposium proceedings giving some applications of supercomputers in a variety of physical problems; and Ortega and Voigt (1985) discuss supercomputers as they are being used to solve partial differential equations. These machines will become increasingly important in all areas of computing, and their architectures are likely to affect smaller mainframe computers of the type that are now more widely used.

*Symbolic mathematics* is a rapidly growing area, and with it one can do analytic rather than numerical mathematics, for example, finding antiderivatives exactly when possible. This area has not significantly affected numerical analysis to date, but that appears to be changing. In many situations, symbolic mathematics are used for part of a calculation, with numerical methods used for the remainder of the calculation. One of the most sophisticated of the programming languages for carrying out symbolic mathematics is MACSYMA, which is described in Rand (1984). For a survey and historical account of programming languages for this area, see Van Hulzen and Calmet (1983).

We conclude by discussing the area of *mathematical software*. This area deals with the implementation of numerical algorithms as computer programs, with careful attention given to questions of accuracy, efficiency, flexibility, portability, and other characteristics that improve the usefulness of the programs. A major journal of the area is the *ACM Transactions on Mathematical Software*. For an extensive survey of the area, including the most important program libraries that have been developed in recent years, see Cowell (1984). In the appendix to this book, we give a further discussion of some currently available numerical analysis computer program packages.

# Bibliography

Alefeld, G., and R. Grigorieff, eds. (1980). *Fundamentals of Numerical Computation (Computer-oriented Numerical Analysis)*. Computing Supplementum 2, Springer-Verlag, Vienna.

Alefeld, G., and J. Herzberger (1983). *Introduction to Interval Computations*. Academic Press, New York.

Bender, E. (1978). *An Introduction to Mathematical Modelling*. Wiley, New York.

Cowell, W., ed. (1984). *Sources and Development of Mathematical Software*. Prentice-Hall, Englewood Cliffs, N.J.

Fadeeva, V. (1959). *Computational Methods of Linear Algebra*. Dover, New York.

Forsythe, G. (1969). What is a satisfactory quadratic equation solver? In B. Dejon and P. Henrici (eds.), *Constructive Aspects of the Fundamental Theorem of Algebra*, pp. 53–61, Wiley, New York.

Forsythe, G., and C. Moler (1967). *Computer Solution of Linear Algebraic Systems*. Prentice-Hall, Englewood Cliffs, N.J.

Forsythe, G., M. Malcolm, and C. Moler (1977). *Computer Methods for Mathematical Computations*. Prentice-Hall, Englewood Cliffs, N.J.

Gautschi, W. (1967). Computational aspects of three term recurrence relations. *SIAM Rev.*, **9**, 24–82.

Groetsch, C. (1984). *The Theory of Tikhonov Regularization for Fredholm Equations of the First Kind*. Pitman, Boston.

Henrici, P. (1962). *Discrete Variable Methods in Ordinary Differential Equations*. Wiley, New York.

Hockney, R., and C. Jesshope (1981). *Parallel Computers: Architecture, Programming, and Algorithms*. Adam Hilger, Bristol, England.

Hogg, R., and A. Craig (1978). *Introduction to Mathematical Statistics*, 4th ed. Macmillan, New York.

Institute of Electrical and Electronics Engineers (1985). Proposed standard for binary floating-point arithmetic. (IEEE Task P754), Draft 10.0. IEEE Society, New York.

Knuth, D. (1981). *The Art of Computer Programming*, vol. 2, *Seminumerical Algorithms*, 2nd ed. Addison-Wesley, Reading, Mass.

Kulisch, U., and W. Miranker (1981). *Computer Arithmetic in Theory and Practice*. Academic Press, New York.

Lavrentiev, M. (1967). *Some Improperly Posed Problems of Mathematical Physics*. Springer-Verlag, New York.

Lin, C., and L. Segal (1974). *Mathematics Applied to Deterministic Problems in the Natural Sciences*. Macmillan, New York.

Maki, D., and M. Thompson (1973). *Mathematical Models and Applications*. Prentice-Hall, Englewood Cliffs, N.J.

Moore, R. (1966). *Interval Analysis*. Prentice-Hall, Englewood Cliffs, N.J.

Moore, R. (1979). *Methods and Applications of Interval Analysis*. Society for Industrial and Applied Mathematics, Philadelphia.

Ortega, J., and R. Voigt (1985). Solution of partial differential equations on vector and parallel computers. *SIAM Rev.*, **27**, 149–240.

Parter, S., ed. (1984). *Large Scale Scientific Computation*. Academic Press, New York.

Palmer, J., and S. Morse (1984). *The 8087 Primer*. Wiley, New York.

Quinn, M. (1987). *Designing Efficient Algorithms for Parallel Computers*. McGraw-Hill, New York.

Rand, R. (1984). *Computer Algebra in Applied Mathematics: An Introduction to MACSYMA*. Pitman, Boston.

Rubinow, S. (1975). *Introduction to Mathematical Biology*. Wiley, New York.

Sterbenz, P. (1974). *Floating-Point Computation*. Prentice-Hall, Englewood Cliffs, N.J.

Tikhonov, A., and V. Arsenin (1977). *Solutions of Ill-posed Problems*. Wiley, New York.

Van Hulzen, J., and J. Calmet (1983). Computer algebra systems. In B. Buchberger, G. Collins, R. Loos (eds.), *Computer Algebra: Symbolic and Algebraic Computation*, 2nd ed. Springer-Verlag, Vienna.

Wahba, G. (1980). Ill-posed problems: Numerical and statistical methods for mildly, moderately, and severely ill-posed problems with noisy data. Tech.

Rep. #595, Statistics Department, Univ. of Wisconsin, Madison. Prepared for the *Proc. Int. Symp. on Ill-Posed Problems*, Newark, Del., 1979.

Wahba, G. (1984). Cross-validated spline methods for the estimation of multi-variate functions from data on functionals, in *Statistics: An Appraisal*, H. A. David and H. T. David (eds.), Iowa State Univ. Press, Ames, pp. 205–235.

Wilkinson, J. (1963). *Rounding Errors in Algebraic Processes*. Prentice-Hall, Englewood Cliffs, N.J.

Wilkinson, J. (1965). *The Algebraic Eigenvalue Problem*. Oxford, England.

Wilkinson, J. (1984). The perfidious polynomial. In Golub (ed.), *Studies in Numerical Analysis*, pp. 1–28, Mathematical Association of America.

## Problems

1.  (a)  Assume $f(x)$ is continuous on $a \le x \le b$, and consider the average

$$S = \frac{1}{n} \sum_{j=1}^{n} f(x_j)$$

with all points $x_j$ in the interval $[a, b]$. Show that

$$S = f(\zeta)$$

for some $\zeta$ in $[a, b]$. *Hint:* Use the intermediate value theorem and consider the range of values of $f(x)$ and thus of $S$.

(b)  Generalize part (a) to the sum

$$S = \sum_{j=1}^{n} w_j f(x_j)$$

with all $x_j$ in $[a, b]$ and all $w_j \ge 0$.

2.  Derive the following inequalities:

(a)  $|e^x - e^z| \le |x - z|$      for all $x, z \le 0$.

(b)  $|x - z| \le |\tan(x) - \tan(z)|$      $-\dfrac{\pi}{2} < x, z < \dfrac{\pi}{2}$

(c)  $py^{p-1}(x - y) \le x^p - y^p \le px^{p-1}(x - y)$      $0 \le y \le x$,   $p \ge 1$.

3.  (a)  Bound the error in the approximation

$$\sin(x) \doteq x \qquad |x| \le \delta$$

**(b)**    For small values of $\delta$, measure the relative error in $\sin(x) \doteq x$ by using

$$\frac{\sin(x) - x}{\sin(x)} \doteq \frac{\sin(x) - x}{x} \qquad x \neq 0$$

Bound this modified relative error for $|x| \leq \delta$. Choose $\delta$ to make this error less than .01, corresponding to a 1 percent error.

**4.**    Assuming $g \in C[a, b]$, show

$$\int_0^h x^2(h - x)^2 g(x)\, dx = \frac{h^5}{30} g(\zeta) \quad \text{some} \quad \zeta \text{ in } [a, b]$$

**5.**    Construct a Taylor series for the following functions, and bound the error when truncating after $n$ terms.

**(a)**    $\dfrac{1}{x} \displaystyle\int_0^x e^{-t^2}\, dt$

**(b)**    $\sin^{-1}(x) \qquad |x| < 1$

**(c)**    $\dfrac{1}{x} \displaystyle\int_0^x \dfrac{\tan^{-1} t\, dt}{t}$

**(d)**    $\cos(x) + \sin(x)$

**(e)**    $\log(1 - x) \qquad -1 < x < 1$

**(f)**    $\log\left[\dfrac{1 + x}{1 - x}\right] \qquad -1 < x < 1$

**6.    (a)**    Using the result (1.1.11), we can show

$$\frac{\pi}{4} = \tan^{-1}(1) = \sum_{j=0}^{\infty} \frac{(-1)^{2j+1}}{2j + 1}$$

and we can obtain $\pi$ by multiplying by 4. Why is this not a practical way to compute $\pi$?

**(b)**    Using a Taylor polynomial approximation, give a practical way to evaluate $\pi$.

**7.**    Using Taylor's theorem for functions of two variables, find linear and quadratic approximations to the following functions $f(x, y)$ for small values of $x$ and $y$. Give the tangent plane function $z = p(x, y)$ whose graph is tangent to that of $z = f(x, y)$ at $(0, 0, f(0, 0))$.

**(a)**    $\sqrt{1 + 2x - y}$

**(b)**    $\dfrac{1 + x}{1 + y}$

**(c)**    $x \cdot \cos(x - y)$

**(d)**    $\cos(x + \sqrt{\pi^2 + y})$

**8.**  Consider the second-order divided difference $f[x_0, x_1, x_2]$ defined in (1.1.13).

**(a)**  Prove the property (1.1.15), that the order of the arguments $x_0, x_1, x_2$ does not affect the value of the divided difference.

**(b)**  Prove formula (1.1.14),

$$f[x_0, x_1, x_2] = \tfrac{1}{2}f''(\zeta)$$

for some $\zeta$ between the minimum and maximum of $x_0$, $x_1$, and $x_2$. *Hint:* From part (a), there is no loss of generality in assuming $x_0 < x_1 < x_2$. Use Taylor's theorem to reduce $f[x_0, x_1, x_2]$, expanding about $x_1$; and then use the intermediate value theorem to simplify the error term.

**(c)**  Assuming $f(x)$ is twice continuously differentiable, show that $f[x_0, x_1, x_2]$ can be extended continuously to the case where some or all of the points $x_0$, $x_1$, and $x_2$ are coincident. For example, show

$$f[x_0, x_1, x_0] \equiv \operatorname*{Limit}_{x_2 \to x_0} f[x_0, x_1, x_2]$$

exists and compute a formula for it.

**9.**  **(a)**  Show that the vector norms (1.1.16) and (1.1.18) satisfy the three general properties of norms that are listed following (1.1.18).

**(b)**  Show $\|x\|_2$ in (1.1.17) is a vector norm, restricting yourself to the $n = 2$ case.

**(c)**  Show that the matrix norm (1.1.19) satisfies (1.1.20) and (1.1.21). For simplicity, consider only matrices of order $2 \times 2$.

**10.**  Convert the following numbers to their decimal equivalents.

**(a)**  $(10101.101)_2$    **(b)**  $(2A3.FF)_{16}$    **(c)**  $(.101010101\ldots)_2$

**(d)**  $(.AAAA\ldots)_{16}$    **(e)**  $(.00011001100110011\ldots)_2$

**(f)**  $(11\ldots1)_2$ with the parentheses enclosing $n$ 1s.

**11.**  To convert a positive decimal integer $x$ to its binary equivalent,

$$x = (a_n a_{n-1} \ldots a_1 a_0)_2$$

begin by writing

$$x = a_n \cdot 2^n + a_{n-1} \cdot 2^{n-1} + \cdots + a_1 \cdot 2^1 + a_0 \cdot 2^0$$

Based on this, use the following algorithm.

(i)    $x_0 := x; \; j := 0$

(ii)   *While* $x_j \neq 0$, *Do* the following:
$\quad a_j :=$ Remainder of integer divide $x_j/2$
$\quad x_{j+1} :=$ Quotient of integer divide $x_j/2$
$\quad j := j + 1$
*End While*

The language of the algorithm should be self-explanatory. Apply it to convert the following integers to their binary equivalents.

(a)   49   (b)   127   (c)   129

12.   To convert a positive decimal fraction $x < 1$ to its binary equivalent

$$x = (.a_1 a_2 a_3 \dots)_2$$

begin by writing

$$x = a_1 \cdot 2^{-1} + a_2 \cdot 2^{-2} + a_3 \cdot 2^{-3} + \cdots$$

Based on this, use the following algorithm.

(i)    $x_1 := x; \; j := 1$

(ii)   *While* $x_j \neq 0$, *Do* the following:
$\quad a_j :=$ Integer part of $2 \cdot x_j$
$\quad x_{j+1} :=$ Fractional part of $2 \cdot x_j$
$\quad j := j + 1$
*End While*

Apply this algorithm to convert the following decimal numbers to their binary equivalents.

(a)   .8125   (b)   12.0625   (c)   .1   (d)   .2   (e)   .4

(f)   $\frac{1}{7} = .142857142857\dots$

13.   Generalize Problems 11 and 12 to the conversion of a decimal integer to its hexadecimal equivalent.

14. Predict the output of the following section of Fortran code if it is run on a binary computer that uses chopped arithmetic.

```
        I = 0
        X = 0.0
        H = .1
10      I = I + 1
        X = X + H
        PRINT *, I, X
        IF (X .LT. 1.0) GO TO 10
```

Would the outcome be any different if the statement "X = X + H" was replaced by "X = I * H"?

15. Derive the bounds (1.2.9) for the relative error in the rounded floating-point representation of (1.2.5)–(1.2.6).

16. Derive the upper bound result $M = \beta^t$ given in (1.2.16).

17. (a) Write a program to create an overflow error on your computer. For example, input a number $x$ and repeatedly square it.

    (b) Write a program to experimentally determine the largest allowable floating-point number.

18. (a) A simple model for population growth is

$$\frac{dN}{dt} = kN \qquad t \geq t_0, \quad N(t_0) = N_0$$

with $N(t)$ the population at time $t$ and $k > 0$. Show that this implies a geometric rate of increase in population:

$$N(t + 1) = CN(t) \qquad t \geq t_0$$

Find a formula for $C$.

    (b) A more sophisticated model for population growth is

$$\frac{dN}{dt} = kN[1 - bN] \qquad t \geq t_0, \quad N(t_0) = N_0$$

with $b, k > 0$ and $1 - bN_0 > 0$. Find the solution to this differential equation problem. Compare its solution to that of part (a). Describe the differences in population growth predicted by the two models, for both large and small values of $t$.

**19.**  On your computer, evaluate the two functions

(a)   $f(x) = x^3 - 3x^2 + 3x - 1$

(b)   $f(x) = x^3 + 2x^2 - x - 2$

Evaluate them for a large sampling of values of $x$ around 1, and try to produce the kind of behavior shown in Figure 1.2. Compare the results for the two functions.

**20.**  Write a program to compute experimentally

$$\text{Limit}_{p \to \infty} (x^p + y^p)^{1/p}$$

where $x$ and $y$ are positive numbers. First do the computation in the form just shown. Second, repeat the computation with the idea used in (1.3.8). Run the program for a variety of large and small values of $x$ and $y$, for example, $x = y = 10^{10}$ and $x = y = 10^{-10}$.

**21.**  For the following numbers $x_A$ and $x_T$, how many significant digits are there in $x_A$ with respect to $x_T$?

(a)   $x_A = 451.023,$      $x_T = 451.01$

(b)   $x_A = -.045113,$      $x_T = -.04518$

(c)   $x_A = 23.4213,$      $x_T = 23.4604$

**22.**  Let all of the following numbers be correctly rounded to the number of digits shown: (a) $1.1062 + .947$, (b) $23.46 - 12.753$, (c) $(2.747)(6.83)$, (d) $8.473/.064$. For each calculation, determine the smallest interval in which the result, using true instead of rounded values, must be located.

**23.**  Prove the formula (1.4.5) for $\text{Rel}(x_A/y_A)$.

**24.**  Given the equation $x^2 - 40x + 1 = 0$, find its roots to five significant digits. Use $\sqrt{399} \doteq 19.975$, correctly rounded to five digits.

**25.**  Give exact ways of avoiding loss-of-significance errors in the following computations.

(a)   $\log(x + 1) - \log(x)$      large $x$

(b)   $\sin(x) - \sin(y)$      $x \doteq y$

(c)   $\tan(x) - \tan(y)$      $x \doteq y$

(d)  $\dfrac{1 - \cos(x)}{x^2}$    $x \doteq 0$

(e)  $\sqrt[3]{1 + x} - 1,$    $x \doteq 0$

26.  Use Taylor approximations to avoid the loss-of-significance error in the following computations.

(a)  $f(x) = \dfrac{e^x - e^{-x}}{2x}$

(b)  $f(x) = \dfrac{\log(1 - x) + xe^{x/2}}{x^3}$

In both cases, what is $\underset{x \to 0}{\text{Limit}} f(x)$?

27.  Consider evaluating $\cos(x)$ for large $x$ by using the Taylor approximation (1.1.5),

$$\cos(x) \doteq 1 - \frac{x^2}{2!} + \cdots + (-1)^n \frac{x^{2n}}{(2n)!}$$

To see the difficulty involved in using this approximation, use it to evaluate $\cos(2\pi) = 1$. Determine $n$ so that the Taylor approximation error is less than .0005. Then repeat the type of computation used in (1.4.17) and Table 1.2. How should $\cos(x)$ be evaluated for larger values of $x$?

28.  Suppose you wish to compute the values of (a) $\cos(1.473)$, (b) $\tan^{-1}(2.621)$, (c) $\ln(1.471)$, (d) $e^{2.653}$. In each case, assume you have only a table of values of the function with the argument $x$ given in increments of .01. Choose the table value whose argument is nearest to your given argument. Estimate the resulting error.

29.  Assume that $x_A = .937$ has three significant digits with respect to $x_T$. Bound the relative error in $x_A$. For $f(x) = \sqrt{1 - x}$, bound the error and relative error in $f(x_A)$ with respect to $f(x_T)$.

30.  The numbers given below are correctly rounded to the number of digits shown. Estimate the errors in the function values in terms of the errors in the arguments. Bound the relative errors.

(a)  $\sin[(3.14)(2.685)]$          (b)  $\ln(1.712)$

(c)  $(1.56)^{3.414}$

31.  Write a computer subroutine to form the sum

$$S = \sum_{1}^{n} a_i$$

in three ways: (1) from smallest to largest, (2) from largest to smallest, and (3) in double precision, with a single precision rounding/chopping error at the conclusion of the summation. Use the double precision result to find the error in the two single precision results. Print the results. Also write a main program to create the following series in single precision, and use the subroutine just given to sum the series. [*Hint:* In writing the subroutine, for simplicity assume the terms of the series are arranged from largest to smallest.]

(a) $\displaystyle\sum_{1}^{n}\frac{1}{j}$ (b) $\displaystyle\sum_{1}^{n}\frac{1}{j^2}$ (c) $\displaystyle\sum_{1}^{n}\frac{1}{j^3}$ (d) $\displaystyle\sum_{1}^{n}\frac{(-1)^j}{j}$

32.  Consider the product $a_0 a_1 \ldots a_m$, where $a_0, a_1, \ldots, a_m$ are $m + 1$ numbers stored in a computer that uses $n$ digit base $\beta$ arithmetic. Define $p_1 = \mathrm{fl}(a_0 a_1)$, $p_2 = \mathrm{fl}(a_2 p_1)$, $p_3 = \mathrm{fl}(a_3 p_2), \ldots, p_m = \mathrm{fl}(a_m p_{m-1})$. If we write $p_m = a_0 a_1 \ldots a_m (1 + w)$, determine an estimate for $w$. Assume that $a_i = \mathrm{fl}(a_i)$, $i = 0, 1, \ldots, m$. What is a rigorous bound for $w$? What is a statistical estimate for the size of $w$?

# TWO

# ROOTFINDING FOR NONLINEAR EQUATIONS

Finding one or more roots of an equation

$$f(x) = 0 \tag{2.0.1}$$

is one of the more commonly occurring problems of applied mathematics. In most cases explicit solutions are not available and we must be satisfied with being able to find a root to any specified degree of accuracy. The numerical methods for finding the roots are called *iterative methods*, and they are the main subject of this chapter.

We begin with iterative methods for solving (2.0.1) when $f(x)$ is any continuously differentiable real valued function of a real variable $x$. The iterative methods for this quite general class of equations will require knowledge of one or more initial guesses $x_0$ for the desired root $\alpha$ of $f(x)$. An initial guess $x_0$ can usually be found by using the context in which the problem first arose; otherwise, a simple graph of $y = f(x)$ will often suffice for estimating $x_0$.

A second major problem discussed in this chapter is that of finding one or more roots of a polynomial equation

$$p(x) \equiv a_0 + a_1 x + \cdots + a_n x^n = 0 \qquad a_n \neq 0 \tag{2.0.2}$$

The methods of the first problem are often specialized to deal with (2.0.2), and that will be our approach. But there is a large literature on methods that have been developed especially for polynomial equations, using their special properties in an essential way. These are the most important methods used in creating automatic computer programs for solving (2.0.2), and we will reference some such methods.

The third class of problems to be discussed is the solution of nonlinear systems of equations. These systems are very diverse in form, and the associated numerical analysis is both extensive and sophisticated. We will just touch on this subject, indicating some successful methods that are fairly general in applicability. An adequate development of the subject requires a good knowledge of both theoretical and numerical linear algebra, and these topics are not taken up until Chapters 7 through 9.

The last class of problems discussed in this chapter are optimization problems. In this case, we seek to maximize or minimize a real valued function $f(x_1, \ldots, x_n)$ and to find the point $(x_1, \ldots, x_n)$ at which the optimum is attained. Such

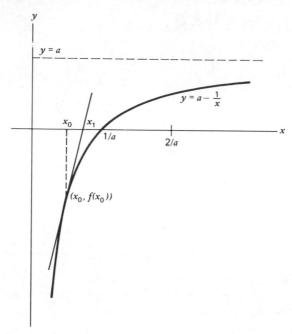

**Figure 2.1**    Iterative solution of $a - (1/x) = 0$.

problems can often be reduced to a system of nonlinear equations, but it is usually better to develop special methods to carry out the optimization directly. The area of optimization is well-developed and extensive. We just briefly introduce and survey the subject.

To illustrate the concept of an iterative method for finding a root of (2.0.1), we begin with an example. Consider solving

$$f(x) \equiv a - \frac{1}{x} = 0 \tag{2.0.3}$$

for a given $a > 0$. This problem has a practical application to computers without a machine divide operation. This was true of some early computers, and some modern-day computers also use the algorithm derived below, as part of their divide operation.

Let $x = 1/a$ be an approximate solution of the equation. At the point $(x_0, f(x_0))$, draw the tangent line to the graph of $y = f(x)$ (see Figure 2.1). Let $x_1$ be the point at which the tangent line intersects the $x$-axis. It should be an improved approximation of the root $\alpha$.

To obtain an equation for $x_1$, match the slopes obtained from the tangent line and the derivative of $f(x)$ at $x_0$

$$f'(x_0) = \frac{f(x_0) - 0}{x_0 - x_1}$$

Substituting from (2.0.3) and manipulating, we obtain

$$x_1 = x_0(2 - ax_0)$$

The general iteration formula is then obtained by repeating the process, with $x_1$ replacing $x_0$, ad infinitum, to get

$$x_{n+1} = x_n(2 - ax_n) \quad n \geq 0 \tag{2.0.4}$$

A form more convenient for theoretical purposes is obtained by introducing the scaled *residual*

$$r_n = 1 - ax_n \tag{2.0.5}$$

Using it,

$$x_{n+1} = x_n(1 + r_n) \quad n \geq 0 \tag{2.0.6}$$

For the error,

$$e_n = \frac{1}{a} - x_n = \frac{r_n}{a} \tag{2.0.7}$$

We will analyze the convergence of this method, its speed, and its dependence on $x_0$. First,

$$r_{n+1} = 1 - ax_{n+1} = 1 - ax_n(1 + r_n) = 1 - (1 - r_n)(1 + r_n)$$

$$r_{n+1} = r_n^2 \tag{2.0.8}$$

Inductively,

$$r_n = r_0^{2^n} \quad n \geq 0 \tag{2.0.9}$$

From (2.0.7), the error $e_n$ converges to zero as $n \to \infty$ if and only if $r_n$ converges to zero. From (2.0.9), $r_n$ converges to zero if and only if $|r_0| < 1$, or equivalently,

$$-1 < 1 - ax_0 < 1$$

$$0 < x_0 < \frac{2}{a} \tag{2.0.10}$$

In order that $x_n$ converge to $1/a$, it is necessary and sufficient that $x_0$ be chosen to satisfy (2.0.10).

To examine the speed of convergence when (2.0.10) is satisfied, we obtain formulas for the error and relative error. For the speed of convergence when (2.0.10) is satisfied,

$$e_{n+1} = \frac{r_{n+1}}{a} = \frac{r_n^2}{a} = \frac{e_n^2 a^2}{a}$$

$$e_{n+1} = ae_n^2 \tag{2.0.11}$$

$$\frac{e_{n+1}}{1/a} = e_n^2 a^2 = \left[\frac{e_n}{1/a}\right]^2$$

$$\text{Rel}(x_{n+1}) = \text{Rel}(x_n)^2 \quad n \geq 0 \tag{2.0.12}$$

The notation $\text{Rel}(x_n)$ denotes the relative error in $x_n$. Based on equation (2.0.11), we say $e_n$ *converges to zero quadratically*. To illustrate how rapidly the error will decrease, suppose that $\text{Rel}(x_0) = 0.1$. Then $\text{Rel}(x_4) = 10^{-16}$. Each iteration doubles the number of significant digits.

This example illustrates the construction of an iterative method for solving an equation; a complete convergence analysis has been given. This analysis included a proof of convergence, a determination of the *interval of convergence* for the choice of $x_0$, and a determination of the speed of convergence. These ideas are examined in more detail in the following sections using more general approaches to solving (2.0.1).

**Definition**    A sequence of iterates $\{x_n | n \geq 0\}$ is said to converge with *order* $p \geq 1$ to a point $\alpha$ if

$$|\alpha - x_{n+1}| \leq c|\alpha - x_n|^p \qquad n \geq 0 \qquad (2.0.13)$$

for some $c > 0$. If $p = 1$, the sequence is said to *converge linearly* to $\alpha$. In that case, we require $c < 1$; the constant $c$ is called the *rate of linear convergence* of $x_n$ to $\alpha$.

Using this definition, the earlier example (2.0.5)–(2.0.6) has order of convergence 2, which is also called *quadratic convergence*. This definition of order is not always a convenient one for some linearly convergent iterative methods. Using induction on (2.0.13) with $p = 1$, we obtain

$$|\alpha - x_n| \leq c^n |\alpha - x_0| \qquad n \geq 0 \qquad (2.0.14)$$

This shows directly the convergence of $x_n$ to $\alpha$. For some iterative methods we can show (2.0.14) directly, whereas (2.0.13) may not be true for any $c < 1$. In such a case, the method will still be said to converge linearly with a rate of $c$.

## 2.1    The Bisection Method

Assume that $f(x)$ is continuous on a given interval $[a, b]$ and that it also satisfies

$$f(a)f(b) < 0 \qquad (2.1.1)$$

Using the intermediate value Theorem 1.1 from Chapter 1, the function $f(x)$ must have at least one root in $[a, b]$. Usually $[a, b]$ is chosen to contain only one root $\alpha$, but the following algorithm for the bisection method will always converge to some root $\alpha$ in $[a, b]$, because of (2.1.1).

**Algorithm**    Bisect $(f, a, b, \text{root}, \epsilon)$

    **1.**   Define $c := (a + b)/2$.

    **2.**   If $b - c \leq \epsilon$, then accept root $:= c$, and exit.

**3.**    If $\text{sign}(f(b)) \cdot \text{sign}(f(c)) \leq 0$, then $a := c$; otherwise, $b := c$.

**4.**    Return to step 1.

The interval $[a, b]$ is halved in size for every pass through the algorithm. Because of step 3, $[a, b]$ will always contain a root of $f(x)$. Since a root $\alpha$ is in $[a, b]$, it must lie within either $[a, c]$ or $[c, b]$; and consequently

$$|c - \alpha| \leq b - c = c - a$$

This is justification for the test in step 2. On completion of the algorithm, $c$ will be an approximation to the root with

$$|c - \alpha| \leq \epsilon$$

***Example***    Find the largest real root $\alpha$ of

$$f(x) \equiv x^6 - x - 1 = 0 \qquad (2.1.2)$$

It is straightforward to show that $1 < \alpha < 2$, and we will use this as our initial interval $[a, b]$. The algorithm *Bisect* was used with $\epsilon = .00005$. The results are shown in Table 2.1. The first two iterates give the initial interval enclosing $\alpha$, and the remaining values $c_n$, $n \geq 1$, denote the successive midpoints found using *Bisect*. The final value $c_{15} = 1.13474$ was accepted as an approximation to $\alpha$ with $|\alpha - c_{15}| \leq .00004$.
    The true solution is

$$\alpha \doteq 1.13472413840152 \qquad (2.1.3)$$

The true error in $c_{15}$ is

$$\alpha - c_{15} = -.000016$$

It is much smaller than the predicted error bound. It might seem as though we could have saved some computation by stopping with an earlier iterate. But there

**Table 2.1    Example of bisection method**

| $n$ | $c_n$ | $f(c_n)$ | $n$ | $c_n$ | $f(c_n)$ |
|---|---|---|---|---|---|
|  | $2.0 = b$ | 61.0 | 8 | 1.13672 | .02062 |
|  | $1.0 = a$ | $-1.0$ | 9 | 1.13477 | .00043 |
| 1 | 1.5 | 8.89063 | 10 | 1.13379 | $-.00960$ |
| 2 | 1.25 | 1.56470 | 11 | 1.13428 | $-.00459$ |
| 3 | 1.125 | $-.09771$ | 12 | 1.13452 | $-.00208$ |
| 4 | 1.1875 | .61665 | 13 | 1.13464 | $-.00083$ |
| 5 | 1.15625 | .23327 | 14 | 1.13470 | $-.00020$ |
| 6 | 1.14063 | .06158 | 15 | 1.13474 | .00016 |
| 7 | 1.13281 | $-.01958$ | | | |

is no way to predict the possibly better accuracy in an earlier iterate, and thus there is no way we can know the iterate is sufficiently accurate. For example, $c_9$ is sufficiently accurate, but there was no way of telling that fact during the computation.

To examine the speed of convergence, let $c_n$ denote the $n$th value of $c$ in the algorithm. Then it is easy to see that

$$\alpha = \lim_{n \to \infty} c_n$$

$$|\alpha - c_n| \le \left[\frac{1}{2}\right]^n (b - a) \tag{2.1.4}$$

where $b - a$ denotes the length of the original interval input into *Bisect*. Using the variant (2.0.14) for defining linear convergence, we say that the bisection method converges linearly with a rate of $\frac{1}{2}$. The actual error may not decrease by a factor of $\frac{1}{2}$ at each step, but the average rate of decrease is $\frac{1}{2}$, based on (2.1.4). The preceding example illustrates the result (2.1.4).

There are several deficiencies in the algorithm *Bisect*. First, it does not take account of the limits of machine precision, as described in Section 1.2 of Chapter 1. A practical program would take account of the unit round on the machine [see (1.2.12)], adjusting the given $\epsilon$ if necessary. The second major problem with *Bisect* is that it converges very slowly when compared with the methods defined in the following sections. The major advantages of the bisection method are: (1) it is guaranteed to converge (provided $f$ is continuous on $[a, b]$ and (2.1.1) is satisfied), and (2) a reasonable error bound is available. Methods that at every step give upper and lower bounds on the root $\alpha$ are called *enclosure methods*. In Section 2.8, we describe an enclosure algorithm that combines the previously stated advantages of the bisection method with the faster convergence of the secant method (described in Section 2.3).

## 2.2   Newton's Method

Assume that an initial estimate $x_0$ is known for the desired root $\alpha$ of $f(x) = 0$. Newton's method will produce a sequence of iterates $\{x_n: n \ge 1\}$, which we hope will converge to $\alpha$. Since $x_0$ is assumed close to $\alpha$, approximate the graph of $y = f(x)$ in the vicinity of its root $\alpha$ by constructing its tangent line at $(x_0, f(x_0))$. Then use the root of this tangent line to approximate $\alpha$; call this new approximation $x_1$. Repeat this process, ad infinitum, to obtain a sequence of iterates $x_n$. As with the example (2.0.3) beginning this chapter, this leads to the iteration formula

$$x_{n+1} = x_n - \frac{f(x_n)}{f'(x_n)} \qquad n \ge 0 \tag{2.2.1}$$

The process is illustrated in Figure 2.2, for the iterates $x_1$ and $x_2$.

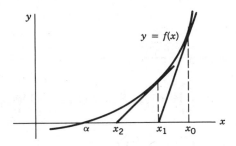

**Figure 2.2**    Newton's method.

Newton's method is the best known procedure for finding the roots of an equation. It has been generalized in many ways for the solution of other, more difficult nonlinear problems, for example, systems of nonlinear equations and nonlinear integral and differential equations. It is not always the best method for a given problem, but its formal simplicity and its great speed often lead it to be the first method that people use in attempting to solve a nonlinear problem.

As another approach to (2.2.1), we use a Taylor series development. Expanding $f(x)$ about $x_n$,

$$f(x) = f(x_n) + (x - x_n)f'(x_n) + \frac{(x - x_n)^2}{2}f''(\xi)$$

with $\xi$ between $x$ and $x_n$. Letting $x = \alpha$ and using $f(\alpha) = 0$, we solve for $\alpha$ to obtain

$$\alpha = x_n - \frac{f(x_n)}{f'(x_n)} - \frac{(\alpha - x_n)^2}{2} \cdot \frac{f''(\xi_n)}{f'(x_n)}$$

with $\xi_n$ between $x_n$ and $\alpha$. We can drop the error term (the last term) to obtain a better approximation to $\alpha$ than $x_n$, and we recognize this approximation as $x_{n+1}$ from (2.2.1). Then

$$\alpha - x_{n+1} = -(\alpha - x_n)^2 \cdot \frac{f''(\xi_n)}{2f'(x_n)} \qquad n \geq 0 \qquad (2.2.2)$$

**Table 2.2 Example of Newton's method**

| $n$ | $x_n$ | $f(x_n)$ | $\alpha - x_n$ | $x_{n+1} - x_n$ |
|---|---|---|---|---|
| 0 | 2.0 | 61.0 | $-8.653\text{E} - 1$ | |
| 1 | 1.680628273 | 19.85 | $-5.459\text{E} - 1$ | $-2.499\text{E} - 1$ |
| 2 | 1.430738989 | 6.147 | $-2.960\text{E} - 1$ | $-1.758\text{E} - 1$ |
| 3 | 1.254970957 | 1.652 | $-1.202\text{E} - 1$ | $-9.343\text{E} - 2$ |
| 4 | 1.161538433 | $2.943\text{E} - 1$ | $-2.681\text{E} - 2$ | $-2.519\text{E} - 2$ |
| 5 | 1.136353274 | $1.683\text{E} - 2$ | $-1.629\text{E} - 3$ | $-1.623\text{E} - 3$ |
| 6 | 1.134730528 | $6.574\text{E} - 5$ | $-6.390\text{E} - 6$ | $-6.390\text{E} - 6$ |
| 7 | 1.134724139 | $1.015\text{E} - 9$ | $-9.870\text{E} - 11$ | $-9.870\text{E} - 11$ |

This formula will be used to show that Newton's method has a quadratic order of convergence, $p = 2$ in (2.0.13).

*Example* We again solve for the largest root of

$$f(x) \equiv x^6 - x - 1 = 0$$

Newton's method (2.2.1) is used, and the results are shown in Table 2.2. The computations were carried out in approximately 16-digit floating-point arithmetic, and the table iterates were rounded from these more accurate computations. The last column, $x_{n+1} - x_n$, is an estimate of $\alpha - x_n$; this is discussed later in the section.

The Newton method converges very rapidly once an iterate is fairly close to the root. This is illustrated in iterates $x_4, x_5, x_6, x_7$. The iterates $x_0, x_1, x_2, x_3$ show the slow initial convergence that is possible with a poor initial guess $x_0$. If the initial guess $x_0 = 1$ had been chosen, then $x_4$ would have been accurate to seven significant digits and $x_5$ to fourteen digits. These results should be compared with those of the bisection method given in Table 2.1. The much greater speed of Newton's method is apparent immediately.

**Convergence analysis** A convergence result will be given, showing the speed of convergence and also an interval from which initial guesses can be chosen.

*Theorem 2.1* Assume $f(x)$, $f'(x)$, and $f''(x)$ are continuous for all $x$ in some neighborhood of $\alpha$, and assume $f(\alpha) = 0$, $f'(\alpha) \neq 0$. Then if $x_0$ is chosen sufficiently close to $\alpha$, the iterates $x_n$, $n \geq 0$, of (2.2.1) will converge to $\alpha$. Moreover,

$$\lim_{n \to \infty} \frac{\alpha - x_{n+1}}{(\alpha - x_n)^2} = -\frac{f''(\alpha)}{2f'(\alpha)} \tag{2.2.3}$$

proving that the iterates have an order of convergence $p = 2$.

***Proof***  Pick a sufficiently small interval $I = [\alpha - \epsilon, \alpha + \epsilon]$ on which $f'(x) \neq 0$ [this exists by continuity of $f'(x)$], and then let

$$M = \frac{\underset{x \in I}{\text{Max}} |f''(x)|}{2 \underset{x \in I}{\text{Min}} |f'(x)|}$$

From (2.2.2),

$$|\alpha - x_1| \leq M|\alpha - x_0|^2$$

$$M|\alpha - x_1| \leq (M|\alpha - x_0|)^2$$

Pick $|\alpha - x_0| \leq \epsilon$ and $M|\alpha - x_0| < 1$. Then $M|\alpha - x_1| < 1$, and $M|\alpha - x_1| \leq M|\alpha - x_0|$, which says $|\alpha - x_1| \leq \epsilon$. We can apply the same argument to $x_1, x_2, \ldots$, inductively, showing that $|\alpha - x_n| \leq \epsilon$ and $M|\alpha - x_n| < 1$ for all $n \geq 1$.

To show convergence, use (2.2.2) to give

$$|\alpha - x_{n+1}| \leq M|\alpha - x_n|^2$$

$$M|\alpha - x_{n+1}| \leq (M|\alpha - x_n|)^2 \tag{2.2.4}$$

and inductively,

$$M|\alpha - x_n| \leq (M|\alpha - x_0|)^{2^n}$$

$$|\alpha - x_n| \leq \frac{1}{M}(M|\alpha - x_0|)^{2^n} \tag{2.2.5}$$

Since $M|\alpha - x_0| < 1$, this shows that $x_n \to \alpha$ as $n \to \infty$.

In formula (2.2.2), the unknown point $\xi_n$ is between $x_n$ and $\alpha$, implying $\xi_n \to \alpha$ as $n \to \infty$. Thus

$$\underset{n \to \infty}{\text{Limit}} \frac{\alpha - x_{n+1}}{(\alpha - x_n)^2} = -\underset{n \to \infty}{\text{Limit}} \frac{f''(\xi_n)}{2f'(x_n)} = \frac{-f''(\alpha)}{2f'(\alpha)} \qquad ∎$$

The error column in Table 2.2 can be used to illustrate (2.2.3). In particular, for that example,

$$-\frac{f''(\alpha)}{2f'(\alpha)} = -2.417 \qquad \frac{\alpha - x_6}{(\alpha - x_5)^2} = -2.41$$

Let $M$ denote the limit on the right side of (2.2.2). Then if $x_n$ is near $\alpha$, (2.2.2) implies

$$M(\alpha - x_{n+1}) \doteq [M(\alpha - x_n)]^2$$

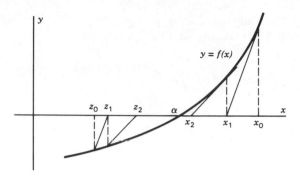

**Figure 2.3**    The Newton–Fourier method.

In order to have convergence of $x_n$ to $\alpha$, this statement says that we should probably have

$$|\alpha - x_0| < \frac{1}{M} \qquad (2.2.6)$$

Thus $M$ is a measure of how close $x_0$ must be chosen to $\alpha$ to ensure convergence to $\alpha$. Some examples with large values of $M$ are given in the problems at the end of the chapter.

Another approach to the error analysis of Newton's method is given by the following construction and theorem. Assume $f(x)$ is twice continuously differentiable on an interval $[a, b]$ containing $\alpha$. Further assume $f(a) < 0$, $f(b) > 0$, and that

$$f'(x) > 0 \qquad f''(x) > 0 \qquad \text{for} \quad a \le x \le b \qquad (2.2.7)$$

Then $f(x)$ is strictly increasing on $[a, b]$, and there is a unique root $\alpha$ in $[a, b]$. Also, $f(x) < 0$ for $a \le x < \alpha$, and $f(x) > 0$ for $\alpha < x \le b$.

Let $x_0 = b$ and define the Newton iterates $x_n$ as in (2.2.1). Next, define a new sequence of iterates by

$$z_{n+1} = z_n - \frac{f(z_n)}{f'(x_n)} \qquad n \ge 0 \qquad (2.2.8)$$

with $z_0 = a$. The resulting iterates are illustrated in Figure 2.3. With the use of $\{z_n\}$, we obtain excellent upper and lower bounds for $\alpha$. The use of (2.2.8) with Newton's method is called the *Newton–Fourier method*.

**Theorem 2.2**    As previously, assume $f(x)$ is twice continuously differentiable on $[a, b]$, $f(a) < 0$, $f(b) > 0$, and condition (2.2.7). Then the iterates $x_n$ are strictly decreasing to $\alpha$, and the iterates $z_n$ are strictly increasing to $\alpha$. Moreover,

$$\underset{n \to \infty}{\text{Limit}} \frac{x_{n+1} - z_{n+1}}{(x_n - z_n)^2} = \frac{f''(\alpha)}{2f'(\alpha)} \qquad (2.2.9)$$

showing that the distance between $x_n$ and $z_n$ decreases quadratically with $n$.

***Proof***   We first show that

$$z_0 < z_1 < \alpha < x_1 < x_0 \qquad (2.2.10)$$

From the definitions (2.2.1) and (2.2.8),

$$x_1 - x_0 = \frac{-f(x_0)}{f'(x_0)} < 0$$

$$z_1 - z_0 = \frac{-f(z_0)}{f'(x_0)} > 0$$

From the error formula (2.2.2),

$$\alpha - x_1 = -(\alpha - x_0)^2 \frac{f''(\xi_0)}{2f'(x_0)} < 0$$

Finally,

$$\alpha - z_1 = \alpha - z_0 + \frac{f(z_0)}{f'(x_0)} = \alpha - z_0 + \frac{f(z_0) - f(\alpha)}{f'(x_0)}$$

$$= \alpha - z_0 - \frac{f'(\zeta_0)(\alpha - z_0)}{f'(x_0)} \qquad \text{some } z_0 < \zeta_0 < \alpha$$

$$= (\alpha - z_0)\left[\frac{f'(x_0) - f'(\zeta_0)}{f'(x_0)}\right] > 0$$

because $f'(x)$ is an increasing function on $[a, b]$. Combining these results proves (2.2.10). This proof can be repeated inductively to prove that

$$z_n < z_{n+1} < \alpha < x_{n+1} < x_n \qquad n \geq 0 \qquad (2.2.11)$$

The sequence $\{x_n\}$ is bounded below by $\alpha$, and thus it has an infimum $\bar{x}$; similarly, the sequence $\{z_n\}$ has a supremum $\bar{z}$:

$$\underset{n \to \infty}{\text{Limit}} \, x_n = \bar{x} \geq \alpha \qquad \underset{n \to \infty}{\text{Limit}} \, z_n = \bar{z} \leq \alpha$$

Taking limits in (2.2.1) and (2.2.8), we obtain

$$\bar{x} = \bar{x} - \frac{f(\bar{x})}{f'(\bar{x})} \qquad \bar{z} = \bar{z} - \frac{f(\bar{z})}{f'(\bar{x})}$$

which leads to $f(\bar{x}) = 0 = f(\bar{z})$. Since $\alpha$ is the unique root of $f(x)$ in $[a, b]$, this proves $\{x_n\}$ and $\{z_n\}$ converge to $\alpha$.

The proof of (2.2.9) is more complicated, and we refer the reader to Ostrowski (1973, p. 70). From Theorem 2.1 and formula (2.2.3), the sequence $\{x_n\}$ converges to $\alpha$ quadratically. The result (2.2.9) shows that

$$|\alpha - x_n| \le |z_n - x_n|$$

is an error bound that decreases quadratically.    ■

The hypotheses of Theorem 2.2 can be reduced to $f(x)$ being twice continuously differentiable in a neighborhood of $\alpha$ and

$$f'(\alpha)f''(\alpha) \ne 0 \qquad (2.2.12)$$

From this, there will be an interval $[a, b]$ about $\alpha$ with $f'(x)$ and $f''(x)$ nonzero on the interval. Then the rootfinding problem $f(x) = 0$ will satisfy (2.2.7) or it can be easily modified to an equivalent problem that will satisfy it. For example, if $f'(\alpha) < 0$, $f''(\alpha) > 0$, then consider the rootfinding problem $g(x) = 0$ with $g(x) \equiv f(-x)$. The root of $g$ will be $-\alpha$, and the conditions in (2.2.7) will be satisfied by $g(x)$ on some interval about $-\alpha$. The numerical illustration of Theorem 2.2 will be left until the Problems section.

**Error estimation**    The preceding procedure gives upper and lower bounds for the root, with the distance $x_n - z_n$ decreasing quadratically. However, in most applications, Newton's method is used alone, without (2.2.8). In that case, we use the following.

Using the mean value theorem,

$$f(x_n) = f(x_n) - f(\alpha) = f'(\xi_n)(x_n - \alpha)$$

$$\alpha - x_n = \frac{f(x_n)}{f'(\xi_n)}$$

with $\xi_n$ between $x_n$ and $\alpha$. If $f'(x)$ is not changing rapidly between $x_n$ and $\alpha$, then we have $f'(\xi_n) \doteq f'(x_n)$, and

$$\alpha - x_n \doteq \frac{-f(x_n)}{f'(x_n)} = x_{n+1} - x_n$$

with the last equality following from the definition of Newton's method. For Newton's method, the standard error estimate is

$$\alpha - x_n \doteq x_{n+1} - x_n \qquad (2.2.13)$$

and this is illustrated in Table 2.2. For relative error, use

$$\frac{\alpha - x_n}{\alpha} \doteq \frac{x_{n+1} - x_n}{x_{n+1}}$$

**The Newton algorithm**    Using the Newton formula (2.2.1) and the error estimate (2.2.13), we give the following algorithm.

***Algorithm***    *Newton* $(f, df, x_0, \epsilon, \text{root}, \text{itmax}, \text{ier})$

1. Remark: *df* is the derivative function $f'(x)$, itmax is the maximum number of iterates to be computed, and ier is an error flag to the user.

2. itnum $:= 1$

3. denom $:= df(x_0)$.

4. If denom $= 0$, then ier $:= 2$ and exit.

5. $x_1 := x_0 - f(x_0)/\text{denom}$

6. If $|x_1 - x_0| \le \epsilon$, then set ier $:= 0$, root $:= x_1$, and exit.

7. If itnum $=$ itmax, set ier $:= 1$ and exit.

8. Otherwise, itnum $:=$ itnum $+ 1$, $x_0 := x_1$, and go to step 3.

As with the earlier algorithm *Bisect*, no account is taken of the limits of the computer arithmetic, although a practical program would need to do such. Also, Newton's method is not guaranteed to converge, and thus a test on the number of iterates (step 7) is necessary.

When Newton's method converges, it generally does so quite rapidly, an advantage over the bisection method. But again, it need not converge. Another source of difficulty in some cases is the necessity of knowing $f'(x)$ explicitly. With some rootfinding problems, this is not possible. The method of the next section remedies this situation, at the cost of a somewhat slower speed of convergence.

## 2.3 The Secant Method

As with Newton's method, the graph of $y = f(x)$ is approximated by a straight line in the vicinity of the root $\alpha$. In this case, assume that $x_0$ and $x_1$ are two initial estimates of the root $\alpha$. Approximate the graph of $y = f(x)$ by the *secant line* determined by $(x_0, f(x_0))$ and $(x_1, f(x_1))$. Let its root be denoted by $x_2$; we hope it will be an improved approximation of $\alpha$. This is illustrated in Figure 2.4.

Using the slope formula with the secant line, we have

$$\frac{f(x_1) - f(x_0)}{x_1 - x_0} = \frac{f(x_1) - 0}{x_1 - x_2}$$

Solving for $x_2$,

$$x_2 = x_1 - f(x_1) \cdot \frac{x_1 - x_0}{f(x_1) - f(x_0)}$$

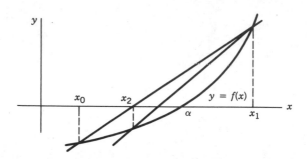

**Figure 2.4**   The secant method.

Using $x_1$ and $x_2$, repeat this process to obtain $x_3$, etc. The general formula based on this is

$$x_{n+1} = x_n - f(x_n) \cdot \frac{x_n - x_{n-1}}{f(x_n) - f(x_{n-1})} \qquad n \geq 1 \qquad (2.3.1)$$

This is the *secant method*. As with Newton's method, it is not guaranteed to converge, but when it does converge, the speed is usually greater than that of the bisection method.

*Example*   Consider again finding the largest root of

$$f(x) \equiv x^6 - x - 1 = 0$$

The secant method (2.3.1) was used, and the iteration continued until the successive differences $x_n - x_{n-1}$ were considered sufficiently small. The numerical results are given in Table 2.3. The calculations were done on a binary machine with approximately 16 decimal digits of accuracy, and the table results are rounded from the computer results.

The convergence is increasingly rapid as $n$ increases. One way of measuring this is to calculate the ratios

$$\frac{\alpha - x_{n+1}}{\alpha - x_n} \qquad n \geq 0$$

For a linear method these are generally constant as $x_n$ converges to $\alpha$. But in this example, these ratios become smaller with increasing $n$. One intuitive explanation is that the straight line connecting $(x_{n-1}, f(x_{n-1}))$ and $(x_n, f(x_n))$ becomes an

**Table 2.3  Example of secant method**

| $n$ | $x_n$ | $f(x_n)$ | $\alpha - x_n$ | $x_n - x_{n-1}$ |
|---|---|---|---|---|
| 0 | 2.0 | 61.0 | 8.65E − 1 | |
| 1 | 1.0 | −1.0 | 1.35E − 1 | 1.61E − 2 |
| 2 | 1.016129032 | − 9.154E − 1 | 1.19E − 1 | 1.74E − 1 |
| 3 | 1.190577769 | 6.575E − 1 | − 5.59E − 2 | − 7.29E − 2 |
| 4 | 1.117655831 | −1.685E − 1 | −1.71E − 2 | 1.49E − 2 |
| 5 | 1.132531550 | − 2.244E − 2 | 2.19E − 3 | 2.29E − 3 |
| 6 | 1.134816808 | 9.536E − 4 | − 9.27E − 5 | − 9.32E − 5 |
| 7 | 1.134723646 | − 5.066E − 6 | 4.92E − 7 | 4.92E − 7 |
| 9 | 1.134724138 | −1.135E − 9 | 1.10E − 10 | 1.10E − 10 |

increasingly accurate approximation to the graph of $y = f(x)$ in the vicinity of $x = \alpha$, and consequently the root $x_{n+1}$ of the straight line is an increasingly improved estimate of $\alpha$. Also note that the iterates $x_n$ move above and below the root $\alpha$ in an apparently random fashion as $n$ increases. An explanation of this will come from the error formula (2.3.3) given below.

**Error analysis**   Multiply both sides of (2.3.1) by $-1$ and then add $\alpha$ to both sides, obtaining

$$\alpha - x_{n+1} = \alpha - x_n + f(x_n) \cdot \frac{x_n - x_{n-1}}{f(x_n) - f(x_{n-1})}$$

The right-hand side can be manipulated algebraically to obtain the formula

$$\alpha - x_{n+1} = -(\alpha - x_{n-1})(\alpha - x_n) \frac{f[x_{n-1}, x_n, \alpha]}{f[x_{n-1}, x_n]} \qquad (2.3.2)$$

The quantities $f[x_{n-1}, x_n]$ and $f[x_{n-1}, x_n, \alpha]$ are first- and second-order Newton divided differences, defined in (1.1.13) of Chapter 1. The reader should check (2.3.2) by substituting from (1.1.13) and then simplifying. Using (1.1.14), formula (2.3.2) becomes

$$\alpha - x_{n+1} = -(\alpha - x_{n-1})(\alpha - x_n) \cdot \frac{f''(\xi_n)}{2f'(\xi_n)} \qquad (2.3.3)$$

with $\xi_n$ between $x_{n-1}$ and $x_n$, and $\zeta_n$ between $x_{n-1}$, $x_n$, and $\alpha$. Using this error formula, we can examine the convergence of the secant method.

***Theorem 2.3***   Assume $f(x)$, $f'(x)$, and $f''(x)$ are continuous for all values of $x$ in some interval containing $\alpha$, and assume $f'(\alpha) \neq 0$. Then if the initial guesses $x_0$ and $x_1$ are chosen sufficiently close to $\alpha$, the iterates $x_n$ of (2.3.1) will converge to $\alpha$. The order of convergence will be $p = (1 + \sqrt{5})/2 \doteq 1.62$.

***Proof***   For the neighborhood $I = [\alpha - \epsilon, \alpha + \epsilon]$ with some $\epsilon > 0$, $f'(x) \neq 0$ everywhere on $I$. Then define

$$M = \frac{\underset{x \in I}{\text{Max}} |f''(x)|}{2 \underset{x \in I}{\text{Min}} |f'(x)|}$$

Then for all $x_0, x_1 \in [\alpha - \epsilon, \alpha + \epsilon]$, using (2.3.3),

$$|e_2| \leq |e_1| \cdot |e_0| M,$$

$$M|e_2| \leq M|e_1| \cdot M|e_0|$$

Further assume that $x_1$ and $x_0$ are so chosen that

$$\delta \equiv \text{Max} \{ M|e_0|, M|e_1| \} < 1 \qquad (2.3.4)$$

Then $M|e_2| < 1$ since

$$M|e_2| \leq \delta^2$$

Also $M|e_2| \leq \delta^2 < \delta$ implies

$$|e_2| < \frac{\delta}{M} = \text{Max} \{ |e_1|, |e_0| \} \leq \epsilon$$

and thus $x_2 \in [\alpha - \epsilon, \alpha + \epsilon]$. We apply this argument inductively to show that $x_n \in [\alpha - \epsilon, \alpha + \epsilon]$ and $M|e_n| \leq \delta$ for $n \geq 2$.
To prove convergence and obtain the order of convergence, continue applying (2.3.3) to get

$$M|e_3| \leq M|e_2| \cdot M|e_1| \leq \delta^2 \cdot \delta = \delta^3$$

$$M|e_4| \leq M|e_3| \cdot M|e_2| \leq \delta^5$$

For

$$M|e_n| \leq \delta^{q_n} \qquad (2.3.5)$$

$$M|e_{n+1}| \leq M|e_n| \cdot M|e_{n-1}| \leq \delta^{q_n + q_{n-1}} = \delta^{q_{n+1}}$$

Thus

$$q_{n+1} = q_n + q_{n-1} \qquad n \geq 1 \qquad (2.3.6)$$

with $q_0 = q_1 = 1$. This is a *Fibonacci sequence* of numbers, and an explicit formula can be given:

$$q_n = \frac{1}{\sqrt{5}} \left[ r_0^{n+1} - r_1^{n+1} \right] \qquad n \geq 0 \qquad (2.3.7)$$

$$r_0 = \frac{1 + \sqrt{5}}{2} \doteq 1.618 \qquad r_1 = \frac{1 - \sqrt{5}}{2} \doteq -.618$$

Thus

$$q_n \doteq \frac{1}{\sqrt{5}} (1.618)^{n+1} \qquad \text{for large} \quad n \qquad (2.3.8)$$

For example $q_5 = 8$, and formula (2.3.8) gives 8.025. Returning to (2.3.5), we obtain the error bound

$$|e_n| \leq \frac{1}{M} \delta^{q_n} \qquad n \geq 0 \qquad (2.3.9)$$

with $q_n$ given by (2.3.7). Since $q_n \to \infty$ as $n \to \infty$, we have $x_n \to \alpha$.

By doing a more careful derivation, we can actually show that the order of convergence is $p = (1 + \sqrt{5})/2$. To simplify the presentation, we instead show that this is the rate at which the bound in (2.3.9) decreases. Let $B_n$ denote the upper bound in (2.3.9). Then

$$\frac{B_{n+1}}{B_n^{r_0}} = \frac{\dfrac{1}{M} \delta^{q_{n+1}}}{\left[ \dfrac{1}{M} \right]^{r_0} \cdot \delta^{r_0 q_n}} = M^{r_0 - 1} \delta^{q_{n+1} - r_0 q_n}$$

$$\leq \delta^{-1} M^{r_0 - 1} \equiv c$$

because $q_{n+1} - r_0 q_n = r_1^{n+1} > -1$. Thus

$$B_{n+1} \leq c B_n^{r_0}$$

which implies an order of convergence $p = r_0 = (1 + \sqrt{5})/2$. A similar result holds for the actual errors $e_n$; moreover,

$$\operatorname*{Limit}_{n \to \infty} \frac{|e_{n+1}|}{|e_n|^{r_0}} = \left| \frac{f''(\alpha)}{2f'(\alpha)} \right|^{(\sqrt{5} - 1)/2} \qquad (2.3.10) \qquad \blacksquare$$

The error formula (2.3.3) can be used to explain the oscillating behavior of the iterates $x_n$ about the root $\alpha$ in the last example. For $x_n$ and $x_{n-1}$ near to $\alpha$, (2.3.3) implies

$$\alpha - x_{n+1} \doteq -(\alpha - x_n)(\alpha - x_{n-1}) \cdot \frac{f''(\alpha)}{2f'(\alpha)} \qquad (2.3.11)$$

The sign of $\alpha - x_{n+1}$ is determined from that of the previous two errors, together with the sign of $f''(\alpha)/f'(\alpha)$.

The condition (2.3.4) gives some information on how close the initial values $x_0$ and $x_1$ should be to $\alpha$ in order to have convergence. If the quantity $M$ is large, or more specifically, if

$$\left| \frac{f''(\alpha)}{2f'(\alpha)} \right|$$

is very large, then $\alpha - x_0$ and $\alpha - x_1$ must be correspondingly smaller. Convergence can occur without (2.3.4), but it is likely to initially be quite haphazard in such a case.

For an error test, use the same error estimate (2.2.13) that was used with Newton's method, namely

$$\alpha - x_n \doteq x_{n+1} - x_n$$

Its use is illustrated in Table 2.3 in the last example. Because the secant method may not converge, programs implementing it should have an upper limit on the number of iterates, as with the algorithm *Newton* in the last section.

A possible problem with the secant method is the calculation of the approximate derivative

$$a_n = \frac{f(x_n) - f(x_{n-1})}{x_n - x_{n-1}} \tag{2.3.12}$$

where the secant method (2.3.1) is then written

$$x_{n+1} = x_n - \frac{f(x_n)}{a_n} \tag{2.3.13}$$

The calculation of $a_n$ involves loss of significance errors, in both the numerator and denominator. Thus it is a less accurate approximation of the derivative of $f$ as $x_n \to \alpha$. Nonetheless, we continue to obtain improvements in the accuracy of $x_n$, until we approach the noise level of $f(x)$ for $x$ near $\alpha$. At that point, $a_n$ may become very different from $f'(\alpha)$, and $x_{n+1}$ can jump rapidly away from the root. For this reason, Dennis and Schnabel (1983, pp. 31–32) recommend the use of (2.3.12) until $x_n - x_{n-1}$ becomes sufficiently small. They then recommend another approximation of $f'(x)$:

$$f'(x_n) \doteq a_n = \frac{f(x_n + h) - f(x_n)}{h},$$

with a fixed $h$. For $h$, they recommend

$$h = \sqrt{\delta} \cdot T_\alpha$$

where $T_\alpha$ is a reasonable nonzero approximation to $\alpha$, say $x_n$, and $\delta$ is the computer's unit round [see (1.2.12)]. They recommend the use of $h$ when $|x_n - x_{n-1}|$ is smaller than $h$. The cost of the secant method in function evaluations will rise slightly, but probably by not more than one or two.

The secant method is well recommended as an efficient and easy-to-use rootfinding procedure for a wide variety of problems. It also has the advantage of not requiring a knowledge of $f'(x)$, unlike Newton's method. In Section 2.8, the secant method will form an important part of another rootfinding algorithm that is guaranteed to converge.

**Comparison of Newton's method and the secant method**   Newton's method and the secant method are closely related. If the approximation

$$f'(x_n) \doteq \frac{f(x_n) - f(x_{n-1})}{x_n - x_{n-1}}$$

is used in the Newton formula (2.2.1), we obtain the secant formula (2.3.1). The conditions for convergence are almost the same [for example, see (2.2.6) and (2.3.4) for conditions on the initial error], and the error formulas are similar [see (2.2.2) and (2.3.3)]. Nonetheless, there are two major differences. Newton's method requires two function evaluations per iterate, that of $f(x_n)$ and $f'(x_n)$, whereas the secant method requires only one function evaluation per iterate, that of $f(x_n)$ [provided the needed function value $f(x_{n-1})$ is retained from the last iteration]. Newton's method is generally more expensive per iteration. On the other hand, Newton's method converges more rapidly [order $p = 2$ vs. the secant method's $p \doteq 1.62$], and consequently it will require fewer iterations to attain a given desired accuracy. An analysis of the effect of these two differences in the secant and Newton methods is given below.

We now consider the expenditure of time necessary to reach a desired root $\alpha$ within a desired tolerance of $\epsilon$. To simplify the analysis, we assume that the initial guesses are quite close to the desired root. Define

$$x_{n+1} = x_n - \frac{f(x_n)}{f'(x_n)} \quad n \geq 0$$

$$\bar{x}_{n+1} = \bar{x}_n - f(\bar{x}_n) \cdot \frac{\bar{x}_n - \bar{x}_{n-1}}{f(\bar{x}_n) - f(\bar{x}_{n-1})} \quad n \geq 1$$

and let $x_0 = \bar{x}_0$. We define $\bar{x}_1$ based on the following convergence formula. From (2.2.3) and (2.3.10), respectively,

$$|\alpha - x_{n+1}| \doteq c|\alpha - x_n|^2 \quad n \geq 0, \quad c = \left| \frac{f''(\alpha)}{2f'(\alpha)} \right|$$

$$|\alpha - \bar{x}_{n+1}| \doteq d|\alpha - \bar{x}_n|^r \quad r = \frac{1 + \sqrt{5}}{2}, \quad d = c^{r-1}$$

Inductively for the error in the Newton iterates,

$$c|\alpha - x_{n+1}| \doteq (c|\alpha - x_n|)^2$$

$$c|\alpha - x_n| \doteq (c|\alpha - x_0|)^{2^n}$$

$$|\alpha - x_n| \doteq \frac{1}{c}(c|\alpha - x_0|)^{2^n} \qquad n \geq 0$$

Similarly for the secant method iterates,

$$|\alpha - \bar{x}_n| \doteq d|\alpha - \bar{x}_{n-1}|^r$$

$$\doteq d^{1 + r + \cdots + r^{n-1}}|\alpha - \bar{x}_0|^{r^n}$$

Using the formula (1.1.9) for a finite geometric series, we obtain

$$d^{1 + r + \cdots + r^{n-1}} = d^{(r^n - 1)/(r-1)} = c^{r^n - 1}$$

and thus

$$|\alpha - \bar{x}_n| \doteq c^{r^n - 1}|\alpha - \bar{x}_0|^{r^n} = \frac{1}{c}[c|\alpha - x_0|]^{r^n}$$

To satisfy $|\alpha - x_n| \leq \epsilon$, for the Newton iterates, we must have

$$(c|\alpha - x_0|)^{2^n} \leq c\epsilon$$

$$n \geq \frac{K}{\log 2} \qquad K = \log\left[\frac{\log \epsilon c}{\log c|\alpha - x_0|}\right]$$

Let $m$ be the time to evaluate $f(x)$, and let $s \cdot m$ be the time to evaluate $f'(x)$. Then the minimum time to obtain the desired accuracy with Newton's method is

$$T_N = (m + ms)n = \frac{(1 + s)mK}{\log 2} \qquad (2.3.14)$$

For the secant method, a similar calculation shows that $|\alpha - \bar{x}_n| \leq \epsilon$ if

$$n \geq \frac{K}{\log r}$$

Thus the minimum time necessary to obtain the desired accuracy is

$$T_S = mn = \frac{mK}{\log r} \qquad (2.3.15)$$

To compare the times for the secant method and Newton's method, we have

$$\frac{T_S}{T_N} = \frac{\log 2}{(1 + s)\log r}$$

The secant method is faster than the Newton method if the ratio is less than one,

$$\frac{T_S}{T_N} < 1$$

$$s > \frac{\log 2}{\log r} - 1 \doteq .44 \tag{2.3.16}$$

If the time to evaluate $f'(x)$ is more than 44 percent of that necessary to evaluate $f(x)$, then the secant method is more efficient. In practice, many other factors will affect the relative costs of the two methods, so that the .44 factor should be used with caution.

The preceding argument is useful in illustrating that the mathematical speed of convergence is not the complete picture. Total computing time, ease of use of an algorithm, stability, and other factors also have a bearing on the relative desirability of one algorithm over another one.

## 2.4   Muller's Method

Muller's method is useful for obtaining both real and complex roots of a function, and it is reasonably straightforward to implement as a computer program. We derive it, discuss its convergence, and give some numerical examples.

Muller's method is a generalization of the approach that led to the secant method. Given three points $x_0$, $x_1$, $x_2$, a quadratic polynomial is constructed that passes through the three points $(x_i, f(x_i))$, $i = 0, 1, 2$; one of the roots of this polynomial is used as an improved estimate for a root $\alpha$ of $f(x)$.

The quadratic polynomial is given by

$$p(x) = f(x_2) + (x - x_2)f[x_2, x_1] + (x - x_2)(x - x_1)f[x_2, x_1, x_0].$$

$$\tag{2.4.1}$$

The divided differences $f[x_2, x_1]$ and $f[x_2, x_1, x_0]$ were defined in (1.1.13) of Chapter 1. To check that

$$p(x_i) = f(x_i) \qquad i = 0, 1, 2$$

just substitute $x_i$ into (2.4.1) and then reduce the resulting expression using (1.1.13). There are other formulas for $p(x)$ given in Chapter 3, but the form shown in (2.4.1) is the most convenient for defining Muller's method. The formula (2.4.1) is called Newton's divided difference form of the interpolating polynomial, and it is developed in general in Section 3.2 of Chapter 3.

To find the zeros of (2.4.1) we first rewrite it in the more convenient form

$$y = f(x_2) + w(x - x_2) + f[x_2, x_1, x_0]](x - x_2)^2$$

$$w = f[x_2, x_1] + (x_2 - x_1)f[x_2, x_1, x_0]$$

$$= f[x_2, x_1] + f[x_2, x_0] - f[x_0, x_1]$$

We want to find the smallest value of $x - x_2$ that satisfies the equation $y = 0$, thus finding the root of (2.4.1) that is closest to $x_2$. The solution is

$$x - x_2 = \frac{-w \pm \sqrt{w^2 - 4f(x_2)f[x_2, x_1, x_0]}}{2f[x_2, x_1, x_0]}$$

with the sign chosen to make the numerator as small as possible. Because of the loss-of-significance errors implicit in this formula, we rationalize the numerator to obtain the new iteration formula

$$x_3 = x_2 - \frac{2f(x_2)}{w \pm \sqrt{w^2 - 4f(x_2)f[x_2, x_1, x_0]}} \tag{2.4.2}$$

with the sign chosen to maximize the magnitude of the denominator.

Repeat (2.4.2) recursively to define a sequence of iterates $\{x_n : n \geq 0\}$. If they converge to a point $\alpha$, and if $f'(\alpha) \neq 0$, then $\alpha$ is a root of $f(x)$. To see this, use (1.1.14) of Chapter 1 and (2.4.2) to give

$$w \to f'(\alpha) \qquad \text{as} \quad n \to \infty$$

$$\alpha = \alpha - \frac{2f(\alpha)}{f'(\alpha) \pm \sqrt{[f'(\alpha)]^2 - 2f(\alpha)f''(\alpha)}}$$

showing that the right-hand fraction must be zero. Since $f'(\alpha) \neq 0$ by assumption, the method of choosing the sign in the denominator implies that the denominator is nonzero. Then the numerator must be zero, showing $f(\alpha) = 0$. The assumption $f'(\alpha) \neq 0$ will say that $\alpha$ is a simple root. (See Section 2.7 for a discussion of simple and multiple roots.)

By an argument similar to that used for the secant method, it can be shown that

$$\underset{n \to \infty}{\text{Limit}} \frac{|\alpha - x_{n+1}|}{|\alpha - x_n|^p} = \left| \frac{f^{(3)}(\alpha)}{6f'(\alpha)} \right|^{(p-1)/2} \qquad p \doteq 1.84 \tag{2.4.3}$$

provided $f(x)$ is three times continuously differentiable in a neighborhood of $\alpha$

and $f'(\alpha) \neq 0$. The order $p$ is the positive root of

$$x^3 - x^2 - x - 1 = 0$$

With the secant method, real choices of $x_0$ and $x_1$ lead to a real value of $x_2$. But with Muller's method, real choices of $x_0, x_1, x_2$ can and do lead to complex roots of $f(x)$. This is an important aspect of Muller's method, being one reason it is used.

The following examples were computed using a commercial program that gives an automatic implementation of Muller's method. With no initial guesses given, it found the roots of $f(x)$ in roughly increasing order. After approximations $z_1, \ldots, z_r$ had been found as roots, the function

$$g(x) = \frac{f(x)}{(x - z_1) \cdots (x - z_r)} \qquad (2.4.4)$$

was used in finding the remaining roots of $f(x)$. [For a discussion of the errors in this use of $g(x)$, see Peters and Wilkinson (1971)]. In order that an approximate root $z$ be acceptable to the program, it had to satisfy one of the following two conditions (specified by the user):

1.  $|f(x)| \leq 10^{-10}$
2.  $z$ has eight significant digits of accuracy.

In Tables 2.4 and 2.5, the roots are given in the order in which they were found. The column IT gives the number of iterates that were calculated for each root. The examples are all for $f(x)$ a polynomial, but the program was designed for general functions $f(x)$, with $x$ allowed to be complex.

**Table 2.4  Muller's method, example 2**

| IT | Root | | $f$(root) | |
|----|------|------|-----------|------|
| 9  | 1.1572211736E − 1 | | 5.96E − 8 | |
| 10 | 6.1175748452E − 1 | + 9.01E − 20$i$ | −2.98E − 7 | +9.06E − 11$i$ |
| 14 | 2.8337513377E0 | − 5.05E − 17$i$ | 2.55E − 5 | −4.78E − 8$i$ |
| 13 | 4.5992276394E0 | − 5.95E − 15$i$ | 7.13E − 5 | +9.37E − 6$i$ |
| 8  | 1.5126102698E0 | + 2.98E − 16$i$ | 3.34E − 6 | −2.35E − 7$i$ |
| 19 | 1.3006054993E1 | + 9.04E − 18$i$ | 2.32E − 1 | +4.15E − 7$i$ |
| 16 | 9.6213168425E0 | − 4.97E − 17$i$ | −3.66E − 2 | −5.38E − 7$i$ |
| 14 | 1.7116855187E1 | − 8.48E − 17$i$ | −1.68E + 0 | +2.40E − 5$i$ |
| 13 | 2.2151090379E1 | + 9.35E − 18$i$ | 8.61E − 1 | +2.60E − 5$i$ |
| 7  | 6.8445254531E0 | − 3.43E − 28$i$ | −4.49E − 3 | −1.22E − 18$i$ |
| 4  | 2.8487967251E1 | + 5.77E − 25$i$ | −6.34E + 1 | −2.96E − 11$i$ |
| 4  | 3.7099121044E1 | + 2.80E − 24$i$ | 2.12E   3 | +7.72E − 9$i$ |

**Table 2.5    Muller's method, example 3**

| IT | Root | $f$(root) |
|---|---|---|
| 41 | 2.9987526  $-6.98E - 4i$ | $-3.33E - 11 + 6.70E - 11i$ |
| 17 | 2.9997591  $-2.68E - 4i$ | $5.68E - 14 - 6.48E - 14i$ |
| 31 | 3.0003095  $-3.17E - 4i$ | $-3.41E - 13 + 3.22E - 14i$ |
| 10 | 3.0003046  $+3.14E - 4i$ | $3.98E - 13 - 3.83E - 14i$ |
| 6 | $5.97E - 15 + 3.000000000i$ | $4.38E - 11 - 1.19E - 11i$ |
| 3 | $5.97E - 15 - 3.000000000i$ | $4.38E - 11 + 1.19E - 11i$ |

*Example 1.*   $f(x) = x^{20} - 1$.   All 20 roots were found with an accuracy of 10 or more significant digits. In all cases, the approximate root $z$ satisfied $|f(z)| < 10^{-10}$, generally much less. The number of iterates ranged from 1 to 18, with an average of 8.5.

**2.**   $f(x) =$ Laguerre polynomial of degree 12.   The real parts of the roots as shown are correct, rounded to the number of places shown, but the imaginary parts should all be zero. The numerical results are given in Table 2.4. Note that $f(x)$ is quite large for many of the approximate roots.

**3.**

$$f(x) = x^6 - 12x^5 + 63x^4 - 216x^3 + 567x^2 - 972x + 729$$

$$= (x^2 + 9)(x - 3)^4$$

The numerical results are given in Table 2.5. Note the inaccuracy in the first four roots, which is inherent due to the noise in $f(x)$ associated with $\alpha = 3$ being a repeated root. See Section 2.7 for a complete discussion of the problems in calculating repeated roots.

The last two examples demonstrate why two error tests are necessary, and they indicate why the routine requests a maximum on the number of iterations to be allowed per root. The form (2.4.2) of Muller's method is due to Traub (1964, pp. 210–213). For a computational discussion, see Whitley (1968).

## 2.5    A General Theory for One-Point Iteration Methods

We now consider solving an equation $x = g(x)$ for a root $\alpha$ by the iteration

$$x_{n+1} = g(x_n) \qquad n \geq 0 \tag{2.5.1}$$

with $x_0$ an initial guess to $\alpha$. The Newton method fits in this pattern with

$$g(x) \equiv x - \frac{f(x)}{f'(x)} \tag{2.5.2}$$

**Table 2.6   Iteration examples for $x^2 - 3 = 0$**

|  | Case (i) | Case (ii) | Case (iii) |
|---|---|---|---|
| $n$ | $x_n$ | $x_n$ | $x_n$ |
| 0 | 2.0 | 2.0 | 2.0 |
| 1 | 3.0 | 1.5 | 1.75 |
| 2 | 9.0 | 2.0 | 1.732143 |
| 3 | 87.0 | 1.5 | 1.732051 |

Each solution of $x = g(x)$ is called a *fixed point* of $g$. Although we are usually interested in solving an equation $f(x) = 0$, there are many ways this can be reformulated as a fixed-point problem. At this point, we just illustrate this reformulation process with some examples.

***Example***   Consider solving $x^2 - a = 0$ for $a > 0$.

**(i)**   $x = x^2 + x - a$, or more generally, $x = x + c(x^2 - a)$ for some $c \neq 0$

**(ii)**   $x = \dfrac{a}{x}$

**(iii)**   $x = \dfrac{1}{2}\left(x + \dfrac{a}{x}\right)$                                    (2.5.3)

We give a numerical example with $a = 3$, $x_0 = 2$, and $\alpha = \sqrt{3} = 1.732051$. With $x_0 = 2$, the numerical results for (2.5.1) in these cases are given in Table 2.6.

It is natural to ask what makes the various iterative schemes behave in the way they do in this example. We will develop a general theory to explain this behavior and to aid in analyzing new iterative methods.

***Lemma 2.4***   Let $g(x)$ be continuous on the interval $a \leq x \leq b$, and assume that $a \leq g(x) \leq b$ for every $a \leq x \leq b$. (We say $g$ sends $[a, b]$ into $[a, b]$, and denote it by $g([a, b]) \subset [a, b]$.) Then $x = g(x)$ has at least one solution in $[a, b]$.

***Proof***   Consider the continuous function $g(x) - x$. At $x = a$, it is positive, and at $x = b$ it is negative. Thus by the intermediate value theorem, it must have a root in the interval $[a, b]$. In Figure 2.5, the roots are the intersection points of $y = x$ and $y = g(x)$.   ■

***Lemma 2.5***   Let $g(x)$ be continuous on $[a, b]$, and assume $g([a, b]) \subset [a, b]$. Furthermore, assume there is a constant $0 < \lambda < 1$, with

$$|g(x) - g(y)| \leq \lambda |x - y| \qquad \text{for all} \quad x, y \in [a, b]  \quad (2.5.4)$$

Then $x = g(x)$ has a unique solution $\alpha$ in $[a, b]$. Also, the iterates

$$x_n = g(x_{n-1}) \qquad n \geq 1$$

will converge to $\alpha$ for any choice of $x_0$ in $[a, b]$, and

$$|\alpha - x_n| \leq \frac{\lambda^n}{1 - \lambda}|x_1 - x_0| \qquad (2.5.5)$$

***Proof***   Suppose $x - g(x)$ has two solutions $\alpha$ and $\beta$ in $[a, b]$. Then

$$|\alpha - \beta| = |g(\alpha) - g(\beta)| \leq \lambda|\alpha - \beta|$$

$$(1 - \lambda)|\alpha - \beta| \leq 0$$

Since $0 < \lambda < 1$, this implies $\alpha = \beta$. Also we know by the earlier lemma that there is at least one root $\alpha$ in $[a, b]$.

To examine the convergence of the iterates $x_n$, first note that they all remain in $[a, b]$. To see this, note that the result

$$x_n \in [a, b] \qquad \text{implies} \qquad x_{n+1} = g(x_n) \in [a, b]$$

can be used with mathematical induction to prove $x_n \in [a, b]$ for all $n$. For the convergence,

$$|\alpha - x_{n+1}| = |g(\alpha) - g(x_n)| \leq \lambda|\alpha - x_n| \qquad (2.5.6)$$

and by induction,

$$|\alpha - x_n| \leq \lambda^n|\alpha - x_0| \qquad n \geq 0 \qquad (2.5.7)$$

As $n \rightarrow \infty$, $\lambda^n \rightarrow 0$; thus, $x_n \rightarrow \alpha$.

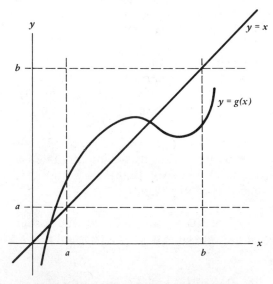

**Figure 2.5**   Example of Lemma 2.4.

To prove the bound (2.5.5), begin with

$$|\alpha - x_0| \le |\alpha - x_1| + |x_1 - x_0| \le \lambda|\alpha - x_0| + |x_1 - x_0|$$

where the last step used (2.5.6). Then solving for $|\alpha - x_0|$, we have

$$|\alpha - x_0| \le \frac{1}{1 - \lambda}|x_1 - x_0| \qquad (2.5.8)$$

Combining this with (2.5.7) will complete the proof.    ∎

The bound (2.5.6) shows that the sequence $\{x_n\}$ is linearly convergent, with the rate of convergence bounded by $\lambda$, based on the definition (2.0.13). Also from the proof, we can devise a possibly more accurate error bound than (2.5.5). Repeating the argument that led to (2.5.8), we obtain

$$|\alpha - x_n| \le \frac{1}{1 - \lambda}|x_{n+1} - x_n|$$

Further, applying (2.5.6) yields the bound

$$|\alpha - x_{n+1}| \le \frac{\lambda}{1 - \lambda}|x_{n+1} - x_n| \qquad (2.5.9)$$

When $\lambda$ is computable, this furnishes a practical bound in most situations. Other error bounds and estimates are discussed in the following section.

If $g(x)$ is differentiable on $[a, b]$, then

$$g(x) - g(y) = g'(\xi)(x - y) \qquad \xi \text{ between } x \text{ and } y$$

for all $x, y \in [a, b]$. Define

$$\lambda = \operatorname*{Max}_{a \le x \le b} |g'(x)|$$

Then

$$|g(x) - g(y)| \le \lambda|x - y| \qquad \text{all} \quad x, y \in [a, b]$$

**Theorem 2.6**    Assume that $g(x)$ is continuously differentiable on $[a, b]$, that $g([a, b]) \subset [a, b]$, and that

$$\lambda = \operatorname*{Max}_{a \le x \le b} |g'(x)| < 1 \qquad (2.5.10)$$

Then

(i)    $x = g(x)$ has a unique solution $\alpha$ in $[a, b]$

(ii)    For any choice of $x_0$ in $[a, b]$, with $x_{n+1} = g(x_n)$, $n \ge 0$,

$$\operatorname*{Limit}_{n \to \infty} x_n = \alpha$$

**(iii)**

$$|\alpha - x_n| \le \lambda^n |\alpha - x_0| \le \frac{\lambda^n}{1 - \lambda}|x_1 - x_0|$$

$$\underset{n \to \infty}{\text{Limit}} \frac{\alpha - x_{n+1}}{\alpha - x_n} = g'(\alpha) \qquad (2.5.11)$$

**Proof**  Every result comes from the preceding lemmas, except for the rate of convergence (2.5.11). For it, use

$$\alpha - x_{n+1} = g(\alpha) - g(x_n) = g'(\xi_n)(\alpha - x_n) \qquad n \ge 0 \quad (2.5.12)$$

with $\xi_n$ an unknown point between $\alpha$ and $x_n$. Since $x_n \to \alpha$, we must have $\xi_n \to \alpha$, and thus

$$\underset{n \to \infty}{\text{Limit}} \frac{\alpha - x_{n+1}}{\alpha - x_n} = \underset{n \to \infty}{\text{Limit}} g'(\xi_n) = g'(\alpha)$$

If $g'(\alpha) \ne 0$, then the sequence $\{x_n\}$ converges to $\alpha$ with order exactly $p = 1$, linear convergence.  ∎

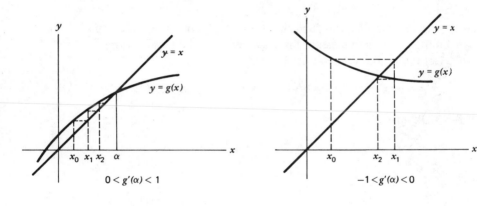

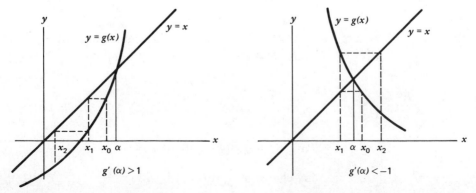

**Figure 2.6**  Examples of convergent and nonconvergent sequences $x_{n+1} = g(x_n)$.

This theorem generalizes to systems of $m$ nonlinear equations in $m$ unknowns. Just regard $x$ as an element of $\mathbf{R}^m$, $g(x)$ as a function from $\mathbf{R}^m$ to $\mathbf{R}^m$, replace the absolute values by vector and matrix norms, and replace $g'(x)$ by the Jacobian matrix for $g(x)$. The assumption $g([a, b]) \subset [a, b]$ must be replaced with a stronger assumption, and care must be exercised in the choice of a region generalizing $[a, b]$. The lemmas generalize, but they are nontrivial to prove. This is discussed further in Section 2.10.

To see the importance of the assumption (2.5.10) on the size of $g'(x)$, suppose $|g'(\alpha)| > 1$. Then if we had a sequence of iterates $x_{n+1} = g(x_n)$ and a root $\alpha = g(\alpha)$, we have (2.5.12). If $x_n$ becomes sufficiently close to $\alpha$, then $|g'(\xi_n)| > 1$ and the error $|\alpha - x_{n+1}|$ will be greater than $|\alpha - x_n|$. Thus convergence is not possible if $|g'(\alpha)| > 1$. We graphically portray the computation of the iterates in four cases (see Figure 2.6).

To simplify the application of the previous theorem, we give the following result.

**Theorem 2.7**   Assume $\alpha$ is a solution of $x = g(x)$, and suppose that $g(x)$ is continuously differentiable in some neighboring interval about $\alpha$ with $|g'(\alpha)| < 1$. Then the results of Theorem 2.6 are still true, provided $x_0$ is chosen sufficiently close to $\alpha$.

**Proof**   Pick a number $\lambda$ satisfying $|g'(\alpha)| < {}^{\lambda} < 1$. Then pick an interval $I = [\alpha - \epsilon, \alpha + \epsilon]$ with

$$\underset{x \in I}{\text{Max}} |g'(x)| \leq \lambda < 1$$

We have $g(I) \subset I$, since $|\alpha - x| \leq \epsilon$ implies

$$|\alpha - g(x)| = |g(\alpha) - g(x)| = |g'(\xi)| \cdot |\alpha - x| \leq \lambda |\alpha - x| \leq \epsilon$$

Now apply the preceding theorem using $[a, b] = [\alpha - \epsilon, \alpha + \epsilon]$.   ∎

**Example**   Referring back to the earlier example in this section, calculate $g'(\alpha)$.

(i)   $g(x) = x^2 + x - 3$   $\qquad g'(\alpha) = g'(\sqrt{3}) = 2\sqrt{3} + 1 > 1$

(ii)   $g(x) = \dfrac{3}{x}$   $\qquad g'(\sqrt{3}) = \dfrac{-3}{(\sqrt{3})^2} = -1$

(iii)   $g(x) = \dfrac{1}{2}(x + \dfrac{3}{x})$   $\qquad g'(x) = \dfrac{1}{2}(1 - \dfrac{3}{x^2})$   $\qquad g'(\sqrt{3}) = 0$

**Example**   For $x = x + c(x^2 - 3)$, pick $c$ to ensure convergence. Since the solution is $\alpha = \sqrt{3}$, and since $g'(x) = 1 + 2cx$, pick $c$ so that

$$-1 < 1 + 2c\sqrt{3} < 1$$

For a good rate of convergence, pick $c$ so that

$$1 + 2c\sqrt{3} \doteq 0$$

**Table 2.7  Numerical example of iteration (2.5.13)**

| $n$ | $x_n$ | $\alpha - x_n$ | Ratio |
|---|---|---|---|
| 0 | 2.0 | $-2.68E - 1$ | |
| 1 | 1.75 | $-1.79E - 2$ | .0668 |
| 2 | 1.7343750 | $-2.32E - 3$ | .130 |
| 3 | 1.7323608 | $-3.10E - 4$ | .134 |
| 4 | 1.7320923 | $-4.15E - 5$ | .134 |
| 5 | 1.7320564 | $-5.56E - 6$ | .134 |
| 6 | 1.7320516 | $-7.45E - 7$ | .134 |
| 7 | 1.7320509 | $-1.00E - 7$ | .134 |

This gives

$$c \doteq \frac{-1}{2\sqrt{3}}$$

Use $c = -\frac{1}{4}$. Then $g'(\sqrt{3}) = 1 - (\sqrt{3}/2) \doteq .134$. This gives the iteration scheme

$$x_{n+1} = x_n - \frac{1}{4}(x_n^2 - 3) \qquad n \geq 0 \qquad (2.5.13)$$

The numerical results are given in Table 2.7. The ratio column gives the values of

$$\frac{\alpha - x_n}{\alpha - x_{n-1}} \qquad n \geq 1$$

The results agree closely with the theoretical value of $g'(\sqrt{3})$.

**Higher order one-point methods**   We complete the development of the theory for one-point iteration methods by considering methods with an order of convergence greater than one, for example, Newtons' method.

*Theorem 2.8*   Assume $\alpha$ is a root of $x = g(x)$, and that $g(x)$ is $p$ times continuously differentiable for all $x$ near to $\alpha$, for some $p \geq 2$. Furthermore, assume

$$g'(\alpha) = \cdots = g^{(p-1)}(\alpha) = 0 \qquad (2.5.14)$$

Then if the initial guess $x_0$ is chosen sufficiently close to $\alpha$, the iteration

$$x_{n+1} = g(x_n) \qquad n \geq 0$$

will have order of convergence $p$, and

$$\underset{n \to \infty}{\text{Limit}} \frac{\alpha - x_{n+1}}{(\alpha - x_n)^p} = (-1)^{p-1} \cdot \frac{g^{(p)}(\alpha)}{p!}$$

***Proof*** Expand $g(x_n)$ about $\alpha$:

$$x_{n+1} = g(x_n) = g(\alpha) + (x_n - \alpha)g'(\alpha) + \cdots + \frac{(x_n - \alpha)^{p-1}}{(p-1)!}g^{(p-1)}(\alpha)$$

$$+ \frac{(x_n - \alpha)^p}{p!}g^{(p)}(\xi_n)$$

for some $\xi_n$ between $x_n$ and $\alpha$. Using (2.5.14) and $\alpha = g(\alpha)$,

$$\alpha - x_{n+1} = -\frac{(x_n - \alpha)^p}{p!}g^{(p)}(\xi_n)$$

Use Theorem 2.7 and $x_n \to \alpha$ to complete the proof.    ∎

The Newton method can be analyzed by this result

$$g(x) = x - \frac{f(x)}{f'(x)} \qquad g'(x) = \frac{f(x)f''(x)}{[f'(x)]^2}$$

$$g'(\alpha) = 0 \qquad g''(\alpha) = \frac{f''(\alpha)}{f'(\alpha)}$$

This and (2.5.14) give the previously obtained convergence result (2.2.3) for Newton's method. For other examples of the application of Theorem 2.8, see the problems at the end of the chapter.

The theory of this section is only for one-point iteration methods, thus eliminating the secant method and Muller's method from consideration. There is a corresponding fixed-point theory for multistep fixed-point methods, which can be found in Traub (1964). We omit it here, principally because only the one-point fixed-point iteration theory will be needed in later chapters.

## 2.6  Aitken Extrapolation for Linearly Convergent Sequences

From (2.5.11) of Theorem 2.6,

$$\underset{x \to \infty}{\text{Limit}} \; \frac{\alpha - x_{n+1}}{\alpha - x_n} = g'(\alpha) \qquad\qquad (2.6.1)$$

for a convergent iteration

$$x_{n+1} = g(x_n) \qquad n \geq 0$$

In this section, we concern ourselves only with the case of linear convergence.

Thus we will assume

$$0 < |g'(\alpha)| < 1 \qquad (2.6.2)$$

We examine estimating the error in the iterates and give a way to accelerate the convergence of $\{x_n\}$.

We begin by considering the ratios

$$\lambda_n = \frac{x_n - x_{n-1}}{x_{n-1} - x_{n-2}} \qquad n \geq 2 \qquad (2.6.3)$$

Claim:

$$\underset{n \to \infty}{\text{Limit}} \, \lambda_n = g'(\alpha) \qquad (2.6.4)$$

To see this, write

$$\lambda_n = \frac{(\alpha - x_{n-1}) - (\alpha - x_n)}{(\alpha - x_{n-2}) - (\alpha - x_{n-1})}$$

Using (2.5.12),

$$\lambda_n = \frac{(\alpha - x_{n-1}) - g'(\xi_{n-1})(\alpha - x_{n-1})}{(\alpha - x_{n-1})/[g'(\xi_{n-2})] - (\alpha - x_{n-1})} = \frac{1 - g'(\xi_{n-1})}{1/[g'(\xi_{n-2})] - 1}$$

$$\underset{n \to \infty}{\text{Limit}} \, \lambda_n = \frac{1 - g'(\alpha)}{1/[g'(\alpha)] - 1} = g'(\alpha)$$

The quantity $\lambda_n$ is computable, and when it converges empirically to a value $\lambda$, we assume $\lambda = g'(\alpha)$.

We use $\lambda_n \doteq g'(\alpha)$ to estimate the error in the iterates $x_n$. Assume

$$\alpha - x_n \doteq \lambda_n(\alpha - x_{n-1})$$

Then

$$\alpha - x_n = (\alpha - x_{n-1}) + (x_{n-1} - x_n)$$

$$\doteq \frac{1}{\lambda_n}(\alpha - x_n) + (x_{n-1} - x_n)$$

$$\alpha - x_n \doteq \frac{\lambda_n}{1 - \lambda_n}(x_n - x_{n-1}) \qquad (2.6.5)$$

This is *Aitken's error formula* for $x_n$, and it is increasingly accurate as $\{\lambda_n\}$ converges to $g'(\alpha)$.

**Table 2.8   Iteration (2.6.6)**

| $n$ | $x_n$ | $x_n - x_{n-1}$ | $\lambda_n$ | $\alpha - x_n$ | Estimate (2.6.5) |
|---|---|---|---|---|---|
| 0 | 6.0000000 | | | 1.55E $-$ 2 | |
| 1 | 6.0005845 | 5.845E $-$ 4 | | 1.49E $-$ 2 | |
| 2 | 6.0011458 | 5.613E $-$ 4 | .9603 | 1.44E $-$ 2 | 1.36E $-$ 2 |
| 3 | 6.0016848 | 5.390E $-$ 4 | .9604 | 1.38E $-$ 2 | 1.31E $-$ 2 |
| 4 | 6.0022026 | 5.178E $-$ 4 | .9606 | 1.33E $-$ 2 | 1.26E $-$ 2 |
| 5 | 6.0027001 | 4.974E $-$ 4 | .9607 | 1.28E $-$ 2 | 1.22E $-$ 2 |
| 6 | 6.0031780 | 4.780E $-$ 4 | .9609 | 1.23E $-$ 2 | 1.17E $-$ 2 |
| 7 | 6.0036374 | 4.593E $-$ 4 | .9610 | 1.18E $-$ 2 | 1.13E $-$ 2 |

*Example*   Consider the iteration

$$x_{n+1} = 6.28 + \sin(x_n) \qquad n \geq 0 \tag{2.6.6}$$

The true root is $\alpha \doteq 6.01550307297$. The results of the iteration are given in Table 2.8, along with the values of $\lambda_n$, $\alpha - x_n$, $x_n - x_{n-1}$, and the error estimate (2.6.5). The values of $\lambda_n$ are converging to

$$g'(\alpha) = \cos(\alpha) \doteq .9644$$

and the estimate (2.6.5) is an accurate indicator of the true error. The size of $g'(\alpha)$ also shows that the iterates will converge very slowly, and in this case, $x_{n+1} - x_n$ is not an accurate indicator of $\alpha - x_n$.

*Aitken's extrapolation formula* is simply (2.6.5), rewritten as an estimate of $\alpha$:

$$\alpha \doteq x_n + \frac{\lambda_n}{1 - \lambda_n}(x_n - x_{n-1}) \tag{2.6.7}$$

We denote this right side by $\hat{x}_n$, for $n \geq 2$. By substituting (2.6.3) into (2.6.7), the formula for $\hat{x}$ can be rewritten as

$$\hat{x}_n = x_n - \frac{(x_n - x_{n-1})^2}{(x_n - x_{n-1}) - (x_{n-1} - x_{n-2})} \qquad n \geq 2 \tag{2.6.8}$$

which is the formula given in many texts.

*Example*   Use the results in Table 2.8 for iteration (2.6.6). With $n = 7$, using either (2.6.7) or (2.6.8),

$$\hat{x}_7 = 6.0149518 \qquad \alpha - \hat{x}_7 = 5.51E - 4$$

Thus the extrapolate $\hat{x}_7$ is significantly more accurate than $x_7$.

We now combine linear iteration and Aitken extrapolation in a simpleminded algorithm.

**Algorithm** Aitken $(g, x_0, \epsilon, \text{root})$

1. Remark: It is assumed that $|g'(\alpha)| < 1$ and that ordinary linear iteration using $x_0$ will converge to $\alpha$.

2. $x_1 := g(x_0)$ $\quad$ $x_2 := g(x_1)$.

3. $\hat{x}_2 := x_2 - \dfrac{(x_2 - x_1)^2}{(x_2 - x_1) - (x_1 - x_0)}$.

4. If $|\hat{x}_2 - x_2| \le \epsilon$, then root $:= \hat{x}_2$ and exit.

5. Set $x_0 := \hat{x}_2$ and go to step 2.

This algorithm will usually converge, provided the assumptions of step 1 are satisfied.

**Example** To illustrate algorithm *Aitken*, we repeat the previous example (2.6.6). The numerical results are given in Table 2.9. The values $x_3$, $x_6$, and $x_9$ are the Aitken extrapolates defined by (2.6.7). The values of $\lambda_n$ are given for only the cases $n = 2$, 5, and 8, since only then do the errors $\alpha - x_n$, $\alpha - x_{n-1}$, and $\alpha - x_{n-2}$ decrease linearly, as is needed for $\lambda_n \doteq g'(\alpha)$.

Extrapolation is often used with slowly convergent linear iteration methods for solving large systems of simultaneous linear equations. The actual methods used are different from that previously described, but they also are based on the general idea of finding the qualitative behavior of the error, as in (2.6.1), and of then using that to produce an improved estimate of the answer. This idea is also pursued in developing numerical methods for integration, solving differential equations, and other mathematical problems.

**Table 2.9** Algorithm *Aitken* applied to (2.6.6)

| $n$ | $x_n$ | $\lambda_n$ | $\alpha - x_n$ |
|---|---|---|---|
| 0 | 6.000000000000 | | 1.55E − 2 |
| 1 | 6.000584501801 | | 1.49E − 2 |
| 2 | 6.001145770761 | .96025 | 1.44E − 2 |
| 3 | 6.014705147543 | | 7.98E − 4 |
| 4 | 6.014733648720 | | 7.69E − 4 |
| 5 | 6.014761128955 | .96418 | 7.42E − 4 |
| 6 | 6.015500802060 | | 2.27E − 6 |
| 7 | 6.015500882935 | | 2.19E − 6 |
| 8 | 6.015500960931 | .96439 | 2.11E − 6 |
| 9 | 6.015503072947 | | 2.05E − 11 |

## 2.7   The Numerical Evaluation of Multiple Roots

We say that the function $f(x)$ has a root $\alpha$ of multiplicity $p > 1$ if

$$f(x) = (x - \alpha)^p h(x) \tag{2.7.1}$$

with $h(\alpha) \neq 0$ and $h(x)$ continuous at $x = \alpha$. We restrict $p$ to be a positive integer, although some of the following is equally valid for nonintegral values. If $h(x)$ is sufficiently differentiable at $x = \alpha$, then (2.7.1) is equivalent to

$$f(\alpha) = f'(\alpha) = \cdots = f^{(p-1)}(\alpha) = 0 \qquad f^{(p)}(\alpha) \neq 0 \tag{2.7.2}$$

When finding a root of any function on a computer, there is always an interval of uncertainty about the root, and this is made worse when the root is multiple. To see this more clearly, consider evaluating the two functions $f_1(x) = x^2 - 3$ and $f_2(x) = 9 + x^2(x^2 - 6)$. Then $\alpha = \sqrt{3}$ has multiplicity one as a root of $f_1$ and multiplicity two as a root of $f_2$. Using four-digit decimal arithmetic, $f_1(x) < 0$ for $x \leq 1.731$, $f_1(1.732) = 0$, and $f_1(x) > 0$ for $x > 1.733$. But $f_2(x) = 0$ for $1.726 \leq x \leq 1.738$, thus limiting the amount of accuracy that can be attained in finding a root of $f_2(x)$. A second example of the effect of noise in the evaluation of a multiple root is illustrated for $f(x) = (x - 1)^3$ in Figures 1.1 and 1.2 of Section 1.3 of Chapter 1. For a final example, consider the following example.

**Example**   Evaluate

$$f(x) = (x - 1.1)^3 (x - 2.1)$$

$$= x^4 - 5.4x^3 + 10.56x^2 - 8.954x + 2.7951 \tag{2.7.3}$$

on an IBM PC microcomputer using double precision arithmetic (in BASIC). The coefficients will not enter exactly because they do not have finite binary expansions (except for the $x^4$ term). The polynomial $f(x)$ was evaluated in its expanded form (2.7.3) and also using the nested multiplication scheme

$$f(x) = 2.7951 + x(-8.954 + x(10.56 + x(-5.4 + x))) \tag{2.7.4}$$

**Table 2.10   Evaluation of $f(x) = (x - 1.1)^3(x - 2.1)$**

| $x$ | $f(x) : (2.7.3)$ | $f(x) : (2.7.4)$ |
|---|---|---|
| 1.099992 | 3.86E − 16 | 5.55E − 16 |
| 1.099994 | 3.86E − 16 | 2.76E − 16 |
| 1.099996 | 2.76E − 16 | 0.0 |
| 1.099998 | − 5.55E − 17 | 1.11E − 16 |
| 1.100000 | 5.55E − 17 | 0.0 |
| 1.100002 | 5.55E − 17 | 5.55E − 17 |
| 1.100004 | − 5.55E − 17 | 0.0 |
| 1.100006 | − 1.67E − 16 | − 1.67E − 16 |
| 1.100008 | − 6.11E − 16 | − 5.00E − 16 |

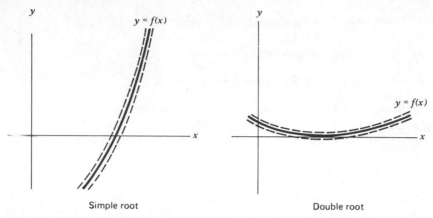

Simple root                                Double root

**Figure 2.7**   Band of uncertainty in evaluation of a function.

The numerical results are given in Table 2.10. Note that the arithmetic being used has about 16 decimal digits in the floating-point representation. Thus, according to the numerical results in the table, no more than 6 digits of accuracy can be expected in calculating the root $\alpha = 1.1$ of $f(x)$. Also note the effect of using the different representations (2.7.3) and (2.7.4).

There is uncertainty in evaluating any function $f(x)$ due to the use of finite precision arithmetic with its resultant rounding or chopping error. This was discussed in Section 3 of Chapter 1, under the name of *noise in function evaluation*. For multiple roots, this leads to considerable uncertainty as to the location of the root. In Figure 2.7, the solid line indicates the graph of $y = f(x)$, and the dotted lines give the region of uncertainty in the evaluation of $f(x)$, which is due to rounding errors and finite-digit arithmetic. The interval of uncertainty in finding the root of $f(x)$ is given by the intersection of the band about the graph of $y = f(x)$ and the $x$-axis. It is clearly greater with the double root than with the simple root, even though the vertical widths of the bands about $y = f(x)$ are the same.

**Newton's method and multiple roots**   Another problem with multiple roots is that the earlier rootfinding methods will not perform as well when the root being sought is multiple. We now investigate this for Newton's method.

We consider Newton's method as a fixed-point method, as in (2.5.2), with $f(x)$ satisfying (2.7.1):

$$x_{n+1} = g(x_n) \qquad g(x) = x - \frac{f(x)}{f'(x)} \qquad x \neq \alpha$$

Before calculating $g'(\alpha)$, we first simplify $g(x)$ using (2.7.1):

$$f'(x) = (x - \alpha)^p h'(x) + p(x - \alpha)^{p-1} h(x)$$

$$g(x) = x - \frac{(x - \alpha)h(x)}{ph(x) + (x - \alpha)h'(x)}$$

**Table 2.11    Newton's method for (2.7.6)**

| $n$ | $x_n$ | $f(x_n)$ | $\alpha - x_n$ | Ratio |
|---|---|---|---|---|
| 0 | 1.22 | $-1.88E - 4$ | $1.00E - 2$ | |
| 1 | 1.2249867374 | $-4.71E - 5$ | $5.01E - 3$ | |
| 2 | 1.2274900222 | $-1.18E - 5$ | $2.51E - 3$ | .502 |
| 3 | 1.2287441705 | $-2.95E - 6$ | $1.26E - 3$ | .501 |
| 4 | 1.2293718746 | $-7.38E - 7$ | $6.28E - 4$ | .501 |
| 5 | 1.2296858846 | $-1.85E - 7$ | $3.14E - 4$ | .500 |
| 18 | 1.2299999621 | $-2.89E - 15$ | $3.80E - 8$ | .505 |
| 19 | 1.2299999823 | $-6.66E - 16$ | $1.77E - 8$ | .525 |
| 20 | 1.2299999924 | $-1.11E - 16$ | $7.58E - 9$ | .496 |
| 21 | 1.2299999963 | $0.0$ | $3.66E - 9$ | .383 |

Differentiating,

$$g'(x) = 1 - \frac{h(x)}{ph(x) + (x - \alpha)h'(x)}$$

$$- (x - \alpha) \cdot \frac{d}{dx}\left[ \frac{h(x)}{ph(x) + (x - \alpha)h'(x)} \right]$$

and

$$g'(\alpha) = 1 - \frac{1}{p} \neq 0 \qquad \text{for} \quad p > 1 \qquad (2.7.5)$$

Thus Newton's method is a linear method with rate of convergence $(p - 1)/p$.

***Example***    Find the smallest root of

$$f(x) = -4.68999 + x(9.1389 + x(-5.56 + x)) \qquad (2.7.6)$$

using Newton's method. The numerical results are shown in Table 2.11. The calculations were done on an IBM PC microcomputer in double precision arithmetic (in BASIC). Only partial results are shown, to indicate the general course of the calculation. The column labeled Ratio is the rate of linear convergence as measured by $\lambda_n$ in (2.6.3).

The Newton method for solving for the root of (2.7.6) is clearly linear in this case, with a linear rate of $g'(x) = \frac{1}{2}$. This is consistent with (2.7.5), since $\alpha = 1.23$ is a root of multiplicity $p = 2$. The final iterates in the table are being affected by the noise in the computer evaluation of $f(x)$. Even though the floating-point representation contains about 16 digits, only about 8 digits of accuracy can be found in this case.

To improve Newton's method, we would like a function $g(x)$ for which $g'(\alpha) = 0$. Based on the derivation of (2.7.5), define

$$g(x) = x - p\frac{f(x)}{f'(x)}$$

Then easily, $g'(\alpha) = 0$; thus,

$$\alpha - x_{n+1} = g(\alpha) - g(x_n)$$

$$= -g'(\alpha)(x_n - \alpha) - \tfrac{1}{2}(x_n - \alpha)^2 g''(\xi_n)$$

with $\xi_n$ between $x_n$ and $\alpha$. Thus

$$\alpha - x_{n+1} = -\tfrac{1}{2}(\alpha - x_n)^2 g''(\xi_n)$$

showing that the method

$$x_{n+1} = x_n - p \cdot \frac{f(x_n)}{f'(x_n)}, \qquad n = 0, 1, 2, \ldots, \qquad (2.7.7)$$

has order of convergence two, the same as the original Newton method for simple roots.

*Example*   Apply (2.7.7) to the preceding example (2.7.6), using $p = 2$ for a double root. The results are given in Table 2.12, using the same computer as previously. The iterates converge rapidly, and then they oscillate around the root. The accuracy (or lack of it) reflects the noise in $f(x)$ and the multiplicity of the root.

Newton's method can be used to determine the multiplicity $p$, as in Table 2.11 combined with (2.7.5), and then (2.7.7) can be used to speed up the convergence. But the inherent uncertainty in the root due to the noise and the multiplicity will remain. This can be removed only by analytically reformulating the rootfinding problem as a new one in which the desired root $\alpha$ is simple. The easiest way to do

**Table 2.12   Modified Newton's method (2.7.7), applied to (2.7.6)**

| $n$ | $x_n$ | $f(x_n)$ | $\alpha - x_n$ |
|---|---|---|---|
| 0 | 1.22 | $-1.88\text{E} - 4$ | $1.00\text{E} - 2$ |
| 1 | 1.2299734748 | $-1.31\text{E} - 9$ | $2.65\text{E} - 5$ |
| 2 | 1.2299999998 | $-1.11\text{E} - 16$ | $1.85\text{E} - 10$ |
| 3 | 1.2300003208 | $-1.92\text{E} - 13$ | $-3.21\text{E} - 7$ |
| 4 | 1.2300000001 | $-1.11\text{E} - 16$ | $-8.54\text{E} - 11$ |
| 5 | 1.2299993046 | $-9.04\text{E} - 13$ | $6.95\text{E} - 7$ |

this is to form the $(p - 1)$st derivative of $f(x)$, and to then solve

$$f^{(p-1)}(x) = 0 \qquad (2.7.8)$$

which will have $\alpha$ as a simple root.

***Example***   The previous example had a root of multiplicity $p = 2$. Then it is a simple root of

$$f'(x) = 3x^2 - 11.12x + 9.1389$$

Using the last iterate in Table 2.11 as an initial guess, and applying Newton's method to finding the root of $f'(x)$ just given, only one iteration was needed to find the value of $\alpha$ to the full precision of the computer.

## 2.8   Brent's Rootfinding Algorithm

We describe an algorithm that combines the advantages of the bisection method and the secant method, while avoiding the disadvantages of each of them. The algorithm is due to Brent (1973, chap. 4), and it is a further development of an earlier algorithm due to Dekker (1969). The algorithm results in a small interval that contains the root. If the function is sufficiently smooth around the desired root $\zeta$, then the order of convergence will be superlinear, as with the secant method.

In describing the algorithm we use the notation of Brent (1973, p. 47). The program is entered with two values, $a_0$ and $b_0$, for which (1) there is at least one root $\zeta$ of $f(x)$ between $a_0$ and $b_0$, and (2) $f(a_0)f(b_0) \leq 0$. The program is also entered with a desired tolerance $t$, from which a stopping tolerance $\delta$ is produced:

$$\delta = t + 2\epsilon|b| \qquad (2.8.1)$$

with $\epsilon$ the unit round for the computer [see (1.2.11) of Chapter 1].

In a typical step of the algorithm, $b$ is the best current estimate of the root $\zeta$, $a$ is the previous value of $b$, and $c$ is a past iterate that has been so chosen that the root $\zeta$ lies between $b$ and $c$ (initially $c = a$). Define $m = \frac{1}{2}(c - b)$.

Stop the algorithm if (1) $f(b) = 0$, or (2) $|m| \leq \delta$. In either case, set the approximate root $\hat{\zeta} = b$. For case (2), because $b$ will usually have been obtained by the secant method, the root $\zeta$ will generally be closer to $b$ than to $c$. Thus usually,

$$|\zeta - \hat{\zeta}| \leq \delta$$

although all that can be guaranteed is that

$$|\zeta - \hat{\zeta}| \leq 2\delta$$

If the error test is not satisfied, set

$$i = b - f(b)\frac{b - a}{f(b) - f(a)} \tag{2.8.2}$$

Then set

$$b'' = \begin{cases} i & \text{if } i \text{ lies between } b \text{ and } b + m = \dfrac{b + c}{2} \\ b + m & \text{otherwise [which is the bisection method]} \end{cases}$$

In the case that $a$, $b$, and $c$ are distinct, the secant method in the definition of $i$ is replaced by an inverse quadratic interpolation method. This results in a very slightly faster convergence for the overall algorithm. Following the determination of $b''$, define

$$b' = \begin{cases} b'' & \text{if } |b - b''| > \delta \\ b + \delta \cdot \text{sign}(m) & \text{if } |b - b''| \le \delta \end{cases} \tag{2.8.3}$$

If you are some distance from the root, then $b' = b''$. With this choice, the method is (1) linear (or quadratic) interpolation, or (2) the bisection method; usually it is (1) for a smooth function $f(x)$. This generally results in a value of $m$ that does not become small. To obtain a small interval containing the root $\zeta$, once we are close to it, we use $b' := b + \delta \cdot \text{sign}(m)$, a step of $\delta$ in the direction of $c$. Because of the way in which a new $c$ is chosen, this will usually result in a new small interval about $\zeta$. Brent makes an additional important, but technical step before choosing a new $b$, usually the $b'$ just given.

Having obtained the new $b'$, we set $b = b'$, $a = $ the old value of $b$. If the sign of $f(b)$, using the new $b$, is the same as with the old $b$, the value of $c$ is unchanged; otherwise, $c$ is set to the old value of $b$, resulting in a smaller interval about $\zeta$. The accuracy of the value of $b$ is now tested, as described earlier.

Brent has taken great care to avoid underflow and overflow difficulties with his method, but the program is somewhat complicated to read as a consequence.

***Example*** Each of the following cases was computed on an IBM PC with 8087 arithmetic coprocessor and single precision arithmetic satisfying the IEEE standard for floating-point arithmetic. The tolerance was $t = 10^{-5}$, and thus

$$\delta = 10^{-5} + 2|b| \times (5.96 \times 10^{-8})$$

$$\doteq 1.01 \times 10^{-5}$$

since the root of $\zeta = 1$ in all cases. The functions were evaluated in the form in which they are given here, and in all cases, the initial interval was $[a, b] = [0, 3]$. The table values for $b$ and $c$ are rounded to seven decimal digits.

**Table 2.13    Example 1 of
Brent's method**

| $b$ | $f(b)$ | $c$ |
|---|---|---|
| 0.0 | $-2.00E + 0$ | 3.0 |
| 0.5 | $-6.25E - 1$ | 3.0 |
| .7139038 | $-3.10E - 1$ | 3.0 |
| .9154507 | $-8.52E - 2$ | 3.0 |
| .9901779 | $-9.82E - 3$ | 3.0 |
| .9998567 | $-1.43E - 4$ | 3.0 |
| .9999999 | $-1.19E - 7$ | 3.0 |
| .9999999 | $-1.19E - 7$ | 1.000010 |

*Case* **(1)**    $f(x) = (x - 1)[1 + (x - 1)^2]$.    The numerical results are given in Table 2.13. This illustrates the necessity of using $b' = b + \delta \cdot \text{sign}(m)$ in order to obtain a small interval enclosing the root $\zeta$.

*Case* **(2)**    $f(x) = x^2 - 1$.    The numerical results are given in Table 2.14.

*Case* **(3)**    $f(x) = -1 + x(3 + x(-3 + x))$.    The root is $\zeta = 1$, of multiplicity three, and it took 50 iterations to converge to the approximate root 1.000001. With the initial values $a = 0$, $b = 3$, the bisection method would use only 19 iterations for the same accuracy. If Brent's method is compared with the bisection method over the class of all continuous functions, then the number of necessary iterates for an error tolerance of $\delta$ is approximately

$$\log_2 \left( \frac{b - a}{\delta} \right) \qquad \text{for bisection method}$$

$$\left[ \log_2 \left( \frac{b - a}{\delta} \right) \right]^2 \qquad \text{for Brent's algorithm}$$

**Table 2.14    Example 2 of
Brent's method**

| $b$ | $f(b)$ | $c$ |
|---|---|---|
| 0.0 | $-1.00$ | 3.0 |
| .3333333 | $-8.89E - 1$ | 3.0 |
| .3333333 | $-8.89E - 1$ | 1.666667 |
| .7777778 | $-4.00E - 1$ | 1.666667 |
| 1.068687 | $1.42E - 1$ | .7777778 |
| .9917336 | $-1.65E - 2$ | 1.068687 |
| .9997244 | $-5.51E - 4$ | 1.068687 |
| 1.000000 | $2.38E - 7$ | .9997244 |
| 1.000000 | $2.38E - 7$ | .9999900 |

**Table 2.15    Example 4 of Brent's method**

| $b$ | $f(b)$ | $c$ |
|---|---|---|
| 0.0 | $-3.68E - 1$ | 3.0 |
| .5731754 | $-1.76E - 3$ | 3.0 |
| .5959331 | $-1.63E - 3$ | 3.0 |
| .6098443 | $-5.47E - 4$ | 3.0 |
| .6136354 | $-4.76E - 4$ | 1.804922 |
| .6389258 | $-1.68E - 4$ | 1.804922 |
| 1.221924 | $3.37E - 10$ | .6389258 |
| 1.221914 | $3.37E - 10$ | .6389258 |
| 1.216585 | $1.20E - 10$ | .6389258 |
| .9277553 | 0.0 | 1.216585 |

Thus there are cases for which bisection is better, as our example shows. But for sufficiently smooth functions with $f'(\alpha) \neq 0$, Brent's algorithm is almost always far faster.

*Case* (4)   $f(x) = (x - 1)\exp[-1/(x - 1)^2]$.   The root $x = 1$ has *infinite multiplicity*, since $f^{(r)}(1) = 0$ for all $r \geq 0$. The numerical results are given in Table 2.15. Note that the routine has found an *exact* root for the machine version of $f(x)$, due to the inherent imprecision in the evaluation of the function; see the preceding section on multiple roots. This root is of course very inaccurate, but this is nothing that the program can treat.

Brent's original program, published in 1973, continues to be very popular and well-used. Nonetheless, improvements and extensions of it continue to be made. For one of them, and for a review of others, see Le (1985).

## 2.9   Roots of Polynomials

We will now consider solving the polynomial equation

$$p(x) \equiv a_0 + a_1 x + \cdots + a_n x^n = 0 \qquad a_n \neq 0 \qquad (2.9.1)$$

This problem arises in many ways, and a large literature has been created to deal with it. Sometimes a particular root is wanted and a good initial guess is known. In that case, the best approach is to modify one of the earlier iterative methods to take advantage of the special form of polynomials. In other cases, little may be known about the location of the roots, and then other methods must be used, of which there are many. In this section, we just give a brief excursion into the area of polynomial rootfinding, without any pretense of completeness. Modifications of the methods of earlier sections will be emphasized, and numerical stability questions will be considered. We begin with a review of some results on bounding or roughly locating the roots of (2.9.1).

**Location theorems**    Because $p(x)$ is a polynomial, many results can be given about the roots of $p(x)$, results that are not true for other functions. The best known of these is the *fundamental theorem of algebra*, which allows us to write $p(x)$ as a unique product (except for order) involving the roots

$$p(x) = a_n(x - z_1) \cdots (x - z_n) \qquad (2.9.2)$$

and $z_1, \ldots, z_n$ are the roots of $p(x)$, repeated according to their multiplicity. We now give some classical results on locating and bounding these roots.

*Descartes's rule of signs* is used to bound the number of positive real roots of $p(x)$, assuming the coefficients $a_0, \ldots, a_n$ are all real.

Let $\nu$ be the number of changes of sign in the coefficients of $p(x)$ in (2.9.1), ignoring the zero terms. Let $k$ denote the number of positive real roots of $p(x)$, counted according to their multiplicity. Then $k \leq \nu$ and $\nu - k$ is even.

A proof of this is given in Henrici (1974, p. 442) and Householder (1970, p. 82).

***Example***    The expression $p(x) = x^6 - x - 1$ has $\nu = 1$ changes of sign. Therefore, $k = 1$; otherwise, $k = 0$, and $\nu - k = 1$ is not an even integer, a contradiction.

Descartes's rule of signs can also be used to bound the number of negative roots of $p(x)$. Apply it to the polynomial

$$q(x) = p(-x)$$

Its positive roots are the negative roots of $p(x)$. Applying this to the last example, $q(x) = x^6 + x - 1$. Again there is one positive real root [of $q(x)$], and thus one negative real root of $p(x)$.

An upper bound for all of the roots of $p(x)$ is given by the following:

$$|z_i| \leq R \equiv 1 + \underset{0 \leq i \leq n-1}{\text{Max}} \left| \frac{a_i}{a_n} \right| \qquad (2.9.3)$$

This is due to Augustin Cauchy, in 1829, and a proof is given in Householder (1970, p. 71). Another such result of Cauchy is based on considering the polynomials

$$|a_n|x^n + |a_{n-1}|x^{n-1} + \cdots + |a_1|x - |a_0| = 0 \qquad (2.9.4)$$

$$|a_n|x^n - |a_{n-1}|x^{n-1} - \cdots - |a_1|x - |a_0| = 0 \qquad (2.9.5)$$

Assume that $a_0 \neq 0$, which is equivalent to assuming $x = 0$ is not a root of $p(x)$. Then by Descartes's law of signs, each of these polynomials has exactly one positive root; call them $\rho_1$ and $\rho_2$, respectively. Then all roots $z_j$ of $p(x)$ satisfy

$$\rho_1 \leq |z_j| \leq \rho_2 \qquad (2.9.6)$$

The proof of the upper bound is given in Henrici (1974, p. 458) and Householder (1970, p. 70). The proof of the lower bound can be based on the following approach, which can also be used in constructing a lower bound for (2.9.3). Consider the polynomial

$$q(x) = x^n p\left(\frac{1}{x}\right) = a_n + a_{n-1}x + \cdots + a_1 x^{n-1} + a_0 x^n \qquad a_0 \neq 0 \quad (2.9.7)$$

Then the roots of $g(x)$ are $1/z$, where $z$ is a root of $p(x)$. If the upper bound result of (2.9.6) is applied to (2.9.7), the lower bound result of (2.9.6) is obtained. We leave this application to be shown in a problem.

Because each of the polynomials (2.9.4), (2.9.5) has a single simple positive root, Newton's method can be easily used to construct $R_1$ and $R_2$. As an initial guess, use the upper bound from (2.9.3) or experiment with smaller positive initial guesses. We leave the illustration of these results to the problems.

There are many other results of the preceding type, and both Henrici (1974, chap. 6) and Householder (1970) wrote excellent treatises on the subject.

**Nested multiplication**    A very efficient way to evaluate the polynomial $p(x)$ given in (2.9.1) is to use nested multiplication:

$$p(x) = a_0 + x\big(a_1 + x(a_2 + \cdots + x(a_{n-1} + a_n x)\cdots)\big) \qquad (2.9.8)$$

With formula (2.9.1), there are $n$ additions and $2n - 1$ multiplications, and with (2.9.8) there are $n$ additions and $n$ multiplications, a considerable saving.

For later work, it is convenient to introduce the following auxiliary coefficients. Let $b_n = a_n$,

$$b_k = a_k + zb_{k+1}, \qquad k = n-1, n-2, \ldots, 0 \qquad (2.9.9)$$

By considering (2.9.8), it is easy to see that

$$p(z) = b_0 \qquad (2.9.10)$$

Introduce the polynomial

$$q(x) = b_1 + b_2 x + \cdots + b_n x^{n-1} \qquad (2.9.11)$$

Then

$$\begin{aligned}
b_0 + (x - z)q(x) &= b_0 + (x - z)\big[b_1 + b_2 x + \cdots + b_n x^{n-1}\big] \\
&= (b_0 - b_1 z) + (b_1 - b_2 z)x + \cdots \\
&\quad + (b_{n-1} - b_n z)x^{n-1} + b_n x^n \\
&= a_0 + a_1 x + \cdots + a_n x^n = p(x) \\
p(x) &= b_0 + (x - z)q(x) \qquad (2.9.12)
\end{aligned}$$

where $q(x)$ is the quotient and $b_0$ the remainder when $p(x)$ is divided by $x - z$. The use of (2.9.9) to evaluate $p(z)$ and to form the quotient polynomial $q(x)$ is also called *Horner's method*.

If $z$ is a root of $p(x)$, then $b_0 = 0$ and $p(x) = (x - z)q(x)$. To find additional roots of $p(x)$, we can restrict our search to the roots of $q(x)$. This reduction process is called *deflation*; it must be used with some caution, a point we will return to later.

**Newton's method**    If we want to apply Newton's method to find a root of $p(x)$, we must be able to evaluate both $p(x)$ and $p'(x)$ at any point $z$. From (2.9.12),

$$p'(x) = (x - z)q'(x) + q(x)$$

$$p'(z) = q(z) \qquad (2.9.13)$$

We use (2.9.10) and (2.9.13) in the following adaption of Newton's method to polynomial rootfinding.

*Algorithm*    Polynew $(a, n, x_0, \epsilon, \text{itmax}, \text{root}, b, \text{ier})$

1.    Remark: $a$ is a vector of coefficients, itmax the maximum number of iterates to be computed, $b$ the vector of coefficients for the deflated polynomial, and ier an error indicator.

2.    itnum $:= 1$

3.    $z := x_0$, $b_n := c := a_n$

4.    For $k = n - 1, \ldots, 1$,   $b_k := a_k + zb_{k+1}$,     $c := b_k + zc$

5.    $b_0 := a_0 + zb_1$

6.    If $c = 0$, ier $:= 2$ and exit.

7.    $x_1 := x_0 - b_0/c$

8.    If $|x_1 - x_0| \le \epsilon$, then ier $:= 0$, root $:= x_1$, and exit.

9.    If itnum $=$ itmax, then ier $:= 1$ and exit.

10.    Otherwise, itnum $:= $ itnum $+ 1$, $x_0 := x_1$, and go to step 3.

**Stability problems**    There are many polynomials in which the roots are quite sensitive to small changes in the coefficients. Some of these are problems with multiple roots, and it is not surprising that these roots are quite sensitive to small changes in the coefficients. But there are many polynomials with only simple roots that appear to be well separated, and for which the roots are still quite sensitive to small perturbations. Formulas are derived below that explain this sensitivity, and numerical examples are also given.

For the theory, introduce

$$p(x) = a_0 + a_1x + \cdots + a_nx^n \qquad a_n \neq 0$$

$$q(x) = b_0 + b_1x + \cdots + b_nx^n \qquad (2.9.14)$$

and define a perturbation of $p(x)$ by

$$p(x; \epsilon) - p(x) + \epsilon q(x) \qquad (2.9.15)$$

Denote the zeros of $p(x; \epsilon)$ by $z_1(\epsilon), \ldots, z_n(\epsilon)$, repeated according to their multiplicity, and let $z_j = z_j(0)$, $i = 1, \ldots, n$, denote the corresponding $n$ zeros of $p(x) = p(x; 0)$. It is well known that the zeros of a polynomial are continuous functions of the coefficients of the polynomial [see, for example, Henrici (1974, p. 281)]. Consequently, $z_j(\epsilon)$ is a continuous function of $\epsilon$. What we want to determine is how rapidly the root $z_j(\epsilon)$ varies with $\epsilon$, for $\epsilon$ near 0.

*Example*  Define

$$p(x; \epsilon) = (x - 1)^3 - \epsilon \qquad p(x) = (x - 1)^3 \qquad \epsilon > 0$$

Then the roots of $p(x)$ are $z_1 = z_1 = z_3 = 1$. The roots of $p(x; \epsilon)$ are

$$z_1(\epsilon) = 1 + \sqrt[3]{\epsilon} \qquad z_2(\epsilon) = 1 + \omega \cdot \sqrt[3]{\epsilon} \qquad z_3 = 1 + \omega^2 \cdot \sqrt[3]{\epsilon}$$

with $\omega = \frac{1}{2}(-1 + i\sqrt{3})$. For all three roots of $p(x; \epsilon)$,

$$|z_j(\epsilon) - 1| = \sqrt[3]{\epsilon}$$

To illustrate this, let $\epsilon = .001$. Then

$$p(x; \epsilon) = x^3 - 3x^2 + 3x - 1.001$$

which is a relatively small change in $p(x)$. But for the roots,

$$|z_j(\epsilon) - 1| = .1$$

a relatively large change in the roots $z_j = 1$.

We now give some more general estimates for $z_j(\epsilon) - z_j$.

*Case* (1)   $z_j$ is a simple root of $p(x)$, and thus $p'(z_j) \neq 0$. Using the theory of functions of a complex variable, it is known that $z_j(\epsilon)$ can be written as a power series:

$$z_j(\epsilon) = z_j + \sum_{l=1}^{\infty} \gamma_l \epsilon^l \qquad (2.9.16)$$

To estimate $z_j(\epsilon) - z_j$, we obtain a formula for the first term $\gamma_1\epsilon$ in the series. To begin, it is easy to see that

$$\gamma_1 = z_j'(0)$$

To calculate $z_j'(\epsilon)$, differentiate the identity

$$p(z_j(\epsilon)) + \epsilon q(z_j(\epsilon)) = 0$$

which holds for all sufficiently small $\epsilon$. We obtain

$$p'(z_j(\epsilon))z_j'(\epsilon) + q(z_j(\epsilon)) + \epsilon q'(z_j(\epsilon))z_j'(\epsilon) = 0$$

$$z_j'(\epsilon) = \frac{-q(z_j(\epsilon))}{p'(z_j(\epsilon)) + \epsilon q'(z_j(\epsilon))} \tag{2.9.17}$$

Substituting $\epsilon = 0$, we obtain

$$\gamma_1 = z_j'(0) = -\frac{q(z_j)}{p'(z_j)}$$

Returning to (2.9.16),

$$z_j(\epsilon) = z_j - \frac{q(z_j)}{p'(z_j)}\epsilon + \sum_{l=2}^{\infty} \gamma_l \epsilon^l$$

$$\left| z_j(\epsilon) - \left[ z_j - \frac{q(z_j)}{p'(z_j)}\epsilon \right] \right| \leq K\epsilon^2 \qquad |\epsilon| \leq \epsilon_0 \tag{2.9.18}$$

for some constants $\epsilon_0 > 0$ and $K > 0$. To estimate $z_j(\epsilon)$ for small $\epsilon$, we use

$$z_j(\epsilon) \doteq z_j - \frac{q(z_j)}{p'(z_j)}\epsilon \tag{2.9.19}$$

The coefficient of $\epsilon$ determines how rapidly $z_j(\epsilon)$ changes relative to $\epsilon$; if it is large, the root $z_j$ is called ill-conditioned.

*Case* (2)    $z_j$ has multiplicity $m > 1$. By using techniques related to those used in 1, we can obtain

$$\left| z_j(\epsilon) - \left[ z_j + \gamma_1 \epsilon^{1/m} \right] \right| \leq K|\epsilon|^{2/m} \qquad |\epsilon| \leq \epsilon_0 \tag{2.9.20}$$

for some $\epsilon_0 > 0$, $K > 0$. There are $m$ possible values to $\gamma_1$, given as the $m$ complex roots of

$$\gamma_1^m = \frac{-m!q(z_j)}{p^{(m)}(z_j)}$$

*Example* Consider the simple polynomial

$$p(x) = (x - 1)(x - 2) \cdots (x - 7)$$

$$= x^7 - 28x^6 + 322x^5 - 1960x^4 + 6769x^3 - 13132x^2$$

$$+ 13068x - 5040 \qquad (2.9.21)$$

For the perturbation, take

$$q(x) = x^6 \qquad \epsilon = -.002$$

Then for the root $z_j = j$,

$$p'(z_j) = \prod_{l \neq j}(j - l) \qquad q(z_j) = j^6$$

From (2.9.19), we have the estimate

$$z_j(\epsilon) \doteq j + \frac{(-1)^{j-1}.002j^6}{(j - 1)!(7 - j)!} \equiv j + \delta(j) \qquad (2.9.22)$$

The numerical values of $\delta(j)$ are given in Table 2.16. The relative error in the coefficient of $x^6$ is $.002/28 = 7.1E - 5$, but the relative errors in the roots are much larger. In fact, the size of some of the perturbations $\delta(j)$ casts doubt on the validity of using the linear estimate (2.9.22). The actual roots of $p(x) + \epsilon q(x)$ are given in Table 2.17, and they correspond closely to the predicted perturbations. The major departure is in the roots for $j = 5$ and $j = 6$. They are complex, which was not predicted by the linear estimate (2.9.22). In these two cases, $\epsilon$ is outside the radius of convergence of the power series (2.9.16), since the latter will only have the real coefficients

$$\gamma_l = \frac{1}{l!} z_j^{(l)}(0)$$

obtained from differentiating (2.9.17).

Table 2.16   Values of $\delta(j)$ from (2.9.22)

| $j$ | $\delta(j)$ |
| --- | --- |
| 1 | $2.78E - 6$ |
| 2 | $-1.07E - 3$ |
| 3 | $3.04E - 2$ |
| 4 | $-2.28E - 1$ |
| 5 | $6.51E - 1$ |
| 6 | $-7.77E - 1$ |
| 7 | $3.27E - 1$ |

**Table 2.17**    **Roots of $p(x;\epsilon)$ for (2.9.21)**

| $j$ | $z_j(\epsilon)$ | $z_j(\epsilon) - z_j(0)$ |
|---|---|---|
| 1 | 1.0000028 | $2.80\mathrm{E} - 6$ |
| 2 | 1.9989382 | $-1.06\mathrm{E} - 3$ |
| 3 | 3.0331253 | $3.31\mathrm{E} - 2$ |
| 4 | 3.8195692 | $-1.80\mathrm{E} - 1$ |
| 5 | $5.4586758 + .54012578i$ | |
| 6 | $5.4586758 - .54012578i$ | |
| 7 | 7.2330128 | $2.33\mathrm{E} - 1$ |

We say that a polynomial whose roots are unstable with respect to small relative changes in the coefficients is ill-conditioned. Many such polynomials occur naturally in applications. The previous example should illustrate the difficulty in determining with only a cursory examination whether or not a polynomial is ill-conditioned.

**Polynomial deflation**    Another problem occurs with deflation of a polynomial to a lower degree polynomial, a process defined following (2.9.12). Since the zeros will not be found exactly, the lower degree polynomial (2.9.11) found by extracting the latest root will generally be in error in all of its coefficients. Clearly from the past example, this can cause a significant perturbation in the roots for some classes of polynomials. Wilkinson (1963) has analyzed the effects of deflation and has recommended the following general strategy: (1) Solve for the roots of smallest magnitude first, ending with those of largest size; (2) after obtaining approximations to all roots, iterate again using the original polynomial and using the previously calculated values as initial guesses. A complete discussion can be found in Wilkinson (1963, pp. 55–65).

*Example*    Consider finding the roots of the degree 6 Laguerre polynomial

$$p(x) = 720 - 4320x + 5400x^2 - 2400x^3 + 450x^4 - 36x^5 + x^6$$

The Newton algorithm of the last section was used to solve for the roots, with deflation following the acceptance of each new root. The roots were calculated in two ways: (1) from largest to smallest, and (2) from smallest to largest. The calculations were in single precision arithmetic on an IBM 360, and the numerical results are given in Table 2.18. A comparison of the columns headed Method (1) and Method (2) shows clearly the superiority of calculating the roots in the order of increasing magnitude. If the results of method (1) are used as initial guesses for further iteration with the original polynomial, then approximate roots are obtained with an accuracy better than that of method (2); see the column headed Method (3) in the table. This table shows the importance of iterating again with the original polynomial to remove the effects of the deflation process.

**Table 2.18**   **Example involving polynomial deflation**

| True | Method (1) | Method (2) | Method (3) |
|------|-----------|-----------|-----------|
| 15.98287 | 15.98287 | 15.98279 | 15.98287 |
| 9.837467 | 9.837471 | 9.837469 | 9.837467 |
| 5.775144 | 5.775764 | 5.775207 | 5.775144 |
| 2.992736 | 2.991080 | 2.992710 | 2.992736 |
| 1.188932 | 1.190937 | 1.188932 | 1.188932 |
| .2228466 | .2219429 | .2228466 | .2228466 |

There are other ways to deflate a polynomial, one of which favors finding roots of largest magnitude first. For a complete discussion see Peters and Wilkinson (1971, sec. 5). An algorithm is given for *composite deflation*, which removes the need to find the roots in any particular order. In that paper, the authors also discuss the use of *implicit deflation*,

$$q(x) = \frac{p(x)}{(x - z_1) \cdots (x - z_r)}$$

to remove the roots $z_1, \ldots, z_r$ that have been computed previously. This was given earlier, in (2.4.4), where it was used in connection with Muller's method.

**General polynomial rootfinding methods**   There are a large number of rootfinding algorithms designed especially for polynomials. Many of these are taken up in detail in the books Dejon and Henrici (1969), Henrici (1974, chap. 6), and Householder (1970). There are far too many types of such methods to attempt to describe them all here.

One large class of important methods uses location theorems related to those described in (2.9.3)–(2.9.6), to iteratively separate the roots into disjoint and ever smaller regions, often circles. The best known of such methods is probably the Lehmer–Schur method [see Householder (1970. sec. 2.7)]. Such methods converge linearly, and for that reason, they are often combined with some more rapidly convergent method, such as Newton's method. Once the roots have been separated into distinct regions, the faster method is applied to rapidly obtain the root within that region. For a general discussion of such rootfinding methods, see Henrici (1974, sec. 6.10).

Other methods that have been developed into widely used algorithms are the method of Jenkins and Traub and the method of Laguerre. For the former, see Householder (1970, p. 173), Jenkins and Traub (1970), (1972). For Laguerre's method, see Householder (1970, sec. 4.5) and Kahan (1967).

Another easy-to-use numerical method is based on being able to calculate the eigenvalues of a matrix. Given the polynomial $p(x)$, it is possible to easily construct a matrix with $p(x)$ as its characteristic polynomial (see Problem 2 of Chapter 9). Since excellent software exists for solving the eigenvalue problem, this software can be used to find the roots of a polynomial $p(x)$.

## 2.10  Systems of Nonlinear Equations

This section and the next are concerned with the numerical solution of systems of nonlinear equations in several variables. These problems are widespread in applications, and they are varied in form. There is a great variety of methods for the solution of such systems, so we only introduce the subject. We give some general theory and some numerical methods that are easily programmed. To do a complete development of the numerical analysis of solving nonlinear systems, we would need a number of results from numerical linear algebra, which is not taken up until Chapters 7–9.

For simplicity of presentation and ease of understanding, the theory is presented for only two equations:

$$f_1(x_1, x_2) = 0 \qquad f_2(x_1, x_2) = 0 \qquad (2.10.1)$$

The generalization to $n$ equations in $n$ variables should be straightforward once the principal ideas have been grasped. As an additional aid, we will simultaneously consider the solution of (2.10.1) in vector notation:

$$\mathbf{f}(\mathbf{x}) = \mathbf{0} \qquad \mathbf{x} = \begin{bmatrix} x_1 \\ x_2 \end{bmatrix} \qquad \mathbf{f}(\mathbf{x}) = \begin{bmatrix} f_1(x_1, x_2) \\ f_2(x_1, x_2) \end{bmatrix} \qquad (2.10.2)$$

The solution of (2.10.1) can be looked upon as a two-step process: (1) Find the zero curves in the $x_1 x_2$-plane of the surfaces $z = f_1(x_1, x_2)$ and $z = f_2(x_1, x_2)$, and (2) find the points of intersection of these zero curves in the $x_1 x_2$-plane. This perspective is used in the next section to generalize Newton's method to solve (2.10.1).

**Fixed-point theory**    We begin by generalizing some of the fixed-point iteration theory of Section 2.5. Assume that the rootfinding problem (2.10.1) has been reformulated in an equivalent form as

$$x_1 = g_1(x_1, x_2) \qquad x_2 = g_2(x_1, x_2) \qquad (2.10.3)$$

Denote its solution by

$$\alpha = \begin{bmatrix} \alpha_1 \\ \alpha_2 \end{bmatrix}$$

We study the fixed-point iteration

$$x_{1,n+1} = g_1(x_{1,n}, x_{2,n}) \qquad x_{2,n+1} = g_2(x_{1,n}, x_{2,n}) \qquad (2.10.4)$$

Using vector notation, we write this as

$$\mathbf{x}_{n+1} = \mathbf{g}(\mathbf{x}_n) \qquad (2.10.5)$$

**Table 2.19    Example (2.10.7) of fixed-point iteration**

| $n$ | $x_{1,n}$ | $x_{2,n}$ | $f_1(x_{1,n}, x_{2,n})$ | $f_2(x_{1,n}, x_{2,n})$ |
|---|---|---|---|---|
| 0 | $-.5$ | .25 | 0.0 | $1.56E - 2$ |
| 1 | $-.497343750$ | .254062500 | $2.43E - 4$ | $5.46E - 4$ |
| 2 | $-.497254794$ | .254077922 | $9.35E - 6$ | $2.12E - 5$ |
| 3 | $-.497251343$ | .254078566 | $3.64E - 7$ | $8.26E - 7$ |
| 4 | $-.497251208$ | .254078592 | $1.50E - 8$ | $3.30E - 8$ |

with

$$\mathbf{x}_n = \begin{bmatrix} x_{1,n} \\ x_{2,n} \end{bmatrix} \qquad \mathbf{g}(\mathbf{x}) = \begin{bmatrix} g_1(x_1, x_2) \\ g_2(x_1, x_2) \end{bmatrix}$$

**Example**    Consider solving

$$f_1 \equiv 3x_1^2 + 4x_2^2 - 1 = 0 \qquad f_2 \equiv x_2^3 - 8x_1^3 - 1 = 0 \qquad (2.10.6)$$

for the solution $\boldsymbol{\alpha}$ near $(x_1, x_2) = (-.5, .25)$. We solve this system iteratively with

$$\begin{bmatrix} x_{1,n+1} \\ x_{2,n+1} \end{bmatrix} = \begin{bmatrix} x_{1,n} \\ x_{2,n} \end{bmatrix} - \begin{bmatrix} .016 & -.17 \\ .52 & -.26 \end{bmatrix} \begin{bmatrix} 3x_{1,n}^2 + 4x_{2,n}^2 - 1 \\ x_{2,n}^3 - 8x_{1,n}^3 - 1 \end{bmatrix} \qquad (2.10.7)$$

The origin of this reformulation of (2.10.6) is given later. The numerical results of (2.10.7) are given in Table 2.19. Clearly the iterates are converging rapidly.

To analyze the convergence of (2.10.5), begin by subtracting the two equations in (2.10.4) from the corresponding equations

$$\alpha_1 = g_1(\alpha_1, \alpha_2) \qquad \alpha_2 = g_2(\alpha_1, \alpha_2)$$

involving the true solution $\boldsymbol{\alpha}$. Apply the mean value theorem for functions of two variables (Theorem 1.5 with $n = 1$) to these differences to obtain

$$\alpha_i - x_{i,n+1} = \frac{\partial g_i\left(\xi_{1,n}^{(i)}, \xi_{2,n}^{(i)}\right)}{\partial x_1}(\alpha_1 - x_{1,n}) + \frac{\partial g_i\left(\xi_{1,n}^{(i)}, \xi_{2,n}^{(i)}\right)}{\partial x_2}(\alpha_2 - x_{2,n})$$

for $i = 1, 2$. The points $\boldsymbol{\xi}_n^{(i)} = (\xi_{1,n}^{(i)}, \xi_{2,n}^{(i)})$ are on the line segment joining $\boldsymbol{\alpha}$ and $\mathbf{x}_n$. In matrix form, these error equations become

$$\begin{bmatrix} \alpha_1 - x_{1,n+1} \\ \alpha_2 - x_{2,n+1} \end{bmatrix} = \begin{bmatrix} \dfrac{\partial g_1\left(\xi_n^{(1)}\right)}{\partial x_1} & \dfrac{\partial g_1\left(\xi_n^{(1)}\right)}{\partial x_2} \\ \dfrac{\partial g_2\left(\xi_n^{(2)}\right)}{\partial x_1} & \dfrac{\partial g_2\left(\xi_n^{(2)}\right)}{\partial x_2} \end{bmatrix} \begin{bmatrix} \alpha_1 - x_{1,n} \\ \alpha_2 - x_{2,n} \end{bmatrix} \qquad (2.10.8)$$

Let $G_n$ denote the matrix in (2.10.8). Then we can rewrite this equation as

$$\alpha - x_{n+1} = G_n(\alpha - x_n) \tag{2.10.9}$$

It is convenient to introduce the Jacobian matrix for the functions $g_1$ and $g_2$:

$$G(x) = \begin{bmatrix} \dfrac{\partial g_1(x)}{\partial x_1} & \dfrac{\partial g_1(x)}{\partial x_2} \\[2mm] \dfrac{\partial g_2(x)}{\partial x_1} & \dfrac{\partial g_2(x)}{\partial x_2} \end{bmatrix} \tag{2.10.10}$$

In (2.10.9), if $x_n$ is close to $\alpha$, then $G_n$ will be close to $G(\alpha)$. This will make the size or norm of $G(\alpha)$ crucial in analyzing the convergence in (2.10.9). The matrix $G(\alpha)$ plays the role of $g'(\alpha)$ in the theory of Section 2.5. To measure the size of the errors $\alpha - x_n$ and of the matrices $G_n$ and $G(\alpha)$, we use the vector and matrix norms of (1.1.16) and (1.1.19) in Chapter 1.

**Theorem 2.9**  Let $D$ be a closed, bounded, and convex set in the plane. (We say $D$ is convex if for any two points in $D$, the line segment joining them is also in $D$.) Assume that the components of $g(x)$ are continuously differentiable at all points of $D$, and further assume

1.  $g(D) \subset D$, $\hspace{4cm}$ (2.10.11)

2.  $\lambda \equiv \underset{x \in D}{\text{Max}} \|G(x)\|_\infty < 1$ $\hspace{2.5cm}$ (2.10.12)

Then

(a)  $x = g(x)$ has a unique solution $\alpha \in D$.

(b)  For any initial point $x_0 \in D$, the iteration (2.10.5) will converge in $D$ to $\alpha$.

(c)  $\|\alpha - x_{n+1}\|_\infty \le (\|G(\alpha)\|_\infty + \epsilon_n)\|\alpha - x_n\|_\infty$

$$\tag{2.10.13}$$

with $\epsilon_n \to 0$ as $n \to \infty$.

**Proof**  (a)  The existence of a fixed point $\alpha$ can be shown by proving that the sequence of iterates $\{x_n\}$ from (2.10.5) are convergent in $D$. We leave that to a problem, and instead just show the uniqueness of $\alpha$. Suppose $\alpha$ and $\beta$ are both fixed points of $g(x)$ in $D$. Then

$$\alpha - \beta = g(\alpha) - g(\beta) \tag{2.10.14}$$

Apply the mean value theorem to component $i$, obtaining

$$g_i(\alpha) - g_i(\beta) = \nabla g_i(\xi^{(i)})(\alpha - \beta) \qquad i = 1, 2 \qquad (2.10.15)$$

with

$$\nabla g_i(x) = \left[ \frac{\partial g_i}{\partial x_1} \cdot \frac{\partial g_i}{\partial x_2} \right]$$

and $\xi^{(i)} \in D$, on the line segment joining $\alpha$ and $\beta$. Since $\|G(x)\|_\infty \le \lambda < 1$, we have from the definition of the norm that

$$\left| \frac{\partial g_i(x)}{\partial x_1} \right| + \left| \frac{\partial g_i(x)}{\partial x_2} \right| \le \lambda < 1, \qquad x \in D, \quad i = 1, 2$$

Combining this with (2.10.15),

$$|g_i(\alpha) - g_i(\beta)| \le \lambda \|\alpha - \beta\|_\infty$$

$$\|g(\alpha) - g(\beta)\|_\infty \le \lambda \|\alpha - \beta\|_\infty \qquad (2.10.16)$$

Combined with (2.10.14), this yields

$$\|\alpha - \beta\|_\infty \le \lambda \|\alpha - \beta\|_\infty$$

which is possible only if $\alpha = \beta$, showing the uniqueness of $\alpha$ in $D$.

**(b)**    Condition (2.10.11) will ensure that all $x_n \in D$ if $x_0 \in D$. Next subtract $x_{n+1} = g(x_n)$ from $\alpha = g(\alpha)$, obtaining

$$\alpha - x_{n+1} = g(\alpha) - g(x_n)$$

The result (2.10.16) applies to any two points in $D$. Applying this,

$$\|\alpha - x_{n+1}\|_\infty \le \lambda \|\alpha - x_n\|_\infty \qquad (2.10.17)$$

Inductively,

$$\|\alpha - x_n\|_\infty \le \lambda^n \|\alpha - x_0\|_\infty \qquad (2.10.18)$$

Since $\lambda < 1$, this shows $x_n \to \alpha$ as $n \to \infty$.

**(c)**    From (2.10.9) and using (1.1.21),

$$\|\alpha - x_{n+1}\|_\infty \le \|G_n\|_\infty \|\alpha - x_n\|_\infty \qquad (2.10.19)$$

As $n \to \infty$, the points $\xi_n^{(i)}$ used in evaluating $G_n$ will all tend to $\alpha$, since they are on the line segment joining $x_n$ and $\alpha$. Then $\|G_n\|_\infty \to \|G(\alpha)\|_\infty$ as $n \to \infty$. Result (2.10.13) follows from (2.10.19) by letting $\epsilon_n = \|G_n\|_\infty - \|G(\alpha)\|_\infty$.    ∎

The preceding theorem is the generalization to two variables of Theorem 2.6 for functions of one variable. The following generalizes Theorem 2.7.

**Corollary 2.10**    Let $\alpha$ be a fixed point of $g(x)$, and assume components of $g(x)$ are continuously differentiable in some neighborhood about $\alpha$. Further assume

$$\|G(\alpha)\|_\infty < 1 \qquad (2.10.20)$$

Then for $x_0$ chosen sufficiently close to $\alpha$, the iteration $x_{n+1} = g(x_n)$ will converge to $\alpha$, and the results of Theorem 2.9 will be valid on some closed, bounded, convex region about $\alpha$.    ∎

We leave the proof of this as a problem. Based on results in Chapter 7, the linear convergence of $x_n$ to $\alpha$ will still be true if all eigenvalues of $G(\alpha)$ are less than one in magnitude, which can be shown to be a weaker assumption than (2.10.20).

**Example**    Continue the earlier example (2.10.7). It is straightforward to compute

$$G(\alpha) \doteq \begin{bmatrix} .038920 & .000401 \\ .008529 & -.006613 \end{bmatrix}$$

and therefore

$$\|G(\alpha)\|_\infty \doteq 0393$$

Thus the condition (2.10.20) of the theorem is satisfied. From (2.10.13), it will be approximately true that

$$\frac{\|\alpha - x_{n+1}\|_\infty}{\|\alpha - x_n\|_\infty} \le \|G_n\|_\infty \doteq .0393$$

for all sufficiently large $n$.

Suppose that $A$ is a constant nonsingular matrix of order $2 \times 2$. We can then reformulate (2.10.1) as

$$x = x + Af(x) \equiv g(x) \qquad (2.10.21)$$

The example (2.10.7) illustrates this procedure. To see the requirements on $A$, we produce the Jacobian matrix. Easily,

$$G(x) = I + AF(x)$$

where $F(x)$ is the Jacobian matrix of $f_1$ and $f_2$,

$$F(x) = \begin{bmatrix} \dfrac{\partial f_1(x)}{\partial x_1} & \dfrac{\partial f_1(x)}{\partial x_2} \\ \dfrac{\partial f_2(x)}{\partial x_1} & \dfrac{\partial f_2(x)}{\partial x_2} \end{bmatrix} \qquad (2.10.22)$$

We want to choose $A$ so that (2.10.20) is satisfied. And for rapid convergence, we want $\|G(\alpha)\|_\infty \doteq 0$, or

$$A \doteq -F(\alpha)^{-1}$$

The matrix in (2.10.7) was chosen in this way using

$$A \doteq -F(x_0)^{-1}$$

This suggests using a continual updating of $A$, say $A = -F(x_n)^{-1}$. The resulting method is

$$x_{n+1} = x_n - F(x_n)^{-1}f(x_n) \qquad n \geq 0 \qquad (2.10.23)$$

We consider this method in the next section.

## 2.11   Newton's Method for Nonlinear Systems

As with Newton's method for a single equation, there is more than one way of viewing and deriving the Newton method for solving a system of nonlinear equations. We begin with an analytic derivation, and then we give a geometric perspective.

Apply Taylor's theorem for functions of two variables to each of the equations $f_i(x_1, x_2) = 0$, expanding $f_i(\alpha)$ about $x_0$: for $i = 1, 2$

$$0 = f_i(\alpha) = f_i(x_0) + (\alpha_1 - x_{1,0})\frac{\partial f_i(x_0)}{\partial x_1} + (\alpha_2 - x_{2,0})\frac{\partial f_i(x_0)}{\partial x_2}$$

$$+ \frac{1}{2}\left[(\alpha_1 - x_{1,0})\frac{\partial}{\partial x_1} + (\alpha_2 - x_{2,0})\frac{\partial}{\partial x_2}\right]^2 f_i(\xi^{(i)}) \quad (2.11.1)$$

with $\xi^{(i)}$ on the line segment joining $x_0$ and $\alpha$. If we drop the second-order terms, we obtain the approximation

$$0 \doteq f_1(x_0) + (\alpha_1 - x_{1,0})\frac{\partial f_1(x_0)}{\partial x_1} + (\alpha_2 - x_{2,0})\frac{\partial f_1(x_0)}{\partial x_2}$$

$$0 \doteq f_2(x_0) + (\alpha_1 - x_{1,0})\frac{\partial f_2(x_0)}{\partial x_1} + (\alpha_2 - x_{2,0})\frac{\partial f_2(x_0)}{\partial x_2} \quad (2.11.2)$$

In matrix form,

$$0 \doteq f(x_0) + F(x_0)(\alpha - x_0) \qquad (2.11.3)$$

with $F(x_0)$ the Jacobian matrix of $f$, given in (2.10.22).

Solving for $\alpha$,

$$\alpha \doteq x_0 - F(x_0)^{-1}f(x_0) \equiv x_1$$

The approximation $x_1$ should be an improvement on $x_0$, provided $x_0$ is chosen sufficiently close to $\alpha$. This leads to the iteration method first obtained at the end of the last section,

$$x_{n+1} = x_n - F(x_n)^{-1}f(x_n) \qquad n \geq 0 \qquad (2.11.4)$$

This is Newton's method for solving the nonlinear system $f(x) = 0$.

In actual practice, we do not invert $F(x_n)$, particularly for systems of more than two equations. Instead we solve a linear system for a correction term to $x_n$:

$$F(x_n)\delta_{n+1} = -f(x_n)$$

$$x_{n+1} = x_n + \delta_{n+1} \qquad (2.11.5)$$

This is more efficient in computation time, requiring only about one-third as many operations as inverting $F(x_n)$. See Sections 8.1 and 8.2 for a discussion of the numerical solution of linear systems of equations.

There is a geometrical derivation for Newton's method, in analogy with the tangent line approximation used with single nonlinear equations in Section 2.2. The graph in space of the equation

$$z = f_i(x_0) + (x_1 - x_{1,0})\frac{\partial f_i(x_0)}{\partial x_1} + (x_2 - x_{2,0})\frac{\partial f_i(x_0)}{\partial x_2} \equiv p_i(x_1, x_2)$$

is a plane that is tangent to the graph of $z = f_i(x_1, x_2)$ at the point $x_0$, for $i = 1, 2$. If $x_0$ is near $\alpha$, then these tangent planes should be good approximations to the associated surfaces of $z = f_i(x_1, x_2)$, for $x = (x_1, x_2)$ near $\alpha$. Then the intersection of the zero curves of the tangent planes $z = p_i(x_1, x_2)$ should be a good approximation to the corresponding intersection $\alpha$ of the zero curves of the original surfaces $z = f_i(x_1, x_2)$. This results in the statement (2.11.2). The intersection of the zero curves of $z = p_i(x_1, x_2)$, $i = 1, 2$, is the point $x_1$.

*Example*    Consider the system

$$f_1 \equiv 4x_1^2 + x_2^2 - 4 = 0 \qquad f_2 \equiv x_1 + x_2 - \sin(x_1 - x_2) = 0$$

There are only two roots, one near $(1, 0)$ and its reflection through the origin near $(-1, 0)$. Using (2.11.4) with $x_0 = (1, 0)$, we obtain the results given in Table 2.20.

**Table 2.20    Example of Newton's method**

| $n$ | $x_{1,n}$ | $x_{2,n}$ | $f_1(x_n)$ | $f_2(x_n)$ |
|---|---|---|---|---|
| 0 | 1.0 | 0.0 | 0.0 | 1.59E − 1 |
| 1 | 1.0 | − .1029207154 | 1.06E − 2 | 4.55E − 3 |
| 2 | .9986087598 | − .1055307239 | 1.46E − 5 | 6.63E − 7 |
| 3 | .9986069441 | − .1055304923 | 1.32E − 11 | 1.87E − 12 |

**Convergence analysis**   For the convergence analysis of Newton's method (2.11.4), regard it as a fixed-point iteration method with

$$g(\mathbf{x}) = \mathbf{x} - F(\mathbf{x})^{-1}\mathbf{f}(\mathbf{x}) \tag{2.11.6}$$

Also assume

$$\text{Determinant } F(\boldsymbol{\alpha}) \neq 0$$

which is the analogue of assuming $\boldsymbol{\alpha}$ is a simple root when dealing with a single equation, as in Theorem 2.1. It can then be shown that the Jacobian $G(\mathbf{x})$ of (2.11.6) is zero at $\mathbf{x} = \boldsymbol{\alpha}$ (see Problem 53); consequently, the condition (2.10.20) is easily satisfied.

Corollary 2.10 then implies that $\mathbf{x}_n$ converges to $\boldsymbol{\alpha}$, provided $\mathbf{x}_0$ is chosen sufficiently close to $\boldsymbol{\alpha}$. In addition, it can be shown that the iteration is quadratic. Specifically, the formulas (2.11.1) and (2.11.4) can be combined to obtain

$$\|\boldsymbol{\alpha} - \mathbf{x}_{n+1}\|_\infty \leq B\|\boldsymbol{\alpha} - \mathbf{x}_n\|_\infty^2 \qquad n \geq 0 \tag{2.11.7}$$

for some constant $B > 0$.

**Variations of Newton's method**   Newton's method has both advantages and disadvantages when compared with other methods for solving systems of nonlinear equations. Among its advantages, it is very simple in form and there is great flexibility in using it on a large variety of problems. If we do not want to bother supplying partial derivatives to be evaluated by a computer program, we can use a difference approximation. For example, we commonly use

$$\frac{\partial f_i(x_1, x_2)}{\partial x_1} \doteq \frac{f_i(x_1 + \epsilon, x_2) - f_i(x_1, x_2)}{\epsilon} \tag{2.11.8}$$

with some very small number $\epsilon$. For a detailed discussion of the choice of $\epsilon$, see Dennis and Schnabel (1983, pp. 94–99).

The first disadvantage of Newton's method is that there are other methods which are (1) less expensive to use, and/or (2) easier to use for some special classes of problems. For a system of $m$ nonlinear equations in $m$ unknowns, each iterate for Newton's method requires $m^2 + m$ function evaluations in general. In addition, Newton's method requires the solution of a system of $m$ linear equations for each iterate, at a cost of about $\frac{2}{3}m^3$ arithmetic operations per linear system. There are other methods that are as fast or almost as fast in their mathematical speed of convergence, but that require fewer function evaluations and arithmetic operations per iteration. These are often referred to as *Newton-like*, *quasi-Newton*, and *modified Newton methods*. For a general presentation of many of these methods, see Dennis and Schnabel (1983).

A simple modification of Newton's method is to fix the Jacobian matrix for several steps, say $k$:

$$\mathbf{x}_{rk+j+1} = \mathbf{x}_{rk+j} - F(\mathbf{x}_{rk})^{-1}\mathbf{f}(\mathbf{x}_{rk+j}) \qquad j = 0, 1, \ldots, k-1 \tag{2.11.9}$$

for $r = 0, 1, 2, \ldots$ . This means the linear system in

$$F(\mathbf{x}_{rk}) \delta_{rk+j+1} = - \mathbf{f}(\mathbf{x}_{rk+j})$$

$$\mathbf{x}_{rk+j+1} = \mathbf{x}_{rk+j} + \delta_{rk+j+1}, \qquad j = 0, 1, \ldots, k - 1, \quad (2.11.10)$$

can be solved much more efficiently than in the original Newton method (2.11.5). The linear system, of order $m$, requires about $\frac{2}{3} m^3$ arithmetic operations for its solution in the first case, when $j = 0$. But each subsequent case, $j = 1, \ldots, k - 1$, will require only $2m^2$ arithmetic operations for its solution. See Section 8.1 for more complete details. The speed of convergence of (2.11.9) will be slower than the original method (2.11.4), but the actual computation time of the modified method will often be much less. For a more detailed examination of this question, see Potra and Ptak (1984, p. 119).

A second problem with Newton's method, and with many other methods, is that often $\mathbf{x}_0$ must be reasonably close to $\alpha$ in order to obtain convergence. There are modifications of Newton's method to force convergence for poor choices of $\mathbf{x}_0$. For example, define

$$\mathbf{x}_{n+1} = \mathbf{x}_n + s\mathbf{d}_n \qquad \mathbf{d}_n = - F(\mathbf{x}_n)^{-1} \mathbf{f}(\mathbf{x}_n) \qquad (2.11.11)$$

and choose $s > 0$ to minimize

$$\| \mathbf{f}(\mathbf{x}_n + s\mathbf{d}_n) \|_2^2 = \sum_{j=1}^{m} \left[ f_j(\mathbf{x}_n + s\mathbf{d}_n) \right]^2 \qquad (2.11.12)$$

The choice $s = 1$ in (2.11.11) yields Newton's method, but it may not be the best choice. In some cases, $s$ may need to be much smaller than 1, at least initially, in order to ensure convergence. For a more detailed discussion, see Dennis and Schnabel (1983, chap. 6).

For an analysis of some current programs for solving nonlinear systems, see Hiebert (1982). He also discusses the difficulties in producing such software.

## 2.12  Unconstrained Optimization

Optimization refers to finding the maximum or minimum of a continuous function $f(x_1, \ldots, x_m)$. This is an extremely important problem, lying at the heart of modern industrial engineering, management science, and other areas. This section discusses some methods and perspectives for calculating the minimum or maximum of a function $f(x_1, \ldots, x_m)$. No formal algorithms are given, since this would require too extensive a development.

Vector notation is used in much of the presentation, to give results for a general number $m$ of variables. We consider only the unconstrained optimization problem, in which there are no limitations on $(x_1, \ldots, x_m)$. For simplicity only, we also assume $f(x_1, \ldots, x_m)$ is defined for all $(x_1, \ldots, x_m)$.

Because the behavior of a function $f(\mathbf{x})$ can be quite varied, the problem must be further limited. A point $\alpha$ is called a *strict local minimum* of $f$ if $f(\mathbf{x}) > f(\alpha)$

for all $\mathbf{x}$ close to $\alpha$, $\mathbf{x} \neq \alpha$. We limit ourselves to finding a strict local minimum of $f(\mathbf{x})$. Generally an initial guess $\mathbf{x}_0$ of $\alpha$ will be known, and $f(\mathbf{x})$ will be assumed to be twice continuously differentiable with respect to its variables $x_1, \ldots, x_m$.

**Reformulation as a nonlinear system**    With the assumption of differentiability, a necessary condition for $\alpha$ to be a strict local minimum is that

$$\frac{\partial f(\alpha)}{\partial x_i} = 0 \qquad i = 1, \ldots, m \tag{2.12.1}$$

Thus the nonlinear system

$$\frac{\partial f(\mathbf{x})}{\partial x_i} = 0 \qquad i = 1, \ldots, m \tag{2.12.2}$$

can be solved, and each calculated solution can be checked as to whether it is a local maximum, minimum, or neither. For notation, introduce the gradient vector

$$\nabla f(\mathbf{x}) = \begin{bmatrix} \dfrac{\partial f}{\partial x_1} \\ \vdots \\ \dfrac{\partial f}{\partial x_m} \end{bmatrix}$$

Using this vector, the system (2.12.2) is written more compactly as

$$\nabla f(\mathbf{x}) = \mathbf{0} \tag{2.12.3}$$

To solve (2.12.3), Newton's method (2.11.4) can be used, as well as other rootfinding methods for nonlinear systems. Using Newton's method leads to

$$\mathbf{x}_{n+1} = \mathbf{x}_n - H(\mathbf{x}_n)^{-1} \nabla f(\mathbf{x}_n) \qquad n \geq 0 \tag{2.12.4}$$

with $H(\mathbf{x})$ the Hessian matrix of $f$,

$$H(\mathbf{x})_{ij} = \frac{\partial^2 f(\mathbf{x})}{\partial x_i \partial x_j}, \qquad 1 \leq i, j \leq m$$

If $\alpha$ is a strict local minimum of $f$, then Taylor's theorem (1.1.12) can be used to show that $H(\alpha)$ is a nonsingular matrix; then $H(\mathbf{x})$ will be nonsingular for $\mathbf{x}$ close to $\alpha$. For convergence, the analysis of Newton's method in the preceding section can be used to prove quadratic convergence of $\mathbf{x}_n$ to $\alpha$ provided $x_0$ is chosen sufficiently close to $\alpha$.

The main drawbacks with the iteration (2.12.4) are the same as those given in the last section for Newton's method for solving nonlinear systems. There are other, more efficient optimization methods that seek to approximate $\alpha$ by using only $f(\mathbf{x})$ and $\nabla f(\mathbf{x})$. These methods may require more iterations, but generally their total computing time will be much less than with Newton's method. In addition, these methods seek to obtain convergence for a larger set of initial values $\mathbf{x}_0$.

**Descent methods**   Suppose we are trying to minimize a function $f(\mathbf{x})$. Most methods for doing so are based on the following general two-step iteration process.

**STEP D1:**   At $\mathbf{x}_n$, pick a direction $\mathbf{d}_n$ such that $f(\mathbf{x})$ will decrease as $\mathbf{x}$ moves away from $\mathbf{x}_n$ in the direction $\mathbf{d}_n$.

**STEP D2:**   Let $\mathbf{x}_{n+1} = \mathbf{x}_n + s\mathbf{d}_n$, with $s$ chosen to minimize

$$\varphi(s) = f(\mathbf{x}_n + s\mathbf{d}_n), \qquad s \geq 0 \qquad (2.12.5)$$

Usually $s$ is chosen as the smallest positive relative minimum of $\varphi(s)$.

Such methods are called *descent methods*. With each iteration,

$$f(\mathbf{x}_{n+1}) < f(\mathbf{x}_n)$$

Descent methods are guaranteed to converge under more general conditions than for Newton's method (2.12.4). Consider the level surface

$$C = \{\mathbf{x} \mid f(\mathbf{x}) = f(\mathbf{x}_0)\}$$

and consider only the connected portion of it, say $C'$, that contains $\mathbf{x}_0$. Then if $C'$ is bounded and contains $\alpha$ in its interior, descent methods will converge under very general conditions. This is illustrated for the two-variable case in Figure 2.8. Several level curves $f(x_1, x_2) = c$ are shown for a set of values $c$ approaching $f(\alpha)$. The vectors $\mathbf{d}_n$ are directions in which $f(\mathbf{x})$ is decreasing.

There are a number of ways for choosing the directions $\mathbf{d}_n$, and the best known are as follows.

1. ***The method of steepest descent.***   Here $\mathbf{d}_n = -\nabla f(\mathbf{x}_n)$. It is the direction in which $f(\mathbf{x})$ decreases most rapidly when moving away from $\mathbf{x}_n$. It is a good strategy near $\mathbf{x}_n$, but it usually turns out to be a poor strategy for rapid convergence to $\alpha$.

2. ***Quasi-Newton methods.***   These methods can be viewed as approximations of Newton's method (2.12.4). They use easily computable approximations of $H(\mathbf{x}_n)$ or $H(\mathbf{x}_n)^{-1}$, and they are also descent methods. The best known examples are the Davidon-Fletcher-Powell method and the Broyden methods.

3. ***The conjugate gradient method.***   This uses a generalization of the idea of an orthogonal basis for a vector space to generate the directions $\mathbf{d}_n$, with the

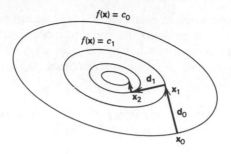

$f(\mathbf{x}) = c_0$

$f(\mathbf{x}) = c_1$

**Figure 2.8**   Illustration of steepest descent method.

directions related in an optimal way to the function $f(\mathbf{x})$ being minimized. In Chapter 8, the conjugate gradient method is used for solving systems of linear equations.

There are many other approaches to minimizing a function, but they are too numerous to include here. As general references to the preceding ideas, see Dennis and Schnabel (1983), Fletcher (1980), Gill et al. (1981), and Luenberger (1984). An important and very different approach to minimizing a function is the simplex method given in Nelder and Mead (1965), with a discussion given in Gill et al. (1981, p. 94) and Woods (1985, chap. 2). This method uses only function values (no derivative values), and it seems to be especially suitable for noisy functions.

An important project to develop programs for solving optimization problems and nonlinear systems is under way at the Argonne National Laboratory. The program package is called MINPACK, and version 1 is available [see Moré et al. (1980) and Moré et al. (1984)]. It contains routines for nonlinear systems and nonlinear least squares problems. Future versions are intended to include programs for both unconstrained and constrained optimization problems.

## Discussion of the Literature

There is a large literature on methods for calculating the roots of a single equation. See the books by Householder (1970), Ostrowski (1973), and Traub (1964) for a more extensive development than has been given here. Newton's method is one of the most widely used methods, and its development is due to many people. For an historical account of contributions to it by Newton, Raphson, and Cauchy, see Goldstine (1977, pp. 64, 278).

For computer programs, most people still use and individually program a method that is especially suitable for their particular application. However, one should strongly consider using one of the general-purpose programs that have been developed in recent years and that are available in the commercial software libraries. They are usually accurate, efficient, and easy to use. Among such general-purpose programs, the ones based on Brent (1973) and Dekker (1969) has been most popular, and further developments of them continue to be made, as in Le (1985). The IMSL and NAG computer libraries include these and other excellent rootfinding programs.

Finding the roots of polynomials is an extremely old area, going back to at least the ancient Greeks. There are many methods and a large literature for them, and many new methods have been developed in the past 2 to 3 decades. As an introduction to the area, see Dejon and Henrici (1969), Henrici (1974, chap. 6), Householder (1970), Traub (1964), and their bibliographies. The article by Wilkinson (1984) shows some of the practical difficulties of solving the polynomial rootfinding problem on a computer. Accurate, efficient, automatic, and reliable computer programs have been produced for finding the roots of polynomials. Among such programs are (a) those of Jenkins (1975), Jenkins and Traub (1970), (1972), and (b) the program ZERPOL of Smith (1967), based on Laguerre's method [see Kahan (1967), Householder (1970, p. 176)]. These automatic programs are much too sophisticated, both mathematically and algorithmically, to discuss in an introductory text such as this one. Nonetheless, they are well worth using. Most people would not be able to write a program that would be as competitive in both speed and accuracy. The latter is especially important, since the polynomial rootfinding problem can be very sensitive to rounding errors, as was shown in examples earlier in the chapter.

The study of numerical methods for solving systems of nonlinear equations and optimization problems is currently a very popular area of research. For introductions to numerical methods for solving nonlinear systems, see Baker and Phillips (1981, pt. 1), Ortega and Rheinboldt (1970), and Rheinboldt (1974). For generalizations of these methods to nonlinear differential and integral equations, see Baker and Phillips (1981), Kantorovich (1948) [a classical paper in this area], Kantorovich and Akilov (1964), and Rall (1969). For a survey of numerical methods for optimization, see Dennis (1984) and Powell (1982). General introductions are given in Dennis and Schnabel (1983), Fletcher (1980), (1981), Gill et al. (1981), and Luenberger (1984). As an example of recent research in optimization theory and in the development of software, see Boggs et al. (1985). For computer programs, see Hiebert (1982) and Moré et al. (1984).

# Bibliography

Baker, C., and C. Phillips, eds. (1981). *The Numerical Solution of Nonlinear Problems*. Clarendon Press, Oxford, England.

Boggs, P., R. Byrd, and R. Schnabel, eds. (1985). *Numerical Optimization 1984*. Society for Industrial and Applied Mathematics, Philadelphia.

Brent, R. (1973). *Algorithms for Minimization Without Derivatives*. Prentice-Hall, Englewood Cliffs, N.J.

Byrne, G., and C. Hall, eds. (1973). *Numerical Solution of Systems of Nonlinear Algebraic Equations*. Academic Press, New York.

Dejon, B., and P. Henrici, eds. (1969). *Constructive Aspects of the Fundamental Theorem of Algebra*. Wiley, New York.

Dekker, T. (1969). Finding a zero by means of successive linear interpolation. In B. Dejon and P. Henrici (eds.), *Constructive Aspects of the Fundamental Theorem of Algebra*, pp. 37–51. Wiley, New York.

Dennis, J. (1984). A user's guide to nonlinear optimization algorithms. *Proc. IEEE*, **72**, 1765–1776.

Dennis, J., and R. Schnabel (1983). *Numerical Methods for Unconstrained Optimization and Nonlinear Equations*. Prentice-Hall, Englewood Cliffs, N.J.

Fletcher, R. (1980). *Practical Methods of Optimization*, Vol. 1, *Unconstrained Optimization*. Wiley, New York.

Fletcher, R. (1981). *Practical Methods of Optimization*, Vol. 2, *Constrained Optimization*. Wiley, New York.

Forsythe, G. (1969). What is a satisfactory quadratic equation solver? In B. Dejon and P. Henrici (eds.), *Constructive Aspects of the Fundamental Theorem of Algebra*, pp. 53–61. Wiley, New York.

Gill, P., W. Murray, and M. Wright (1981). *Practical Optimization*. Academic Press, New York.

Goldstine, H. (1977). *A History of Numerical Analysis*. Springer-Verlag, New York.

Henrici, P. (1974). *Applied and Computational Complex Analysis*, Vol. 1. Wiley, New York.

Hiebert, K. (1982). An evaluation of mathematical software that solve systems of nonlinear equations. *ACM Trans. Math. Softw.*, **11**, 250–262.

Householder, A. (1970). *The Numerical Treatment of a Single Nonlinear Equation*. McGraw-Hill, New York.

Jenkins, M. (1975). Algorithm 493: Zeroes of a real polynomial. *ACM Trans. Math. Softw.*, **1**, 178–189.

Jenkins, M., and J. Traub (1970). A three state algorithm for real polynomials using quadratic iteration. *SIAM J. Numer. Anal.*, **7**, 545–566.

Jenkins, M., and J. Traub (1972). Algorithm 419—Zeros of a complex polynomial. *Commun. ACM*, **15**, 97–99.

Kahan, W. (1967). Laguerre's method and a circle which contains at least one zero of a polynomial. *SIAM J. Numer. Anal.*, **4**, 474–482.

Kantorovich, L. (1948). Functional analysis and applied mathematics. *Usp. Mat. Nauk*, **3**, 89–185.

Kantorovich, L., and G. Akilov (1964). *Functional Analysis in Normed Spaces*. Pergamon, London.

Le, D. (1985). An efficient derivative-free method for solving nonlinear equations. *ACM Trans. Math. Softw.*, **11**, 250–262.

Luenberger, D. (1984). *Linear and Nonlinear Programming*, 2nd ed. Wiley, New York.

Moré, J., B. Garbow, and K. Hillstrom (1980). *User Guide for MINPACK-1*. Argonne Nat. Lab. Rep. ANL-80-74.

Moré, J., and D. Sorenson. Newton's method. In *Studies in Numerical Analysis*, G. Golub (ed.), pp. 29–82. Math. Assoc. America, Washington, D.C.

Moré, J., D. Sorenson, B. Garbow, and K. Hillstrom (1984). The MINPACK project. In *Sources and Development of Mathematical Software*, Cowell (ed.), pp. 88–111. Prentice-Hall, Englewood Cliffs, N.J.

Nelder, A., and R. Mead (1965). A simplex method for function minimization. *Comput. J.*, **7**, 308–313.

Ortega, J., and W. Rheinboldt (1970). *Iterative Solution of Nonlinear Equations in Several Variables*. Academic Press, New York.

Ostrowski, A. (1973). *Solution of Equations in Euclidean and Banach Spaces*. Academic Press, New York.

Peters, G., and J. Wilkinson (1971). Practical problems arising in the solution of polynomial equations. *J. Inst. Math. Its Appl.* **8**, 16–35.

Potra, F., and V. Ptak (1984). *Nondiscrete Induction and Iterative Processes*. Pitman, Boston.

Powell, M., ed. (1982). *Nonlinear Optimization 1981*. NATO Conf. Ser. Academic Press, New York.

Rall, L. (1969). *Computational Solution of Nonlinear Operator Equations*. Wiley, New York.

Rheinboldt, W. (1974). *Methods for Solving Systems of Nonlinear Equations*. Society for Industrial and Applied Mathematics, Philadelphia.

Smith, B. (1967). ZERPOL: A zero finding algorithm for polynomials using Laguerre's method. Dept. of Computer Science, Univ. Toronto, Toronto, Ont., Canada.

Traub, J. (1964). *Iterative Methods for the Solution of Equations*. Prentice-Hall, Englewood Cliffs, N.J.

Whitley, V. (1968). Certification of algorithm 196: Muller's method for finding roots of an arbitrary function. *Commun. ACM* **11**, 12–14.

Wilkinson, J. (1963). *Rounding Errors in Algebraic Processes*. Prentice-Hall, Englewood Cliffs, N.J.

Wilkinson, J. (1984). The perfidious polynomial. In *Studies in Numerical Analysis*, G. Golub (ed.). Math. Assoc. America, Washington, D.C.

Woods, D. (1985). An interactive approach for solving multi-objective optimization problems. Ph.D. dissertation, William Marsh Rice Univ., Houston, Tex.

## Problems

1. The introductory examples for $f(x) = a - (1/x)$ is related to the infinite product

$$\prod_{j=0}^{\infty} (1 + r^{2^j}) \equiv \underset{n \to \infty}{\text{Limit}} \left[ (1 + r)(1 + r^2)(1 + r^4) \cdots (1 + r^{2^n}) \right]$$

By using formula (2.0.6) and (2.0.9), we can calculate the value of the infinite product. What is this value, and what condition on $r$ is required for the infinite product to converge? *Hint:* Let $r = r_0$, and write $x_n$ in terms of $x_0$ and $r_0$.

2.  Write a program implementing the algorithm *Bisect* given in Section 2.1. Use the program to calculate the real roots of the following equations. Use an error tolerance of $\epsilon = 10^{-5}$.

    (a)  $e^x - 3x^2 = 0$    (b)  $x^3 = x^2 + x + 1$    (c)  $e^x = \dfrac{1}{.1 + x^2}$

    (d)  $x = 1 + .3\cos(x)$

3.  Use the program from Problem 2 to calculate (a) the smallest positive root of $x - \tan(x) = 0$, and (b) the root of this equation that is closest to $x = 100$.

4.  Implement the algorithm *Newton* given in Section 2.2. Use it to solve the equations in Problem 2.

5.  Use Newton's method to calculate the roots requested in Problem 3. Attempt to explain the differences in finding the roots of parts (a) and (b).

6.  Use Newton's method to calculate the unique root of

$$x + e^{-Bx^2}\cos(x) = 0$$

with $B > 0$ a parameter to be set. Use a variety of increasing values of $B$, for example, $B = 1, 5, 10, 25, 50$. Among the choices of $x_0$ used, choose $x_0 = 0$ and explain any anomalous behavior. Theoretically, the Newton method will converge for any value of $x_0$ and $B$. Compare this with actual computations for larger values of $B$.

7.  An interesting polynomial rootfinding problem occurs in the computation of annuities. An amount of $P_1$ dollars is put into an account at the beginning of years $1, 2, \ldots, N_1$. It is compounded annually at a rate of $r$ (e.g., $r = .05$ means a 5 percent rate of interest). At the beginning of years $N_1 + 1, \ldots, N_1 + N_2$, a payment of $P_2$ dollars is removed from the account. After the last payment, the account is exactly zero. The relationship of the variables is

$$P_1\left[(1 + r)^{N_1} - 1\right] = P_2\left[1 - (1 + r)^{-N_2}\right]$$

If $N_1 = 30$, $N_2 = 20$, $P_1 = 2000$, $P_2 = 8000$, then what is $r$? Use a rootfinding method of your choice.

8.  Use the Newton–Fourier method to solve the equations in Problems 2 and 6.

9. Use the secant method to solve the equations given in Problem 2.

10. Use the secant method to solve the equation of Problem 6.

11. Show the error formula (2.3.2) for the secant method,

$$\alpha - c = -(\alpha - b)(\alpha - a)\frac{f[a, b, \alpha]}{f[a, b]}$$

12. Consider Newton's method for finding the positive square root of $a > 0$. Derive the following results, assuming $x_0 > 0$, $x_0 \neq \sqrt{a}$.

   (a) $x_{n+1} = \frac{1}{2}\left(x_n + \frac{a}{x_n}\right)$

   (b) $x_{n+1}^2 - a = \left[\frac{x_n^2 - a}{2x_n}\right]^2$   $n \geq 0$, and thus $x_n > \sqrt{a}$ for all $n > 0$.

   (c) The iterates $\{x_n\}$ are a strictly decreasing sequence for $n \geq 1$. *Hint:* Consider the sign of $x_{n+1} - x_n$.

   (d) $e_{n+1} = -e_n^2/(2x_n)$, with $e_n = \sqrt{a} - x_n$,

   $$\text{Rel}(x_{n+1}) = \frac{-\sqrt{a}}{2x_n}[\text{Rel}(x_n)]^2 \quad n \geq 0$$

   with $\text{Rel}(x_n)$ the relative error in $x_n$.

   (e) If $x_0 \geq \sqrt{a}$ and $|\text{Rel}(x_0)| \leq 0.1$, bound $\text{Rel}(x_4)$.

13. Newton's method is the commonly used method for calculating square roots on a computer. To use Newton's method to calculate $\sqrt{a}$, an initial guess $x_0$ must be chosen, and it would be most convenient to use a fixed number of iterates rather than having to test for convergence. For definiteness, suppose that the computer arithmetic is binary and that the mantissa contains 48 binary bits. Write

$$a = \hat{a} \cdot 2^e \quad \tfrac{1}{2} \leq \hat{a} < 1$$

This can be easily modified to the form

$$a = b \cdot 2^f \quad \tfrac{1}{4} \leq b < 1$$

with $f$ an even integer. Then

$$\sqrt{a} = \sqrt{b} \cdot 2^{f/2} \quad \tfrac{1}{2} \leq \sqrt{b} < 1$$

and the number $\sqrt{a}$ will be in standard floating-point form, once $\sqrt{b}$ is known.

This reduces the problem to that of calculating $\sqrt{b}$ for $\frac{1}{4} \le b < 1$. Use the linear interpolating formula

$$x_0 = \tfrac{1}{3}(2b + 1) \qquad \tfrac{1}{4} \le b \le 1$$

as an initial guess for the Newton iteration for calculating $\sqrt{b}$. Bound the error $\sqrt{b} - x_0$. Estimate how many iterates are necessary in order that

$$0 \le x_n - \sqrt{b} \le 2^{-48}$$

which is the limit of machine precision for $b$ on a particular computer. [Note that the effect of rounding errors is being ignored]. How might the choice of $x_0$ be improved?

14. Numerically calculate the Newton iterates for solving $x^2 - 1 = 0$, and use $x_0 = 100,000$. Identify and explain the resulting speed of convergence.

15. (a) Apply Newton's method to the function

$$f(x) = \begin{cases} \sqrt{x} & x \ge 0 \\ -\sqrt{-x} & x < 0 \end{cases}$$

with the root $\alpha = 0$. What is the behavior of the iterates? Do they converge, and if so, at what rate?

(b) Do the same as in (a), but with

$$f(x) = \begin{cases} \sqrt[3]{x^2} & x \ge 0 \\ -\sqrt[3]{x^2} & x < 0 \end{cases}$$

16. A sequence $\{x_n\}$ is said to converge superlinearly to $\alpha$ if

$$|\alpha - x_{n+1}| \le c_n |\alpha - x_n| \qquad n \ge 0$$

with $c_n \to 0$ as $n \to \infty$. Show that in this case,

$$\operatorname*{Limit}_{n \to \infty} \frac{|\alpha - x_n|}{|x_{n+1} - x_n|} = 1$$

Thus $|\alpha - x_n| \doteq |x_{n+1} - x_n|$ is increasingly valid as $n \to \infty$.

17. Newton's method for finding a root $\alpha$ of $f(x) = 0$ sometimes requires the initial guess $x_0$ to be quite close to $\alpha$ in order to obtain convergence. Verify that this is the case for the root $\alpha = \pi/2$ of

$$f(x) = \cos(x) + \sin^2(50x)$$

Give a rough estimate of how small $|x_0 - \alpha|$ should be in order to obtain convergence to $\alpha$. *Hint:* Consider (2.2.6).

18.  Write a program to implement Muller's method. Apply it to the rootfinding problems in Problems 2, 3, and 6.

19.  Show that $x = 1 + \tan^{-1}(x)$ has a solution $\alpha$. Find an interval $[a, b]$ containing $\alpha$ such that for every $x_0 \in [a, b]$, the iteration

$$x_{n+1} = 1 + \tan^{-1}(x_n) \qquad n \geq 0$$

will converge to $\alpha$. Calculate the first few iterates and estimate the rate of convergence.

20.  Do the same as in Problem 19, but with the iteration

$$x_{n+1} = 3 - 2\log(1 + e^{-x_n}) \qquad n \geq 0$$

21.  To find a root for $f(x) = 0$ by iteration, rewrite the equation as

$$x = x + cf(x) \equiv g(x)$$

for some constant $c \neq 0$. If $\alpha$ is a root of $f(x)$ and if $f'(\alpha) \neq 0$, how should $c$ be chosen in order that the sequence $x_{n+1} = g(x_n)$ converges to $\alpha$?

22.  Consider the equation

$$x = d + hf(x)$$

with $d$ a given constant and $f(x)$ continuous for all $x$. For $h = 0$, a root is $\alpha = d$. Show that for all sufficiently small $h$, this equation has a root $\alpha(h)$. What condition is needed, if any, in order to ensure the uniqueness of the root $\alpha(h)$ in some interval about $d$?

23.  The iteration $x_{n+1} = 2 - (1 + c)x_n + cx_n^3$ will converge to $\alpha = 1$ for some values of $c$ [provided $x_0$ is chosen sufficiently close to $\alpha$]. Find the values of $c$ for which this is true. For what value of $c$ will the convergence be quadratic?

24.  Which of the following iterations will converge to the indicated fixed point $\alpha$ (provided $x_0$ is sufficiently close to $\alpha$)? If it does converge, give the order of convergence; for linear convergence, give the rate of linear convergence.

(a)  $x_{n+1} = -16 + 6x_n + \dfrac{12}{x_n} \qquad \alpha = 2$

(b)  $x_{n+1} = \dfrac{2}{3}x_n + \dfrac{1}{x_n^2} \qquad \alpha = 3^{1/3}$

(c)  $x_{n+1} = \dfrac{12}{1 + x_n} \qquad \alpha = 3$

**25.** Show that

$$x_{n+1} = \frac{x_n(x_n^2 + 3a)}{3x_n^2 + a} \qquad n \geq 0$$

is a third-order method for computing $\sqrt{a}$. Calculate

$$\underset{n \to \infty}{\text{Limit}} \frac{\sqrt{a} - x_{n+1}}{\left(\sqrt{a} - x_n\right)^3}$$

assuming $x_0$ has been chosen sufficiently close to $\alpha$.

**26.** Using Theorem 2.8, show that formula (2.4.11) is a cubically convergent iteration method.

**27.** Define an iteration formula by

$$x_{n+1} = z_{n+1} - \frac{f(z_{n+1})}{f'(x_n)}, \qquad z_{n+1} = x_n - \frac{f(x_n)}{f'(x_n)}$$

Show that the order of convergence of $\{x_n\}$ to $\alpha$ is at least 3. *Hint:* Use theorem 2.8, and let

$$g(x) = h(x) - \frac{f(h(x))}{f'(x)} \qquad h(x) = x - \frac{f(x)}{f'(x)}$$

**28.** There is another modification of Newton's method, similar to the secant method, but using a different approximation to the derivative $f'(x_n)$. Define

$$x_{n+1} = x_n - \frac{f(x_n)}{D(x_n)} \qquad D(x_n) = \frac{f(x_n + f(x_n)) - f(x_n)}{f(x_n)} \qquad n \geq 0$$

This one-point method is called *Steffenson's method*. Assuming $f'(\alpha) \neq 0$, show that this is a second-order method. *Hint:* Write the iteration as $x_{n+1} = g(x_n)$. Use $f(x) = (x - \alpha)h(x)$ with $h(\alpha) \neq 0$, and then compute the formula for $g(x)$ in terms of $h(x)$. Having done so, apply Theorem 2.8.

**29.** Given below is a table of iterates from a linearly convergent iteration $x_{n+1} = g(x_n)$. Estimate (a) the rate of linear convergence, (b) the fixed point $\alpha$, and (c) the error $\alpha - x_5$.

| $n$ | $x_n$ |
|---|---|
| 0 | 1.0949242 |
| 1 | 1.2092751 |
| 2 | 1.2807917 |
| 3 | 1.3254943 |
| 4 | 1.3534339 |
| 5 | 1.3708962 |

**30.**   The algorithm *Aitken*, given in Section 2.6, can be shown to be second order in its speed of convergence. Let the original iteration be $x_{n+1} = g(x_n)$, $n \geq 0$. The formula (2.6.8) can be rewritten in the equivalent form

$$\alpha \doteq \hat{x}_{n-2} = x_{n-2} + \frac{(x_{n-1} - x_{n-2})^2}{(x_{n-1} - x_{n-2}) - (x_n - x_{n-1})} \qquad n \geq 2$$

To examine the speed of convergence of the Aitken extrapolates, we consider the associated sequence

$$z_{n+1} = z_n + \frac{[g(z_n) - z_n]^2}{[g(z_n) - z_n] - [g(g(z_n)) - g(z_n)]} \qquad n \geq 0$$

The values $z_n$ are the successive values of $\hat{x}_n$ produced in the algorithm *Aitken*.

For $g'(\alpha) \neq 0$ or 1, show that $z_n$ converges to $\alpha$ quadratically. This is true even if $|g'(\alpha)| > 1$ and the original iteration is divergent. *Hint:* Do not attempt to use Theorem 2.8 directly, as it will be too complicated. Instead write

$$g(x) = (x - \alpha)h(x) \qquad h(\alpha) = g'(\alpha) \neq 0$$

Use this to show that

$$\alpha - z_{n+1} = H(z_n)(\alpha - z_n)^2$$

for some function $H(x)$ bounded about $x = \alpha$.

**31.**   Consider the sequence

$$x_n = \alpha + \beta \rho^n + \gamma \rho^{2n}, \qquad n \geq 0, \quad |\rho| < 1$$

with $\beta, \gamma \neq 0$, which converges to $\alpha$ with a linear rate of $\rho$. Let $\hat{x}_{n-2}$ be the Aitken extrapolate:

$$\hat{x}_{n-2} = x_n - \frac{(x_n - x_{n-1})^2}{(x_n - x_{n-1}) - (x_{n-1} - x_{n-2})} \qquad n \geq 0$$

Show that

$$\hat{x}_{n-2} = \alpha + a\rho^{2n} + b\rho^{4n} + c_n \rho^{6n}$$

where $c_n$ is bounded as $n \to \infty$. Derive expressions for $a$ and $b$. The sequence $\{\hat{x}_n\}$ converges to $\alpha$ with a linear rate of $\rho^2$.

**32.**   Let $f(x)$ have a multiple root $\alpha$, say of multiplicity $m \geq 1$. Show that

$$K(x) = \frac{f(x)}{f'(x)}$$

has $\alpha$ as a simple root. Why does this not help with the fundamental difficulty in numerically calculating multiple roots, namely that of the large interval of uncertainty in $\alpha$?

33.    Use Newton's method to calculate the real roots of the following polynomials as accurately as possible. Estimate the multiplicity of each root, and then if necessary, try an alternative way of improving your calculated values.

(a)    $x^4 - 3.2x^3 + .96x^2 + 4.608x - 3.456$

(b)    $x^5 + .9x^4 - 1.62x^3 - 1.458x^2 + .6561x + .59049$

34.    Use the program from Problem 2 to solve the following equations for the root $\alpha = 1$. Use the initial interval $[0, 3]$, and in all cases use $\epsilon = 10^{-5}$ as the stopping tolerance. Compare the results with those obtained in Section 2.8 using Brent's algorithm.

(i)    $(x - 1)[1 + (x - 1)^2] = 0$

(ii)    $x^2 - 1 = 0$

(iii)    $-1 + x(3 + x(-3 + x)) = 0$

(iv)    $(x - 1)\exp(-1/(x - 1)^2) = 0$

35.    Prove the lower bound in (2.9.6), using the upper bound in (2.9.6) and the suggestion in (2.9.7).

36.    Let $p(x)$ be a polynomial of degree $n$. Let its distinct roots be denoted by $\alpha_1, \ldots, \alpha_r$, of respective multiplicities $m_1, \ldots, m_r$.

(a)    Show that

$$\frac{p'(x)}{p(x)} = \sum_{j=1}^{r} \frac{m_j}{x - \alpha_j}$$

(b)    Let $c$ be a number for which $p'(c) \neq 0$. Show there exists a root $\alpha$ of $p(x)$ satisfying

$$|\alpha - c| \leq n \left| \frac{p(c)}{p'(c)} \right|$$

37.    For the polynomial

$$p(x) = a_0 + a_1 x + \cdots + a_n x^n \qquad a_n \neq 0$$

define

$$R = \frac{|a_0| + |a_1| + \cdots + |a_{n-1}|}{|a_n|}$$

Show that every root $x$ of $p(x) = 0$ satisfies

$$|x| \leq \text{Max}\left\{ R, \sqrt[n]{R} \right\}$$

**38.** Write a computer program to evaluate the following polynomials $p(x)$ for the given values of $x$ and to evaluate the *noise* in the values $p(x)$. For each $x$, evaluate $p(x)$ in both single and double precision arithmetic; use their difference as the noise in the single precision value, due to rounding errors in the evaluation of $p(x)$. Use both the ordinary formula (2.9.1) and Horner's rule (2.9.8) to evaluate each polynomial; this should show that the noise is different in the two cases.

**(a)** $p(x) = x^4 - 5.7x^3 - .47x^2 + 29.865x - 26.1602$ $\qquad -3 \leq x \leq 5$
with steps of $h = 0.1$ for $x$.

**(b)** $p(x) = x^4 - 5.4x^3 + 10.56x^2 - 8.954x + 2.7951$ $\qquad 1 \leq x \leq 1.2$
in steps of $h = .001$ for $x$.

*Note:* Smaller or larger values of $h$ may be appropriate on different computers. Also, before using double precision, enter the coefficients in single precision, for a more valid comparison.

**39.** Use complex arithmetic and Newton's method to calculate a complex root of

$$p(z) = z^4 - 3z^3 + 20z^2 + 44z + 54$$

located near to $z_0 = 2.5 + 4.5i$.

**40.** Write a program to find the roots of the following polynomials as accurately as possible.

**(a)** $676039x^{12} - 1939938x^{10} + 2078505x^8 - 1021020x^6 + 225225x^4$
$- 18018x^2 + 231$

**(b)** $x^4 - 4.096152422706631x^3 + 3.284232335022705x^2$
$+ 4.703847577293368x - 5.715767664977294$

**41.** Use a package rootfinding program for polynomials to find the roots of the polynomials in Problems 38, 39, and 40.

**42.** For the example $f(x) = (x - 1)(x - 2) \cdots (x - 7)$, (2.9.21) in Section 2.9, consider perturbing the coefficient of $x^i$ by $\epsilon_i x^i$, in which $\epsilon_i$ is chosen so that the relative perturbation in the coefficient of $x^i$ is the same as that of the example in the text in the coefficient of $x^6$. What does the linearized theory (2.9.19) predict for the perturbations in the roots? A change in which coefficient will lead to the greatest changes in the roots?

43.   The polynomial $f(x) = x^5 - 300x^2 - 126x + 5005$ has $\alpha = 5$ as a root. Estimate the effect on $\alpha$ of changing the coefficient of $x^5$ from 1 to $1 + \epsilon$.

44.   The stability result (2.9.19) for polynomial roots can be generalized to general functions. Let $\alpha$ be a simple root of $f(x) = 0$, and let $f(x)$ and $g(x)$ be continuously differentiable about $\alpha$. Define $F_\epsilon(x) = f(x) + \epsilon g(x)$. Let $\alpha(\epsilon)$ denote a root of $F_\epsilon(x) = 0$, corresponding to $\alpha = \alpha(0)$ for small $\epsilon$. To see that there exists such an $\alpha(\epsilon)$, and to prove that it is continuously differentiable, use the *implicit function theorem* for functions of one variable. From this, generalize (2.9.19) to the present situation.

45.   Using the stability result in Problem 44, estimate the root $\alpha(\epsilon)$ of

$$x - \tan(x) + \epsilon = 0$$

Consider two cases explicitly for roots $\alpha$ of $x - \tan(x) = 0$: (1) $\alpha \in (.5\pi, 1.5\pi)$, (2) $\alpha \in (31.5\pi, 32.5\pi)$.

46.   Consider the system

$$x = \frac{.5}{1 + (x + y)^2} \qquad y = \frac{.5}{1 + (x - y)^2}$$

Find a bounded region $D$ for which the hypotheses of Theorem 2.9 are satisfied. *Hint:* What will be the sign of the components of the root $\alpha$? Also, what are the maximum possible values for $x$ and $y$ in the preceding formulas?

47.   Consider the system

$$x = 1 + h \cdot \frac{e^{-x^2}}{1 + y^2} \qquad y = .5 + h \cdot \tan^{-1}(x^2 + y^2)$$

Show that if $h$ is chosen sufficiently small, then this system has a unique solution $\alpha$ within some rectangular region. Moreover, show that simple iteration of the form (2.10.4) will converge to this solution.

48.   Prove Corollary 2.10. *Hint:* Use the continuity of the partial derivatives of the components of $\mathbf{g}(\mathbf{x})$.

49.   Prove that the iterates $\{\mathbf{x}_n\}$ in Theorem 2.9 will converge to a solution of $\mathbf{x} = \mathbf{g}(\mathbf{x})$. *Hint:* Consider the infinite sum

$$\mathbf{x}_0 + \sum_{n=0}^{\infty} [\mathbf{x}_{n+1} - \mathbf{x}_n]$$

Its partial sums are

$$\mathbf{x}_0 + \sum_{n=0}^{N-1} [\mathbf{x}_{n+1} - \mathbf{x}_n] = \mathbf{x}_N$$

Thus if the infinite series converges, say to $\alpha$, then $\mathbf{x}_N$ converges to $\alpha$. Show the infinite series converges absolutely by showing and using the result

$$\|\mathbf{x}_{n+1} - \mathbf{x}_n\|_\infty \le \lambda \|\mathbf{x}_n - \mathbf{x}_{n-1}\|_\infty$$

Following this, show that $\alpha$ is a fixed point of $\mathbf{g}(\mathbf{x})$.

**50.** Using Newton's method for nonlinear systems, solve the nonlinear system

$$x^2 + y^2 = 4 \qquad x^2 - y^2 = 1$$

The true solutions are easily determined to be $(\pm \sqrt{2.5}, \pm \sqrt{1.5})$. As an initial guess, use $(x_0, y_0) = (1.6, 1.2)$.

**51.** Solve the system

$$x^2 + xy^3 = 9 \qquad 3x^2 y - y^3 = 4$$

using Newton's method for nonlinear systems. Use each of the initial guesses $(x_0, y_0) = (1.2, 2.5)$, $(-2, 2.5)$, $(-1.2, -2.5)$, $(2, -2.5)$. Observe which root to which the method converges, the number of iterates required, and the speed of convergence.

**52.** Using Newton's method for nonlinear systems, solve for all roots of the following nonlinear system. Use graphs to estimate initial guesses.

(a)   $x^2 + y^2 - 2x - 2y + 1 = 0 \qquad x + y - 2xy = 0$

(b)   $x^2 + 2xy + y^2 - x + y - 4 = 0$
       $5x^2 - 6xy + 5y^2 + 16x - 16y + 12 = 0$

**53.** Prove that the Jacobian of

$$\mathbf{g}(\mathbf{x}) = \mathbf{x} - F(\mathbf{x})^{-1}\mathbf{f}(\mathbf{x})$$

is zero at any root $\alpha$ of $\mathbf{f}(\mathbf{x}) = 0$, provided $F(\alpha)$ is nonsingular. Combined with Corollary 2.10 of Section 2.10, this will prove the convergence of Newton's method.

**54.** Use Newton's method (2.12.4) to find the minimum value of the function

$$f(\mathbf{x}) = x_1^4 + x_1 x_2 + (1 + x_2)^2$$

Experiment with various initial guesses and observe the pattern of convergence.

# THREE

# INTERPOLATION THEORY

The concept of interpolation is the selection of a function $p(x)$ from a given class of functions in such a way that the graph of $y = p(x)$ passes through a finite set of given data points. In most of this chapter we limit the interpolating function $p(x)$ to being a polynomial.

Polynomial interpolation theory has a number of important uses. In this text, its primary use is to furnish some mathematical tools that are used in developing methods in the areas of approximation theory, numerical integration, and the numerical solution of differential equations. A second use is in developing means for working with functions that are stored in tabular form. For example, almost everyone is familiar from high school algebra with linear interpolation in a table of logarithms. We derive computationally convenient forms for polynomial interpolation with tabular data and analyze the resulting error. It is recognized that with the widespread use of calculators and computers, there is far less use for table interpolation than in the recent past. We have included it because the resulting formulas are still useful in other connections and because table interpolation provides us with convenient examples and exercises.

The chapter concludes with introductions to two other topics. These are (1) piecewise polynomial interpolating functions, spline functions in particular; and (2) interpolation with trigonometric functions.

## 3.1 Polynomial Interpolation Theory

Let $x_0, x_1, \ldots, x_n$ be distinct real or complex numbers, and let $y_0, y_1, \ldots, y_n$ be associated function values. We now study the problem of finding a polynomial $p(x)$ that interpolates the given data:

$$p(x_i) = y_i \qquad i = 0, 1, \ldots, n \qquad (3.1.1)$$

Does such a polynomial exist, and if so, what is its degree? Is it unique? What is a formula for producing $p(x)$ from the given data?

By writing

$$p(x) = a_0 + a_1 x + \cdots + a_m x^m$$

for a general polynomial of degree $m$, we see there are $m + 1$ independent parameters $a_0, a_1, \ldots, a_m$. Since (3.1.1) imposes $n + 1$ conditions on $p(x)$, it is reasonable to first consider the case when $m = n$. Then we want to find $a_0, a_1, \ldots, a_n$ such that

$$a_0 + a_1 x_0 + a_2 x_0^2 + \cdots + a_n x_0^n = y_0$$

$$\vdots$$

$$a_0 + a_1 x_n + a_2 x_n^2 + \cdots + a_n x_n^n = y_n \qquad (3.1.2)$$

This is a system of $n + 1$ linear equations in $n + 1$ unknowns, and solving it is completely equivalent to solving the polynomial interpolation problem. In vector and matrix notation, the system is

$$Xa = y$$

with

$$X = \left[ x_i^j \right] \qquad i, j = 0, 1, \ldots, n \qquad (3.1.3)$$

$$a = \left[ a_0, a_1, \ldots, a_n \right]^T \qquad y = \left[ y_0, \ldots, y_n \right]^T$$

The matrix $X$ is called a *Vandermonde matrix*.

**Theorem 3.1**    Given $n + 1$ distinct points $x_0, \ldots, x_n$ and $n + 1$ ordinates $y_0, \ldots, y_n$, there is a polynomial $p(x)$ of degree $\leq n$ that interpolates $y_i$ at $x_i$, $i = 0, 1, \ldots, n$. This polynomial $p(x)$ is unique among the set of all polynomials of degree at most $n$.

**Proof**    Three proofs of this important result are given. Each will furnish some needed information and has important uses in other interpolation problems.

(i)    It can be shown that for the matrix $X$ in (3.1.3),

$$\det(X) = \prod_{0 \leq j < i \leq n} (x_i - x_j) \qquad (3.1.4)$$

(see Problem 1). This shows that $\det(X) \neq 0$, since the points $x_i$ are distinct. Thus $X$ is nonsingular and the system $Xa = y$ has a unique solution $a$. This proves the existence and uniqueness of an interpolating polynomial of degree $\leq n$.

(ii)    By a standard theorem of linear algebra (see Theorem 7.2 of Chapter 7), the system $Xa = y$ has a unique solution if and only if the homogeneous system $Xb = 0$ has only the trivial solution $b = 0$. Therefore, assume $Xb = 0$ for some $b$. Using $b$, define

$$p(x) = b_0 + b_1 x + \cdots + b_n x^n$$

From the system $Xb = 0$, we have

$$p(x_i) = 0 \qquad i = 0, 1, \ldots, n$$

The polynomial $p(x)$ has $n + 1$ zeros and degree $p(x) \leq n$. This is not possible unless $p(x) \equiv 0$. But then all coefficients $b_i = 0$, $i = 0, 1, \ldots, n$, completing the proof.

**(iii)**   We now exhibit explicitly the interpolating polynomial. To begin, we consider the special interpolation problem in which

$$y_i = 1 \qquad y_j = 0 \qquad \text{for} \quad j \neq i$$

for some $i$, $0 \leq i \leq n$. We want a polynomial of degree $\leq n$ with the $n$ zeros $x_j$, $j \neq i$. Then

$$p(x) = c(x - x_0) \cdots (x - x_{i-1})(x - x_{i+1}) \cdots (x - x_n)$$

for some constant $c$. The condition $p(x_i) = 1$ implies

$$c = [(x_i - x_0) \cdots (x_i - x_{i-1})(x_i - x_{i+1}) \cdots (x_i - x_n)]^{-1}$$

This special polynomial is written as

$$l_i(x) = \prod_{j \neq i} \left( \frac{x - x_j}{x_i - x_j} \right) \qquad i = 0, 1, \ldots, n \qquad (3.1.5)$$

To solve the general interpolation problem (3.1.1), write

$$p(x) = y_0 l_0(x) + y_1 l_1(x) + \cdots + y_n l_n(x)$$

With the special properties of the polynomials $l_i(x)$, $p(x)$ easily satisfies (3.1.1). Also, degree $p(x) \leq n$, since all $l_i(x)$ have degree $n$.

To prove uniqueness, suppose $q(x)$ is another polynomial of degree $\leq n$ that satisfies (3.1.1). Define

$$r(x) = p(x) - q(x)$$

Then degree $r(x) \leq n$, and

$$r(x_i) = p(x_i) - q(x_i) = y_i - y_i = 0 \qquad i = 0, 1, \ldots, n$$

Since $r(x)$ has $n + 1$ zeros, we must have $r(x) \equiv 0$. This proves $p(x) \equiv q(x)$, completing the proof.  ■

Uniqueness is a property that is of practical use in much that follows. We derive other formulas for the interpolation problem (3.1.1), and uniqueness says they are the same polynomial. Also, without uniqueness the linear system (3.1.2) would not be uniquely solvable; from results in linear algebra, this would imply the existence of data vectors $y$ for which there is no interpolating polynomial of degree $\leq n$.

The formula

$$p_n(x) = \sum_{i=0}^{n} y_i l_i(x) \tag{3.1.6}$$

is called *Lagrange's formula* for the interpolating polynomial.

*Example*

$$p_1(x) = \frac{x - x_1}{x_0 - x_1} y_0 + \frac{x - x_0}{x_1 - x_0} y_1 = \frac{(x_1 - x) y_0 + (x - x_0) y_1}{x_1 - x_0}$$

$$p_2(x) = \frac{(x - x_1)(x - x_2)}{(x_0 - x_1)(x_0 - x_2)} y_0 + \frac{(x - x_0)(x - x_2)}{(x_1 - x_0)(x_1 - x_2)} y_1$$

$$+ \frac{(x - x_0)(x - x_1)}{(x_2 - x_0)(x_2 - x_1)} y_2$$

The polynomial of degree $\leq 2$ that passes through the three points $(0, 1)$, $(-1, 2)$, and $(1, 3)$ is

$$p_2(x) = \frac{(x + 1)(x - 1)}{(0 + 1)(0 - 1)} \cdot 1 + \frac{(x - 0)(x - 1)}{(-1 - 0)(-1 - 1)} \cdot 2 + \frac{(x - 0)(x + 1)}{(1 - 0)(1 + 1)} \cdot 3$$

$$= 1 + \frac{1}{2} x + \frac{3}{2} x^2$$

If a function $f(x)$ is given, then we can form an approximation to it using the interpolating polynomial

$$p_n(x; f) \equiv p_n(x) = \sum_{i=0}^{n} f(x_i) l_i(x) \tag{3.1.7}$$

This interpolates $f(x)$ at $x_0, \ldots, x_n$. For example, we later consider $f(x) = \log_{10} x$ with linear interpolation. The basic result used in analyzing the error of interpolation is the following theorem. As a notation, $\mathcal{H}\{a, b, c, \ldots\}$ denotes the smallest interval containing all of the real numbers $a, b, c, \ldots$.

**Theorem 3.2**   Let $x_0, x_1, \ldots, x_n$ be distinct real numbers, and let $f$ be a given real valued function with $n + 1$ continuous derivatives on the interval $I_t = \mathcal{H}\{t, x_0, \ldots, x_n\}$, with $t$ some given real number.

Then there exists $\xi \in I_t$ with

$$f(t) - \sum_{j=0}^{n} f(x_j) l_j(t) = \frac{(t - x_0) \cdots (t - x_n)}{(n + 1)!} f^{(n+1)}(\xi) \quad (3.1.8)$$

***Proof*** Note that the result is trivially true if $t$ is any node point, since then both sides of (3.1.8) are zero. Assume $t$ does not equal any node point. Then define

$$E(x) = f(x) - p_n(x) \qquad p_n(x) = \sum_{j=0}^{n} f(x_j) l_j(x)$$

$$G(x) = E(x) - \frac{\Psi(x)}{\Psi(t)} E(t) \qquad \text{for all} \quad x \in I_t \qquad (3.1.9)$$

with

$$\Psi(x) = (x - x_0) \cdots (x - x_n)$$

The function $G(x)$ is $n + 1$ times continuously differentiable on the interval $I_t$, as are $E(x)$ and $\Psi(x)$. Also,

$$G(x_i) = E(x_i) - \frac{\Psi(x_i)}{\Psi(t)} E(t) = 0 \qquad i = 0, 1, \ldots, n$$

$$G(t) = E(t) - E(t) = 0$$

Thus $G$ has $n + 2$ distinct zeros in $I_t$. Using the mean value theorem, $G'$ has $n + 1$ distinct zeros. Inductively, $G^{(j)}(x)$ has $n + 2 - j$ zeros in $I_t$, for $j = 0, 1, \ldots, n + 1$. Let $\xi$ be a zero of $G^{(n+1)}(x)$,

$$G^{(n+1)}(\xi) = 0$$

Since

$$E^{(n+1)}(x) = f^{(n+1)}(x)$$

$$\Psi^{(n+1)}(x) = (n + 1)!$$

we obtain

$$G^{(n+1)}(x) = f^{(n+1)}(x) - \frac{(n + 1)!}{\Psi(t)} E(t)$$

Substituting $x = \xi$ and solving for $E(t)$,

$$E(t) = \frac{\Psi(t)}{(n + 1)!} \cdot f^{(n+1)}(\xi)$$

the desired result.

This may seem a "tricky" derivation, but it is a commonly used technique for obtaining some error formulas. ∎

**Example** For $n = 1$, using $x$ in place of $t$,

$$f(x) - \frac{(x_1 - x)f(x_0) + (x - x_0)f(x_1)}{x_1 - x_0} = \frac{(x - x_0)(x - x_1)}{2}f''(\xi_x)$$

$$(3.1.10)$$

for some $\xi_x \in \mathcal{H}\{x_0, x_1, x\}$. The subscript $x$ on $\xi_x$ shows explicitly that $\xi$ depends on $x$; usually we omit the subscript, for convenience.

We now apply the $n = 1$ case to the common high school technique of linear interpolation in a logarithm table. Let

$$f(x) = \log_{10} x$$

Then $f''(x) = -\log_{10} e/x^2$, $\log_{10} e \doteq 0.434$. In a table, we generally would have $x_0 < x < x_1$. Then

$$E(x) = \frac{(x - x_0)(x_1 - x)}{2} \cdot \frac{\log_{10} e}{\xi^2} \qquad x_0 \le \xi \le x_1$$

This gives the upper and lower bounds

$$\frac{\log_{10} e}{x_1^2} \cdot \frac{(x - x_0)(x_1 - x)}{2} \le E(x) \le \frac{\log_{10} e}{x_0^2} \cdot \frac{(x - x_0)(x_1 - x)}{2}$$

This shows that the error function $E(x)$ looks very much like a quadratic polynomial, especially if the distance $h = x_1 - x_0$ is reasonably small. For a uniform bound on $[x_0, x_1]$,

$$\underset{x_0 \le x \le x_1}{\text{Max}} (x_1 - x)(x - x_0) = \frac{h^2}{4}$$

$$|\log_{10} x - p_1(x)| \le \frac{h^2}{8} \frac{.434}{x_0^2} = \frac{.0542 h^2}{x_0^2} \le .0542 h^2 \qquad (3.1.11)$$

for $x_0 \ge 1$, as is usual in a logarithm table. Note that the interpolation error in a standard table is much less for $x$ near 10 than near 1. Also, the maximum error is near the midpoint of $[x_0, x_1]$.

For a four-place table, $h = .01$,

$$|\log_{10} x - p_1(x)| \le 5.42 \times 10^{-6} \qquad 1 \le x_0 < x_1 \le 10$$

Since the entries in the table are given to four digits (e.g., $\log_{10} 2 = .3010$), this result is sufficiently accurate. Why do we need a more accurate five-place table if the preceding is so accurate? Because we have neglected to include the effects of the rounding errors present in the table entries. For example, with $\log_{10} 2 \doteq .3010$.

$$|\log_{10} 2 - .3010| \leq .00005$$

and this will dominate the interpolation error if $x_0$ or $x_1 = 2$.

**Rounding error analysis for linear interpolation**   Let

$$f(x_0) = f_0 + \epsilon_0 \qquad f(x_1) = f_1 + \epsilon_1$$

with $f_0$ and $f_1$ the table entries and $\epsilon_0, \epsilon_1$ the rounding errors. We will assume

$$|\epsilon_0|, |\epsilon_1| \leq \epsilon$$

for a known $\epsilon$. In the case of the four-place logarithm table, $\epsilon = .00005$.
   We want to bound

$$\mathcal{E}(x) = f(x) - \frac{(x_1 - x)f_0 + (x - x_0)f_1}{x_1 - x_0} \qquad x_0 \leq x \leq x_1 \quad (3.1.12)$$

Using $f_i = f(x_i) - \epsilon_i$,

$$\mathcal{E}(x) = f(x) - \frac{(x_1 - x)f(x_0) + (x - x_0)f(x_1)}{x_1 - x_0}$$

$$+ \frac{(x_1 - x)\epsilon_0 + (x - x_0)\epsilon_1}{x_1 - x_0}$$

$$\equiv E(x) + R(x) \qquad\qquad (3.1.13)$$

$$E(x) = \frac{(x - x_0)(x - x_1)}{2}f''(\xi) \qquad \xi \in [x_0, x_1]$$

The error $\mathcal{E}(x)$ is the sum of the theoretical interpolation error $E(x)$ and the function $R(x)$, which depends on $\epsilon_0, \epsilon_1$. Since $R(x)$ is a straight line, its maximum on $[x_0, x_1]$ is attained at an endpoint,

$$\underset{x_0 \leq x \leq x_1}{\text{Max}} |R(x)| = \text{Max}\left\{|\epsilon_0|, |\epsilon_1|\right\} \leq \epsilon \qquad (3.1.14)$$

With $x_1 = x_0 + h$, $x_0 \leq x \leq x_1$,

$$|\mathcal{E}(x)| \leq \frac{h^2}{8}\underset{x_0 \leq t \leq x_1}{\text{Max}} |f''(t)| + \text{Max}\left\{|\epsilon_0|, |\epsilon_1|\right\} \qquad (3.1.15)$$

*Example*   For the earlier logarithm example using a four-place table,

$$|\mathcal{E}(x)| \leq 5.42 \times 10^{-6} + 5 \times 10^{-5} \doteq 5.5 \times 10^{-5}$$

For a five-place table, $h = .001$, $\epsilon = .000005$, and

$$|\mathcal{E}(x)| \leq 5.42 \times 10^{-8} + 5 \times 10^{-6} \doteq 5.05 \times 10^{-6} \qquad x_0 \leq x \leq x_1$$

The rounding error is the only significant error in using linear interpolation in a five-place logarithm table. In fact, it would seem worthwhile to increase the five-place table to a six-place table, without changing the mesh size $h$. Then we would have a maximum error for $\mathcal{E}(x)$ of $5.5 \times 10^{-7}$, without any significant increase in computation. These arguments on rounding error generalize to higher degree polynomial interpolation, although the result on Max $|R(x)|$ is slightly more complicated (see Problem 8).

None of the results of this section take into account new rounding errors that occur in the evaluation of $p_n(x)$. These are minimized by results given in the next section.

## 3.2   Newton Divided Differences

The Lagrange form of the interpolation polynomial can be used for interpolation to a function given in tabular form; tables in Abramowitz and Stegun (1964, chap. 25) can be used to evaluate the functions $l_i(x)$ more easily. But there are other forms that are much more convenient, and they are developed in this and the following section. With the Lagrange form, it is inconvenient to pass from one interpolation polynomial to another of degree one greater. Such a comparison of different degree interpolation polynomials is a useful technique in deciding what degree polynomial to use. The formulas developed in this section are for nonevenly spaced grid points $\{x_i\}$. As such they are convenient for inverse interpolation in a table, a point we illustrate later. These formulas are specialized in Section 3.3 to the case of evenly spaced grid points.

We would like to write

$$p_n(x) = p_{n-1} + C(x) \qquad C(x) = \text{correction term} \qquad (3.2.1)$$

Then, in general, $C(x)$ is a polynomial of degree $n$, since usually degree $(p_{n-1})$ $= n - 1$ and degree $(p_n) = n$. Also we have

$$C(x_i) = p_n(x_i) - p_{n-1}(x_i) = f(x_i) - f(x_i) = 0 \qquad i = 0, \ldots, n - 1$$

Thus

$$C(x) = a_n(x - x_0) \cdots (x - x_{n-1})$$

Since $p_n(x_n) = f(x_n)$, we have from (3.2.1) that

$$a_n = \frac{f(x_n) - p_{n-1}(x_n)}{(x_n - x_0) \cdots (x_n - x_{n-1})}$$

For reasons derived below, this coefficient $a_n$ is called the n*th-order Newton divided difference* of $f$, and it is denoted by

$$a_n \equiv f[x_0, x_1, \ldots, x_n]$$

Thus our interpolation formula becomes

$$p_n(x) = p_{n-1}(x) + (x - x_0) \cdots (x - x_{n-1}) f[x_0, \ldots, x_n] \quad (3.2.2)$$

To obtain more information on $a_n$, we return to the Lagrange formula (3.1.7) for $p_n(x)$. Write

$$\Psi_n(x) = (x - x_0) \cdots (x - x_n) \quad (3.2.3)$$

Then

$$\Psi'_n(x_i) = (x_i - x_0) \cdots (x_i - x_{i-1})(x_i - x_{i+1}) \cdots (x_i - x_n)$$

and if $x$ is not a node point

$$p_n(x) = \sum_{j=0}^{n} \frac{\Psi_n(x)}{(x - x_j)\Psi'_n(x_j)} \cdot f(x_j) \quad (3.2.4)$$

Since $a_n$ is the coefficient of $x^n$ in $p_n(x)$, we use the Lagrange formula to obtain the coefficient of $x^n$. By looking at each n th-degree term in the formula (3.2.4), we obtain

$$f[x_0, x_1, \ldots, x_n] = \sum_{j=0}^{n} \frac{f(x_j)}{\Psi'_n(x_j)} \quad (3.2.5)$$

From this formula, we obtain an important property of the divided difference. Let $(i_0, i_1, \ldots, i_n)$ be some permutation of $(0, 1, \ldots, n)$. Then easily

$$\sum_{j=0}^{n} \frac{f(x_j)}{\Psi'_n(x_j)} = \sum_{j=0}^{n} \frac{f(x_{i_j})}{\Psi'_n(x_{i_j})}$$

since the second sum is merely a rearrangement of the first one. But then

$$f[x_0, x_1, \ldots, x_n] = f[x_{i_0}, x_{i_1}, \ldots, x_{i_n}] \quad (3.2.6)$$

for any permutation $(i_0, \ldots, i_n)$ of $(0, 1, \ldots, n)$.

Another useful formula for computing $f[x_0, \ldots, x_n]$ is

$$f[x_0, x_1, \ldots, x_n] = \frac{f[x_1, \ldots, x_n] - f[x_0, \ldots, x_{n-1}]}{x_n - x_0} \quad (3.2.7)$$

which also explains the name of *divided difference*. This result can be proved

**Table 3.1   Format for constructing divided differences of $f(x)$**

| $x_i$ | $f(x_i)$ | $f[x_i, x_{i+1}]$ | $f[x_i, x_{i+1}, x_{i+2}] \cdots$ |
|---|---|---|---|
| $x_0$ | $f_0$ | | |
| | | $f[x_0, x_1]$ | |
| $x_1$ | $f_1$ | | $f[x_0, x_1, x_2] \cdots$ |
| | | $f[x_1, x_2]$ | |
| $x_2$ | $f_2$ | | $f[x_1, x_2, x_3] \cdots$ |
| | | $f[x_2, x_3]$ | |
| $x_3$ | $f_3$ | | $f[x_2, x_3, x_4] \cdots$ |
| | | $f[x_3, x_4]$ | |
| $x_4$ | $f_4$ | | $f[x_3, x_4, x_5] \cdots$ |
| | | $f[x_4, x_5]$ | |
| $x_5$ | $f_5$ | | |
| $\vdots$ | $\vdots$ | $\vdots$ | $\vdots$ |

from (3.2.5) or from the following alternative formula for $p_n(x)$:

$$p_n(x) = \frac{(x_n - x)p_{n-1}^{(0, n-1)}(x) + (x - x_0)p_{n-1}^{(1, n)}(x)}{x_n - x_0} \tag{3.2.8}$$

with $p_{n-1}^{(0, n-1)}(x)$ the polynomial of degree $\leq n - 1$ interpolating $f(x)$ at $\{x_0, \ldots, x_{n-1}\}$ and $p_{n-1}^{(1, n)}(x)$ the polynomial interpolating $f(x)$ at $\{x_1, \ldots, x_n\}$. The proofs of (3.2.7) and (3.2.8) appear in Problem 13.

Returning to the formula (3.2.2), we have the formulas

$$p_0(x) = f(x_0)$$

$$p_1(x) = f(x_0) + (x - x_0)f[x_0, x_1]$$

$$\vdots$$

$$p_n(x) = f(x_0) + (x - x_0)f[x_0, x_1] + (x - x_0)(x - x_1)f[x_0, x_1, x_2]$$

$$+ \cdots + (x - x_0) \cdots (x - x_{n-1})f[x_0, x_1, \ldots, x_n] \tag{3.2.9}$$

This is called *Newton's divided difference formula for the interpolation polynomial.* It is much better for computation than the Lagrange formula (although there are variants of the Lagrange formula that are more efficient than the Lagrange formula).

To construct the divided differences, use the format shown in Table 3.1. Each numerator of a difference is obtained by differencing the two adjacent entries in the column to the left of the column you are constructing.

*Example*   We construct a divided difference table for $f(x) = \sqrt{x}$, shown in Table 3.2. We have used the notation $D^r f(x_i) = f[x_i, x_{i+1}, \ldots, x_{i+r}]$. Note that the table entries for $f(x_i)$ have rounding errors in the seventh place, and that this

**Table 3.2  Example of constructing divided differences**

| $x_i$ | $f(x_i)$ | $f[x_i, x_{i+1}]$ | $D^2 f[x_i]$ | $D^3 f[x_i]$ | $D^4 f[x_i]$ |
|-------|----------|-------------------|--------------|--------------|--------------|
| 2.0   | 1.414214 |                   |              |              |              |
|       |          | .34924            |              |              |              |
| 2.1   | 1.449138 |                   | $-.04110$    |              |              |
|       |          | .34102            |              | .009167      |              |
| 2.2   | 1.483240 |                   | $-.03835$    |              | $-.002084$   |
|       |          | .33335            |              | .008333      |              |
| 2.3   | 1.516575 |                   | $-.03585$    |              |              |
|       |          | .32618            |              |              |              |
| 2.4   | 1.549193 |                   |              |              |              |

affects the accuracy of the resulting divided differences. A discussion of the effects of rounding error in evaluating $p_n(x)$ in both its Lagrange and Newton finite difference form is given in Powell (1981, p. 51).

A simple algorithm can be given for constructing the differences

$$f(x_0), f[x_0, x_1], f[x_0, x_1, x_2], \ldots, f[x_0, x_1, \ldots, x_n]$$

which are necessary for evaluating the Newton form (3.2.9).

***Algorithm***   *Divdif* $(d, x, n)$

1.  Remark: $d$ and $x$ are vectors with entries $f(x_i)$ and $x_i$, $i = 0, 1, \ldots, n$, respectively. On exit, $d_i$ will contain $f[x_0, \ldots, x_i]$.

2.  Do through step 4 for $i = 1, 2, \ldots, n$.

3.  Do through step 4 for $j = n, n - 1, \ldots, i$.

4.  $d_j := (d_j - d_{j-1})/(x_j - x_{j-i})$.

5.  Exit from the algorithm.

To evaluate the Newton form of the interpolating polynomial (3.2.9), we give a simple variant of the nested polynomial multiplication (2.9.8) of Chapter 2.

***Algorithm***   Interp $(d, x, n, t, p)$

1.  Remark: On entrance, $d$ and $x$ are vectors containing $f[x_0, \ldots, x_i]$ and $x_i$, $i = 0, 1, \ldots, n$, respectively. On exit, $p$ will contain the value $p_n(t)$ of the $n$th-degree polynomial interpolating $f$ on $x$.

2.   $p := d_n$

3.   Do through step 4 for $i = n - 1, n - 2, \ldots, 0$.

4.   $p := d_i + (t - x_i)p$

5.   Exit the algorithm.

***Example***   For $f(x) = \sqrt{x}$, we give in Table 3.3 the values of $p_n(x)$ for various values of $x$ and $n$. The highest degree polynomial $p_4(x)$ uses function values at the grid points $x_0 = 2.0$ through $x_4 = 2.4$. The necessary divided differences are given in the last example, Table 3.2.

When a value of $x$ falls outside $\mathscr{H}\{x_0, x_1, \ldots, x_n\}$, we often say that $p_n(x)$ extrapolates $f(x)$. In the last example, note the greater inaccuracy of the extrapolated value $p_3(2.45)$ as compared with $p_3(2.05)$ and $p_3(2.15)$. In this text, however, the word *interpolation* always includes the possibility that $x$ falls outside the interval $\mathscr{H}\{x_0, \ldots, x_n\}$.

Often we know the value of the function $f(x)$, and we want to compute the corresponding value of $x$. This is called *inverse interpolation*. It is commonly known to users of logarithm tables as computing the antilog of a number. To compute $x$, we treat it as the dependent variable and $y = f(x)$ as the independent variable. Given table values $(x_i, y_i)$, $i = 0, \ldots, n$, we produce a polynomial $p_n(y)$ that interpolates $x_i$ at $y_i$, $i = 0, \ldots, n$. In effect, we are interpolating the inverse function $g(y) \equiv f^{-1}(y)$; in the error formula of Theorem 3.2, with $x = f^{-1}(y)$,

$$x - p_n(y) = \frac{(y - y_0) \cdots (y - y_n)}{(n + 1)!} g^{(n+1)}(\zeta) \qquad (3.2.10)$$

for some $\zeta \in \mathscr{H}\{y, y_0, y_1, \ldots, y_n\}$. If they are needed, the derivatives of $g(y)$ can be computed by differentiating the composite formula

$$g(f(x)) = x$$

for example,

$$g'(y) = \frac{1}{f'(x)} \qquad \text{for} \quad y = f(x)$$

***Example***   Consider the Table 3.4 of values of the Bessel function $J_0(x)$, taken

**Table 3.3    Example of use of Newton's formula (3.20)**

| $x$ | $p_1(x)$ | $p_2(x)$ | $p_3(x)$ | $p_4(x)$ | $\sqrt{x}$ |
|------|----------|----------|----------|----------|----------|
| 2.05 | 1.431676 | 1.431779 | 1.431782 | 1.431782 | 1.431782 |
| 2.15 | 1.466600 | 1.466292 | 1.466288 | 1.466288 | 1.466288 |
| 2.45 | 1.571372 | 1.564899 | 1.565260 | 1.565247 | 1.565248 |

**Table 3.4    Values of Bessel function $J_0(x)$**

| $x$ | $J_0(x)$ |
|---|---|
| 2.0 | .2238907791 |
| 2.1 | .1666069803 |
| 2.2 | .1103622669 |
| 2.3 | .0555397844 |
| 2.4 | .0025076832 |
| 2.5 | −.0483837764 |
| 2.6 | −.0968049544 |
| 2.7 | −.1424493700 |
| 2.8 | −.1850360334 |
| 2.9 | −.2243115458 |

**Table 3.5    Example of inverse interpolation**

| $n$ | $p_n(y)$ | $p_n(y) - p_{n-1}(y)$ | $g[y_0, \ldots, y_n]$ |
|---|---|---|---|
| 0 | 2.0 | | 2.0 |
| 1 | 2.216275425 | 2.16E − 1 | −1.745694282 |
| 2 | 2.218619608 | 2.34E − 3 | .2840748405 |
| 3 | 2.218686252 | 6.66E − 5 | −.7793711812 |
| 4 | 2.218683344 | −2.91E − 6 | .7648986704 |
| 5 | 2.218683964 | 6.20E − 7 | −1.672357264 |
| 6 | 2.218683773 | −1.91E − 7 | 3.477333126 |

from Abramowitz and Stegun (1964, chap. 9). We will calculate the value of $x$ for which $J_0(x) = 0.1$. Table 3.5 gives values of $p_n(y)$ for $n = 0, 1, \ldots, 6$, with $x_0 = 2.0$. The polynomial $p_n(y)$ is interpolating the inverse function of $J_0(x)$, call it $g(y)$. The answer $x = 2.2186838$ is correct to eight significant digits.

**An interpolation error formula using divided differences**    Let $t$ be a real number, distinct from the node points $x_0, x_1, \ldots, x_n$. Construct the polynomial interpolating to $f(x)$ at $x_0, \ldots, x_n$, and $t$:

$$p_{n+1}(x) = f(x_0) + (x - x_0)f[x_0, x_1] + \cdots$$

$$+ (x - x_0) \cdots (x - x_{n-1})f[x_0, \ldots, x_n]$$

$$+ (x - x_0) \cdots (x - x_n)f[x_0, x_1, \ldots, x_n, t]$$

$$= p_n(x) + (x - x_0) \cdots (x - x_n)f[x_0, \ldots, x_n, t]$$

Since $p_{n+1}(t) = f(t)$, let $x = t$ to obtain

$$f(t) - p_n(t) = (t - x_0) \cdots (t - x_n)f[x_0, \ldots, x_n, t] \qquad (3.2.11)$$

where $p_n(t)$ was moved to the left side of the equation. This gives us another formula for the error $f(t) - p_n(t)$, one that is very useful in many situations. Comparing this with the earlier error formula (3.1.8) and canceling the multiplying polynomial $\Psi_n(t)$, we have

$$f[x_0, x_1, \ldots, x_n, t] = \frac{f^{(n+1)}(\xi)}{(n+1)!}$$

for some $\xi \in \mathcal{H}\{x_0, x_1, \ldots, x_n, t\}$. To make this result symmetric in the arguments, we generally let $t = x_{n+1}$, $n = m - 1$, and obtain

$$f[x_0, x_1, \ldots, x_m] = \frac{f^{(m)}(\xi)}{m!} \qquad \text{some} \quad \xi \in \mathcal{H}\{x_0, \ldots, x_m\} \quad (3.2.12)$$

With this result, the Newton formula (3.2.9) looks like a truncated Taylor series for $f(x)$, expanded about $x_0$, provided the size $x_n - x_0$ is not too large.

***Example***   From Table 3.2, of divided differences for $f(x) = \sqrt{x}$,

$$f[2.0, 2.1, \ldots, 2.4] = -.002084$$

Since $f^{(4)}(x) = -15/(16x^3\sqrt{x})$, it is easy to show that

$$\frac{f^{(4)}(2.3103)}{4!} \doteq -.002084$$

so $\xi \doteq 2.31$ in (3.2.12) for this case.

***Example***   If the formula (3.2.12) is used to estimate the derivatives of the inverse function $g(y)$ of $J_0(x)$, given in a previous example, then the derivatives $g^{(n)}(y)$ are growing rapidly with $n$. For example,

$$g[y_0, \ldots, y_6] \doteq 3.48$$

$$g^{(6)}(\xi) \doteq (6!)(3.48) \doteq 2500$$

for some $\xi$ in $[-.0968, .2239]$. Similar estimates can be computed for the other derivatives.

To extend the definition of Newton divided difference to the case in which some or all of the nodes coincide, we introduce another formula for the divided difference.

***Theorem* 3.3** (Hermite–Gennochi)   Let $x_0, x_1, \ldots, x_n$ be distinct, and let $f(x)$ be $n$ times continuously differentiable on the interval $\mathcal{H}\{x_0, x_1, \ldots, x_n\}$. Then

$$f[x_0, x_1, \ldots, x_n] = \int \cdots \int_{\tau_n} f^{(n)}(t_0 x_0 + \cdots + t_n x_n)\, dt_1 \ldots dt_n \quad (3.2.13)$$

in which

$$T_n = \left\{ (t_1, t_2, \ldots, t_n)| \quad \text{all} \quad t_i \ge 0, \sum_1^n t_i \le 1 \right\} \quad (3.2.14)$$

$$t_0 = 1 - \sum_1^n t_i$$

Note that $t_0 \ge 0$ and $\sum_0^n t_i = 1$.

***Proof***   We show that (3.2.13) is true for $n = 1$ and 2, and these two cases should suggest the general induction proof.

1.   $n = 1$.   Then $T_1 = [0, 1]$.

$$\int_0^1 f'(t_0 x_0 + t_1 x_1)\, dt_1 = \int_0^1 f'(x_0 + t_1(x_1 - x_0))\, dt_1$$

$$= \frac{1}{x_1 - x_0} f(x_0 + t_1(x_1 - x_0)) \Bigg|_{t_1=0}^{t_1=1}$$

$$= \frac{f(x_1) - f(x_0)}{x_1 - x_0} = f[x_0, x_1]$$

2.   $n = 2$.   Then $T_2$ is the triangle with vertices $(0, 0)$, $(0, 1)$, and $(1, 0)$, shown in Figure 3.1.

$$\int\int_{T_n} f''(t_0 x_0 + t_1 x_1 + t_2 x_2)\, dt_1\, dt_2$$

$$= \int_0^1 \int_0^{1-t_1} f''(x_0 + t_1(x_1 - x_0) + t_2(x_2 - x_0))\, dt_2\, dt_1$$

$$= \int_0^1 \frac{1}{x_2 - x_0}[f'(x_0 + t_1(x_1 - x_0) + t_2(x_2 - x_0))]_{t_2=0}^{t_2=1-t_1}\, dt_1$$

$$= \frac{1}{x_2 - x_0}\left[ \int_0^1 f'(x_2 + t_1(x_1 - x_2))\, dt_1 \right.$$

$$\left. - \int_0^1 f'(x_0 + t_1(x_1 - x_0))\, dt_1 \right]$$

$$= \frac{1}{x_2 - x_0}\{ f[x_1, x_2] - f[x_0, x_1] \} = f[x_0, x_1, x_2]$$

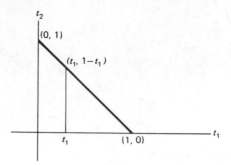

**Figure 3.1**   Region $\tau_2$.

Do the general case by a similar procedure. Integrate once and reduce to one lower dimension. Then invoke the induction hypothesis and use (3.2.7) to complete the proof.   ∎

We can now look at $f[x_0, x_1, \ldots, x_n]$ using (3.2.13). Doing so, we see that if $f(x)$ is $n$ times continuously differentiable on $\mathscr{H}\{x_0, \ldots, x_n\}$, then $f[x_0, \ldots, x_n]$ is a continuous function of the $n$ variables $x_0, x_1, \ldots, x_n$, regardless of whether they are distinct or not. For example, if we let all points coalesce to $x_0$, then for the $n$th-order divided difference,

$$f[x_0, \ldots, x_0] = \int \cdots \int_{\tau_n} f^{(n)}(x_0)\, dt_1 \ldots dt_n$$

$$= f^{(n)}(x_0) \cdot \text{Vol}(\tau_n).$$

From Problem 15, $\text{Vol}(\tau_n) = 1/n!$, and thus

$$f[x_0, \ldots, x_0] = \frac{f^{(n)}(x_0)}{n!} \tag{3.2.15}$$

This could have been predicted directly from (3.2.12). But if only some of the nodes coalesce, we must use (3.2.13) to justify the existence of $f[x_0, \ldots, x_n]$.

In applications to numerical integration, we need to know whether

$$\frac{d}{dx} f[x_0, \ldots, x_n, x] \tag{3.2.16}$$

exists. If $f$ is $n + 2$ times continuously differentiable, then we can apply Theorem 3.3. By applying theorems on differentiating an integral with respect to a parameter in the integrand, we can conclude the existence of (3.2.16). More directly,

$$\frac{d}{dx} f[x_0, \ldots, x_n, x] = \operatorname*{Limit}_{h \to 0} \frac{f[x_0, \ldots, x_n, x + h] - f[x_0, \ldots, x_n, x]}{h}$$

$$= \operatorname*{Limit}_{h \to 0} \frac{f[x_0, \ldots, x_n, x + h] - f[x, x_0, \ldots, x_n]}{h}$$

$$= \operatorname*{Limit}_{h \to 0} f[x, x_0, \ldots, x_n, x + h]$$

$$= f[x, x_0, x_1, \ldots, x_n, x]$$

$$\frac{d}{dx} f[x_0, x_1, \ldots, x_n, x] = f[x_0, x_1, \ldots, x_n, x, x] \tag{3.2.17}$$

The existence and continuity of the right-hand side is guaranteed using (3.2.13).

There is a rich theory involving polynomial interpolation and divided differences, but we conclude at this point with one final straightforward result. If $f(x)$ is a polynomial of degree $m$, then

$$f[x_0, \ldots, x_n, x] = \begin{cases} \text{polynomial of degree } m - n - 1 & n < m - 1 \\ a_m & n = m - 1 \\ 0 & n > m - 1 \end{cases} \tag{3.2.18}$$

where $f(x) = a_m x^m +$ lower degree terms. For the proof, see Problem 14.

## 3.3   Finite Differences and Table-Oriented Interpolation Formulas

In this section, we introduce forward and backward differences, along with interpolation formulas that use them. These differences are referred to collectively as *finite differences*, and they are used to produce interpolation formulas for tables in which the abscissae $\{x_i\}$ are evenly spaced. Such interpolation formulas are also used in the numerical solution of ordinary and partial differential equations. In addition, finite differences can be used to determine the maximum degree of interpolation polynomial that can be used safely, based on the accuracy of the table entries. And finite differences can be used to detect *noise* in data, when the noise is large with respect to the rounding errors or uncertainty errors of physical measurement. This idea is developed in Section 3.4.

For a given $h > 0$, define

$$\Delta_h f(z) = f(z + h) - f(z)$$

Generally the $h$ is understood from the context, and we write

$$\Delta f(z) = f(z + h) - f(z) \tag{3.3.1}$$

This quantity is called the *forward difference* of $f$ at $z$, and $\Delta$ is called the *forward difference operator*. We will always be working with evenly spaced node points $x_i = x_0 + ih$, $i = 0, 1, 2, 3, \ldots$ . Then we write

$$\Delta f(x_i) = f(x_{i+1}) - f(x_i)$$

or more concisely,

$$\Delta f_i = f_{i+1} - f_i \qquad f(x_i) = f_i \qquad (3.3.2)$$

For $r \geq 0$, define

$$\Delta^{r+1} f(z) = \Delta^r f(z + h) - \Delta^r f(z) \qquad (3.3.3)$$

with $\Delta^0 f(z) = f(z)$. The term $\Delta^r f(z)$ is the $r$th-order forward difference of $f$ at $z$. Forward differences are quite easy to compute, and examples are given later in connection with an interpolation formula.

We first derive results for the forward difference operator by applying the results on the Newton divided difference.

**Lemma 1**   For $k \geq 0$,

$$f[x_0, x_1, \ldots, x_k] = \frac{1}{k!h^k} \Delta^k f_0 \qquad (3.3.4)$$

**Proof**   For $k = 0$, the result is trivially true. For $k = 1$,

$$f[x_0, x_1] = \frac{f_1 - f_0}{x_1 - x_0} = \frac{1}{h} \Delta f_0$$

which shows (3.3.4). Assume the result (3.3.4) is true for all forward differences of order $k \leq r$. Then for $k = r + 1$, using (3.2.7),

$$f[x_0, x_1, \ldots, x_{r+1}] = \frac{f[x_1, \ldots, x_{r+1}] - f[x_0, \ldots, x_r]}{x_{r+1} - x_0}$$

Applying the induction hypothesis, this equals

$$\frac{1}{(r+1)h} \left[ \frac{1}{r!h^r} \Delta^r f_1 - \frac{1}{r!h^r} \Delta^r f_0 \right] = \frac{1}{(r+1)!h^{r+1}} \Delta^{r+1} f_0 \qquad \blacksquare$$

We now modify the Newton interpolation formula (3.2.9) to a formula involving forward differences in place of divided differences. For a given value of $x$ at which we will evaluate the interpolating polynomial, define

$$\mu = \frac{x - x_0}{h}$$

to indicate the position of $x$ relative to $x_0$. For example, $\mu = 1.6$ means $x$ is $6/10$ of the distance from $x_1$ to $x_2$. We need a formula for

$$(x - x_0) \cdots (x - x_k)$$

with respect to the variable $\mu$:

$$x - x_j = x_0 + \mu h - (x_0 + jh) = (\mu - j)h$$

$$(x - x_0) \cdots (x - x_k) = \mu(\mu - 1) \cdots (\mu - k)h^{k+1} \qquad (3.3.5)$$

Combining (3.3.4) and (3.3.5) with the divided difference interpolation formula (3.2.9), we obtain

$$p_n(x) = f_0 + \mu h \cdot \frac{\Delta f_0}{h} + \mu(\mu - 1)h^2 \cdot \frac{\Delta^2 f_0}{2! h^2}$$

$$+ \cdots + \mu(\mu - 1) \cdots (\mu - n + 1)h^n \cdot \frac{\Delta^n f_0}{n! h^n}$$

Define the *binomial coefficients*,

$$\binom{\mu}{k} = \frac{\mu(\mu - 1) \cdots (\mu - k + 1)}{k!} \qquad k > 0 \qquad (3.3.6)$$

and $\binom{\mu}{0} = 1$. Then

$$p_n(x) = \sum_{j=0}^{n} \binom{\mu}{j} \Delta^j f_0 \qquad \mu = \frac{x - x_0}{h} \qquad (3.3.7)$$

This is the *Newton forward difference form of the interpolating polynomial.*

**Table 3.6    Format for constructing forward differences**

| $x_i$ | $f_i$ | $\Delta f_i$ | $\Delta^2 f_i$ | $\Delta^3 f_i$ | $\cdots$ |
|-------|-------|--------------|----------------|----------------|----------|
| $x_0$ | $f_0$ | | | | |
| | | $\Delta f_0$ | | | |
| $x_1$ | $f_1$ | | $\Delta^2 f_0$ | | |
| | | $\Delta f_1$ | | $\Delta^3 f_0$ | |
| $x_2$ | $f_2$ | | $\Delta^2 f_1$ | | |
| | | $\Delta f_2$ | | $\Delta^3 f_1$ | $\vdots$ |
| $x_3$ | $f_3$ | | $\Delta^2 f_2$ | | |
| | | $\Delta f_3$ | | $\Delta^3 f_2$ | |
| $x_4$ | $f_4$ | | $\Delta^2 f_3$ | | |
| | | $\Delta f_4$ | | | |
| $x_5$ | $f_5$ | | | | |
| $\vdots$ | $\vdots$ | | | | |

**Table 3.7** Forward difference table for $f(x) = \sqrt{x}$

| $x_i$ | $f_i$ | $\Delta f_i$ | $\Delta^2 f_i$ | $\Delta^3 f_i$ | $\Delta^4 f_i$ |
|---|---|---|---|---|---|
| 2.0 | 1.414214 | | | | |
| | | .034924 | | | |
| 2.1 | 1.449138 | | −.000822 | | |
| | | .034102 | | .000055 | |
| 2.2 | 1.483240 | | −.000767 | | −.000005 |
| | | .033335 | | .000050 | |
| 2.3 | 1.516575 | | −.000717 | | |
| | | .032618 | | | |
| 2.4 | 1.549193 | | | | |

*Example* For $n = 1$,

$$p_1(x) = f_0 + \mu \Delta f_0 \qquad (3.3.8)$$

This is the formula that most people use when doing linear interpolation in a table.

For $n = 2$,

$$p_2(x) = f_0 + \mu \Delta f_0 + \frac{\mu(\mu - 1)}{2} \Delta^2 f_0 \qquad (3.3.9)$$

which is an easily computable form of the quadratic interpolating polynomial.

The forward differences are constructed in a pattern like that for divided differences, but now there are no divisions (see Table 3.6).

*Example* The forward differences for $f(x) = \sqrt{x}$ are given in Table 3.7. The values of the interpolating polynomial will be the same as those obtained using the Newton divided difference formula, but the forward difference formula (3.3.7) is much easier to compute.

*Example* Evaluate $p_n(x)$ using Table 3.7 for $n = 1, 2, 3, 4$, with $x = 2.15$. Note that $\sqrt{2.15} = 1.4662878$; and $\mu = 1.5$.

$$p_1(x) = 1.414214 + 1.5(.034924) = 1.414214 + 0.52386 = 1.4666$$

$$p_2(x) = p_1(x) + \frac{(1.5)(.5)}{2}(-.000822) = 1.4666 - .00030825 = 1.466292$$

$$p_3(x) = p_2(x) + \frac{(1.5)(.5)(-.5)}{6}(.000055) = 1.466292 - .0000034$$

$$= 1.466288$$

$$p_4(x) = p_3(x) + \frac{(1.5)(.5)(-.5)(-1.5)}{24}(-.000005) = 1.466288 - .00000012$$

$$= 1.466288$$

The correction terms are easily computed; and by observing their size, you obtain a generally accurate idea of when the degree $n$ is sufficiently large. Note that the seven-place accuracy in the table values of $\sqrt{x}$ leads to at most one place of accuracy in the forward difference $\Delta^4 f_0$. The forward differences of order greater than three are almost entirely the result of differencing the rounding errors in the table entries; consequently, interpolation in this table should be limited to polynomials of degree less than four. This idea is given further theoretical justification in the next section.

There are other forms of differences and associated interpolation formulas. Define the *backward difference* by

$$\nabla f(z) = f(z) - f(z - h)$$

$$\nabla^{r+1} f(z) = \nabla^r f(z) - \nabla^r f(z - h) \qquad r \geq 1 \qquad (3.3.10)$$

Completely analogous results to those for forward differences can be derived. And we obtain the Newton backward difference interpolation formula,

$$p_n(x) = f_0 + \binom{-\nu}{1} \nabla f_0 + \binom{-\nu+1}{2} \nabla^2 f_0 + \cdots + \binom{-\nu+n-1}{n} \nabla^n f_0$$

$$(3.3.11)$$

In this formula, the interpolation nodes are $x_0, x_{-1}, x_{-2}, \ldots, x_{-n}$, with $x_{-j} = x_0 - jh$, as before. The value $\nu$ is given by

$$\nu = \frac{x_0 - x}{h}$$

reflecting the fact that $x$ will generally be less than $x_0$ when using this formula. A backward difference diagram can be constructed in an analogous way to that for forward differences. The backward difference formula is used in Chapter 6 to develop the Adams family of formulas (named after John Couch Adams, a nineteenth-century astronomer) for the numerical solution of differential equations.

Other difference formulas and associated interpolation formulas can be given. Since they are used much less than the preceding formula, we just refer the reader to Hildebrand (1956).

## 3.4   Errors in Data and Forward Differences

We can use a forward difference table to detect *noise* in physical data, as long as the noise is large relative to the usual limits of experimental error. We must begin with some preliminary lemmas.

**Lemma 2**   $\Delta^r f(x_i) = h^r f^{(r)}(\xi_i)$, for some $x_i \leq \xi_i \leq x_{i+r}$.

*Proof*

$$\Delta^r f_i = h^r r! f[x_i, \ldots, x_{i+r}] = h^r r! \frac{f^{(r)}(\xi_i)}{r!} = h^r f^{(r)}(\xi_i)$$

using Lemma 1 and (3.2.12).    ∎

**Lemma 3**   For any two functions $f$ and $g$, and for any two constants $\alpha$ and $\beta$,

$$\Delta^r(\alpha f(x) + \beta g(x)) = \alpha \Delta^r f(x) + \beta \Delta^r g(x) \qquad r \geq 0$$

*Proof*   The result is trivial if $r = 0$ or $r = 1$. Assume the result is true for all $r \leq n$, and prove it for $r = n + 1$:

$$\Delta^{n+1}[\alpha f(x) + \beta g(x)] = \Delta^n[\alpha f(x + h) + \beta g(x + h)]$$

$$- \Delta^n[\alpha f(x) + \beta g(x)]$$

$$= \alpha \Delta^n f(x + h) + \beta \Delta^n g(x + h)$$

$$- \alpha \Delta^n f(x) - \beta \Delta^n g(x)$$

using the definition (3.3.3) of $\Delta^{n+1}$ and the induction hypothesis. Then by recombining, we obtain

$$\alpha[\Delta^n f(x + h) - \Delta^n f(x)] + \beta[\Delta^n g(x + h) - \Delta^n g(x)]$$

$$= \alpha \Delta^{n+1} f(x) + \beta \Delta^{n+1} g(x)$$    ∎

Lemma 2 says that if the derivatives of $f(x)$ are bounded, or if they do not increase rapidly compared with $h^{-n}$, then the forward differences $\Delta^n f(x)$ should become smaller as $n$ increases. We next look at the effect of rounding errors and other errors of a larger magnitude than rounding errors. Let

$$f(x_i) = \tilde{f}_i + e(x_i) \qquad i = 0, 1, 2, \ldots \tag{3.4.1}$$

with $\tilde{f}_i$ a table value that we use in constructing the forward difference table. Then

$$\Delta^r \tilde{f}_i = \Delta^r f(x_i) - \Delta^r e(x_i)$$

$$= h^r f^{(r)}(\xi_i) - \Delta^r e(x_i) \tag{3.4.2}$$

The first term becomes smaller as $r$ increases, as illustrated in the earlier forward difference table for $f(x) = \sqrt{x}$.

To better understand the behavior of $\Delta^r e(x_i)$, consider the simple case in which

$$e(x_i) = \begin{cases} 0 & i \neq k \\ \epsilon & i = k \end{cases} \tag{3.4.3}$$

**Table 3.8    Forward differences of error function $e(x)$**

| $x_i$ | $e(x_i)$ | $\Delta e(x_i)$ | $\Delta^2 e(x_i)$ | $\Delta^3 e(x_i)$ |
|---|---|---|---|---|
| $\vdots$ | $\vdots$ | 0 | $\vdots$ | 0 |
| $x_{k-2}$ | 0 | $\vdots$ | 0 | $\vdots$ |
| | | 0 | | $\epsilon$ |
| $x_{k-1}$ | 0 | | $\epsilon$ | |
| | | $\epsilon$ | | $-3\epsilon$ |
| $x_k$ | $\epsilon$ | | $-2\epsilon$ | |
| | | $-\epsilon$ | | $3\epsilon$ |
| $x_{k+1}$ | 0 | | $\epsilon$ | |
| | | 0 | | $-\epsilon$ |
| $x_{k+2}$ | 0 | | 0 | |
| | | | | 0 |
| $\vdots$ | $\vdots$ | $\vdots$ | $\vdots$ | $\vdots$ |

The forward difference of this function are given in Table 3.8. It can be proved that the column for $\Delta^r e(x_i)$ will look like

$$0,\ldots,0,\epsilon,-\binom{r}{1}\epsilon,\binom{r}{2}\epsilon,-\binom{r}{3}\epsilon,\ldots,(-1)^{r+1}\epsilon,0,\ldots \qquad (3.4.4)$$

Thus the effect of a single rounding error will propagate and increase in value as larger order differences are formed.

With rounding errors defining a general error function as in (3.4.1), their effect can be looked upon as a sum of functions of the form (3.4.3). Since the values of $e(x_i)$ in general will vary in sign and magnitude, their effects will overlap in a seemingly random manner. But their differences will still grow in size, and the higher order differences of table values $\tilde{f}_i$ will eventually become useless. When differences $\Delta^r \tilde{f}_i$ begin to increase in size with increasing $r$, then these differences are most likely dominated by rounding errors and should not be used. An interpolation formula of degree less than $r$ should be used.

**Detecting noise in data**    This same analysis can be used to detect and correct isolated errors that are large relative to rounding error. Since (3.4.2) says that the effect of the errors will eventually dominate, we look for a pattern like (3.4.4). The general technique is illustrated in Table 3.9.

From (3.4.2)

$$\Delta^r e(x_i) = \Delta^r f(x_i) - \Delta^r \tilde{f}_i$$

Using $r = 3$ and one of the error entries chosen arbitrarily, say the first,

$$\epsilon = -.00002 - (.00006) = -.00008$$

Try this to see how it will alter the column of $\Delta^r \tilde{f}_i$ (see Table 3.10). This will not be improved on by another choice of $\epsilon$, say $\epsilon = -.00007$, although the results

**Table 3.9   Example of detecting an isolated error in data**

| $\tilde{f}_i$ | $\Delta \tilde{f}_i$ | $\Delta^2 \tilde{f}_i$ | $\Delta^3 \tilde{f}_i$ | Error Guess | Guess $\Delta^3 f(x_i)$ |
|---|---|---|---|---|---|
| .10396 | | | | | |
| | .01700 | | | | |
| .12096 | | −.00014 | | | |
| | .01686 | | −.00003 | 0 | −.00003 |
| .13782 | | −.00017 | | | |
| | .01669 | | −.00002 | 0 | −.00002 |
| .15451 | | −.00019 | | | |
| | .01650 | | .00006 | $\epsilon$ | −.00002 |
| .17101 | | −.00013 | | | |
| | .01637 | | −.00025 | $-3\epsilon$ | −.00002 |
| .18738 | | −.00038 | | | |
| | .01599 | | .00021 | $3\epsilon$ | −.00002 |
| .20337 | | −.00017 | | | |
| | .01582 | | −.00010 | $-\epsilon$ | −.00002 |
| .21919 | | −.00027 | | | |
| | .01555 | | | | |
| .23474 | | | | | |

**Table 3.10   Correcting a data error**

| $\Delta^3 \tilde{f}_i$ | $\Delta^3 e(x_i)$ | $\Delta^3 f(x_i)$ |
|---|---|---|
| −.00002 | 0.0 | −.00002 |
| .00006 | −.00008 | −.00002 |
| −.00025 | .00024 | −.00001 |
| .00021 | −.00024 | −.00003 |
| −.00010 | .00008 | −.00002 |

may be equally good. Tracing backwards, the entry $\tilde{f}_i = .18738$ should be

$$f(x_i) = \tilde{f}_i + e(x_i) = .18738 + (-.00008) = .18730$$

In a table in which there are two or three isolated errors, their higher order differences may overlap, making it more difficult to discover the errors (see Problem 22).

## 3.5   Further Results on Interpolation Error

Consider again the error formula

$$f(x) - p_n(x) = \frac{(x - x_0) \cdots (x - x_n)}{(n + 1)!} f^{(n+1)}(\xi_x) \qquad \xi_x \in \mathscr{H}\{x_0, \ldots, x_n, x\}$$

$$(3.5.1)$$

We assume that $f(x)$ is $n + 1$ times continuously differentiable on an interval $I$ that contains $\mathcal{H}\{x_0, \ldots, x_n, x\}$ for all values $x$ of interest. Since $\xi_x$ is unknown, we must replace $f^{(n+1)}(\xi_x)$ by

$$c_{n+1} = \underset{t \in I}{\text{Max}} |f^{(n+1)}(t)| \tag{3.5.2}$$

in order to evaluate (3.5.1). We will concentrate our attention on bounding the polynomial

$$\Psi_n(x) = (x - x_0) \cdots (x - x_n) \tag{3.5.3}$$

Then from (3.5.1), for

$$\underset{x \in I}{\text{Max}} |f(x) - p_n(x)| \leq \frac{c_{n+1}}{(n + 1)!} \cdot \underset{x \in I}{\text{Max}} |\Psi_n(x)| \tag{3.5.4}$$

We will consider only the use of evenly spaced nodes: $x_j = x_0 + jh$ for $j = 0, 1, \ldots, n$. We first consider cases of specific values of $n$, and later we comment on the case of general $n$.

**Case 1**    $n = 1$.    $\Psi_1(x) = (x - x_0)(x - x_1)$. Then easily

$$\underset{x_0 \leq x \leq x_1}{\text{Max}} |\Psi_1(x)| = \frac{h^2}{4}$$

See Figure 3.2 for an illustration.

**Case 2**    $n = 2$.    To bound $\Psi_2(x)$ on $[x_0, x_2]$, shift it along the $x$-axis to obtain an equivalent polynomial

$$\hat{\Psi}_2(x) = (x + h)x(x - h)$$

whose graph is shown in Figure 3.3. Its shape and size are exactly the same as with the original polynomial $\Psi_2(x)$, but it is easier to bound analytically. Using

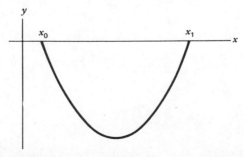

**Figure 3.2**    $y = \Psi_1(x)$.

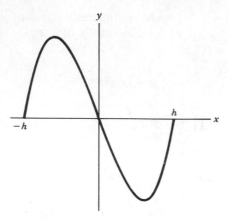

**Figure 3.3**   $y = \Psi_2(x)$.

$\hat{\Psi}(x)$, we easily obtain

$$\operatorname*{Max}_{x_1 - h/2 \le x \le x_1 + h/2} |\Psi_2(x)| = .375h^3$$

$$\operatorname*{Max}_{x_0 \le x \le x_2} |\Psi_2(x)| = \frac{2\sqrt{3}}{9}h^3 \doteq .385h^3 \tag{3.5.5}$$

Thus it doesn't matter if $x$ is located near $x_1$ in the interval $[x_0, x_2]$, although it will make a difference for higher degree interpolation. Combining (3.5.5) with (3.5.4),

$$\operatorname*{Max}_{x_0 \le x \le x_2} |f(x) - p_2(x)| \le \frac{\sqrt{3}}{27}h^3 \cdot \operatorname*{Max}_{x_0 \le x \le x_2} |f^{(3)}(t)| \tag{3.5.6}$$

and $\dfrac{\sqrt{3}}{27} \doteq .064$.

***Case 3*** $n = 3$.   As previously, shift the polynomial to make the nodes symmetric about the origin, obtaining

$$\hat{\Psi}_3(x) = \left(x^2 - \tfrac{9}{4}h^2\right)\left(x^2 - \tfrac{1}{4}h^2\right)$$

The graph of $\hat{\Psi}_3(x)$ is shown in Figure 3.4. Using this modification,

$$\operatorname*{Maximum}_{x_1 \le x \le x_2} |\Psi_3(x)| = \tfrac{9}{16}h^4 \doteq 0.56h^4$$

$$\operatorname*{Max}_{x_0 \le x \le x_3} |\Psi_3(x)| = h^4$$

Thus in interpolation to $f(x)$ at $x$, the nodes should be so chosen that

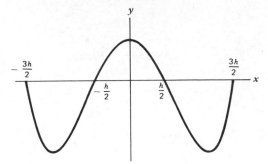

**Figure 3.4**    $y = \hat{\Psi}_3(x)$.

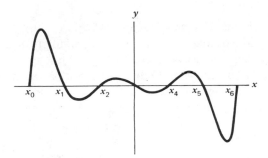

**Figure 3.5**    $y = \Psi_6(x)$.

$x_1 < x < x_2$. Then from (3.5.1),

$$\underset{x_1 \le x \le x_2}{\text{Max}} |f(x) - p_3(x)| \le \frac{3h^4}{128} \cdot \underset{x_0 \le t \le x_3}{\text{Max}} |f^{(4)}(t)| \qquad (3.5.7)$$

*Case* **4**    For a general $n > 3$, the behavior just exhibited with $n = 3$ is accentuated. For example, consider the graph of $\Psi_6(x)$ in Figure 3.5. As earlier, we can show

$$\underset{x_2 \le x \le x_4}{\text{Max}} |\Psi_6(x)| \doteq 12.36h^7$$

$$\underset{x_0 \le x \le x_6}{\text{Max}} |\Psi_6(x)| \doteq 95.8h^7$$

To minimize the interpolation error, the interpolation nodes should be chosen so that the interpolation point $x$ is as near as possible to the midpoint of $[x_0, x_n]$. As the degree $n$ increases, for interpolation on evenly spaced nodes, it becomes increasingly advisable to choose the nodes so that $x$ is near the middle of $[x_0, x_n]$.

*Example*    Consider interpolation of degree five to $J_0(x)$ at $x = 2.45$, with the values of $J_0(x)$ given in Table 3.4 in Section 3.2. Using first $x_0 = 2.4$, $x_5 = 2.9$, we obtain

$$p_5(2.45) = -0.232267384 \qquad \text{Error} = -4.9 \times 10^{-9}$$

Second, use $x_0 = 2.2$, $x_5 = 2.7$. Then

$$p_5(2.45) = -0.232267423 \qquad \text{Error} = 1.0 \times 10^{-9}$$

The error is about five times smaller, due to positioning $x$ near the middle of $[x_0, x_5]$.

The tables in Abramowitz and Stegun (1964) are given to many significant digits, and the grid spacing $h$ is not correspondingly small. Consequently, high-degree interpolation must be used. Although this results in more work for the user of the tables, it allows the table to be compacted into a much smaller space; more tables of more functions can then be included in the volume. When interpolating with these tables and using high-degree interpolation, the nodes should be so chosen that $x$ is near $(x_0 + x_n)/2$, if possible.

**The approximation problem**    In using a computer, we generally prefer to store an analytic approximation to a function rather than a table of values from which we interpolate. Consider approximating a given function $f(x)$ on a given interval $[a, b]$ by using interpolating polynomials. In particular, consider the polynomial $p_n(x)$ produced by interpolating $f(x)$ on an evenly spaced subdivision of $[a, b]$.

For each $n \geq 1$, define $h = (b - a)/n$, $x_j = a + jh$, $j = 0, 1, \ldots, n$. Let $p_n(x)$ be the polynomial interpolating to $f(x)$ at $x_0, \ldots, x_n$. Then we ask whether

$$\underset{a \leq x \leq b}{\text{Max}} |f(x) - p_n(x)| \tag{3.5.8}$$

tend to zero as $n \to \infty$? The answer is *not necessarily*. For many functions, for example, $e^x$ on $[0, 1]$, the error in (3.5.8) does converge to zero as $n \to \infty$ (see Problem 24). But there are other functions, that are quite well behaved, for which convergence does not occur.

The most famous example of failure to converge is one due to Carl Runge. Let

$$f(x) = \frac{1}{1 + x^2} \qquad -5 \leq x \leq 5 \tag{3.5.9}$$

In Isaacson and Keller (1966, pp. 275–279), it is shown that for any $3.64 < |x| < 5$,

$$\underset{n \geq k}{\text{Supremum}} |f(x) - p_n(x)| = \infty \qquad \text{and} \quad k \geq 0 \tag{3.5.10}$$

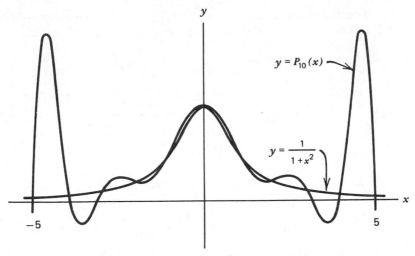

**Figure 3.6**   Interpolation to $1/(1 + x^2)$.

Thus $p_n(x)$ does not converge to $f(x)$ as $n \to \infty$ for any such $x$. This is at first counterintuitive, but it is based on the behavior of the polynomials $y = \Psi_n(x)$ near the endpoints of $[a, b] = [x_0, x_n]$. This is further illustrated with the graphs of $f(x)$ and $p_{10}(x)$, given in Figure 3.6. Although interpolation on an even grid may not produce a convergent sequence of interpolation polynomials, there are suitable sets of grid points $\{x_j\}$ that do result in good approximations for all continuously differentiable functions. This grid is developed in Section 4.7 of Chapter 4, which is on approximation theory.

## 3.6   Hermite Interpolation

For a variety of applications, it is convenient to consider polynomials $p(x)$ that interpolate a function $f(x)$, and in addition have the derivative polynomial $p'(x)$ interpolate the derivative function $f'(x)$. In this text, the primary application of this is as a mathematical tool in the development of Gaussian numerical integration in Chapter 5. But such interpolation is also convenient in developing numerical methods for solving some differential equation problems.

We begin by considering an existence theorem for the basic interpolation problem

$$p(x_i) = y_i \qquad p'(x_i) = y_i' \qquad i = 1, \dots, n \qquad (3.6.1)$$

in which $x_1, \dots, x_n$ are distinct nodes (real or complex) and $y_1, \dots, y_n, y_1', \dots, y_n'$ are given data (The notation has been changed from $n + 1$ nodes $\{x_0, \dots, x_n\}$ to $n$ nodes $\{x_1, \dots, x_n\}$ in line with the eventual application in Chapter 5.) There are $2n$ conditions imposed in (3.6.1); thus we look for a polynomial $p(x)$ of at most degree $2n - 1$.

To deal with the existence and uniqueness for $p(x)$, we generalize the third proof of Theorem 3.1. In line with previous notation in Section 3.1, let

$$\Psi_n(x) = (x - x_1) \cdots (x - x_n)$$

$$l_i(x) = \frac{(x - x_1) \cdots (x - x_{i-1})(x - x_{i+1}) \cdots (x - x_n)}{(x_i - x_1) \cdots (x_i - x_{i-1})(x_i - x_{i+1}) \cdots (x - x_n)} = \frac{\Psi_n(x)}{(x - x_i)\Psi_n'(x_i)}$$

$$\tilde{h}_i(x) = (x - x_i)[l_i(x)]^2$$

$$h_i(x) = [1 - 2l_i'(x_i)(x - x_i)][l_i(x)]^2 \qquad (3.6.2)$$

Then for $i, j = 1, \ldots, n$,

$$h_i'(x_j) = \tilde{h}_i(x_j) = 0 \qquad 1 \le i, j \le n$$

$$h_i(x_j) = \tilde{h}_i'(x_j) = \begin{cases} 0 & i \ne j \\ 1 & i = j \end{cases} \qquad (3.6.3)$$

The interpolating polynomial for (3.6.1) is given by

$$H_n(x) = \sum_{i=1}^{n} y_i h_i(x) + \sum_{i=1}^{n} y_i' \tilde{h}_i(x) \qquad (3.6.4)$$

To show the uniqueness of $H_n(x)$, suppose there is a second polynomial $G(x)$ that satisfies (3.6.1) with degree $\le 2n - 1$. Define $R = H - G$. Then from (3.6.1)

$$R(x_i) = R'(x_i) = 0 \qquad i = 1, 2, \ldots, n$$

where $R$ is a polynomial of degree $\le 2n - 1$, with $n$ double roots, $x_1, x_2, \ldots, x_n$. This can be true only if

$$R(x) = q(x)(x - x_1)^2 \cdots (x - x_n)^2$$

for some polynomial $q(x)$. If $q(x) \ne 0$, then degree $R(x) \ge 2n$, a contradiction. Therefore we must have $R(x) \equiv 0$.

To obtain a more computable form than (3.6.4) and an error term, first consider the polynomial interpolating $f(x)$ at nodes $z_1, z_2, \ldots, z_{2n}$, written in the Newton divided difference form:

$$p_{2n-1}(x) = f(z_1) + (x - z_1)f[z_1, z_2] + \cdots$$

$$+ (x - z_1) \cdots (x - z_{2n-1})f[z_1, \ldots, z_{2n}] \qquad (3.6.5)$$

For the error,

$$f(x) - p_{2n-1}(x) = (x - z_1) \cdots (x - z_{2n})f[z_1, \ldots, z_{2n}, x] \qquad (3.6.6)$$

In the formula (3.6.5), we can let nodes coincide and the formula will still exist. In particular, let

$$z_1, z_2 \to x_1, \qquad z_3, z_4 \to x_2, \ldots, z_{2n-1}, z_{2n} \to x_n$$

to obtain

$$p_{2n-1}(x) = f(x_1) + (x - x_1)f[x_1, x_1] + (x - x_1)^2 f[x_1, x_1, x_2] + \cdots$$

$$+ (x - x_1)^2 \cdots (x - x_{n-1})^2 (x - x_n) f[x_1, x_1, \ldots, x_n, x_n] \quad (3.6.7)$$

This is a polynomial of degree $\leq 2n - 1$. For its error, take limits in (3.6.6) as $z_1, z_2 \to x_1, \ldots, z_{2n-1}, z_{2n} \to x_n$. By the continuity of the divided difference, assuming $f$ is sufficiently differentiable,

$$f(x) - p_{2n-1}(x) = (x - x_1)^2 \cdots (x - x_n)^2 f[x_1, x_1, \ldots, x_n, x_n, x] \quad (3.6.8)$$

Claim: $p_{2n-1}(x) = H_n(x)$. To prove this, assume $f(x)$ is $2n + 1$ times continuously differentiable. Then note that

$$f(x_i) - p_{2n-1}(x_i) = 0 \qquad i = 1, 2, \ldots, n$$

Also,

$$f'(x) - p'_{2n-1}(x) = (x - x_1)^2 \cdots (x - x_n)^2 \frac{d}{dx} f[x_1, x_1, \ldots, x_n, x_n, x]$$

$$+ 2f[x_1, x_1, \ldots, x_n, x_n, x] \sum_{i=1}^{n} \left[ (x - x_i) \prod_{\substack{j=1 \\ j \neq i}}^{n} (x - x_j)^2 \right]$$

and

$$f'(x_i) - p'_{2n-1}(x_i) = 0 \qquad i = 1, \ldots, n$$

Thus degree $(p_{2n-1}) \leq 2n - 1$ and it satisfies (3.6.1) relative to the data $y_i = f(x_i)$, $y'_i = f'(x_i)$. By the uniqueness of the Hermite interpolating polynomial, we have $p_{2n-1} = H_n$. Thus (3.6.7) gives a divided difference formula for calculating $H_n(x)$, and (3.6.8) gives an error formula:

$$f(x) - H_n(x) = [\Psi_n(x)]^2 f[x_1, x_1, \ldots, x_n, x_n, x] \qquad (3.6.9)$$

Using (3.2.13) to generalize (3.2.12), we obtain

$$f(x) - H_n(x) = [\Psi_n(x)]^2 \frac{f^{(2n)}(\xi_x)}{(2n)!} \qquad \xi_x \in \mathscr{H}\{x_1, \ldots, x_n, x\} \quad (3.6.10)$$

***Example***    The most widely used form of Hermite interpolation is probably the cubic Hermite polynomial, which solves

$$p(a) = f(a) \qquad p'(a) = f'(a)$$

$$p(b) = f(b) \qquad p'(b) = f'(b) \tag{3.6.11}$$

The formula (3.6.4) becomes

$$H_2(x) = \left[1 + 2\frac{x-a}{b-a}\right]\left[\frac{b-x}{b-a}\right]^2 \cdot f(a) + \left[1 + 2\frac{b-x}{b-a}\right]\left[\frac{x-a}{b-a}\right]^2 \cdot f(b)$$

$$+ \frac{(x-a)(b-x)^2}{(b-a)^2}f'(a) - \frac{(x-a)^2(b-x)}{(b-a)^2}f'(b) \tag{3.6.12}$$

The divided difference formula (3.6.7) becomes

$$H_2(x) = f(a) + (x-a)f'(a) + (x-a)^2 f[a, a, b]$$

$$+ (x-a)^2(x-b)f[a, a, b, b] \tag{3.6.13}$$

in which

$$f[a, a, b] = \frac{f[a, b] - f'(a)}{b - a}$$

$$f[a, a, b, b] = \frac{f'(b) - 2f[a, b] + f'(a)}{(b - a)^2}$$

The formula (3.6.13) can be evaluated by a nested multiplication algorithm analogous to *Interp* in Section 3.2.

The error formula for (3.6.12) or (3.6.13) is

$$f(x) - H_2(x) = (x-a)^2(x-b)^2 f[a, a, b, b, x] \tag{3.6.14}$$

$$= \frac{(x-a)^2(x-b)^2}{24}f^{(4)}(\xi_x) \quad \xi_x \in \mathcal{H}\{a, b, x\}$$

$$\underset{a \le x \le b}{\text{Max}} |f(x) - H_2(x)| \le \frac{(b-a)^4}{384} \cdot \underset{a \le t \le b}{\text{Max}} |f^{(4)}(t)| \tag{3.6.15}$$

Further use will be made of the cubic Hermite polynomial in the next section, and a numerical example is given in Table 3.12 of that section.

**The general Hermite interpolation problem**    We generalize the simple Hermite problem (3.6.1) to the following: Find a polynomial $p(x)$ to satisfy

$$p^{(i)}(x_1) = y_1^{(i)} \qquad i = 0, 1, \ldots, \alpha_1 - 1$$

$$\vdots \qquad\qquad\qquad\qquad (3.6.16)$$

$$p^{(i)}(x_n) = y_n^{(i)} \qquad i = 0, 1, \ldots, \alpha_n - 1$$

The numbers $y_j^{(i)}$ are given data, and the number of conditions on $p(x)$ at $x_j$ is $\alpha_j$, $j = 1, \ldots, n$.

Define

$$N = \alpha_1 + \cdots + \alpha_n$$

Then there is a polynomial $p(x)$, unique among those of degree $\leq N - 1$, which satisfies (3.6.16). The proof is left as Problem 25. All of the earlier results, such as (3.6.4) and (3.6.9), can be generalized. As an interesting special case, consider $\alpha_1 = N$, $n = 1$. This means that $p(x)$ is to satisfy

$$p^{(i)}(x_1) = f^{(i)}(x_1) \qquad i = 0, 1, \ldots, N - 1$$

We have replaced $y_1^{(i)}$ by $f^{(i)}(x_1)$ for notational convenience. Then using (3.2.15), the Newton divided difference form of the Hermite interpolating polynomial is

$$p(x) = f(x_1) + (x - x_1)f[x_1, x_1] + (x - x_1)^2 f[x_1, x_1, x_1] + \cdots$$

$$+ (x - x_1)^{N-1} f[x_1, \ldots, x_1]$$

$$= f(x_1) + (x - x_1)f'(x_1) + \frac{(x - x_1)^2}{2!} f''(x_1) + \cdots$$

$$+ \frac{(x - x_1)^{N-1}}{(N - 1)!} f^{(N-1)}(x_1)$$

which is also the Taylor polynomial of $f$ about $x_1$.

## 3.7    Piecewise Polynomial Interpolation

Since the early 1960s, the subject of piecewise polynomial functions has become increasingly popular, especially spline functions. These polynomial functions been used in a large variety of ways in approximation theory, computer graphics, data fitting, numerical integration and differentiation, and the numerical solution of integral, differential, and partial differential equations. We look at piecewise polynomial functions from only the viewpoint of interpolation theory, but much of their useful application occurs in some other area, with interpolation occurring only in a peripheral way.

For a piecewise polynomial function $p(x)$, there is an associated grid:

$$-\infty < x_0 < x_1 < \cdots < x_n < \infty \qquad (3.7.1)$$

The points $x_j$ are sometimes called *knots*, *breakpoints*, or *nodes*. The function $p(x)$ is a polynomial on each of the subintervals

$$(-\infty, x_0], [x_0, x_1], \ldots, [x_n, \infty) \qquad (3.7.2)$$

although often the intervals $(-\infty, x_0]$ and $[x_n, \infty)$ are not included. We say $p(x)$ is a *piecewise polynomial of order r* if the degree of $p(x)$ is less than $r$ on each of the subintervals in (3.7.2). No restrictions of continuity need be placed on $p(x)$ or its derivatives, although usually $p(x)$ is continuous. In this section, we mostly restrict ourselves to piecewise cubic polynomial functions (of order four). This is the most common case in applications, and it simplifies the presentation to be definite as to order.

One way of classifying piecewise polynomial interpolation problems is as *local* or *global*. For the local type of problem, the polynomial $p(x)$ on each subinterval $[x_{i-1}, x_i]$ is completely determined by the interpolation data at node points inside and neighboring $[x_{i-1}, x_i]$. But for a global problem, the choice of $p(x)$ on each $[x_{i-1}, x_i]$ is dependent on all of the interpolation data. The global problems are somewhat more complicated to study; the best known examples are the spline functions, to be defined later.

**Local interpolation problems**    Consider that we want to approximate $f(x)$ on an interval $[a, b]$. We begin by choosing a grid

$$a = x_0 < x_1 < \cdots < x_n = b \qquad (3.7.3)$$

often evenly spaced. Our first case of a piecewise polynomial interpolation function is based on using ordinary polynomial interpolation on each subinterval $[x_{i-1}, x_i]$.

Let four interpolation nodes be given on each subinterval $[x_{i-1}, x_i]$,

$$x_{i-1} \le z_{i,1} < z_{i,2} < z_{i,3} < z_{i,4} \le x_i \qquad i = 1, \ldots, n \qquad (3.7.4)$$

Define $p(x)$ on $x_{i-1} < x < x_i$ by letting it be the polynomial of degree $\le 3$ that interpolates $f(x)$ at $z_{i,1}, \ldots, z_{i,4}$. If

$$z_{i,1} = x_{i-1} \qquad z_{i,4} = x_i \qquad i = 1, 2, \ldots, n \qquad (3.7.5)$$

then $p(x)$ is continuous on $[a, b]$. For the nodes in (3.7.4), we call this interpolating function the *Lagrange piecewise polynomial function*, and it will be denoted by $L_n(x)$.

From the error formula (3.1.8) for polynomial interpolation, the error formula for $L_n(x)$ satisfies

$$f(x) - L_n(x) = \frac{(x - z_{i,1})(x - z_{i,2})(x - z_{i,3})(x - z_{i,4})}{4!} f^{(4)}(\xi_i) \qquad (3.7.6)$$

for $x_{i-1} \leq x \leq x_i$, $i = 1, \ldots, n$, with $x_{i-1} < \xi_x < x_i$. Consider the special case of even spacing. Let

$$\delta_i = \frac{(x_i - x_{i-1})}{3} \qquad z_{i,j} = x_{i-1} + (j-1)\delta_i \qquad j = 1, 2, 3, 4 \quad (3.7.7)$$

Then (3.5.7) and (3.7.6) yield

$$|f(x) - L_n(x)| \leq \frac{\delta_i^4}{24} \cdot \underset{x_{i-1} \leq x \leq x_i}{\text{Max}} |f^{(4)}(t)| \qquad x_{i-1} < x < x_i \quad (3.7.8)$$

for $i = 1, 2, \ldots, n$. From this, we can see that to maintain an equal level of error throughout $a \leq x \leq b$, the spacing $\delta_i$ should be chosen based on the size of the derivative $f^{(4)}(t)$ at points of $[x_{i-1}, x_i]$. Thus if the function $f(x)$ has varying behavior in the interval $[a, b]$, the piecewise polynomial function $L_n(x)$ can be chosen to mimic this behavior by suitably adjusting the grid in (3.7.3). Ordinary polynomial interpolation on $[a, b]$ is not this flexible, which is one reason for using piecewise polynomial interpolation. For cases where we use an evenly spaced grid (3.7.3), the result (3.7.8) guarantees convergence where ordinary polynomial interpolation may fail, as with Runge's example (3.5.9).

***Example*** For $f(x) = e^x$ on $[0, 1]$, suppose we want even spacing and a maximum error less than $10^{-6}$. Using $\delta_i = \delta$ in (3.7.8), we require that

$$\frac{\delta^4}{24} e \leq 10^{-6}$$

$$\delta \leq .055 \qquad n = \frac{1}{3\delta} \geq 6.12$$

It will probably be sufficient to use $n = 6$ because of the conservative nature of the bound (3.7.8). There will will be six subintervals of cubic interpolation.

For the storage requirements of Lagrange piecewise polynomial interpolation, assuming (3.7.7), we need to save four pieces of information on each subinterval $[x_{i-1}, x_i]$. This gives a total storage requirement of $4n$ memory locations for $L_n(x)$, as well as an additional $n - 1$ locations for the breakpoints of (3.7.3). The choice of information to be saved depends on how $L_n(x)$ is to be evaluated. If derivatives of $L_n(x)$ are desired, then it is most convenient to store $L_n(x)$ in its Taylor polynomial form on each subinterval $[x_{j-1}, x_j]$:

$$L_n(x) = a_j + b_j(x - x_{j-1}) + c_j(x - x_{j-1})^2 + d_j(x - x_{j-1})^3$$

$$x_{j-1} \leq x \leq x_j \quad (3.7.9)$$

This should be evaluated in nested form. The coefficients $\{a_j, b_j, c_j, d_j\}$ are easily produced from the standard forms of the cubic interpolation polynomial.

A second widely used piecewise polynomial interpolation function is based on the cubic Hermite interpolation polynomial of (3.6.12)–(3.6.15). On each subinterval $[x_{i-1}, x_i]$, let $Q_n(x)$ be the cubic Hermite polynomial interpolating $f(x)$ at $x_{i-1}$ and $x_i$. The function $Q_n(x)$ is piecewise cubic on the grid (3.7.3), and because of the interpolation conditions, both $Q_n(x)$ and $Q_n'(x)$ are continuous on $[a, b]$. Thus $Q_n(x)$ will generally be smoother than $L_n(x)$.

For the error in $Q_n(x)$ on $[x_{i-1}, x_i]$, use (3.6.15) to get

$$|f(x) - Q_n(x)| \le \frac{h_i^4}{384} \cdot \underset{x_{i-1} \le x \le x_i}{\text{Max}} |f^{(4)}(t)| \qquad x_{i-1} \le x \le x_i \quad (3.7.10)$$

with $h_i = x_i - x_{i-1}$, $1 \le i \le n$. When this is compared to (3.7.8) for the error in $L_n(x)$, it might seem that piecewise cubic Hermite interpolation is superior. This is deceptive. To see this more clearly, let the grid (3.7.3) be evenly spaced, with $x_i - x_{i-1} = h$, all $i$. Let $L_n(x)$ be based on (3.7.7), and let $\delta = h/3$ in (3.7.8). Note that $Q_n(x)$ is based on $2n + 2$ pieces of data about $f(x)$, namely $\{f(x_i), f'(x_i)| \ i = 0, 1, \ldots, n\}$, and $L_n(x)$ is based on $3n + 1$ pieces of data about $f(x)$. Equalize this by comparing the error for $L_{n_1}(x)$ and $Q_{n_2}(x)$ with $n_2 \doteq 1.5n_1$. Then the resultant error bounds from (3.7.8) and (3.7.10) will be exactly the same.

Since there is no difference in error, the form of piecewise polynomial function used will depend on the application for which it is to be used. In numerical integration applications, the piecewise Lagrange function is most suitable; it is also used in solving some singular integral equations, by means of the product integration methods of Section 5.6 in Chapter 5. The piecewise Hermite function is useful for solving some differential equation problems. For example, it is a popular function used with the finite element method for solving boundary value problems for second-order differential equations; see Strang and Fix (1973, chap. 1). Numerical examples comparing $L_n(x)$ and $Q_n(x)$ are given in Tables 3.11 and 3.12, following the introduction of spline functions.

**Spline functions** As before, consider a grid

$$a = x_0 < x_1 < \cdots < x_n = b$$

We say $s(x)$ is a *spline function* of order $m \ge 1$ if it satisfies the following two properties:

**(P1)** $s(x)$ is a polynomial of degree $< m$ on each subinterval $[x_{i-1}, x_i]$.
**(P2)** $s^{(r)}(x)$ is continuous on $[a, b]$, for $0 \le r \le m - 2$.

The derivative of a spline of order $m$ is a spline of order $m - 1$, and similarly for antiderivatives. If the continuity in P1 is extended to $s^{(m-1)}(x)$, then it can be proved that $s(x)$ is a polynomial of degree $\le m - 1$ on $[a, b]$ (see Problem 33).

Cubic spline functions (order $m = 4$) are the most popular spline functions, for a variety of reasons. They are smooth functions with which to fit data, and when used for interpolation, they do not have the oscillatory behavior that is characteristic of high-degree polynomial interpolation. Some further motivation for particular forms of cubic spline interpolation is given in (3.7.27) and in Problem 38.

For our interpolation problem, we wish to find a cubic spline $s(x)$ for which

$$s(x_i) = y_i \qquad i = 0, 1, \ldots, n \tag{3.7.11}$$

We begin by investigating how many degrees of freedom are left in the choice of $s(x)$, once it satisfies (3.7.11). The technique used does not lead directly to a practical means for calculating $s(x)$, but it does furnish additional insight. Write

$$s(x) = a_i + b_i x + c_i x^2 + d_i x^3 \qquad x_{i-1} \leq x \leq x_i \qquad i = 1, \ldots, n \tag{3.7.12}$$

There are $4n$ unknown coefficients $\{a_i, b_i, c_i, d_i\}$. The constraints on $s(x)$ are (3.7.11) and the continuity restrictions from P2,

$$s^{(j)}(x_i + 0) = s^{(j)}(x_i - 0) \qquad i = 1, \ldots, n-1 \qquad j = 0, 1, 2 \tag{3.7.13}$$

Together this gives

$$n + 1 + 3(n - 1) = 4n - 2$$

constraints, as compared with $4n$ unknowns. Thus there are at least two degrees of freedom in choosing the coefficients of (3.7.12). We should expect to impose extra conditions on $s(x)$ in order to obtain a unique interpolating spline $s(x)$.

We will now give a method for constructing $s(x)$. Introduce the notation

$$M_i = s''(x_i) \qquad i = 0, 1, \ldots, n \tag{3.7.14}$$

Since $s(x)$ is cubic on $[x_i, x_{i+1}]$, $s''(x)$ is linear and thus

$$s''(x) = \frac{(x_{i+1} - x)M_i + (x - x_i)M_{i+1}}{h_i} \qquad i = 0, 1, \ldots, n-1 \tag{3.7.15}$$

where $h_i = x_{i+1} - x_i$. With this formula, $s''(x)$ is continuous on $[x_0, x_n]$. Integrate twice to get

$$s(x) = \frac{(x_{i+1} - x)^3 M_i + (x - x_i)^3 M_{i+1}}{6h_i} + C(x_{i+1} - x) + D(x - x_i)$$

with $C$ and $D$ arbitrary. The interpolating condition (3.7.11) implies

$$C = \frac{y_i}{h_i} - \frac{h_i M_i}{6} \qquad D = \frac{y_{i+1}}{h_i} - \frac{h_i M_{i+1}}{6}$$

$$s(x) = \frac{(x_{i+1} - x)^3 M_i + (x - x_i)^3 M_{i+1}}{6h_i} + \frac{(x_{i+1} - x)y_i + (x - x_i)y_{i+1}}{h_i}$$

$$- \frac{h_i}{6}\left[(x_{i+1} - x)M_i + (x - x_i)M_{i+1}\right]$$

$$x_i \leq x \leq x_{i+1} \qquad 0 \leq i \leq n - 1 \tag{3.7.16}$$

This formula implies the continuity on $[a, b]$ of $s(x)$, as well as the interpolating condition (3.7.11). To determine the constants $M_0, \ldots, M_n$, we require $s'(x)$ to be continuous at $x_1, \ldots, x_{n-1}$:

$$\text{Limit}_{x \searrow x_i} s'(x) = \text{Limit}_{x \nearrow x_i} s'(x) \qquad i = 1, \ldots, n-1 \qquad (3.7.17)$$

On $[x_i, x_{i+1}]$,

$$s'(x) = \frac{-(x_{i+1} - x)^2 M_i + (x - x_i)^2 M_{i+1}}{2h_i} + \frac{y_{i+1} - y_i}{h_i} - \frac{(M_{i+1} - M_i)h_i}{6}$$

$$(3.7.18)$$

and on $[x_{i-1}, x_i]$,

$$s'(x) = \frac{-(x_i - x)^2 M_{i-1} + (x - x_{i-1})^2 M_i}{2h_{i-1}} + \frac{y_i - y_{i-1}}{h_{i-1}} - \frac{(M_i - M_{i-1})h_{i-1}}{6}$$

Using (3.7.17) and some manipulation, we obtain

$$\frac{h_{i-1}}{6} M_{i-1} + \frac{h_i + h_{i-1}}{3} M_i + \frac{h_i}{6} M_{i+1} = \frac{y_{i+1} - y_i}{h_i} - \frac{y_i - y_{i-1}}{h_{i-1}} \qquad (3.7.19)$$

for $i = 1, \ldots, n-1$. This gives $n-1$ equations for the $n+1$ unknowns $M_0, \ldots, M_n$. We generally specify endpoint conditions, at $x_0$ and $x_n$, to remove the two degrees of freedom present in (3.7.19).

*Case 1*    Endpoint derivative conditions.    Require that $s(x)$ satisfy

$$s'(x_0) = y_0' \qquad s'(x_n) = y_n' \qquad (3.7.20)$$

with $y_0', y_n'$ given constants. Using these conditions with (3.7.18), for $i = 0$ and $i = n - 1$, we obtain the additional equations

$$\frac{h_0}{3} M_0 + \frac{h_0}{6} M_1 = \frac{y_1 - y_0}{h_0} - y_0'$$

$$\frac{h_{n-1}}{6} M_{n-1} + \frac{h_{n-1}}{3} M_n = y_n' - \frac{y_n - y_{n-1}}{h_{n-1}}$$

Combined with (3.7.19), we have a system of linear equations

$$AM = D \qquad (3.7.21)$$

with

$$D^T = \left[ \frac{y_1 - y_0}{h_0} - y_0', \frac{y_2 - y_1}{h_1} - \frac{y_1 - y_0}{h_0}, \ldots, \right.$$

$$\left. \frac{y_n - y_{n-1}}{h_{n-1}} - \frac{y_{n-1} - y_{n-2}}{h_{n-2}}, y_n' - \frac{y_n - y_{n-1}}{h_{n-1}} \right]$$

$$M^T = [M_0, M_1, \ldots, M_n]$$

$$A = \begin{bmatrix}
\dfrac{h_0}{3} & \dfrac{h_0}{6} & 0 & 0 & & \cdots & & 0 \\[2mm]
\dfrac{h_0}{6} & \dfrac{h_0 + h_1}{3} & \dfrac{h_1}{6} & & & & & \\[2mm]
0 & \dfrac{h_1}{6} & \dfrac{h_1 + h_2}{3} & \dfrac{h_2}{6} & & & & \vdots \\[2mm]
0 & & & \ddots & & & & \\[2mm]
\vdots & & & & \dfrac{h_{n-2}}{6} & \dfrac{h_{n-2} + h_{n-1}}{3} & \dfrac{h_{n-1}}{6} \\[2mm]
0 & & \cdots & & & \dfrac{h_{n-1}}{6} & \dfrac{h_{n-1}}{3}
\end{bmatrix}$$

$$(3.7.22)$$

This matrix is symmetric, positive definite, and diagonally dominant, and the linear system $AM = D$ is uniquely solvable. This system can be solved easily and rapidly, using about $8n$ arithmetic operations. (See the material on tridiagonal systems in Section 8.3.)

The resulting cubic spline function $s(x)$ is sometimes called the *complete cubic spline interpolant*, and we denote it by $s_c(x)$. An error analysis of it would require too extensive a development, so we just quote results from de Boor (1978, pp. 68–69).

**Theorem 3.4**    Let $f(x)$ be four times continuously differentiable for $a \leq x \leq b$. Let a sequence of partitions be given,

$$\tau_n: a = x_0^{(n)} < x_1^{(n)} < \cdots < x_n^{(n)} = b$$

and define

$$|\tau_n| = \operatorname*{Max}_{1 \leq i \leq n} \left( x_i^{(n)} - x_{i-1}^{(n)} \right)$$

Let $s_{c,n}(x)$ be the complete cubic spline interpolant of $f(x)$ on the partition $\tau_n$:

$$s_{c,n}\left( x_i^{(n)} \right) = f\left( x_i^{(n)} \right) \qquad i = 0, 1, \ldots, n$$

$$s_{c,n}'(a) = f'(a) \qquad s_{c,n}'(b) = f'(b)$$

Then for suitable constants $c_j$,

$$\underset{a \leq x \leq b}{\text{Max}} |f^{(j)}(x) - s_{c,n}^{(j)}(x)| \leq c_j |\tau_n|^{4-j} \cdot \underset{a \leq x \leq b}{\text{Max}} |f^{(4)}(x)| \quad (3.7.23)$$

for $j = 0, 1, 2$. With the additional assumption

$$\underset{n}{\text{Supremum}} \left\{ \frac{|\tau_n|}{\underset{1 \leq i \leq n}{\text{Min}} \left( x_i^{(n)} - x_{i-1}^{(n)} \right)} \right\} < \infty$$

the result (3.7.23) also holds for $j = 3$. Acceptable constants are

$$c_0 = \frac{5}{384} \qquad c_1 = \frac{1}{24} \qquad c_2 = \frac{3}{8} \qquad (3.7.24)$$

**Proof**   The proofs of most of these results can be found in de Boor (1978, pp. 68–69), along with other results on $s_{c,n}(x)$.    ∎

Letting $j = 0$ in (3.7.23), we see that for a uniform grid $\tau_n$, the rate of convergence is proportional to $1/n^4$. This is the same as for piecewise cubic Lagrange and Hermite interpolation, but the multiplying constant $c_0$ is smaller by a factor of about 3. Thus the complete spline interpolant should be a somewhat superior approximation, as the results in Tables 3.11–3.14 bear out.

Another motivation for using $s_c(x)$ is the following optimality property. Let $g(x)$ be any function that is twice continuously differentiable on $[a, b]$, and moreover, let it satisfy the interpolating conditions (3.7.11) and (3.7.20). Then

$$\int_a^b |s_c''(x)|^2 \, dx \leq \int_a^b |g''(x)|^2 \, dx \qquad (3.7.25)$$

with equality only if $g(x) = s_c(x)$. Thus $s_c(x)$ "oscillates least" of all smooth functions satisfying the interpolating conditions (3.7.11) and (3.7.20). To prove the result, let $k(x) = s_c(x) - g(x)$, and write

$$\int_a^b |g''(x)|^2 \, dx = \int_a^b |s_c''(x) - k''(x)|^2 \, dx$$

$$= \int_a^b |s_c''(x)|^2 \, dx - 2 \int_a^b s_c''(x) k''(x) \, dx + \int_a^b |k''(x)|^2 \, dx$$

By integration by parts, and using the interpolating conditions and the properties of $s_c(x)$, we can show

$$\int_a^b s_c''(x) k''(x) \, dx = 0 \qquad (3.7.26)$$

and thus

$$\int_a^b |g''(x)|^2 \, dx = \int_a^b |s_c''(x)|^2 \, dx + \int_a^b |s_c''(x) - g''(x)|^2 \, dx \qquad (3.7.27)$$

This proves (3.7.25). Equality in (3.7.25) occurs only if $s_c''(x) - g''(x) \equiv 0$ on $[a,b]$, or equivalently $s_c(x) - g(x)$ is linear. The interpolating conditions then imply $s_c(x) - g(x) \equiv 0$. We leave a further discussion of this topic to Problem 38.

*Case 2*    The "not-a-knot" condition.   When the derivative values $f'(a)$ and $f'(b)$ are not available, we need other end conditions on $s(x)$ in order to complete the system of equations (3.7.19). This is accomplished by requiring $s^{(3)}(x)$ to be continuous at $x_1$ and $x_{n-1}$. This is equivalent to requiring that $s(x)$ be a cubic spline function with knots $\{x_0, x_2, x_3, \ldots, x_{n-2}, x_n\}$, while still requiring interpolation at all node points in $\{x_0, x_1, x_2, \ldots, x_{n-1}, x_n\}$. This reduces system (3.7.19) to $n - 3$ equations, and the interpolation at $x_1$ and $x_{n-1}$ introduces two new equations (we leave their derivation to Problem 34). Again we obtain a tridiagonal linear system $AM = D$, although the matrix $A$ does not possess some of the nice properties of that in (3.7.22). The resulting spline function will be denoted here by $s_{nk}(x)$, with the subscript indicating the "not-a-knot" condition. A convergence analysis can be given for $s_{nk}(x)$, similar to that given in Theorem 3.4. For a discussion of this, see de Boor (1978, p. 211), (1985).

There are other ways of introducing endpoint conditions when $f'(a)$ and $f'(b)$ are unknown. A discussion of some of these can be found in de Boor (1978, p. 56). However, the preceding scheme is the simplest to apply, and it is widely used. In special cases, there are simpler endpoint conditions that can be used than those discussed here, and we take up one of these in Problem 38. In general, however, the preceding type of endpoint conditions are needed in order to preserve the rates of convergence given in Theorem 3.4.

**Numerical examples**   Let $f(x) = \tan^{-1} x$, $0 \le x \le 5$. Table 3.11 gives the errors

$$E_i = \underset{0 \le x \le 5}{\text{Max}} |f^{(i)}(x) - L_n^{(i)}(x)| \qquad i = 0, 1, 2, 3 \qquad (3.7.28)$$

where $L_n(x)$ is the Lagrange piecewise cubic function interpolating $f(x)$ on the nodes $x_j = a + jh$, $j = 0, 1, \ldots, n$, $h = (b - a)/n$. The columns labeled Ratio

**Table 3.11    Lagrange piecewise cubic interpolation: $L_n(x)$**

| $n$ | $E_0$ | Ratio | $E_1$ | Ratio | $E_2$ | Ratio | $E_3$ | Ratio |
|---|---|---|---|---|---|---|---|---|
| 2 | 1.20E − 2 | | 1.22E − 1 | | 7.81E − 1 | | 2.32 | |
| | | 3.3 | | 2.1 | | 1.5 | | 1.2 |
| 4 | 3.62E − 3 | | 5.83E − 2 | | 5.24E − 1 | | 1.95 | |
| | | 11.4 | | 6.1 | | 3.2 | | 1.6 |
| 8 | 3.18E − 4 | | 9.57E − 3 | | 1.64E − 1 | | 1.19 | |
| | | 16.9 | | 8.1 | | 3.9 | | 1.7 |
| 16 | 1.88E − 5 | | 1.11E − 3 | | 4.21E − 2 | | .682 | |
| | | 14.5 | | 7.3 | | 3.7 | | 1.9 |
| 32 | 1.30E − 6 | | 1.61E − 4 | | 1.14E − 2 | | .359 | |

**Table 3.12   Hermite piecewise cubic interpolation:** $Q_n(x)$

| $n$ | $E_0$ | Ratio | $E_1$ | Ratio | $E_2$ | Ratio | $E_3$ | Ratio |
|---|---|---|---|---|---|---|---|---|
| 3 | 2.64E − 2 | | 5.18E − 2 | | 4.92E − 1 | | 2.06 | |
| | | 5.6 | | 3.0 | | 2.3 | | 1.5 |
| 6 | 4.73E − 3 | | 1.74E − 2 | | 2.14E − 1 | | 1.33 | |
| | | 16.0 | | 8.0 | | 3.6 | | 1.5 |
| 12 | 2.95E − 4 | | 2.17E − 3 | | 5.91E − 2 | | .891 | |
| | | 13.1 | | 6.7 | | 3.6 | | 1.9 |
| 24 | 2.26E − 5 | | 3.25E − 4 | | 1.66E − 2 | | .475 | |
| | | 16.0 | | 8.0 | | 4.0 | | 2.0 |
| 48 | 1.41E − 6 | | 4.06E − 5 | | 4.18E − 3 | | .241 | |

give the rate of decrease in the error when $n$ is doubled. Note that the rate of convergence for $L_n^{(i)}$ to $f^{(i)}$ is proportional to $h^{4-i}$, $i = 0, 1, 2, 3$. This can be rigorously proved, and an indication of the proof is given in Problem 32.

Table 3.12 gives the analogous errors for the Hermite piecewise cubic function $Q_n(x)$ interpolating $f(x)$. Note that again the errors agree with

$$\underset{a \le x \le b}{\text{Max}} |f^{(i)}(x) - Q_n^{(i)}(x)| \le ch^{4-i} \qquad i = 0, 1, 2, 3$$

which can also be proved, for some $c > 0$.

As was stated earlier following (3.7.10), the functions $L_n(x)$ and $Q_m(x)$, $m = 1.5n$, are of comparable accuracy in approximating $f(x)$, and Tables 3.11 and 3.12 confirm this. In contrast, the derivative $Q_m'(x)$ is a more accurate approximation to $f'(x)$ than is $L_n'(n)$. An explanation is given in Problem 32.

In Table 3.13, we give the results of using the complete cubic spline interpolant $s_c(x)$. To compare with $L_n(x)$ and $Q_m(x)$ for comparable amounts of given data on $f(x)$, we use the same number of evenly spaced interpolation points as used in $L_n(x)$.

***Example*** Another informative example is to take $f(x) = x^4$, $0 \le x \le 1$. All of the preceding interpolation formulas have $f^{(4)}(x)$ as a multiplier in their error

**Table 3.13   Complete cubic spline interpolation:** $s_{c,n}(x)$

| $n$ | $E_0$ | Ratio | $E_1$ | Ratio | $E_2$ | Ratio | $E_3$ | Ratio |
|---|---|---|---|---|---|---|---|---|
| 6 | 7.09E − 3 | | 2.45E − 2 | | 1.40E − 1 | | 1.06E0 | |
| | | 21.9 | | 10.7 | | 4.8 | | 2.6 |
| 12 | 3.24E − 4 | | 2.28E − 3 | | 2.90E − 2 | | 4.09E − 1 | |
| | | 10.6 | | 5.6 | | 2.9 | | 1.6 |
| 24 | 3.06E − 5 | | 4.09E − 4 | | 9.84E − 3 | | 2.53E − 1 | |
| | | 20.7 | | 9.7 | | 4.6 | | 2.1 |
| 48 | 1.48E − 6 | | 4.22E − 5 | | 2.13E − 3 | | 1.22E − 1 | |
| | | 16.4 | | 8.1 | | 4.0 | | 2.0 |
| 96 | 9.04E − 8 | | 5.19E − 6 | | 5.30E − 4 | | 6.09E − 2 | |

**Table 3.14    Comparison of three forms of piecewise cubic interpolation**

| Method | $E_0$ | $E_1$ | $E_2$ | $E_3$ |
|---|---|---|---|---|
| Lagrange $n = 32$ | 1.10E − 8 | 6.78E − 6 | 2.93E − 3 | .375 |
| Hermite $n = 48$ | 1.18E − 8 | 1.74E − 6 | 8.68E − 4 | .250 |
| Spline $n = 96$ | 7.36E − 10 | 2.12E − 7 | 1.09E − 4 | .0625 |

formulas. Since $f^{(4)}(x) = 24$, a constant, the error for all three forms of interpolation satisfy

$$\underset{0 \le x \le 1}{\text{Max}} |x^4 - f_n(x)| = c_j h^{4-j} \qquad j = 0, 1, 2, 3 \qquad (3.7.29)$$

The constants $c_j$ will vary with the form of interpolation being used. In the actual computations, the errors behaved exactly like (3.7.29), thus providing another means for comparing the methods. We give the results in Table 3.14 for only the most accurate case.

These examples show that the complete cubic interpolating spline is more accurate, significantly so in some cases. But the examples also show that all of the methods are probably adequate in terms of accuracy, and that they all converge at the same rate. Therefore, the decision as to which method of interpolation to use should depend on other factors, usually arising from the intended area of application. Spline functions have proved very useful with data fitting problems and curve fitting, and Lagrange and Hermite functions are more useful for analytic approximations in solving integral and differential equations, respectively. All of these forms of piecewise polynomial approximation are useful with all of these applications, and one should choose the form of approximation based on the needs of the problem being solved.

**B-splines**    One way of representing cubic spline functions is given in (3.7.12)–(3.7.13), in which a cubic polynomial is given on each subinterval. This is satisfactory for interpolation problems, as given in (3.7.16), but for most applications, there are better ways to represent cubic spline functions. As before, we look at cubic splines with knots $\{x_0, x_1, \ldots, x_n\}$.
    Define

$$x_+^r = \begin{cases} 0 & x < 0 \\ x^r & x \ge 0 \end{cases} \qquad (3.7.30)$$

This is a spline of order $r + 1$, and it has only the one knot $x = 0$. This can be used to give a second representation of spline functions. Let $s(x)$ be a spline function of order $m$ with knots $\{x_0, \ldots, x_n\}$. Then for $x_0 \le x \le x_n$,

$$s(x) = p_{m-1}(x) + \sum_{j=1}^{n-1} \beta_j (x - x_j)_+^{m-1} \qquad (3.7.31)$$

with $p_{m-1}(x)$ a uniquely chosen polynomial of degree $\leq m - 1$ and $\beta_1, \ldots, \beta_{n-1}$ uniquely determined coefficients. The proof of this result is left as Problem 37. There are several unsatisfactory features to this representation when applying it to the solution of other problems. The most serious problem is that it often leads to numerical schemes that are ill-conditioned. For this reason, we introduce another numerical representation of $s(x)$, one that is much better in its numerical properties. To simplify the presentation, we consider only cubic splines.

We begin by augmenting the knots $\{x_0, \ldots, x_n\}$. Choose additional knots

$$x_{-3} < x_{-2} < x_{-1} < x_0 \qquad x_n < x_{n+1} < x_{n+2} < x_{n+3} \qquad (3.7.32)$$

in some arbitrary manner. For $i = -3, -2, \ldots, n - 1$, define

$$B_i(x) = (x_{i+4} - x_i) f_x[x_i, x_{i+1} x_{i+2}, x_{i+3}, x_{i+4}] \qquad (3.7.33)$$

a fourth-order divided difference of

$$f_x(t) = (t - x)_+^3 \qquad (3.7.34)$$

The function $B_i(x)$ is called a *B-spline*, which is short for basic spline function. As an alternative to (3.7.33), apply the formula (3.2.5) for divided differences, obtaining

$$B_i(x) = (x_{i+4} - x_i) \sum_{j=i}^{i+4} \frac{(x_j - x)_+^3}{\Psi_i'(x_j)}$$

$$\Psi_i(x) = (x - x_i)(x - x_{i+1})(x - x_{i+2})(x - x_{i+3})(x - x_{i+4}) \qquad (3.7.35)$$

This shows $B_i(x)$ is a cubic spline with knots $x_i, \ldots, x_{i+4}$. A graph of a typical B-spline is shown in Figure 3.7. We summarize some important properties of B-splines as follows.

***Theorem 3.5*** The cubic B-splines satisfy

(a) $B_i(x) = 0$ outside of $x_i < x < x_{i+4}$; $\qquad (3.7.36)$

(b) $0 \leq B_i(x) \leq 1 \qquad$ all $x$; $\qquad (3.7.37)$

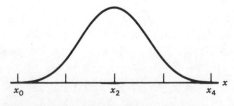

**Figure 3.7** The B-spline $B_0(x)$.

(c) $\displaystyle\sum_{i=-3}^{n-1} B_i(x) = 1 \qquad x_0 \le x \le x_n;$ (3.7.38)

(d) $\displaystyle\int_{x_i}^{x_{i+4}} B_i(x)\, dx = \frac{(x_{i+4} - x_i)}{4};$ (3.7.39)

(e) If $s(x)$ is a cubic spline function with knots $\{x_0, \ldots, x_n\}$, then for $x_0 \le x \le x_n$,

$$s(x) = \sum_{i=-3}^{n-1} \alpha_i B_i(x) \qquad (3.7.40)$$

with the choice of $\alpha_{-3}, \ldots, \alpha_{n-1}$ unique.

***Proof*** (a)  For $x \le x_i$, the function $f_x(t)$ is a cubic polynomial for the interval $x_i \le t \le x_{i+4}$. Thus its fourth-order divided difference is zero. For $x \ge x_{i+4}$, the function $f_x(t) \equiv 0$ for $x_i \le t \le x_{i+4}$, and thus $B_i(x) = 0$.

(b)  See de Boor (1978, p. 131).

(c)  Using the recursion relation for divided differences,

$$B_i(x) = f_x[x_{i+1}, x_{i+2}, x_{i+3}, x_{i+4}] - f_x[x_i, x_{i+1}, x_{i+2}, x_{i+3}] \quad (3.7.41)$$

Next, assume $x_k \le x \le x_{k+1}$. Then the only $B$-splines that can be nonzero at $x$ are $B_{k-3}(x), B_{k-2}(x), \ldots, B_k(x)$. Using (3.7.41),

$$\sum_{i=-3}^{n-1} b_i(x) = \sum_{i=k-3}^{k} B_i(x)$$

$$= \sum_{i=k-3}^{k} \left( f_x[x_{i+1}, x_{i+2}, x_{i+3}, x_{i+4}] - f[x_i, x_{i+1}, x_{i+2}, x_{i+3}] \right)$$

$$= f_x[x_{k+1}, x_{k+2}, x_{k+3}, x_{k+4}] - f_x[x_{k-3}, x_{k-2}, x_{k-1}, x_k]$$

$$= 1 - 0 = 1$$

The last step uses (1) the fact that $f_x(t)$ is cubic on $[x_{k+1}, x_{k+4}]$, so that the divided difference equals 1, from (3.2.18); and (2) $f_x(t) \equiv 0$ on $[x_{k-3}, x_k]$.

(d)  See de Boor (1978, p. 151).

(e)  The concept of B-splines originated with I. J. Schoenberg, and the result (3.7.40) is due to him. For a proof, see de Boor (1978, p. 113).

■

Because of (3.7.36), the sum in (3.7.40) involves at most four nonzero terms. For $x_k \leq x < x_{k+1}$,

$$s(x) = \sum_{i=k-3}^{k} \alpha_i B_i(x) \qquad (3.7.42)$$

In addition, using (3.7.37) and (3.7.38),

$$\text{Min } \{\alpha_{k-3}, \alpha_{k-2}, \alpha_{k-1}, \alpha_k\} \leq s(x) \leq \text{Max } \{\alpha_{k-3}, \alpha_{k-2}, \alpha_{k-1}, \alpha_k\}$$

showing that the value of $s(x)$ is bounded by coefficients for B-splines near to $x$. In this sense, (3.7.40) is a local representation of $s(x)$, at each $x \in [x_0, x_n]$.

A more general treatment of B-splines is given in de Boor (1978, chaps. 9–11), along with further properties omitted here. Programs are also given for computing with B-splines.

An important generalization of splines arises when the knots are allowed to coalesce. In particular, let some of the nodes in (3.7.33) become coincident. Then, so as long as $x_i < x_{i+4}$, the function $B_i(x)$ will be a cubic piecewise polynomial. Letting two knots coalesce will reduce from two to one the number of continuous derivatives at the multiple knot. Letting three knots coalesce will mean that $B_i(x)$ will only be continuous. Doing this, (3.7.40) becomes a representation for all cubic piecewise polynomials. In this scheme of things, all piecewise polynomial functions are spline functions, and vice versa. This is fully explored in de Boor (1978).

## 3.8    Trigonometric Interpolation

An extremely important class of functions are the periodic functions. A function $f(t)$ is said to be *periodic* with period $\tau$ if

$$f(t + \tau) = f(t) \qquad -\infty < t < \infty$$

and this is not to be true for any smaller positive value of $\tau$. The best known periodic functions are the trigonometric functions. Periodic functions occur widely in applications, and this motivates our consideration of interpolation suitable for data derived from such functions. In addition, we use this topic to introduce the fast Fourier transform (FFT), which is used in solving many problems that involve data from periodic functions.

By suitably scaling the independent variable, it is always possible to let $\tau = 2\pi$ be the period:

$$f(t + 2\pi) = f(t) \qquad -\infty < t < \infty \qquad (3.8.1)$$

We approximate such functions $f(t)$ by using trigonometric polynomials,

$$p_n(t) = a_0 + \sum_{j=1}^{n} a_j \cos(jt) + b_j \sin(jt) \qquad (3.8.2)$$

If $|a_n| + |b_n| \neq 0$, then this function $p_n(t)$ is called a *trigonometric polynomial of degree n*. It can be shown by using trigonometric addition formulas that an equivalent formulation is

$$p_n(t) = \alpha_0 + \sum_{j=1}^{n} \alpha_j [\cos(t)]^j + \beta_j [\sin(t)]^j \qquad (3.8.3)$$

thus partially explaining our use of the word polynomial for such a function. The polynomial $p_n(t)$ has period $2\pi$ or integral fraction thereof.

To study interpolation problems with $p_n(t)$ as a solution, we must impose $2n + 1$ interpolating conditions, since $p_n(t)$ contains $2n + 1$ coefficients $a_j, b_j$. Because of the periodicity of the function $f(t)$ and the polynomial $p_n(t)$, we also require the interpolation nodes to lie in the interval $0 \leq t < 2\pi$ (or equivalently, $-\pi \leq t < \pi$ or $0 < t \leq 2\pi$). Thus we assume the existence of the interpolation nodes

$$0 \leq t_0 < t_1 < \cdots < t_{2n} < 2\pi \qquad (3.8.4)$$

and we require $p_n(t)$ to be chosen to satisfy

$$p_n(t_i) = f(t_i) \qquad i = 0, 1, \ldots, 2n \qquad (3.8.5)$$

It is shown later that this problem has a unique solution.

This interpolation problem has an explicit solution, comparable to the Lagrange formula (3.1.6) for polynomial interpolation; this is dealt with in Problem 41. Rather than proceeding with such a development, we first convert (3.8.4)–(3.8.5) to an equivalent problem involving polynomials and functions of a complex variable. This new formulation is the more natural mathematical setting for trigonometric polynomial interpolation.

Using *Euler's formula*

$$e^{i\theta} = \cos(\theta) + i \cdot \sin(\theta) \qquad i = \sqrt{-1} \qquad (3.8.6)$$

we obtain

$$\cos(\theta) = \frac{e^{i\theta} + e^{-i\theta}}{2} \qquad \sin(\theta) = \frac{e^{i\theta} - e^{-i\theta}}{2i} \qquad (3.8.7)$$

Using these in (3.8.2), we obtain

$$p_n(t) = \sum_{j=-n}^{n} c_j e^{ijt} \qquad (3.8.8)$$

The coefficients are related by

$$c_0 = a_0 \qquad c_{-j} = \tfrac{1}{2}(a_j - ib_j) \qquad c_j = \tfrac{1}{2}(a_j + ib_j) \qquad 1 \leq j \leq n$$

Given $\{c_j\}$, the coefficients $\{a_j, b_j\}$ are easily obtained by solving these latter equations. Letting $z = e^{it}$, we can rewrite (3.8.8) as the complex function

$$P_n(z) = \sum_{j=-n}^{n} c_j z^j \qquad (3.8.9)$$

The function $z^n P_n(z)$ is a polynomial of degree $\leq 2n$.

To reformulate the interpolation problem (3.8.4)–(3.8.5), let $z_j = e^{it_j}$, $j = 0, \ldots, 2n$. With the restriction in (3.8.4), the numbers $z_j$ are distinct points on the unit circle $|z| = 1$ in the complex plane. The interpolation problem is

$$P_n(z_j) = f(t_j) \qquad j = 0, \ldots, 2n \qquad (3.8.10)$$

To see that this is always uniquely solvable, note that it is equivalent to

$$Q(z_j) = z_j^n f(t_j) \qquad j = 0, \ldots, 2n$$

with $Q(z) = z^n P_n(z)$. This is a polynomial interpolation problem, with $2n + 1$ distinct node points $z_0, \ldots, z_{2n}$; Theorem 3.1 shows there is a unique solution. Also, the Lagrange formula (3.1.6) generalizes to $Q(z)$, and thence to $P(z)$.

There are a number of reasons, both theoretical and practical, for converting to the complex variable form of trigonometric interpolation. The most important in our view is that interpolation and approximation by trigonometric polynomials are intimately connected to the subject of differentiable functions of a complex variable, and much of the theory is better understood from this perspective. We do not develop this theory, but a complete treatment is given in Henrici (1986, chap. 13) and Zygmund (1959, chap. 10).

**Evenly spaced interpolation**   The case of interpolation that is of most interest in applications is to use evenly spaced nodes $t_j$. More precisely, define

$$t_j = j\frac{2\pi}{2n + 1} \qquad j = 0, \pm 1, \pm 2, \ldots \qquad (3.8.11)$$

The points $t_0, \ldots, t_{2n}$ satisfy (3.8.4), and the points $z_j = e^{it_j}$, $j = 0, \ldots, 2n$, are evenly spaced points on the unit circle $|z| = 1$. Note also that the points $z_j$ repeat as $j$ increases by $2n + 1$.

We now develop an alternative to the Lagrange form for $p_n(t)$ when the nodes $\{t_j\}$ satisfy (3.8.11). We begin with the following lemma.

**Lemma 4**   For all integers $k$,

$$\sum_{j=0}^{2n} e^{ikt_j} = \begin{cases} 2n + 1 & e^{it_k} = 1 \\ 0 & e^{it_k} \neq 1 \end{cases} \qquad (3.8.12)$$

The condition $e^{it_k} = 1$ is equivalent to $k$ being an integer multiple of $2n + 1$.

***Proof***    Let $z = e^{it_k}$. Then using (3.8.11), $e^{ikt_j} = e^{ijt_k}$, and the sum in (3.8.12) becomes

$$S = \sum_{j=0}^{2n} z^j$$

If $z = 1$, then this sums to $2n + 1$. If $z \neq 1$, then the geometric series formula (1.1.8) implies

$$S = \frac{z^{2n+1} - 1}{z - 1}$$

Using (3.8.11), $z^{2n+1} = e^{2\pi ki} = 1$; thus, $S = 0$.    ■

The interpolation conditions (3.8.10) can be written as

$$\sum_{k=-n}^{n} c_k e^{ikt_j} = f(t_j) \qquad j = 0, 1, \ldots, 2n \tag{3.8.13}$$

To find the coefficients $c_k$, we use Lemma 4. Multiply equation $j$ by $e^{-ilt_j}$, then sum over $j$, restricting $l$ to satisfy $-n \leq l \leq n$. This yields

$$\sum_{j=0}^{2n} \sum_{k=-n}^{n} c_k e^{i(k-l)t_j} = \sum_{j=0}^{2n} e^{-ilt_j} f(t_j) \tag{3.8.14}$$

Reverse the order of summation, and then use Lemma 4 to obtain

$$\sum_{j=0}^{2n} e^{i(k-l)t_j} = \begin{cases} 0 & k \neq l \\ 2n + 1 & k = l \end{cases}$$

Using this in (3.8.14), we obtain

$$c_l = \frac{1}{2n+1} \sum_{j=0}^{2n} e^{-ilt_j} f(t_j) \qquad l = -n, \ldots, n \tag{3.8.15}$$

The coefficients $\{c_{-n}, \ldots, c_n\}$ are called the *finite Fourier transform* of the data $\{f(t_0), \ldots, f(t_{2n})\}$. They yield an explicit formula for the trigonometric interpolating polynomial $p_n(t)$ of (3.8.8). The formula (3.8.15) is related to the Fourier coefficients of $f(t)$:

$$\gamma_l = \frac{1}{2\pi} \int_0^{2\pi} e^{-ilt} f(t) \, dt \qquad -\infty < l < \infty \tag{3.8.16}$$

If the trapezoidal numerical integration rule [see Section 5.1] is applied to these integrals, using $2n + 1$ subdivisions of $[0, 2\pi]$, then (3.8.15) is the result, provided $f(t)$ is periodic on $[0, 2\pi]$. We next discuss the convergence of $p_n(t)$ to $f(t)$.

**Theorem 3.6**   Let $f(t)$ be a continuous, periodic function, and let $2\pi$ be an integer multiple of its period. Define

$$\rho_n(f) = \operatorname*{Infimum}_{\deg(q) \le n} \left[ \operatorname*{Max}_{0 \le t \le 2\pi} |f(t) - q(t)| \right] \qquad (3.8.17)$$

with $q(t)$ a trigonometric polynomial. Then the interpolating function $p_n(t)$ from (3.8.8) and (3.8.15) satisfies

$$\operatorname*{Max}_{0 \le t \le 2\pi} |f(t) - p_n(t)| \le c[\ln(n + 2)] \rho_n(f) \qquad n \ge 0 \quad (3.8.18)$$

The constant $c$ is independent of $f$ and $n$.

**Proof**   See Zygmund (1959, chap. 10, p. 19), since the proof is fairly complicated. ∎

The quantity $\rho_n(f)$ is called a minimax error (see Chapter 4), and it can be estimated in a variety of ways. The most important bound on $\rho_n(f)$ is probably that of D. Jackson. Assume that $f(t)$ is $k$ times continuously differentiable on $[0, 2\pi]$, $k \ge 0$, and further assume $f^{(k)}(t)$ satisfies the condition

$$|f^{(k)}(t_1) - f^{(k)}(t_2)| \le c_f |t_1 - t_2|^\alpha \qquad 0 \le t_1, t_2 \le 2\pi$$

for some $0 < \alpha \le 1$. (This is called a Hölder condition.) Then

$$\rho_n(f) \le \frac{c_k(f)}{n^{k+\alpha}} \qquad n \ge 1 \qquad (3.8.19)$$

with $c_k(f)$ independent of $n$. For a proof, see Meinardus (1967, p. 55).

An alternative error formula to that of (3.8.18) is given in Henrici (1986, cor. 13.6c), using the Fourier series coefficients (3.8.16) for $f(t)$.

**Example**   Consider approximating $f(t) = e^{\sin(t)}$, using the interpolating function $p_n(t)$. The maximum error

$$E_n = \operatorname*{Max}_{0 \le t \le 2\pi} |f(t) - p_n(t)|$$

for various values of $n$, is given in Table 3.15. The convergence is rapid.

**Table 3.15   Error in trigonometric polynomial interpolation**

| $n$ | $E_n$ | $n$ | $E_n$ |
|---|---|---|---|
| 1 | 5.39E − 1 | 6 | 6.46E − 6 |
| 2 | 9.31E − 2 | 7 | 4.01E − 7 |
| 3 | 1.10E − 2 | 8 | 2.22E − 8 |
| 4 | 1.11E − 3 | 9 | 1.10E − 9 |
| 5 | 9.11E − 5 | 10 | 5.00E − 11 |

**The fast Fourier transform**   The approximation of $f(t)$ by $p_n(t)$ in the preceding example was very accurate for small values of $n$. In contrast, the calculation of the finite Fourier transform (3.8.15) in other applications will often require large values of $n$. We introduce a method that is very useful in reducing the cost of calculating the coefficients $\{c_j\}$ when $n$ is large.

Rather than using formula (3.8.15), we consider the equivalent formula

$$d_k = \frac{1}{m} \sum_{j=0}^{m-1} w_m^{jk} \cdot f_j \qquad w_m = e^{-2\pi i/m} \qquad k = 0, 1, \ldots, m - 1 \quad (3.8.20)$$

with given data $\{f_0, \ldots, f_{m-1}\}$. This is called a finite Fourier transform of *order* $m$. For formula (3.8.15), let $m = 2n + 1$. We can allow $k$ to be any integer, noting that

$$d_{k+m} = d_k \qquad -\infty < k < \infty \qquad (3.8.21)$$

Thus it is sufficient to compute $d_0, \ldots, d_{m-1}$ or any other $m$ consecutive coefficients $d_k$.

To contrast the formula (3.8.20) with the alternative presented below, we calculate the cost of evaluating $d_0, \ldots, d_{m-1}$ using (3.8.20). To evaluate $d_k$, let $z_k = w_m^k$. Then

$$d_k = \frac{1}{m} \sum_{j=0}^{m-1} f_j z_k^j \qquad (3.8.22)$$

Using nested multiplication, this requires $m - 1$ multiplications and $m - 1$ additions. We ignore the division by $m$, since often other factors are used. The evaluation of $z_k$ requires only 1 multiplication, since $z_k = w_m z_{k-1}$, $k \geq 2$. The total cost of evaluating $d_0, \ldots, d_{m-1}$ is $m^2$ multiplications and $m(m - 1)$ additions.

To introduce the main idea behind the *fast Fourier transform*, let $m = pq$ with $p$ and $q$ positive integers greater than 1. Rewrite the definition (3.8.20) in the equivalent form

$$d_k = \frac{1}{p} \sum_{l=0}^{p-1} \frac{1}{q} \sum_{g=0}^{q-1} w_m^{k(l+pg)} f_{l+pg}$$

Use $w_m^p = \exp(-2\pi i/q) = w_q$. Then

$$d_k = \frac{1}{p} \sum_{l=0}^{p-1} w_m^{kl} \left[ \frac{1}{q} \sum_{g=0}^{q-1} w_q^{kg} f_{l+pg} \right] \qquad k = 0, 1, \ldots, m - 1$$

Write

$$e_k^{(l)} = \frac{1}{q} \sum_{g=0}^{q-1} w_q^{kg} f_{l+pq} \qquad 0 \leq l \leq p - 1 \qquad (3.8.23)$$

$$d_k = \frac{1}{p} \sum_{l=0}^{p-1} w_m^{kl} e_k^{(l)} \qquad 0 \leq k \leq m - 1 \qquad (3.8.24)$$

Once $\{e_k^{(l)}\}$ is known, each value of $d_k$ will require $p$ multiplications, using a nested multiplication scheme as in (3.8.22). The evaluation of (3.8.24) will require $mp$ multiplications, assuming all $e_k^{(l)}$ have been computed previously. There will be a comparable number of additions.

We turn our attention to the computation of the quantities $e_k^{(l)}$. The index $k$ ranges from 0 to $m - 1$, but not all of these need be computed. Note that

$$e_{k+q}^{(l)} = \frac{1}{q} \sum_{g=0}^{q-1} w_q^{(k+q)g} f_{l+pq} = e_k^{(l)}$$

because $w_q^q = 1$. Thus $e_k^{(l)}$ needs to be calculated for only $k = 0, 1, \ldots, q - 1$, and then it repeats itself. For each $l$, $\{e_0^{(l)}, \ldots, e_{q-1}^{(l)}\}$ is the finite Fourier transform of the data $\{f_l, f_{l+p}, \ldots, f_{l+p(q-1)}\}$, $0 \le l \le p - 1$. Thus the computation of $\{e_k^{(l)}\}$ amounts to the computation of $p$ finite Fourier transforms of order $q$ (i.e., for data of length $q$).

The fast Fourier transform amounts to a repeated use of this idea, to reduce the computation to finite Fourier transforms of smaller and smaller order. To be more specific, suppose $m = 2^r$ for some integer $r$. As a first step, let $p = 2$, $q = 2^{r-1}$. Then the computation of (3.8.24) requires $2m$ multiplications plus the cost of evaluating two finite Fourier transforms of order $q = 2^{r-1}$. For each of these, repeat the process recursively. There will be $r$ levels in this process, resulting eventually in the evaluation of finite Fourier transforms of order one. The total number of multiplications is given by

$$2m + 2\left[2\left(\frac{m}{2}\right)\right] + 4\left[2\left(\frac{m}{4}\right)\right] + \cdots + 2^r\left[2\left(\frac{m}{2^r}\right)\right]$$

which sums to

$$2rm = 2m \cdot \log_2 m \tag{3.8.25}$$

Thus the number of operations is proportional to $m \cdot \log_2 m$, in contrast to the $m^2$ operations of the nested multiplication algorithm of (3.8.22). When $m$ is large, say $2^{10}$ or larger, this results in an enormous savings in calculation time. For the particular case of $m = 2^r$, a more careful accounting will show that only $m \cdot \log_2 m$ multiplications are actually needed, and there is a variant procedure that requires only half of this.

For other values of $m$, there are modifications of the preceding procedure that will still yield an operations count proportional to $m \cdot \log m$, just as previously shown, but the case of $m = 2^r$ leads to the greatest savings. For a further discussion of this topic, we refer the reader to Henrici (1986, chap. 13). Henrici also contains a discussion of the stability of the algorithm when rounding errors are taken into account. The use of the fast Fourier transform has revolutionized many subjects, making computationally feasible many calculations that previously were not practical.

# Discussion of the Literature

As noted in the introduction, interpolation theory is a foundation for the development of methods in numerical integration and differentiation, approximation theory, and the numerical solution of differential equations. Each of these topics is developed in the following chapters, and the associated literature is discussed at that point. Additional results on interpolation theory are given in de Boor (1978), Davis (1963), Henrici (1982, chaps. 5 and 7), and Hildebrand (1956). For an historical account of many of the topics of this chapter, see Goldstine (1977).

The introduction of digital computers produced a revolution in numerical analysis, including interpolation theory. Before the use of digital computers, hand computation was necessary, which meant that numerical methods were used that minimized the need for computation. Such methods were often more complicated than the methods now used on computers, taking special advantage of the unique mathematical characteristics of each problem. These methods also made extensive use of tables, to avoid repeating calculations done by others; interpolation formulas based on finite differences were used extensively. A large subject was created, called the *finite difference calculus*, and it was used in solving problems in several areas of numerical analysis and applied mathematics. For a general introduction to this approach to numerical analysis, see Hildebrand (1956) and the references contained therein.

The use of digital computers has changed the needs of other areas for interpolation theory, vastly reducing the need for finite difference based interpolation formulas. But there is still an important place for both hand computation and the use of mathematical tables, especially for the more complicated functions of mathematical physics. Everyone doing numerical work should possess an elementary book of tables such as the well-known CRC tables. The National Bureau of Standards tables of Abramowitz and Stegun (1964) are an excellent reference for nonelementary functions. The availability of sophisticated hand calculators and microcomputers makes possible a new level of hand (or personal) calculation.

Piecewise polynomial approximation theory has been very popular since the early 1960s, and it is finding use in a variety of fields. For example, see Strang and Fix (1973, chap. 1) for applications to the solution of boundary value problems for ordinary differential equations, and see Pavlidis (1982, chaps. 10–12) for applications in computer graphics. Most of the interest in piecewise polynomial functions has centered on spline functions. The beginning of the theory of spline functions is generally credited to Schoenberg in his 1946 papers, and he has been prominent in helping to develop the subject [e.g., see Schoenberg (1973)]. There is now an extensive literature on spline functions, involving many individuals and groups. For general surveys, see Ahlberg et al. (1967), de Boor (1978), and Schumaker (1981). Some of the most widely used computer software for using spline functions is based on the programs in de Boor (1978). Versions of these are available in the IMSL and NAG numerical analysis libraries.

Finite Fourier transforms, trigonometric interpolation, and associated topics are quite old topics; for example, see Goldstine (1977, p. 238) for a discussion of

Gauss' work on trigonometric interpolation. Since Fourier series and Fourier transforms are important tools in much of applied mathematics, it is not surprising that there is a great deal of interest in their discrete approximations. Following the famous paper by Cooley and Tukey (1965) on the fast Fourier transform, there has been a large increase in the use of finite Fourier transforms and associated topics. For example, this has led to very fast methods for solving Laplace's partial differential equation on rectangular regions, which we discuss further in Chapter 8. For a classical account of trigonometric interpolation, see Zygmund (1959, chap. 10), and for a more recent survey of the entire area of finite Fourier analysis, see Henrici (1986, chap. 13).

Multivariate polynomial interpolation theory is a rapidly developing area, which for reasons of space has been omitted in this text. The finite element method for solving partial differential equations makes extensive use of multivariate interpolation theory, and some of the better presentations of this theory are contained in books on the finite element method. For example, see Jain (1984), Lapidus and Pinder (1982), Mitchell and Wait (1977), and Strang and Fix (1973). More recently, work in computer graphics has led to new developments [see Barnhill (1977) and Pavlidis (1982, chap. 13)].

# Bibliography

Abramowitz, M., and I. Stegun, eds. (1964). *Handbook of Mathematical Functions*. National Bureau of Standards, Washington, D.C. (Available now from Dover, New York.)

Ahlberg, J., E. Nilson, and J. Walsh (1967). *The Theory of Splines and Their Applications*. Academic Press, New York.

Barnhill, R. (1977). Representation and approximation of surfaces. In *Mathematical Software III*, J. Rice, ed., pp. 69–120. Academic Press, New York.

De Boor, C. (1978). *A Practical Guide to Splines*. Springer-Verlag, New York.

De Boor, C. (1985). Convergence of cubic spline interpolation with the not-a-knot condition. Math. Res. Ctr. Tech. Rep. 1876, Madison, Wis.

*CRC Standard Mathematical Tables*. Chem. Rubber Publ. Co., Cleveland. (Published yearly).

Cooley, J., and J. Tukey (1965). An algorithm for the machine calculation of complex Fourier series. *Math. Comput.*, **19**, 297–301.

Davis, P. (1963). *Interpolation and Approximation*. Ginn (Blaisdell), Boston.

Goldstine, H. (1977). *A History of Numerical Analysis from the 16th through the 19th Century*. Springer-Verlag, New York.

Henrici, P. (1982). *Essentials of Numerical Analysis*. Wiley, New York.

Henrici, P. (1986). *Applied and Computational Complex Analysis*, Vol. 3. Wiley, New York.

Hildebrand, F. (1956). *Introduction to Numerical Analysis*. McGraw-Hill, New York.

Isaacson, E., and H. Keller (1966). *Analysis of Numerical Methods*. Wiley, New York.

Jain, M. (1984). *Numerical Solution of Differential Equations*, 2nd ed., Wiley, New Delhi.

Lapidus, L., and G. Pinder (1982). *Numerical Solution of Partial Differential Equations in Science and Engineering*. Wiley, New York.

Lorentz, G., K. Jetter, and S. Riemenschneider (1983). *Birkhoff Interpolation*. In *Encyclopedia of Mathematics and Its Applications*, Vol. 19. Addison-Wesley, Reading, Mass.

Meinardus, G. (1967). *Approximation of Functions: Theory and Numerical Methods* (transl. L. Schumaker). Springer-Verlag, New York.

Mitchell, A., and R. Wait (1977). *The Finite Element Method in Partial Differential Equations*. Wiley, London.

Pavlidis, T. (1982). *Algorithms for Graphics and Image Processing*. Computer Science Press, Rockville, Md.

Powell, M. (1981). *Approximation Theory and Methods*. Cambridge Univ. Press, Cambridge, England.

Schoenberg, I. (1946). Contributions to the approximation of equidistant data by analytic functions. *Quart. Appl. Math*, **4**(Part A), 45–99; (Part B), 112–141.

Schoenberg, I. (1973). *Cardinal Spline Interpolation*. Society for Industrial and Applied Mathematics, Philadelphia.

Schumaker, L. (1981). *Spline Functions: Basic Theory*. Wiley, New York.

Stoer, J., and R. Bulirsch (1980). *Introduction to Numerical Analysis*. Springer-Verlag, New York.

Strang, G., and G. Fix (1973). *An Analysis of the Finite Element Method*. Prentice-Hall, Englewood Cliffs, N.J.

Zygmund, A. (1959). *Trigonometric Series*, Vols. 1 and 2. Cambridge Univ. Press, Cambridge.

## Problems

**1.**  Recall the Vandermonde matrix $X$ given in (3.1.3), and define

$$
V_n(x) = \det
\begin{bmatrix}
1 & x_0 & x_0^2 & \cdots & x_0^n \\
\vdots & & & & \vdots \\
1 & x_{n-1} & x_{n-1}^2 & & x_{n-1}^n \\
1 & x & x^2 & \cdots & x^n
\end{bmatrix}
$$

(a)   Show that $V_n(x)$ is a polynomial of degree $n$, and that its roots are $x_0, \ldots, x_{n-1}$. Obtain the formula

$$V_n(x) = (x - x_0) \cdots (x - x_{n-1})V_{n-1}(x_{n-1})$$

*Hint:* Expand the last row of $V_n(x)$ by minors to show that $V_n(x)$ is a polynomial of degree $n$ and to find the coefficient of the term $x^n$.

(b)   Show

$$\det(X) \equiv V_n(x_n) = \prod_{0 \leq j < i \leq n} (x_i - x_j)$$

2.   For the basis functions $l_{j,n}(x)$ given in (3.1.5), prove that for any $n \geq 1$,

$$\sum_{j=0}^{n} l_{j,n}(x) = 1 \qquad \text{for all} \quad x$$

3.   Recall the Lagrange functions $l_0(x), \ldots, l_n(x)$, defined in (3.1.5) and then rewritten in a slightly different form in (3.2.4), using

$$\Psi_n(x) = (x - x_0) \cdots (x - x_n)$$

Let $w_j = [\Psi_n'(x_j)]^{-1}$. Show that the polynomial $p_n(x)$ interpolating $f(x)$ can be written as

$$p_n(x) = \frac{\displaystyle\sum_{j=0}^{n} [w_j f(x_j)]/(x - x_j)}{\displaystyle\sum_{j=0}^{n} w_j/(x - x_j)}$$

provided $x$ is not a node point. This is called the *barycentric* representation of $p_n(x)$, giving it as a weighted sum of the values $\{f(x_0), \ldots, f(x_n)\}$. For a discussion of the use of this representation, see Henrici (1982, p. 237).

4.   Consider linear interpolation in a table of values of $e^x$, $0 \leq x \leq 2$, with $h = .01$. Let the table values be given to five significant digits, as in the CRC tables. Bound the error of linear interpolation, including that part due to the rounding errors in the table entries.

5.   Consider linear interpolation in a table of $\cos(x)$ with $x$ given in degrees, $0 \leq x \leq 90°$, with a stepsize $h = 1' = \frac{1}{60}$ degree. Assuming that the table entries are given to five significant digits, bound the total error of interpolation.

6.   Suppose you are to make a table of values of $\sin(x)$, $0 \leq x \leq \pi/2$, with a stepsize of $h$. Assume linear interpolation is to be used with the table, and suppose the total error, including the effects due to rounding in table entries, is to be at most $10^{-6}$. What should $h$ equal (choose it in a

convenient size for actual use) and to how many significant digits should the table entries be given?

7.   Repeat Problem 6, but using $e^x$ on $0 \le x \le 1$.

8.   Generalize to quadratic interpolation the material on the effect of rounding errors in table entries, given in Section 3.1. Let $\epsilon_i = f(x_i) - \tilde{f_i}$, $i = 0, 1, 2$, and $\epsilon \ge \text{Max}\{|\epsilon_0|, |\epsilon_1|, |\epsilon_2|\}$. Show that the effect of these rounding errors on the quadratic interpolation error is bounded by $1.25\epsilon$, assuming $x_0 \le x \le x_2$ and $x_1 - x_0 = x_2 - x_1 = h$.

9.   Repeat Problem 6, but use quadratic interpolation and the result of Problem 8.

10.  Consider producing a table of values for $f(x) = \log_{10} x$, $1 \le x \le 10$, and assume quadratic interpolation is to be used. Let the total interpolation error, including the effect of rounding in table entries, be less than $10^{-6}$. Choose an appropriate grid spacing $h$ and the number of digits to which the entries should be given. Would it be desirable to vary the spacing $h$ as $x$ varies in $[1, 10]$? If so, suggest a suitable partition of $[1, 10]$ with corresponding values of $h$. Use the result of Problem 8 on the effect of rounding error.

11.  Let $x_0, \ldots, x_n$ be distinct real points, and consider the following interpolation problem. Choose a function

$$P_n(x) = \sum_{j=0}^{n} c_j e^{jx}$$

such that

$$P_n(x_i) = y_i \qquad i = 0, 1, \ldots, n$$

with the $\{y_i\}$ given data. Show there is a unique choice of $c_0, \ldots, c_n$. *Hint:* The problem can be reduced to that of ordinary polynomial interpolation.

12.  Consider finding a rational function $p(x) = (a + bx)/(1 + cx)$ that satisfies

$$p(x_i) = y_i \qquad i = 1, 2, 3$$

with $x_1, x_2, x_3$ distinct. Does such a function $p(x)$ exist, or are additional conditions needed to ensure existence and uniqueness of $p(x)$? For a general theory of rational interpolation, see Stoer and Bulirsch (1980, p. 58).

13.  (a)  Prove the recursive form of the interpolation formula, as given in (3.2.8).

     (b)  Prove the recursion formula (3.2.7) for divided differences.

14.    Prove the relations (3.2.18), pertaining to the variable divided difference of a polynomial.

15.    Prove that $\text{Vol}(\tau_n) = 1/n!$, where $\tau_n$ is the simplex in $\mathbf{R}^n$ defined in (3.2.14) of Theorem 3.3 in Section 3.2. *Hint:* Use (3.2.13) and other results on divided differences, along with a special choice for $f(x)$.

16.    Let $p_2(x)$ be the quadratic polynomial interpolating $f(x)$ at the evenly spaced points $x_0$, $x_1 = x_0 + h$, $x_2 = x_0 + 2h$. Derive formulas for the errors $f'(x_i) - p_2'(x_i)$, $i = 0, 1, 2$. Assuming $f(x)$ is three times continuously differentiable, give computable bounds for these errors. *Hint:* Use the error formula (3.2.11).

17.    Produce computer subroutine implementations of the algorithms *Divdif* and *Interp* given in Section 3.2, and then write a main driver program to use them in doing table interpolation. Choose a table from Abramowitz and Stegun (1964) to test the program, considering several successive degrees of $n$ for the interpolation polynomial.

18.    Do an inverse interpolation problem using the table for $J_0(x)$ given in Section 3.2. Find the value of $x$ for which $J_0(x) = 0$, that is, calculate an accurate estimate of the root. Estimate your accuracy, and compare this with the actual value $x = 2.4048255577$.

19.    Derive the analogue of Lemma 1, given in Section 3.3, for backward differences. Use this and Newton's divided form of the interpolating polynomial (3.2.9) to derive the backward difference interpolation formula (3.3.11).

20.    Consider the following table of values for

$$j_0(x) = \sqrt{\frac{\pi}{2x}} \cdot J_{1/2}(x)$$

taken from Abramowitz and Stegun (1964, chap. 10).

| $x$ | $j_0(x)$ | $x$ | $j_0(x)$ |
|-----|----------|-----|----------|
| 0.0 | 1.00000  | 0.7 | .92031   |
| 0.1 | .99833   | 0.8 | .89670   |
| 0.2 | .99335   | 0.9 | .87036   |
| 0.3 | .98507   | 1.0 | .84147   |
| 0.4 | .97355   | 1.1 | .81019   |
| 0.5 | .95885   | 1.2 | .77670   |
| 0.6 | .94107   | 1.3 | .74120   |

Based on the rounding errors in the table entries, what should be the maximum degree of polynomial interpolation used with the table? *Hint:* Use the forward difference table to detect the influence of the rounding errors.

21. The following data are taken from a polynomial of degree $\leq 5$. What is the degree of the polynomial?

| $x$ | $-2$ | $-1$ | $0$ | $1$ | $2$ | $3$ |
|---|---|---|---|---|---|---|
| $p(x)$ | $-5$ | $1$ | $1$ | $1$ | $7$ | $25$ |

22. The following data have noise in them that is large relative to rounding error. Find the noise and change the data appropriately. Only the function values are given, since the node points are unnecessary for computing the forward difference table.

| | | | |
|---|---|---|---|
| 304319 | 419327 | 545811 | 683100 |
| 326313 | 443655 | 572433 | 711709 |
| 348812 | 468529 | 599475 | 740756 |
| 371806 | 493852 | 626909 | 770188 |
| 395285 | 519615 | 654790 | 800000 |

23. For $f(x) = 1/(1 + x^2)$, $-5 \leq x \leq 5$, produce $p_n(x)$ using $n + 1$ evenly spaced nodes on $[-5, 5]$. Calculate $p_n(x)$ at a large number of points, and graph it or its error on $[-5, 5]$, as in Figure 3.6.

24. Consider the function $e^x$ on $[0, b]$ and its approximation by an interpolating polynomial. For $n \geq 1$, let $h = b/n$, $x_j = jh$, $j = 0, 1, \ldots, n$, and let $p_n(x)$ be the $n$th-degree polynomial interpolating $e^x$ on the nodes $x_0, \ldots, x_n$. Prove that

$$\underset{0 \leq x \leq b}{\text{Max}} |e^x - p_n(x)| \to 0 \quad \text{as} \quad n \to \infty$$

*Hint:* Show $|\Psi_n(x)| \leq n! h^{n+1}$, $0 \leq x \leq b$; look separately at each subinterval $[x_{j-1}, x_j]$.

25. Prove that the general Hermite problem (3.6.16) has a unique solution $p(x)$ among all polynomials of degree $\leq N - 1$. *Hint:* Show that the homogeneous problem for the associated linear system has only the zero solution.

26. Consider the Hermite problem

$$p^{(j)}(x_i) = y_i^{(j)} \quad i = 1, 2 \quad j = 0, 1, 2$$

with $p(x)$ a polynomial of degree $\leq 5$.

(a) Give a Lagrange type of formula for $p(x)$, generalizing (3.6.12) for cubic Hermite interpolation. *Hint:* For the basis functions satisfying

$l(x_2) = l'(x_2) = l''(x_2) = 0$, use $l(x) = (x - x_2)^3 g(x)$, with $g(x)$ of degree $\leq 2$. Find $g(x)$.

**(b)**   Give a Newton divided difference formula, generalizing (3.6.13).

**(c)**   Derive an error formula generalizing (3.6.14).

**27.**   Let $p(x)$ be a polynomial solving the Hermite interpolation problem

$$p^{(j)}(a) = f^{(j)}(a) \qquad p^{(j)}(b) = f^{(j)}(b) \qquad j = 0, 1, \ldots, n - 1$$

Its existence is guaranteed by the argument in Problem 25. Assuming $f(x)$ has $2n$ continuous derivatives on $[a, b]$, show that for $a \leq x \leq b$,

$$f(x) - p_n(x) = \frac{(x - a)^n (x - b)^n}{(2n)!} f^{(2n)}(\xi_x)$$

with $a \leq \xi_x \leq b$. *Hint:* Generalize the argument used in Theorem 3.2.

**28.**   **(a)**   Find a polynomial $p(x)$ of degree $\leq 2$ that satisfies

$$p(x_0) = y_0 \qquad p'(x_0) = y_0' \qquad p'(x_1) = y_1'$$

Give a formula in the form

$$p(x) = y_0 l_0(x) + y_0' l_1(x) + y_1' l_2(x)$$

**(b)**   Find a formula for the following polynomial interpolation problem. Let $x_i = x_0 + ih$, $i = 0, 1, 2$. Find a polynomial $p(x)$ of degree $\leq 4$ for which

$$p(x_i) = y_i \qquad i = 0, 1, 2$$

$$p'(x_0) = y_0' \qquad p'(x_2) = y_2'$$

with the $y$ values given.

**29.**   Consider the problem of finding a quadratic polynomial $p(x)$ for which

$$p(x_0) = y_0 \qquad p'(x_1) = y_1' \qquad p(x_2) = y_2$$

with $x_0 \neq x_2$ and $\{y_0, y_1', y_2\}$ the given data. Assuming that the nodes $x_0, x_1, x_2$ are real, what conditions must be satisfied for such a $p(x)$ to exist and be unique? This problem, Problem 28(a), and the following problem are examples of *Hermite–Birkhoff interpolation* problems [see Lorentz et al. (1983)].

**30.** **(a)** Show there is a unique cubic polynomial $p(x)$ for which

$$p(x_0) = f(x_0) \qquad p(x_2) = f(x_2)$$

$$p'(x_1) = f'(x_1) \qquad p''(x_1) = f''(x_1)$$

where $f(x)$ is a given function and $x_0 \neq x_2$. Derive a formula for $p(x)$.

**(b)** Let $x_0 = -1$, $x_1 = 0$, $x_2 = 1$. Assuming $f(x)$ is four times continuously differentiable on $[-1, 1]$, show that for $-1 \leq x \leq 1$,

$$f(x) - p(x) = \frac{x^4 - 1}{4!} f^{(4)}(\xi_x)$$

for some $\xi_x \in [-1, 1]$. *Hint:* Mimic the proof of Theorem 3.2.

**31.** For the function $f(x) = \sin(x)$, $0 \leq x \leq \pi/2$, find the piecewise cubic Hermite function $Q_n(x)$ and the piecewise cubic Lagrange function $L_m(x)$, for $n = 3, 6, 12$, $m = \frac{2}{3}n$. Evaluate the maximum errors $f(x) - L_m(x)$, $f'(x) - L'_m(x)$, $f(x) - Q_n(x)$, and $f'(x) - Q'_n(x)$. This can be done with reasonable accuracy by evaluating the errors at $8n$ evenly spaced points on $[0, \pi/2]$.

**32.** **(a)** Let $p_3(x)$ denote the cubic polynomial interpolating $f(x)$ at the evenly spaced points $x_j = x_0 + jh$, $j = 0, 1, 2, 3$. Assuming $f(x)$ is sufficiently differentiable, bound the error in using $p_3'(x)$ as an approximation to $f'(x)$, $x \leq x \leq x_3$.

**(b)** Let $H_2(x)$ denote the cubic Hermite polynomial interpolating $f(x)$ and $f'(x)$ at $x_0$ and $x_1 = x_0 + h$. Bound the error $f'(x) - H_2'(x)$ for $x_0 \leq x \leq x_1$.

**(c)** Consider the piecewise polynomial functions $L_m(x)$ and $Q_n(x)$ of Section 3.7, $m = 2n/3$, and bound the errors $f'(x) - L'_m(x)$ and $f'(x) - Q'_n(x)$. Apply it to the specific case of $f(x) = x^4$ and compare your answers with the numerical results given in Table 3.14.

**33.** Let $s(x)$ be a spline function of order $m$. Let $b$ be a knot, and let $s(x)$ be a polynomial of degree $\leq m - 1$ on $[a, b]$ and $[b, c]$. Show that if $s^{(m-1)}(x)$ is continuous at $x = b$, then $s(x)$ is a polynomial of degree $\leq m - 1$ for $a \leq x \leq c$.

**34.** Derive the conditions on the cubic interpolating spline $s(x)$ that are implied by the "not-a-knot" endpoint conditions. Refer to the discussion of case 2, following (3.7.27).

**35.** Consider finding a cubic spline interpolating function for the data

| $x$ | 0 | 1 | 2 | 2.5 | 3 | 4 |
|---|---|---|---|---|---|---|
| $y$ | 1.4 | 0.6 | 1.0 | .65 | .6 | 1.0 |

Use the "not-a-knot" condition to obtain boundary conditions supplementing (3.7.19). Graph the resulting function $s(x)$. Compare it to the use of piecewise linear interpolation, connecting the successive points $(x_i, y_i)$ by line segments.

**36.** Write a program to investigate the rate of convergence of cubic spline interpolation (as in Table 3.13) with various boundary conditions. Many computer centers will have a package to produce such an interpolant, with the user allowed to specify the boundary conditions. Otherwise, use a linear systems solver with (3.7.19) and the additional boundary conditions. Investigate the following boundary conditions: (a) derivatives given as in (3.7.20); (b) the "not-a-knot" condition, given in Problem 34; and (c) the natural spline conditions, $M_0 = M_n = 0$, of Problem 38. Apply this program to studying the convergence of $s(x)$ to $f(x)$ in the following cases: (1) $f(x) = e^x$ on $[0, 1]$, (2) $f(x) = \sin(x)$ on $[0, \pi/2]$, and (3) $f(x) = x\sqrt{x}$ on $[0, 1]$. Note and compare the behavior of the error near the endpoints.

**37. (a)** Let $q(x)$ be a cubic spline with a single knot $x = a$. In addition, suppose that $q(x) \equiv 0$ for $x \leq a$. Show that $q(x) = c(x - a)_+^3$ for some $c$. *Hint:* For $x \geq a$, write the cubic polynomial $s(x)$ as

$$s(x) = c_0 + c_1(x - a) + c_2(x - a)^2 + c_3(x - a)^3.$$

Then apply the assumptions about $s(x)$.

**(b)** Using part (a), prove the representation (3.7.31) for cubic spline functions $(m = 4)$.

**38.** Consider calculating a cubic interpolating spline based on (3.7.19) with the additional boundary conditions

$$s(x_0) = M_0 = 0 \qquad s(x_n) = M_n = 0$$

This has a unique solution $\hat{s}(x)$. Show that

$$\int_{x_0}^{x_n} [\hat{s}''(x)]^2 \, dx \leq \int_{x_0}^{x_n} [g''(x)]^2 \, dx$$

where $g(x)$ is any twice continuously differentiable function that satisfies the interpolating conditions $g(x_i) = y_i$, $i = 0, 1, \ldots, n$. *Hint:* Show (3.7.27) is valid for $\hat{s}(x)$. [This interpolating spline $\hat{s}(x)$ is called the *natural cubic interpolating spline*. It is a smooth interpolant to the data, but it usually converges slowly near the endpoints. For more information on it, see de Boor (1978, p. 55).]

**39.** To define a B-spline of order $m$, with $[x_i, x_{i+m}]$ the interval on which it is nonzero, define

$$B_i^{(m)}(x) = (x_{i+m} - x_i) f_x[x_i, x_{i+1}, \ldots, x_{i+m}]$$

with $f_x(t) = (t - x)_+^{m-1}$. Derive a recursion relation for $B_i^{(m)}(x)$ in terms of B-splines of order $m - 1$.

**40.** Use the definition (3.7.33) to investigate the behavior of cubic $B$-splines when nodes are allowed to coincide. Show that if two of the nodes in $\{x_i, x_{i+1}, \ldots, x_{i+4}\}$ coincide, then $B_i(x)$ has only one continuous derivative at the coincident node. Similarly, if three of them coincide, show that $B_i(x)$ is continuous at that point, but is not differentiable.

**41.** Let $0 \le t_0 < t_1 < \cdots < t_{2n} < 2\pi$, and consider the trigonometric polynomial interpolation problem (3.8.5). Define

$$l_j(t) = \prod_{\substack{k=0 \\ k \ne j}}^{2n} \frac{\sin \frac{1}{2}(t - t_k)}{\sin \frac{1}{2}(t_j - t_k)}$$

for $j = 0, 1, \ldots, 2n$. Easily, $l_j(t_i) = \delta_{ij}$, $0 \le i$, $j \le 2n$. Show that $l_j(t)$ is a trigonometric polynomial of degree $\le n$. Then the solution to (3.8.5) is given by

$$p_n(t) = \sum_{j=0}^{2n} f(t_j) l_j(t)$$

*Hint:* Use induction on $n$, along with standard trigonometric identities.

**42. (a)** Prove the following formulas:

$$\sum_{j=1}^{m-1} \sin\left(\frac{2\pi jk}{m}\right) = 0, \; m \ge 2, \text{ all integers } k$$

$$\sum_{j=0}^{m-1} \cos\left(\frac{2\pi jk}{m}\right) = \begin{cases} m & k \text{ a multiple of } m \\ 0 & k \text{ not a multiple of } m \end{cases}$$

**(b)** Use these formulas to derive formulas for the following:

$$\sum_{j=0}^{m-1} \cos\left(\frac{2\pi jk}{m}\right) \cos\left(\frac{2\pi jl}{m}\right) \qquad \sum_{j=1}^{m-1} \sin\left(\frac{2\pi jk}{m}\right) \sin\left(\frac{2\pi jl}{m}\right)$$

$$\sum_{j=0}^{m-1} \cos\left(\frac{2\pi jk}{m}\right) \sin\left(\frac{2\pi jl}{m}\right)$$

for $0 \leq k$, $l \leq m - 1$. The formulas obtained are referred to as *discrete orthogonality relations*, and they are the analogues of integral orthogonality relations for $\{\cos(kx), \sin(kx)\}$.

43.  Calculate the finite Fourier transform of order $m$ of the following sequences.

(a)   $x_k = 1$      $0 \leq k \leq m - 1$

(b)   $x_k = (-1)^k$      $0 \leq k \leq m - 1$      $m$ even

(c)   $x_k = k$      $0 \leq k \leq m - 1$

# FOUR

# APPROXIMATION
# OF FUNCTIONS

To evaluate most mathematical functions, we must first produce computable approximations to them. Functions are defined in a variety of ways in applications, with integrals and infinite series being the most common types of formulas used for the definition. Such a definition is useful in establishing the properties of the function, but it is generally not an efficient way to evaluate the function. In this chapter we examine the use of polynomials as approximations to a given function. Various means of producing polynomial approximations are described, and they are compared as to their relative accuracy.

For evaluating a function $f(x)$ on a computer, it is generally more efficient of space and time to have an analytic approximation to $f(x)$ rather than to store a table and use interpolation. It is also desirable to use the lowest possible degree of polynomial that will give the desired accuracy in approximating $f(x)$. The following sections give a number of methods for producing an approximation, and generally the better approximations are also the more complicated to produce. The amount of time and effort expended on producing an approximation should be directly proportional to how much the approximation will be used. If it is only to be used a few times, a truncated Taylor series will often suffice. But if an approximation is to be used millions of times by many people, then much care should be used in producing the approximation.

There are forms of approximating functions other than polynomials. Rational functions are quotients of polynomials, and they are usually a somewhat more efficient form of approximation. But because polynomials furnish an adequate and efficient form of approximation, and because the theory for rational function approximation is more complicated than that of polynomial approximation, we have chosen to consider only polynomials. The results of this chapter can also be used to produce piecewise polynomial approximations, somewhat analogous to the piecewise polynomial interpolating functions of Section 3.7 of the preceding chapter.

## 4.1 The Weierstrass Theorem and Taylor's Theorem

To justify using polynomials to approximate continuous functions, we present the following theorem.

**Theorem 4.1** (Weierstrass)    Let $f(x)$ be continuous for $a \leq x \leq b$ and let $\epsilon > 0$. Then there is a polynomial $p(x)$ for which

$$|f(x) - p(x)| \leq \epsilon \qquad a \leq x \leq b$$

**Proof**    There are many proofs of this result and of generalizations of it. Since this is not central to our numerical analysis development, we just indicate a constructive proof. For other proofs, see Davis (1963, chap. 6).

Assume $[a, b] = [0, 1]$ for simplicity: by an appropriate change of variables, we can always reduce to this case if necessary. Define

$$p_n(x) = \sum_{k=0}^{n} \binom{n}{k} f\left(\frac{k}{n}\right) x^k (1 - x)^{n-k} \qquad 0 \leq x \leq 1$$

Let $f(x)$ be bounded on $[0, 1]$. Then

$$\underset{n \to \infty}{\text{Limit}}\, p_n(x) = f(x)$$

at any point $x$ at which $f$ is continuous. If $f(x)$ is continuous at every $x$ in $[0, 1]$, then the convergence of $p_n$ to $f$ is uniform on $[0, 1]$, that is,

$$\underset{0 \leq x \leq 1}{\text{Max}}\, |f(x) - p_n(x)| \to 0 \qquad \text{as} \quad n \to \infty \qquad (4.1.1)$$

This gives an explicit way of finding a polynomial that satisfies the conclusion of the theorem. The proof of these results can be found in Davis (1963, pp. 108–118), along with additional properties of the approximating polynomials $p_n(x)$, which are called the *Bernstein polynomials*. They mimic extremely well the qualitative behavior of the function $f(x)$. For example, if $f(x)$ is $r$ times continuously differentiable on $[0, 1]$, then

$$\underset{0 \leq x \leq 1}{\text{Max}}\, |f^{(r)}(x) - p_n^{(r)}(x)| \to 0 \qquad \text{as} \quad n \to \infty$$

But such an overall approximating property has its price, and in this case the convergence in (4.1.1) is generally very slow. For example, if $f(x) = x^2$, then

$$\underset{n \to \infty}{\text{Limit}}\, n[p_n(x) - f(x)] = x(1 - x)$$

and thus

$$p_n(x) - x^2 \doteq \frac{1}{n} x(1 - x)$$

for large values of $n$. The error does not decrease rapidly, even for approximating such a trivial case as $f(x) = x^2$.    ∎

**Taylor's theorem**   Taylor's theorem was presented earlier, in Theorem 1.4 of Section 1.1, Chapter 1. It is the first important means for the approximation of a function, and it is often used as a preliminary approximation for computing some more efficient approximation. To aid in understanding why the Taylor approximation is not particularly efficient, consider the following example.

**Example**   Find the error of approximating $e^x$ using the third-degree Taylor polynomial $p_3(x)$ on the interval $[-1, 1]$, expanding about $x = 0$. Then

$$p_3(x) = 1 + x + \tfrac{1}{2}x^2 + \tfrac{1}{6}x^3$$

$$e^x - p_3(x) = R_4(x) = \tfrac{1}{24}x^4 e^\xi \qquad (4.1.2)$$

with $\xi$ between $x$ and 0.

To examine the error carefully, we bound it from above and below:

$$\frac{1}{24}x^4 \le e^x - p_3(x) \le \frac{e}{24}x^4 \qquad 0 \le x \le 1$$

$$\frac{e^{-1}}{24}x^4 \le e^x - p_3(x) \le \frac{1}{24}x^4 \qquad -1 \le x \le 0$$

The error increases for increasing $|x|$, and by direct calculation,

$$\underset{-1 \le x \le 1}{\text{Max}} |e^x - p_3| \doteq .0516 \qquad (4.1.3)$$

The error is not distributed evenly through the interval $[-1, 1]$ (see Figure 4.1). It is much smaller near the origin than near the endpoints $-1$ and $1$. This uneven distribution of the error, which is typical of the Taylor remainder, means that there are usually much better approximating polynomials of the same degree. Further examples are given in the next section.

**The function space $C[a, b]$**   The set $C[a, b]$ of all continuous real valued functions on the interval $[a, b]$ was introduced earlier in Section 1.1 of Chapter 1.

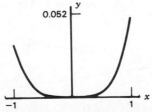

**Figure 4.1**   Error curve for $p_3(x) \doteq e^x$.

With it we will generally use the norm

$$\|f\|_\infty = \underset{a \leq x \leq b}{\text{Max}} |f(x)| \qquad f \in C[a, b] \tag{4.1.4}$$

It is called variously the *maximum norm*, Chebyshev norm, infinity norm, and uniform norm. It is the natural measure to use in approximation theory, since we will want to determine and compare

$$\|f - p\|_\infty = \underset{a \leq x \leq b}{\text{Max}} |f(x) - p(x)| \tag{4.1.5}$$

for various polynomials $p(x)$. Another norm for $C[a, b]$ is introduced in Section 4.3, one that is also useful in measuring the size of $f(x) - p(x)$.

As noted earlier in Section 1.1, the maximum norm satisfies the following characteristic properties of a norm:

$$\|f\| = 0 \qquad \text{if and only if} \quad f \equiv 0 \tag{4.1.6}$$

$$\|\alpha f\| = |\alpha| \|f\| \quad \text{for all } f \in C[a, b] \text{ and all scalars } \alpha \tag{4.1.7}$$

$$\|f + g\| \leq \|f\| + \|g\| \qquad \text{all } f, g \in C[a, b] \tag{4.1.8}$$

The proof of these properties for (4.1.4) is quite straightforward. And the properties show that the norm should be thought of as a generalization of the absolute value of a number.

Although we will not make any significant use of the idea, it is often useful to regard $C[a, b]$ as a vector space. The vectors are the functions $f(x)$, $a \leq x \leq b$. We define the distance from a vector $f$ to a vector $g$ by

$$D(f, g) = \|f - g\| \tag{4.1.9}$$

which is in keeping with our intuition about the concept of distance in simpler vector spaces. This is illustrated in Figure 4.2.

Using the inequality (4.1.8),

$$\|f - g\| = \|(f - h) + (h - g)\| \leq \|f - h\| + \|h - g\|$$

$$D(f, g) \leq D(f, h) + D(h, g) \tag{4.1.10}$$

This is called the *triangle inequality*, because of its obvious interpretation in measuring the lengths of sides of a triangle whose vertices are $f$, $g$, and $h$. The name "triangle inequality" is also applied to the equivalent formulation (4.1.8).

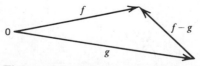

**Figure 4.2**   Illustration for defining distance $D(f, g)$.

Another useful result is the *reverse triangle inequality*

$$\left| \|f\| - \|g\| \right| \le \|f - g\| \tag{4.1.11}$$

To prove it, use (4.1.8) to give

$$\|f\| \le \|f - g\| + \|g\|$$

$$\|f\| - \|g\| \le \|f - g\|$$

By the symmetric argument,

$$\|g\| - \|f\| \le \|g - f\| = \|f - g\|$$

with the last equality following from (4.1.7) with $\alpha = -1$. Combining these two inequalities proves (4.1.11).

A more complete introduction to vector spaces (although only finite dimensional) and to vector norms is given in Chapter 7. And some additional geometry for $C[a, b]$ is given in Sections 4.3 and 4.4 with the introduction of another norm, different from (4.1.4). For cases where we want to talk about functions that have several continuous derivatives, we introduce the function space $C'[a, b]$, consisting of functions $f(x)$ that have $r$ continuous derivatives on $[a, b]$. This function space is of independent interest, but we regard it as just a simplifying notational device.

## 4.2    The Minimax Approximation Problem

Let $f(x)$ be continuous on $[a, b]$. To compare polynomial approximations $p(x)$ to $f(x)$, obtained by various methods, it is natural to ask what is the best possible accuracy that can be attained by using polynomials of each degree $n \ge 0$. Thus we are lead to introduce the *minimax error*

$$\rho_n(f) = \underset{\deg(q) \le n}{\text{Infimum}} \|f - q\|_\infty \tag{4.2.1}$$

There does not exist a polynomial $q(x)$ of degree $\le n$ that can approximate $f(x)$ with a smaller maximum error than $\rho_n(f)$.

Having introduced $\rho_n(f)$, we seek whether there is a polynomial $q_n^*(x)$ for which

$$\rho_n(f) = \|f - q_n^*\|_\infty \tag{4.2.2}$$

And if so, is it unique, what are its characteristics, and how can it be constructed? The approximation $q_n^*(x)$ is called the *minimax approximation* to $f(x)$ on $[a, b]$, and its theory is developed in Section 4.6.

*Example*    Compute the minimax polynomial approximation $q_1^*(x)$ to $e^x$ on $-1 \le x \le 1$. Let $q_1^*(x) = a_0 + a_1 x$. To find $a_0$ and $a_1$, we have to use some geometric insight. Consider the graph of $y = e^x$ with that of a possible ap-

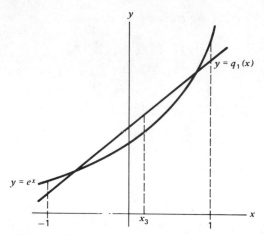

**Figure 4.3** Linear minimax approximation to $e^x$.

proximation $y = q_1(x)$, as in Figure 4.3. Let

$$\epsilon(x) = e^x - [a_0 + a_1 x] \qquad (4.2.3)$$

Clearly, $q_1^*(x)$ and $e^x$ must be equal at two points in $[-1,1]$, say at $-1 < x_1 < x_2 < 1$. Otherwise, we could improve on the approximation by moving the graph of $y = q_1^*(x)$ appropriately. Also

$$\rho_1 = \underset{-1 \le x \le 1}{\text{Max}} |\epsilon(x)|$$

and $\epsilon(x_1) = \epsilon(x_2) = 0$. By another argument based on shifting the graph of $y = q_1^*(x)$, we conclude that the maximum error $\rho_1$ is attained at exactly three points.

$$\epsilon(-1) = \rho_1 \qquad \epsilon(1) = \rho_1 \qquad \epsilon(x_3) = -\rho_1 \qquad (4.2.4)$$

where $x_1 < x_3 < x_2$. Since $\epsilon(x)$ has a relative minimum at $x_3$, we have $\epsilon'(x_3) = 0$. Combining these four equations, we have

$$e^{-1} - [a_0 - a_1] = \rho_1 \qquad e - [a_0 + a_1] = \rho_1$$

$$e^{x_3} - [a_0 + a_1 x_3] = -\rho_1 \qquad e^{x_3} - a_1 = 0 \qquad (4.2.5)$$

These have the solution

$$a_1 = \frac{e - e^{-1}}{2} \doteq 1.1752 \qquad x_3 = \ln_e (a_1) \doteq .1614$$

$$\rho_1 = \frac{1}{2} e^{-1} + \frac{x_3}{4} (e - e^{-1}) \doteq .2788$$

$$a_0 = \rho_1 + (1 - x_3) a_1 \doteq 1.2643$$

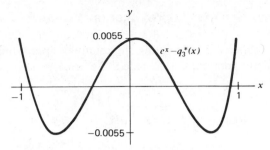

**Figure 4.4** Error for cubic minimax approximation to $e^x$.

Thus

$$q_1^*(x) = 1.2643 + 1.1752x \qquad (4.2.6)$$

and $\rho_1 \doteq .2788$.

By using what is called the Remes algorithm, we can also construct $q_3^*(x)$ for $e^x$ on $[-1, 1]$:

$$q_3^*(x) = .994579 + .995668x + .542973x^2 + .179533x^3 \qquad (4.2.7)$$

The graph of its error is given in Figure 4.4, and in contrast to the Taylor approximations (see Figure 4.1), the error is evenly distributed throughout the interval of approximation.

We conclude the example by giving the errors in the minimax approximation $q_n^*(x)$ and the Taylor approximations $p_n(x)$ for $f(x) = e^x$, $-1 \leq x \leq 1$. The maximum errors for various $n$ are given in Table 4.1.

The accuracy of $q_n^*(x)$ is significantly better than that of $p_n(x)$, and the disparity increases as $n$ increases. It should be noted that $e^x$ is a function with a rapidly convergent Taylor series, say in comparison to $\log x$ and $\tan^{-1} x$. With these latter functions, the minimax would look even better in comparison to the Taylor series.

**Table 4.1   Taylor and minimax errors for $e^x$**

| $n$ | $\|f - p_n\|_\infty$ | $\|f - q_n^*\|_\infty$ |
|---|---|---|
| 1 | 7.18E − 1 | 2.79E − 1 |
| 2 | 2.18E − 1 | 4.50E − 2 |
| 3 | 5.16E − 2 | 5.53E − 3 |
| 4 | 9.95E − 3 | 5.47E − 4 |
| 5 | 1.62E − 3 | 4.52E − 5 |
| 6 | 2.26E − 4 | 3.21E − 6 |
| 7 | 2.79E − 5 | 2.00E − 7 |
| 8 | 3.06E − 6 | 1.11E − 8 |
| 9 | 3.01E − 7 | 5.52E − 10 |

## 4.3   The Least Squares Approximation Problem

Because of the difficulty in calculating the minimax approximation, we often go to an intermediate approximation called the *least squares approximation*. As notation, introduce

$$\|g\|_2 = \sqrt{\int_a^b |g(x)|^2 \, dx} \qquad g \in C[a, b] \qquad (4.3.1)$$

This is a function norm, satisfying the same properties (4.1.6)–(4.1.8) as the maximum norm of (4.1.4). It is a generalization of the ordinary Euclidean norm for $R^n$, defined in (1.1.17). We return to the proof of the triangle inequality (4.1.8) in the next section when we generalize the preceding definition.

For a given $f \in C[a, b]$ and $n \geq 0$, define

$$M_n(f) = \underset{\deg(r) \leq n}{\text{Infimum}} \|f - r\|_2 \qquad (4.3.2)$$

As before, does there exist a polynomial $r_n^*$ that minimizes this expression:

$$M_n(f) = \|f - r_n^*\|_2 \qquad (4.3.3)$$

Is it unique? Can we calculate it?

To further motivate (4.3.2), consider calculating an average error in the approximation of $f(x)$ by $r(x)$. For an integer $m \geq 1$, define the nodes $x_j$ by

$$x_j = a + \left(j - \frac{1}{2}\right)\left(\frac{b - a}{m}\right) \qquad j = 1, 2, \ldots, m$$

These are the midpoints of $m$ evenly spaced subintervals of $[a, b]$. Then an *average error* of approximating $f(x)$ by $r(x)$ on $[a, b]$ is

$$E = \underset{m \to \infty}{\text{Limit}} \left\{ \frac{1}{m} \sum_{j=1}^m \left[f(x_j) - r(x_j)\right]^2 \right\}^{1/2}$$

$$= \underset{m \to \infty}{\text{Limit}} \left\{ \frac{1}{b - a} \sum_{j=1}^m \left[f(x_j) - r(x_j)\right]^2 \left(\frac{b - a}{m}\right) \right\}^{1/2}$$

$$= \frac{1}{\sqrt{b - a}} \sqrt{\int_a^b |f(x) - r(x)|^2 \, dx}$$

$$E = \frac{\|f - r\|_2}{\sqrt{b - a}} \qquad (4.3.4)$$

Thus the least squares approximation should have a small average error on $[a, b]$. The quantity $E$ is called the *root-mean-square error* in the approximation of $f(x)$ by $r(x)$.

***Example*** Let $f(x) = e^x$, $-1 \le x \le 1$, and let $r_1(x) = b_0 + b_1x$. Minimize

$$\|f - r_1\|_2^2 = \int_{-1}^{1} [e^x - b_0 - b_1x]^2 \, dx \equiv F(b_0, b_1) \tag{4.3.5}$$

If we expand the integrand and break the integral apart, then $F(b_0, b_1)$ is a quadratic polynomial in the two variables $b_0, b_1$,

$$F = \int_{-1}^{1} \left\{ e^{2x} + b_0^2 + b_1^2x^2 - 2b_0xe^x + 2b_0b_1x \right\} dx$$

To find a minimum, we set

$$\frac{\partial F}{\partial b_0} = 0 \qquad \frac{\partial F}{\partial b_1} = 0$$

which is a necessary condition at a minimal point. Rather than differentiating in the previously given integral, we merely differentiate through the integral in (4.3.5),

$$0 = \frac{\partial F}{\partial b_0} = \int_{-1}^{1} \frac{\partial}{\partial b_0} [e^x - b_0 - b_1x]^2 \, dx = 2 \int_{-1}^{1} [e^x - b_0 - b_1x](-1) \, dx$$

$$0 = \frac{\partial F}{\partial b_1} = 2 \int_{-1}^{1} [e^x - b_0 - b_1x](-x) \, dx$$

Then

$$b_0 = \frac{1}{2} \int_{-1}^{1} e^x \, dx = \sinh(1) \doteq 1.1752$$

$$b_1 = \frac{3}{2} \int_{-1}^{1} xe^x \, dx = 3e^{-1} \doteq 1.1036$$

$$r_1^*(x) = 1.1752 + 1.1036x \tag{4.3.6}$$

By direct examination,

$$\|e^x - r_1^*\|_\infty \doteq .44$$

This is intermediate to the approximations $q_1^*$ and $p_1(x)$ derived earlier. Usually the least squares approximation is a fairly good uniform approximation, superior to the Taylor series approximations.

As a further example, the cubic least squares approximation to $e^x$ on $[-1, 1]$ is given by

$$r_3^*(x) = .996294 + .997955x + .536722x^2 + .176139x^3 \tag{4.3.7}$$

and

$$\|e^x - r_3^*\|_\infty = .0112$$

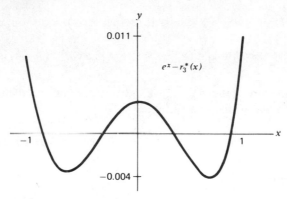

**Figure 4.5**   Error in cubic least squares approximation to $e^x$.

The graph of the error is given in Figure 4.5. Note that the error is not distributed as evenly in $[-1, 1]$ as was true of the minimax $q_3^*(x)$ in Figure 4.4.

**The general least squares problem**    We give a generalization of the least squares problem (4.3.2), allowing weighted average errors in the approximation of $f(x)$ by a polynomial $r(x)$. The general theory of the existence, uniqueness, and construction of least squares approximations is given in Section 4.5. It uses the theory of orthogonal polynomials, which is given in the next section.

Let $w(x)$ be a nonnegative *weight function* on the interval $(a, b)$, which may be infinite, and assume the following properties:

1.

$$\int_a^b |x|^n w(x)\, dx \qquad (4.3.8)$$

is integrable and finite for all $n \geq 0$;

2.    Suppose that

$$\int_a^b w(x) g(x)\, dx = 0 \qquad (4.3.9)$$

for some nonnegative continuous function $g(x)$; then the function $g(x) \equiv 0$ on $(a, b)$.

*Example*    The following are the weight functions of most interest in the developments of this text:

$$w(x) \equiv 1 \qquad a \leq x \leq b$$

$$w(x) = \frac{1}{\sqrt{1 - x^2}} \qquad -1 \leq x \leq 1$$

$$w(x) = e^{-x} \qquad 0 \leq x < \infty$$

$$w(x) = e^{-x^2} \qquad -\infty < x < \infty$$

For a finite interval $[a, b]$, the general least squares problem can now be stated. *Given* $g \in C[a, b]$, *does there exist a polynomial* $r_n^*(x)$ *of degree* $\leq n$ *that minimizes*

$$\int_a^b w(x)[f(x) - r(x)]^2 \, dx \qquad (4.3.10)$$

*among all polynomials* $r(x)$ *of degree* $\leq n$? The function $w(x)$ allows different degrees of importance to be given to the error at different points in the interval $[a, b]$. This will prove useful in developing *near minimax approximations*.

Define

$$F(a_0, a_1, \ldots, a_n) = \int_a^b w(x)\left[f(x) - \sum_{j=0}^n a_j x^j\right]^2 dx \qquad (4.3.11)$$

in order to compute (4.3.10) for an arbitrary polynomial $r(x)$ of degree $\leq n$. We want to minimize $F$ as the coefficients $\{a_i\}$ range over all real numbers. A necessary condition for a point $(a_0, \ldots, a_n)$ to be a minimizing point is

$$\frac{\partial F}{\partial a_i} = 0 \qquad i = 0, 1, \ldots, n \qquad (4.3.12)$$

By differentiating through the integral in (4.3.11) and using (4.3.12), we obtain the linear system

$$\sum_{j=0}^n a_j \int_a^b w(x) x^{i+j} \, dx = \int_a^b w(x) f(x) x^i \, dx \qquad i = 0, 1, \ldots, n \quad (4.3.13)$$

To see why this solution of the least squares problem is unsatisfactory, consider the special case $w(x) \equiv 1$, $[a, b] = [0, 1]$. Then the linear system becomes

$$\sum_{j=0}^n \frac{a_j}{i + j + 1} = \int_0^1 f(x) x^i \, dx \qquad i = 0, 1, \ldots, n \qquad (4.3.14)$$

The matrix of coefficients is the Hilbert matrix of order $n + 1$, which was introduced in (1.6.9). The solution of linear system (4.3.14) is extremely sensitive to small changes in the coefficients or right-hand constants. Thus this is not a good way to approach the least squares problem. In single precision arithmetic on an IBM 3033 computer, the cases $n \geq 4$ will be completely unsatisfactory.

## 4.4  Orthogonal Polynomials

As is evident from graphs of $x^n$ on $[0, 1]$, $n \geq 0$, these monomials are very nearly *linearly dependent*, looking much alike, and this results in the instability of the linear system (4.3.14). To avoid this problem, we consider an alternative basis for the polynomials, based on polynomials that are orthogonal in a function space sense that is given below. These results are of fundamental importance in the approximation of functions and for much work in classical applied mathematics.

Let $w(x)$ be the same as in (4.3.8) and (4.3.9), and define the *inner product* of two continuous functions $f$ and $g$ by

$$(f, g) = \int_a^b w(x)f(x)g(x)\, dx \qquad f, g \in C[a, b] \qquad (4.4.1)$$

Then the following simple properties are easily shown.

1. $(\alpha f, g) = (f, \alpha g) = \alpha(f, g)$      for all scalars $\alpha$
2. $(f_1 + f_2, g) = (f_1, g) + (f_2, g)$
   $(f, g_1 + g_2) = (f, g_1) + (f, g_2)$
3. $(f, g) = (g, f)$
4. $(f, f) \geq 0$ for all $f \in C[a, b]$, and $(f, f) = 0$ if and only if $f(x) = 0$, $a \leq x \leq b$

Define the *two norm* or *Euclidean norm* by

$$\|f\|_2 = \sqrt{\int_a^b w(x)[f(x)]^2\, dx} = \sqrt{(f, f)} \qquad (4.4.2)$$

This definition will satisfy the norm properties (4.1.6)–(4.1.8). But the proof of the triangle inequality (4.1.8) is no longer obvious, and it depends on the following well-known inequality.

*Lemma* (Cauchy–Schwartz inequality)    For $f, g \in C[a, b]$,

$$|(f, g)| \leq \|f\|_2 \|g\|_2 \qquad (4.4.3)$$

*Proof*    If $g = 0$, the result is trivially true. For $g \neq 0$, consider the following. For any real number $\alpha$,

$$0 \leq (f + \alpha g, f + \alpha g) = (f, f) + 2\alpha(f, g) + \alpha^2(g, g)$$

The polynomial on the right has at most one real root, and thus the discriminant cannot be positive,

$$4|(f, g)|^2 - 4(f, f)(g, g) \leq 0$$

This implies (4.4.3). Note that we have equality in (4.4.3) only if the discriminant is exactly zero. But that implies there is an $\alpha^*$ for which the polynomial is zero. Then

$$(f + \alpha^* g, f + \alpha^* g) = 0$$

and thus $f = -\alpha^* g$. Thus equality holds in (4.3.4) if and only if either (1) $f$ is a multiple of $g$, or (2) $f$ or $g$ is identically zero. ∎

To prove the triangle inequality (4.1.8),

$$\|f + g\|_2^2 = (f + g, f + g) = (f, f) + 2(f, g) + (g, g)$$

$$\leq \|f\|_2^2 + 2\|f\|_2\|g\|_2 + \|g\|_2^2 = (\|f\|_2 + \|g\|_2)^2$$

Take square roots of each side of the inequality to obtain

$$\|f + g\|_2 \leq \|f\|_2 + \|g\|_2 \tag{4.4.4}$$

We are interested in obtaining a basis for the polynomials other than the ordinary basis, the monomials $\{1, x, x^2, \ldots\}$. We produce what is called an orthogonal basis, a generalization of orthogonal basis in the space $R^n$ (see Section 7.1). We say that $f$ and $g$ are *orthogonal* if

$$(f, g) = 0 \tag{4.4.5}$$

The following is a constructive existence result for orthogonal polynomials.

**Theorem 4.2** (Gram–Schmidt)  There exists a sequence of polynomials $\{\varphi_n(x) | n \geq 0\}$ with degree $(\varphi_n) = n$, for all $n$, and

$$(\varphi_n, \varphi_m) = 0 \qquad \text{for all} \quad n \neq m \qquad n, m \geq 0 \tag{4.4.6}$$

In addition, we can construct the sequence with the additional properties: (1) $(\varphi_n, \varphi_n) = 1$, for all $n$; (2) the coefficient of $x^n$ in $\varphi_n(x)$ is positive. With these additional properties, the sequence $\{\varphi_n\}$ is unique.

**Proof**  We show a constructive and recursive method of obtaining the members of the sequence; it is called the *Gram–Schmidt process*. Let

$$\varphi_0(x) = c$$

a constant. Pick it such that $\|\varphi_0\|_2 = 1$ and $c > 0$. Then

$$(\varphi_0, \varphi_0) = c^2 \int_a^b w(x)\, dx = 1$$

$$c = \left[\int_a^b w(x)\, dx\right]^{-1/2}$$

For constructing $\varphi_1(x)$, begin with

$$\psi_1(x) = x + a_{1,0}\varphi_0(x)$$

Then

$$(\psi_1, \varphi_0) = 0 \quad \text{implies} \quad 0 = (x, \varphi_0) + a_{1,0}(\varphi_0, \varphi_0)$$

$$a_{1,0} = -(x, \varphi_0) = \frac{-\int_a^b xw(x)\,dx}{\left[\int_a^b w(x)\,dx\right]^{1/2}}$$

Define

$$\varphi_1(x) = \frac{\psi_1(x)}{\|\psi_1\|_2}$$

and note that

$$\|\varphi_1\|_2 = 1 \qquad (\varphi_1, \varphi_0) = 0$$

and the coefficient of $x$ is positive.
  To construct $\varphi_n(x)$, first define

$$\psi_n(x) = x^n + a_{n,n-1}\varphi_{n-1}(x) + \cdots + a_{n,0}\varphi_0(x) \qquad (4.4.7)$$

and choose the constants to make $\psi_n$ orthogonal to $\varphi_j$ for $j = 0, \ldots,$
$n-1$. Then

$$(\psi_n, \varphi_j) = 0 \quad \text{implies} \quad a_{n,j} = -(x^n, \varphi_j) \qquad j = 0, 1, \ldots, n-1 \quad (4.4.8)$$

The desired $\varphi_n(x)$ is

$$\varphi_n(x) = \frac{\psi_n(x)}{\|\psi_n\|_2} \qquad (4.4.9)$$

Continue the derivation inductively.  ∎

**Example**   For the special case of $w(x) \equiv 1$, $[a, b] = [-1, 1]$, we have

$$\varphi_0(x) = \sqrt{\tfrac{1}{2}} \qquad \varphi_1(x) = \sqrt{\tfrac{3}{2}}\,x \qquad \varphi_2(x) = \tfrac{1}{2}\sqrt{\tfrac{5}{2}}(3x^2 - 1)$$

and further polynomials can be constructed by the same process.

  There is a very large literature on these polynomials, including a variety of formulas for them. Of necessity we only skim the surface. The polynomials are usually given in a form for which $\|\varphi_n\|_2 \neq 1$.

**Particular Cases 1.**   The Legendre polynomials.   Let $w(x) \equiv 1$ on $[-1, 1]$.
Define

$$P_n(x) = \frac{(-1)^n}{2^n n!} \cdot \frac{d^n}{dx^n}\left[(1 - x^2)^n\right] \qquad n \geq 1 \qquad (4.4.10)$$

with $P_0(x) \equiv 1$. These are orthogonal on $[-1, 1]$, degree $P_n(x) = n$, and $P_n(1) = 1$ for all $n$. Also,

$$(P_n, P_n) = \frac{2}{2n + 1}$$

$$\varphi_n(x) = \sqrt{\frac{2n + 1}{2}} P_n(x) \qquad (4.4.11)$$

**2.** The Chebyshev polynomials. Let $w(x) = 1/\sqrt{1 - x^2}$, $-1 \le x \le 1$. Then

$$T_n(x) = \cos(n \cos^{-1} x) \qquad n \ge 0 \qquad (4.4.12)$$

is an orthogonal family of polynomials with $\deg(T_n) = n$. To see that $T_n(x)$ is a polynomial, let $\cos^{-1} x = \theta$, $0 \le \theta \le \pi$. Then

$$T_{n \pm 1}(x) = \cos(n \pm 1)\theta = \cos(n\theta)\cos\theta \mp \sin(n\theta)\sin\theta$$

$$T_{n+1}(x) + T_{n-1}(x) = 2\cos(n\theta)\cos\theta = 2T_n(x)x$$

$$T_{n+1}(x) = 2xT_n(x) - T_{n-1}(x) \qquad n \ge 1 \qquad (4.4.13)$$

Also by direct calculation in (4.4.12),

$$T_0(x) \equiv 1 \qquad T_1(x) = x$$

Using (4.4.13)

$$T_2(x) = 2x^2 - 1 \qquad T_3(x) = 2x(2x^2 - 1) - x = 4x^3 - 3x$$

The polynomials also satisfy $T_n(1) = 1$, $n \ge 1$,

$$(T_n, T_m) = \begin{cases} 0 & n \ne m \\ \pi & m = m = 0 \\ \dfrac{\pi}{2} & n = m > 0 \end{cases} \qquad (4.4.14)$$

The Chebyshev polynomials are extremely important in approximation theory, and they also arise in many other areas of applied mathematics. For a more complete discussion of them, see Rivlin (1974), and Fox and Parker (1968). We give further properties of the Chebyshev polynomials in the following sections.

**3.** The Laguerre polynomials. Let $w(x) = e^{-x}$, $[a, b] = [0, \infty)$. Then

$$L_n(x) = \frac{1}{n!e^{-x}} \cdot \frac{d^n}{dx^n}\{x^n e^{-x}\} \qquad n \ge 0 \qquad (4.4.15)$$

$\|L_n\|_2 = 1$ for all $n$, and $\{L_n\}$ are orthogonal on $[0, \infty)$ relative to the weight function $e^{-x}$.

We say that a family of functions is an *orthogonal family* if each member is orthogonal to every other member of the family. We call it an *orthonormal family* if it is an orthogonal family and if every member has length one, that is, $\|f\|_2 = 1$. For other examples of orthogonal polynomials, see Abramowitz and Stegun (1964, chap. 22), Davis (1963, app.), Szego (1968).

**Some properties of orthogonal polynomials**    These results will be useful later in this chapter and in the next chapter.

**Theorem 4.3**    Let $\{\varphi_n(x) | n \geq 0\}$ be an orthogonal family of polynomials on $(a, b)$ with weight function $w(x)$. With such a family we always assume implicitly that degree $\varphi_n = n$, $n \geq 0$. If $f(x)$ is a polynomial of degree $m$, then

$$f(x) = \sum_{n=0}^{m} \frac{(f, \varphi_n)}{(\varphi_n, \varphi_n)} \varphi_n(x) \qquad (4.4.16)$$

**Proof**    We begin by showing that every polynomial can be written as a combination of orthogonal polynomials of no greater degree. Since degree $(\varphi_0) = 0$, we have $\varphi_0(x) = c$, a constant, and thus

$$1 \equiv \frac{1}{c} \varphi_0(x)$$

Since degree $(\varphi_1) = 1$, we have from the construction in the Gram–Schmidt process,

$$\varphi_1(x) = c_{1,1}x + c_{1,0}\varphi_0(x) \qquad c_{1,1} \neq 0$$

$$x = \frac{1}{c_{1,1}} \left[ \varphi_1(x) - c_{1,0}\varphi_0(x) \right]$$

By induction in the Gram–Schmidt process,

$$\varphi_r(x) = c_{r,r}x^r + c_{r,r-1}\varphi_{r-1}(x) + \cdots + c_{r,0}\varphi_0(x) \qquad c_{r,r} \neq 0$$

and

$$x^r = \frac{1}{c_{r,r}} \left[ \varphi_r(x) - c_{r,r-1}\varphi_{r-1}(x) - \cdots - c_{r,0}\varphi_0(x) \right]$$

Thus every monomial can be rewritten as a combination of orthogonal polynomials of no greater degree. From this it follows easily that an arbitrary polynomial $f(x)$ of degree $m$ can be written as

$$f(x) = b_m \varphi_m(x) + \cdots + b_0 \varphi_0(x)$$

for some choice of $b_0, \ldots, b_m$. To calculate each $b_i$, multiply both sides by $w(x)$ and $\varphi_i(x)$, and integrate over $(a, b)$. Then

$$(f, \varphi_i) = \sum_{j=0}^{m} b_j(\varphi_j, \varphi_i) = b_i(\varphi_i, \varphi_i)$$

$$b_i = \frac{(f, \varphi_i)}{(\varphi_i, \varphi_i)}$$

which proves (4.4.16) and the theorem.    ∎

**Corollary**    If $f(x)$ is a polynomial of degree $\leq m - 1$, then

$$(f, \varphi_m) = 0 \tag{4.4.17}$$

and $\varphi_m(x)$ is orthogonal to $f(x)$.

**Proof**    It follows easily from (4.4.16) and the orthogonality of the family $\{\varphi_n(x)\}$.    ∎

The following result gives some intuition as to the shape of the graphs of the orthogonal polynomials. It is also crucial to the work on Gaussian quadrature in Chapter 5.

**Theorem 4.4**    Let $\{\varphi_n(x) | n \geq 0\}$ be an orthogonal family of polynomials on $(a, b)$ with weight function $w(x) \geq 0$. Then the polynomial $\varphi_n(x)$ has exactly $n$ distinct real roots in the open interval $(a, b)$.

**Proof**    Let $x_1, x_2, \ldots, x_m$ be all of the zeros of $\varphi_n(x)$ for which

1.    $a < x_i < b$

2.    $\varphi_n(x)$ changes sign at $x_i$

Since degree $(\varphi_n) = n$, we trivially have $m \leq n$. We assume $m < n$ and then derive a contradiction.
Define

$$B(x) = (x - x_1) \cdots (x - x_m)$$

By the definition of the points $x_1, \ldots, x_m$, the polynomial

$$\varphi_n(x)B(x) = (x - x_1) \cdots (x - x_m)\varphi_n(x)$$

does not change sign in $(a, b)$. To see this more clearly, the assumptions on $x_1, \ldots, x_m$ imply

$$\varphi_n(x) = h(x)(x - x_1)^{r_1} \cdots (x - x_m)^{r_m}$$

with each $r_i$ odd and with $h(x)$ not changing sign in $(a, b)$. Then

$$\varphi_n(x)B(x) = (x - x_1)^{r_1+1} \cdots (x - x_m)^{r_m+1}h(x)$$

and the conclusion follows. Consequently,

$$\int_a^b w(x)B(x)\varphi_n(x)\,dx \neq 0$$

since clearly $B(x)\varphi_n(x) \not\equiv 0$. But since degree$(B) = m < n$, the corollary to Theorem 4.3 implies

$$\int_a^b w(x)B(x)\varphi_n(x)\,dx = (B, \varphi_n) = 0$$

This is a contradiction, and thus we must have $m = n$. But then the conclusion of the theorem will follow, since $\varphi_n(x)$ can have at most $n$ roots, and the assumptions on $x_1, \ldots, x_n$ imply that they must all be simple, that is, $\varphi'(x_i) \neq 0$. ∎

As previously, let $\{\varphi_n(x) | n \geq 0\}$ be an orthogonal family on $(a, b)$ with weight function $w(x) \geq 0$. Define $A_n$ and $B_n$ by

$$\varphi_n(x) = A_n x^n + B_n x^{n-1} + \cdots \qquad (4.4.18)$$

Also, write

$$\varphi_n(x) = A_n(x - x_{n,1})(x - x_{n,2}) \cdots (x - x_{n,n}) \qquad (4.4.19)$$

Let

$$a_n = \frac{A_{n+1}}{A_n} \qquad \gamma_n = (\varphi_n, \varphi_n) > 0 \qquad (4.4.20)$$

**Theorem 4.5** (Triple Recursion Relation)   Let $\{\varphi_n\}$ be an orthogonal family of polynomials on $(a, b)$ with weight function $w(x) \geq 0$. Then for $n \geq 1$,

$$\varphi_{n+1}(x) = (a_n x + b_n)\varphi_n(x) - c_n\varphi_{n-1}(x) \qquad (4.4.21)$$

with

$$b_n = a_n \cdot \left[ \frac{B_{n+1}}{A_{n+1}} - \frac{B_n}{A_n} \right] \qquad c_n = \frac{A_{n+1}A_{n-1}}{A_n^2} \cdot \frac{\gamma_n}{\gamma_{n-1}} \qquad (4.4.22)$$

**Proof**   First note that the triple recursion relation (4.4.13) for Chebyshev polynomials is an example of (4.4.21). To derive (4.4.21), we begin by considering the polynomial

$$G(x) = \varphi_{n+1}(x) - a_n x \varphi_n(x)$$

$$= \left[ A_{n+1} x^{n+1} + B_{n+1} x^n + \cdots \right]$$

$$- \frac{A_{n+1}}{A_n} x \left[ A_n x^n + B_n x^{n-1} + \cdots \right]$$

$$= \left[ B_{n+1} - \frac{A_{n+1} B_n}{A_n} \right] x^n + \cdots$$

and degree $(G) \leq n$. By Theorem 4.3, we can write

$$G(x) = d_n \varphi_n(x) + \cdots + d_0 \varphi_0(x)$$

for an appropriate set of constants $d_0, \ldots, d_n$. Calculating $d_i$,

$$d_i = \frac{(G, \varphi_i)}{(\varphi_i, \varphi_i)} = \frac{1}{\gamma_i} \left[ (\varphi_{n+1}, \varphi_i) - a_n (x \varphi_n, \varphi_i) \right] \qquad (4.4.23)$$

We have $(\varphi_{n+1}, \varphi_i) = 0$ for $i \leq n$, and

$$(x \varphi_n, \varphi_i) = \int_a^b w(x) \varphi_n(x) x \varphi_i(x)\, dx = 0$$

for $i \leq n - 2$, since then degree $(x \varphi_i(x)) \leq n - 1$. Combining these results,

$$d_i = 0 \qquad 0 \leq i \leq n - 2$$

and therefore

$$G(x) = d_n \varphi_n(x) + d_{n-1} \varphi_{n-1}(x)$$

$$\varphi_{n+1}(x) = (a_n x + d_n) \varphi_n(x) + d_{n-1} \varphi_{n-1}(x) \qquad (4.4.24)$$

This shows the existence of a triple recursion relation, and the remaining work is manipulation of the formulas to obtain $d_n$ and $d_{n-1}$, given as $b_n$ and $c_n$ in (4.4.22). These constants are not derived here, but their values are important for some applications (see Problem 18). ∎

*Example* 1.   For Laguerre polynomials,

$$L_{n+1}(x) = \frac{1}{n+1} [2n + 1 - x] L_n(x) - \frac{n}{n+1} L_{n-1}(x) \qquad (4.4.25)$$

2.   For Legendre polynomials,

$$P_{n+1}(x) = \frac{2n+1}{n+1} x P_n(x) - \frac{n}{n+1} P_{n-1}(x) \qquad (4.4.26)$$

**Theorem 4.6** (Christoffel–Darboux Identity)    For $\{\varphi_n\}$ an orthogonal family of
polynomials with weight function $w(x) \geq 0$,

$$\sum_{k=0}^{n} \frac{\varphi_k(x)\varphi_k(y)}{\gamma_k} = \frac{\varphi_{n+1}(x)\varphi_n(y) - \varphi_n(x)\varphi_{n+1}(y)}{a_n\gamma_n(x-y)} \qquad x \neq y$$

(4.4.27)

**Proof**    The proof is based on manipulating the triple recursion relation [see
Szego (1967, p. 43)].    ∎

## 4.5    The Least Squares Approximation Problem (continued)

We now return to the general least squares problem, of minimizing (4.3.10)
among all polynomials of degree $\leq n$. Assume that $\{\varphi_k(x)|k \geq 0\}$ is an orthogo-
nal family of polynomials with weight function $w(x) \geq 0$, that is,

$$(\varphi_n, \varphi_m) = \delta_{n,m} = \begin{cases} 1 & n = m \\ 0 & n \neq m \end{cases}$$

Then an arbitrary polynomial $f(x)$ of degree$\leq n$ can be written as

$$r(x) = b_0\varphi_0(x) + \cdots + b_n\varphi_n(x) \tag{4.5.1}$$

For a given $f \in C[a, b]$,

$$\|f - r\|_2^2 = \int_a^b w(x)\left[f(x) - \sum_{j=0}^{n} b_j\varphi_j(x)\right]^2 dx \equiv G(b_0, \ldots, b_n) \quad (4.5.2)$$

We solve the least squares problem by minimizing $G$.

As before, we could set

$$\frac{\partial G}{\partial b_i} = 0 \qquad i = 0, 1, \ldots, n$$

But to obtain a more complete result, we proceed in another way. For any choice
of $b_0, \ldots, b_n$,

$$0 \leq G(b_0, \ldots, b_n) = \left(f - \sum_{j=0}^{n} b_j\varphi_j, f - \sum_{i=0}^{n} b_i\varphi_i\right)$$

$$= (f, f) - 2\sum_{j=0}^{n} b_j(f, \varphi_j) + \sum_i\sum_j b_ib_j(\varphi_i, \varphi_j)$$

$$= \|f\|_2^2 - 2\sum_{j=0}^{n} b_j(f, \varphi_j) + \sum_{j=0}^{n} b_j^2$$

$$= \|f\|_2^2 - \sum_{j=0}^{n} (f, \varphi_j)^2 + \sum_{j=0}^{n} \left[(f, \varphi_j) - b_j\right]^2 \quad (4.5.3)$$

which can be checked by expanding the last term. Thus $G$ is a minimum if and only if

$$b_j = (f, \varphi_j) \qquad j = 0, 1, \ldots, n$$

Then the least squares approximation exists, is unique, and is given by

$$r_n^*(x) = \sum_{j=0}^{n} (f, \varphi_j) \varphi_j(x)$$

Moreover, from (4.5.2) and (4.5.3),

$$\|f - r_n^*\|_2 = \left[ \|f\|_2^2 - \sum_{j=0}^{n} (f, \varphi_j)^2 \right]^{1/2}$$

$$= \sqrt{\|f\|_2^2 - \|r_n^*\|_2^2} \qquad (4.5.4)$$

$$\|f\|_2^2 = \|r_n^*\|_2^2 + \|f - r_n^*\|_2^2 \qquad (4.5.5)$$

As a practical note in obtaining $r_{n+1}^*(x)$, use

$$r_{n+1}^*(x) = r_n^*(x) + (f, \varphi_{n+1}) \varphi_{n+1}(x) \qquad (4.5.6)$$

**Theorem 4.7**   Assuming $[a, b]$ is finite,

$$\lim_{n \to \infty} \|f - r_n^*\|_2 = 0 \qquad (4.5.7)$$

**Proof**   By definition of $r_n^*$ as a minimizing polynomial for $\|f - r_n\|_2$, we have

$$\|f - r_1^*\|_2 \geq \|f - r_2^*\|_2 \geq \cdots \geq \|f - r_n^*\|_2 \geq \cdots \qquad (4.5.8)$$

Let $\epsilon > 0$ be arbitrary. Then by the Weierstrass theorem, there is a polynomial $Q(x)$ of some degree, say $m$, for which

$$\max_{a \leq x \leq b} |f(x) - Q(x)| \leq \frac{\epsilon}{c}, \qquad c = \sqrt{\int_a^b w(x)\, dx}$$

By the definition of $r_m^*(x)$,

$$\|f - r_m^*\|_2 \leq \|f - Q\|_2 = \left[ \int_a^b w(x)[f(x) - Q(x)]^2\, dx \right]^{1/2}$$

$$\leq \left[ \int_a^b w(x) \frac{\epsilon^2}{c^2}\, dx \right]^{1/2} = \epsilon$$

Combining this with (4.5.8),

$$\|f - r_n^*\|_2 \leq \epsilon$$

for all $n \geq m$. Since $\epsilon$ was arbitrary, this proves (4.5.7).  ∎

Using (4.5.5) and a straightforward computation of $\|r_n^*\|_2$, we have *Bessel's inequality*:

$$\|r_n^*\|_2^2 = \sum_{j=0}^{n} (f, \varphi_j)^2 \leq \|f\|_2^2 \qquad (4.5.9)$$

and using (4.5.7) in (4.5.5), we obtain *Parseval's equality*,

$$\|f\|_2 = \left[ \sum_{j=0}^{\infty} (f, \varphi_j)^2 \right]^{1/2} \qquad (4.5.10)$$

Theorem 4.7 does not say that $\|f - r_n^*\|_\infty \to 0$. But if additional differentiability assumptions are placed on $f(x)$, results on the uniform convergence of $r_n^*$ to $f$ can be proved. An example is given later.

**Legendre Polynomial Expansions**   To solve the least squares problem on a finite interval $[a, b]$ with $w(x) \equiv 1$, we can convert it to a problem on $[-1, 1]$. The change of variable

$$x = \frac{b + a + (b - a)t}{2} \qquad (4.5.11)$$

converts the interval $-1 \leq t \leq 1$ to $a \leq x \leq b$. Define

$$F(t) = f\left( \frac{b + a + (b - a)t}{2} \right) \qquad -1 \leq t \leq 1 \qquad (4.5.12)$$

for a given $f \in C[a, b]$. Then

$$\int_a^b [f(x) - r_n(x)]^2 \, dx = \left( \frac{b - a}{2} \right) \int_{-1}^{1} [F(t) - R_n(t)]^2 \, dt$$

with $R_n(t)$ obtained from $r_n(x)$ using (4.5.11). The change of variable (4.5.11) gives a one-to-one correspondence between polynomials of degree $m$ on $[a, b]$ and of degree $m$ on $[-1, 1]$, for every $m \geq 0$. Thus minimizing $\|f - r_n\|_2$ on $[a, b]$ is equivalent to minimizing $\|\varphi - R_n\|_2$ on $[-1, 1]$. We therefore restrict our interest to the least squares problem on $[-1, 1]$.

Given $f \in [-1, 1]$, the orthonormal family described in Theorem 4.2 is $\varphi_0(x) \equiv 1/\sqrt{2}$,

$$\varphi_n(x) = \sqrt{\frac{2n + 1}{2}} \cdot \frac{(-1)^n}{2^n n!} \cdot \frac{d^n}{dx^n}[(1 - x^2)^n] \qquad n \geq 1 \quad (4.5.13)$$

The least squares approximation is

$$r_n^*(x) = \sum_{j=0}^{n} (f, \varphi_j)\varphi_j(x) \qquad (f, \varphi_j) = \int_{-1}^{1} f(x)\varphi_j(x) \, dx \quad (4.5.14)$$

**Table 4.2    Legendre expansion coefficients for $e^x$**

| $j$ | $(f, \varphi_j)$ | $j$ | $(f, \varphi_j)$ |
|-----|------------------|-----|------------------|
| 0 | 1.661985 | 3 | .037660 |
| 1 | .901117 | 4 | .004698 |
| 2 | .226302 | 5 | .000469 |

the solution of the original least squares problem posed in Section 4.3. The coefficients $(f, \varphi_j)$ are called Legendre coefficients.

***Example***    For $f(x) = e^x$ on $[-1, 1]$, the expansion coefficients $(f, \varphi_j)$ of (4.5.14) are given in Table 4.2. The approximation $r_3^*(x)$ was given earlier in (4.3.7), written in standard polynomial form. For the average error $E$ in $r_3^*(x)$, combine (4.3.4), (4.5.4), (4.5.9), and the table coefficients to give

$$E = \frac{1}{\sqrt{2}} \| e^x - r_3^*(x) \|_2 \doteq .0034$$

**Chebyshev polynomial expansions**    The weight function is $w(x) = 1/\sqrt{1 - x^2}$, and

$$\varphi_0(x) = \frac{1}{\sqrt{\pi}}, \qquad \varphi_n(x) = \sqrt{\frac{2}{\pi}} T_n(x) \qquad n \geq 1 \qquad (4.5.15)$$

The least squares solution is

$$C_n(x) = \sum_{j=0}^{n} (f, \varphi_j) \varphi_j(x) \qquad (f, \varphi_j) = \int_{-1}^{1} \frac{f(x)\varphi_j(x)\, dx}{\sqrt{1 - x^2}} \qquad (4.5.16)$$

Using the definition of $\varphi_n(x)$ in terms of $T_n(x)$,

$$C_n(x) = {\sum_{j=0}^{n}}' c_j T_j(x) \qquad c_j = \frac{2}{\pi} \int_{-1}^{1} \frac{f(x)T_j(x)\, dx}{\sqrt{1 - x^2}} \qquad (4.5.17)$$

The prime on the summation symbol means that the first term should be halved before beginning to sum.

The Chebyshev expansion is closely related to Fourier cosine expansions. Using $x = \cos\theta$, $0 \leq \theta \leq \pi$,

$$C_n(\cos\theta) = {\sum_{j=0}^{n}}' c_j \cos(j\theta) \qquad (4.5.18)$$

$$c_j = \frac{2}{\pi} \int_{0}^{\pi} \cos(j\theta) f(\cos\theta)\, d\theta \qquad (4.5.19)$$

Thus $C_n(\cos\theta)$ is the truncation after $n + 1$ terms of the Fourier cosine expansion

$$f(\cos\theta) = \sum_{0}^{\infty}{}' c_j \cos(j\theta)$$

If the Fourier cosine expansion on $[0, \pi]$ of $f(\cos\theta)$ is known, then by substituting $\theta = \cos^{-1} x$, we have the Chebyshev expansion of $f(x)$.

For reasons to be given later, the Chebyshev least squares approximation is more useful than the Legendre least squares approximation. For this reason, we give a more detailed convergence theorem for (4.5.17).

**Theorem 4.8**    Let $f(x)$ have $r$ continuous derivatives on $[-1, 1]$, with $r \geq 1$. Then for the Chebyshev least squares approximation $C_n(x)$ defined in (4.5.17),

$$\|f - C_n\|_\infty \leq \frac{B \ln n}{n^r} \qquad n \geq 2 \qquad (4.5.20)$$

for a constant $B$ dependent on $f$ and $r$. Thus $C_n(x)$ converges uniformly to $f(x)$ as $n \to \infty$, provided $f(x)$ is continuously differentiable.

**Proof**    Combine Rivlin (1974, theorem 3.3, p. 134) and Meinardus (1967, theorem 45, p. 57).    ∎

**Example**    We illustrate the Chebyshev expansion by again considering approximations to $f(x) = e^x$. For the coefficients $c_j$ of (4.5.17),

$$c_j = \frac{2}{\pi} \int_{-1}^{1} \frac{e^x T_j(x)\, dx}{\sqrt{1 - x^2}}$$

$$= \frac{2}{\pi} \int_{0}^{\pi} e^{\cos\theta} \cos(j\theta)\, d\theta \qquad (4.5.21)$$

The latter formula is better for numerical integration, since there are no singularities in the integrand. Either the midpoint rule or the trapezoidal rule is an

**Table 4.3    Chebyshev expansion coefficients for $e^x$**

| $j$ | $c_j$ |
|---|---|
| 0 | 2.53213176 |
| 1 | 1.13031821 |
| 2 | .27149534 |
| 3 | .04433685 |
| 4 | .00547424 |
| 5 | .00054293 |

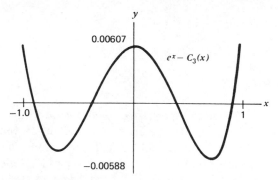

**Figure 4.6**  Error in cubic Chebyshev least squares approximation to $e^x$.

excellent integration method because of the periodicity of the integrand (see the Corollary 1 to Theorem 5.5 in Section 5.4). Using numerical integration we obtain the values given in Table 4.3. Using (4.5.17) and the formulas for $T_j(x)$, we obtain

$$C_1(x) = 1.266 + 1.130x$$

$$C_3(x) = .994571 + .997308x + .542991x^2 + .177347x^3$$

$$\|e^x - C_1(x)\|_\infty = .32 \qquad \|e^x - C_3(x)\|_\infty = .00607 \qquad (4.5.22)$$

The graph of $e^x - C_3(x)$ is given in Figure 4.6, and it is very similar to Figure 4.4, the graph for the minimax error. The maximum errors for these Chebyshev least squares approximations are quite close to the minimax errors, and are generally adequate for most practical purposes.

The polynomial $C_n(x)$ can be evaluated accurately and rapidly using the form (4.5.17), rather than converting it to the ordinary form using the monomials $x^j$, as was done previously for the example in (4.5.22). We make use of the triple recursion relation (4.4.13) for the Chebyshev polynomials $T_n(x)$. The following algorithm is due to C. W. Clenshaw, and our presentation follows Rivlin (1974, p. 125).

**Algorithm**  *Chebeval*   $(a, n, x$ value$)$

    **1.**   Remark: This algorithm evaluates

$$\text{Value} := \sum_{j=0}^{n} a_j T_j(x)$$

    **2.**   $b_{n+1} = b_{n+2} := 0 \qquad z = 2x$

    **3.**   Do through step 5 for $j = n, n-1, \ldots, 0$.

4.   $b_j = zb_{j+1} - b_{j+2} + a_j$

5.   Next $j$.

6.   Value $:= (b_0 - b_2)/2$.

This is almost as efficient as the nested multiplication algorithm (2.9.8) of Section 2.9. We leave the detailed comparison to Problem 25. Similar algorithms are available for other orthogonal polynomial expansions, again using their associated triple recursion relation. For an analysis of the effect of rounding errors in *Chebeval*, see Rivlin (1974, p. 127), and Fox and Parker (1968, p. 57).

## 4.6   Minimax Approximations

For a good uniform approximation to a given function $f(x)$, it seems reasonable to expect the error to be fairly uniformly distributed throughout the interval of approximation. The earlier examples with $f(x) = e^x$ further illustrate this point, and they show that the maximum error will oscillate in sign. Table 4.4 summarizes the statistics for the various forms of approximation to $e^x$ on $[-1, 1]$, including some methods given in Section 4.7. To show the importance of a uniform distribution of the error, with the error function oscillating in sign, we present two theorems. The first one is useful for estimating $\rho_n(f)$, the minimax error, without having to calculate the minimax approximation $q_n^*(x)$.

***Theorem 4.9*** (de la Vallee–Poussin)    Let $f \in C[a, b]$ and $n \geq 0$. Suppose we have a polynomial $Q(x)$ of degree $\leq n$ which satisfies

$$f(x_j) - Q(x_j) = (-1)^j e_j \qquad j = 0, 1, \ldots, n+1 \quad (4.6.1)$$

with all $e_j$ nonzero and of the same sign, and with

$$a \leq x_0 < x_1 < \cdots < x_{n+1} \leq b$$

Then

$$\underset{0 \leq j \leq n+1}{\text{Min}} |e_j| \leq \rho_n(f) \equiv \|f - q_n^*\|_\infty \leq \|f - Q\|_\infty \quad (4.6.2)$$

**Table 4.4   Comparison of various linear and cubic approximations to $e^x$**

| | Maximum Error | |
|---|---|---|
| Method of Approximation | Linear | Cubic |
| Taylor polynomial $p_n(x)$ | .718 | .0516 |
| Legendre least squares $r_n^*(x)$ | .439 | .0112 |
| Chebyshev least squares $C_n(x)$ | .322 | .00607 |
| Chebyshev node interpolation formula $I_n(x)$ | .372 | .00666 |
| Chebyshev forced oscillation formula $F_n(x)$ | .286 | .00558 |
| Minimax $q_n^*(x)$ | .279 | .00553 |

***Proof*** The upper inequality in (4.6.2) follows from the definition of $\rho_n(f)$. To prove the lower bound, we assume it is false and produce a contradiction. Assume

$$\rho_n(f) < \underset{0 \le j \le n+1}{\text{Min}} |e_j| \qquad (4.6.3)$$

Then by the definition of $\rho_n(f)$, there is a polynomial $P(x)$ of degree $\le n$ for which

$$\rho_n(f) \le \|f - P\|_\infty < \text{Min}|e_j| \qquad (4.6.4)$$

Define

$$R(x) = Q(x) - P(x)$$

a polynomial of degree $\le n$. For simplicity, let all $e_j > 0$; an analogous argument works when all $e_j < 0$.

Evaluate $R(x_j)$ for each $j$ and observe the sign of $R(x_j)$. First,

$$R(x_0) = Q(x_0) - P(x_0) = [f(x_0) - P(x_0)] - [f(x_0) - Q(x_0)]$$

$$= [f(x_0) - P(x_0)] - e_0 < 0$$

by using (4.6.4). Next,

$$R(x_1) = Q(x_1) - P(x_1) = [f(x_1) - P(x_1)] + e_1 > 0$$

Inductively, the sign of $R(x_j)$ is $(-1)^{j+1}$, $j = 0, 1, \ldots, n + 1$. This gives $R(x)$ $n + 2$ changes of sign and implies $R(x)$ has $n + 1$ zeros. Since degree$(R) \le n$, this is not possible unless $R \equiv 0$. Then $P \equiv Q$, contrary to (4.6.1) and (4.6.4). ∎

***Example*** Recall the cubic Chebyshev least squares approximation $C_n(x)$ to $e^x$ on $[-1, 1]$, given in (4.5.22). It has the maximum errors on the interval $[-1, 1]$ given in Table 4.5. These errors satisfy the hypotheses of the Theorem 4.9, and thus

$$.00497 \le \rho_3(f) \le .00607$$

**Table 4.5   Relative maxima of $|e^x - C_3(x)|$**

| $x$ | $e^x - C_3(x)$ |
|---|---|
| $-1.0$ | .00497 |
| $-.6919$ | $-.00511$ |
| .0310 | .00547 |
| .7229 | $-.00588$ |
| 1.0 | .00607 |

From this, we could conclude that $C_3(x)$ was quite close to the best possible approximation. Note that $\rho_3(f) = .00553$ from direct calculations using (4.2.7).

**Theorem 4.10** (Chebyshev Equioscillation Theorem)   Let $f \in C[a, b]$ and $n \geq 0$. Then there exists a unique polynomial $q_n^*(x)$ of degree $\leq n$ for which

$$\rho_n(f) = \|f - q_n^*\|_\infty$$

This polynomial is uniquely characterized by the following property: There are at least $n + 2$ points,

$$a \leq x_0 < x_1 < \cdots < x_{n+1} \leq b$$

for which

$$f(x_j) - q_n^*(x_j) = \sigma(-1)^j \rho_n(f) \qquad j = 0, 1, \ldots, n + 1 \quad (4.6.5)$$

with $\sigma = \pm 1$, depending only on $f$ and $n$.

**Proof**   The proof is quite technical, amounting to a complicated and manipulatory proof by contradiction. For that reason, it is omitted. For a complete development, see Davis (1963, chap. 7).   ∎

**Example**   The cubic minimax approximation $q_3^*(x)$ to $e^x$, given in (4.2.7), satisfies the conclusions of this theorem, as can be seen from the graph of the error in Figure 4.4 of Section 4.2.

From this theorem we can see that the Taylor series is always a poor uniform approximation. The Taylor series error,

$$f(x) - p_n(x) = \frac{(x - x_0)^{n+1}}{(n + 1)!} f^{(n+1)}(\xi_x) \qquad (4.6.6)$$

does not vary uniformly through the interval of approximation, nor does it oscillate much in sign.

To give a better idea of how well $q_n^*(x)$ approximates $f(x)$ with increasing $n$, we give the following theorem of D. Jackson.

**Theorem 4.11** (Jackson)   Let $f(x)$ have $k$ continuous derivatives for some $k \geq 0$. Moreover, assume $f^{(k)}(x)$ satisfies

$$\text{Supremum}_{a \leq x, y \leq b} |f^{(k)}(x) - f^{(k)}(y)| \leq M|x - y|^\alpha \quad (4.6.7)$$

for some $M > 0$ and some $0 < \alpha \leq 1$. [We say $f^{(k)}(x)$ satisfies a Hölder condition with exponent $\alpha$.] Then there is a constant $d_k$, independent of $f$ and $n$, for which

$$\rho_n(f) \leq \frac{Md_k}{n^{k+\alpha}} \qquad n \geq 1 \qquad (4.6.8)$$

**Table 4.6    Comparison of $\rho_n$ and $M_n$ for $f(x) = e^x$**

| $n$ | 2 | 3 | 4 | 5 | 6 | 7 |
|---|---|---|---|---|---|---|
| $M_n(f)$ | 1.13E − 1 | 1.42E − 2 | 1.42E − 3 | 1.18E − 4 | 8.43E − 6 | 5.27E − 7 |
| $\rho_n(f)$ | 4.50E − 2 | 5.53E − 3 | 5.47E − 4 | 4.52E − 5 | 3.21E − 6 | 2.00E − 7 |

***Proof***    See Meinardus (1967, Theorem 45, p. 57). Note that if we wish to ignore (4.6.7) with a $k$-times continuously differentiable function $f(x)$, then just use $k - 1$ in place of $k$ in the theorem, with $\alpha = 1$ and $M = \|f^{(k)}\|_\infty$. This will then yield

$$\rho_n(f) \le \frac{d_{k-1}}{n^k}\|f^{(k)}\|_\infty \tag{4.6.9}$$

Also note that if $f(x)$ is infinitely differentiable, then $q_n^*(x)$ converges to $f(x)$ uniformly on $[a, b]$, faster than any power $1/n^k$, $k \ge 1$.    ∎

From Theorem 4.12 in the next section, we are able to prove the following result. If $f(x)$ is $n + 1$ times continuously differentiable on $[a, b]$, then

$$\rho_n(f) \le \frac{[(b - a)/2]^{n+1}}{(n + 1)!\,2^n}\|f^{(n+1)}\|_\infty \equiv M_n(f) \tag{4.6.10}$$

The proof is left as Problem 38. There are infinitely differentiable functions for which $M_n(f) \to \infty$. However, for most of the widely used functions, the bound in (4.6.10) seems to be a fairly accurate estimate of the magnitude of $\rho_n(f)$. This estimate is illustrated in Table 4.6 for $f(x) = e^x$ on $[-1, 1]$. For other estimates and bounds for $\rho_n(f)$, see Meinardus (1967, sec. 6.2).

## 4.7    Near-Minimax Approximations

In the light of the Chebyshev equioscillation theorem, we can deduce methods that often give a good estimate of the minimax approximation. We begin with the least squares approximation $C_n(x)$ of (4.5.17). It is often a good estimate of $q_n^*(x)$, and our other near-minimax approximations are motivated by properties of $C_n(x)$.

From (4.5.17),

$$C_n(x) = \sum_{j=0}^{n}{}' c_j T_j(x) \qquad c_j = \frac{2}{\pi}\int_{-1}^{1}\frac{f(x)T_j(x)\,dx}{\sqrt{1 - x^2}} \tag{4.7.1}$$

with the prime on the summation meaning that the first term ($j = 0$) should be halved before summing the series. Using (4.5.7), if $f \in C[-1, 1]$, then

$$f(x) = \sum_{j=0}^{\infty}{}' c_j T_j(x) \tag{4.7.2}$$

with convergence holding in the sense that

$$\underset{n \to \infty}{\text{Limit}} \int_{-1}^{1} \frac{1}{\sqrt{1 - x^2}} \left[ f(x) - \sum_{j=0}^{n} {}' c_j T_j(x) \right]^2 dx = 0$$

For uniform convergence, we have the quite strong result that

$$\rho_n(f) \leq \|f - C_n\|_\infty \leq \left( 4 + \frac{4}{\pi^2} \ln(n) \right) \rho_n(f) \tag{4.7.3}$$

For a proof, see Rivlin (1974, p. 134). Combining this with Jackson's result (4.6.9) implies the earlier convergence bound (4.5.20) of Theorem 4.8.

If $f \in C'[a, b]$, it can be proved that there is a constant $c$, dependent on $f$ and $r$, for which

$$|c_j| \leq \frac{c}{j^r} \qquad j \geq 1 \tag{4.7.4}$$

This is proved by considering the $c_j$ as the Fourier coefficients of $f(\cos \theta)$, and then by using results from the theory of Foruier series on the rate of decrease of such coefficients. Thus as $r$ becomes larger, the coefficients $c_j$ decrease more rapidly.

For the truncated expansion $C_n(x)$,

$$f(x) - C_n(x) = \sum_{n+1}^{\infty} c_j T_j(x) \doteq c_{n+1} T_{n+1}(x) \tag{4.7.5}$$

if $c_{n+1} \neq 0$ and if the coefficients $c_j$ are rapidly convergent to zero. From the definition of $T_{n+1}(x)$,

$$|T_{n+1}(x)| \leq 1 \qquad -1 \leq x \leq 1 \tag{4.7.6}$$

Also, for the $n + 2$ points

$$x_j = \cos \left( \frac{j\pi}{n+1} \right) \qquad j = 0, 1, \ldots, n + 1 \tag{4.7.7}$$

we have

$$T_{n+1}(x_j) = (-1)^j \tag{4.7.8}$$

The bound in (4.7.6) is attained at exactly $n + 2$ points, the maximum possible number. Applying this to (4.7.5), the term $c_{n+1} T_{n+1}(x)$ has exactly $n + 2$ relative maxima and minima, all of equal magnitude. From the Chebyshev equioscillation theorem, we would therefore expect $C_n(x)$ to be nearly equal to the minimax approximation $q_n^*(x)$.

***Example*** Recall the example (4.5.22) for $f(x) = e^x$ near the end of Section 4.5. In it, the coefficients $c_j$ decreased quite rapidly, and

$$\|e^x - C_3\|_\infty = .00607 \qquad c_4 T_4(x) = .00547 T_4(x) \qquad \|e^x - q_3^*\|_\infty = .00553$$

This illustrates the comments of the last paragraph.

***Example*** It can be shown that

$$\tan^{-1} x = 2\left[\alpha T_1(x) - \frac{\alpha^3}{3} T_3(x) + \frac{\alpha^5}{5} T_5(x) - \cdots\right] \qquad (4.7.9)$$

converging uniformly for $-1 \le x \le 1$, with $\alpha = \sqrt{2} - 1 \doteq 0.414$. Then

$$C_{2n+1}(x) = 2\left[\alpha T_1(x) - \frac{\alpha^3}{3} T_3(x) + \cdots + \frac{(-1)^n}{2n+1}\alpha^{2n+1} T_{2n+1}(x)\right] \quad (4.7.10)$$

For the error,

$$E_{2n+1}(x) \equiv \tan^{-1} x - C_{2n+1}(x) = \frac{2}{2n+3}(-1)^{n+1}\alpha^{2n+3} T_{2n+3}(x)$$

$$+ 2 \sum_{j=n+2}^{\infty} \frac{(-1)^j \alpha^{2j+1}}{2j+1} T_{2j+1}(x)$$

We bound these terms to estimate the error in $C_{2n+1}(x)$ and the minimax error $\rho_{2n+1}(f)$.

By taking upper bounds,

$$\left|2\sum_{n+2}^{\infty} \frac{(-1)^j \alpha^{2j+1}}{2j+1} T_{2j+1}(x)\right| \le 2 \sum_{n+2}^{\infty} \frac{\alpha^{2j+1}}{2j+1} < \frac{2}{2n+5} \cdot \frac{\alpha^{2n+5}}{1-\alpha^2}$$

Thus we obtain upper and lower bounds for $E_{2n+1}(x)$,

$$E_{2n+1}(x) \le \frac{2}{2n+3}(-1)^{n+1}\alpha^{2n+3} T_{2n+3}(x) + \frac{2\alpha^{2n+5}}{(2n+5)(1-\alpha^2)}$$

$$\le \frac{2\alpha^{2n+3}}{2n+3}\left[(-1)^{n+1} T_{2n+3}(x) + .207\right]$$

using $\alpha^2/(1-\alpha^2) \doteq .207$. Similarly,

$$E_{2n+1}(x) \ge \frac{\alpha^{2n+3}}{2n+3}\left[(-1)^{n+1} T_{2n+3}(x) - .207\right]$$

Thus

$$E_{2n+1}(x) \doteq \frac{\alpha^{2n+3}}{2n+3}(-1)^{n+1}T_{2n+3}(x) \tag{4.7.11}$$

which has $2n + 4$ relative maxima and minima, of alternating sign. By Theorem 4.9 and the preceding inequalities,

$$\frac{2(.793)\alpha^{2n+3}}{2n+3} \le \rho_{2n+1}(f) = \rho_{2n+2}(f) \le \frac{2(1.207)\alpha^{2n+3}}{2n+3} \tag{4.7.12}$$

For the practical evaluation of the coefficients $c_j$ in (4.7.1) use the formula

$$c_j = \frac{2}{\pi}\int_0^\pi \cos(j\theta)f(\cos\theta)\,d\theta \tag{4.7.13}$$

and the midpoint or the trapezoidal numerical integration rule. This was illustrated earlier in (4.5.21).

**Interpolation at the Chebyshev zeros**    If the error in the minimax approximations is nearly $c_{n+1}T_{n+1}(x)$, as derived in (4.7.5), then the error should be nearly zero at the roots of $T_{n+1}(x)$ on $[-1,1]$. These are the points

$$x_j = \cos\left[\frac{2j+1}{2n+2}\pi\right] \qquad j = 0,1,\ldots,n \tag{4.7.14}$$

Let $I_n(x)$ be the polynomial of degree $\le n$ that interpolates $f(x)$ at these nodes $\{x_j\}$. Since it has zero error at the Chebyshev zeros $x_j$ of $T_{n+1}(x)$, the continuity of an interpolation polynomial with respect to the function values defining it suggests that $I_n(x)$ should approximately equal $C_n(x)$, and hence also $q_n^*(x)$. More precisely, write $I_n(x)$ in its Lagrange form and manipulate it as follows:

$$I_n(x) = \sum_{j=0}^{n} f(x_j)l_j(x)$$

$$= \sum_{j=0}^{n} C_n(x_j)l_j(x) + \sum_{j=0}^{n}\left[f(x_j) - C_n(x_j)\right]l_j(x)$$

$$\doteq C_n(x)$$

$$I_n(x) \doteq C_n(x) \tag{4.7.15}$$

because $f(x_j) - C_n(x_j) \doteq c_{n+1}T_{n+1}(x_j) = 0$.

The term $I_n(x)$ can be calculated using the algorithms *Divdif* and *Interp* of Section 3.2. A note of caution: If $c_{n+1} = 0$, then the error $f(x) - q_n^*(x)$ is likely to have as approximate zeros those of a higher degree Chebyshev polynomial, usually $T_{n+2}(x)$. This is likely to happen if $f(x)$ is either odd or even on $[-1, 1]$ (see Problem 29). [A function $f(x)$ is even {odd} on an interval $[-a, a]$ if $f(-x) = f(x)$ {$f(-x) = -f(x)$} for all $x$ in $[-a, a]$.] The preceding example with $f(x) = \tan^{-1} x$ is an apt illustration.

A further motivation for considering $I_n(x)$ as a near-minimax approximation is based on the following important theorem about Chebyshev polynomials.

**Theorem 4.12**   For a fixed integer $n > 0$, consider the minimization problem:

$$\tau_n = \operatorname*{Infimum}_{\deg(Q) \leq n-1} \left[ \operatorname*{Max}_{-1 \leq x \leq 1} |x^n + Q(x)| \right] \qquad (4.7.16)$$

with $Q(x)$ a polynomial. The minimum $\tau_n$ is attained uniquely by letting

$$x^n + Q(x) = \frac{1}{2^{n-1}} T_n(x) \qquad (4.7.17)$$

defining $Q(x)$ implicitly. The minimum is

$$\tau_n = \frac{1}{2^{n-1}} \qquad (4.7.18)$$

**Proof**   We begin by considering some facts about the Chebyshev polynomials. From the definition, $T_0(x) \equiv 1$, $T_1(x) = x$. The triple recursion relation

$$T_{n+1}(x) = 2xT_n(x) - T_{n-1}(x)$$

is the basis of an induction proof that

$$T_n(x) = 2^{n-1}x^n + \text{lower degree terms} \qquad n \geq 1 \qquad (4.7.19)$$

Thus

$$\frac{1}{2^{n-1}} T_n(x) = x^n + \text{lower degree terms} \qquad n \geq 1 \qquad (4.7.20)$$

Since $T_n(x) = \cos(n\theta)$, $x = \cos\theta$, $0 \leq \theta \leq \pi$, the polynomial $T_n(x)$ attains relative maxima and minima at $n + 1$ points in $[-1, 1]$:

$$x_j = \cos\left(\frac{j\pi}{n}\right) \qquad j = 0, 1, \ldots, n \qquad (4.7.21)$$

Additional values of $j$ do not lead to new values of $x_j$ because of the periodicity of the cosine function. For these points,

$$T_n(x_j) = (-1)^j \qquad j = 0, 1, \ldots, n \qquad (4.7.22)$$

and

$$-1 = x_n < x_{n-1} < \cdots < x_1 < x_0 = 1$$

The polynomial $T_n(x)/2^{n-1}$ has leading coefficient 1 and

$$\underset{-1\leq x\leq 1}{\text{Max}}\left|\frac{1}{2^{n-1}}T_n(x)\right| = \frac{1}{2^{n-1}} \qquad (4.7.23)$$

Thus $\tau_n \leq 1/2^{n-1}$. Suppose that

$$\tau_n < \frac{1}{2^{n-1}} \qquad (4.7.24)$$

We show that this leads to a contradiction. The assumption (4.7.24) and the definition (4.7.16) imply the existence of a polynomial

$$M(x) = x^n + Q(x) \qquad \text{degree}(Q) \leq n - 1$$

with

$$\tau_n \leq \underset{-1\leq x\leq 1}{\text{Max}}|M(x)| < \frac{1}{2^{n-1}} \qquad (4.7.25)$$

Define

$$R(x) = \frac{1}{2^{n-1}}T_n(x) - M(x)$$

which has degree $\leq n - 1$. We examine the sign of $R(x_j)$ at the points of (4.7.21). Using (4.7.22) and (4.7.24),

$$R(x_0) = R(1) = \frac{1}{2^{n-1}} - M(1) > 0$$

$$R(x_1) = -\frac{1}{2^{n-1}} - M(x_1) = -\left[\frac{1}{2^{n-1}} + M(x_1)\right] < 0$$

and the sign of $R(x_j)$ is $(-1)^j$. Since $R$ has $n + 1$ changes of sign, $R$ must have at least $n$ zeros. But then degree$(R) < n$ implies that $R \equiv 0$; thus $M \equiv (1/2^{n-1})T_n$.

To prove that no polynomial other than $(1/2^{n-1})T_n(x)$ will minimize (4.7.16), a variation of the preceding proof is used. We omit it.    ■

Consider now the problem of determining $n + 1$ nodes $x_j$ in $[-1, 1]$, to be used in constructing an interpolating polynomial $p_n(x)$ that is to approximate the given function $f(x)$ on $[-1, 1]$. The error in $p_n(x)$ is

$$f(x) - p_n(x) = \frac{(x - x_0) \cdots (x - x_n)}{(n + 1)!}f^{(n+1)}(\xi_x) \qquad (4.7.26)$$

The value $f^{(n+1)}(\xi_x)$ depends on $\{x_j\}$, but the dependence is not one that can be dealt with explicitly. Thus to try to make $\|f - p_n\|_\infty$ as small as possible, we

**Table 4.7    Interpolation data for $f(x) = e^x$**

| $i$ | $x_i$ | $f(x_i)$ | $f[x_0, \ldots, x_i]$ |
|---|---|---|---|
| 0 | .923880 | 2.5190442 | 2.5190442 |
| 1 | .382683 | 1.4662138 | 1.9453769 |
| 2 | − .382683 | .6820288 | .7047420 |
| 3 | .923880 | .3969760 | .1751757 |

consider only the quantity

$$\underset{-1 \le x \le 1}{\text{Max}} |(x - x_0) \cdots (x - x_n)| \tag{4.7.27}$$

We choose $\{x_j\}$ to minimize this quantity. The polynomial in (4.7.27) is of degree $n + 1$ and has leading coefficient 1. From the preceding theorem, (4.7.27) is minimized by taking this polynomial to be $T_{n+1}(x)/2^n$, and the minimum value of (4.7.27) is $1/2^n$. The nodes $\{x_j\}$ are the zeros of $T_{n+1}(x)$, and these are given in (4.7.14). With this choice of nodes, $p_n = I_n$ and

$$\|f - I_n\|_\infty \le \frac{1}{(n + 1)! \, 2^n} \|f^{(n+1)}\|_\infty \tag{4.7.28}$$

***Example***    Let $f(x) = e^x$ and $x = 3$. We use the Newton divided difference form of the interpolating polynomial. The nodes, function values, and needed divided differences are given in Table 4.7. By direct computation,

$$\underset{-1 \le x \le 1}{\text{Max}} |e^x - I_3(x)| \doteq .00666 \tag{4.7.29}$$

whereas the bound in (4.7.28) is .0142. A graph of $e^x - I_3(x)$ is shown in Figure 4.7.

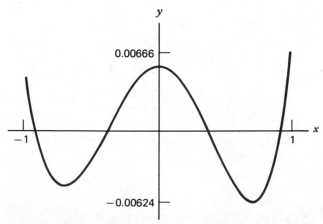

**Figure 4.7**    $e^x - I_3(x)$.

The error $\|f - I_n\|_\infty$ is generally not much worse than $\rho_n(f)$. A precise result is that

$$\|f - I_n\|_\infty \leq \left[\frac{2}{\pi}\log(n+1) + 2\right]\rho_n(f) \qquad n \geq 0 \qquad (4.7.30)$$

For a proof, see Rivlin (1974, p. 13). Actual numerical results are generally better than that predicted by the bound in (4.7.29), as is illustrated by (4.7.29) in the preceding examples of $f(x) = e^x$.

**Forced oscillation of the error**   Let $f(x) \in C[-1, 1]$, and define

$$F_n(x) = \sum_{k=0}^{n}{}' c_{n,k} T_k(x) \qquad -1 \leq x \leq 1 \qquad (4.7.31)$$

Let

$$-1 \leq x_{n+1} < x_n < \cdots < x_1 < x_0 \leq 1$$

be nodes, the choice of which is discussed below. Determine the coefficients $c_{n,k}$ by forcing the error $f(x) - F_n(x)$ to oscillate in the manner specified as necessary by Theorem 4.10:

$$f(x_i) - F_n(x_i) = (-1)^i E_n \qquad i = 0, 1, \ldots, n+1 \qquad (4.7.32)$$

We have introduced another unknown, $E_n$, which we hope will be nonzero. There are $n+2$ unknowns, the coefficients $c_{n,0}, \ldots, c_{n,n}$ and $E_n$, and there are $n+2$ equations in (4.7.32). Thus there is a reasonable chance that a solution exists. If a solution exists, then by Theorem 4.9,

$$|E_n| \leq \rho_n(f) \leq \|f - F_n\|_\infty \qquad (4.7.33)$$

To pick the nodes, note that if $c_{n+1}T_{n+1}(x)$ in (4.7.5) is nearly the minimax error, then the relative maxima and minima in the minimax error, $f(x) - q_n^*(x)$, occur at the relative maxima and minima of $T_{n+1}(x)$. These maxima and minima occur when $T_{n+1}(x) = \pm 1$, and they are given by

$$x_i = \cos\left(\frac{i\pi}{n+1}\right) \qquad i = 0, 1, \ldots, n+1 \qquad (4.7.34)$$

These seem an excellent choice to use in (4.7.32) if $F_n(x)$ is to look like the minimax approximation $q_n^*(x)$.

The system (4.7.32) becomes

$$\sum_{k=0}^{n}{}' c_{n,k} T_k(x_i) + (-1)^i E_n = f(x_i) \qquad i = 0, 1, \ldots, n+1 \qquad (4.7.35)$$

Note that

$$T_k(x_i) = \cos\left(k \cdot \cos^{-1}(x_i)\right) = \cos\left(\frac{ki\pi}{n+1}\right)$$

Introduce $E_n = c_{n,n+1}/2$. The system (4.7.35) becomes

$$\sum_{k=0}^{n+1}{}'' c_{n,k}\cos\left(\frac{ki\pi}{n+1}\right) = f(x_i) \qquad i = 0,1,\ldots,n+1 \qquad (4.7.36)$$

since

$$\cos\left(\frac{ki\pi}{n+1}\right) = (-1)^i \qquad \text{for} \quad k = n+1$$

The notation $\sum''$ means that the first and last terms are to be halved before summation begins.

To solve (4.7.36), we need the following relations:

$$\sum_{i=0}^{n+1}{}'' \cos\left(\frac{ij\pi}{n+1}\right)\cos\left(\frac{ik\pi}{n+1}\right) = \begin{cases} n+1 & j=k=0 \quad \text{or} \quad n+1 \\ \dfrac{n+1}{2} & 0 < j = k < n+1 \qquad (4.7.37) \\ 0 & j \neq k \quad 0 \leq j \quad k \leq n+1 \end{cases}$$

The proof of these relations is closely related to the proof of the relations in Problem 42 of Chapter 3.

Multiply equation $i$ in (4.7.36) by $\cos[(ij\pi)/(n+1)]$ for some $0 \leq j \leq n+1$. Then sum on $i$, halving the first and last terms. This gives

$$\sum_{k=0}^{n+1}{}'' c_{n,k} \sum_{i=0}^{n+1}{}'' \cos\left(\frac{ij\pi}{n+1}\right)\cos\left(\frac{ik\pi}{n+1}\right) = \sum_{i=0}^{n+1}{}'' f(x_i)\cos\left(\frac{ij\pi}{n+1}\right)$$

Using the relations (4.7.37), all but one of the terms in the summation on $k$ will be zero. By checking the two cases $j = 0$, $n+1$, and $0 < j < n+1$, the same formula is obtained:

$$c_{n,j} = \frac{2}{n+1}\sum_{i=0}^{n+1}{}'' f(x_i)\cos\left(\frac{ij\pi}{n+1}\right) \qquad 0 \leq j \leq n+1 \qquad (4.7.38)$$

The formula for $E_n$ is

$$E_n = \frac{1}{n+1}\sum_{i=0}^{n+1}{}'' (-1)^i f(x_i) \qquad (4.7.39)$$

There are a number of connections between this approximation $F_n(x)$ and the Chebyshev expansion $C_n(x)$. Most importantly, the coefficients $c_{n,j}$ are approximations to the coefficients $c_j$ in $C_n(x)$. Evaluate formula (4.7.13) for $c_j$ by using the trapezoidal rule (5.1.5) with $n + 1$ subdivisions:

$$c_j = \frac{2}{\pi} \int_0^\pi f(\cos \theta) \cos(j\theta) \, d\theta$$

$$\doteq \frac{2}{\pi} \sum_{i=0}^{n+1}{}'' f\left(\cos\left(\frac{i\pi}{n+1}\right)\right) \cdot \cos\left(\frac{ji\pi}{n+1}\right) \cdot \frac{\pi}{n+1} = c_{n,j} \qquad (4.7.40)$$

It is well-known that for periodic integrands, the trapezoidal numerical integration rule is especially accurate (see Theorem 5.5 in Section 5.4). In addition, it can be shown that

$$c_{nj} = c_j + c_{2(n+1)-j} + c_{2(n+1)+j} + c_{4(n+1)-j} + \cdots \qquad (4.7.41)$$

If the Chebyshev coefficients $c_n$ in (4.7.2) decrease rapidly, then the approximation $F_n(x)$ nearly equals $C_n(x)$, and it is easier to calculate than $C_n(x)$.

***Example***   Use $f(x) = e^x$, $-1 \le x \le 1$, as before. For $n = 1$, the nodes are

$$\{x_i\} = \{-1, 0, 1\}$$

$E_1 = .272$, and

$$F_1(x) = 1.2715 + 1.1752x \qquad (4.7.42)$$

For the error, the relative maxima for $e^x - F_1(x)$ are given in Table 4.8. For $n = 3$, $E_3 = .00547$ and

$$F_3(x) = .994526 + .995682x + .543081x^2 + .179519x^3 \qquad (4.7.43)$$

The points of maximum error are given in Table 4.9. From Theorem 4.9

$$.00547 \le \rho_3(f) \le .00558$$

and this says that $F_3(x)$ is an excellent approximation to $q_3^*(x)$.

To see that $F_3(x)$ is an approximation to $C_3(x)$, compare the coefficients $c_{3,j}$ with the coefficients $c_j$ given in Table 4.3 in example (4.5.22) at the end of Section 4.5. The results are given in Table 4.10.

**Table 4.8   Relative maxima of $|e^x - F_1(x)|$**

| $x$ | $e^x - F_1(x)$ |
|---|---|
| $-1.0$ | .272 |
| .1614 | $-.286$ |
| 1.0 | .272 |

**Table 4.9   Relative maxima of $|e^x - F_3(x)|$**

| $x$ | $e^x - F_3(x)$ |
|---|---|
| $-1.0$ | .00547 |
| $-.6832$ | $-.00552$ |
| .0493 | .00558 |
| .7324 | $-.00554$ |
| 1.0 | .00547 |

**Table 4.10   Expansion coefficients for $C_3(x)$ and $F_3(x)$ to $e^x$**

| $j$ | $c_j$ | $c_{n,j}$ |
|---|---|---|
| 0 | 2.53213176 | 2.53213215 |
| 1 | 1.13031821 | 1.13032142 |
| 2 | .27149534 | .27154032 |
| 3 | .04433685 | .04487978 |
| 4 | .00547424 | $E_4 = .00547424$ |

As with the interpolatory approximation $I_n(x)$, care must be taken if $f(x)$ is odd or even in $[-1, 1]$. In such a case, choose $n$ as follows:

$$\text{If } f \text{ is } \left\{ \begin{array}{c} \text{even} \\ \text{odd} \end{array} \right\}, \text{ then chooses } n \text{ to be } \left\{ \begin{array}{c} \text{odd} \\ \text{even} \end{array} \right\} \qquad (4.7.44)$$

This ensures $c_{n+1} \neq 0$ in (4.7.5), and thus the nodes chosen will be the correct ones.

An analysis of the convergence of $F_n(x)$ to $f(x)$ is given in Shampine (1970), resulting in a bound similar to the bound (4.7.30) for $I_n(x)$:

$$\|f - F_n\|_\infty \leq w(n)\rho_n(f), \qquad (4.7.45)$$

with $w(n)$ empirically nearly equal to the bounding coefficient in (4.7.30). Both $I_n(x)$ and $F_n(x)$ are practical near-minimax approximations.

We now give an algorithm for computing $F_n(x)$, which can then be evaluated using the algorithm *Chebeval* of Section 4.5.

***Algorithm*** *Approx*   $(c, E, f, n)$

1.   Remark: This algorithm calculates the coefficients $c_j$ in

$$F_n(x) = \sum_{j=0}^{n}{}' c_j T_j(x) \qquad -1 \leq x \leq 1$$

according to the formula (4.7.38), and $E$ is calculated from (4.7.39). The term $c_0$ should be halved before using algorithm *Chebeval*.

**2.** Create $x_i := \cos(i\pi/(n+1))$

$$f_i := f(x_i) \qquad i = 0, 1, \ldots, n+1$$

**3.** Do through step 8 for $j = 0, 1, \ldots, n+1$.

**4.** sum $:= [f_0 + (-1)^j f_{n+1}]/2$

**5.** Do through step 6 for $i = 1, \ldots, n$.

**6.** sum $:=$ sum $+ f_i \cos(ij\pi/(n+1))$

**7.** End loop on $i$.

**8.** $c_j \doteq 2\,\mathrm{sum}/(n+1)$

**9.** End loop on $j$.

**10.** $E := c_{n+1}/2$ and exit.

The cosines in step 6 can be evaluated more efficiently by using the trigonometric addition formulas for the sine and cosine functions, but we have opted for pedagogical simplicity, since the computer running time will still be quite small with our algorithm. For the same reason, we have not used the FFT techniques of Section 3.8.

## Discussion of the Literature

Approximation theory is a classically important area of mathematics, and it is also an increasingly important tool in studying a wide variety of modern problems in applied mathematics, for example, in mathematical physics and combinatorics. The variety of problems and approaches in approximation theory can be seen in the books of Achieser (1956), Askey (1975b), Davis (1963), Lorentz (1966), Meinardus (1967), Powell (1981), and Rice (1964), (1968). The classic work on orthogonal polynomials is Szego (1967), and a survey of more recent work is given in Askey (1975a). For Chebyshev polynomials and their many uses throughout applied mathematics and numerical analysis, see Fox and Parker (1968) and Rivlin (1974). The related subject of Fourier series and approximation by trigonometric polynomials was only alluded to in the text, but it is of central importance in a large number of applications. The classical reference is Zygmund (1959). There are many other areas of approximation theory that we have not even defined. For an excellent survey of these areas, including an excellent bibliography, see Gautschi (1975). A major area omitted in the present text is approximation by rational functions. For the general area, see Meinardus (1967, chap. 9) and Rice (1968, chap. 9). The generalization to rational functions of the Taylor polynomial is called the *Pade approximation*; for introductions, see Baker (1975) and Brezinski (1980). For the related area of *continued fraction expansions* of functions, see Wall (1948). Many of the functions that are of practical interest are examples of what are called the *special functions* of mathematical physics. These include the basic transcendental functions (sine, log, exp, square root), and in addition, orthogonal polynomials, the Bessel functions, gamma function, and hypergeometric function. There is an extensive literature on special functions, and special approximations have been devised for most of them. The most important references for special functions are in Abramowitz and Stegun (1964), a handbook produced under the auspices of the U.S. National Bureau of Standards, and Erdélyi et al. (1953), a three volume set that is often referred to as the "Bateman project." For a general overview and survey of the techniques for approximating special functions, see Gautschi (1975). An extensive compendium of theoretical results for special functions and of methods for their numerical evaluation is given in Luke (1969), (1975), (1977). For a somewhat more current sampling of trends in the study of special functions, see the symposium proceedings in Askey (1975b).

From the advent of large-scale use of computers in the 1950s, there has been a need for high-quality polynomial or rational function approximations of the basic mathematical functions and other special functions. As pointed out previously, the approximation of these functions requires a knowledge of their properties. But it also requires an intimate knowledge of the arithmetic of digital computers, as surveyed in Chapter 1. A general survey of numerical methods for producing polynomial approximations is given in Fraser (1965), which has influenced the organization of this chapter. For a very complete discussion of approximation of the elementary functions, together with detailed algorithms, see Cody and Waite (1980); a discussion of the associated programming project is discussed in Cody (1984). For a similar presentation of approximations, but one that also includes some of the more common special functions, see Hart et al. (1968). For an extensive set of approximations for special functions, see Luke (1975), (1977). For general functions, a successful and widely used program for generating minimax approximations is given in Cody et al. (1968). General programs for computing minimax approximations are available in the IMSL and NAG libraries.

# Bibliography

Abramowitz, M., and I. Stegun (eds.) (1964). *Handbook of Mathematical Functions*. National Bureau of Standards, U.S. Government Printing Office, Washington, D.C. (It is now published by Dover, New York.)

Achieser, N. (1956). *Theory of Approximation* (transl. C. Hyman). Ungar, New York.

Askey, R. (1975a). *Orthogonal Polynomials and Special Functions*. Society for Industrial and Applied Mathematics, Philadelphia.

Askey, R. (ed.) (1975b). *Theory and Application of Special Functions*. Academic Press, New York.

Baker, G., Jr. (1975). *Essentials of Padé Approximants*. Academic Press, New York.

Brezinski, C. (1980). *Padé-Type Approximation and General Orthogonal Polynomials*. Birkhäuser, Basel.

Cody, W. (1984). FUNPACK—A package of special function routines. In *Sources and Development of Mathematical Software*, W. Cowell (ed.), pp. 49–67. Prentice-Hall, Englewood Cliffs, N.J.

Cody, W. and W. Waite (1980). *Software Manual for the Elementary Functions*. Prentice-Hall, Englewood Cliffs, N.J.

Cody, W., W. Fraser, and J. Hart (1968). Rational Chebyshev approximation using linear equations. *Numer. Math.*, **12**, 242–251.

Davis, P. (1963). *Interpolation and Approximation*. Ginn (Blaisdell), Boston.

Erdélyi, A., W. Magnus, F. Oberhettinger, and F. Tricomi (1953). *Higher Transcendental Functions*, Vols. I, II, and III. McGraw-Hill, New York.

Fox, L., and I. Parker (1968). *Chebyshev Polynomials in Numerical Analysis*. Oxford Univ. Press, Oxford, England.

Fraser, W. (1965). A survey of methods of computing minimax and near-minimax polynomial approximations for functions of a single independent variable. *J. ACM*, **12**, 295–314.

Gautschi, W. (1975). Computational methods in special functions—A survey. In R. Askey (ed.), *Theory and Application of Special Functions*, pp. 1–98. Academic Press, New York.

Hart, J., E. Cheney, C. Lawson, H. Maehly, C. Mesztenyi, J. Rice, H. Thacher, and C. Witzgall (1968). *Computer Approximations*. Wiley, New York. (Reprinted in 1978, with corrections, by Krieger, Huntington, N.Y.)

Isaacson, E., and H. Keller (1966). *Analysis of Numerical Methods*. Wiley, New York.

Lorentz, G. (1966). *Approximation of Functions*. Holt, Rinehart & Winston, New York.

Luke, Y. (1969). *The Special Functions and Their Applications*, Vols. I and II. Academic Press, New York.

Luke, Y. (1975). *Mathematical Functions and Their Approximations*. Academic Press, New York.

Luke, Y. (1977). *Algorithms for the Computation of Mathematical Functions*. Academic Press, New York.

Meinardus, G. (1967). *Approximation of Functions: Theory and Numerical Methods* (transl. L. Schumaker). Springer-Verlag, New York.

Powell, M. (1981). *Approximation Theory and Methods*. Cambridge Univ. Press, Cambridge, England.

Rice, J. (1964). *The Approximation of Functions: Linear Theory*. Addison-Wesley, Reading, Mass.

Rice, J. (1968). *The Approximation of Functions: Advanced Topics.* Addison-Wesley, Reading, Mass.

Rivlin, T. (1974). *The Chebyshev Polynomials.* Wiley, New York.

Shampine, L. (1970). Efficiency of a procedure for near-minimax approximation. *J. ACM*, **17**, 655–660.

Szego, G. (1967). *Orthogonal Polynomials*, 3rd ed. Amer. Math. Soc., Providence, R.I.

Wall, H. (1948). *Analytic Theory of Continued Fractions.* Van Nostrand, New York.

Zygmund, A. (1959). *Trigonometric Series*, Vols. I and II. Cambridge Univ. Press, Cambridge, England.

## Problems

1.  To illustrate that the Bernstein polynomials $p_n(x)$ in Theorem 4.1 are poor approximations, calculate the fourth-degree approximation $p_n(x)$ for $f(x) = \sin(\pi x)$, $0 \le x \le 1$. Compare it with the fourth-degree Taylor polynomial approximation, expanded about $x = \frac{1}{2}$.

2.  Let $S = \sum_{1}^{\infty} (-1)^j a_j$ be a convergent series, and assume that all $a_j \ge 0$ and

    $$a_1 \ge a_2 \ge \cdots \ge a_n \ge \cdots$$

    Prove that

    $$\left| S - \sum_{1}^{n} (-1)^j a_j \right| \le a_{n+1}$$

3.  Using Problem 2, examine the convergence of the following series. Bound the error when truncating after $n$ terms, and note the dependence on $x$. Find the value of $n$ for which the error is less than $10^{-5}$. This problem illustrates another common technique for bounding the error in Taylor series.

    **(a)** $J_0(x) = \sum_{j=0}^{\infty} \dfrac{(-1)^j \left(\frac{1}{4}x^2\right)^j}{(j!)^2}$

    **(b)** $\sum_{j=1}^{\infty} \dfrac{(-1)^j x^{2j}}{j^2}$

4.  Graph the errors of the Taylor series approximations $p_n(x)$ to $f(x) = \sin\left[(\pi/2)x\right]$ on $-1 \le x \le 1$, for $n = 1, 3, 5$. Note the behavior of the error both near the origin and near the endpoints.

5.  Let $f(x)$ be three times continuously differentiable on $[-\alpha, \alpha]$ for some $\alpha > 0$, and consider approximating it by the rational function

$$R(x) = \frac{a + bx}{1 + cx}$$

To generalize the idea of the Taylor series, choose the constants $a$, $b$, and $c$ so that

$$R^{(j)}(0) = f^{(j)}(0) \qquad j = 0, 1, 2$$

Is it always possible to find such an approximation $R(x)$? The function $R(x)$ is an example of a *Pade approximation* to $f(x)$. See Baker (1975) and Brezinski (1980).

6.  Apply the results of Problem 5 to the case $f(x) = e^x$, and give the resulting approximation $R(x)$. Analyze its error on $[-1, 1]$, and compare it with the error for the quadratic Taylor polynomial.

7.  By means of various identities, it is often possible to reduce the interval on which a function needs to be approximated. Show how to reduce each of the following functions from $-\infty < x < \infty$ to the given interval. Usually a few additional, but simple, operations will be needed.

    (a)  $e^x$    $0 \le x \le 1$

    (b)  $\cos(x)$    $0 \le x \le \pi/4$

    (c)  $\tan^{-1}(x)$    $0 \le x \le 1$

    (d)  $\ln(x)$    $1 \le x \le 2$. Reduce from $\ln(x)$ on $0 < x < \infty$.

8.  (a)  Let $f(x)$ be continuously differentiable on $[a, b]$. Let $p(x)$ be a polynomial for which

    $$\|f' - p\|_\infty \le \epsilon$$

    and define

    $$q(x) = f(a) + \int_a^x p(t)\, dt \qquad a \le x \le b$$

    Show that $q(x)$ is a polynomial and satisfies

    $$\|f - q\|_\infty \le \epsilon(b - a)$$

    (b)  Extend part (a) to the case where $f(x)$ is $N$ times continuously differentiable on $[a, b]$, $N \ge 2$, and $p(x)$ is a polynomial satisfying

    $$\|f^{(N)} - p\|_\infty \le \epsilon$$

Obtain a formula for a polynomial $q(x)$ approximating $f(x)$, with the formula for $q(x)$ involving a single integral.

(c)   Assume that $f(x)$ is infinitely differentiable on $[a, b]$, that is, $f^{(j)}(x)$ exists and is continuous on $[a, b]$, for all $j \geq 0$. [This does not imply that $f(x)$ has a convergent Taylor series on $[a, b]$.] Prove there exists a sequence of polynomials $\{\, p_n(x) | n \geq 1 \}$ for which

$$\underset{n \to \infty}{\text{Limit}} \| f^{(j)} - p_n^{(j)} \|_\infty = 0$$

for all $j \geq 0$. *Hint:* Use the Weierstrass theorem and part (b).

9.   Prove the following result. Let $f \in C^2[a, b]$ with $f''(x) > 0$ for $a \leq x \leq b$. If $q_1^*(x) = a_0 + a_1 x$ is the linear minimax approximation to $f(x)$ on $[a, b]$, then

$$a_1 = \frac{f(b) - f(a)}{b - a} \qquad a_0 = \frac{f(a) + f(c)}{2} - \left( \frac{a + c}{2} \right) \left[ \frac{f(b) - f(a)}{b - a} \right]$$

where $c$ is the unique solution of

$$f'(c) = \frac{f(b) - f(a)}{b - a}$$

What is $\rho$?

10.   (a)   Produce the linear Taylor polynomials to $f(x) = \ln(x)$ on $1 \leq x \leq 2$, expanding about $x_0 = \frac{3}{2}$. Graph the error.

(b)   Produce the linear minimax approximation to $f(x) = \ln(x)$ on $[1, 2]$. Graph the error, and compare it with the Taylor approximation.

11.   (a)   Show that the linear minimax approximation to $\sqrt{1 + x^2}$ on $[0, 1]$ is

$$q_1^*(x) = .955 + 414x$$

(b)   Using part (a), derive the approximation

$$\sqrt{y^2 + z^2} \doteq .955z + .414y \qquad 0 \leq y \leq z$$

and determine the error.

12.   Find the linear least squares approximation to $f(x) = \ln(x)$ on $[1, 2]$. Compare the error with the results of Problem 10.

13.   Find the value of $\alpha$ that minimizes

$$\int_0^1 |e^x - \alpha| \, dx$$

What is the minimum? This is a simple illustration of yet another way to measure the error of an approximation and of the resulting best approximation.

14. Solve the following minimization problems and determine whether there is a unique value of $\alpha$ that gives the minimum. In each case, $\alpha$ is allowed to range over all real numbers. We are approximating the function $f(x) = x$ with polynomials of the form $\alpha x^2$.

(a)    $\displaystyle \text{Min}_{\alpha} \int_{-1}^{1} [x - \alpha x^2]^2 \, dx$

(b)    $\displaystyle \text{Min}_{\alpha} \int_{-1}^{1} |x - \alpha x^2| \, dx$

(c)    $\displaystyle \text{Min}_{\alpha} \ \text{Max}_{-1 \le x \le 1} \ |x - \alpha x^2|$

15. Using (4.4.10), show that $\{P_n(x)\}$ is an orthogonal family and that $\|P_n\|_2 = \sqrt{2/(2n + 1)}$, $n \ge 0$.

16. Verify that

$$\varphi_n(x) = \frac{(-1)^n}{n!} e^x \frac{d^n}{dx^n} (x^n e^{-x})$$

for $n \ge 0$ are orthogonal on the interval $[0, \infty)$ with respect to the weight function $w(x) = e^{-x}$. (Note: $\int_0^{\infty} e^{-x} x^m \, dx = m!$ for $m = 0, 1, 2 \dots$)

17. (a)    Find the relative maxima and minima of $T_n(x)$ on $[-1, 1]$, obtaining (4.7.21).

(b)    Find the zeros of $T_n(x)$, obtaining (4.7.14).

18. Derive the formulas for $b_n$ and $c_n$ given in the triple recursion relation in (4.4.21) of Theorem 4.5.

19. Modify the Gram–Schmidt procedure of Theorem 4.2, to avoid the normalization step $\varphi_n = \psi_n / \|\psi_n\|_2$:

$$\psi_n(x) = x^n + b_{n, n-1} \psi_{n-1}(x) + \cdots + b_{n, 0} \psi_0(x)$$

and find the coefficients $b_{n, j}$, $0 \le j \le n - 1$.

20. Using Problem 19, find $\psi_0, \psi_1, \psi_2$ for the following weight functions $w(x)$ on the indicated intervals $[a, b]$.

(a)    $w(x) = \ln(x)$,    $0 \le x \le 1$

Hint: $\displaystyle \int_0^1 x^n \ln(x) \, dx = [-1/(n + 1)^2]$    $n \ge 0$

**(b)**   $w(x) = x$     $0 \le x \le 1$

**(c)**   $w(x) = \sqrt{1 - x^2}$     $-1 \le x \le 1$

**21.**   Let $\{\varphi_n(x) | n \ge 1\}$ be orthogonal on $(a, b)$ with weight function $w(x) \ge 0$. Denote the zeros of $\varphi_n(x)$ by

$$a < z_{n, n} < z_{n-1, n} < \cdots < z_{1, n} < b$$

Prove that the zeros of $\varphi_{n+1}(x)$ are separated by those of $\varphi_n(x)$, that is,

$$a < z_{n+1, n+1} < z_{n, n} < z_{n, n+1} < \cdots < z_{2, n+1} < z_{1, n} < z_{1, n+1} < b$$

*Hint:* Use induction on the degree $n$. Write $\varphi_n(x) = A_n x^n + \cdots$, with $A_n > 0$, and use the triple recursion relation (4.4.21) to evaluate the polynomials at the zeros of $\varphi_n(x)$. Observe that the sign changes for $\varphi_{n-1}(x)$ and $\varphi_{n+1}(x)$.

**22.**   Extend the Christoffel–Darboux identity (4.4.27) to the case with $x = y$, obtaining a formula for

$$\sum_{k=0}^{n} \frac{[\varphi_k(x)]^2}{\gamma_k}$$

*Hint:* Consider the limit in (4.4.27) as $y \to x$.

**23.**   Let $f(x) = \cos^{-1}(x)$ for $-1 \le x \le 1$ (the principal branch $0 \le f \le \pi$). Find the polynomial of degree two,

$$p(x) = a_0 + a_1 x + a_2 x^2$$

which minimizes

$$\int_{-1}^{1} \frac{[f(x) - p(x)]^2}{\sqrt{1 - x^2}} \, dx$$

**24.**   Define $S_n(x) = \dfrac{1}{n+1} T'_{n+1}(x)$, $n \ge 0$, with $T_{n+1}(x)$ the Chebyshev polynomial of degree $n + 1$. The polynomials $S_n(x)$ are called *Chebyshev polynomials of the second kind*.

**(a)**   Show that $\{S_n(x) | n \ge 0\}$ is an orthogonal family on $[-1, 1]$ with respect to the weight function $w(x) = \sqrt{1 - x^2}$.

**(b)**   Show that the family $\{S_n(x)\}$ satisfies the same triple recursion relation (4.4.13) as the family $\{T_n(x)\}$.

(c)    Given $f \in C[-1, 1]$, solve the problem

$$\text{Min} \int_{-1}^{1} \sqrt{1 - x^2} \, [f(x) - P_n(x)]^2 \, dx$$

where $p_n(x)$ is allowed to range over all polynomials of degree $\leq n$.

25.    Do an operations count for algorithm *Chebeval* of Section 4.5. Give the number of multiplications and the number of additions. Compare this to the ordinary nested multiplication algorithm.

26.    Show that the framework of Sections 4.4 and 4.5 also applies to the trigonometric polynomials of degree $\leq n$. Show that the family $\{1, \sin(x), \cos(x), \ldots, \sin(nx), \cos(nx)\}$ is orthogonal on $[0, 2\pi]$. Derive the least squares approximation to $f(x)$ on $[0, 2\pi]$ using such polynomials. [Letting $n \to \infty$, we obtain the well-known Fourier series (see Zygmund (1959))].

27.    Let $f(x)$ be a continuous even (odd) function on $[-a, a]$. Show that the minimax approximation $q_n^*(x)$ to $f(x)$ will be an even (odd) function on $[-a, a]$, regardless of whether $n$ is even or odd. *Hint:* Use Theorem 4.10, including the uniqueness result.

28.    Using (4.6.10), bound $\rho_n(f)$ for the following functions $f(x)$ on the given interval, $n = 1, 2, \ldots, 10$.

(a)    $\sin(x)$      $0 \leq x \leq \pi/2$

(b)    $\ln(x)$      $1 \leq x \leq e$

(c)    $\tan^{-1}(x)$      $0 \leq x \leq \pi/4$

(d)    $e^x$      $0 \leq x \leq 1$

29.    For the Chebyshev expansion (4.7.2), show that if $f(x)$ is even (odd) on $[-1, 1]$, then $c_j = 0$ if $j$ is odd (even).

30.    For $f(x) = \sin[(\pi/2)x]$, $-1 \leq x \leq 1$, find both the Legendre and Chebyshev least squares approximations of degree three to $f(x)$. Determine the error in each approximation and graph them. Use Theorem 4.9 to bound the minimax error $\rho_3(f)$. *Hint:* Use numerical integration to simplify constructing the least squares coefficients. Note the comments following (4.7.13), for the Chebyshev least squares approximation.

31.    Produce the interpolatory near-minimax approximation $I_n(x)$ for the following basic mathematical functions $f$ on the indicated intervals, for $n = 1, 2, \ldots, 8$. Using the standard routines of your computer, compute the error. Graph the error, and using Theorem 4.9, give upper and lower bounds for $\rho_n(f)$.

(a)    $e^x$      $0 \leq x \leq 1$          (b)    $\sin(x)$      $0 \leq x \leq \pi/2$

(c)    $\tan^{-1}(x)$      $0 \leq x \leq 1$          (d)    $\ln(x)$      $1 \leq x \leq 2$

**32.** Repeat Problem 31 with the near-minimax approximation $F_n(x)$.

**33.** Repeat Problem 31 and 32 for

$$f(x) = \frac{1}{x} \int_0^x \frac{\sin(t)}{t} \, dt \qquad 0 \le x \le \frac{\pi}{2}$$

*Hint:* Find a Taylor approximation of high degree for evaluating $f(x)$, then use the transformation (4.5.12) to obtain an approximation problem on $[-1, 1]$.

**34.** For $f(x) = e^x$ on $[-1, 1]$, consider constructing $I_n(x)$. Derive the error result

$$\alpha_n |T_{n+1}(x)| \le |f(x) - I_n(x)| \le \beta_n |T_{n+1}(x)| \qquad -1 \le x \le 1$$

for appropriate constants $\alpha_n$, $\beta_n$. Find nonzero upper and lower bounds for $\rho_n(f)$.

**35. (a)** The function $\sin(x)$ vanishes at $x = 0$. In order to better approximate it in a relative error sense, we consider the function $f(x) = \sin(x)/x$. Calculate the near-minimax approximation $I_n(x)$ on $0 \le x \le \pi/2$ for $n = 1, 2, \ldots, 7$, and then compare $\sin(x) - xI_n(x)$ with the results of Problem 31(b).

**(b)** Repeat part (a) for $f(x) = \tan^{-1}(x)$, $\qquad 0 \le x \le 1$.

**36.** Let $f(x) = a_n x^n + \cdots + a_1 x + a_0$, $a_n \ne 0$. Find the minimax approximation to $f(x)$ on $[-1, 1]$ by a polynomial of degree $\le n - 1$, and also find $\rho_{n-1}(f)$.

**37.** Let $\alpha = \operatorname{Min} \left[ \operatorname{Max}_{|x| \le 1} |x^6 - x^3 - p_5(x)| \right]$, where the minimum is taken over all polynomials of degree $\le 5$.

**(a)** Find $\alpha$.

**(b)** Find the polynomial $p_5(x)$ for which the minimum $\alpha$ is attained.

**38.** Prove the result (4.6.10). *Hint:* Consider the near-minimax approximation $I_n(x)$.

**39. (a)** For $f(x) = e^x$ on $[-1, 1]$, find the Taylor polynomial $p_4(x)$ of degree four, expanded about $x = 0$.

**(b)** Using Problem 36, find the minimax polynomial $m_{4,3}(x)$ of degree three that approximates $p_4(x)$. Graph the error $e^x - m_{4,3}(x)$ and compare it to the Taylor error $e^x - p_3(x)$ shown in Figure 4.1. The process of reducing a Taylor polynomial to a lower degree one by this process is called *economization* or *telescoping*. It is usually used several times in succession, to reduce a high-degree Taylor polynomial to a polynomial approximation of much lower degree..

40. Using a standard program from your computer center for computing minimax approximations, calculate the minimax approximation $q_n^*(x)$ for the following given functions $f(x)$ on the given interval. Do this for $n = 1, 2, 3, \ldots, 8$. Compare the results with those for problem 31.

(a) $e^x \quad 0 \le x \le 1$ 　　　　　　(b) $\sin(x) \quad 0 \le x \le \pi/2$

(c) $\tan^{-1}(x) \quad 0 \le x \le 1$ 　　　(d) $\ln(x) \quad 1 \le x \le 2$

41. Produce the minimax approximations $q_n^*(x)$, $n = 1, 3, 5, 7, 9$, for

$$f(x) = \frac{1}{x} \int_0^x e^{-t^2} \, dt \qquad -1 \le x \le 1$$

*Hint:* First produce a Taylor approximation of high accuracy, and then use it with the program of Problem 40.

42. Repeat Problem 41 for

$$f(x) = \frac{1}{x} \int_0^x \frac{\sin(t)}{t} \, dt \qquad |x| \le \pi$$

# FIVE

# NUMERICAL INTEGRATION

In this chapter we derive and analyze numerical methods for evaluating definite integrals. The integrals are mainly of the form

$$I(f) = \int_a^b f(x)\, dx \tag{5.0.1}$$

with $[a, b]$ finite. Most such integrals cannot be evaluated explicitly, and with many others, it is often faster to integrate them numerically rather than evaluating them exactly using a complicated antiderivative of $f(x)$. The approximation of $I(f)$ is usually referred to as *numerical integration* or *quadrature*.

There are many numerical methods for evaluating (5.0.1), but most can be made to fit within the following simple framework. For the integrand $f(x)$, find an approximating family $\{ f_n(x) | n \geq 1 \}$ and define

$$I_n(f) = \int_a^b f_n(x)\, dx = I(f_n) \tag{5.0.2}$$

We usually require the approximations $f_n(x)$ to satisfy

$$\| f - f_n \|_\infty \to 0 \qquad \text{as} \quad n \to \infty \tag{5.0.3}$$

And the form of each $f_n(x)$ should be chosen such that $I(f_n)$ can be evaluated easily. For the error,

$$E_n(f) = I(f) - I_n(f) = \int_a^b [f(x) - f_n(x)]\, dx$$

$$|E_n(f)| \leq \int_a^b |f(x) - f_n(x)|\, dx \leq (b - a)\| f - f_n \|_\infty \tag{5.0.4}$$

Most numerical integration methods can be viewed within this framework, although some of them are better studied from some other perspective. The one class of methods that does not fit within the framework are those based on extrapolation using asymptotic estimates of the error. These are examined in Section 5.4.

Most numerical integrals $I_n(f)$ will have the following form when they are evaluated:

$$I_n(f) = \sum_{j=1}^{n} w_{j,n} f(x_{j,n}) \qquad n \geq 1 \qquad (5.0.5)$$

The coefficients $w_{j,n}$ are called the integration *weights* or quadrature weights; and the points $x_{j,n}$ are the integration *nodes*, usually chosen in $[a, b]$. The dependence on $n$ is usually suppressed, writing $w_j$ and $x_j$, although it will be understood implicitly. Standard methods have nodes and weights that have simple formulas or else they are tabulated in tables that are readily available. Thus there is usually no need to explicitly construct the functions $f_n(x)$ of (5.0.2), although their role in defining $I_n(f)$ may be useful to keep in mind.

The following example is a simple illustration of (5.0.2)–(5.0.4), but it is not of the form (5.0.5).

***Example*** Evaluate

$$I = \int_0^1 \frac{e^x - 1}{x} \, dx \qquad (5.0.6)$$

This integrand has a removable singularity at the origin. Use a Taylor series for $e^x$ [see (1.1.4) of Chapter 1] to define $f_n(x)$, and then define

$$I_n = \int_0^1 \sum_{j=1}^{n} \frac{x^{j-1}}{j!} \, dx$$

$$= \sum_{j=1}^{n} \frac{1}{(j!)(j)} \qquad (5.0.7)$$

For the error in $I_n$, use the Taylor formula (1.1.4) to obtain

$$f(x) - f_n(x) = \frac{x^n}{(n+1)!} e^{\xi_x}$$

for some $0 \leq \xi_x \leq x$. Then

$$I - I_n = \int_0^1 \frac{x^n}{(n+1)!} e^{\xi_x} \, dx$$

$$\frac{1}{(n+1)!(n+1)} \leq I - I_n \leq \frac{e}{(n+1)!(n+1)} \qquad (5.0.8)$$

The sequence in (5.0.7) is rapidly convergent, and (5.0.8) allows us to estimate the error very accurately. For example, with $n = 6$

$$I_6 = 1.31787037$$

and from (5.0.8)

$$2.83 \times 10^{-5} \le I - I_6 \le 7.70 \times 10^{-5}$$

The true error is $3.18 \times 10^{-5}$.

For integrals in which the integrand has some kind of bad behavior, for example, an infinite value at some point, we often will consider the integrand in the form

$$I(f) = \int_a^b w(x) f(x)\, dx \qquad (5.0.9)$$

The bad behavior is assumed to be located in $w(x)$, called the *weight function*, and the function $f(x)$ will be assumed to be well-behaved. For example, consider evaluating

$$\int_0^1 (\ln x) f(x)\, dx$$

for arbitrary continuous functions $f(x)$. The framework (5.0.2)–(5.0.4) generalizes easily to the treatment of (5.0.9). Methods for such integrals are considered in Sections 5.3 and 5.6.

Most numerical integration formulas are based on defining $f_n(x)$ in (5.0.2) by using polynomial or piecewise polynomial interpolation. Formulas using such interpolation with evenly spaced node points are derived and discussed in Sections 5.1 and 5.2. The Gaussian quadrature formulas, which are optimal in a certain sense and which have very rapid convergence, are given in Section 5.3. They are based on defining $f_n(x)$ using polynomial interpolation at carefully selected node points that need not be evenly spaced.

Asymptotic error formulas for the methods of Sections 5.1 and 5.2 are given and discussed in Section 5.4, and some new formulas are derived based on extrapolation with these error formulas. Some methods that control the integration error in an automatic way, while remaining efficient, are given in Section 5.5. Section 5.6 surveys methods for integrals that are singular or ill-behaved in some sense, and Section 5.7 discusses the difficult task of numerical differentiation.

## 5.1  The Trapezoidal Rule and Simpson's Rule

We begin our development of numerical integration by giving two well-known numerical methods for evaluating

$$I(f) = \int_a^b f(x)\, dx \qquad (5.1.1)$$

We analyze and illustrate these methods very completely, and they serve as an introduction to the material of later sections. The interval $[a, b]$ is always finite in this section.

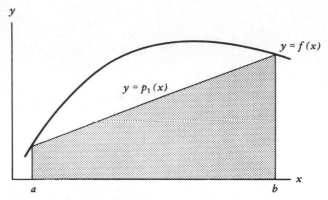

**Figure 5.1**    Illustration of trapezoidal rule.

**The trapezoidal rule**    The simple trapezoidal rule is based on approximating $f(x)$ by the straight line joining $(a, f(a))$ and $(b, f(b))$. By integrating the formula for this straight line, we obtain the approximation

$$I_1(f) = \left(\frac{b-a}{2}\right)[f(a) + f(b)] \tag{5.1.2}$$

This is of course the area of the trapezoid shown in Figure 5.1. To obtain an error formula, we use the interpolation error formula (3.1.10):

$$f(x) - \frac{(b-x)f(a) + (x-a)f(b)}{b-a} = (x-a)(x-b)f[a, b, x]$$

We also assume for all work with the error for the trapezoidal rule in this section that $f(x)$ is twice continuously differentiable on $[a, b]$. Then

$$E_1(f) = \int_a^b f(x)\, dx - \frac{(b-a)}{2}[f(a) + f(b)]$$

$$= \int_a^b (x-a)(x-b)f[a, b, x]\, dx \tag{5.1.3}$$

Using the integral mean value theorem [Theorem 1.3 of Chapter 1],

$$E_1(f) = f[a, b, \xi]\int_a^b (x-a)(x-b)\, dx \quad \text{some} \quad a \le \xi \le b$$

$$= \left[\frac{1}{2}f''(\eta)\right]\left[-\frac{1}{6}(b-a)^3\right] \quad \text{some} \quad \eta \in [a, b]$$

using (3.2.12). Thus

$$E_1(f) = -\frac{(b-a)^3}{12}f''(\eta) \qquad \eta \in [a,b] \qquad (5.1.4)$$

If $b - a$ is not sufficiently small, the trapezoidal rule (5.1.2) is not of much use. For such an integral, we break it into a sum of integrals over small subintervals, and then we apply (5.1.2) to each of these smaller integrals. Let $n \geq 1$, $h = (b-a)/n$, and $x_j = a + jh$ for $j = 0, 1, \ldots, n$. Then

$$I(f) = \int_a^b f(x)\,dx = \sum_{j=1}^n \int_{x_{j-1}}^{x_j} f(x)\,dx$$

$$= \sum_{j=1}^n \left\{ \left(\frac{h}{2}\right)[f(x_{j-1}) + f(x_j)] - \frac{h^3}{12}f''(\eta_j) \right\}$$

with $x_{j-1} \leq \eta_j \leq x_j$. There is no reason why the subintervals $[x_{j-1}, x_j]$ must all have equal length, but it is customary to first introduce the general principles involved in this way. Although this is also the customary way in which the method is applied, there are situations in which it is desirable to vary the spacing of the nodes.

The first terms in the sum can be combined to give the *composite trapezoidal rule*,

$$I_n(f) = h\left[\tfrac{1}{2}f_0 + f_1 + f_2 + \cdots + f_{n-1} + \tfrac{1}{2}f_n\right] \qquad n \geq 1 \qquad (5.1.5)$$

with $f(x_j) \equiv f_j$. For the error in $I_n(f)$,

$$E_n(f) = I(f) - I_n(f) = \sum_{j=1}^n -\frac{h^3}{12}f''(\eta_j)$$

$$= -\frac{h^3 n}{12}\left[\frac{1}{n}\sum_{j=1}^n f''(\eta_j)\right] \qquad (5.1.6)$$

For the term in brackets,

$$\min_{a \leq x \leq b} f''(x) \leq M \equiv \frac{1}{n}\sum_{j=1}^n f''(\eta_j) \leq \max_{a \leq x \leq b} f''(x)$$

Since $f''(x)$ is continuous for $a \leq x \leq b$, it must attain all values between its minimum and maximum at some point of $[a, b]$; thus $f''(\eta) = M$ for some $\eta \in [a, b]$. Thus we can write

$$E_n(f) = -\frac{(b-a)h^2}{12}f''(\eta) \qquad \text{some} \quad \eta \in [a,b] \qquad (5.1.7)$$

Another error estimate can be derived using this analysis. From (5.1.6)

$$\underset{n \to \infty}{\text{Limit}} \frac{E_n(f)}{h^2} = \underset{n \to \infty}{\text{Limit}} \left[ -\frac{h}{12} \sum_{j=1}^{n} f''(\eta_j) \right]$$

$$= -\frac{1}{12} \underset{n \to \infty}{\text{Limit}} \sum_{j=1}^{n} f''(\eta_j) h$$

Since $x_{j-1} \le \eta_j \le x_j$, $j = 1, \ldots, n$, the last sum is a Riemann sum; thus

$$\underset{n \to \infty}{\text{Limit}} \frac{E_n(f)}{h^2} = -\frac{1}{12} \int_a^b f''(x) \, dx = -\frac{1}{12} [f'(b) - f'(a)] \quad (5.1.8)$$

$$E_n(f) \doteq -\frac{h^2}{12} [f'(b) - f'(a)] \equiv \tilde{E}_n(f) \quad (5.1.9)$$

The term $\tilde{E}_n(f)$ is called an asymptotic error estimate for $E_n(f)$, and is valid in the sense of (5.1.8).

**Definition**    Let $E_n(f)$ be an exact error formula, and let $\tilde{E}_n(f)$ be an estimate of it. We say that $\tilde{E}_n(f)$ is an *asymptotic error estimate* for $E_n(f)$ if

$$\underset{n \to \infty}{\text{Limit}} \frac{\tilde{E}_n(f)}{E_n(f)} = 1 \quad (5.1.10)$$

or equivalently,

$$\underset{n \to \infty}{\text{Limit}} \frac{E_n(f) - \tilde{E}_n(f)}{E_n(f)} = 0$$

The estimate in (5.1.9) meets this criteria, based on (5.1.8).

The composite trapezoidal rule (5.1.5) could also have been obtained by replacing $f(x)$ by a piecewise linear interpolating function $f_n(x)$ interpolating $f(x)$ at the nodes $x_0, x_1, \ldots, x_n$. From here on, we generally refer to the composite trapezoidal rule as simply the trapezoidal rule.

**Example**    We use the trapezoidal rule (5.1.9) to calculate

$$I = \int_0^\pi e^x \cos(x) \, dx \quad (5.1.11)$$

The true value is $I = -(e^\pi + 1)/2 \doteq -12.0703463164$. The values of $I_n$ are given in Table 5.1, along with the true errors $E_n$ and the asymptotic estimates $\tilde{E}_n$, obtained from (5.1.9). Note that the errors decrease by a factor of 4 when $n$ is doubled (and hence $h$ is halved). This result was predictable from the multiplying

**Table 5.1    Trapezoidal rule for evaluating (5.1.11)**

| $n$ | $I_n$ | $E_n$ | Ratio | $\tilde{E}_n$ |
|---|---|---|---|---|
| 2 | $-17.389259$ | 5.32 | | 4.96 |
| | | | 4.20 | |
| 4 | $-13.336023$ | 1.27 | | 1.24 |
| | | | 4.06 | |
| 8 | $-12.382162$ | $3.12\mathrm{E}-1$ | | $3.10\mathrm{E}-1$ |
| | | | 4.02 | |
| 16 | $-12.148004$ | $7.77\mathrm{E}-2$ | | $7.76\mathrm{E}-2$ |
| | | | 4.00 | |
| 32 | $-12.089742$ | $1.94\mathrm{E}-2$ | | $1.94\mathrm{E}-2$ |
| | | | 4.00 | |
| 64 | $-12.075194$ | $4.85\mathrm{E}-3$ | | $4.85\mathrm{E}-3$ |
| | | | 4.00 | |
| 128 | $-12.071558$ | $1.21\mathrm{E}-3$ | | $1.21\mathrm{E}-3$ |
| | | | 4.00 | |
| 256 | $-12.070649$ | $3.03\mathrm{E}-4$ | | $3.03\mathrm{E}-4$ |
| | | | 4.00 | |
| 512 | $-12.070422$ | $7.57\mathrm{E}-5$ | | $7.57\mathrm{E}-5$ |

factor of $h^2$ present in (5.1.7) and (5.1.9); when $h$ is halved, $h^2$ decreases by a factor of 4. This example also shows that the trapezoidal rule is relatively inefficient when compared with other methods to be developed in this chapter.

Using the error estimate $\tilde{E}_n(f)$, we can define an improved numerical integration rule:

$$CT_n(f) \equiv I_n(f) + \tilde{E}_n(f)$$

$$= h\left[\frac{1}{2}f_0 + f_1 + \cdots + f_{n-1} + \frac{1}{2}f_n\right] - \frac{h^2}{12}[f'(b) - f'(a)] \quad (5.1.12)$$

This is called the *corrected trapezoidal rule*. The accuracy of $\tilde{E}_n(f)$ should make $CT_n(f)$ much more accurate than the trapezoidal rule. Another derivation of (5.1.12) is suggested in Problem 4, one showing that (5.1.12) will fit into the approximation theoretic framework (5.0.2)–(5.0.4). The major difficulty of using $CT_n(f)$ is that $f'(a)$ and $f'(b)$ are required.

***Example***    Apply $CT_n(f)$ to the earlier example (5.1.11). The results are shown in Table 5.2, together with the errors for the trapezoidal rule, for comparison. Empirically, the error in $CT_n(f)$ is proportional to $h^4$, whereas it was proportional to $h^2$ with the trapezoidal rule. A proof of this is suggested in Problem 4.

**Table 5.2    The corrected trapezoidal rule for (5.1.11)**

| $n$ | $CT_n(f)$ | Error | Ratio | Trap Error |
|---|---|---|---|---|
| 2 | $-12.425528367$ | $3.55\mathrm{E}-1$ | | 5.32 |
| | | | 14.4 | |
| 4 | $-12.095090106$ | $2.47\mathrm{E}-2$ | | 1.27 |
| | | | 15.6 | |
| 8 | $-12.071929245$ | $1.58\mathrm{E}-3$ | | $3.12\mathrm{E}-1$ |
| | | | 15.9 | |
| 16 | $-12.070445804$ | $9.95\mathrm{E}-5$ | | $7.77\mathrm{E}-2$ |
| | | | 16.0 | |
| 32 | $-12.070352543$ | $6.23\mathrm{E}-6$ | | $1.94\mathrm{E}-2$ |
| | | | 16.0 | |
| 64 | $-12.070346706$ | $3.89\mathrm{E}-7$ | | $4.85\mathrm{E}-3$ |
| | | | 16.0 | |
| 128 | $-12.070346341$ | $2.43\mathrm{E}-8$ | | $1.21\mathrm{E}-3$ |

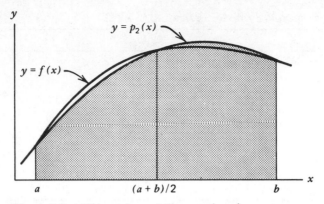

**Figure 5.2**   Illustration of Simpson's rule.

**Simpson's rule**   To improve upon the simple trapezoidal rule (5.1.2), we use a quadratic interpolating polynomial $p_2(f)$ to approximate $f(x)$ on $[a, b]$. Let $c = (a + b)/2$, and define

$$I_2(f) = \int_a^b \left[ \frac{(x - c)(x - b)}{(a - c)(a - b)} f(a) + \frac{(x - a)(x - b)}{(c - a)(c - b)} f(c) \right.$$

$$\left. + \frac{(x - a)(x - c)}{(b - a)(b - c)} f(b) \right] dx$$

Carrying out the integration, we obtain

$$I_2(f) = \frac{h}{3} \left[ f(a) + 4f\left(\frac{a + b}{2}\right) + f(b) \right] \qquad h = \frac{b - a}{2} \qquad (5.1.13)$$

This is called *Simpson's rule*. An illustration is given in Figure 5.2, with the shaded region denoting the area under the graph of $y = p_2(x)$.

For the error, we begin with the interpolation error formula (3.2.11) to obtain

$$E_2(f) = I(f) - I_2(f)$$

$$= \int_a^b (x - a)(x - c)(x - b) f[a, b, c, x] \, dx \qquad (5.1.14)$$

We cannot apply the integral mean value theorem since the polynomial in the integrand changes sign at $x = c = (a + b)/2$. We will assume $f(x)$ is four times continuously differentiable on $[a, b]$ for the work of this section on Simpson's rule. Define

$$w(x) = \int_a^x (t - a)(t - c)(t - b) \, dt$$

It is not hard to show that

$$w(a) = w(b) = 0 \qquad w(x) > 0 \qquad \text{for} \quad a < x < b$$

Integrating by parts,

$$E_2(f) = \int_a^b w'(x) f[a, b, c, x] \, dx$$

$$= [w(x) f[a, b, c, x]]_{x=a}^{x=b} - \int_a^b w(x) \frac{d}{dx} f[a, b, c, x] \, dx$$

$$= - \int_a^b w(x) f[a, b, c, x, x] \, dx$$

The last equality used (3.2.17). Applying the integral mean value theorem and (3.2.12),

$$E_2(f) = -f[a, b, c, \xi, \xi] \int_a^b w(x) \, dx \qquad \text{some} \quad a \le \xi \le b$$

$$= -\frac{f^{(4)}(\eta)}{24} \left[ \frac{4}{15} h^5 \right] \cdot \quad h = \frac{b-a}{2} \quad \text{some} \quad \eta \in [a, b].$$

Thus

$$E_2(f) = -\frac{h^5}{90} f^{(4)}(\eta) \qquad \eta \in [a, b] \tag{5.1.15}$$

From this we see that $E_2(f) = 0$ if $f(x)$ is a polynomial of degree $\le 3$, even though quadratic interpolation is exact only if $f(x)$ is a polynomial of degree at most two. The additional fortuitous cancellation of errors is suggested in Figure 5.2. This results in Simpson's rule being much more accurate than the trapezoidal rule.

Again we create a composite rule. For $n \ge 2$ and even, define $h = (b - a)/n$, $x_j = a + jh$ for $j = 0, 1, \ldots, n$. Then

$$I(f) = \int_a^b f(x) \, dx = \sum_{j=1}^{n/2} \int_{x_{2j-2}}^{x_{2j}} f(x) \, dx$$

$$= \sum_{j=1}^{n/2} \left\{ \frac{h}{3} [f_{2j-2} + 4f_{2j-1} + f_{2j}] - \frac{h^5}{90} f^{(4)}(\eta_j) \right\}$$

with $x_{2j-2} \le \eta_j \le x_{2j}$. Simplifying the first terms in the sum, we obtain the composite Simpson rule:

$$I_n(f) = \frac{h}{3} [f_0 + 4f_1 + 2f_2 + 4f_3 + 2f_4 + \cdots + 2f_{n-2} + 4f_{n-1} + f_n] \tag{5.1.16}$$

As before, we will simply call this Simpson's rule. It is probably the most well-used numerical integration rule. It is simple, easy to use, and reasonably accurate for a wide variety of integrals.

For the error, as with the trapezoidal rule,

$$E_n(f) = I(f) - I_n(f) = -\frac{h^5(n/2)}{90} \cdot \frac{2}{n} \sum_{j=1}^{n/2} f^{(4)}(\eta_j)$$

$$E_n(f) = -\frac{h^4(b-a)}{180} f^{(4)}(\eta) \quad \text{some} \quad \eta \in [a, b] \qquad (5.1.17)$$

We can also derive the asymptotic error formula

$$E_n(f) = -\frac{h^4}{180} \left[ f^{(3)}(b) - f^{(3)}(a) \right] \equiv \tilde{E}_n(f) \qquad (5.1.18)$$

The proof is essentially the same as was used to obtain (5.1.9).

**Example**   We use Simpson's rule (5.1.16) to evaluate the integral (5.1.11),

$$I = \int_0^{\pi} e^x \cos(x)\, dx$$

used earlier as an example for the trapezoidal rule. The numerical results are given in Table 5.3. Again, the rate of decrease in the error confirms the results given by (5.1.17) and (5.1.18). Comparing with the earlier results in Table 5.1 for the trapezoidal rule, it is clear that Simpson's rule is superior. Comparing with Table 5.2, Simpson's rule is slightly inferior, but the speed of convergence is the same. Simpson's rule has the advantage of not requiring derivative values.

**Peano kernel error formulas**   There is another approach to deriving the error formulas, and it does not result in the derivative being evaluated at an unknown point $\eta$. We first consider the trapezoidal rule. Assume $f' \in C[a, b]$ and that $f''(x)$ is integrable on $[a, b]$. Then using Taylor's theorem [Theorem 1.4 in

**Table 5.3   Simpson's rule for evaluating (5.1.11)**

| $n$ | $I_n$ | $E_n$ | Ratio | $\tilde{E}_n$ |
|-----|-------|-------|-------|---------------|
| 2 | $-11.5928395534$ | $-4.78E-1$ | | $-1.63$ |
| 4 | $-11.9849440198$ | $-8.54E-2$ | 5.59 | $-1.02E-1$ |
| 8 | $-12.0642089572$ | $-6.14E-3$ | 14.9 | $-6.38E-3$ |
| 16 | $-12.0699513233$ | $-3.95E-4$ | 15.5 | $-3.99E-4$ |
| 32 | $-12.0703214561$ | $-2.49E-5$ | 15.9 | $-2.49E-5$ |
| 64 | $-12.0703447599$ | $-1.56E-6$ | 16.0 | $-1.56E-6$ |
| 128 | $-12.0703462191$ | $-9.73E-8$ | 16.0 | $-9.73E-8$ |

Chapter 1],

$$f(x) = p_1(x) + R_2(x) \qquad p_1(x) = f(a) + (x - a)f'(a)$$

$$R_2(x) = \int_a^x (x - t)f''(t)\, dt$$

Note that from (5.1.3),

$$E_1(F + G) = E_1(F) + E_1(G) \tag{5.1.19}$$

for any two functions $F, G \in C[a, b]$. Thus

$$E_1(f) = E_1(p_1) + E_1(R_2) = E_1(R_2)$$

since $E_1(p_1) = 0$ from (5.1.4). Substituting,

$$E_1(R_2) = \int_a^b R_2(x)\, dx - \left(\frac{b - a}{2}\right)[R_2(a) + R_2(b)]$$

$$= \int_a^b \int_a^x (x - t)f''(t)\, dt - \left(\frac{b - a}{2}\right)\int_a^b (b - t)f''(t)\, dt$$

In general for any integrable function $G(x, t)$,

$$\int_a^b \int_a^x G(x, t)\, dt\, dx = \int_a^b \int_t^b G(x, t)\, dx\, dt \tag{5.1.20}$$

Thus

$$E_1(R_2) = \int_a^b f''(t)\int_t^b (x - t)\, dx\, dt - \left(\frac{b - a}{2}\right)\int_a^b (b - t)f''(t)\, dt$$

Combining integrals and simplifying the results,

$$E_1(f) = \frac{1}{2}\int_a^b f''(t)(t - a)(t - b)\, dt \tag{5.1.21}$$

For the composite trapezoidal rule (5.1.5),

$$E_n(f) = \int_a^b K(t)f''(t)\, dt \tag{5.1.22}$$

$$K(t) = \frac{1}{2}(t - t_{j-1})(t - t_j) \qquad t_{j-1} \le t \le t_j \qquad j = 1, 2, \ldots, n \tag{5.1.23}$$

The formulas (5.1.21) and (5.1.22) are called the Peano kernel formulation of the error, and $K(t)$ is called the *Peano kernel*. For a more general presentation, see Davis (1963, chap. 3).

As a simple illustration of its use, take bounds in (5.1.22) to obtain

$$|E_n(f)| \leq \|K\|_\infty \int_a^b |f''(t)| \, dt = \frac{h^2}{8} \int_a^b |f''(t)| \, dt \qquad (5.1.24)$$

If $f''(t)$ is very peaked, this may give a better bound on the error than (5.1.7), because in (5.1.7) we generally must replace $|f''(\eta)|$ by $\|f''\|_\infty$.

For Simpson's rule, use Taylor's theorem to write

$$f(x) = p_3(x) + R_4(x)$$

$$R_4(x) = \frac{1}{6} \int_a^x (x - t)^3 f^{(4)}(t) \, dt$$

As before

$$E_2(f) = E_2(p_3) + E_2(R_4) = E_2(R_4)$$

and we then calculate $E_2(R_4)$ by direct substitution and simplification:

$$E_2(f) = \int_a^b R_4(x) \, dx - \frac{h}{3}\left[ R_4(a) + 4R_4\left(\frac{a+b}{2}\right) + R_4(b) \right]$$

This yields

$$E_2(f) = \int_a^b K(t) f^{(4)}(t) \, dt \qquad (5.1.25)$$

$$K(t) = \begin{cases} \dfrac{1}{72}(t - a)^3(3t - a - 2b) & a \leq t \leq \dfrac{a+b}{2} \\ \dfrac{1}{72}(b - t)^3(b + 2a - 3t) & \dfrac{a+b}{2} \leq t \leq b \end{cases} \qquad (5.1.26)$$

A graph of $K(t)$ is given in Figure 5.3. By direct evaluation,

$$\|K\|_\infty = \frac{h^4}{72} \qquad \int_a^b K(t) \, dt = -\frac{h^5}{90} \qquad h = \frac{b-a}{2}$$

As with the composite trapezoidal method, these results extend easily to the composite Simpson rule.

The following examples are intended to describe more fully the behavior of Simpson's and trapezoidal rules.

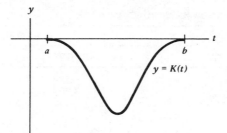

**Figure 5.3**    The Peano kernel for Simpson's rule.

*Example*    **1.**

$$f(x) = x^3 \sqrt{x}, \ [a, b] = [0, 1], \ I = \frac{2}{9}$$

Table 5.4 gives the error for increasing values of $n$. The derivative $f^{(4)}(x)$ is singular at $x = 0$, and thus the formula (5.1.17) cannot be applied to this case. As an alternative, we use the generalization of (5.1.25) to the composite Simpson rule to obtain

$$|E_n(f)| \leq \|K\|_\infty \int_a^b |f^{(4)}(t)| \, dt = \left(\frac{h^4}{72}\right)\left(\frac{105}{8}\right) = \frac{35}{192} h^4$$

Thus the error should decrease by a factor of 16 when $h$ is halved (i.e., $n$ is doubled). This also gives a fairly realistic bound on the error. Note the close agreement of the empirical values of ratio with the theoretically predicted values of 4 and 16, respectively.

**2.**

$$f(x) = \frac{1}{1 + (x - \pi)^2} \qquad [a, b] = [0, 5] \qquad I \doteq 2.33976628367$$

**Table 5.4    Trapezoidal, Simpson integration: case (1)**

| | Trapezoidal Rule | | Simpson's Rule | |
|---|---|---|---|---|
| $n$ | Error | Ratio | Error | Ratio |
| 2 | $-7.197 - 2$ | | $-3.370 - 3$ | |
| 4 | $-1.817 - 2$ | 3.96 | $-2.315 - 4$ | 14.6 |
| 8 | $-4.553 - 3$ | 3.99 | $-1.543 - 5$ | 15.0 |
| 16 | $-1.139 - 3$ | 4.00 | $-1.008 - 6$ | 15.3 |
| 32 | $-2.848 - 4$ | 4.00 | $-6.489 - 8$ | 15.5 |
| 64 | $-7.121 - 5$ | 4.00 | $-4.141 - 9$ | 15.7 |
| 128 | $-1.780 - 5$ | 4.00 | $-2.626 - 10$ | 15.8 |

**Table 5.5  Trapezoidal, Simpson integration: case (2)**

| | Trapezoidal Rule | | Simpson's Rule | |
|---|---|---|---|---|
| $n$ | Error | Ratio | Error | Ratio |
| 2 | 1.731 − 1 | | −2.853 − 1 | |
| | | 2.43 | | −7.69 |
| 4 | 7.110 − 2 | | 3.709 − 2 | |
| | | 9.48 | | −2.71 |
| 8 | 7.496 − 3 | | −1.371 − 2 | |
| | | 3.84 | | −130 |
| 16 | 1.953 − 3 | | 1.059 − 4 | |
| | | 3.99 | | 9.81 |
| 32 | 4.892 − 4 | | 1.080 − 6 | |
| | | 4.00 | | 16.0 |
| 64 | 1.223 − 4 | | 6.743 − 8 | |
| | | 4.00 | | 16.0 |
| 128 | 3.059 − 5 | | 4.217 − 9 | |

According to theory, the infinite differentiability of $f(x)$ implies a value for ratio of 4.0 and 16.0 for the trapezoidal and Simpson rules, respectively. But these need not hold for the first several values of $I_n$, as Table 5.5 shows. The integrand is relatively peaked, especially its higher derivatives, and this affects the speed of convergence.

**3.**

$$f(x) = \sqrt{x} \qquad [a, b] = [0, 1] \qquad I = \frac{2}{3}$$

Since $f'(x)$ has an infinite value at $x = 0$, none of the theoretical results given previously apply to this case. The numerical results are in Table 5.6; note that there is still a regular behavior to the error. In fact, the errors of the two methods decrease by the same ratio as $n$ is doubled. This ratio of $2^{1.5} \doteq 2.83$ is explained in Section 5.4, formula (5.4.24).

**4.**

$$f(x) = e^{\cos(x)} \qquad [a, b] = [0, 2\pi] \qquad I \doteq 7.95492652101284$$

The results are shown in Table 5.7, and they are extremely good. Both methods

**Table 5.6  Trapezoidal, Simpson integration: case (3)**

| | Trapezoidal Rule | | Simpson's Rule | |
|---|---|---|---|---|
| $n$ | Error | Ratio | Error | Ratio |
| 2 | 6.311 − 2 | | 2.860 − 2 | |
| | | 2.70 | | 2.82 |
| 4 | 2.338 − 2 | | 1.012 − 2 | |
| | | 2.74 | | 2.83 |
| 8 | 8.536 − 3 | | 3.587 − 3 | |
| | | 2.77 | | 2.83 |
| 16 | 3.085 − 3 | | 1.268 − 3 | |
| | | 2.78 | | 2.83 |
| 32 | 1.108 − 3 | | 4.485 − 4 | |
| | | 2.80 | | 2.83 |
| 64 | 3.959 − 4 | | 1.586 − 4 | |
| | | 2.81 | | 2.83 |
| 128 | 1.410 − 4 | | 5.606 − 5 | |

**Table 5.7    Trapezoidal, Simpson integration: case (4)**

| | Trapezoidal Rule | | Simpson's Rule | |
|---|---|---|---|---|
| $n$ | Error | Ratio | Error | Ratio |
| 2 | $-1.74$ | | $7.21E - 1$ | |
| 4 | $-3.44E - 2$ | $5.06E + 1$ | $5.34E - 1$ | $1.35$ |
| 8 | $-1.25E - 6$ | $2.75E + 4$ | $1.15E - 2$ | $4.64E + 1$ |
| 16 | $< 1.00E - 14$ | $> 1.25E + 8$ | $4.17E - 7$ | $2.76E + 4$ |
| | | | $< 1.00E - 14$ | $> 4.17E + 7$ |

are very rapidly convergent, with the trapezoidal rule superior to Simpson's rule. This illustrates the excellent convergence of the trapezoidal rule for periodic integrands; this is analyzed in Section 5.4. An indication of this behavior can be seen from the asymptotic error terms (5.1.9) and (5.1.18), since both estimates are zero in this case of $f(x)$.

## 5.2    Newton–Cotes Integration Formulas

The simple trapezoidal rule (5.1.2) and Simpson's rule (5.1.13) are the first two cases of the Newton–Cotes integration formula. For $n \geq 1$, let $h = (b - a)/n$, $x_j = a + jh$ for $j = 0, 1, \ldots, n$. Define $I_n(f)$ by replacing $f(x)$ by its interpolating polynomial $p_n(x)$ on the nodes $x_0, x_1, \ldots, x_n$:

$$I(f) = \int_a^b f(x)\, dx \doteq I_n(f) = \int_a^b p_n(x)\, dx \qquad (5.2.1)$$

Using the Lagrange formula (3.1.6) for $p_n(x)$,

$$I_n(f) = \int_a^b \sum_{j=0}^n l_{j,n}(x) f(x_j)\, dx = \sum_{j=0}^n w_{j,n} f(x_j) \qquad (5.2.2)$$

with

$$w_{j,n} = \int_a^b l_{j,n}(x)\, dx \quad j = 0, 1, \ldots, n \qquad (5.2.3)$$

Usually we suppress the subscript $n$ and write just $w_j$. We have already calculated the cases $n = 1$ and 2. To illustrate the calculation of the weights, we give the case of $w_0$ for $n = 3$.

$$w_0 = \int_a^b l_0(x)\, dx = \int_{x_0}^{x_3} \frac{(x - x_1)(x - x_2)(x - x_3)\, dx}{(x_0 - x_1)(x_0 - x_2)(x_0 - x_3)}$$

A change of variable simplifies the calculations. Let $x = x_0 + \mu h$, $0 \leq \mu \leq 3$.

Then

$$w_0 = -\frac{1}{6h^3}\int_{x_0}^{x_3}(x-x_1)(x-x_2)(x-x_3)\,dx$$

$$= -\frac{1}{6h^3}\int_0^3(\mu-1)h(\mu-2)h(\mu-3)h\cdot h\,d\mu$$

$$= -\frac{h}{6}\int_0^3(\mu-1)(\mu-2)(\mu-3)\,d\mu$$

$$w_0 = \frac{3h}{8}$$

The complete formula for $n = 3$ is

$$I_3(f) = \frac{3h}{8}[f(x_0) + 3f(x_1) + 3f(x_2) + f(x_3)] \qquad (5.2.4)$$

and is called the *three-eighths rule*.

For the error, we give the following theorem.

***Theorem 5.1*** **(a)**   For $n$ even, assume $f(x)$ is $n + 2$ times continuously differentiable on $[a, b]$. Then

$$I(f) - I_n(f) = C_n h^{n+3} f^{(n+2)}(\eta) \qquad \text{some} \quad \eta \in [a, b] \quad (5.2.5)$$

with

$$C_n = \frac{1}{(n+2)!}\int_0^n \mu^2(\mu-1)\cdots(\mu-n)\,d\mu \qquad (5.2.6)$$

**(b)**   For $n$ odd, assume $f(x)$ is $n + 1$ times continuously differentiable on $[a, b]$. Then

$$I(f) - I_n(f) = C_n h^{n+2} f^{(n+1)}(\eta) \qquad \text{some} \quad \eta \in [a, b] \quad (5.2.7)$$

with

$$C_n = \frac{1}{(n+1)!}\int_0^n \mu(\mu-1)\cdots(\mu-n)\,d\mu$$

***Proof***   We sketch the proof for part (a), the most important case. For complete proofs of both cases, see Isaacson and Keller (1966, pp. 308–314). From (3.2.11),

$$E_n(f) = I(f) - I_n(f)$$

$$= \int_a^b(x-x_0)(x-x_1)\cdots(x-x_n)f[x_0, x_1, \ldots, x_n, x]\,dx$$

Define

$$w(x) = \int_a^x (t - x_0) \cdots (t - x_n) \, dt$$

Then

$$w(a) = w(b) = 0 \qquad w(x) > 0 \qquad \text{for} \quad a < x < b$$

The proof that $w(x) > 0$ can be found in Isaacson and Keller (1966, p. 309). It is easy to show $w(b) = 0$, since the integrand $(t - x_0) \cdots (t - x_n)$ is an odd function with respect to the middle node $x_{n/2} = (a + b)/2$.

Using integration by parts and (3.2.17),

$$E_n(f) = \int_a^b w'(x) f[x_0, \ldots, x_n, x] \, dx$$

$$= [w(x) f[x_0, \ldots, x_n, x]]_a^b - \int_a^b w(x) \frac{d}{dx} f[x_0, \ldots, x_n, x] \, dx$$

$$E_n(f) = -\int_a^b w(x) f[x_0, \ldots, x_n, x, x] \, dx$$

Using the integral mean value theorem and (3.2.12),

$$E_n(f) = -f[x_0, \ldots, x_n, \xi, \xi] \int_a^b w(x) \, dx$$

$$= -\frac{f^{(n+2)}(\eta)}{(n + 2)!} \int_a^b \int_a^x (t - x_0) \cdots (t - x_n) \, dt \, dx \quad (5.2.8)$$

We change the order of integration and then use the change of variable $t = x_0 + \mu h$, $0 \le \mu \le n$:

$$\int_a^b w(x) \, dx = \int_a^b \int_t^b (t - x_0) \cdots (t - x_n) \, dx \, dt$$

$$= \int_{x_0}^{x_n} (t - x_0) \cdots (t - x_{n-1})(t - x_n)(x_n - t) \, dt$$

$$= -h^{n+3} \int_0^n \mu(\mu - 1) \cdots (\mu - n + 1)(\mu - n)^2 \, d\mu$$

Use the change of variable $\nu = n - \mu$ to give the result

$$\int_a^b w(x) \, dx = -h^{n+3} \int_0^n (n - \nu) \cdots (1 - \nu)\nu^2 \, d\nu$$

Use the fact that $n$ is even and combine the preceding with (5.2.8), to obtain the result (5.2.5)–(5.2.6).   ■

**Table 5.8    Commonly used Newton–Cotes formulas**

$n = 1$    $\int_a^b f(x)\, dx = \dfrac{h}{2}[f(a) + f(b)] - \dfrac{h^3}{12} f''(\xi)$    trapezoidal rule

$n = 2$    $\int_a^b f(x)\, dx = \dfrac{h}{3}\left[ f(a) + 4f\left(\dfrac{a+b}{2}\right) + f(b) \right] - \dfrac{h^5}{90} f^{(4)}(\xi)$    Simpson's rule

$n = 3$    $\int_a^b f(x)\, dx = \dfrac{3h}{8}[f(a) + 3f(a+h) + 3f(b-h) + f(b)] - \dfrac{3h^5}{80} f^{(4)}(\xi)$

$n = 4$    $\int_a^b f(x)\, dx = \dfrac{2h}{45}\left[ 7f(a) + 32f(a+h) + 12f\left(\dfrac{a+b}{2}\right) + 32f(b-h) + 7f(b) \right] - \dfrac{8h^7}{945} f^{(6)}(\xi)$

For easy reference, the most commonly used Newton–Cotes formulas are given in Table 5.8. For $n = 4$, $I_4(f)$ is often called *Boole's rule*. As previously, let $h = (b - a)/n$ in the table.

**Definition**    A numerical integration formula $\tilde{I}(f)$ that approximates $I(f)$ is said to have *degree of precision m* if

1.    $\tilde{I}(f) = I(f)$ for all polynomials $f(x)$ of degree $\le m$.

2.    $\tilde{I}(f) \ne I(f)$ for some polynomial $f$ of degree $m + 1$.

**Example**    With $n = 1, 3$ in Table 5.8, the degrees of precision are also $m = n = 1, 3$, respectively. But with $n = 2, 4$, the degrees of precision are ($m = n + 1 = 3, 5$, respectively. This illustrates the general result that Newton–Cotes formulas with an even index $n$ gain an extra degree of precision as compared with those of an odd index [see formulas (5.2.5) and (5.2.7)].

Each Newton–Cotes formula can be used to construct a composite rule. The most useful remaining one is probably that based on Boole's rule (see Problem 7). We omit any further details.

**Convergence discussion**    The next question of interest is whether $I_n(f)$ converges to $I(f)$ as $n \to \infty$. Given the lack of convergence of the interpolation polynomials on evenly spaced nodes for some choices of $f(x)$ [see (3.5.10)], we should expect some difficulties. Table 5.9 gives the results for a well-known example,

$$I = \int_{-4}^4 \frac{dx}{1 + x^2} = 2 \cdot \tan^{-1}(4) \doteq 2.6516 \qquad (5.2.9)$$

These Newton–Cotes numerical integrals are diverging; and this illustrates the fact that the Newton–Cotes integration formulas $I_n(f)$ in (5.2.2), need not converge to $I(f)$.

To understand the implications of the lack of convergence of Newton–Cotes quadrature for (5.2.9), we first give a general discussion of the convergence of numerical integration methods.

**Table 5.9   Newton–Cotes example
(5.2.9)**

| $n$ | $I_n$ |
|---|---|
| 2 | 5.4902 |
| 4 | 2.2776 |
| 6 | 3.3288 |
| 8 | 1.9411 |
| 10 | 3.5956 |

. *Definition*   Let $\mathscr{F}$ be a family of continuous functions on a given interval $[a, b]$. We say $\mathscr{F}$ is *dense* in $C[a, b]$ if for every $f \in C[a, b]$ and every $\epsilon > 0$, there is a function $f_\epsilon$ in $\mathscr{F}$ for which

$$\underset{a \le x \le b}{\text{Max}} |f(x) - f_\epsilon(x)| \le \epsilon \qquad (5.2.10)$$

*Example* 1.   From the Weierstrass theorem [see Theorem 4.1], the set of all polynomials is dense in $C[a, b]$.

2.   Let $n \ge 1$, $h = (b - a)/n$, $x_j = a + jh$, $0 \le j \le n$. Let $f(x)$ be linear on each of the subintervals $[x_{j-1}, x_j]$. Define $\mathscr{F}$ to be the set of all such piecewise linear functions $f(x)$ for all $n$. We leave to Problem 11 the proof that $\mathscr{F}$ is dense in $C[a, b]$.

*Theorem 5.2*   Let

$$I_n(f) = \sum_{j=0}^{n} w_{j,n} f(x_{j,n}) \qquad n \ge 1$$

be a sequence of numerical integration formulas that approximate

$$I(f) = \int_a^b f(x) \, dx$$

Let $\mathscr{F}$ be a family dense in $C[a, b]$. Then

$$I_n(f) \to I(f) \qquad \text{all} \quad f \in C[a, b] \qquad (5.2.11)$$

if and only if

1.   $\qquad\qquad I_n(f) \to I(f) \qquad \text{all} \quad f \in \mathscr{F} \qquad (5.2.12)$

and

2.   $\qquad\qquad B \equiv \underset{n \ge 1}{\text{Supremum}} \sum_{j=0}^{n} |w_{j,n}| < \infty \qquad (5.2.13)$

***Proof***   **(a)**   Trivially, (5.2.11) implies (5.2.12). But the proof that (5.2.11) implies (5.2.13) is much more difficult. It is an example of the *principle of uniform boundedness*, and it can be found in almost any text on functional analysis; for example, see Cryer (1982, p. 121).

**(b)**   We now prove that (5.2.12) and (5.2.13) implies (5.2.11). Let $f \in C[a, b]$ be given, and let $\epsilon > 0$ be arbitrary. Using the assumption that $\mathscr{F}$ is dense in $C[a, b]$, pick $f_\epsilon \in \mathscr{F}$ such that

$$\underset{a \leq x \leq b}{\text{Max}} |f(x) - f_\epsilon(x)| \leq \frac{\epsilon}{[2(b - a + B)]} \qquad (5.2.14)$$

Then write

$$I(f) - I_n(f) = [I(f) - I(f_\epsilon)] + [I(f_\epsilon) - I_n(f_\epsilon)]$$
$$+ [I_n(f_\epsilon) - I_n(f)]$$

It is straightforward to derive, using (5.2.13) and (5.2.14), that

$$|I(f) - I_n(f)| \leq |I(f) - I(f_\epsilon)| + |I(f_\epsilon) - I_n(f_\epsilon)|$$
$$+ |I_n(f_\epsilon) - I_n(f)|$$

$$\leq \frac{\epsilon}{2} + |I(f_\epsilon) - I_n(f_\epsilon)|$$

Using (5.2.12), $I_n(f_\epsilon) \to I(f_\epsilon)$ as $n \to \infty$. Thus for all sufficiently large $n$, say $n \geq n_\epsilon$,

$$|I(f) - I_n(f)| \leq \epsilon$$

Since $\epsilon$ was arbitrary, this shows $I_n(f) \to I(f)$ as $n \to \infty$.   ∎

Since the Newton–Cotes numerical integrals $I_n(f)$ do not converge to $I(f)$ for $f(x) = 1/(1 + x^2)$ on $[-4, 4]$, it must follow that either condition (5.2.12) or (5.2.13) is violated. If we choose $\mathscr{F}$ as the polynomials, then (5.2.12) is satisfied, since $I_n(p) = I(p)$ for any polynomial $p$ of degree $\leq n$. Thus (5.2.13) must be false. For the Newton–Cotes formulas (5.2.2),

$$\underset{n}{\text{Supremum}} \sum_{j=0}^{n} |w_{j,n}| = \infty \qquad (5.2.15)$$

Since $I(f) = I_n(f)$ for the special case $f(x) \equiv 1$, for any $n$, we have

$$\sum_{j=0}^{n} w_{j,n} = b - a \qquad n \geq 1 \qquad (5.2.16)$$

Combining these results, the weights $w_{j,n}$ must vary in sign as $n$ becomes sufficiently large. For example, using $n = 8$,

$$\int_{x_0}^{x_8} f(x)\, dx \doteq I_8(f) = \frac{4h}{14,175} [989(f_0 + f_8) + 5,888(f_1 + f_7)$$

$$- 928(f_2 + f_6) + 10,496(f_3 + f_5) - 4,540 f_4]$$

Such formulas can cause loss-of-significance errors, although it is unlikely to be a serious problem until $n$ is larger. But because of this problem, people have generally avoided using Newton–Cotes formulas for $n \geq 8$, even in forming composite formulas.

The most serious problem of the Newton–Cotes method (5.2.2) is that it may not converge for perfectly well-behaved integrands, as in (5.2.9).

**The midpoint rule**    There are additional Newton–Cotes formulas in which one or both of the endpoints of integration are deleted from the interpolation (and integration) node points. The best known of these is also the simplest, the *midpoint rule*. It is based on interpolation of the integrand $f(x)$ by the constant $f((a + b)/2)$; and the resulting integration formula is

$$\int_a^b f(x)\, dx = (b - a)f\left(\frac{a + b}{2}\right) + \frac{(b - a)^3}{24} f''(\eta) \qquad \text{some} \quad \eta \in [a, b]$$

$$(5.2.17)$$

For its composite form, define

$$x_j = a + \left(j - \tfrac{1}{2}\right)h \qquad j = 1, 2, \ldots, n$$

the midpoints of the intervals $[a + (j - 1)h, a + jh]$. Then

$$\int_a^b f(x)\, dx = I_n(f) + E_n(f)$$

$$I_n(f) = h[f_1 + f_2 + \cdots + f_n] \qquad\qquad (5.2.18)$$

$$E_n(f) = \frac{h^2(b - a)}{24} f''(\eta) \qquad \text{some} \quad \eta \in [a, b] \qquad (5.2.19)$$

The proof of these results is left as Problem 10.

These integration formulas in which one or both endpoints are missing are called *open Newton–Cotes formulas*, and the previous formulas are called *closed formulas*. The open formulas of higher order were used classically in deriving numerical formulas for the solution of ordinary differential equations.

## 5.3   Gaussian Quadrature

The composite trapezoidal and Simpson rules are based on using a low-order polynomial approximation of the integrand $f(x)$ on subintervals of decreasing size. In this section, we investigate a class of methods that use polynomial approximations of $f(x)$ of increasing degree. The resulting integration formulas are extremely accurately in most cases, and they should be considered seriously by anyone faced with many integrals to evaluate.

For greater generality, we will consider formulas

$$I_n(f) = \sum_{j=1}^{n} w_{j,n} f(x_{j,n}) \doteq \int_a^b w(x) f(x) \, dx = I(f) \qquad (5.3.1)$$

The weight function $w(x)$ is assumed to be nonnegative and integrable on $[a, b]$, and it is to also satisfy the hypotheses (4.3.8) and (4.3.9) of Section 4.3. The nodes $\{x_{j,n}\}$ and weights $\{w_{j,n}\}$ are to be chosen so that $I_n(f)$ equals $I(f)$ exactly for polynomials $f(x)$ of as large a degree as possible. It is hoped that this will result in a formula $I_n(f)$ that is nearly exact for integrands $f(x)$ that are well approximable by polynomials. In Section 5.2, the Newton–Cotes formulas have an increasing degree of precision as $n$ increased, but nonetheless they do not converge for many well-behaved integrands. The difficulty with the Newton–Cotes formulas is that the nodes $\{x_{j,n}\}$ must be evenly spaced. By omitting this restriction, we will be able to obtain new formulas $I_n(f)$ that converge for all $f \in C[a, b]$.

To obtain some intuition for the determination of $I_n(f)$, consider the special case

$$\int_{-1}^{1} f(x) \, dx \doteq \sum_{j=1}^{n} w_j f(x_j) \qquad (5.3.2)$$

where $w(x) \equiv 1$ and the explicit dependence of $\{w_j\}$ and $\{x_j\}$ on $n$ has been dropped. The weights $\{w_j\}$ and nodes $\{x_j\}$ are to be determined to make the error

$$E_n(f) = \int_{-1}^{1} f(x) \, dx - \sum_{j=1}^{n} w_j f(x_j) \qquad (5.3.3)$$

equal zero for as high a degree polynomial $f(x)$ as possible. To derive equations for the nodes and weights, we first note that

$$E_n(a_0 + a_1 x + \cdots + a_m x^m) = a_0 E_n(1) + a_1 E_n(x) + \cdots + a_m E_n(x^m) \quad (5.3.4)$$

Thus $E_n(f) = 0$ for every polynomial of degree $\leq m$ if and only if

$$E_n(x^i) = 0 \qquad i = 0, 1, \ldots, m \qquad (5.3.5)$$

***Case 1.*** $n = 1$. Since there are two parameters, $w_1$ and $x_1$, we consider requiring

$$E_n(1) = 0 \qquad E_n(x) = 0$$

This gives

$$\int_{-1}^{1} 1 \, dx - w_1 = 0 \qquad \int_{-1}^{1} x \, dx - w_1 x_1 = 0$$

This implies $w_1 = 2$ and $x_1 = 0$. Thus the formula (5.3.2) becomes

$$\int_{-1}^{1} f(x) \, dx \doteq 2f(0)$$

the midpoint rule.

***Case 2.*** $n = 2$. There are four parameters, $w_1, w_2, x_1, x_2$, and thus we put four constraints on these parameters:

$$E_n(x^i) = \int_{-1}^{1} x^i \, dx - \left[ w_1 x_1^i + w_2 x_2^i \right] = 0 \qquad i = 0,1,2,3$$

or

$$w_1 + w_2 = 2$$

$$w_1 x_1 + w_2 x_2 = 0$$

$$w_1 x_1^2 + w_2 x_2^2 = \tfrac{2}{3}$$

$$w_1 x_1^3 + w_2 x_2^3 = 0$$

These equations lead to the unique formula

$$\int_{-1}^{1} f(x) \, dx \doteq f\left( -\frac{\sqrt{3}}{3} \right) + f\left( \frac{\sqrt{3}}{3} \right) \qquad (5.3.6)$$

which has degree of precision three. Compare this with Simpson's rule (5.1.13), which uses three nodes to attain the same degree of precision.

***Case 3.*** For a general $n$ there are $2n$ free parameters $\{x_i\}$ and $\{w_i\}$, and we would guess that there is a formula (5.3.2) that uses $n$ nodes and gives a degree of precision of $2n - 1$. The equations to be solved are

$$E_n(x^i) = 0 \qquad i = 0,1,\ldots,2n-1$$

or

$$\sum_{j=1}^{n} w_j x_j^i = \begin{cases} 0 & i = 1,3,\ldots,2n-1 \\ \dfrac{2}{i+1} & i = 0,2,\ldots,2n-2 \end{cases} \qquad (5.3.7)$$

These are nonlinear equations, and their solvability is not at all obvious. Because of the difficulty in working with this nonlinear system, we use another approach to the theory for (5.3.2), one that is somewhat circuitous.

Let $\{\varphi_n(x)|n \geq 0\}$ be the orthogonal polynomials on $(a, b)$ with respect to the weight function $w(x) \geq 0$. Denote the zeros of $\varphi_n(x)$ by

$$a < x_1 < \cdots < x_n < b \tag{5.3.8}$$

Also, recall the notation from (4.4.18)–(4.4.20):

$$\varphi_n(x) = A_n x^n + \cdots \qquad a_n = \frac{A_{n+1}}{A_n}$$

$$\gamma_n = \int_a^b w(x)[\varphi_n(x)]^2 \, dx \tag{5.3.9}$$

**Theorem 5.3**    For each $n \geq 1$, there is a unique numerical integration formula (5.3.1) of degree of precision $2n - 1$. Assuming $f(x)$ is $2n$ times continuously differentiable on $[a, b]$, the formula for $I_n(f)$ and its error is given by

$$\int_a^b w(x)f(x) \, dx = \sum_{j=1}^n w_j f(x_j) + \frac{\gamma_n}{A_n^2(2n)!} f^{(2n)}(\eta) \tag{5.3.10}$$

for some $a < \eta < b$. The nodes $\{x_j\}$ are the zeros of $\varphi_n(x)$, and the weights $\{w_j\}$ are given by

$$w_j = \frac{-a_n \gamma_n}{\varphi_n'(x_j)\varphi_{n+1}(x_j)} \qquad j = 1, \ldots, n \tag{5.3.11}$$

**Proof**    The proof is divided into three parts. We first obtain a formula with degree of precision $2n - 1$, using the nodes (5.3.8). We then show that it is unique. Finally, we sketch the derivation of the error formula and the weights.

**(a)**    Construction of the formula. Hermite interpolation is used as the vehicle for the construction (see Section 3.6 to review the notation and results). For the nodes in (5.3.8), the Hermite polynomial interpolating $f(x)$ and $f'(x)$ is

$$H_n(x) = \sum_{j=1}^n f(x_j)h_j(x) + \sum_{j=1}^n f'(x_j)\tilde{h}_j(x) \tag{5.3.12}$$

with $h_j(x)$ and $\tilde{h}_j(x)$ defined in (3.6.2) of Section 3.6. The interpolation error is given by

$$\mathcal{E}_n(x) = f(x) - H_n(x) = [\psi_n(x)]^2 f[x_1, x_1, \ldots, x_n, x_n, x]$$

$$= \frac{[\psi_n(x)]^2}{(2n)!} f^{(2n)}(\xi) \qquad \xi \in [a, b] \tag{5.3.13}$$

with

$$\psi_n(x) = (x - x_1) \cdots (x - x_n)$$

Note that

$$\psi_n(x) = \frac{\varphi_n(x)}{A_n} \qquad (5.3.14)$$

since both $\varphi_n(x)$ and $\psi_n(x)$ are of degree $n$ and have the same zeros.

Using (5.3.12), if $f(x)$ is continuously differentiable, then

$$\int_a^b w(x)f(x)\, dx = \int_a^b w(x)H_n(x)\, dx + \int_a^b w(x)\mathscr{E}_n(x)\, dx$$

$$\equiv I_n(f) + E_n(f) \qquad (5.3.15)$$

The degree of precision is at least $2n - 1$, since $\mathscr{E}_n(x) = 0$ if $f(x)$ is a polynomial of degree $< 2n$, from (5.3.13). Also from (5.3.13),

$$E_n(x^{2n}) = \int_a^b w(x)\mathscr{E}_n(x)\, dx = \int_a^b w(x)[\psi_n(x)]^2\, dx > 0 \quad (5.3.16)$$

Thus the degree of precision of $I_n(f)$ is exactly $2n - 1$.

To derive a simpler formula for $I_n(f)$,

$$I_n(f) = \sum_{j=1}^n f(x_j) \int_a^b w(x)h_j(x)\, dx + \sum_{j=1}^n f'(x_j) \int_a^b w(x)\tilde{h}_j(x)\, dx$$

$$(5.3.17)$$

we show that all of the integrals in the second sum are zero. Recall that from (3.6.2),

$$\tilde{h}_j(x) = (x - x_j)\left[l_j(x)\right]^2$$

$$l_j(x) = \frac{\psi_n(x)}{(x - x_j)\psi_n'(x_j)} = \frac{\varphi_n(x)}{(x - x_j)\varphi_n'(x_j)}$$

The last step uses (5.3.14). Thus

$$\tilde{h}_j(x) = (x - x_j)l_j(x)l_j(x) = \frac{\varphi_n(x)l_j(x)}{\varphi_n'(x_j)} \qquad (5.3.18)$$

Since degree $(l_j) = n - 1$, and since $\varphi_n(x)$ is orthogonal to all polynomials of degree $< n$, we have

$$\int_a^b w(x)\tilde{h}_j(x)\, dx = \frac{1}{\varphi_n'(x_j)} \int_a^b w(x)\varphi_n(x)l_j(x)\, dx = 0 \qquad j = 1, \ldots, n$$

$$(5.3.19)$$

The integration formula (5.3.15) becomes

$$\int_a^b w(x)f(x)\, dx = \sum_{j=1}^n w_j f(x_j) + E_n(f)$$

$$w_j = \int_a^b w(x) h_j(x)\, dx \qquad j = 1, \dots, n \qquad (5.3.20)$$

**(b)**   Uniqueness of formula (5.3.19). Suppose that we have a numerical integration formula

$$\int_a^b w(x)f(x)\, dx \doteq \sum_{j=1}^n v_j f(z_j) \qquad (5.3.21)$$

that has degree of precision $\geq 2n - 1$. Construct the Hermite interpolation formula to $f(x)$ at the nodes $z_1, \dots, z_n$. Then for any polynomial $f(x)$ of degree $\leq 2n - 1$,

$$f(x) = \sum_{j=1}^n f(z_j) h_j(x) + \sum_{j=1}^n f'(z_j)\tilde{h}_j(x) \qquad \deg(f) \leq 2n-1 \quad (5.3.22)$$

where $h_j(x)$ and $\tilde{h}_j(x)$ are defined using $\{z_j\}$. Multiply (5.3.22) by $w(x)$, use the assumption on the degree of precision of (5.3.21), and integrate to get

$$\sum_{j=1}^n v_j f(z_j) = \sum_{j=1}^n f(z_j) \int_a^b w(x) h_j(x)\, dx + \sum_{j=1}^n f'(z_j) \int_a^b w(x)\tilde{h}_j(x)\, dx$$

$$(5.3.23)$$

for any polynomial $f(x)$ of degree $\leq 2n - 1$.

Let $f(x) = \tilde{h}_i(x)$. Use the properties (3.6.3) of $\tilde{h}_i(x)$ to obtain from (5.3.23) that

$$0 = \int_a^b w(x)\tilde{h}_i(x)\, dx \qquad i = 1, \dots, n \qquad (5.3.24)$$

As before in (5.3.18), we can write

$$\tilde{h}_i(x) = (x - z_i)[l_i(x)]^2 = \frac{l_i(x)\omega_n(x)}{\omega_n'(z_i)}$$

$$\omega(x) = (x - z_1) \cdots (x - z_n)$$

Then (5.3.24) becomes

$$\int_a^b w(x)\omega_n(x) l_i(x)\, dx = 0 \qquad i = 1, 2, \dots, n$$

Since all polynomials of degree $\leq n - 1$ can be written as a combination of $l_1(x), \ldots, l_n(x)$, we have that $\omega_n(x)$ is orthogonal to every polynomial of degree $\leq n - 1$. Using the uniqueness of the orthogonal polynomials [from Theorem 4.2], $\omega_n(x)$ must be a constant multiple of $\varphi_n(x)$. Thus they must have the same zeros, and

$$z_i = x_i \qquad i = 1, \ldots, n$$

To complete the proof of the uniqueness, we must show that $w_i = v_i$, where $v_i$ is the weight in (5.3.21) and $w_i$ in (5.3.10). Use (5.3.23) with (5.3.24) and $f(x) = h_i(x)$. The result will follow immediately, since $h_i(x)$ is now constructed using $\{x_i\}$.

**(c)** The error formula. We begin by deducing some further useful properties about the weights $\{w_i\}$ in (5.3.10). Using the definition (3.6.2) of $h_i(x)$,

$$w_i = \int_a^b w(x) h_i(x)\, dx = \int_a^b w(x)[1 - 2l_i'(x_i)(x - x_i)][l_i(x)]^2\, dx$$

$$= \int_a^b w(x)[l_i(x)]^2\, dx - 2l_i'(x_i) \int_a^b w(x)(x - x_i)[l_i(x)]^2\, dx$$

The last integral is zero from (5.3.19), since $\tilde{h}_i(x) = (x - x_i)[l_i(x)]^2$. Thus

$$w_i = \int_a^b w(x)[l_i(x)]^2\, dx > 0 \qquad i = 1, 2, \ldots, n \qquad (5.3.25)$$

and all the weights are positive, for all $n$.

To construct $w_i$, begin by substituting $f(x) = l_i(x)$ into (5.3.20), and note that $E_n(f) = 0$, since degree $(l_i) = n - 1$. Then using $l_i(x_j) = \delta_{ij}$, we have

$$w_i = \int_a^b w(x) l_i(x)\, dx \qquad i = 1, \ldots, n \qquad (5.3.26)$$

To further simplify the formula, the Christoffel–Darboux identity (Theorem 4.6) can be used, followed by much manipulation, to give the formula (5.3.11). For the details, see Isaacson and Keller (1966, pp. 333–334).

For the integration error, if $f(x)$ is $2n$ times continuously differentiable on $[a, b]$, then

$$E_n(f) = \int_a^b w(x) \mathscr{E}_n(x)\, dx$$

$$= \int_a^b w(x)[\psi_n(x)]^2 f[x_1, x_1, \ldots, x_n, x_n, x]\, dx$$

$$= f[x_1, x_1, \ldots, x_n, x_n, \xi] \int_a^b w(x)[\psi_n(x)]^2\, dx \qquad \text{some } \xi \in [a, b]$$

the last step using the integral mean value theorem. Using (5.3.14) in the last integral, and replacing the divided difference by a derivative, we have

$$E_n(f) = \frac{f^{(2n)}(\eta)}{(2n)!} \int_a^b w(x) \frac{[\varphi_n(x)]^2}{A_n^2} \, dx$$

which gives the error formula in (5.3.10). ∎

**Gauss–Legendre quadrature**  For $w(x) \equiv 1$, the Gaussian formula on $[-1, 1]$ is given by

$$\int_{-1}^1 f(x) \, dx \doteq \sum_{j=1}^n w_j f(x_j) \qquad (5.3.27)$$

with the nodes equal to the zeros of the degree $n$ Legendre polynomial $P_n(x)$ on $[-1, 1]$. The weights are

$$w_i = \frac{-2}{(n+1)P_n'(x_i)P_{n+1}(x_i)} \qquad i = 1, 2, \ldots, n \qquad (5.3.28)$$

and

$$E_n(f) = \frac{2^{2n+1}(n!)^4}{(2n+1)[(2n)!]^2} \cdot \frac{f^{(2n)}(\eta)}{(2n)!} \equiv e_n \frac{f^{(2n)}(\eta)}{(2n)!}, \qquad (5.3.29)$$

**Table 5.10   Gauss–Legendre nodes and weights**

| $n$ | $x_i$ | $w_i$ |
|---|---|---|
| 2 | $\pm.5773502692$ | 1.0 |
| 3 | $\pm.7745966692$ | .5555555556 |
|   | 0.0 | .8888888889 |
| 4 | $\pm.8611363116$ | .3478546451 |
|   | $\pm.3399810436$ | .6521451549 |
| 5 | $\pm.9061798459$ | .2369268851 |
|   | $\pm.5384693101$ | .4786286705 |
|   | 0.0 | .5688888889 |
| 6 | $\pm.9324695142$ | .1713244924 |
|   | $\pm.6612093865$ | .3607615730 |
|   | $\pm.2386191861$ | .4679139346 |
| 7 | $\pm.9491079123$ | .1294849662 |
|   | $\pm.7415311856$ | .2797053915 |
|   | $\pm.4058451514$ | .3818300505 |
|   | 0.0 | .4179591837 |
| 8 | $\pm.9602898565$ | .1012285363 |
|   | $\pm.7966664774$ | .2223810345 |
|   | $\pm.5255324099$ | .3137066459 |
|   | $\pm.1834346425$ | .3626837834 |

**Table 5.11    Gaussian quadrature for (5.1.11)**

| $n$ | $I_n$ | $I - I_n$ |
|---|---|---|
| 2 | $-12.33621046570$ | $2.66E - 1$ |
| 3 | $-12.12742045017$ | $5.71E - 2$ |
| 4 | $-12.07018949029$ | $-1.57E - 4$ |
| 5 | $-12.07032853589$ | $-1.78E - 5$ |
| 6 | $-12.07034633110$ | $1.47E - 8$ |
| 7 | $-12.07034631753$ | $1.14E - 9$ |
| 8 | $-12.07034631639$ | $-4.25E - 13$ |

for some $-1 < \eta < 1$. For integrals on other finite intervals with weight function $w(x) \equiv 1$, use the following linear change of variables:

$$\int_a^b f(t)\, dt = \left( \frac{b - a}{2} \right) \int_{-1}^1 f\left( \frac{a + b + x(b - a)}{2} \right) dx \qquad (5.3.30)$$

reducing the integral to the standard interval $[-1, 1]$.

For convenience, we include Table 5.10, which gives the nodes and weights for formula (5.3.27) with small values of $n$. For larger values of $n$, see the very complete tables in Stroud and Secrest (1966), which go up to $n = 512$.

***Example***    Evaluate the integral (5.1.11),

$$I = \int_0^\pi e^x \cos(x)\, dx \doteq -12.0703463164$$

which was used previously in Section 5.1 as an example for the trapezoidal rule (see Table 5.1) and Simpson's rule (see Table 5.3). The results given in Table 5.11 show the marked superiority of Gaussian quadrature.

**A general error result**    We give a useful result in trying to explain the excellent convergence of Gaussian quadrature. In the next subsection, we consider in more detail the error in Gauss–Legendre quadrature.

***Theorem 5.4***    Assume $[a, b]$ is finite. Then the error in Gaussian quadrature,

$$E_n(f) = \int_a^b w(x) f(x)\, dx - \sum_{j=1}^n w_j f(x_j)$$

satisfies

$$|E_n(f)| \le 2 \left[ \int_a^b w(x)\, dx \right] \rho_{2n-1}(f) \qquad n \ge 1 \quad (5.3.31)$$

with $\rho_{2n-1}(f)$ the minimax error from (4.2.1).

**Proof**  $E_n(p) = 0$ for any polynomial $p(x)$ of degree $\leq 2n - 1$. Also, the error function $E_n$ satisfies

$$E_n(F + G) = E_n(F) + E_n(G)$$

for all $F, G \in C[a, b]$. Let $p(x) = q^*_{2n-1}(x)$, the minimax approximation of degree $\leq 2n - 1$ to $f(x)$ on $[a, b]$. Then

$$E_n(f) = E_n(f) - E_n(q^*_{2n-1}) = E_n(f - q^*_{2n-1})$$

$$= \int_a^b w(x)[f(x) - q^*_{2n-1}(x)]\, dx - \sum_{j=1}^{n} w_j[f(x_j) - q^*_{2n-1}(x_j)]$$

$$|E_n(f)| \leq \|f - q^*_{2n-1}\|_\infty \left[ \int_a^b w(x) + \sum_{j=1}^{n} |w_j| \right]$$

From (5.3.25), all $w_j > 0$. Also, since $p(x) \equiv 1$ is of degree 0,

$$\sum_{j=1}^{n} w_j = \int_a^b w(x)\, dx$$

This completes the proof of (5.3.31). ∎

From the results in Sections 4.6 and 4.7, the speed of convergence to zero of $\rho_m(f)$ increases with the smoothness of the integrand. From (5.3.31), the same is true of Gaussian quadrature. In contrast, the composite trapezoidal rule will usually not converge faster than order $h^2$ [in particular, if $f'(b) - f'(a) \neq 0$], regardless of the smoothness of $f(x)$. Gaussian quadrature takes advantage of additional smoothness in the integrand, in contrast to most composite rules.

**Example**  Consider using Gauss–Legendre quadrature to integrate

$$I = \int_0^1 e^{-x^2}\, dx \doteq .746824132812 \tag{5.3.32}$$

Table 5.12 contains error bounds based on (5.3.31),

$$|E_n(f)| \leq 2\rho_{2n-1}(f) \tag{5.3.33}$$

**Table 5.12  Gaussian quadrature of (5.3.32)**

| $n$ | $E_n(f)$ | (5.3.33) |
|---|---|---|
| 1 | $-3.20\mathrm{E} - 2$ | $1.06\mathrm{E} - 1$ |
| 2 | $2.29\mathrm{E} - 4$ | $1.33\mathrm{E} - 3$ |
| 3 | $9.55\mathrm{E} - 6$ | $3.24\mathrm{E} - 5$ |
| 4 | $-3.35\mathrm{E} - 7$ | $9.24\mathrm{E} - 7$ |
| 5 | $6.05\mathrm{E} - 9$ | $1.61\mathrm{E} - 8$ |

along with the true error. The error bound is of approximately the same magnitude as the true error.

**Discussion of Gauss–Legendre quadrature**  We begin by trying to make the error term (5.3.29) more understandable. First define

$$M_m = \operatorname*{Max}_{-1 \le x \le 1} \frac{|f^{(m)}(x)|}{m!} \qquad m \ge 0 \qquad (5.3.34)$$

For a large class of infinitely differentiable functions $f$ on $[-1,1]$, we have Supremum$_{m \ge 0} M_m < \infty$. For example, this will be true if $f(z)$ is analytic on the region $R$ of the complex plane defined by

$$R = \{ z : |z - x| \le 1 \text{ for some } x, -1 \le x \le 1 \}$$

With many functions, $M_m \to 0$ as $m \to \infty$, for example, $f(x) = e^x$ and $\cos(x)$. Combining (5.3.29) and (5.3.34), we obtain

$$|E_n(f)| \le e_n M_{2n} \qquad n \ge 1 \qquad (5.3.35)$$

and the size of $e_n$ is essential in examining the speed of convergence.

The term $e_n$ can be made more understandable by estimating it using Stirling's formula,

$$n! \doteq e^{-n} n^n \sqrt{2\pi n}$$

which is true in a relative error sense as $n \to \infty$. Then we obtain

$$e_n \doteq \frac{\pi}{4^n} \qquad \text{as} \quad n \to \infty \qquad (5.3.36)$$

This is quite a good estimate. For example, $e_5 = .00293$, and (5.3.36) gives the estimate .00307. Combined with (5.3.35), this implies

$$|E_n(f)| \le \frac{\pi}{4^n} \cdot M_{2n} \qquad (5.3.37)$$

which is a correct bound in an asymptotic sense as $n \to \infty$. This shows that $E_n(f) \to 0$ with an exponential rate of decrease as a function of $n$. Compare this with the polynomial rates of $1/n^2$ and $1/n^4$ for the trapezoidal and Simpson rules, respectively.

In order to consider integrands that are not infinitely differentiable, we can use the Peano kernel form of the error, just as in Section 5.1 for Simpson's and the trapezoidal rules. If $f(x)$ is $r$ times differentiable on $[-1,1]$, with $f^{(r)}(x)$ integrable on $[-1,1]$, then

$$E_n(f) = \int_{-1}^{1} K_{n,r}(t) f^{(r)}(t) \, dt \qquad n > \frac{r}{2} \qquad (5.3.38)$$

**Table 5.13   Error constants $e_{n,r}$ for (5.3.39)**

| $n$ | $e_{n,2}$ | Ratio | $e_{n,4}$ | Ratio |
|---|---|---|---|---|
| 2 | .162 | | .178 | |
| 4 | .437E − 1 | 3.7 | .647E − 2 | 27.5 |
| 8 | .118E − 1 | 3.7 | .417E − 3 | 15.5 |
| 16 | .311E − 2 | 3.8 | .279E − 4 | 14.9 |
| 32 | .800E − 3 | 3.9 | .183E − 5 | 15.3 |
| 64 | .203E − 3 | 3.9 | | |
| 128 | .511E − 4 | 4.0 | | |

for an appropriate Peano kernel $K_{n,r}(t)$. The procedure for constructing $K_{n,r}(t)$ is exactly the same as with the Peano kernels (5.1.21) and (5.1.25) in Section 5.1. From (5.3.38)

$$|E_n(f)| \le e_{n,r}M_r$$

$$e_{n,r} = r!\int_{-1}^{1}|K_{n,r}(t)|\,dt \qquad (5.3.39)$$

The values of $e_{n,r}$ given in Table 5.13 are taken from Stroud–Secrest (1966, pp. 152–153). The table shows that for $f$ twice continuously differentiable, Gaussian quadrature converges at least as rapidly as the trapezoidal rule (5.1.5). Using (5.3.39) and the table, we can construct the empirical bound

$$|E_n(f)| \le \frac{.42}{n^2}\left[\ \underset{-1\le x\le 1}{\text{Max}}\ |f''(x)|\right] \qquad (5.3.40)$$

The corresponding formula (5.1.7) for the trapezoidal rule on $[-1,1]$ gives

$$|\text{Trapezoidal error}| \le \frac{.67}{n^2}\left[\ \underset{-1\le x\le 1}{\text{Max}}\ |f''(x)|\right]$$

which is slightly larger than (5.3.40). In actual computation, Gaussian quadrature appears to always be superior to the trapezoidal rule, except for the case of periodic integrands with the integration interval an integer multiple of the period of the integrand, as in Table 5.7. An analogous discussion, using Table 5.13 with $e_{n,4}$, can be carried out for integrands $f(x)$, which are four times differentiable (see Problem 20).

***Example***   We give three further examples that are not as well behaved as the ones in Tables 5.11 and 5.12. Consider

$$I^{(1)} = \int_0^1 \sqrt{x}\ dx = \tfrac{2}{3} \qquad I^{(2)} = \int_0^5 \frac{dx}{1+(x-\pi)^2} \doteq 2.33976628367$$

$$I^{(3)} = \int_0^{2\pi} e^{-x}\sin(50x)\ dx \doteq .019954669278 \qquad (5.3.41)$$

**Table 5.14    Gaussian quadrature examples (5.3.41)**

| $n$ | $I^{(1)} - I_n^{(1)}$ | Ratio | $I^{(2)} - I_n^{(2)}$ | $I^{(3)} - I_n^{(3)}$ |
|---|---|---|---|---|
| 2 | $-7.22\text{E} - 3$ | | $3.50\text{E} - 1$ | $3.48\text{E} - 1$ |
| | | 6.2 | | |
| 4 | $-1.16\text{E} - 3$ | | $-9.19\text{E} - 2$ | $-1.04\text{E} - 1$ |
| | | 6.9 | | |
| 8 | $-1.69\text{E} - 4$ | | $-4.03\text{E} - 3$ | $-1.80\text{E} - 2$ |
| | | 7.4 | | |
| 16 | $-2.30\text{E} - 5$ | | $-6.24\text{E} - 7$ | $-3.34\text{E} - 1$ |
| | | 7.6 | | |
| 32 | $-3.00\text{E} - 6$ | | $-2.98\text{E} - 11$ | $1.16\text{E} - 1$ |
| | | 7.8 | | |
| 64 | $-3.84\text{E} - 7$ | | — | $1.53\text{E} - 1$ |
| | | 7.9 | | |
| 128 | $-4.85\text{E} - 8$ | | — | $6.69\text{E} - 15$ |

The values in Table 5.14 show that Gaussian quadrature is still very effective, in spite of the bad behavior of the integrand.

Compare the results for $I^{(1)}$ with those in Table 5.6 for the trapezoidal and Simpson rules. Gaussian quadrature is converging with an error proportional to $1/n^3$, whereas in Table 5.6, the errors converged with a rate proportional to $1/n^{1.5}$. Consider the integral

$$I(f) = \int_0^1 x^\alpha f(x)\, dx \qquad (5.3.42)$$

with $\alpha > -1$ and nonintegral, and $f(x)$ smooth with $f(0) \neq 0$. It has been shown by Donaldson and Elliott (1972) that the error in Gauss–Legendre quadrature for (5.3.42) will have the asymptotic estimate

$$E_n(f) \doteq \frac{c(f, \alpha)}{n^{2(1+\alpha)}} \qquad (5.3.43)$$

This agrees with $I_n^{(1)}$ in Table 5.14, using $\alpha = \frac{1}{2}$. Other important results on Gauss–Legendre quadrature are also given in the Donaldson and Elliott paper.

The initial convergence of $I_n^{(2)}$ to $I^{(2)}$ is quite slow, but as $n$ increases, the speed increases dramatically. For $n \geq 64$, $I_n^{(2)} = I^{(2)}$ within the limits of the machine arithmetic. Also compare these results with those of Table 5.5, for the trapezoidal and Simpson rules.

The approximations in Table 5.14 for $I^{(3)}$ are quite poor because the integrand is so oscillatory. There are 101 zeros of the integrand in the interval of integration. To obtain an accurate value $I_n^{(3)}$, the degree of the approximating polynomial underlying Gaussian quadrature must be very large. With $n = 128$, $I_n^{(3)}$ is a very accurate approximation of $I^{(3)}$.

**General comments**    Gaussian quadrature has a number of strengths and weaknesses.

1.  Because of the form of the nodes and weights and the resulting need to use a table, many people prefer a simpler formula, such as Simpson's rule. This shouldn't be a problem when doing integration using a computer. Programs should be written containing these weights and nodes for standard values of $n$, for example, $n = 2, 4, 8, 16, \ldots, 512$ [taken from Stroud and Secrest (1966)].

In addition, there are a number of very rapid programs for calculating the nodes and weights for a variety of commonly used weight functions. Among the better known algorithms is that in Golub and Welsch (1969).

2.  It is difficult to estimate the error, and thus we usually take

$$I - I_n \doteq I_m - I_n \qquad (5.3.44)$$

for some $m > n$, for example, $m = n + 2$ with well-behaved integrands, and $m = 2n$ otherwise. This results in greater accuracy than necessary, but even with the increased number of function evaluations, Gaussian quadrature is still faster than most other methods.

3.  The nodes for each formula $I_n$ are distinct from those of preceding formulas $I_m$, and this results in some inefficiency. If $I_n$ is not sufficiently accurate, based on an error estimate like (5.3.44), then we must compute a new value of $I_n$. However, none of the previous values of the integrand can be reused, resulting in wasted effort. This is discussed more extensively in the last part of this section, resulting in some new methods without this drawback. Nonetheless, in many situations, the resulting inefficiency in Gaussian quadrature is usually not significant because of its rapid rate of convergence.

4.  If a large class of integrals of a similar nature are to be evaluated, then proceed as follows. Pick a few representative integrals, including some with the worst behavior in the integrand that is likely to occur. Determine a value of $n$ for which $I_n(f)$ will have sufficient accuracy among the representative set. Then fix that value of $n$, and use $I_n(f)$ as the numerical integral for all members of the original class of integrals.

5.  Gaussian quadrature can handle many near-singular integrands very effectively, as is shown in (5.3.43) for (5.3.42). But all points of singular behavior must occur as endpoints of the integration interval. Gaussian quadrature is very poor on an integral such as

$$\int_0^1 \sqrt{|x - .7|} \, dx$$

which contains a singular point in the interval of integration. (Most other numerical integration methods will also perform poorly on this integral.) The integral should be decomposed and evaluated in the form

$$\int_0^{.7} \sqrt{.7 - x} \, dx + \int_{.7}^1 \sqrt{x - .7} \, dx$$

**Extensions that reuse node points**    Suppose we have a quadrature formula

$$I_n(f) = \sum_{k=1}^n w_k f(x_k) \doteq \int_a^b w(x) f(x) \, dx \qquad (5.3.45)$$

We want to produce a new quadrature formula that uses the $n$ nodes $x_1, \ldots, x_n$ and $m$ new nodes $x_{n+1}, \ldots, x_{n+m}$:

$$I_{n+m}(f) = \sum_{k=1}^{n+m} v_k f(x_k) \doteq \int_a^b w(x) f(x) \, dx \qquad (5.3.46)$$

These $n + 2m$ unspecified parameters, namely the nodes $x_{n+1}, \ldots, x_{n+m}$ and the weights $v_1, \ldots, v_{n+m}$, are to be chosen to give (5.3.46) as large a degree of precision as is possible. We seek a formula of degree of precision $n + 2m - 1$. Whether such a formula can be determined with the new nodes $x_{n+1}, \ldots, x_{n+m}$ located in $[a, b]$ is in general unknown.

In the case that (5.3.45) is a Gauss formula, Kronrod studied extensions (5.3.46) with $m = n + 1$. Such pairs of formulas give a less expensive way of producing an error estimate for a Gauss rule (as compared with using a Gauss rule with $2n + 1$ node points). And the degree of precision is high enough to produce the kind of accuracy associated with the Gauss rules.

A variation on the preceding theme was introduced in Patterson (1968). For $w(x) \equiv 1$, he started with a Gauss–Legendre rule $I_{n_0}(f)$. He then produced a sequence of formulas by repeatedly constructing formulas (5.3.46) from the preceding member of the sequence, with $m = n + 1$. A paper by Patterson (1973) contains an algorithm based on a sequence of rules $I_3, I_7, I_{15}, I_{31}, I_{63}, I_{127}, I_{255}$; the formula $I_3$ is the three-point Gauss rule. Another such sequence $\{I_{10}, I_{21}, I_{43}, I_{87}\}$ is given in Piessens et al. (1983, pp. 19, 26, 27), with $I_{10}$ the ten-point Gauss rule. All such Patterson formulas to date have had all nodes located inside the interval of integration and all weights positive.

The degree of precision of the Patterson rules increases with the number of points. For the sequence $I_3, I_7, \ldots, I_{255}$ previously referred to, the respective degrees of precision are $d = 5, 11, 23, 47, 95, 191, 383$. Since the weights are positive, the proof of Theorem 5.4 can be repeated to show that the Patterson rules are rapidly convergent.

A further discussion of the Patterson and Kronrod rules, including programs, is given in Piessens et al. (1983, pp. 15–27); they also give reference to much of the literature on this subject.

***Example***    Let (5.3.45) be the three-point Gauss rule on $[-1, 1]$:

$$I_3(f) = \frac{8}{9}f(0) + \frac{5}{9}\left[f(-\sqrt{.6}) + f(\sqrt{.6})\right] \qquad (5.3.47)$$

The Kronrod rule for this is

$$I_7(f) = \alpha_0 f(0) + \alpha_1\left[f(-\sqrt{.6}) + f(\sqrt{.6})\right]$$
$$+ \alpha_2\left[f(-\beta_1) + f(\beta_1)\right] + \alpha_3\left[f(-\beta_2) + f(\beta_2)\right] \qquad (5.3.48)$$

with $\beta_1^2$ and $\beta_2^2$ the smallest and largest roots, respectively, of

$$x^2 - \frac{10}{9}x + \frac{155}{891} = 0$$

The weights $\alpha_0, \alpha_1, \alpha_2, \alpha_3$ come from integrating over $[-1, 1]$ the Lagrange polynomial $p_7(x)$ that interpolates $f(x)$ at the nodes $\{0, \pm\sqrt{.6}, \pm\beta_1, \pm\beta_2\}$. Approximate values are

$$\alpha_0 = .450916538658 \qquad \alpha_1 = .268488089868$$

$$\alpha_2 = .401397414776 \qquad \alpha_3 = .104656226026$$

## 5.4   Asymptotic Error Formulas and Their Applications

Recall the definition (5.1.10) of an asymptotic error formula for a numerical integration formula: $\tilde{E}_n(f)$ is an asymptotic error formula for $E_n(f) = I(f) - I_n(f)$ if

$$\underset{n \to \infty}{\text{Limit}} \frac{\tilde{E}_n(f)}{E_n(f)} = 1 \qquad (5.4.1)$$

or equivalently,

$$\underset{n \to \infty}{\text{Limit}} \frac{E_n(f) - \tilde{E}_n(f)}{E_n(f)} = 0$$

Examples are (5.1.9) and (5.1.18) from Section 5.1.

By obtaining an asymptotic error formula, we are obtaining the form or structure of the error. With this information, we can either estimate the error in $I_n(f)$, as in Tables 5.1 and 5.3, or we can develop a new and more accurate formula, as with the corrected trapezoidal rule in (5.1.12). Both of these alternatives are further illustrated in this section, concluding with the rapidly convergent Romberg integration method. We begin with a further development of asymptotic error formulas.

**The Bernoulli polynomials**   For use in the next theorem, we introduce the *Bernoulli polynomials* $B_n(x)$, $n \geq 0$. These are defined implicitly by the *generating function*

$$\frac{t(e^{xt} - 1)}{e^t - 1} = \sum_{j=1}^{\infty} B_j(x) \frac{t^j}{j!} \qquad (5.4.2)$$

The first few polynomials are

$$B_0(x) = 1 \qquad B_1(x) = x \qquad B_2(x) = x^2 - x$$

$$B_3(x) = x^3 - \frac{3x^2}{2} + \frac{x}{2} \qquad B_4(x) = x^2(1 - x)^2 \qquad (5.4.3)$$

With these polynomials,

$$B_k(0) = 0 \qquad k \geq 1 \qquad (5.4.4)$$

There are easily computable recursion relations for calculating these polynomials (see Problem 23).

Also of interest are the *Bernoulli numbers*, defined implicitly by

$$\frac{t}{e^t - 1} = \sum_{j=0}^{\infty} B_j \cdot \frac{t^j}{j!} \qquad (5.4.5)$$

The first few numbers are

$$B_0 = 1 \quad B_1 = -\frac{1}{2} \quad B_2 = \frac{1}{6} \quad B_4 = \frac{-1}{30} \quad B_6 = \frac{1}{42} \quad B_8 = \frac{-1}{30} \quad (5.4.6)$$

and for all odd integers $j \geq 3$, $B_j = 0$. To obtain a relation to the Bernoulli polynomials $B_j(x)$, integrate (5.4.2) with respect to $x$ on $[0, 1]$. Then

$$1 - \frac{t}{e^t - 1} = \sum_{1}^{\infty} \frac{t^j}{j!} \int_0^1 B_j(x) \, dx$$

and thus

$$B_j = -\int_0^1 B_j(x) \, dx \qquad j \geq 1 \qquad (5.4.7)$$

We will also need to define a periodic extension of $B_j(x)$,

$$\overline{B}_j(x) = \begin{cases} B_j(x) & 0 \leq x < 1 \\ \overline{B}_j(x - 1) & x \geq 1 \end{cases} \qquad (5.4.8)$$

**The Euler–MacLaurin formula**  The following theorem gives a very detailed asymptotic error formula for the trapezoidal rule. This theorem is at the heart of much of the asymptotic error analysis of this section. The connection with some other integration formulas appears later in the section.

**Theorem 5.5**  (Euler–MacLaurin Formula)  Let $m \geq 0$, $n \geq 1$, and define $h = (b - a)/n$, $x_j = a + jh$ for $j = 0, 1, \ldots, n$. Further assume $f(x)$ is $2m + 2$ times continuously differentiable on $[a, b]$ for some $m \geq 0$. Then for the error in the trapezoidal rule,

$$E_n(f) = \int_a^b f(x) \, dx - h \sum_{j=0}^{n} {}'' f(x_j)$$

$$= -\sum_{i=1}^{m} \frac{B_{2i}}{(2i)!} h^{2i} \left[ f^{(2i-1)}(b) - f^{(2i-1)}(a) \right]$$

$$+ \frac{h^{2m+2}}{(2m+2)!} \int_a^b \overline{B}_{2m+2}\left( \frac{x - a}{h} \right) f^{(2m+2)}(x) \, dx \qquad (5.4.9)$$

*Note:* The double prime notation on the summation sign means that the first and last terms are to be halved before summing.

***Proof***  A complete proof is given in Ralston (1965, pp. 131–133), and a more general development is sketched in Lyness and Puri (1973, sec. 2). The proof in Ralston is short and correct, making full use of the special properties of the Bernoulli polynomials. We give a simpler, but less general, version of that proof, showing it to be based on integration by parts with a bit of clever algebraic manipulation.

The proof of (5.4.9) for general $n > 1$ is based on first proving the result for $n = 1$. Thus we concentrate on

$$E_1(f) = \int_0^h f(x)\, dx - \frac{h}{2}[f(0) + f(h)]$$

$$= \frac{1}{2} \int_0^h f''(x) x (x - h)\, dx \qquad (5.4.10)$$

the latter formula coming from (5.1.21). Since we know the asymptotic formula

$$E_1(f) \doteq -\frac{h^2}{12}[f'(h) - f'(0)]$$

we attempt to manipulate (5.4.10) to obtain this. Write

$$E_1(f) = \int_0^h f''(x)\left[-\frac{h^2}{12}\right] dx + \int_0^h f''(x)\left[\frac{x(x - h)}{2} + \frac{h^2}{12}\right] dx$$

Then

$$E_1(f) = -\frac{h^2}{12}[f'(h) - f'(0)] + \int_0^h f''(x)\left[\frac{x^2}{2} - \frac{xh}{2} + \frac{h^2}{12}\right] dx$$

Using integration by parts,

$$E_1(f) = -\frac{h^2}{12}[f'(h) - f'(0)] + \left[f''(x)\left(\frac{x^3}{6} - \frac{x^2 h}{4} + \frac{h^2 x}{12}\right)\right]_0^h$$

$$- \int_0^h f^{(3)}(x)\left[\frac{x^3}{6} - \frac{x^2 h}{4} + \frac{h^2 x}{12}\right] dx$$

The evaluation of the quantity in brackets at $x = 0$ and $x = h$ gives zero. Integrate by parts again; the parts outside the integral will again be zero. The result will be

$$E_1(f) = -\frac{h^2}{12}[f'(h) - f'(0)] + \frac{1}{24} \int_0^h f^{(4)}(x) x^2 (x - h)^2\, dx \quad (5.4.11)$$

which is (5.4.9) with $m = 1$. To obtain the $m = 2$ case, first note that

$$\frac{1}{24}\int_0^h x^2(x-h)^2\,dx = \frac{h^5}{720}$$

Then as before, write

$$\frac{1}{24}\int_0^h f^{(4)}(x)x^2(x-h)^2\,dx = \int_0^h f^{(4)}(x)\left[\frac{h^4}{720}\right]dx$$

$$+ \int_0^h f^{(4)}(x)\left[\frac{x^2(x-h)^2}{24} - \frac{h^4}{720}\right]dx$$

$$= \frac{h^4}{720}\left[f^{(3)}(h) - f^{(3)}(0)\right]$$

$$+ \int_0^h f^{(4)}(x)\left[\frac{x^4 - 2x^3h + x^2h^2}{24} - \frac{h^4}{720}\right]dx$$

Integrate by parts twice to obtain the $m = 2$ case of (5.4.9). This can be continued indefinitely. The proof in Ralston uses integration by parts, taking advantage of special relations for the Bernoulli polynomials (see Problem 23).

For the proof for general $n > 1$, write

$$E_n(f) = \sum_{j=1}^n \left\{\int_{x_{j-1}}^{x_j} f(x)\,dx - \frac{h}{2}\left[f(x_{j-1}) + f(x_j)\right]\right\}$$

For the $m = 1$ case, using (5.4.11),

$$E_n(f) = \sum_{j=1}^n \left\{-\frac{h^2}{12}\left[f'(x_j) - f'(x_{j-1})\right]\right\}$$

$$+ \sum_{j=1}^n \frac{1}{24}\int_{x_{j-1}}^{x_j} f^{(4)}(x)(x-x_{j-1})^2(x-x_j)^2\,dx$$

$$= -\frac{h^2}{12}\left[f'(b) - f'(a)\right] + \frac{h^4}{24}\int_a^b f^{(4)}(x)\bar{B}_4\left(\frac{x-a}{h}\right)dx \quad (5.4.12)$$

The proof for $m > 1$ is essentially the same.  ∎

The error term in (5.4.9) can be simplified using the integral mean value theorem. It can be shown that

$$B_{2j}(x) > 0 \qquad 0 < x < 1 \qquad\qquad (5.4.13)$$

and consequently the error term satisfies

$$\int_a^b \overline{B}_{2m+2}\left(\frac{x-a}{h}\right) f^{(2m+2)}(x)\, dx = f^{(2m+2)}(\xi) \int_a^b \overline{B}_{2m+2}\left(\frac{x-a}{h}\right) dx$$

$$= nf^{(2m+2)}(\xi) \int_a^{a+h} B_{2m+2}\left(\frac{x-a}{h}\right) dx$$

$$= nhf^{(2m+2)}(\xi) \int_0^1 B_{2m+2}(u)\, du$$

$$= -(b-a) B_{2m+2} f^{(2m+2)}(\xi) \quad \text{some } a \le \xi \le b$$

Thus (5.4.9) becomes

$$E_n(f) = -\sum_{i=1}^m \frac{B_{2i}}{(2i)!} h^{2i} \left[ f^{(2i-1)}(b) - f^{(2i-1)}(a) \right]$$

$$-\frac{h^{2m+2}(b-a) B_{2m+2}}{(2m+2)!} f^{(2m+2)}(\xi) \qquad (5.4.14)$$

for some $a \le \xi \le b$.

As an important corollary of (5.4.9), we can show that the trapezoidal rule performs especially well when applied to periodic functions.

**Corollary 1**   Suppose $f(x)$ is infinitely differentiable for $a \le x \le b$, and suppose that all of its odd ordered derivatives are periodic with $b - a$ an integer multiple of the period. Then the order of convergence of the trapezoidal rule $I_n(f)$ applied to

$$I(f) = \int_a^b f(x)\, dx$$

is greater than any power of $h$.   ■

**Proof**   Directly from the assumptions on $f(x)$,

$$f^{(2j-1)}(b) = f^{(2j-1)}(a) \qquad j \ge 1 \qquad (5.4.15)$$

Consequently for any $m \ge 0$, with $h = (b - a)/n$, (5.4.14) implies

$$I(f) - I_n(f) = -\frac{h^{2m+2}}{(2m+2)!}(b-a) B_{2m+2} f^{(2m+2)}(\xi) \qquad a \le \xi \le b \quad (5.4.16)$$

Thus as $n \to \infty$ (and $h \to 0$) the rate of convergence is proportional to $h^{2m+2}$. But $m$ was arbitrary, which shows the desired result.    ∎

This result is illustrated in Table 5.7 for $f(x) = \exp(\cos(x))$. The trapezoidal rule is often the best numerical integration rule for smooth periodic integrands of the type specified in the preceding corollary. For a comparison of the Gauss–Legendre formula and the trapezoidal rule for a one-parameter family of periodic functions, of varying behavior, see Donaldson and Elliott (1972, p. 592). They suggest that the trapezoidal rule is superior, even for very peaked integrands. This conclusion improves on an earlier analysis that seemed to indicate that Gaussian quadrature was superior for peaked integrands.

**The Euler–MacLaurin summation formula**    Although it doesn't involve numerical integration, an important application of (5.4.9) or (5.4.14) is to the summation of series.

***Corollary 2*** (Euler–MacLaurin summation formula)    Assume $f(x)$ is $2m + 2$ times continuously differentiable for $0 \le x < \infty$, for some $m \ge 0$. Then for all $n \ge 1$,

$$\sum_{j=0}^{n} f(j) = \int_{0}^{n} f(x)\, dx + \frac{1}{2}[f(0) + f(n)]$$

$$+ \sum_{i=1}^{m} \frac{B_{2i}}{(2i)!} \left[ f^{(2i-1)}(n) - f^{(2i-1)}(0) \right] \qquad (5.4.17)$$

$$- \frac{1}{(2m+2)!} \int_{0}^{n} \overline{B}_{2m+2}(x) f^{(2m+2)}(x)\, dx \qquad ∎$$

***Proof***    Merely substitute $a = 0$, $b = n$ into (5.4.9), and then rearrange the terms appropriately.    ∎

***Example***    In a later example we need the sum

$$S = \sum_{1}^{\infty} \frac{1}{n^{3/2}} \qquad (5.4.18)$$

If we use the obvious choice $f(x) = (x+1)^{-3/2}$ in (5.4.17), the results are disappointing. By letting $n \to \infty$, we obtain

$$S = \int_{0}^{\infty} \frac{dx}{(x+1)^{3/2}} + \frac{1}{2} - \sum_{1}^{m} \frac{B_{2i}}{(2i)!} f^{(2i-1)}(0)$$

$$- \frac{1}{(2m+2)!} \int_{0}^{\infty} \overline{B}_{2m+2}(x) f^{(2m+2)}(x)\, dx$$

and the error term does not become small for any choice of $m$. But if we divide the series $S$ into two parts, we are able to treat it very accurately.

Let $f(x) = (x + 10)^{-3/2}$. Then with $m = 1$,

$$\sum_{10}^{\infty} \frac{1}{n^{3/2}} = \sum_{0}^{\infty} f(j) = \int_0^{\infty} \frac{dx}{(x+10)^{3/2}} + \frac{1}{2(10)^{3/2}} - \frac{(1/6)}{2} \cdot \frac{(-\frac{3}{2})}{(10)^{5/2}} + E$$

$$E = -\frac{1}{24} \int_0^{\infty} \overline{B}_4(x) f^{(4)}(x)\, dx$$

Since $\overline{B}_4(x) \geq 0$, $f^{(4)}(x) > 0$, we have $E < 0$. Also

$$0 < -E < \frac{1}{24} \int_0^{\infty} \left(\frac{1}{16}\right) f^{(4)}(x)\, dx = \frac{35}{(1024)(10)^{9/2}} \doteq 1.08 \times 10^{-6}$$

Thus

$$\sum_{10}^{\infty} \frac{1}{n^{3/2}} = .648662205 + E$$

By directly summing $\sum_{1}^{9} (1/n^{3/2}) \doteq 1.963713717$, we obtain

$$\sum_{1}^{\infty} \frac{1}{n^{3/2}} = 2.6123759 + E \quad 0 < -E < 1.08 \times 10^{-6} \qquad (5.4.19)$$

See Ralston (1965, pp. 134–138) for more information on summation techniques. To appreciate the importance of the preceding summation method, it would have been necessary to have added $3.43 \times 10^{12}$ terms in $S$ to have obtained comparable accuracy.

**A generalized Euler–MacLaurin formula**   For integrals in which the integrand is not differentiable at some points, it is still often possible to obtain an asymptotic error expansion. For the trapezoidal rule and other numerical integration rules applied to integrands with algebraic and/or logarithmic singularities, see the article Lyness and Ninham (1967). In the following paragraphs, we specialize their results to the integral

$$I = \int_0^1 x^{\alpha} f(x)\, dx \qquad \alpha > 0 \qquad (5.4.20)$$

with $f \in C^{m+1}[0, 1]$, using the trapezoidal numerical integration rule.

Assuming $\alpha$ is not an integer,

$$E_n(f) = \sum_{j=0}^{m-1} \frac{c_j}{n^{\alpha+j+1}} + \sum_{j=1}^{m-1} \frac{d_j}{n^{j+1}} + O\left(\frac{1}{n^{m+1}}\right) \qquad (5.4.21)$$

The term $O(1/n^{m+1})$ denotes a quantity whose size is proportional to $1/n^{m+1}$, or possibly smaller. The constants are given by

$$c_j = \frac{2\Gamma(\alpha + j + 1) \sin\left[(\pi/2)(\alpha + j)\right]\zeta(\alpha + j + 1)}{(2\pi)^{\alpha+j+1}j!} f^{(j)}(0)$$

$$d_j = 0 \quad \text{for } j \text{ even}$$

$$d_j = (-1)^{(j-1)/2} \cdot \frac{2\zeta(j + 1)}{(2\pi)^{j+1}} g^{(j)}(1) \quad j \text{ odd}$$

with $g(x) = x^\alpha f(x)$, $\Gamma(x)$ the gamma function, and $\zeta(p)$ the zeta function,

$$\zeta(p) = \sum_{j=1}^{\infty} \frac{1}{j^p} \qquad p > 1 \tag{5.4.22}$$

For $0 < \alpha < 1$ with $m = 1$, we obtain the asymptotic error estimate

$$E_n(f) = \frac{2\Gamma(\alpha + 1) \sin\left[(\pi/2)\alpha\right]\zeta(\alpha + 1)f(0)}{(2\pi)^{\alpha+1}n^{\alpha+1}} + O\left(\frac{1}{n^2}\right) \qquad f \in C^2[0,1].$$

$$\tag{5.4.23}$$

For example with $I = \int_0^1 \sqrt{x} f(x)\, dx$, and using (5.4.19) for evaluating $\zeta(\frac{3}{2})$,

$$E_n(f) = \frac{c}{n\sqrt{n}} f(0) + O\left(\frac{1}{n^2}\right) \qquad c = \frac{\zeta(\frac{3}{2})}{4\pi} \doteq .208 \tag{5.4.24}$$

This is confirmed numerically in the example given in Table 5.6 in Section 5.1.

For logarithmic endpoint singularities, the results of Lyness and Ninham (1967) imply an asymptotic error formula

$$E_n(f) \doteq \frac{c \cdot \log n}{n^p} \tag{5.4.25}$$

for some $p > 0$ and some constant $c$. For numerical purposes, this is essentially $O(1/n^p)$. To justify this, calculate the following limit using L'Hospital's rule, with $p > q$:

$$\lim_{n \to \infty} \frac{\log(n)/n^p}{1/n^q} = \lim_{n \to \infty} \frac{\log(n)}{n^{p-q}} = 0$$

This means that $\log(n)/n^p$ decreases more rapidly than $1/n^q$ for any $q < p$. And it clearly decreases less rapidly than $1/n^p$, although not by much.

For practical computation, (5.4.25) is essentially $O(1/n^p)$. For example, calculate the limit of successive errors:

$$\underset{n \to \infty}{\text{Limit}} \frac{I - I_n}{I - I_{2n}} = \underset{n \to \infty}{\text{Limit}} \frac{c \cdot \log(n)/n^p}{c \cdot \log(2n)/2^p n^p} = \underset{n \to \infty}{\text{Limit}} 2^p \cdot \frac{\log(n)}{\log(2n)}$$

$$= \underset{n \to \infty}{\text{Limit}} 2^p \cdot \frac{1}{1 + (\log 2/\log n)} = 2^p$$

This is the same limiting ratio as would occur if the error were just $O(1/n^p)$.

**Aitken extrapolation**    Motivated by the preceding, we assume that the integration formula has an asymptotic error formula

$$I - I_n \doteq \frac{c}{n^p} \qquad p > 0 \qquad (5.4.26)$$

This is not always valid. For example, Gaussian quadrature does not usually satisfy (5.4.26), and the trapezoidal rule applied to periodic integrands does not satisfy it. Nonetheless, many numerical integration rules do satisfy it, for a wide variety of integrands. Using this assumed form for the error, we attempt to estimate the error. An analogue of this work is that on Aitken extrapolation in Section 2.6.

First we estimate $p$. Using (5.4.26),

$$\frac{I_{2n} - I_n}{I_{4n} - I_{2n}} = \frac{(I - I_n) - (I - I_{2n})}{(I - I_{2n}) - (I - I_{4n})} \doteq \frac{(c/n^p) - (c/2^p n^p)}{(c/2^p n^p) - (c/4^p n^p)} = 2^p$$

$$R_{4n} \equiv \frac{I_{2n} - I_n}{I_{4n} - I_{2n}} \doteq 2^p \qquad (5.4.27)$$

This gives a simple way of computing $p$.

**Example**    Consider the use of Simpson's rule with $\int_0^1 x\sqrt{x}\, dx = 0.4$. In Table 5.15, column $R_n$ should approach $2^{2.5} \doteq 5.66$, a theoretical result from Lyness

**Table 5.15    Simpson integration errors for $\int_0^1 x\sqrt{x}\, dx$**

| $n$ | $I_n$ | $I - I_n$ | $I_n - I_{n/2}$ | $R_n$ |
|---|---|---|---|---|
| 2 | .402368927062 | $-2.369 - 3$ | | |
| 4 | .400431916045 | $-4.319 - 4$ | $-1.937 - 3$ | |
| 8 | .400077249447 | $-7.725 - 5$ | $-3.547 - 4$ | 5.46 |
| 16 | .400013713469 | $-1.371 - 5$ | $-6.354 - 5$ | 5.58 |
| 32 | .400002427846 | $-2.428 - 6$ | $-1.129 - 5$ | 5.63 |
| 64 | .400000429413 | $-4.294 - 7$ | $-1.998 - 6$ | 5.65 |
| 128 | .400000075924 | $-7.592 - 8$ | $-3.535 - 7$ | 5.65 |
| 256 | .400000013423 | $-1.342 - 8$ | $-6.250 - 8$ | 5.66 |
| 512 | .400000002373 | $-2.373 - 9$ | $-1.105 - 8$ | 5.66 |

and Ninham (1967) for the order of convergence. Clearly the numerical results confirm the theory.

To estimate the integral $I$ with increased accuracy, suppose that $I_n$, $I_{2n}$, and $I_{4n}$ have been computed. Using (5.4.26),

$$\frac{I - I_n}{I - I_{2n}} \doteq 2^p \doteq \frac{I - I_{2n}}{I - I_{4n}}$$

and thus

$$(I - I_n)(I - I_{4n}) \doteq (I - I_{2n})^2$$

Solving for $I$, and manipulating to obtain a desirable form,

$$I \doteq \tilde{I}_{4n} \equiv I_{4n} - \frac{(I_{4n} - I_{2n})^2}{(I_{4n} - I_{2n}) - (I_{2n} - I_n)} \tag{5.4.28}$$

***Example***   Using the previous example for $f(x) = x\sqrt{x}$ and Table 5.15, we obtain the difference table in Table 5.16. Then

$$I \doteq \tilde{I}_{64} = .399999999387$$

$$I - \tilde{I}_{64} = 6.13 \times 10^{-10} \qquad I - I_{64} = -4.29 \times 10^{-7}$$

Thus $\tilde{I}_{64}$ is a considerable improvement on $I_{64}$. Also note that $\tilde{I}_{64} - I_{64}$ is an excellent approximation to $I - I_{64}$.

Summing up, given a numerical integration rule satisfying (5.4.26) and given three values $I_n$, $I_{2n}$, $I_{4n}$, calculate the Aitken extrapolate $\tilde{I}_{4n}$ of (5.4.28). It is usually a significant improvement on $I_{4n}$ as an approximation to $I$; and based on this,

$$I - I_{4n} \doteq \tilde{I}_{4n} - I_{4n} \tag{5.4.29}$$

With Simpson's rule, or any other composite closed Newton–Cotes formula, the expense of evaluating $I_n$, $I_{2n}$, $I_{4n}$ is no more than that of $I_{4n}$ alone, namely $4n + 1$ function evaluations. And when the algorithm is designed correctly, there is no need for temporary storage of large numbers of function values $f(x_j)$. For

**Table 5.16   Difference table for Simpson integration**

| $m$ | $I_m$ | $\Delta I_m = I_{2m} - I_m$ | $\Delta^2 I_m$ |
|---|---|---|---|
| 16 | .400013713469 | | |
| | | $-1.1285623E - 5$ | |
| 32 | .400002427846 | | $9.28719E - 6$ |
| | | $-1.998433E - 6$ | |
| 64 | .400000429413 | | |

that reason, one should never use Simpson's rule with just one value of the index $n$. With no extra expenditure of time, and with only a slightly more complicated algorithm, an Aitken extrapolate and an error estimate can be produced.

**Richardson extrapolation**    If we assume sufficient smoothness for the integrand $f(x)$ in our integral $I(f)$, then we can write the trapezoidal error term (5.4.9) as

$$I - I_n = \frac{d_2^{(0)}}{n^2} + \frac{d_4^{(0)}}{n^4} + \cdots + \frac{d_{2m}^{(0)}}{n^{2m}} + F_{n,m} \tag{5.4.30}$$

where $I_n$ denotes the trapezoidal rule, and

$$F_{n,m} = \frac{(b-a)^{2m+2}}{(2m+2)!n^{2m+2}} \int_a^b \overline{B}_{2m+2}\left(\frac{x-a}{h}\right) f^{(2m+2)}(x)\,dx$$

$$d_{2j}^{(0)} = -\frac{B_{2j}}{(2j)!}(b-a)^{2j}\left[f^{(2j-1)}(b) - f^{(2j-1)}(a)\right] \tag{5.4.31}$$

Although the series dealt with are always finite and have an error term, we will usually not directly concern ourselves with it.

For $n$ even,

$$I - I_{n/2} = \frac{4d_2^{(0)}}{n^2} + \frac{16d_4^{(0)}}{n^4} + \frac{64d_6^{(0)}}{n^6} + \cdots \tag{5.4.32}$$

Multiply (5.4.30) by 4 and subtract from it (5.4.32):

$$4(I - I_n) - (I - I_{n/2}) = \frac{-12d_4^{(0)}}{n^4} - \frac{60d_6^{(0)}}{n^6} - \cdots$$

$$I = \frac{4I_n - I_{n/2}}{3} - \frac{4d_4^{(0)}}{n^4} - \frac{20d_6^{(0)}}{n^6} - \cdots$$

Define

$$I_n^{(1)} = \frac{1}{3}\left[4I_n^{(0)} - I_{n/2}^{(0)}\right] \qquad n \text{ even} \qquad n \geq 2 \tag{5.4.33}$$

and $I_m^{(0)} \equiv I_m$. We call $\{I_n^{(1)}\}$ the Richardson extrapolate of $\{I_n^{(0)}\}$.

The sequence

$$I_2^{(1)}, I_4^{(1)}, I_6^{(1)}, \ldots$$

is a new numerical integration rule. For the error,

$$I - I_n^{(1)} = \frac{d_4^{(1)}}{n^4} + \frac{d_6^{(1)}}{n^6} + \cdots \tag{5.4.34}$$

$$d_4^{(1)} = -4d_4^{(0)}, \; d_6^{(1)} = -20d_6^{(0)}, \ldots \tag{5.4.35}$$

To see the explicit formula for $I_n^{(1)}$, let $h = (b - a)/n$ and $x_j = a + jh$ for $j = 0, 1, \ldots, n$. Then using (5.4.33) and the definition of the trapezoidal rule,

$$I_n^{(1)} = \frac{4h}{3}\left[\frac{1}{2}f_0 + f_1 + f_2 + f_3 + \cdots + f_{n-1} + \frac{1}{2}f_n\right]$$

$$- \frac{2h}{3}\left[\frac{1}{2}f_0 + f_2 + f_4 + f_6 + \cdots + f_{n-2} + \frac{1}{2}f_n\right]$$

$$I_n^{(1)} = \frac{h}{3}[f_0 + 4f_1 + 2f_2 + 4f_3 + \cdots + 2f_{n-2} + 4f_{n-1} + f_n] \quad (5.4.36)$$

which is Simpson's rule with $n$ subdivisions. For the error, using (5.4.35) and (5.4.31),

$$I - I_n^{(1)} = -\frac{h^4}{180}\left[f^{(3)}(b) - f^{(3)}(a)\right] + \frac{h^6}{1512}\left[f^{(5)}(b) - f^{(5)}(a)\right] + \cdots$$

$$(5.4.37)$$

This means that the work on the Euler–MacLaurin formula transfers to Simpson's rule by means of some simple algebraic manipulations. We omit any numerical examples since they would just be Simpson's rule, due to (5.4.36).

The preceding argument, which led to $I_n^{(1)}$, can be continued to produce other new formulas. As before, if $n$ is a multiple of 4, then

$$I - I_{n/2}^{(1)} = \frac{16d_4^{(1)}}{n^4} + \frac{64d_6^{(1)}}{n^6} + \cdots$$

$$16\left(I - I_n^{(1)}\right) - \left(I - I_{n/2}^{(1)}\right) = \frac{-48d_6^{(1)}}{n^6} + \cdots$$

$$I = \frac{16I_n^{(1)} - I_{n/2}^{(1)}}{15} - \frac{48d_6^{(1)}}{15n^6} + \cdots \qquad (5.4.38)$$

Then

$$I - I_n^{(2)} = \frac{d_6^{(2)}}{n^6} + \frac{d_8^{(2)}}{n^8} + \cdots \qquad (5.4.39)$$

with

$$I_n^{(2)} = \frac{16I_n^{(1)} - I_{n/2}^{(1)}}{15} \qquad n \geq 4 \qquad (5.4.40)$$

and $n$ divisible by 4. We call $\{I_n^{(2)}\}$ the Richardson extrapolate of $\{I_n^{(1)}\}$. If we derive the actual integration weights of $I_n^{(2)}$, in analogy with (5.4.36), we will find that $I_n^{(2)}$ is simply the composite Boole's rule.

Using the preceding formulas, we can obtain useful estimates of the error. Using (5.4.39),

$$I - I_n^{(1)} = \frac{16I_n^{(1)} - I_{n/2}^{(1)}}{15} - I_n^{(1)} + \frac{d_6^{(2)}}{n^6} + \cdots$$

$$= \frac{I_n^{(1)} - I_{n/2}^{(1)}}{15} + \frac{d_6^{(2)}}{n^6} + \cdots$$

Using $h = (b - a)/n$,

$$I - I_n^{(1)} = \frac{1}{15}\left[I_n^{(1)} - I_{n/2}^{(1)}\right] + O(h^6) \tag{5.4.41}$$

and thus

$$I - I_n^{(1)} \doteq \frac{1}{15}\left[I_n^{(1)} - I_{n/2}^{(1)}\right] \tag{5.4.42}$$

since both terms are $O(h^4)$ and the remainder term is $O(h^6)$. This is called *Richardson's error estimate* for Simpson's rule.

This extrapolation process can be continued inductively. Define

$$I_n^{(k)} = \frac{4^k I_n^{(k-1)} - I_{n/2}^{(k-1)}}{4^k - 1} \qquad n \geq 2^k \tag{5.4.43}$$

with $n$ a multiple of $2^k$, $k \geq 1$. It can be shown that the error has the form

$$I - I_n^{(k)} = \frac{d_{2k+2}^{(k)}}{n^{2k+2}} + \cdots$$

$$= A_k(b - a)h^{2k+2}f^{(2k+2)}(\zeta_n) \qquad a < \zeta_n < b \tag{5.4.44}$$

with $A_k$ a constant independent of $f$ and $h$, and

$$d_{2k+2}^{(k)} = A_k(b - a)^{2k+2}\left[f^{(2k+1)}(b) - f^{(2k+1)}(a)\right]$$

Finally, it can be shown that for any $f \in C[a, b]$,

$$\underset{n \to \infty}{\text{Limit}} I_n^{(k)}(f) = I(f) \tag{5.4.45}$$

The rules $I_n^{(k)}(f)$ for $k > 2$ bear no direct relation to the composite Newton–Cotes rules. See Bauer et al. (1963) for complete details.

**Romberg integration**  Define

$$J_k(f) = I_{2^k}^{(k)} \qquad k = 0, 1, 2, \ldots \tag{5.4.46}$$

$$I_1^{(0)}$$

$$I_2^{(0)} \quad I_2^{(1)}$$

$$I_4^{(0)} \quad I_4^{(1)} \quad I_4^{(2)}$$

$$I_8^{(0)} \quad I_8^{(1)} \quad I_8^{(2)} \quad I_8^{(3)}$$

$$I_{16}^{(0)} \quad I_{16}^{(1)} \quad I_{16}^{(2)} \quad I_{16}^{(3)} \quad I_{16}^{(4)}$$

$$\vdots \quad \vdots \quad \vdots \quad \vdots \quad \vdots$$

**Figure 5.4**   Romberg integration table.

This is the Romberg integration rule. Consider the diagram in Figure 5.4 for the Richardson extrapolates of the trapezoidal rule, with the number of subdivisions a power of 2. The first column denotes the trapezoidal rule, the second Simpson's rule, etc. By (5.4.45), each column converges to $I(f)$. Romberg integration is the rule of taking the diagonal. Since each column converges more rapidly than the preceding column, assuming $f(x)$ is infinitely differentiable, it could be expected that $J_k(f)$ would converge more rapidly than $\{I_n^{(k)}\}$ for any $k$. This is usually the case, and consequently the method has been very popular since the late 1950s. Compared with Gaussian quadrature, Romberg integration has the advantage of using evenly spaced abscissas. For a more complete analysis of Romberg integration, see Bauer et al. (1963).

*Example*   Using Romberg integration, evaluate

$$I = \int_0^\pi e^x \cos(x)\, dx = -\tfrac{1}{2}(e^\pi + 1)$$

This was used previously as an example, in Tables 5.1, 5.3, and 5.11, for the trapezoidal, Simpson, and Gauss–Legendre rules, respectively. The Romberg results are given in Table 5.17. They show that Romberg integration is superior to Simpson's rule, but Gaussian quadrature is still more rapidly convergent.

To compute $J_k(f)$ for a particular $k$, having already computed $J_1(f)$, $\ldots, J_{k-1}(f)$, the row

$$I_n^{(0)}, I_n^{(1)}, \ldots, I_n^{(k-1)} \qquad n = 2^{k-1} \tag{5.4.47}$$

**Table 5.17   Example of Romberg integration**

| $k$ | Nodes | $J_k(f)$ | Error |
|---|---|---|---|
| 0 | 2 | −34.77851866026 | 2.27E + 1 |
| 1 | 3 | −11.59283955342 | −4.78E − 1 |
| 2 | 5 | −12.01108431754 | −5.93E − 2 |
| 3 | 9 | −12.07042041287 | 7.41E − 5 |
| 4 | 17 | −12.07034720873 | 8.92E − 7 |
| 5 | 33 | −12.07034631632 | −6.82E − 11 |
| 6 | 65 | −12.07034631639 | < 5.00E − 12 |

should have been saved in temporary storage. Then compute $I_{2^k}^{(0)}$ from $I_{2^{k-1}}^{(0)}$ and $2^{k-1}$ new function values. Using (5.4.33), (5.4.40), and (5.4.43) compute the next row in the table, including $J_k(f)$. Compare $J_k(f)$ and $J_{k-1}(f)$ to see if there is sufficient accuracy to accept $J_k(f)$ as an accurate approximation to $I(f)$.

We give this procedure in a formal way in the following algorithm. It is included for pedagogical reasons, and it should not be considered as a serious program unless some improvements are included. For example, the error test is primitive and much too conservative, and a safety check needs to be included for the numerical integrals associated with small $k$, when not enough function values have yet been sampled.

**Algorithm** *Romberg* $(f, a, b, \epsilon, \text{int})$

1. *Remark:* Use Romberg integration to calculate int, an estimate of the integral

$$I = \int_a^b f(x)\, dx$$

Stop when $|I - \text{int}| \le \epsilon$.

2. Initialize:

$k := 0, \; n := 1,$

$T_0 := R_0 := \alpha_0 := (b - a)[f(a) + f(b)]/2$

3. Begin the main loop:

$n := 2n \qquad k := k + 1 \qquad h = (b - a)/n$

4. $\text{sum} := \displaystyle\sum_{j=1}^{n/2} f(a + (2j - 1)h)$

5. $T_k := h \cdot \text{sum} + \dfrac{1}{2} T_{k-1}$

6. $\beta_j := \alpha_j \qquad j = 0, 1, \ldots, k - 1$

7. $\alpha_0 := T_k \qquad m := 1$

8. Do through step 10 for $j = 1, 2, \ldots, k$

9. $m := 4m$

10. $\alpha_j := \dfrac{m \cdot \alpha_{j-1} - \beta_{j-1}}{m - 1}$

11.   $R_k := \alpha_k$

12.   If $|R_k - R_{k-1}| > \epsilon$, then go to step 3

13.   Since $|R_k - R_{k-1}| \leq \epsilon$, accept int $= R_{k-1}$ and return.

There are many variants of Romberg integration. For example, other ways of increasing the number of nodes have been studied. For a very complete survey of the literature on Romberg integration, see Davis and Rabinowitz (1984, pp. 434–446). They also give a Fortran program for Romberg integration.

## 5.5   Automatic Numerical Integration

An automatic numerical integration program calculates an approximate integral to within an accuracy specified by the user of the program. The user does not need to specify either the method or the number of nodes to be used. There are some excellent automatic integration programs, and many people use them. Such a program saves you the time of writing your own program, and for many people, it avoids having to understand the needed numerical integration theory. Nonetheless, it is almost always possible to improve upon an automatic program, although it usually requires a good knowledge of the numerical integration needed for your particular problem. When doing only a small number of numerical integrations, automatic integration is often a good way to save time. But for problems involving many integrations, it is probably better to invest the time to find a less expensive numerical integration procedure.

An automatic numerical integration program functions as a "black box," without the user being able to see the intermediate steps of the computation. Because of this, the most important characteristic of such a program is that it be reliable: The approximate integral that is returned by the program and that the program says satisfies the user's error tolerance must, in fact, be that accurate. In theory, no such algorithm exists, as we explain in the next paragraph. But for the type of integrands that one usually considers in practice, there are programs that have a high order of reliability. This reliability will be improved if the user reads the program description, to see the restrictions and assumptions of the program.

To understand the theoretical impossibility of a perfectly reliable automatic integration program, note that the program will evaluate the integrand $f(x)$ at only a finite number of points, say $x_1, \ldots, x_n$. Then there are an infinity of continuous functions $\hat{f}(x)$ for which

$$\hat{f}(x_i) = f(x_i) \qquad i = 1, \ldots, n$$

and

$$\int_a^b \hat{f}(x)\, dx \neq \int_a^b f(x)\, dx$$

In fact, there are an infinity of such functions $\hat{f}(x)$ that are infinitely differentiable. For practical problems, it is unlikely that a well-constructed automatic integration program will be unreliable, but it is possible. An automatic integration program can be made more reliable by increasing the stringency of its error tests, but this also makes the program less efficient. Generally there is a tradeoff between reliability and efficiency. For a further discussion of the questions of reliability and efficiency of automatic quadrature programs, see Lyness and Kaganove (1976).

**Adaptive quadrature**   Automatic programs can be divided into (1) those using a global rule, such as Gaussian quadrature or the trapezoidal rule with even spacing, and (2) those using an *adaptive* strategy, in which the integration rule varies its placement of node points and even its definition to reflect the varying local behavior of the integrand. Global strategies use the type of error estimation that we have discussed in previous sections. We now discuss the concept and practice of an adaptive strategy.

Many integrands vary in their smoothness or differentiability at different points of the interval of integration $[a, b]$. For example, with

$$I = \int_0^1 \sqrt{x} \, dx$$

the integrand has infinite slope at $x = 0$, but the function is well behaved at points $x$ near 1. Most numerical methods use a uniform grid of node points, that is, the density of node points is about equal throughout the integration interval. This includes composite Newton–Cotes formulas, Gaussian quadrature, and Romberg integration. When the integrand is badly behaved at some point $\alpha$ in the interval $[a, b]$, many node points must be placed near $\alpha$ to compensate for this. But this forces many more node points than necessary to be used at all other parts of $[a, b]$. Adaptive integration attempts to place node points according to the behavior of the integrand, with the density of node points being greater near points of bad behavior.

We now explain the basic concept of adaptive integration using a simplified *adaptive Simpson's rule*. To see more precisely why variable spacing is necessary, consider Simpson's rule with such a spacing of the nodes:

$$I(f) = \sum_{j=1}^{n/2} \int_{x_{2j-2}}^{x_{2j}} f(x) \, dx \doteq I_n(f) = \sum_{j=1}^{n/2} \left( \frac{x_{2j} - x_{2j-2}}{6} \right) (f_{2j-2} + 4f_{2j-1} + f_{2j})$$

$$(5.5.1)$$

with $x_{2j-1} = (x_{2j-2} + x_{2j})/2$. Using (5.1.15),

$$I(f) - I_n(f) = -\frac{1}{2880} \sum_{j=1}^{n/2} (x_{2j} - x_{2j-2})^5 f^{(4)}(\xi_j) \qquad (5.5.2)$$

with $x_{2j-2} \le \xi_j \le x_{2j}$. Clearly, you want to choose $x_{2j} - x_{2j-2}$ according to the size of $f^{(4)}(\xi)$, which is unknown in general. If $f^{(4)}(x)$ varies greatly in magnitude, you do not want even spacing of the node points.

As notation, introduce

$$I_{\alpha, \beta} = \int_{\alpha}^{\beta} f(x)\, dx$$

$$I^{(1)}_{\alpha, \beta} = \frac{h}{3}\left[ f(\alpha) + 4f\left(\frac{\alpha + \beta}{2}\right) + f(\beta)\right] \qquad h = \frac{\beta - \alpha}{2} \qquad (5.5.3)$$

$$I^{(2)}_{\alpha, \beta} = I^{(1)}_{\alpha, \gamma} + I^{(1)}_{\gamma, \beta} \qquad \gamma = \frac{\beta + \alpha}{2}$$

To describe the adaptive algorithm for computing

$$I = \int_{a}^{b} f(x)\, dx$$

we use a recursive definition. Suppose that $\epsilon > 0$ is given, and that we want to find an approximate integral $\tilde{I}$ for which

$$|I - \tilde{I}| \le \epsilon \qquad (5.5.4)$$

Begin by setting $\alpha = a$, $\beta = b$. Compute $I^{(1)}_{\alpha, \beta}$ and $I^{(2)}_{\alpha, \beta}$. If

$$|I^{(2)}_{\alpha, \beta} - I^{(1)}_{\alpha, \beta}| \le \epsilon \qquad (5.5.5)$$

then accept $I^{(2)}_{\alpha, \beta}$ as the adaptive integral approximation to $I_{\alpha, \beta}$. Otherwise let $\epsilon := \epsilon/2$, and set the adaptive integral for $I_{\alpha, \beta}$ equal to the sum of the adaptive integrals for $I_{\alpha, \gamma}$ and $I_{\gamma, \beta}$, $\gamma = (\alpha + \beta)/2$, each to be computed with an error tolerance of $\epsilon$.

In an actual implementation as a computer program, many extra limitations are included as safeguards; and the error estimation is usually much more sophisticated. All function evaluations are handled carefully in order to ensure that the integrand is never evaluated twice at the same point. This requires a clever stacking procedure for those values of $f(x)$ that must be temporarily stored because they will be needed again later in the computation. There are many small modifications that can be made to improve the performance of the program, but generally a great deal of experience and empirical investigation is first necessary. For that and other reasons, it is recommended that standard well-tested adaptive procedures be used [e.g., de Boor (1971), Piessens et al. (1983)]. This is discussed further at the end of the section.

**Table 5.18    Adaptive Simpson's example (5.5.6)**

| $[\alpha, \beta]$ | $I^{(2)}$ | $I - I^{(2)}$ | $I - I^{(1)}$ | $|I^{(2)} - I^{(1)}|$ | $\epsilon$ |
|---|---|---|---|---|---|
| [0.0, .0625] | .010258 | 1.6E − 4 | 4.5E − 4 | 2.9E − 4 | .0003125 |
| [.0625, .0125] | .019046 | 1.2E − 7 | 1.1E − 6 | 1.0E − 6 | .0003125 |
| [.125, .25] | .053871 | 4.5E − 7 | 3.6E − 6 | 4.0E − 6 | .000625 |
| [.25, .5] | .152368 | 9.3E − 7 | 1.1E − 5 | 1.0E − 5 | .00125 |
| [.5, 1.0] | .430962 | 2.4E − 6 | 3.0E − 5 | 2.8E − 5 | .0025 |

*Example*   Consider using the preceding simpleminded adaptive Simpson procedure to evaluate

$$I = \int_0^1 \sqrt{x} \; dx \qquad (5.5.6)$$

with $\epsilon = .005$ on $[0, 1]$. The final intervals $[\alpha, \beta]$ and integrals $I_{\alpha, \beta}^{(2)}$ are given in Table 5.18. The column labeled $\epsilon$ gives the error tolerance used in the test (5.5.5), which estimates the error in $I_{\alpha, \beta}^{(1)}$. The error estimated for $I_{\alpha, \beta}^{(1)}$ on $[0, .0625]$ was inaccurate, but it was accurate for the remaining subintervals. The value used to estimate $I_{\alpha, \beta}$ is actually $I_{\alpha, \beta}^{(2)}$, and it is sufficiently accurate on all subintervals. The total integral, obtained by summing all $I_{\alpha, \beta}^{(2)}$, is

$$\tilde{I} = .666505 \qquad I - \tilde{I} = 1.6E - 4$$

and the calculated bound is

$$|I - \tilde{I}| \le 3.3E - 4$$

obtained by summing the column labeled $|I^{(2)} - I^{(1)}|$. Note that the error is concentrated on the first subinterval, as could have been predicted from the behavior of the integrand near $x = 0$. For an example where the test (5.5.5) is not adequate, see Problem 32.

**Some automatic integration programs**   One of the better known automatic integration programs is the adaptive program CADRE (Cautious Adaptive Romberg Extrapolation), given in de Boor (1971). It includes a means of recognizing algebraic singularities at the endpoints of the integration interval. The asymptotic error formulas of Lyness and Ninham (1967), given in (5.4.21) in a special case, are used to produce a more rapidly convergent integration method, again based on repeated Richardson extrapolation. The routine CADRE has been found empirically to be both quite reliable and efficient.

**Table 5.19   Integration examples for CADRE**

| | Desired Error | | | | | |
|---|---|---|---|---|---|---|
| | $10^{-2}$ | | $10^{-5}$ | | $10^{-8}$ | |
| Integral | Error | $N$ | Error | $N$ | Error | $N$ |
| $I_1$ | $A = 2.49E - 6$ | 76 | $A = 1.40E - 7$ | 73 | $A = 4.60E - 11$ | 225 |
| | $P = 5.30E - 4$ | | $P = 4.45E - 6$ | | $P = 2.48E - 9$ | |
| $I_2$ | $A = 1.18E - 5$ | 9 | $A = 3.96E - 7$ | 17 | $P = 2.73E - 10$ | 129 |
| | $P = 3.27E - 3$ | | $P = 3.56E - 6$ | | $P = 2.81E - 9$ | |
| $I_3$ | $A = 1.03E - 4$ | 17 | $A = 3.23E - 8$ | 33 | $A = 1.98E - 9$ | 65 |
| | $P = 2.98E - 3$ | | $P = 4.43E - 8$ | | $P = 2.86E - 9$ | |
| $I_4$ | $A = 6.57E - 5$ | 105 | $A = 6.45E - 8$ | 209 | $A = 4.80E - 9$ | 281 |
| | $P = 4.98E - 3$ | | $P = 9.22E - 6$ | | $P = 1.55E - 8$ | |
| $I_5$ | $A = 2.77E - 5$ | 226 | $A = 7.41E - 8$ | 418 | $A = 5.89E - 9$ | 562 |
| | $P = 3.02E - 3$ | | $P = 1.00E - 5$ | | $P = 1.11E - 8$ | |
| $I_6$ | $A = 8.49E - 6$ | 955 | $A = 2.37E - 8$ | 1171 | $A = 4.30E - 11$ | 2577 |
| | $P = 8.48E - 3$ | | $P = 1.67E - 5$ | | $P = 2.07E - 8$ | |
| $I_7$ | $A = 4.54E - 4$ | 98 | $A = 7.72E - 7$ | 418 | $A = **$ | 1506 |
| | $P = 1.30E - 3$ | | $P = 8.02E - 6$ | | $P = **$ | |

A more recently developed package is QUADPACK, some of whose programs are general purpose, while others deal with special classes of integrals. The package was a colloborative effort, and a complete description of it is given in Piessens et al. (1983). The package is well tested and appears to be an excellent collection of programs.

We illustrate the preceding by calculating numerical approximations to the following integrals:

$$I_1 = \int_0^1 \frac{4\,dx}{1 + 256(x - .375)^2} = \frac{1}{4}\left[\tan^{-1}(10) + \tan^{-1}(6)\right]$$

$$I_2 = \int_0^1 x^2\sqrt{x}\,dx = \frac{2}{7}$$

$$I_3 = \int_0^1 \sqrt{x}\,dx = \frac{2}{3}$$

$$I_4 = \int_0^1 \log(x)\,dx = -1$$

$$I_5 = \int_0^1 \log|x - .7|\,dx = .3\log(.3) + .7\log(.7) - 1$$

$$I_6 = \int_0^{2\pi} e^{-x}\sin(50x)\,dx = \frac{50}{2501}(1 - e^{-2\pi})$$

$$I_7 = \int_{-9}^{10000} \frac{dx}{\sqrt{|x|}} = 206$$

From QUADPACK, we chose DQAGP. It too contains ways to recognize algebraic singularities at the endpoints and to compensate for their presence. To improve performance, it allows the user to specify points interior to the integration interval at which the integrand is singular.

We used both CADRE and DQAGP to calculate the preceding integrals, with error tolerances of $10^{-2}$, $10^{-5}$, and $10^{-8}$. The results are shown in Tables 5.19 and 5.20. To more fairly compare DQAGP and CADRE, we applied CADRE to two integrals in both $I_5$ and $I_7$, to have the singularities occur at endpoints. For example, we used CADRE for each of the integrals in

$$I_7 = \int_{-9}^0 \frac{dx}{\sqrt{-x}} + \int_0^{10000} \frac{dx}{\sqrt{x}} \tag{5.5.7}$$

In the tables, $P$ denotes the error bound predicted by the program and $A$ denotes the actual absolute error in the calculated answer. Column $N$ gives the number of integrand evaluations. At all points at which the integrand was undefined, it was arbitrarily set to zero. The examples were computed in double precision on a Prime 850, with a unit round of $2^{-46} \doteq 1.4 \times 10^{-14}$.

In Table 5.19, CADRE failed for $I_7$ with the tolerance $10^{-8}$, even though (5.5.7) was used. Otherwise, it performed quite well. When the decomposition

**Table 5.20   Integration examples for DQAGP**

| Integral | Desired Error | | | | | |
| | $10^{-2}$ | | $10^{-5}$ | | $10^{-8}$ | |
| | Error | $N$ | Error | $N$ | Error | $N$ |
|---|---|---|---|---|---|---|
| $I_1$ | $A = 2.88\text{E} - 9$ | 105 | $A = 5.40\text{E} - 13$ | 147 | $A = 5.40\text{E} - 13$ | 147 |
| | $P = 2.96\text{E} - 3$ | | $P = 5.21\text{E} - 10$ | | $P = 5.21\text{E} - 10$ | |
| $I_2$ | $A = 1.17\text{E} - 11$ | 21 | $A = 1.17\text{E} - 11$ | 21 | $A = 1.17\text{E} - 11$ | 21 |
| | $P = 7.46\text{E} - 9$ | | $P = 7.46\text{E} - 9$ | | $P = 7.46\text{E} - 9$ | |
| $I_3$ | $A = 4.79\text{E} - 6$ | 21 | $A = 4.62\text{E} - 13$ | 189 | $A = 4.62\text{E} - 13$ | 189 |
| | $P = 4.95\text{E} - 3$ | | $P = 4.77\text{E} - 14$ | | $P = 4.77\text{E} - 14$ | |
| $I_4$ | $A = 5.97\text{E} - 13$ | 231 | $A = 5.97\text{E} - 13$ | 231 | $A = 5.97\text{E} - 13$ | 231 |
| | $P = 7.15\text{E} - 14$ | | $P = 7.15\text{E} - 14$ | | $P = 7.15\text{E} - 14$ | |
| $I_5$ | $A = 8.67\text{E} - 13$ | 462 | $A = 8.67\text{E} - 13$ | 462 | $A = 8.67\text{E} - 13$ | 462 |
| | $P = 1.15\text{E} - 13$ | | $P = 1.15\text{E} - 13$ | | $P = 1.15\text{E} - 13$ | |
| $I_6$ | $A = 1.00\text{E} - 3$ | 525 | $A = 6.33\text{E} - 14$ | 861 | $A = 5.33\text{E} - 14$ | 1239 |
| | $P = 4.36\text{E} - 3$ | | $P = 8.13\text{E} - 6$ | | $P = 7.12\text{E} - 9$ | |
| $I_7$ | $A = 1.67\text{E} - 10$ | 462 | $A = 1.67\text{E} - 10$ | 462 | $A = 1.67\text{E} - 10$ | 462 |
| | $P = 1.16\text{E} - 10$ | | $P = 1.16\text{E} - 10$ | | $P = 1.16\text{E} - 10$ | |

(5.5.7) is not used and CADRE is called only once for the single interval $[-9, 10000]$, it fails for all three error tolerances.

In Table 5.20, the predicted error is in some cases smaller than the actual error. This difficulty appears to be due to working at the limits of the machine arithmetic precision, and in all cases the final error was well within the limits requested.

In comparing the two programs, DQAGP and CADRE are both quite reliable and efficient. Also, both programs perform relatively poorly for the highly oscillatory integral $I_6$, showing that $I_6$ should be evaluated using a program designed for oscillatory integrals (such as DQAWO in QUADPACK, for Fourier coefficient calculations). From the tables, DQAGP is somewhat more able to deal with difficult integrals, while remaining about equally efficient compared to CADRE. Much more detailed examples for CADRE are given in Robinson (1979).

Automatic quadrature programs can be easily misused in large calculations, resulting in erroneous results and great inefficiency. For comments on the use of such programs in large calculations and suggestions for choosing when to use them, see Lyness (1983). The following are from his concluding remarks.

*The Automatic Quadrature Rule* (AQR) is an impressive and practical item of numerical software. Its main advantage for the user is that it is convenient. He can take it from the library shelf, plug it in, and feel confident that it will work. For this convenience, there is in general a modest charge in CPU time, this surcharge being a factor of about 3. The *Rule Evaluation Quadrature Routine* (REQR) [non-automatic quadrature rule] does not carry this surcharge, but to code and check out an REQR might take a couple of hours of the user's time. So unless the expected CPU time is high, many user's willingly pay the surcharge in order to save themselves time and trouble.

However there are certain—usually large scale—problems for which the AQR is not designed and in which its uncritical use can lead to CPU time surcharges by factors of 100 or more. ... These are characterized by the circumstances that a large number of separate quadratures are involved, and that the results of these quadratures are subsequently used as input to some other numerical process. In order to recognize this situation, it is necessary to examine the subsequent numerical process to see whether it requires a smooth input function. ... For some of these problems, an REQR is quite suitable while an AQR may lead to a numerical disaster.

## 5.6  Singular Integrals

We discuss the approximate evaluation of integrals for which methods of the type discussed in Sections 5.1 through 5.4 do not perform well: these methods include the composite Newton–Cotes rules (e.g., the trapezoidal rule), Gauss–Legendre quadrature, and Romberg integration. The integrals discussed here lead to poorly convergent numerical integrals when evaluated using the latter integration rules, for a variety of reasons. We discuss (1) integrals whose integrands contain a singularity in the interval of integration $(a, b)$, and (2) integrals with an infinite interval of integration. Adaptive integration methods can be used for these integrals, but it is usually possible to obtain more rapidly convergent approximations by carefully examining the nature of the singular behavior and then compensating for it.

**Change of the variable of integration**  We illustrate the importance of this idea with several examples. For

$$I = \int_0^b \frac{f(x)\, dx}{\sqrt{x}} \tag{5.6.1}$$

with $f(x)$ a function with several continuous derivatives, let $x = u^2, 0 \le u \le \sqrt{b}$. Then

$$I = 2 \int_0^{\sqrt{b}} f(u^2)\, du$$

This integral has a smooth integrand and standard techniques can be applied to it.

Similarly,

$$\int_0^1 \sin(x)\sqrt{1 - x^2}\, dx = 2 \int_0^1 u^2 \sqrt{2 - u^2}\, \sin(1 - u^2)\, du$$

using the change of variable $u = \sqrt{1 - x}$. The right-hand integrand has an infinite number of continuous derivatives on $[0, 1]$, whereas the derivative of the first integrand was singular at $x = 1$.

For an infinite interval of integration, the change of variable technique is also useful. Suppose

$$I = \int_1^\infty \frac{f(x)}{x^p} \, dx \qquad p > 1 \tag{5.6.2}$$

with $\text{Limit}_{x \to \infty} f(x)$ existing. Also assume $f(x)$ is smooth on $[1, \infty)$. Then use the change of variable

$$x = \frac{1}{u^\alpha} \qquad dx = \frac{-\alpha}{u^{1+\alpha}} \, du \qquad \text{for some} \quad \alpha > 0$$

Then

$$I = \alpha \int_0^1 u^{p\alpha} f\left(\frac{1}{u^\alpha}\right) \frac{du}{u^{1+\alpha}} = \alpha \int_0^1 u^{(p-1)\alpha - 1} f\left(\frac{1}{u^\alpha}\right) du \tag{5.6.3}$$

Maximize the smoothness of the new integrand at $u = 0$ by picking $\alpha$ to produce a large value for the exponent $(p - 1)\alpha - 1$. For example, with

$$I = \int_1^\infty \frac{f(x) \, dx}{x\sqrt{x}}$$

the change of variable $x = 1/u^4$ leads to

$$I = 4 \int_0^1 uf\left(\frac{1}{u^4}\right) du \tag{5.6.4}$$

If we assume a behavior at $x = \infty$ of

$$f(x) = c_0 + \frac{c_1}{x} + \frac{c_2}{x^2} + \cdots$$

then

$$uf\left(\frac{1}{u^4}\right) = c_0 u + c_1 u^5 + c_2 u^9 + \cdots$$

and (5.6.4) has a smooth integrand at $u = 0$.

An interesting idea has been given in Iri et al. (1970) to deal with endpoint singularities in the integral

$$I = \int_a^b f(x) \, dx \tag{5.6.5}$$

Define

$$\psi(t) = \exp\left(\frac{-c}{1 - t^2}\right) \tag{5.6.6}$$

$$\varphi(t) = a + \frac{b - a}{\gamma} \int_{-1}^t \psi(u) \, du \qquad -1 \le t \le 1 \tag{5.6.7}$$

where $c$ is a positive constant and

$$\gamma = \int_{-1}^{1} \psi(u) \, du$$

As $t$ varies from $-1$ to $1$, $\varphi(t)$ varies from $a$ to $b$. Using $x = \varphi(t)$ as a change of variable in (5.6.5), we obtain

$$I = \int_{-1}^{1} f(\varphi(t))\varphi'(t) \, dt \qquad (5.6.8)$$

The function $\varphi'(t) = ((b - a)/\gamma)\psi(t)$ is infinitely differentiable on $[-1, 1]$, and it and all of derivatives are zero at $t = \pm 1$. In (5.6.8), the integrand and all of derivatives will vanish at $t = \pm 1$ for virtually all functions $f(x)$ of interest. Using the error formula (5.4.9) for the trapezoidal rule on $[-1, 1]$, it can be seen that the trapezoidal rule will converge very rapidly when applied to (5.6.8). We will call this method the IMT method.

This method has been implemented in de Doncker and Piessens (1976), and in the general comparisons of Robinson (1979), it is rated as an extremely reliable and quite efficient way of handling integrals (5.6.4) that have endpoint singularities. De Doncker and Piessens (1976) also treat integrals over $[0, \infty)$ by first using the change of variable $x = (1 + u)/(1 - u)$, $-1 \le u < 1$, followed by the change of variable $u = \varphi(t)$.

***Example*** Use the preceding method (5.6.5)–(5.6.8) with the trapezoidal rule, to evaluate

$$I = \int_{0}^{1} \sqrt{-\ln x} \, dx = \frac{\sqrt{\pi}}{2} \doteq .8862269 \qquad (5.6.9)$$

Note that the integrand has singular behavior at both endpoints, although it is different in the two cases. The constant in (5.6.6) is $c = 4$, and the evaluation of (5.6.7) is taken from Robinson and de Doncker (1981). The results are shown in Table 5.21. The column labeled nodes gives the number of nodes interior to $[0, 1]$.

**Table 5.21    Example of the IMT method**

| Nodes | Error |
|---|---|
| 2 | $-6.54\text{E} - 2$ |
| 4 | $5.82\text{E} - 3$ |
| 8 | $-1.30\text{E} - 4$ |
| 16 | $7.42\text{E} - 6$ |
| 32 | $1.17\text{E} - 8$ |
| 64 | $1.18\text{E} - 12$ |

**Gaussian quadrature**   In Section 5.3, we developed a general theory for Gaussian quadrature formulas

$$\int_a^b w(x)f(x)\, dx \doteq \sum_{j=1}^n w_{j,n} f(x_{j,n}) \qquad n \geq 1$$

that have degree of precision $2n - 1$. The construction of the nodes and weights, and the form of the error, are given in Theorem 5.3. For our work in this section we note that (1) the interval $(a, b)$ is allowed to be infinite, and (2) $w(x)$ can have singularities on $(a, b)$, provided it is nonnegative and satisfies the assumptions (4.3.8) and (4.3.9) of Section 4.3. For rapid convergence, we would also expect that $f(x)$ would need to be a smooth function, as was illustrated with Gauss–Legendre quadrature in Section 5.3.

The weights and nodes for a wide variety of weight functions $w(x)$ and intervals $(a, b)$ are known. The tables of Stroud and Secrest (1966) include the integrals

$$\int_0^\infty x^\alpha e^{-x} f(x)\, dx \qquad \int_{-\infty}^\infty e^{-x^2} f(x)\, dx \qquad \int_0^1 \ln\left(\frac{1}{x}\right) f(x)\, dx \quad (5.6.10)$$

and others. The constant $\alpha > -1$. There are additional books containing tables for integrals other than those in (5.6.10). In addition, the paper by Golub and Welsch (1969) describes a procedure for constructing the nodes and weights in (5.6.10), based on solving a matrix eigenvalue problem. A program is given, and it includes most of the more popular weighted integrals to which Gaussian quadrature is applied. For an additional discussion of Gaussian quadrature, with references to the literature (including tables and programs), see Davis and Rabinowitz (1984, pp. 95–132, 222–229).

***Example***   We illustrate the use of Gaussian quadrature for evaluating integrals

$$I = \int_0^\infty g(x)\, dx$$

We use Gauss–Laguerre quadrature, in which $w(x) = e^{-x}$. Then write $I$ as

$$I = \int_0^\infty e^{-x} [e^x g(x)]\, dx = \int_0^\infty e^{-x} f(x)\, dx \qquad (5.6.11)$$

We give results for three integrals:

$$I^{(1)} = \int_0^\infty \frac{x\, dx}{e^x - 1} = \frac{\pi^2}{6}$$

$$I^{(2)} = \int_0^\infty \frac{x\, dx}{(1 + x^2)^5} = \frac{1}{8}$$

$$I^{(3)} = \int_0^\infty \frac{dx}{1 + x^2} = \frac{\pi}{2}$$

**Table 5.22    Examples of Gauss–Laguerre quadrature**

| Nodes | $I^{(1)} - I_n^{(1)}$ | $I^{(2)} - I_n^{(2)}$ | $I^{(3)} - I_n^{(3)}$ |
|---|---|---|---|
| 2 | $1.01\text{E} - 4$ | $-8.05\text{E} - 2$ | $7.75\text{E} - 2$ |
| 4 | $1.28\text{E} - 5$ | $-4.20\text{E} - 2$ | $6.96\text{E} - 2$ |
| 8 | $-9.48\text{E} - 8$ | $1.27\text{E} - 2$ | $3.70\text{E} - 2$ |
| 16 | $3.16\text{E} - 11$ | $-1.39\text{E} - 3$ | $1.71\text{E} - 2$ |
| 32 | $7.11\text{E} - 14$ | $3.05\text{E} - 5$ | $8.31\text{E} - 3$ |
| 64 | — | $1.06\text{E} - 7$ | $4.07\text{E} - 3$ |

Gauss–Laguerre quadrature is best for integrands that decrease exponentially as $x \to \infty$. For integrands that are $O(1/x^p)$, $p > 1$, as $x \to \infty$, the convergence rate becomes quite poor as $p \to 1$. These comments are illustrated in Table 5.22. For a formal discussion of the convergence of Gauss–Laguerre quadrature, see Davis and Rabinowitz (1984, p. 227).

One especially easy case of Gaussian quadrature is for the singular integral

$$I(f) = \int_{-1}^{1} \frac{f(x)\, dx}{\sqrt{1 - x^2}} \qquad (5.6.12)$$

With this weight function, the orthogonal polynomials are the Chebyshev polynomials $\{T_n(x), n \geq 0\}$. Thus the integration nodes in (5.6.10) are given by

$$x_{j,n} = \cos\left(\frac{2j - 1}{2n}\pi\right) \qquad j = 1, \ldots, n \qquad (5.6.13)$$

and from (5.3.11), the weights are

$$w_{j,n} = \frac{\pi}{n} \qquad j = 1, \ldots, n$$

Using the formula (5.3.10) for the error, the Gaussian quadrature formula for (5.6.12) is given by

$$\int_{-1}^{1} \frac{f(x)\, dx}{\sqrt{1 - x^2}} = \frac{\pi}{n} \sum_{j=1}^{n} f(x_{j,n}) + \frac{2\pi}{2^{2n}(2n)!} f^{(2n)}(\eta) \qquad (5.6.14)$$

for some $-1 < \eta < 1$.

This formula is related to the composite midpoint rule (5.2.18). Make the change of variable $x = \cos\theta$ in (5.6.14) to obtain

$$\int_{0}^{\pi} f(\cos\theta)\, d\theta = \frac{\pi}{n} \sum_{j=1}^{n} f(\cos\theta_{j,n}) + E \qquad (5.6.15)$$

where $\theta_{j,n} = (2j - 1)\pi/2n$. Thus Gaussian quadrature for (5.6.12) is equivalent to the composite midpoint rule applied to the integral on the left in (5.6.15). Like the trapezoidal rule, the midpoint rule has an error expansion very similar to that given in (5.4.9) using the Euler–MacLaurin formula. The Corollary 1 to Theorem 5.5 also is valid, showing the composite midpoint rule to be highly accurate for periodic functions. This is reflected in the high accuracy of (5.6.14). Thus Gaussian quadrature for (5.6.12) results in a formula that would have been reasonable from the asymptotic error expansion for the composite midpoint rule applied to the integral on the left of (5.6.15).

**Analytic treatment of singularity**    Divide the interval of integration into two parts, one containing the singular point, which is to be treated analytically. For example, consider

$$I = \int_0^b f(x) \log (x) \, dx = \left[ \int_0^\epsilon + \int_\epsilon^b \right] f(x) \log (x) \, dx \equiv I_1 + I_2 \quad (5.6.16)$$

Assuming $f(x)$ is smooth on $[\epsilon, b]$, apply a standard technique to the evaluation of $I_2$. For $f(x)$ about zero, assume it has a convergent Taylor series on $[0, \epsilon]$. Then

$$I_1 = \int_0^\epsilon f(x) \log (x) \, dx = \int_0^\epsilon \left( \sum_0^\infty a_j x^j \right) \log (x) \, dx$$

$$= \sum_0^\infty a_j \frac{\epsilon^{j+1}}{j+1} \left[ \log (\epsilon) - \frac{1}{j+1} \right] \quad (5.6.17)$$

For example, with

$$I = \int_0^{4\pi} \cos (x) \log (x) \, dx$$

define

$$I_1 = \int_0^{.1} \cos (x) \log (x) \, dx \qquad \epsilon = .1$$

$$= \epsilon [\log (\epsilon) - 1] - \frac{\epsilon^3}{6} \left[ \log (\epsilon) - \frac{1}{3} \right] + \frac{\epsilon^5}{600} \left[ \log (\epsilon) - \frac{1}{5} \right] - \cdots$$

This is an alternating series, and thus it is clear that using the first three terms will give a very accurate value for $I_1$. A standard method can be applied to $I_2$ on $[.1, 4\pi]$.

Similar techniques can be used with infinite intervals of integration $[a, \infty)$, discarding the integral over $[b, \infty)$ for some large value of $b$. This is not developed here.

**Product integration**    Let $I(f) = \int_a^b w(x) f(x) \, dx$ with a near-singular or singular weight function $w(x)$ and a smooth function $f(x)$. The main idea is to produce a sequence of functions $f_n(x)$ for which

**1.** $\|f - f_n\|_\infty = \underset{a \le x \le b}{\text{Max}} |f(x) - f_n(x)| \to 0$ as $n \to \infty$

**2.** The integrals

$$I_n(f) \equiv \int_a^b w(x) f_n(x)\, dx \qquad (5.6.18)$$

can be fairly easily evaluated.

This generalizes the schema (5.0.2) of the introduction. For the error,

$$|I(f) - I_n(f)| \le \int_a^b |w(x)|\,|f(x) - f_n(x)|\, dx$$

$$\le \|f - f_n\|_\infty \int_a^b |w(x)|\, dx \qquad (5.6.19)$$

Thus $I_n(f) \to I(f)$ as $n \to \infty$, and the rate of convergence is at least as rapid as that of $f_n(x)$ to $f(x)$ on $[a, b]$.

Within the preceding framework, the product integration methods are usually defined by using piecewise polynomial interpolation to define $f_n(x)$ from $f(x)$. To illustrate the main ideas, while keeping the algebra simple, we will define the *product trapezoidal method* for evaluating

$$I(f) = \int_0^b f(x) \log(x)\, dx \qquad (5.6.20)$$

Let $n \ge 1$, $h = b/n$, $x_j = jh$ for $j = 0, 1, \ldots, n$. Define $f_n(x)$ as the piecewise linear function interpolating to $f(x)$ on the nodes $x_0, x_1, \ldots, x_n$. For $x_{j-1} \le x \le x_j$, define

$$f_n(x) = \frac{1}{h}\big[(x_j - x)f(x_{j-1}) + (x - x_{j-1})f(x_j)\big] \qquad (5.6.21)$$

for $j = 1, 2, \ldots, n$. From (3.1.10), it is straightforward to show

$$\|f - f_n\|_\infty \le \frac{h^2}{8}\|f''\|_\infty \qquad (5.6.22)$$

provided $f(x)$ is twice continuously differentiable for $0 \le x \le b$. From (5.6.19) we obtain the error bound

$$|I(f) - I_n(f)| \le \frac{h^2}{8}\|f''\|_\infty \int_0^b |\log(x)|\, dx \qquad (5.6.23)$$

This method of defining $I_n(f)$ is similar to the regular trapezoidal rule (5.1.5). The rule (5.1.5) could also have been obtained by integrating the preceding function $f_n(x)$, but the weight function would have been simply $w(x) \equiv 1$. We can easily generalize the preceding by using higher degree piecewise polynomial interpolation. The use of piecewise quadratic interpolation to define $f_n(x)$ leads to a formula $I_n(f)$ called the *product Simpson's rule*. And using the same reasoning as led to (5.6.23), it can be shown that

$$|I(f) - I_n(f)| \le \frac{\sqrt{3}}{27} h^3 \|f^{(3)}\|_\infty \int_0^b |\log(x)|\, dx \qquad (5.6.24)$$

Higher order formulas can be obtained by using even higher degree interpolation. For the computation of $I_n(f)$ using (5.6.21),

$$I_n(f) = \sum_{j=1}^{n} \int_{x_{j-1}}^{x_j} \log(x) \left[ \frac{(x_j - x)f(x_{j-1}) + (x - x_{j-1})f(x_j)}{h} \right] dx$$

$$= \sum_{k=0}^{n} w_k f(x_k) \qquad (5.6.25)$$

$$w_0 = \frac{1}{h} \int_{x_0}^{x_1} (x_1 - x) \log(x)\, dx \qquad w_n = \frac{1}{h} \int_{x_{n-1}}^{x_n} (x - x_{n-1}) \log(x)\, dx,$$

$$w_j = \frac{1}{h} \int_{x_{j-1}}^{x_j} (x - x_{j-1}) \log(x)\, dx$$

$$+ \frac{1}{h} \int_{x_j}^{x_{j+1}} (x_{j+1} - x) \log(x)\, dx \qquad j = 1, \ldots, n-1 \qquad (5.6.26)$$

The calculation of these weights can be simplified considerably. Making the change of variable $x - x_{j-1} = uh$, $0 \le u \le 1$, we have

$$\frac{1}{h} \int_{x_{j-1}}^{x_j} (x - x_{j-1}) \log(x)\, dx = h \int_0^1 u \log[(j-1+u)h]\, du$$

$$= \frac{h}{2} \log(h) + h \int_0^1 u \log(j-1+u)\, du$$

and

$$\frac{1}{h} \int_{x_{j-1}}^{x_j} (x_j - x) \log(x)\, dx = h \int_0^1 (1-u) \log[(j-1+u)h]\, du$$

$$= \frac{h}{2} \log(h) + h \int_0^1 (1-u) \log(j-1+u)\, du$$

Define

$$\psi_1(k) = \int_0^1 u \log(u+k)\, du \qquad \psi_2(k) = \int_0^1 (1-u) \log(u+k)\, du \qquad (5.6.27)$$

**Table 5.23    Weights for product trapezoidal rule**

| $k$ | $\psi_1(k)$ | $\psi_2(k)$ |
|---|---|---|
| 0 | $-.250$ | $-.750$ |
| 1 | .250 | .1362943611 |
| 2 | .4883759281 | .4211665768 |
| 3 | .6485778545 | .6007627239 |
| 4 | .7695705457 | .7324415720 |
| 5 | .8668602747 | .8365069785 |
| 6 | .9482428376 | .9225713904 |
| 7 | 1.018201652 | .9959596385 |

for $k = 0, 1, 2, \ldots$ . Then

$$w_0 = \frac{h}{2} \log(h) + h\psi_2(0) \qquad w_n = \frac{h}{2} \log(h) + h\psi_1(n-1)$$

$$w_j = h \log(h) + h[\psi_1(j-1) + \psi_2(j)] \qquad j = 1, 2, \ldots, n-1 \quad (5.6.28)$$

The functions $\psi_1(k)$ and $\psi_2(k)$ do not depend on $h$, $b$, or $n$. They can be calculated and stored in a table for use with a variety of values of $b$ and $n$. For example, a table of $\psi_1(k)$ and $\psi_2(k)$ for $k = 0, 1, \ldots, 99$ can be used with any $b > 0$ and with any $n \leq 100$. Once the table of the values $\psi_1(k)$ and $\psi_2(k)$ has been calculated, the cost of using the product trapezoidal rule is no greater than the cost of any other integration rule.

The integrals $\psi_1(k)$ and $\psi_2(k)$ in (5.6.27) can be evaluated explicitly; some values are given in Table 5.23.

***Example***    Compute $I = \int_0^1 (1/(x+2)) \log(x)\, dx \doteq -.4484137$. The computed values are given in Table 5.24. The computed rate of convergence is in agreement with the order of convergence of (5.6.23).

Many types of interpolation may be used to define $f_n(x)$, but most applications to date have used piecewise polynomial interpolation on evenly spaced node points. Other weight functions may be used, for example,

$$w(x) = x^\alpha \qquad \alpha > -1 \qquad x \geq 0 \qquad (5.6.29)$$

and again the weights can be reduced to a fairly simple formula similar to

**Table 5.24    Example of product trapezoidal rule**

| $n$ | $I_n$ | $I - I_n$ | Ratio |
|---|---|---|---|
| 1 | $-.4583333$ | .00992 | |
| 2 | $-.4516096$ | .00320 | 3.10 |
| 4 | $-.4493011$ | .000887 | 3.61 |
| 8 | $-.4486460$ | .000232 | 3.82 |

(5.6.28). For an irrational value of $\alpha$, say

$$w(x) = \frac{1}{x^{\sqrt{2}-1}}$$

a change of variables can no longer be used to remove the singularity in the integral. Also, one of the major applications of product integration is to integral equations in which the kernel function has an algebraic and/or logarithmic singularity. For such equations, changes of variables are no longer possible, even with square root singularities. For example, consider the equation

$$\lambda\varphi(x) - \int_a^b \frac{\varphi(y)\,dy}{|x-y|^{1/2}} = f(x) \qquad a \le x \le b$$

with $\lambda$, $a$, $b$, and $f$ given and $\varphi$ the desired unknown function. Product integration leads to efficient procedures for such equations, provided $\varphi(y)$ is a smooth function [see Atkinson (1976), p. 106].

For complicated weight functions in which the weights $w_j$ can no longer be calculated, it is often possible to modify the problem to one in which product integration is still easily applicable. This will be examined using an example.

***Example***   Consider $I = \int_0^\pi f(x)\log(\sin x)\,dx$. The integrand has a singularity at both $x = 0$ and $x = \pi$. Use

$$\log(\sin x) = \log\left[\frac{\sin(x)}{x(\pi - x)}\right] + \log(x) + \log(\pi - x)$$

and this gives

$$I = \int_0^\pi f(x)\log\left[\frac{\sin x}{x(\pi - x)}\right]dx + \int_0^\pi f(x)\log(x)\,dx$$

$$+ \int_0^\pi f(x)\log(\pi - x)\,dx$$

$$\equiv I_1 + I_2 + I_3 \tag{5.6.30}$$

Integral $I_1$ has an infinitely differentiable integrand, and any standard numerical method will perform well. Integral $I_2$ has already been discussed, with $w(x) = \log(x)$. For $I_3$, use a change of variable to write

$$I_3 = \int_0^\pi f(x)\log(\pi - x)\,dx = \int_0^\pi f(\pi - z)\log(z)\,dz$$

Combining with $I_2$,

$$I_2 + I_3 = \int_0^\pi \log(x)[f(x) + f(\pi - x)]\,dx$$

to which the preceding work applies. By such manipulations, the applicability of the cases $w(x) = \log(x)$ and $w(x) = x^\alpha$ is much greater than might first be imagined.

For an asymptotic error analysis of product integration, see the work of de Hoog and Weiss (1973), in which some generalizations of the Euler–MacLaurin expansion are derived. Using their results, it can be shown that the error in the product Simpson rule is $O(h^4 \log(h))$. Thus the bound (5.6.24) based on the interpolation error $f(x) - f_n(x)$ does not predict the correct rate of convergence. This is similar to the result (5.1.17) for the Simpson rule error, in which the error was smaller than the use of quadratic interpolation would lead us to believe.

## 5.7  Numerical Differentiation

Numerical approximations to derivatives are used mainly in two ways. First, we are interested in calculating derivatives of given data that are often obtained empirically. Second, numerical differentiation formulas are used in deriving numerical methods for solving ordinary and partial differential equations. We begin this section by deriving some of the most commonly used formulas for numerical differentiation.

The problem of numerical differentiation is in some ways more difficult than that of numerical integration. When using empirically determined function values, the error in these values will usually lead to instability in the numerical differentiation of the function. In contrast, numerical integration is stable when faced with such errors (see Problem 13).

**The classical formulas**    One of the main approaches to deriving a numerical approximation to $f'(x)$ is to use the derivative of a polynomial $p_n(x)$ that interpolates $f(x)$ at a given set of node points. Let $x_0, x_1, \ldots, x_n$ be given, and let $p_n(x)$ interpolate $f(x)$ at these nodes. Usually $\{x_i\}$ are evenly spaced. Then use

$$f'(x) \doteq p_n'(x) \tag{5.7.1}$$

From (3.1.6), (3.2.4), and (3.2.11):

$$p_n(x) = \sum_{j=0}^{n} f(x_j) l_j(x)$$

$$l_j(x) = \frac{\Psi_n(x)}{(x - x_j)\Psi_n'(x_j)}$$

$$= \frac{(x - x_0) \cdots (x - x_{j-1})(x - x_{j+1}) \cdots (x - x_n)}{(x_j - x_0) \cdots (x_j - x_{j-1})(x_j - x_{j+1}) \cdots (x_j - x_n)}$$

$$\Psi_n(x) = (x - x_0) \cdots (x - x_n)$$

$$f(x) - p_n(x) = \Psi_n(x) f[x_0, \ldots, x_n, x] \tag{5.7.2}$$

Thus

$$f'(x) \doteq p_n'(x) = \sum_{j=0}^{n} f(x_j) l_j'(x) \equiv D_h f(x) \qquad (5.7.3)$$

$$f'(x) - D_h f(x) = \Psi_n'(x) f[x_0, \ldots, x_n, x]$$

$$+ \Psi_n(x) f[x_0, \ldots, x_n, x, x] \qquad (5.7.4)$$

with the last step using (3.2.17). Applying (3.2.12),

$$f'(x) - D_h f(x) = \Psi_n'(x) \frac{f^{(n+1)}(\xi_1)}{(n+1)!} + \Psi_n(x) \frac{f^{(n+2)}(\xi_2)}{(n+2)!} \qquad (5.7.5)$$

with $\xi_1, \xi_2 \in \mathcal{H}\{x_0, \ldots, x_n, x\}$. Higher order differentiation formulas and their error can be obtained by further differentiation of (5.7.3) and (5.7.4).

The most common application of the preceding is to evenly spaced nodes $\{x_i\}$. Thus let

$$x_i = x_0 + ih \qquad i \geq 0$$

with $h > 0$. In this case, it is straightforward to show that

$$\Psi_n(x) = O(h^{n+1}) \qquad \Psi_n'(x) = O(h^n) \qquad (5.7.6)$$

Thus

$$f'(x) - p_n'(x) = \begin{cases} O(h^n) & \Psi_n'(x) \neq 0 \\ O(h^{n+1}) & \Psi_n'(x) = 0 \end{cases} \qquad (5.7.7)$$

We now derive examples of each case.

Let $n = 1$, so that $p_n(x)$ is just the linear interpolate of $(x_0, f(x_0))$ and $(x_1, f(x_1))$. Then (5.7.3) yields

$$f'(x_0) \doteq D_h f(x_0) \equiv \frac{1}{h} [f(x_0 + h) - f(x_0)] \qquad (5.7.8)$$

From (5.7.5),

$$f'(x_0) - D_h f(x_0) = \frac{h}{2} f''(\xi_1) \qquad x_0 \leq \xi_1 \leq x_1 \qquad (5.7.9)$$

since $\Psi(x_0) = 0$.

To improve on this with linear interpolation, choose $x = m \equiv (x_0 + x_1)/2$. Then

$$f'(m) \doteq \frac{1}{h} [f(x_1) - f(x_0)]$$

We usually rewrite this by letting $\delta = h/2$, to obtain

$$f'(m) \doteq D_\delta f(m) = \frac{1}{2\delta} [f(m + \delta) - f(m - \delta)] \qquad (5.7.10)$$

For the error, using (5.7.5) and $\Psi_1'(m) = 0$,

$$f'(m) - D_\delta f(m) = \frac{-\delta^2}{6} f^{(3)}(\xi_2) \qquad m - \delta \le \xi_2 \le m + \delta \quad (5.7.11)$$

In general, to obtain the higher order case in (5.7.7), we want to choose the nodes $\{x_i\}$ to have $\Psi_n'(x) = 0$. This will be true if $n$ is odd and the nodes are placed symmetrically about $x$, as in (5.7.10).

To obtain higher order formulas in which the nodes all lie on one side of $x$, use higher values of $n$ in (5.7.3). For example, with $x = x_0$ and $n = 2$,

$$f'(x_0) \doteq D_h f(x_0) \equiv \frac{1}{2h} [-3f(x_0) + 4f(x_1) - f(x_2)] \qquad (5.7.12)$$

$$f'(x_0) - D_h f(x_0) = \frac{h^2}{3} f^{(3)}(\xi_1) \qquad x_0 \le \xi_1 \le x_2 \qquad (5.7.13)$$

**The method of undetermined coefficients**   Another method to derive formulas for numerical integration, differentiation, and interpolation is called the method of undetermined coefficients. It is often equivalent to the formulas obtained from a polynomial interpolation formula, but sometimes it results in a simpler derivation. We will illustrate the method by deriving a formula for $f''(x)$.

Assume

$$f''(x) \doteq D_h^{(2)} f(x) = A f(x + h) + B f(x) + C f(x - h) \qquad (5.7.14)$$

with $A$, $B$, and $C$ unspecified. Replace $f(x + h)$ and $f(x - h)$ by the Taylor expansions

$$f(x \pm h) = f(x) \pm h f'(x) + \frac{h^2}{2} f''(x) \pm \frac{h^3}{6} f^{(3)}(x) + \frac{h^4}{24} f^{(4)}(\xi_\pm)$$

with $x - h \le \xi_- \le x \le \xi_+ \le x + h$. Substitute into (5.7.14) and rearrange into a polynomial in powers of $h$:

$$A f(x + h) + B f(x) + C f(x - h)$$

$$= (A + B + C) f(x) + h(A - C) f(x) + \frac{h^2}{2} (A + C) f''(x)$$

$$+ \frac{h^3}{6} (A - C) f^{(3)}(x) + \frac{h^4}{24} [A f^{(4)}(\xi_+) + B f^{(4)}(\xi_-)] \qquad (5.7.15)$$

In order for this to equal $f''(x)$, we set

$$A + B + C = 0 \qquad A - C = 0 \qquad A + C = \frac{2}{h^2}$$

The solution of this system is

$$A = C = \frac{1}{h^2} \qquad B = \frac{-2}{h^2} \tag{5.7.16}$$

This yields the formula

$$D_h^{(2)}f(x) = \frac{f(x + h) - 2f(x) + f(x - h)}{h^2} \tag{5.7.17}$$

For the error, substitute (5.7.16) into (5.7.15) and use (5.7.17). This yields

$$f''(x) - D_h^{(2)}f(x) = \frac{h^2}{24}\left[f^{(4)}(\xi_+) + f^{(4)}(\xi_-)\right]$$

Using Problem 1 of Chapter 1, and assuming $f(x)$ is four times continuously differentiable,

$$f''(x) - D_h^{(2)}f(x) = -\frac{h^2}{12}f^{(4)}(\xi) \tag{5.7.18}$$

for some $x - h \le \xi \le x + h$. Formulas (5.7.17) and (5.7.18) could have been derived by calculating $p_2''(x)$ for the quadratic polynomial interpolating $f(x)$ at $x - h$, $x$, $x + h$, but the preceding is probably simpler.

The general idea of the method of undetermined coefficients is to choose the Taylor coefficients in an expansion in $h$ so as to obtain the desired derivative (or integral) as closely as possible.

**Effect of error in function values**   The preceding formulas are useful when deriving methods for solving ordinary and partial differential equations, but they can lead to serious errors when applied to function values that are obtained empirically. To illustrate a method for analyzing the effect of such errors, we consider the second derivative approximation (5.7.17).

Begin by rewriting (5.7.17) as

$$f''(x_1) \doteq D_h^{(2)}f(x_1) = \frac{f(x_2) - 2f(x_1) + f(x_0)}{h^2}$$

with $x_j = x_0 + jh$. Let the actual values used be $\tilde{f}_i$ with

$$f(x_i) = \tilde{f}_i + \epsilon_i \qquad i = 0, 1, 2 \tag{5.7.19}$$

The actual numerical derivative computed is

$$\tilde{D}_h^{(2)}f(x_1) = \frac{\tilde{f}_2 - 2\tilde{f}_1 + \tilde{f}_0}{h^2} \tag{5.7.20}$$

For its error, substitute (5.7.19) into (5.7.20), obtaining

$$f''(x_1) - \tilde{D}_h^{(2)}f(x_1) = f''(x_1) - \frac{f(x_2) - 2f(x_1) + f(x_0)}{h^2}$$

$$+ \frac{\epsilon_2 - 2\epsilon_1 + \epsilon_0}{h^2}$$

$$= \frac{-h^2}{12}f^{(4)}(\xi) + \frac{\epsilon_2 - 2\epsilon_1 + \epsilon_0}{h^2} \tag{5.7.21}$$

For the term involving $\{\epsilon_i\}$, assume these errors are random within some interval $-E \le \epsilon \le E$. Then

$$\left| f''(x_1) - \tilde{D}_h^{(2)}f(x_1) \right| \le \frac{h^2}{12}\left| f^{(4)}(\xi) \right| + \frac{4E}{h^2} \tag{5.7.22}$$

and the last bound would be attainable in many situations. An example of such errors would be rounding errors, with $E$ a bound on their magnitude.

The error bound in (5.7.22) will initially get smaller as $h$ decreases, but for $h$ sufficiently close to zero, the error will begin to increase again. There is an optimal value of $h$, call it $h^*$, to minimize the right side of (5.7.22), and presumably there is a similar value for the actual error $f''(x_1) - \tilde{D}_h^{(2)}f(x_1)$.

**Example** Let $f(x) = -\cos(x)$, and compute $f''(0)$ using the numerical approximation (5.7.17). In Table 5.25, we give the errors in (1) $D_h^{(2)}f(0)$, computed exactly, and (2) $\tilde{D}_h^{(2)}f(0)$, computed using 8-digit rounded decimal arithmetic. In

**Table 5.25    Example of $D_h^{(2)}f(0)$ and $\tilde{D}_h^{(2)}f(0)$**

| $h$ | $f''(0) - D_h^{(2)}f(0)$ | Ratio | $f''(0) - \tilde{D}_h^{(2)}f(0)$ |
|---|---|---|---|
| .5 | 2.07E − 2 | | 2.07E − 2 |
| .25 | 5.20E − 3 | 3.98 | 5.20E − 3 |
| .125 | 1.30E − 3 | 3.99 | 1.30E − 3 |
| .0625 | 3.25E − 4 | 4.00 | 3.25E − 4 |
| .03125 | 8.14E − 5 | 4.00 | 8.45E − 5 |
| .015625 | 2.03E − 5 | 4.00 | 2.56E − 6 |
| .0078125 | 5.09E − 6 | 4.00 | −7.94E − 5 |
| .00390625 | 1.27E − 6 | 4.00 | −7.94E − 5 |
| .001953125 | 3.18E − 7 | 4.00 | −1.39E − 3 |

this last case,

$$\left| f''(0) - \tilde{D}_h^{(2)}f(0) \right| \le \frac{h^2}{12} + \frac{2 \times 10^{-8}}{h^2} \tag{5.7.23}$$

This bound is minimized at $h^* = .0022$, which is consistent with the errors $f''(0) - \tilde{D}_h^{(2)}f(0)$ given in the table. For the exactly computed $D_h^{(2)}f(0)$, note that the errors decrease by four whenever $h$ is halved, consistent with the error formula (5.7.18).

## Discussion of the Literature

Even though the topic of numerical integration is one of the oldest in numerical analysis and there is a very large literature, new papers continue to appear at a fairly high rate. Many of these results give methods for special classes of problems, for example, oscillatory integrals, and others are a response to changes in computers, for example, the use of vector pipeline architectures. The best survey of numerical integration is the large and detailed work of Davis and Rabinowitz (1984). It contains a comprehensive survey of most quadrature methods, a very extensive bibliography, a set of computer programs, and a bibliography of published quadrature programs. It also contains the article "On the practical evaluation of integrals" by Abramowitz, which gives some excellent suggestions on analytic approaches to quadrature. Other important texts in numerical integration are Engels (1980), Krylov (1962), and Stroud (1971). For a history of the classical numerical integration methods, see Goldstine (1977).

For reasons of space, we have had to omit some important ideas. Chief among these are (1) Clenshaw–Curtis quadrature, and (2) multivariable quadrature. The former is based on integrating a Chebyshev expansion of the integrand; empirically the method has proved excellent for a wide variety of integrals. The original method is presented in Clenshaw and Curtis (1960); a current account of the method is given in Piessens et al. (1983, pp. 28–39). The area of multivariable quadrature is an active area of research, and the texts of Engels (1980) and Stroud (1971) are the best introductions to the area. Because of the widespread use of multivariable quadrature in the finite element method for solving partial differential equations, texts on the finite element method will often contain integration formulas for triangular and rectangular regions.

Automatic numerical integration was a very active area of research in the 1960s and 1970s, when it was felt that most numerical integrations could be done in this way. Recently, there has been a return to a greater use of nonautomatic quadrature especially adapted to the integral at hand. An excellent discussion of the relative advantages and disadvantages of automatic quadrature is given in Lyness (1983). The most powerful and flexible of the current automatic programs are probably those given in QUADPACK, which is discussed and illustrated in Piessens et al. (1983). Versions of QUADPACK are included in the IMSL and NAG libraries.

For microcomputers and hand computation, Simpson's rule is still popular because of its simplicity. Nonetheless, serious consideration should be given to Gaussian quadrature because of its much greater accuracy. The nodes and weights are readily available, in Abramowitz and Stegun (1964) and Stroud and Secrest (1966), and programs for their calculation are also available.

Numerical differentiation is an ill-posed problem in the sense of Section 1.6. Numerical differentiation procedures that account for this have been developed in the past ten to fifteen years. In particular, see Anderssen and Bloomfield (1974a), (1974b), Cullum (1971), Wahba (1980), and Woltring (1986).

## Bibliography

Abramowitz, M., and I. Stegun, Eds. (1964). *Handbook of Mathematical Functions*. National Bureau of Standards, U.S. Government Printing Office, Washington, D.C.

Anderssen, R., and P. Bloomfield (1974a). Numerical differentiation procedures for non-exact data, *Numer. Math.*, **22**, 157–182.

Anderssen, R., and P. Bloomfield (1974b). A time series approach to numerical differentiation, *Technometrics* **16**, 69–75.

Atkinson, K. (1976). *A Survey of Numerical Methods for the Solution of Fredholm Integral Equations of the Second Kind*. Society for Industrial and Applied Mathematics, Philadelphia.

Atkinson, K. (1982). Numerical integration on the sphere, *J. Austr. Math. Soc.* (*Ser. B*) **23**, 332–347.

Bauer, F., H. Rutishauser, and E. Stiefel (1963). New aspects in numerical quadrature. In *Experimental Arithmetic, High Speed Computing, and Mathematics*, pp. 199–218. Amer. Math. Soc., Providence, R.I.

de Boor, C. (1971). CADRE: An algorithm for numerical quadrature. In *Mathematical Software*, pp. 201–209. Academic Press, New York.

Clenshaw, C., and A. Curtis (1960). A method for numerical integration on an automatic computer, *Numer. Math.* **2**, 197–205.

Cryer, C. (1982). *Numerical Functional Analysis*. Oxford Univ. Press (Clarendon), Oxford, England.

Cullum, J. (1971). Numerical differentiation and regularization, *SIAM J. Numer. Anal.* **8**, 254–265.

Davis, P. (1963). *Interpolation and Approximation*. Ginn (Blaisdell), Boston.

Davis, P., and P. Rabinowitz (1984). *Methods of Numerical Integration*, 2nd ed. Academic Press, New York.

Dixon, V. (1974). Numerical quadrature: A survey of the available algorithms. In *Software for Numerical Mathematics*, D. Evans, Ed., pp. 105–137. Academic Press, London.

Donaldson, J., and D. Elliott (1972). A unified approach to quadrature rules with asymptotic estimates of their remainders, *SIAM J. Numer. Anal.* **9**, 573–602.

de Doncker, E., and R. Piessens (1976). Algorithm 32: Automatic computation of integrals with singular integrand, over a finite or an infinite interval, *Computing* **17**, 265–279.

Engels, H. (1980). *Numerical Quadrature and Cubature*. Academic Press, New York.

Goldstine, H. (1977). *A History of Numerical Analysis*. Springer-Verlag, New York.

Golub, G., and J. Welsch (1969). Calculation of Gauss quadrature rules, *Math. Comput.* **23**, 221–230.

de Hoog, F., and R. Weiss (1973). Asymptotic expansions for product integration, *Math. Comput.* **27**, 295–306.

Iri, M., S. Moriguti, and Y. Takasawa (1970). On a numerical integration formula (in Japanese), J. Res. Inst. *Math. Sci.*, **91**, 82. Kyoto Univ., Kyoto, Japan.

Isaacson, E., and H. Keller (1966). *Analysis of Numerical Methods*. Wiley, New York.

Kronrod, A. (1965). *Nodes and Weights of Quadrature Formulas*, Consultants Bureau, New York.

Krylov, V. (1962). *Approximate Calculation of Integrals*. Macmillan, New York.

Lyness, J. (1983). When not to use an automatic quadrature routine, *SIAM Rev.* **25**, 63–88.

Lyness, J., and C. Moler (1967). Numerical differentiation of analytic functions, *SIAM J. Num. Anal.* **4**, 202–210.

Lyness, J., and J. Kaganove (1976). Comments on the nature of automatic quadratic routines, *ACM Trans. Math. Softw.* **2**, 65–81.

Lyness, J., and K. Puri (1973). The Euler-MacLaurin expansion for the simplex, *Math. Comput.* **27**, 273–293.

Lyness, J., and B. Ninham (1967). Numerical quadrature and asymptotic expansions, *Math. Comput.* **21**, 162–178.

Patterson, T. (1968). The optimum addition of points to quadrature formulae, *Math. Comput.* **22**, 847–856. (Includes a microfiche enclosure; a microfiche correction is in **23**, 1969.)

Patterson, T. (1973) Algorithm 468: Algorithm for numerical integration over a finite interval. *Commun. ACM* **16**, 694–699.

Piessens, R., E. deDoncker-Kapenga, C. Überhuber, and D. Kahaner (1983). *QUADPACK: A Subroutine Package for Automatic Integration*. Springer-Verlag, New York.

Ralston, A. (1965). *A First Course in Numerical Analysis*. McGraw-Hill, New York.

Robinson, I. (1979). A comparison of numerical integration programs, *J. Comput. Appl. Math.* **5**, 207–223.

Robinson, I., and E. de Doncker (1981). Automatic computation of improper integrals over a bounded or unbounded planar region, *Computing* **27**, 253–284.

Stenger, F. (1981). Numerical methods based on Whittaker cardinal or sinc functions, *SIAM Rev.* **23**, 165–224.

Stroud, A. (1971). *Approximate Calculation of Multiple Integrals.* Prentice-Hall, Englewood Cliffs, N.J.

Stroud, A., and D. Secrest (1966). *Gaussian Quadrature Formulas.* Prentice-Hall, Englewood Cliffs, N.J.

Wahba, G. (1980). Ill-posed problems: Numerical and statistical methods for mildly, moderately, and severely ill-posed problems with noisy data, Tech. Rep. #595, Statistics Dept., Univ. of Wisconsin, Madison. (Prepared for the Proc. Int. Symp. on Ill-posed Problems, Newark, Del., 1979.)

Woltring, H. (1986). A Fortran package for generalized, cross-validatory spline smoothing and differentiation, *Adv. Eng. Softw.* **8**, 104–113.

## Problems

1.  Write a program to evaluate $I = \int_a^b f(x)\, dx$ using the trapezoidal rule with $n$ subdivisions, calling the result $I_n$. Use the program to calculate the following integrals with $n = 2, 4, 8, 16, \ldots, 512$.

    (a)  $\displaystyle\int_0^1 e^{-x^2}\, dx$    (b)  $\displaystyle\int_0^1 x^{5/2}\, dx$    (c)  $\displaystyle\int_{-4}^4 \frac{dx}{1 + x^2}$

    (d)  $\displaystyle\int_0^{2\pi} \frac{dx}{2 + \cos(x)}$    (e)  $\displaystyle\int_0^\pi e^x \cos(4x)\, dx$

    Analyze empirically the rate of convergence of $I_n$ to $I$ by calculating the ratios of (5.4.27):

    $$R_n = \frac{I_{2n} - I_n}{I_{4n} - I_{2n}}$$

2.  Repeat Problem 1 using Simpson's rule.

3.  Apply the corrected trapezoidal rule (5.1.12) to the integrals in Problem 1. Compare the results with those of Problem 2 for Simpson's rule.

**4.** As another approach to the corrected trapezoidal rule (5.1.12), use the cubic Hermite interpolation polynomial to $f(x)$ to obtain

$$\int_a^b f(x)\, dx \doteq \left(\frac{b-a}{2}\right)[f(a) + f(b)] - \frac{(b-a)^2}{12}[f'(b) - f'(a)]$$

Use the error formula for Hermite interpolation to obtain an error formula for the preceding approximation. Generalize these results to (5.1.12), with $n$ subdivisions of $[a, b]$.

**5. (a)** Assume that $f(x)$ is continuous and that $f'(x)$ is integrable on $[0, 1]$. Show that the error in the trapezoidal rule for calculating $\int_0^1 f(x)\, dx$ has the form

$$E_n(f) = \int_0^1 K(t) f'(t)\, dt$$

$$K(t) = \frac{t_{j-1} + t_j}{2} - t \qquad t_{j-1} \le t \le t_j \qquad j = 1, \ldots, n$$

This contrasts with (5.1.22) in which $f'(x)$ is continuous and $f''(x)$ is integrable.

**(b)** Apply the result to $f(x) = x^\alpha$ and to $f(x) = x^\alpha \log(x)$, $0 < \alpha < 1$. This gives an order of convergence, although it is less than the true order. [See Problem 6 and (5.4.23).]

**6.** Using the program of Problem 1, determine empirically the rate of convergence of the trapezoidal rule applied to $\int_0^1 x^\alpha \ln(x)\, dx$, $0 \le \alpha \le 1$, with a range of values of $\alpha$, say $\alpha = .25, .5, .75, 1.0$.

**7.** Derive the composite form of Boole's rule, which is given as the $n = 4$ entry in Table 5.8. Develop error formulas analogous to those given in (5.1.17) and (5.1.18) for Simpson's rule.

**8.** Repeat Problem 1 using the composite Boole's rule obtained in Problem 7.

**9.** Let $p_2(x)$ be the quadratic polynomial interpolating $f(x)$ at $x = 0, h, 2h$. Use this to derive a numerical integration formula $I_h$ for $I = \int_0^{3h} f(x)\, dx$. Use a Taylor series expansion of $f(x)$ to show

$$I - I_h = \frac{3}{8} h^4 f^{(3)}(0) + O(h^5)$$

10. For the midpoint integration formula (5.2.17), derive the Peano kernel error formula

$$\int_0^h f(x)\,dx - hf\left(\frac{h}{2}\right) = \int_0^h K(t)f''(t)\,dt$$

$$K(t) = \begin{cases} \dfrac{1}{2}t^2 & 0 \le t \le \dfrac{h}{2} \\[2ex] \dfrac{1}{2}(h-t)^2 & \dfrac{h}{2} \le t \le h. \end{cases}$$

Use this to derive the error term in (5.2.17).

11. Consider functions defined in the following way. Let $n > 0$, $h = (b - a)/n$, $t_j = a + jh$ for $j = 0, 1, \ldots, n$. Let $f(x)$ be linear on each subinterval $[t_{j-1}, t_j]$, for $j = 1, \ldots, n$. Show that the set of all such $f$, for all $n \ge 1$, is dense in $C[a, b]$.

12. Let $w(x)$ be an integrable function on $[a, b]$,

$$\int_a^b |w(x)|\,dx < \infty$$

and let

$$\int_a^b w(x)f(x)\,dx \doteq \sum_{j=1}^n w_{j,n} f(x_{j,n})$$

be a sequence of numerical integration rules. Generalize Theorem 5.2 to this case.

13. Assume the numerical integration rule

$$\int_a^b f(x)\,dx \doteq \sum_{j=1}^n w_{j,n} f(x_{j,n})$$

is convergent for all continuous functions. Consider the effect of errors in the function values. Assume we use $\tilde{f}_i \doteq f(x_i)$, with

$$|f(x_i) - \tilde{f}_i| \le \epsilon \qquad 1 \le i \le n$$

What is the effect on the numerical integration of these errors in the function values?

14. Apply Gauss–Legendre quadrature to the integrals in Problem 1. Compare the results with those for the trapezoidal and Simpson methods.

15. Use Gauss–Legendre quadrature to evaluate $\int_{-4}^{4} dx/(1 + x^2)$ with the $n = 2, 4, 6, 8$ node-point formulas. Compare with the results in Table 5.9, obtained using the Newton–Cotes formula.

**16.** Prove that the Gauss–Legendre nodes and weights on $[-1, 1]$ are symmetrically distributed about the origin $x = 0$.

**17.** Derive the two-point Gaussian quadrature formula for

$$I(f) = \int_0^1 f(x) \log\left(\frac{1}{x}\right) dx$$

in which the weight function is $w(x) = \log(1/x)$. *Hint:* See Problem 20(a) of Chapter 4. Also, use the analogue of (5.3.7) to compute the weights, not formula (5.3.11).

**18.** Derive the one- and two-point Gaussian quadrature formulas for

$$I = \int_0^1 xf(x)\, dx \doteq \sum_{j=1}^n w_j f(x_j)$$

with weight function $w(x) = x$.

**19.** For the integral $I = \int_{-1}^1 \sqrt{1 - x^2} f(x)\, dx$ with weight $w(x) = \sqrt{1 - x^2}$, find explicit formulas for the nodes and weights of the Gaussian quadrature formula. Also give the error formula. *Hint:* See Problem 24 of Chapter 4.

**20.** Using the column $e_{n,4}$ of Table 5.13, produce the fourth-order error formula analogous to the second-order formula (5.3.40) for $e_{n,2}$. Compare it with the fourth-order Simpson error formula (5.1.17).

**21.** The weights in the Kronrod formula (5.3.48) can be calculated as the solution of four simultaneous linear equations. Find that system and then solve it to verify the values given following (5.3.48). *Hint:* Use the approach leading to (5.3.7).

**22.** Compare the seven-point Gauss–Legendre formula with the seven-point Kronrod formula of (5.3.48). Use each of them on a variety of integrals, and then compare their respective errors.

**23.** **(a)** Derive the relation (5.4.7) for the Bernoulli polynomials $B_j(x)$ and Bernoulli numbers $B_j$. Show that $B_j = 0$ for all odd integers $j \geq 3$.

**(b)** Derive the identities

$$B_j'(x) = jB_{j-1}(x) \qquad j \geq 4 \text{ and even}$$

$$B_j'(x) = j\left[B_{j-1}(x) + B_{j-1}\right] \qquad j \geq 3 \text{ and odd}$$

These can be used to give a general proof of the Euler–MacLaurin formula (5.4.9).

**24.** Using the Euler–MacLaurin summation formula (5.4.17), obtain an estimate of $\zeta(\frac{5}{4})$, accurate to three decimal places. The zeta function $\zeta(p)$ is
defined in (5.4.22).

**25.** Obtain the asymptotic error formula for the trapezoidal rule applied to
$\int_0^1 \sqrt[4]{x} f(x)\, dx$. Use the estimate from Problem 24.

**26.** Consider the following table of approximate integrals $I_n$ produced using
Simpson's rule. Predict the order of convergence of $I_n$ to $I$:

| $n$ | $I_n$ |
|---|---|
| 2 | .28451779686 |
| 4 | .28559254576 |
| 8 | .28570248748 |
| 16 | .28571317731 |
| 32 | .28571418363 |
| 64 | .28571427643 |

That is, if $I - I_n \doteq c/n^p$, then what is $p$? Does this appear to be a valid
form for the error for these data? Predict a value of $c$ and the error in $I_{64}$.
How large should $n$ be chosen if $I_n$ is to be in error by less than $10^{-11}$?

**27.** Assume that the error in an integration formula has the asymptotic expansion

$$I - I_n = \frac{C_1}{n\sqrt{n}} + \frac{C_2}{n^2} + \frac{C_3}{n^2\sqrt{n}} + \frac{C_4}{n^3} + \cdots$$

Generalize the Richardson extrapolation process of Section 5.4 to obtain
formulas for $C_1$ and $C_2$. Assume that three values $I_n$, $I_{2n}$, and $I_{4n}$ have
been computed, and use these to compute $C_1$, $C_2$, and an estimate of $I$,
with an error of order $1/n^2\sqrt{n}$.

**28.** For the trapezoidal rule (denoted by $I_n^{(T)}$) for evaluating $I = \int_a^b f(x)\, dx$, we
have the asymptotic error formula

$$I - I_n^{(T)} = -\frac{h^2}{12}[f'(b) - f'(a)] + O(h^4)$$

and for the midpoint formula $I_n^{(M)}$, we have

$$I - I_n^{(M)} = \frac{h^2}{24}[f'(b) - f'(a)] + O(h^4)$$

provided $f$ is sufficiently differentiable on $[a, b]$. Using these results, obtain
a new numerical integration formula $\tilde{I}_n$ combining $I_n^{(T)}$ and $I_n^{(M)}$, with a
higher order of convergence. Write out the weights to the new formula $\tilde{I}_n$.

**29.**   Obtain an asymptotic error formula for Simpson's rule, comparable to the Euler–MacLaurin formula (5.4.9) for the trapezoidal rule. Use (5.4.9), (5.4.33), and (5.4.36), as in (5.4.37).

**30.**   Show that the formula (5.4.40) for $I_n^{(2)}$ is the composite Boole's rule. See Problem 7.

**31.**   Implement the algorithm *Romberg* of Section 5.4, and then apply it to the integrals of Problem 1. Compare the results with those for the trapezoidal and Simpson rules.

**32.**   Consider evaluating $I = \int_0^1 x^\alpha \, dx$ by using the adaptive Simpson's rule that was described following (5.5.3). To see that the error test (5.5.5) may fail, consider the case of $-1 < \alpha < 0$, with the integrand arbitrarily set to zero when $x = 0$. Show that for sufficiently small $\epsilon$, the test (5.5.5) will never be satisfied for the subinterval $[0, h]$. Note that for $\epsilon$ specified on $[0, 1]$, the error tolerance for $[0, h]$ will be $\epsilon h$.

**33.**   Use Simpson's rule with even spacing to integrate $I = \int_0^1 \log(x) \, dx$. For $x = 0$, set the integrand to zero. Compare the results with those in Tables 5.19 and 5.20, for integral $I_4$.

**34.**   Use an adaptive integration program [e.g., DQAGP from Piessens et al. (1983)] to calculate the integrals in Problem 1. Compare the results with those of Problems 1, 2, 14, and 31.

**35.**   Decrease the singular behavior of the integrand in

$$I = \int_0^1 f(x) \log(x) \, dx$$

by using the change of variable $x = t^r$, $r > 0$. Analyze the smoothness of the resulting integrand. Also explore the empirical behavior of the trapezoidal and Simpson rules for various $r$.

**36.**   Apply the IMT method, described following (5.6.5), to the calculation of $I = \int_0^\infty f(x) \, dx$, using some change of variable from $[0, \infty)$ to a finite interval. Apply this to the calculation of the integrals following (5.6.11).

**37.**   Use Gauss–Laguerre quadrature with $n = 2, 4, 6,$ and $8$ node points to evaluate the following integrals. Use (5.6.11) to put the integrals in proper form.

    **(a)** $\displaystyle\int_0^\infty e^{-x^2} \, dx = \frac{\sqrt{\pi}}{2}$     **(b)** $\displaystyle\int_0^\infty \frac{x \, dx}{(1 + x^2)^2} = \frac{1}{2}$

    **(c)** $\displaystyle\int_0^\infty \frac{\sin(x)}{x} \, dx = \frac{\pi}{2}$

**38.** Evaluate the following singular integrals within the prescribed error tolerance. Do not use an automatic program.

(a) $\displaystyle\int_0^{4\pi} \cos(u)\log(u)\,du \qquad \epsilon = 10^{-8}$

(b) $\displaystyle\int_0^{2/\pi} x^2 \sin\left(\frac{1}{x}\right) dx \qquad \epsilon = 10^{-3}$

(c) $\displaystyle\int_0^1 \frac{\cos(x)\,dx}{\sqrt[3]{x^2}} \qquad \epsilon = 10^{-8}$

(d) $\displaystyle\int_0^1 \frac{\sqrt{1-x^4}\,dx}{x^{\alpha}} \qquad \epsilon = 10^{-5} \quad \alpha = 1 - \frac{1}{\pi}$

**39.** Use an adaptive program (e.g., DQAGP) to evaluate the integrals of
Problem 38.

**40.** (a) Develop the product trapezoidal integration rule for calculating

$$I(f) = \int_0^b \frac{f(x)\,dx}{x^{\alpha}} \qquad 0 < \alpha \le 1$$

Put the weights into a convenient form, analogous to (5.6.28) for the
product trapezoidal rule when $w(x) = \log(x)$. Also, give an error
formula analogous to (5.6.24).

(b) Write a simple program to evaluate the following integrals using the
results of part (a).

(i) $\displaystyle\int_0^1 \frac{e^x}{x^{1/\pi}}\,dx$ (ii) $\displaystyle\int_0^1 \frac{dx}{\sin(x^{2/\pi})}$

*Hint:* For part (ii), first let $u = x^{2/\pi}$.

**41.** To show the ill-posedness of the differentiation of a function $y(t)$ on an
interval $[0, 1]$, consider calculating $x(t) = y'(t)$ and $x_n(t) = y_n'(t)$, with

$$y_n(t) = y(t) + \frac{1}{n}t^n \qquad n \ge 1$$

Recall the definition of ill-posed and well-posed problems from Section 1.6.
Use the preceding construction to show that the evaluation of $x(t)$ is
unstable relative to changes in $y$. For measuring changes in $x$ and $y$, use
the maximum norm of (1.1.8) and (4.1.4).

**42.** Repeat the numerical differentiation example of (5.7.23) and Table 5.25,
using your own computer or hand calculator.

**43.** Derive error results for $D_8 f(m)$ in (5.7.10) relative to the effects of
rounding errors, in analogy with the results (5.7.21)–(5.7.22) for $\tilde{D}_n^{(2)} f(x)$.
Apply it to $f(x) = e^x$ at $x = 0$, and compare the results with the actual
results of your own hand computations.

# SIX

# NUMERICAL METHODS FOR ORDINARY DIFFERENTIAL EQUATIONS

Differential equations are one of the most important mathematical tools used in modeling problems in the physical sciences. In this chapter we derive and analyze numerical methods for solving problems for ordinary differential equations. The main form of problem that we study is the *initial value problem*:

$$y' = f(x, y) \qquad y(x_0) = Y_0 \tag{6.0.1}$$

The function $f(x, y)$ is to be continuous for all $(x, y)$ in some domain $D$ of the $xy$-plane, and $(x_0, Y_0)$ is a point in $D$. The results obtained for (6.0.1) will generalize in a straightforward way to both systems of differential equations and higher order equations, provided appropriate vector and matrix notation is used. These generalizations are discussed and illustrated in the next two sections.

We say that a function $Y(x)$ is a solution on $[a, b]$ of (6.0.1) if for all $a \le x \le b$,

1.  $(x, Y(x)) \in D$.
2.  $Y(x_0) = Y_0$.
3.  $Y'(x)$ exists and $Y'(x) = f(x, Y(x))$.

As notation throughout this chapter, $Y(x)$ will denote the true solution of whatever differential equation problem is being considered.

***Example* 1.** The general first-order linear differential equation is

$$y' = a_0(x)y + g(x) \qquad a \le x \le b$$

in which the coefficients $a_0(x)$ and $g(x)$ are assumed to be continuous on $[a, b]$. The domain $D$ for this problem is

$$D = \{(x, y) | a \le x \le b, -\infty < y < \infty\}$$

The exact solution of this equation can be found in any elementary differential equations textbook [e.g., Boyce and Diprima (1986), p. 13]. As a special case, consider

$$y' = \lambda y + g(x) \qquad 0 \le x < \infty \tag{6.0.2}$$

with $g(x)$ continuous on $[0, \infty)$. The solution that satisfies $Y(0) = Y_0$ is given by

$$Y(x) = Y_0 e^{\lambda x} + \int_0^x e^{\lambda(x-t)} g(t)\, dt \qquad 0 \le x < \infty \qquad (6.0.3)$$

This is used later to illustrate various theoretical results.

**2.** The equation $y' = -y^2$ is nonlinear. One of its solutions is $Y(x) \equiv 0$, and the remaining solutions have the form

$$Y(x) = \frac{1}{x + c}$$

with $c$ an arbitrary constant. Note that $|Y(-c)| = \infty$. Thus the global smoothness of $f(x, y) = -y^2$ does not guarantee a similar behavior in the solutions.

To obtain some geometric insight for the solutions of a single first-order differential equation, we can look at the *direction field* induced by the equation on the $xy$-plane. If $Y(x)$ is a solution that passes through $(x_0, Y_0)$, then the slope of $Y(x)$ at $(x_0, Y_0)$ is $Y'(x_0) = f(x_0, Y_0)$. Within the domain $D$ of $f(x, y)$, pick a representative set of points $(x, y)$ and then draw a short line segment with slope $f(x, y)$ through each $(x, y)$.

*Example* Consider the equation $y' = -y$. The direction field is given in Figure 6.1, and several representative solutions have been drawn in. It is clear from the graph that all solutions $Y(x)$ satisfy

$$\underset{x \to \infty}{\text{Limit }} Y(x) = 0$$

To make it easier to draw the direction field of $y' = f(x, y)$, look for those curves in the $xy$-plane along which $f(x, y)$ is constant. Solve

$$f(x, y) = c$$

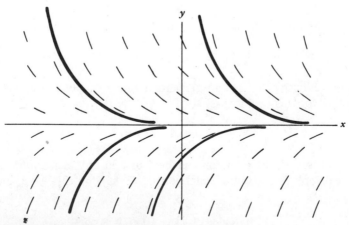

**Figure 6.1**   Direction field for $y' = -y$.

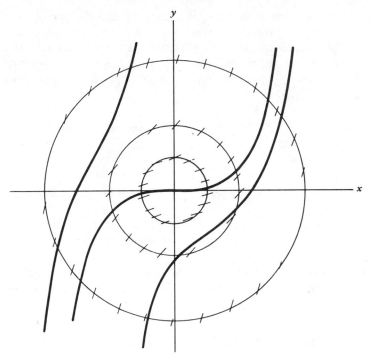

**Figure 6.2**    Direction field for $y' = x^2 + y^2$.

for various choices of $c$. Then each solution $Y(x)$ of $y' = f(x, y)$ that intersects the curve $f(x, y) = c$ satisfies $Y'(x) = c$ at the point of intersection. The curves $f(x, y) = c$ are called *level curves* of the differential equation.

***Example***    Sketch the qualitative behavior of solutions of

$$y' = x^2 + y^2$$

The level curves of the equation are given by

$$x^2 + y^2 = c \qquad c > 0$$

which are circles with center $(0, 0)$ and radius $\sqrt{c}$. The direction field with some representative solutions is given in Figure 6.2. Three circles are drawn, with $c = \frac{1}{4}, 1, 4$. This gives some qualitative information about the solutions $Y(x)$, and occasionally this can be useful.

## 6.1    Existence, Uniqueness, and Stability Theory

From the examples of the introduction it should be intuitive that in most cases the initial value problem (6.0.1) has a unique solution. Before beginning the numerical analysis for (6.0.1), we present some theoretical results for it. Conditions are given to ensure that it has a unique solution, and we consider the stability of the solution when the initial data $Y_0$ and the derivative $f(x, y)$ are changed by small amounts. These results are necessary in order to better understand the numerical methods presented later. The domain $D$ will henceforth be assumed to satisfy the following minor technical requirement: If two points $(x, y_1)$ and $(x, y_2)$ both belong to $D$, then the vertical line segment joining them is also contained in $D$.

**Theorem 6.1**   Let $f(x, y)$ be a continuous function of $x$ and $y$, for all $(x, y)$ in $D$, and let $(x_0, Y_0)$ be an interior point of $D$. Assume $f(x, y)$ satisfies the *Lipschitz condition*

$$|f(x, y_1) - f(x, y_2)| \le K|y_1 - y_2| \qquad \text{all } (x, y_1), (x, y_2) \text{ in } D \quad (6.1.1)$$

for some $K \ge 0$. Then for a suitably chosen interval $I = [x_0 - \alpha, x_0 + \alpha]$, there is a unique solution $Y(x)$ on $I$ of (6.0.1).

**Proof**   The proof of this result can be found in most texts on the theory of ordinary differential equations [e.g., see Boyce and Diprima (1986), p. 95]. For that reason, we omit it.   ∎

Showing the Lipschitz is straightforward if $\partial f(x, y)/\partial y$ exists and is bounded on $D$. Simply let

$$K = \underset{(x, y) \in D}{\text{Max}} \left| \frac{\partial f(x, y)}{\partial y} \right| \qquad (6.1.2)$$

Then using the mean value theorem

$$f(x, y_1) - f(x, y_2) = \frac{\partial f(x, \xi)}{\partial y}(y_1 - y_2)$$

for some $\xi$ between $y_1$ and $y_2$. Result (6.1.1) follows immediately using (6.1.2).

**Example**   The following two examples illustrate the theorem.

1.   Consider $y' = 1 + \sin(xy)$ with

$$D = \{(x, y) | 0 \le x \le 1, \ -\infty < y < \infty\}$$

To compute the Lipschitz constant $K$, use (6.1.2). Then

$$\frac{\partial f(x, y)}{\partial y} = x \cdot \cos(xy) \qquad K = 1.$$

Thus for any $(x_0, Y_0)$ with $0 < x_0 < 1$, there is a solution $Y(x)$ to the associated initial value problem on some interval $[x_0 - \alpha, x_0 + \alpha] \subset [0, 1]$.

**2.**    Consider the problem

$$y' = \frac{2x}{a^2}y^2 \qquad y(0) = 1$$

for any constant $a > 0$. The solution is

$$Y(x) = \frac{a^2}{a^2 - x^2} \qquad -a < x < a$$

The solution exists on only a small interval if $a$ is small. To determine the Lipschitz constant, we must calculate

$$\frac{\partial f(x, y)}{\partial y} = \frac{4xy}{a^2}$$

To have a finite Lipschitz constant on $D$, the region $D$ must be bounded in $x$ and $y$, say $-c \le x \le c$ and $-b \le y \le b$. The preceding theorem then states that there is a solution $Y(x)$ on some interval $-\alpha \le x \le \alpha$, with $\alpha \le c$.

Lest it be thought that we can evaluate the partial derivative at the initial point $(x_0, Y_0) = (0, 1)$ and obtain sufficient information to estimate the Lipschitz constant $K$, note that

$$\frac{\partial f(0, 1)}{\partial y} = 0$$

for any $a > 0$.

**Stability of the solution**    The stability of the solution $Y(x)$ is examined when the initial value problem is changed by a small amount. This is related to the discussion of stability in Section 1.6 of Chapter 1. We consider the perturbed problem

$$y' = f(x, y) + \delta(x)$$
$$y(x_0) = Y_0 + \epsilon \qquad\qquad (6.1.3)$$

with the same hypotheses for $f(x, y)$ as in Theorem 6.1. Furthermore, we assume $\delta(x)$ is continuous for all $x$ such that $(x, y) \in D$ for some $y$. The problem (6.1.3) can then be shown to have a unique solution, denoted by $Y(x; \delta, \epsilon)$.

***Theorem 6.2***    Assume the same hypotheses as in Theorem 6.1. Then the problem (6.1.3) will have a unique solution $Y(x; \delta, \epsilon)$ on an interval $[x_0 - \alpha, x_0 + \alpha]$, some $\alpha > 0$, uniformly for all perturbations $\epsilon$ and $\delta(x)$ that satisfy

$$|\epsilon| \le \epsilon_0 \qquad \|\delta\|_\infty \le \epsilon_0$$

for $\epsilon_0$ sufficiently small. In addition, if $Y(x)$ is the solution of the unperturbed problem, then

$$\underset{|x-x_0|\le a}{\text{Max}}\ |Y(x)-Y(x;\delta,\epsilon)|\le k\big[|\epsilon|+\alpha\|\delta\|_\infty\big]\quad (6.1.4)$$

with $k=1/(1-\alpha K)$, using the Lipschitz constant $K$ of (6.1.1).

***Proof***    The derivation of (6.1.4) is much the same as the proof of Theorem 6.1, and it can be found in most graduate texts on ordinary differential equations.    ∎

Using this result, we can say that the initial value problem (6.0.1) is well-posed or stable, in the sense of Section 1.6 in Chapter 1. If small changes are made in the differential equation or in the initial value, then the solution will also change by a small amount. The solution $Y$ depends continuously on the data of the problem, namely the function $f$ and the initial data $Y_0$.

It was pointed out in Section 1.6 that a problem could be stable but ill-conditioned with respect to numerical computation. This is true with differential equations, although it does not occur often in practice. To better understand when this may happen, we estimate the perturbation in $Y$ due to perturbations in the problem. To simplify our discussion, we consider only perturbations $\epsilon$ in the initial value $Y_0$; perturbations $\delta(x)$ in the equation enter into the final answer in much the same way, as indicated in (6.1.4).

Perturbing the initial value $Y_0$ as in (6.1.3), let $Y(x;\epsilon)$ denote the perturbed solution. Then

$$Y'(x;\epsilon)=f(x,Y(x;\epsilon))\qquad x_0-\alpha\le x\le x_0+\alpha$$

$$Y(x_0;\epsilon)=Y_0+\epsilon \tag{6.1.5}$$

Subtract the corresponding equations of (6.0.1) for $Y(x)$, and let $Z(x)=Y(x;\epsilon)-Y(x)$. Then

$$Z'(x;\epsilon)=f(x,Y(x;\epsilon))-f(x,Y(x))$$

$$\doteq \frac{\partial f(x,Y(x))}{\partial y}Z(x;\epsilon) \tag{6.1.6}$$

and $Z(x_0;\epsilon)=\epsilon$. The approximation (6.1.6) is valid when $Y(x;\epsilon)$ is sufficiently close to $Y(x)$, which it is for small values of $\epsilon$ and small intervals $[x_0-\alpha,x_0+\alpha]$.

We can easily solve the approximate differential equation of (6.1.6), obtaining

$$Z(x;\epsilon)\doteq \epsilon\cdot\exp\left[\int_{x_0}^x \frac{\partial f(t,Y(t))}{\partial y}\,dt\right] \tag{6.1.7}$$

If the partial derivative satisfies

$$\frac{\partial f(t,Y(t))}{\partial y}\le 0\qquad |x_0-t|\le\alpha \tag{6.1.8}$$

then we have that $Z(x, \epsilon)$ probably remains bounded by $\epsilon$ as $x$ increases. In this case, we say the initial value problem is well-conditioned.

As an example of the opposite behavior, suppose we consider

$$y' = \lambda y + g(x) \qquad y(0) = Y_0 \tag{6.1.9}$$

with $\lambda > 0$. Then $\partial f / \partial y = \lambda$, and we can calculate exactly

$$Z(x; \epsilon) = \epsilon e^{\lambda x}$$

Thus the change in $Y(x)$ becomes increasing large as $x$ increases.

*Example*    The equation

$$y' = 100y - 101e^{-x} \qquad y(0) = 1 \tag{6.1.10}$$

has the solution $Y(x) = e^{-x}$. The perturbed problem

$$y' = 100y - 101e^{-x} \qquad y(0) = 1 + \epsilon$$

has the solution

$$Y(x; \epsilon) = e^{-x} + \epsilon e^{100x}$$

which rapidly departs from the true solution. We say (6.1.10) is an ill-conditioned problem.

For a problem to be well-conditioned, we want the integral

$$\int_{x_0}^{x} \frac{\partial f(t, Y(t))}{\partial y} \, dt$$

to be bounded from above by zero or a small positive number, as $x$ increases. Then the perturbation $Z(x; \epsilon)$ will be bounded by some constant times $\epsilon$, with the constant not too large.

In the case that (6.1.8) is satisfied, but the partial derivative is large in magnitude, we will have that $Z(x; \epsilon) \to 0$ rapidly as $x$ increases. Such equations are considered well-conditioned, but they may also be troublesome for most of the numerical methods of this chapter. These equations are called *stiff differential equations*, and we will return to them later in Section 6.9.

**Systems of differential equations**    The material of this chapter generalizes to a system of $m$ first-order equations, written

$$y_1' = f_1(x, y_1, \ldots, y_m) \qquad y_1(x_0) = Y_{1,0}$$

$$\vdots \tag{6.1.11}$$

$$y_m' = f_m(x, y_1, \ldots, y_m) \qquad y_m(x_0) = Y_{m,0}$$

This is often made to look like a first-order equation by introducing vector notation. Let

$$\mathbf{y}(x) = \begin{bmatrix} y_1(x) \\ \vdots \\ y_m(x) \end{bmatrix} \qquad \mathbf{f}(x,\mathbf{y}) = \begin{bmatrix} f_1(x,\mathbf{y}) \\ \vdots \\ f_m(x,\mathbf{y}) \end{bmatrix} \qquad \mathbf{Y}_0 = \begin{bmatrix} Y_{1,0} \\ \vdots \\ Y_{m,0} \end{bmatrix} \qquad (6.1.12)$$

The system (6.1.11) can then be written as

$$\mathbf{y}' = \mathbf{f}(x,\mathbf{y}) \qquad \mathbf{y}(x_0) = \mathbf{Y}_0 \qquad (6.1.13)$$

By introducing vector norms to replace absolute values, virtually all of the preceding results generalize to this vector initial value problem.

For higher order equations, the initial value problem

$$y^{(m)} = f(x, y, y', \dots, y^{(m-1)})$$

$$y(x_0) = Y_0, \dots, y^{(m-1)}(x_0) = Y_0^{(m-1)} \qquad (6.1.14)$$

can be converted to a first-order system. Introduce the new unknown functions

$$y_1 = y \qquad y_2 = y', \dots, y_m = y^{(m-1)}.$$

These functions satisfy the system

$$y_1' = y_2 \qquad\qquad y_1(x_0) = Y_0$$

$$y_2' = y_3 \qquad\qquad y_2(x_0) = Y_0'$$

$$\vdots \qquad\qquad\qquad\qquad \vdots \qquad\qquad (6.1.15)$$

$$y_{m-1}' = y_m$$

$$y_m' = f(x, y_1, \dots, y_m) \qquad y_m(x_0) = Y_0^{(m-1)}$$

*Example*   The linear second-order equation

$$y'' = a_1(x)y' + a_0(x)y + g(x) \qquad y(x_0) = \alpha \qquad y'(x_0) = \beta$$

becomes

$$\begin{bmatrix} y_1 \\ y_2 \end{bmatrix}' = \begin{bmatrix} 0 & 1 \\ a_0(x) & a_1(x) \end{bmatrix} \begin{bmatrix} y_1 \\ y_2 \end{bmatrix} + \begin{bmatrix} 0 \\ g(x) \end{bmatrix} \qquad \begin{bmatrix} y_1(x_0) \\ y_2(x_0) \end{bmatrix} = \begin{bmatrix} \alpha \\ \beta \end{bmatrix}$$

In vector form with $A(x)$ denoting the coefficient matrix,

$$\mathbf{y}' = A(x)\mathbf{y} + \mathbf{G}(x) \qquad \mathbf{y}(x_0) = \mathbf{Y}_0 = \begin{bmatrix} \alpha \\ \beta \end{bmatrix}$$

a linear system of first-order differential equations.

There are special numerical methods for $m$th order equations, but these have been developed to a significant extent only for $m = 2$, which arises in applications of Newtonian mechanics from Newton's second law of mechanics. Most high-order equations are solved by first converting them to an equivalent first-order system, as was just described.

## 6.2  Euler's Method

The most popular numerical methods for solving (6.0.1) are called *finite difference methods*. Approximate values are obtained for the solution at a set of grid points

$$x_0 < x_1 < x_2 < \cdots < x_n < \cdots \qquad (6.2.1)$$

and the approximate value at each $x_n$ is obtained by using some of the values obtained in previous steps. We begin with a simple but computationally inefficient method attributed to Leonhard Euler. The analysis of it has many of the features of the analyses of the more efficient finite difference methods, but without their additional complexity. First we give several derivations of Euler's method, and follow with a complete convergence and stability analysis for it. We given an asymptotic error formula, and conclude the section by generalizing the earlier results to systems of equations.

As before, $Y(x)$ will denote the true solution to (6.0.1):

$$Y'(x) = f(x, Y(x)) \qquad Y(x_0) = Y_0 \qquad (6.2.2)$$

The approximate solution will be denoted by $y(x)$, and the values $y(x_0)$, $y(x_1), \ldots, y(x_n), \ldots$ will often be denoted by $y_0, y_1, \ldots, y_n, \ldots$. An equal grid size $h > 0$ will be used to define the node points,

$$x_j = x_0 + jh \qquad j = 0, 1, \ldots$$

When we are comparing numerical solutions for various values of $h$, we will also use the notation $y_h(x)$ to refer to $y(x)$ with stepsize $h$. The problem (6.0.1) will be solved on a fixed finite interval, which will always be denoted by $[x_0, b]$. The notation $N(h)$ will denote the largest index $N$ for which

$$x_N \le b \qquad x_{N+1} > b$$

In later sections, we discuss varying the stepsize at each $x_n$, in order to control the error.

**Derivation of Euler's method**   Euler's method is defined by

$$y_{n+1} = y_n + hf(x_n, y_n) \qquad n = 0, 1, 2, \ldots \qquad (6.2.3)$$

with $y_0 \doteq Y_0$. Four viewpoints of it are given.

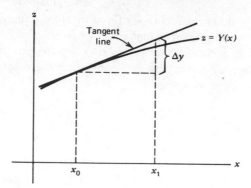

**Figure 6.3**    Geometric interpretation
of Euler's method.

1.  **A geometric viewpoint.**    Consider the graph of the solution $Y(x)$ in Figure
    6.3. Form the tangent line to the graph of $Y(x)$ at $x_0$, and use this line as an
    approximation to the curve for $x_0 \le x \le x_1$. Then

$$\frac{\Delta y}{h} = Y'(x_0) = f(x_0, Y_0)$$

$$Y(x_1) - Y(x_0) \doteq \Delta y = hY'(x_0)$$

$$Y(x_1) \doteq Y(x_0) + hf(x_0, Y(x_0))$$

By repeating this argument on $[x_1, x_2], [x_2, x_3], \ldots,$ we obtain the general
formula (6.2.3).

2.  **Taylor series.**    Expand $Y(x_{n+1})$ about $x_n$,

$$Y(x_{n+1}) = Y(x_n) + hY'(x_n) + \frac{h^2}{2}Y''(\xi_n) \qquad x_n \le \xi_n \le x_{n+1} \quad (6.2.4)$$

By dropping the error term, we obtain the Euler method (6.2.3). The term

$$T_n = \frac{h^2}{2}Y''(\xi_n) \tag{6.2.5}$$

is called the *truncation error* or *discretization error* at $x_{n+1}$. We use the
former name in this text.

3.  **Numerical differentiation.**    From the definition of a derivative,

$$\frac{Y(x_{n+1}) - Y(x_n)}{h} \doteq Y'(x_n) = f(x_n, Y(x_n))$$

$$Y(x_{n+1}) \doteq Y(x_n) + hf(x_n, Y(x_n))$$

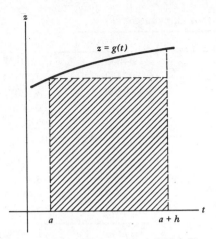

**Figure 6.4**    Illustration of (6.2.7).

4.    *Numerical integration.*    Integrate $Y'(t) = f(t, Y(t))$ over $[x_n, x_{n+1}]$:

$$Y(x_{n+1}) = Y(x_n) + \int_{x_n}^{x_{n+1}} f(t, Y(t))\, dt \qquad (6.2.6)$$

Consider the simple numerical integration method

$$\int_a^{a+h} g(t)\, dt \doteq hg(a) \qquad (6.2.7)$$

called the *left-hand rectangular rule*. Figure 6.4 shows the numerical integral as the crosshatched area. Applying this to (6.2.6), we obtain

$$Y(x_{n+1}) \doteq Y(x_n) + hf(x_n, Y(x_n))$$

as before.

Of the three analytical derivations (2)–(4), both (2) and (4) are the simplest cases of a set of increasingly accurate methods. Approach (2) leads to the *single-step methods*, particularly the Runge–Kutta formulas. Approach (4) leads to *multistep methods*, especially the predictor–corrector methods. Perhaps surprisingly, method (3) often does not lead to other successful methods. The first example given of (3) is the midpoint method in Section 6.4, and it leads to problems of numerical instability. In contrast in Section 6.9, numerical differentiation is used to derive a class of methods for solving stiff differential equations. The bulk of this chapter is devoted to multistep methods, partly because they are generally the most efficient class of methods and partly because they are more complex to analyze than are the Runge–Kutta methods. The latter are taken up in Section 6.10.

Before analyzing the Euler method, we give some numerical examples. They also serve as illustrations for some of the theory that is presented.

**Table 6.1   Euler's method, example (1)**

|            | $x$  | $y_h(x)$ | $Y(x)$  | $Y(x) - y_h(x)$ |
|------------|------|----------|---------|------------------|
| $h = .2$   | .40  | 1.44000  | 1.49182 | .05182           |
|            | .80  | 2.07360  | 2.22554 | .15194           |
|            | 1.20 | 2.98598  | 3.32012 | .33413           |
|            | 1.60 | 4.29982  | 4.95303 | .65321           |
|            | 2.00 | 6.19174  | 7.38906 | 1.19732          |
| $h = .1$   | .40  | 1.46410  | 1.49182 | .02772           |
|            | .80  | 2.14359  | 2.22554 | .08195           |
|            | 1.20 | 3.13843  | 3.32012 | .18169           |
|            | 1.60 | 4.59497  | 4.95303 | .35806           |
|            | 2.00 | 6.72750  | 7.38906 | .66156           |
| $h = .05$  | .40  | 1.47746  | 1.49182 | .01437           |
|            | .80  | 2.18287  | 2.22554 | .04267           |
|            | 1.20 | 3.22510  | 3.32012 | .09502           |
|            | 1.60 | 4.76494  | 4.95303 | .18809           |
|            | 2.00 | 7.03999  | 7.38906 | .34907           |

***Example 1.***   Consider the equation $y' = y$, $y(0) = 1$. Its true solution is $Y(x) = e^x$. Numerical results are given in Table 6.1 for several values of $h$. The answers $y_h(x_n)$ are given at only a few points, rather than at all points at which they were calculated. Note that the error at each point $x$ decreases by about half when $h$ is halved.

**2.**   Consider the equation

$$y' = \frac{1}{1 + x^2} - 2y^2 \quad y(0) = 0 \quad\quad (6.2.8)$$

The true solution is

$$Y(x) = \frac{x}{1 + x^2}$$

and the results are given in Table 6.2. Again note the behavior of the error as $h$ is decreased.

**Convergence analysis**   At each step of the Euler method, an additional truncation error (6.2.5) is introduced. We analyze the cumulative effect of these errors. The total error $Y(x) - y(x)$ is called the *global error*, and the last columns of Tables 6.1 and 6.2 are examples of global error.

To obtain some intuition about the behavior of the global error for Euler's method, we consider the very simple problem

$$y' = 2x \quad y(0) = 0 \quad\quad (6.2.9)$$

**Table 6.2    Euler's method, example (2)**

|  | $x$ | $y_h(x)$ | $Y(x)$ | $Y(x) - y_h(x)$ |
|---|---|---|---|---|
| $h = .2$ | 0.00 | 0.0 | 0.0 | 0.0 |
|  | .40 | .37631 | .34483 | $-.03148$ |
|  | .80 | .54228 | .48780 | $-.05448$ |
|  | 1.20 | .52709 | .49180 | $-.03529$ |
|  | 1.60 | .46632 | .44944 | $-.01689$ |
|  | 2.00 | .40682 | .40000 | $-.00682$ |
| $h = .1$ | .40 | .36085 | .34483 | $-.01603$ |
|  | .80 | .51371 | .48780 | $-.02590$ |
|  | 1.20 | .50961 | .49180 | $-.01781$ |
|  | 1.60 | .45872 | .44944 | $-.00928$ |
|  | 2.00 | .40419 | .40000 | $-.00419$ |
| $h = .05$ | .40 | .35287 | .34483 | $-.00804$ |
|  | .80 | .50049 | .48780 | $-.01268$ |
|  | 1.20 | .50073 | .49180 | $-.00892$ |
|  | 1.60 | .45425 | .44944 | $-.00481$ |
|  | 2.00 | .40227 | .40000 | $-.00227$ |

Its solution is $Y(x) = x^2$. Euler's method for this problem is

$$y_{n+1} = y_n + 2hx_n \qquad y_0 = 0$$

Then it can be verified by the induction that

$$y_n = x_{n-1}x_n \qquad n \geq 1$$

For the error,

$$Y(x_n) - y_n = x_n^2 - x_n x_{n-1} = hx_n \qquad (6.2.10)$$

Thus the global error at each fixed value of $x$ is proportional to $h$. This agrees with the behavior of the examples in Table 6.1 and 6.2, in which the error decreases by about half when $h$ is halved.

For the complete error analysis, we begin with the following lemma, which is quite useful in the analysis of finite difference methods.

**Lemma 1**   For any real $x$,

$$1 + x \leq e^x$$

and for any $x \geq -1$,

$$0 \leq (1 + x)^m \leq e^{mx} \qquad (6.2.11)$$

**Proof**   Using Taylor's theorem,

$$e^x = 1 + x + \frac{x^2}{2}e^\xi$$

with $\xi$ between 0 and $x$. Since the remainder is always positive, the first result is proved. Formula (6.2.11) follows easily. ∎

For the remainder of this chapter, it is assumed that the function $f(x, y)$ satisfies the following stronger Lipschitz condition:

$$|f(x, y_1) - f(x, y_2)| \le K|y_1 - y_2| \qquad -\infty < y_1, y_2 < \infty \qquad x_0 \le x \le b$$

$$(6.2.12)$$

for some $K \ge 0$. Although stronger than necessary, it will simplify the proofs. And given a function $f(x, y)$ satisfying the weaker condition (6.1.1) and a solution $Y(x)$ to the initial value problem (6.0.1), the function $f$ can be modified to satisfy (6.2.12) without changing the solution $Y(x)$ or the essential character of the problem (6.0.1) and its numerical solution. [See Shampine and Gordon (1975), p. 24, for the details].

**Theorem 6.3**   Assume that the solution $Y(x)$ of (6.0.1) has a bounded second derivative on $[x_0, b]$. Then the solution $\{y_h(x_n)|x_0 \le x_n \le b\}$ obtained by Euler's method (6.2.3) satisfies

$$\underset{x_0 \le x_n \le b}{\text{Max}} |Y(x_n) - y_h(x_n)| \le e^{(b-x_0)K}|e_0|$$

$$+ \left[ \frac{e^{(b-x_0)K} - 1}{K} \right] \tau(h) \quad (6.2.13)$$

where

$$\tau(h) = \frac{h}{2}\|Y''\|_\infty \qquad (6.2.14)$$

and $e_0 = Y_0 - y_h(x_0)$.
    If in addition,

$$|Y_0 - y_h(x_0)| \le c_1 h \qquad \text{as } h \to 0 \qquad (6.2.15)$$

for some $c_1 \ge 0$ (e.g., if $Y_0 = y_0$ for all $h$), then there is a constant $B \ge 0$ for which

$$\underset{x_0 \le x_n \le b}{\text{Max}} |Y(x_n) - y_h(x_n)| \le Bh \qquad (6.2.16)$$

**Proof**   Let $e_n = Y(x_n) - y(x_n)$, $n \ge 0$, and recall the definition of $N(h)$ from the beginning of the section. Define

$$\tau_n = \frac{h}{2} Y''(\xi_n) \qquad 0 \le n \le N(h) - 1$$

based on the truncation error in (6.2.5). Easily

$$\underset{0 \le n \le N-1}{\text{Max}} |\tau_n| \le \tau(h)$$

using (6.2.14). From (6.2.4), (6.2.2), and (6.2.3), we have

$$Y_{n+1} = Y_n + hf(x_n, Y_n) + h\tau_n \qquad (6.2.17)$$

$$y_{n+1} = y_n + hf(x_n, y_n) \qquad 0 \le n \le N(h) - 1 \qquad (6.2.18)$$

We are using the common notation $Y_n \equiv Y(x_n)$. Subtracting (6.2.18) from (6.2.17),

$$e_{n+1} = e_n + h[f(x_n, Y_n) - f(x_n, y_n)] + h\tau_n \qquad (6.2.19)$$

and taking bounds, using (6.2.12),

$$|e_{n+1}| \le |e_n| + hK|Y_n - y_n| + h|\tau_n|$$

$$|e_{n+1}| \le (1 + hK)|e_n| + h\tau(h) \qquad 0 \le n \le N(h) - 1 \qquad (6.2.20)$$

Apply (6.2.20) recursively to obtain

$$|e_n| \le (1 + hK)^n |e_0| + \{1 + (1 + hK) + \cdots + (1 + hK)^{n-1}\} h\tau(h)$$

Using the formula for the sum of a finite geometric series,

$$1 + r + r^2 + \cdots + r^{n-1} = \frac{r^n - 1}{r - 1} \qquad r \ne 1 \qquad (6.2.21)$$

we obtain

$$|e_n| \le (1 + hK)^n |e_0| + \left[\frac{(1 + hK)^n - 1}{K}\right] \tau(h) \qquad (6.2.22)$$

Using Lemma 1,

$$(1 + hK)^n \le e^{nhK} = e^{(x_n - x_0)K} \le e^{(b - x_0)K}$$

and this with (6.2.22) implies the main result (6.2.13).

The remaining result (6.2.16) is a trivial corollary of (6.2.13) with the constant $B$ given by

$$B = c_1 e^{(b - x_0)K} + \left[\frac{e^{(b - x_0)K} - 1}{K}\right] \cdot \frac{\|Y''\|_\infty}{2}$$

This completes the proof.    ∎

The result (6.2.16) implies that the error should decrease by at least one-half when the stepsize $h$ is halved. This is confirmed in the examples of Table 6.1 and 6.2. It is shown later in the section that (6.2.16) gives exactly the correct rate of convergence (also see Problem 7).

The bound (6.2.16) gives the correct speed of convergence for Euler's method, but the multiplying constant $B$ is much too large for most equations. For example, with the earlier example (6.2.8), the formula (6.2.13) will predict that the error grows with $b$. But clearly from Table 6.2, the error decreases with increasing $x$. We give the following improvement of (6.2.13) to handle many cases such as (6.2.8).

**Corollary**    Assume the same hypotheses as in Theorem 6.3; in addition, assume that

$$\frac{\partial f(x, y)}{\partial y} \le 0 \tag{6.2.23}$$

for $x_0 \le x \le b$, $-\infty < y < \infty$. Then for all sufficiently small $h$,

$$|Y(x_n) - y_h(x_n)| \le |e_0| + \frac{h}{2}(x_n - x_0) \underset{x_0 \le x_n \le b}{\text{Max}} |Y''(x)| \tag{6.2.24}$$

for $x_0 \le x_n \le b$.    ■

**Proof**    Apply the mean value theorem to the error equation (6.2.19):

$$e_{n+1} = \left[1 + h\frac{\partial f(x_n, \zeta_n)}{\partial y}\right]e_n + \frac{h^2}{2}Y''(\xi_n) \tag{6.2.25}$$

with $\zeta_n$ between $y_h(x_n)$ and $Y(x_n)$. From the convergence of $y_h(x)$ to $Y(x)$ on $[x_0, b]$, we know that the partial derivatives $\partial f(x_n, \zeta_n)/\partial y$ approach $\partial f(x, Y(x))/\partial y$, and thus they must be bounded in magnitude over $[x_0, b]$. Pick $h_0 > 0$ so that

$$1 + h\frac{\partial f(x_n, \zeta_n)}{\partial y} \ge -1 \qquad x_0 \le x_n \le b \tag{6.2.26}$$

for all $h \le h_0$. From (6.2.23), we know the left side is also bounded above by 1, for all $h$. Apply these results to (6.2.25) to get

$$|e_{n+1}| \le |e_n| + \frac{h^2}{2}|Y''(\xi_n)| \tag{6.2.27}$$

By induction, we can show

$$|e_n| \le |e_0| + \frac{h^2}{2}\big[|Y''(\xi_0)| + \cdots + |Y''(\xi_{n-1})|\big]$$

which easily leads to (6.2.14).    ■

Result (6.2.24) is a considerable improvement over the earlier bound (6.2.13); the exponential $\exp(K(b - x_0))$ is replaced by $b - x_0$ (bounding $x_n - x_0$), which increases less rapidly with $b$. The theorem does not apply directly to the earlier example (6.2.8), but a careful examination of the proof in this case will show that the proof is still valid.

**Stability analysis**   Recall the stability analysis for the initial value problem, given in Theorem 6.2. To consider a similar idea for Euler's method, we consider the numerical method

$$z_{n+1} = z_n + h[f(x_n, z_n) + \delta(x_n)] \qquad 0 \le n \le N(h) - 1 \quad (6.2.28)$$

with $z_0 = y_0 + \epsilon$. This is in analogue to the comparison of (6.1.5) with (6.0.1), showing the stability of the initial value problem. We compare the two numerical solutions $\{z_n\}$ and $\{y_n\}$ as $h \to 0$.

Let $e_n = z_n - y_n$, $n \ge 0$. Then $e_0 = \epsilon$, and subtracting (6.2.3) from (6.2.28),

$$e_{n+1} = e_n + h[f(x_n, z_n) - f(x_n, y_n)] + h\delta(x_n)$$

This has exactly the same form as (6.2.19). Using the same procedure as that following (6.2.19), we have

$$\underset{0 \le n \le N(h)}{\text{Max}} |z_n - y_n| \le e^{(b-x_0)K}|\epsilon| + \left[\frac{e^{(b-x_0)K} - 1}{K}\right]\|\delta\|_\infty$$

Consequently, there are constants $k_1$, $k_2$, independent of $h$, with

$$\underset{0 \le n \le N(h)}{\text{Max}} |z_n - y_n| \le k_1|\epsilon| + k_2\|\delta\|_\infty \qquad (6.2.29)$$

This is the analogue to the result (6.1.4) for the original problem (6.0.1). This says that Euler's method is a stable numerical method for the solution of the initial value problem (6.0.1). We insist that all numerical methods for initial value problems possess this form of stability, imitating the stability of the original problem (6.0.1). In addition, we require other forms of stability as well, which are introduced later. In the future we take $\delta(x) \equiv 0$ and consider only the effect of perturbing the initial value $Y_0$. This simplifies the analysis, and the results are equally useful.

**Rounding error analysis**   Introduce an error into each step of the Euler method, with each error derived from the rounding errors of the operations being performed. This number, denoted by $\rho_n$, is called the *local rounding error*. Calling the resultant numerical values $\tilde{y}_n$, we have

$$\tilde{y}_{n+1} = \tilde{y}_n + hf(x_n, \tilde{y}_n) + \rho_n \qquad n = 0, 1, \ldots, N(h) - 1 \quad (6.2.30)$$

The values $\tilde{y}_n$ are the finite-place numbers actually obtained in the computer, and $y_n$ is the value that would be obtained if exact arithmetic were being used. Let

$\rho(h)$ be a bound on the rounding errors,

$$\rho(h) = \max_{0 \le n \le N(h)-1} |\rho_n| \qquad (6.2.31)$$

In a practical situation, using a fixed word length computer, the bound $\rho(h)$ does not decrease as $h \to 0$. Instead, it remains approximately constant, and $\rho(h)/\|Y\|_\infty$ will be proportional to the unit roundoff $u$ on the computer, that is, the smallest number $u$ for which $1 + u > 1$ [see (1.2.12) in Chapter 1].

To see the effect of the roundoff errors in (6.2.30), subtract it from the true solution in (6.2.4), to obtain

$$\tilde{e}_{n+1} = \tilde{e}_n + h\big[f(x_n, Y_n) - f(x_n, \tilde{y}_n)\big] + h\tau_n - \rho_n,$$

where $\tilde{e}_n = Y(x_n) - \tilde{y}_n$. Proceed as in the proof of Theorem 6.3, but identify $\tau_n - \rho_n/h$ with $\tau_n$ in the earlier proof. Then we obtain

$$|\tilde{e}_n| \le e^{(b-x_0)K}|Y_0 - \tilde{y}_0| + \left[\frac{e^{(b-x_0)K} - 1}{K}\right]\left[\tau(h) + \frac{\rho(h)}{h}\right] \qquad (6.2.32)$$

To further examine this bound, let $\rho(h)/\|Y\|_\infty \doteq u$, as previously discussed. Then

$$|\tilde{e}_n| \le c\left[\frac{h}{2}\|Y''\|_\infty + \frac{u\|Y\|_\infty}{h}\right] \equiv E(h) \qquad (6.2.33)$$

The qualitative behavior of $E(h)$ is shown in the graph of Figure 6.5. At $h^*$, $E(h)$ is a minimum, and any further decrease will cause an increased error bound $E(h)$.

This derivation gives the worst possible case for the effect of the rounding errors. In practice the rounding errors vary in both size and sign. The resultant cancellation will cause $\tilde{e}_n$ to increase less rapidly than is implied by the $1/h$ term in $E(h)$ and (6.2.32). But there will still be an optimum value of $h^*$, and below it the error will again increase. For a more complete analysis, see Henrici (1962, pp. 35–59).

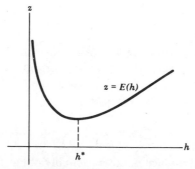

**Figure 6.5**   Error curve for (6.2.33).

Usually rounding error is not a serious problem. However, if the desired accuracy is close to the best that can be attained because of the computer word length, then greater attention must be given to the effects due to rounding. On an IBM mainframe computer (see Table 1.1) with double precision arithmetic, if $h \geq .001$, then the maximum of $u/h$ is $2.2 \times 10^{-13}$, where $u$ is the unit round. Thus the rounding error will usually not present a significant problem unless very small error tolerances are desired. But in single precision with the same restriction on $h$, the maximum of $u/h$ is $0.5 \times 10^{-4}$, and with an error tolerance of this magnitude (not an unreasonable one), the rounding error will be a more significant factor.

*Example*  We solve the problem

$$y' = -y + 2\cos(x) \qquad y(0) = 1$$

whose true solution is $Y(x) = \sin(x) + \cos(x)$. We solve it using Euler's method, with three different forms of arithmetic: (1) four-digit decimal floating-point arithmetic with chopping; (2) four-digit decimal floating-point arithmetic with rounding; and (3) exact, or very high-precision, arithmetic. In the first two cases, the unit rounding errors are $u = .001$ and $u = .0005$, respectively. The bound (6.2.32) applies to cases 1 and 2, whereas case 3 satisfies the theoretical bound (6.2.24). The errors for the three forms of Euler's method are given in Table 6.3. The errors for the answers obtained using decimal arithmetic are based on the true answers $Y(x)$ rounded to four digits.

For the case of chopped decimal arithmetic, the errors are beginning to be affected with $h = .02$; with $h = .01$, the chopping error has a significant effect on

**Table 6.3  Example of rounding effects in Euler's method**

| $h$ | $x$ | Chopped Decimal | Rounded Decimal | Exact Arithmetic |
|-----|-----|-----------------|-----------------|------------------|
| .04 | 1 | $-1.00E - 2$ | $-1.70E - 2$ | $-1.70E - 2$ |
|     | 2 | $-1.17E - 2$ | $1.83E - 2$ | $-1.83E - 2$ |
|     | 3 | $-1.20E - 3$ | $-2.80E - 3$ | $-2.78E - 3$ |
|     | 4 | $1.00E - 2$ | $1.60E - 2$ | $1.53E - 2$ |
|     | 5 | $1.13E - 2$ | $1.96E - 2$ | $1.94E - 2$ |
| .02 | 1 | $7.00E - 3$ | $-9.00E - 3$ | $-8.46E - 3$ |
|     | 2 | $4.00E - 3$ | $-9.10E - 3$ | $-9.13E - 3$ |
|     | 3 | $2.30E - 3$ | $-1.40E - 3$ | $-1.40E - 3$ |
|     | 4 | $-6.00E - 3$ | $8.00E - 3$ | $7.62E - 3$ |
|     | 5 | $-6.00E - 3$ | $8.50E - 3$ | $9.63E - 3$ |
| .01 | 1 | $2.80E - 2$ | $-3.00E - 3$ | $-4.22E - 3$ |
|     | 2 | $2.28E - 2$ | $-4.30E - 3$ | $-4.56E - 3$ |
|     | 3 | $7.40E - 3$ | $-4.00E - 4$ | $-7.03E - 4$ |
|     | 4 | $-2.30E - 2$ | $3.00E - 3$ | $3.80E - 3$ |
|     | 5 | $-2.41E - 2$ | $4.60E - 3$ | $4.81E - 3$ |

the total error. In contrast, the errors using rounded arithmetic are continuing to decrease, although the $h = .1$ case is affected slightly. With this problem, as with most others, the use of rounded arithmetic is far superior to that of chopped arithmetic.

**Asymptotic error analysis**  An asymptotic estimate of the error in Euler's method is derived, ignoring any effects due to rounding. Before beginning, some special notation is necessary to simplify the algebra in the analysis. If $B(x, h)$ is a function defined for $x_0 \leq x \leq b$ and for all sufficiently small $h$, then the notation

$$B(x, h) = O(h^p)$$

for some $p > 0$, means there is a constant $c$ such that

$$|B(x, h)| \leq ch^p \qquad x_0 \leq x \leq b \qquad (6.2.34)$$

for all sufficiently small $h$. If $B$ depends on $h$ only, the same kind of bound is implied.

**Theorem 6.4**  Assume $Y(x)$ is the solution of the initial value problem (6.0.1) and that it is three times continuously differentiable. Assume

$$f_y(x, y) \equiv \frac{\partial f(x, y)}{\partial y} \qquad f_{yy}(x, y) \equiv \frac{\partial^2 f(x, y)}{\partial y^2}$$

are continuous and bounded for $x_0 \leq x \leq b$, $-\infty < y < \infty$. Let the initial value $y_h(x_0)$ satisfy

$$Y_0 - y_h(x_0) = \delta_0 h + O(h^2) \qquad (6.2.35)$$

Usually this error is zero and thus $\delta_0 = 0$.

Then the error in Euler's method (6.2.3) satisfies

$$Y(x_n) - y_h(x_n) = D(x_n)h + O(h^2) \qquad (6.2.36)$$

where $D(x)$ is the solution of the linear initial value problem

$$D'(x) = f_y(x, Y(x))D(x) + \tfrac{1}{2}Y''(x) \qquad D(x_0) = \delta_0 \quad (6.2.37)$$

**Proof**  Using Taylor's theorem,

$$Y(x_{n+1}) = Y(x_n) + hY'(x_n) + \frac{h^2}{2}Y''(x_n) + \frac{h^3}{6}Y^{(3)}(\xi_n)$$

for some $x_n \leq \xi_n \leq x_{n+1}$. Subtract (6.2.3) and use (6.2.2) to obtain

$$e_{n+1} = e_n + h[f(x_n, Y_n) - f(x_n, y_n)]$$

$$+ \frac{h^2}{2}Y''(x_n) + \frac{h^3}{6}Y^{(3)}(\xi_n) \qquad (6.2.38)$$

Using Taylor's theorem on $f(x_n, y_n)$, regarded as a function of $y_n$,

$$f(x_n, y_n) = f(x_n, Y_n) + (y_n - Y_n)f_y(x_n, Y_n) + \tfrac{1}{2}(y_n - Y_n)^2 f_{yy}(x_n, \zeta_n)$$

for some $\zeta_n$ between $y_n$ and $Y_n$. Using this in (6.2.38),

$$e_{n+1} = \left[1 + hf_y(x_n, Y_n)\right]e_n + \frac{h^2}{2}Y''(x_n) + B_n$$

$$B_n = \frac{h^3}{6}Y^{(3)}(\xi_n) - \frac{1}{2}hf_{yy}(x_n, \zeta_n)e_n^2 \tag{6.2.39}$$

Using (6.2.16)

$$B_n = O(h^3) \tag{6.2.40}$$

Because $B_n$ is small relative to the remaining terms in (6.2.39), we find the dominant part of the error by neglecting $B_n$. Let $g_n$ represent the dominant part of the error. It is defined implicitly by

$$g_{n+1} = \left[1 + hf_y(x_n, Y_n)\right]g_n + \frac{h^2}{2}Y''(x_n) \tag{6.2.41}$$

with

$$g_0 = \delta_0 h \tag{6.2.42}$$

the dominant part of the initial error (6.2.35). Since we expect $g_n \doteq e_n$, and since $e_n = O(h)$, we introduce the sequence $\{\delta_n\}$ implicitly by

$$g_n = h\delta_n \tag{6.2.43}$$

Substituting into (6.2.41), canceling $h$ and rearranging, we obtain

$$\delta_{n+1} = \delta_n + h\left[f_y(x_n, Y_n)\delta_n + \tfrac{1}{2}Y''(x_n)\right] \qquad x_0 \le x_n \le b \tag{6.2.44}$$

The initial value $\delta_0$ from (6.2.35) was defined as independent of $h$.

The equation (6.2.44) is Euler's method for solving the initial value problem (6.2.37); thus from Theorem 6.3,

$$D(x_n) - \delta_n = O(h) \qquad x_0 \le x_n \le b \tag{6.2.45}$$

Combining this with (6.2.43),

$$g_n = D(x_n)h + O(h^2) \tag{6.2.46}$$

To complete the proof, it must be shown that $g_n$ is indeed the principal part of the error $e_n$. Introduce

$$k_n = e_n - g_n \tag{6.2.47}$$

Then $k_0 = e_0 - g_0 = O(h^2)$ from (6.2.35) and (6.2.42). Subtracting (6.2.41) from (6.2.39), and using (6.2.40),

$$k_{n+1} = \left[1 + hf_y(x_n, Y_n)\right]k_n + B_n$$

$$|k_{n+1}| \le (1 + hK)|k_n| + O(h^3)$$

This is of the form of (6.2.20) in the proof of Theorem 6.3, with the term $h\tau(h)$ replaced by $O(h^3)$. Using the same derivation,

$$|k_n| = O(h^2) \tag{6.2.48}$$

Combining (6.2.46)–(6.2.48),

$$e_n = g_n + k_n = \left[hD(x_n) + O(h^2)\right] + O(h^2)$$

which proves (6.2.36).    ∎

The function $D(x)$ is rarely produced explicitly, but the form of the error (6.2.36) furnishes useful qualitative information. It is often used as the basis for extrapolation procedures, some of which are discussed in later sections.

*Example*    Consider the problem

$$y' = -y \qquad y(0) = 1$$

with the solution $Y(x) = e^{-x}$. The equation for $D(x)$ is

$$D'(x) = -D(x) + \tfrac{1}{2}e^{-x} \qquad D(0) = 0$$

and its solution is

$$D(x) = \tfrac{1}{2}xe^{-x}$$

This gives the following asymptotic formula for the error in Euler's method.

$$Y(x_n) - y_h(x_n) \doteq \frac{h}{2}x_n e^{-x_n} \tag{6.2.49}$$

Table 6.4 contains the actual errors and the errors predicted by (6.2.49) for $h = .05$. Note then the error decreases with increasing $x$, just as with the solution $Y(x)$. But the relative error increases linearly with $x$,

$$\frac{Y(x_n) - y_h(x_n)}{Y(x_n)} \doteq \frac{h}{2}x_n$$

Also, the estimate (6.2.49) is much better than the bound given by (6.2.13) in

**Table 6.4    Example (6.2.49)
of theorem 6.4**

| $x_n$ | $Y_n - y_n$ | $hD(x_n)$ |
|-------|-------------|-----------|
| .4    | .00689      | .00670    |
| .8    | .00920      | .00899    |
| 1.2   | .00921      | .00904    |
| 1.6   | .00819      | .00808    |
| 2.0   | .00682      | .00677    |

Theorem 6.3. That bound is

$$|Y(x_n) - y_h(x_n)| \le \frac{h}{2}(e^{x_n} - 1)$$

and it increases exponentially with $x_n$.

**Systems of equations**    To simplify the presentation, we will consider only the following system of order 2:

$$y_1' = f_1(x, y_1, y_2) \qquad y_1(x_0) = Y_{1,0}$$
$$y_2' = f_2(x, y_1, y_2) \qquad y_2(x_0) = Y_{2,0} \tag{6.2.50}$$

The generalization to higher order systems is straightforward. Euler's method for solving (6.2.50) is

$$y_{1, n+1} = y_{1, n} + h f_1(x_n, y_{1, n}, y_{2, n})$$
$$y_{2, n+1} = y_{2, n} + h f_2(x_n, y_{1, n}, y_{2, n}) \tag{6.2.51}$$

a clear generalization of (6.2.3).

All of the preceding results of this section will generalize to (6.2.51), and the form of the generalization is clear if we use the vector notation for (6.2.50) and (6.2.51), as given in (6.1.13) in the last section. Use

$$\mathbf{y}' = \mathbf{f}(x, \mathbf{y}) \qquad \mathbf{y}(x_0) = \mathbf{Y}_0$$

In place of absolute values, use the norm (1.1.16) from Chapter 1:

$$\|\mathbf{y}\| = \underset{i}{\text{Max}}|y_i|$$

To generalize the Lipschitz condition (6.2.12), use Taylor's theorem (Theorem 1.5) for functions of several variables to obtain

$$\|\mathbf{f}(x, \mathbf{z}) - \mathbf{f}(x, \mathbf{y})\| \le K\|\mathbf{z} - \mathbf{y}\| \tag{6.2.52}$$

$$K = \underset{i}{\text{Max}} \sum_j \underset{\substack{x_0 \le x \le b \\ -\infty < w_1, w_2 < \infty}}{\text{Max}} \left| \frac{\partial f_i(x, w_1, w_2)}{\partial w_j} \right| \tag{6.2.53}$$

The role of $\partial f(x, y)/\partial y$ is replaced by the Jacobian matrix

$$\mathbf{f}_y(x,y) = \begin{bmatrix} \dfrac{\partial f_1}{\partial y_1} & \dfrac{\partial f_1}{\partial y_2} \\ \dfrac{\partial f_2}{\partial y_1} & \dfrac{\partial f_2}{\partial y_2} \end{bmatrix} \tag{6.2.54}$$

As an example, the asymptotic error formula (6.2.36) becomes

$$\mathbf{Y}(x_n) - \mathbf{y}_h(x_n) = h\mathbf{D}(x_n) + \mathbf{R}_n \qquad \|\mathbf{R}_n\| = O(h^2) \tag{6.2.55}$$

with $\mathbf{D}(x)$ the solution of the linear system

$$\mathbf{D}'(x) = \mathbf{f}_y(x,\mathbf{Y}(x))\mathbf{D}(x) + \tfrac{1}{2}\mathbf{Y}''(x) \qquad \mathbf{D}(x_0) = \delta_0 \tag{6.2.56}$$

using the preceding matrix $\mathbf{f}_y(x,\mathbf{y})$.

*Example*   Solve the pendulum equation,

$$\theta''(t) = -\sin(\theta(t)) \qquad \theta(0) = \frac{\pi}{2} \qquad \theta'(0) = 0$$

Convert this to a system by letting $y_1 = \theta$, $y_2 = \theta'$, and replace the variable $t$ by $x$. Then

$$y_1' = y_2 \qquad\qquad y_1(0) = \frac{\pi}{2}$$

$$y_2' = -\sin(y_1) \qquad y_2(0) = 0 \tag{6.2.57}$$

The numerical results are given in Table 6.5. Note that the error decreases by about half when $h$ is halved.

**Table 6.5   Euler's method for example (6.2.57)**

| $h$ | $x_n$ | $y_{1,n}$ | $Y_1(x_n)$ | Error | $y_{2,n}$ | $Y_2(x_n)$ | Error |
|-----|-------|-----------|------------|-------|-----------|------------|-------|
| 0.2 | .2 | 1.5708 | 1.5508 | −.0200 | −.20000 | −.199999 | .000001 |
|     | .6 | 1.4508 | 1.3910 | −.0598 | −.59984 | −.59806 | .00178 |
|     | 1.0 | 1.1711 | 1.0749 | −.0962 | −.99267 | −.97550 | .01717 |
| 0.1 | .2 | 1.5608 | 1.5508 | −.0100 | −.20000 | −.199999 | .000001 |
|     | .6 | 1.4208 | 1.3910 | −.0298 | −.59927 | −.59806 | .00121 |
|     | 1.0 | 1.1223 | 1.0749 | −.0474 | −.98568 | −.97550 | .01018 |

## 6.3   Multistep Methods

This section contains an introduction to the theory of multistep methods. Some particular methods are examined in greater detail in the following sections. A more complete theory is given in Section 6.8.

As before, let $h > 0$ and define the nodes by $x_n = x_0 + nh$, $n \geq 0$. The general form of the multistep methods to be considered is

$$y_{n+1} = \sum_{j=0}^{p} a_j y_{n-j} + h \sum_{j=-1}^{p} b_j f(x_{n-j}, y_{n-j}) \qquad n \geq p \qquad (6.3.1)$$

The coefficients $a_0, \ldots, a_p, b_{-1}, b_0, \ldots, b_p$ are constants, and $p \geq 0$. If either $a_p \neq 0$ or $b_p \neq 0$, the method is called a $p + 1$ step method, because $p + 1$ previous solution values are being used to compute $y_{n+1}$. The values $y_1, \ldots, y_p$ must be obtained by other means; this is discussed in later sections. Euler's method is an example of a one-step method, with $p = 0$ and

$$a_0 = 1 \qquad b_0 = 1 \qquad b_{-1} = 0$$

If $b_{-1} = 0$, then $y_{n+1}$ occurs on only the left side of equation (6.3.1). Such formulas are called *explicit methods*. If $b_{-1} \neq 0$, then $y_{n+1}$ is present on both sides of (6.3.1) and the formula is called an *implicit method*. The existence of the solution $y_{n+1}$, for all sufficiently small $h$, can be shown by using the fixed-point theory of Section 2.5. Implicit methods are generally solved by iteration, which are discussed in detail for the trapezoidal method in Section 6.5.

***Example 1.***   The midpoint method is defined by

$$y_{n+1} = y_{n-1} + 2hf(x_n, y_n) \qquad n \geq 1 \qquad (6.3.2)$$

and it is an explicit two-step method. It is examined in Section 6.4.

**2.**   The trapezoidal method is defined by

$$y_{n+1} = y_n + \frac{h}{2}[f(x_n, y_n) + f(x_{n+1}, y_{n+1})] \qquad n \geq 0 \qquad (6.3.3)$$

It is an implicit one-step method, and it is discussed in Sections 6.5 and 6.6.

For any differentiable function $Y(x)$, define the truncation error for integrating $Y'(x)$ by

$$T_n(Y) = Y(x_{n+1}) - \left[ \sum_{j=0}^{p} a_j Y(x_{n-j}) + h \sum_{j=-1}^{p} b_j Y'(x_{n-j}) \right] \qquad n \geq p \quad (6.3.4)$$

Define the function $\tau_n(Y)$ by

$$\tau_n(Y) = \frac{1}{h} T_n(Y) \qquad (6.3.5)$$

In order to prove the convergence of the approximate solution $\{y_n | x_0 \le x_n \le b\}$ of (6.3.1) to the solution $Y(x)$ of the initial value problem (6.0.1), it is necessary to have

$$\tau(h) \equiv \underset{x_p \le x_n \le b}{\text{Max}} |\tau_n(Y)| \to 0 \quad \text{as} \quad h \to 0 \quad (6.3.6)$$

This is often called the *consistency condition* for method (6.3.1). The speed of convergence of the solution $\{y_n\}$ to the true solution $Y(x)$ is related to the speed of convergence in (6.3.6), and thus we need to know the conditions under which

$$\tau(h) = O(h^m) \quad (6.3.7)$$

for some desired choice of $m \ge 1$. We now examine the implications of (6.3.6) and (6.3.7) for the coefficients in (6.3.1). The convergence result for (6.3.1) is given later as Theorem 6.6.

**Theorem 6.5**  Let $m \ge 1$ be a given integer. In order that (6.3.6) hold for all continuously differentiable functions $Y(x)$, that is, that the method (6.3.1) be consistent, it is necessary and sufficient that

$$\sum_{j=0}^{p} a_j = 1 \quad - \sum_{j=0}^{p} j a_j + \sum_{j=-1}^{p} b_j = 1 \quad (6.3.8)$$

And for (6.3.7) to be valid for all functions $Y(x)$ that are $m + 1$ times continuously differentiable, it is necessary and sufficient that (6.3.8) hold and that

$$\sum_{j=0}^{p} (-j)^i a_j + i \sum_{j=-1}^{p} (-j)^{i-1} b_j = 1 \quad i = 2, \ldots, m \quad (6.3.9)$$

**Proof**  Note that

$$T_n(\alpha Y + \beta W) = \alpha T_n(Y) + \beta T_n(W) \quad (6.3.10)$$

for all constants $\alpha, \beta$ and all differentiable functions $Y, W$. To examine the consequences of (6.3.6) and (6.3.7) expand $Y(x)$ about $x_n$, using Taylor's theorem 1.4, to obtain

$$Y(x) = \sum_{i=0}^{m} \frac{1}{i!} (x - x_n)^i Y^{(i)}(x_n) + R_{m+1}(x) \quad (6.3.11)$$

assuming $Y(x)$ is $m + 1$ times continuously differentiable. Substituting into (6.3.4) and using (6.3.10),

$$T_n(Y) = \sum_{i=0}^{m} \frac{1}{i!} Y^{(i)}(x_n) T_n\big((x - x_n)^i\big) + T_n(R_{m+1})$$

It is necessary to calculate $T_n((x - x_n)^i)$ for $i \ge 0$.

For $i = 0$,

$$T_n(1) = c_0 \equiv 1 - \sum_{j=0}^{p} a_j \qquad (6.3.12)$$

For $i \geq 1$,

$$T_n\left((x - x_n)^i\right)$$

$$= (x_{n+1} - x_n)^i - \left[ \sum_{j=0}^{p} a_j (x_{n-j} - x_n)^i + h \sum_{j=-1}^{p} b_j i (x_{n-j} - x_n)^{i-1} \right]$$

$$= c_i h^i \qquad (6.3.13)$$

$$c_i = 1 - \left[ \sum_{j=0}^{p} (-j)^i a_j + i \sum_{j=-1}^{p} (-j)^{i-1} b_j \right] \qquad i \geq 1$$

This gives

$$T_n(Y) = \sum_{i=0}^{m} \frac{c_i}{i!} h^i Y^{(i)}(x_n) + T_n(R_{m+1}) \qquad (6.3.14)$$

If we write the remainder $R_{m+1}(x)$ as

$$R_{m+1}(x) = \frac{1}{(m+1)!} (x - x_n)^{m+1} Y^{(m+1)}(x_n) + \cdots$$

then

$$T_n(R_{m+1}) = \frac{c_{m+1}}{(m+1)!} h^{m+1} Y^{(m+1)}(x_n) + O(h^{m+2}) \qquad (6.3.15)$$

assuming $Y$ is $m + 2$ times differentiable.

To obtain the consistency condition (6.3.6), we need $\tau(h) = O(h)$, and this requires $T_n(Y) = O(h^2)$. Using (6.3.14) with $m = 1$, we must have $c_0, c_1 = 0$, which gives the set of equations (6.3.8). In some texts, these equations are referred to as the consistency conditions. To obtain (6.3.7) for some $m \geq 1$, we must have $T_n(Y) = O(h^{m+1})$. From (6.3.14) and (6.3.13), this will be true if and only if $c_i = 0$, $i = 0, 1, \ldots, m$. This proves the conditions (6.3.9) and completes the proof. ∎

The largest value of $m$ for which (6.3.7) holds is called the *order* or *order of convergence* of the method (6.3.1). In Section 6.7, we examine the deriving of methods of any desired order.

We now give a convergence result for the solution of (6.3.1). Although the theorem will not include all the multistep methods that are convergent, it does include most methods of current interest. Moreover, the proof is much easier than that of the more general Theorem 6.8 of Section 6.8.

**Theorem 6.6**    Consider solving the initial value problem

$$y' = f(x, y) \qquad y(x_0) = Y_0 \qquad x_0 \leq x \leq b$$

using the multistep method (6.3.1). Let the initial errors satisfy

$$\eta(h) \equiv \underset{0 \leq i \leq p}{\text{Max}} |Y(x_i) - y_h(x_i)| \to 0 \qquad \text{as} \qquad h \to 0 \quad (6.3.16)$$

Assume the method is consistent, that is, it satisfies (6.3.6). And finally, assume that the coefficients $a_j$ are all nonnegative,

$$a_j \geq 0 \qquad j = 0, 1, \ldots, p \qquad (6.3.17)$$

Then the method (6.3.1) is convergent, and

$$\underset{x0 \leq x_n \leq b}{\text{Max}} |Y(x_n) - y_h(x_n)| \leq c_1 \eta(h) + c_2 \tau(h) \quad (6.3.18)$$

for suitable constants $c_1, c_2$. If the method (6.3.1) is of order $m$, and if the initial errors satisfy $\eta(h) = O(h^m)$, then the speed of convergence of the method is $O(h^m)$.

**Proof**    Rewrite (6.3.4) and use $Y'(x) = f(x, Y(x))$ to get

$$Y(x_{n+1}) = \sum_{j=0}^{p} a_j Y(x_{n-j}) + h \sum_{j=-1}^{p} b_j f\left(x_{n-j}, Y(x_{n-j})\right) + h\tau_n(Y)$$

Subtracting (6.3.1), and using the notation $e_i = Y(x_i) - y_i$,

$$e_{n+1} = \sum_{j=0}^{p} a_j e_{n-j} + h \sum_{j=-1}^{p} b_j \left[ f(x_{n-j}, Y_{n-j}) \right.$$

$$\left. - f(x_{n-j}, y_{n-j}) \right] + h\tau_n(Y)$$

Apply the Lipschitz condition and the assumption (6.3.17) to obtain

$$|e_{n+1}| \leq \sum_{j=0}^{p} a_j |e_{n-j}| + hK \sum_{j=-1}^{p} |b_j| \, |e_{n-j}| + h\tau(h)$$

Introduce the following error bounding function,

$$f_n = \underset{0 \leq i \leq n}{\text{Max}} |e_i| \qquad n = 0, 1, \ldots, N(h)$$

Using this function,

$$|e_{n+1}| \leq \sum_{j=0}^{p} a_j f_n + hK \sum_{j=-1}^{p} |b_j| f_{n+1} + h\tau(h)$$

and applying (6.3.8),

$$|e_{n+1}| \leq f_n + hcf_{n+1} + h\tau(h) \qquad c = K \sum_{j=-1}^{p} |b_j|$$

The right side is trivially a bound for $f_n$, and thus

$$f_{n+1} \leq f_n + hcf_{n+1} + h\tau(h)$$

For $hc \leq \frac{1}{2}$, which must be true as $h \to 0$,

$$f_{n+1} \leq \frac{f_n}{1 - hc} + \frac{h}{1 - hc}\tau(h)$$

$$\leq (1 + 2hc)f_n + 2h\tau(h)$$

Noting that $f_p = \eta(h)$, proceed as in the proof of Theorem 6.3 following (6.2.20). Then

$$f_n \leq e^{2c(b-x_0)}\eta(h) + \left[\frac{e^{2c(b-x_0)} - 1}{c}\right]\tau(h) \qquad x_0 \leq x_n \leq b \quad (6.3.19)$$

This completes the proof.  ∎

The conclusions of the theorem can be proved under weaker assumptions; in particular, (6.3.17) can be replaced by a much weaker assumption. These results are given in Section 6.8. To obtain a rate of convergence of $O(h^m)$ for the method (6.3.1), it is necessary that each step have an error

$$T_n(Y) = O(h^{m+1})$$

But the initial values $y_0, \ldots, y_p$ need to be computed with only an accuracy of $O(h^m)$, since $\eta(h) = O(h^m)$ is sufficient in (6.3.18). Examples illustrating the use of a lower order method for generating the initial values $y_0, \ldots, y_p$ are given in the following sections.

The result (6.3.19) can be improved somewhat for particular cases, but the speed of convergence will remain the same. Examples of the theorem are given in the following sections. As with Euler's method, a complete stability analysis can be given, including a result of the form (6.2.29). The proof is a straightforward modification of the proof of Theorem 6.6. Also, an asymptotic error analysis can be given; examples are given in the next two sections.

## 6.4  The Midpoint Method

We will define and analyze the midpoint method, using it to illustrate some ideas not possible with Euler's method. We can derive the midpoint method in several ways, as with Euler's method, and we use numerical differentiation here.

From (5.7.11) of Chapter 5, we have

$$g'(a) = \frac{g(a+h) - g(a-h)}{2h} - \frac{h^2}{6}g^{(3)}(\xi)$$

for some $a - h \le \xi \le a + h$. Applying this to

$$Y'(x_n) = f(x_n, Y(x_n))$$

we have

$$\frac{Y(x_{n+1}) - Y(x_{n-1})}{2h} - \frac{h^2}{6}Y^{(3)}(\xi_n) = f(x_n, Y(x_n))$$

with $x_{n-1} \le \xi_n \le x_{n+1}$. Solving for $Y(x_{n+1})$,

$$Y(x_{n+1}) = Y(x_{n-1}) + 2hf(x_n, Y(x_n)) + \tfrac{1}{3}h^3 Y^{(3)}(\xi_n) \qquad (6.4.1)$$

The *midpoint method* is obtained by dropping the last term:

$$y_{n+1} = y_{n-1} + 2hf(x_n, y_n) \qquad n \ge 1 \qquad (6.4.2)$$

It is an explicit two-step method, and the order of convergence is two. The value of $y_1$ must be obtained by another method.

The midpoint method could also have been obtained by applying the midpoint numerical integration rule (5.2.17) to the following integral reformulation of the differential equation (6.0.1):

$$Y(x_{n+1}) = Y(x_{n-1}) + \int_{x_{n-1}}^{x_{n+1}} f(t, Y(t)) \, dt \qquad (6.4.3)$$

We omit the details. This approach is used in Section 6.7, to obtain other multistep methods.

To analyze the convergence of (6.4.2), we use Theorem 6.6. For the truncation error, we easily obtain from (6.4.1) that

$$\tau_n(Y) = \tfrac{1}{3}h^2 Y^{(3)}(\xi_n) \qquad x_{n-1} \le \xi_n \le x_{n-1} \qquad (6.4.4)$$

An improved version of the proof of Theorem 6.6, for the midpoint method, yields

$$\underset{x_0 \le x_n \le b}{\text{Max}} |Y(x_n) - y_h(x_n)| \le e^{2K(b-x_0)}\eta(h) + \left[\frac{e^{2K(b-x_0)} - 1}{2K}\right]\left[\frac{1}{3}h^2\|Y^{(3)}\|_\infty\right]$$

$$(6.4.5)$$

$$\eta(h) = \text{Max}\{|Y_0 - y_0|, |Y(x_1) - y_h(x_1)|\}$$

Assuming $y_0 = Y_0$ for all $h$, we need to have $Y(x_1) - y_h(x_1) = O(h^2)$ in order to have a global order of convergence $O(h^2)$ in (6.4.5). From (6.2.4), a single step of Euler's method has the desired property:

$$y_1 = y_0 + hf(x_0, y_0) \qquad y_0 = Y_0 \tag{6.4.6}$$

$$Y(x_1) - y_1 = \frac{h^2}{2} Y''(\zeta) \qquad x_0 \le \zeta \le x_1 \tag{6.4.7}$$

With this initial value $y_h(x_1) = y_1$, the error result (6.4.5) implies

$$\underset{x_0 \le x_n \le b}{\text{Max}} |Y(x_n) - y_h(x_n)| = O(h^2) \tag{6.4.8}$$

A complete stability analysis can be given for the midpoint method, paralleling that given for Euler's method. If we assume for simplicity that $\eta(h) = O(h^3)$, then we have the following asymptotic error formula:

$$Y(x_n) - y_h(x_n) = D(x_n)h^2 + O(h^3) \qquad x_0 \le x_n \le b$$

$$D' = f_y(x, Y(x))D + \tfrac{1}{6}Y^{(3)}(x) \qquad D(x_0) = 0 \tag{6.4.9}$$

There is little that is different in the proofs of these results, and we dispense with them.

**Weak stability**   As previously noted, the midpoint method possesses the same type of stability as that shown for Euler's method in (6.2.28) and (6.2.29). This is, however, not sufficient for practical purposes. We show that the midpoint method is unsatisfactory with respect to another sense of stability yet to be defined.

We consider the numerical solution of the problem

$$y' = \lambda y \qquad y(0) = 1 \tag{6.4.10}$$

which has the solution $Y(x) = e^{\lambda x}$. This is used as a *model problem* for the more general problem (6.0.1), an idea that we explain in Section 6.8. At this point, it is sufficient to note that if a numerical method performs badly for a problem as simple as (6.4.10), then such a method is unlikely to perform well for other more complicated differential equations.

The midpoint method for (6.4.10) is

$$y_{n+1} = y_{n-1} + 2h\lambda y_n \qquad n \ge 1 \tag{6.4.11}$$

We calculate the exact solution of this equation and compare it to the solution $Y(x) = e^{\lambda x}$. The equation (6.4.11) is an example of a *linear difference equation* of order 2. There is a general theory for $p$th-order linear difference equations, paralleling the theory for $p$th-order linear differential equations. Most methods for solving the differential equations have an analogue in solving difference equations, and this is a guide in solving (6.4.11). We begin by looking for linearly independent solutions of the difference equation. These are then combined to form the general solution. For a general theory of linear difference equations, see Henrici (1962, pp. 210–215).

In analogy with the exponential solutions of linear differential equations, we assume a solution for (6.4.11) of the form

$$y_n = r^n \qquad n \geq 0 \tag{6.4.12}$$

for some unknown $r$. Substituting into (6.4.11) to find necessary conditions on $r$,

$$r^{n+1} = r^{n-1} + 2h\lambda r^n$$

Canceling $r^{n-1}$,

$$r^2 = 1 + 2h\lambda r \tag{6.4.13}$$

This argument is reversible. If $r$ satisfies the quadratic equation (6.4.13), then (6.4.12) satisfies (6.4.11).

The equation (6.4.13) is called the *characteristic equation* for the midpoint method. Its roots are

$$r_0 = h\lambda + \sqrt{1 + h^2\lambda^2} \qquad r_1 = h\lambda - \sqrt{1 + h^2\lambda^2} \tag{6.4.14}$$

The general solution to (6.4.11) is then

$$y_n = \beta_0 r_0^n + \beta_1 r_1^n \qquad n \geq 0 \tag{6.4.15}$$

The coefficients $\beta_0$ and $\beta_1$ are chosen so that the values of $y_0$ and $y_1$ that were given originally will agree with the values calculated using (6.4.15):

$$\beta_0 + \beta_1 = y_0$$

$$\beta_0 r_0 + \beta_1 r_1 = y_1$$

The general solution is

$$\beta_0 = \frac{y_1 - r_1 y_0}{r_0 - r_1} \qquad \beta_1 = \frac{y_0 r_0 - y_1}{r_0 - r_1}$$

To gain some intuition for these formulas, consider taking the exact initial values

$$y_0 = 1 \qquad y_1 = e^{\lambda h}$$

Then using Taylor's theorem,

$$\beta_0 = \frac{e^{\lambda h} - r_1}{2\sqrt{1 + h^2\lambda^2}} = 1 + O(h^2\lambda^2)$$

$$\beta_1 = \frac{r_0 - e^{\lambda h}}{2\sqrt{1 + h^2\lambda^2}} = O(h^3\lambda^3) \tag{6.4.16}$$

From these values, $\beta_0 \to 1$ and $\beta_1 \to 0$ as $h \to 0$. Therefore in the formula (6.4.15), the term $\beta_0 r_0^n$ should correspond to the true solution $e^{\lambda x_n}$, since the term $\beta_1 r_1^n \to 0$ as $h \to 0$. In fact,

$$r_0^n = e^{\lambda x_n}[1 + O(h)] \tag{6.4.17}$$

whose proof we leave to the reader.

To see the difficulty in the numerical solution of $y' = \lambda y$ using (6.4.15), examine carefully the relative sizes of $r_0$ and $r_1$. We consider only the case of real $\lambda$. For $0 < \lambda < \infty$, for all $h$,

$$r_0 > |r_1| > 0$$

Thus the term $r_1^n$ will increase less rapidly than $r_0^n$, and the correct term in the general solution (6.4.15) will dominate, namely $\beta_0 r_0^n$.

However, for $-\infty < \lambda < 0$, we will have

$$0 < r_0 < 1 \qquad r_1 < -1 \qquad h > 0$$

As a consequence, $\beta_1 r_1^n$ will eventually dominate $\beta_0 r_0^n$ as $n$ increases, for fixed $h$, no matter how small $h$ is chosen initially. The term $\beta_0 r_0^n \to 0$ as $n \to \infty$; whereas, the term $\beta_1 r_1^n$ increases in magnitude, alternating in sign as $n$ increases.

The term $\beta_1 r_1^n$ is called a *parasitic solution* of the numerical method (6.4.11), since it does not correspond to any solution of the original differential equation $y' = \lambda y$. This original equation has a one-parameter family of solutions, depending on the initial value $Y_0$, but the approximation (6.4.11) has the two-parameter family (6.4.15), which is dependent on $y_0$ and $y_1$. The new solution $\beta_1 r_1^n$ is a creation of the numerical method; for problem (6.4.10) with $\lambda < 0$, it will cause the numerical solution to diverge from the true solution as $x_n \to \infty$. Because of this behavior, we say that the midpoint method is only *weakly stable*.

We return to this topic in Section 6.8, after some necessary theory has been introduced. We generalize the applicability of the model problem (6.4.10) by considering the sign of $\partial f(x, Y(x))/\partial y$. If it is negative, then the weak instability of the midpoint method will usually appear in solving the associated initial value problem. This is illustrated in the second example below.

***Example* 1.**    Consider the model problem (6.4.10) with $\lambda = -1$. The numerical results are given in Table 6.6 for $h = .25$. The value $y_1$ was obtained using Euler's method, as in (6.4.6). From the values in the table, the parasitic solution is clearly growing in magnitude. For $x_n = 2.25$, the numerical solution $y_n$ becomes negative, and it alternates in sign with each successive step.

**2.**    Consider the problem

$$y' = x - y^2 \qquad y(0) = 0$$

The solution $Y(x)$ is strictly increasing for $x \geq 0$; for large $x$, $Y(x) \doteq \sqrt{x}$.

**Table 6.6    Example 1 of midpoint method instability**

| $x_n$ | $y_n$ | $Y(x_n)$ | Error |
|-------|-------|----------|-------|
| .25 | .7500 | .7788 | .0288 |
| .50 | .6250 | .6065 | − .0185 |
| .75 | .4375 | .4724 | .0349 |
| 1.00 | .4063 | .3679 | − .0384 |
| 1.25 | .2344 | .2865 | .0521 |
| 1.50 | .2891 | .2231 | − .0659 |
| 1.75 | .0898 | .1738 | .0839 |
| 2.00 | .2441 | .1353 | − .1088 |
| 2.25 | − .0322 | .1054 | .1376 |

**Table 6.7    Example 2 of midpoint method instability**

| $x_n$ | $y_n$ | $Y(x_n)$ | Error |
|-------|-------|----------|-------|
| .25 | 0.0 | .0312 | .0312 |
| .50 | .1250 | .1235 | − .0015 |
| .75 | .2422 | .2700 | .0278 |
| 1.00 | .4707 | .4555 | − .0151 |
| 1.25 | .6314 | .6585 | .0271 |
| 1.50 | .8963 | .8574 | − .0389 |
| 1.75 | .9797 | 1.0376 | .0579 |
| 2.00 | 1.2914 | 1.1936 | − .0978 |
| 2.25 | 1.1459 | 1.3264 | .181 |
| 2.50 | 1.7599 | 1.4405 | − .319 |
| 2.75 | .8472 | 1.5404 | .693 |
| 3.00 | 2.7760 | 1.6302 | − 1.15 |
| 3.25 | − 1.5058 | 1.7125 | 3.22 |

Although it is increasing,

$$\frac{\partial f(x, y)}{\partial y} = -2y < 0 \quad \text{for} \quad y > 0$$

Therefore, we would expect that the midpoint method would exhibit some instability. This is confirmed with the results given in Table 6.7, obtained with a stepsize of $h = .25$. By $x_n = 2.25$, the numerical solution begins to decrease, and at $x_n = 3.25$, the solution $y_n$ becomes negative.

## 6.5    The Trapezoidal Method

We use the trapezoidal method to introduce implicit methods and ideas associated with them. In addition, the trapezoidal method is of independent interest because of a special stability property it possesses. To introduce the trapezoidal rule, we use numerical integration.

Integrate the differential equation $Y'(t) = f(t, Y(t))$ over $[x_n, x_{n+1}]$ to obtain

$$Y(x_{n+1}) = Y(x_n) + \int_{x_n}^{x_{n+1}} f(t, Y(t)) \, dt$$

Apply the simple trapezoidal rule, (5.1.2) and (5.1.4), to obtain

$$Y(x_{n+1}) = Y(x_n) + \frac{h}{2}\left[ f(x_n, Y(x_n)) + f(x_{n+1}, Y(x_{n+1})) \right]$$

$$- \frac{h^3}{12} Y^{(3)}(\xi_n) \tag{6.5.1}$$

for some $x_n \le \xi_n \le x_{n+1}$. By dropping the remainder term, we obtain the trapezoidal method,

$$y_{n+1} = y_n + \frac{h}{2}\left[ f(x_n, y_n) + f(x_{n+1}, y_{n+1}) \right] \qquad n \ge 0 \tag{6.5.2}$$

It is a one-step method with an $O(h^2)$ order of convergence. It is also a simple example of an implicit method, since $y_{n+1}$ occurs on both sides of (6.5.2). A numerical example is given at the end of this section.

**Iterative solution**  The formula (6.5.2) is a nonlinear equation with root $y_{n+1}$, and any of the general techniques of Chapter 2 can be used to solve it. Simple linear iteration (see Section 2.5) is most convenient and it is usually sufficient. Let $y_{n+1}^{(0)}$ be a good initial guess of the solution $y_{n+1}$, and define

$$y_{n+1}^{(j+1)} = y_n + \frac{h}{2}\left[ f(x_n, y_n) + f(x_{n+1}, y_{n+1}^{(j)}) \right] \qquad j = 0, 1, \dots \tag{6.5.3}$$

The initial guess is usually obtained using an explicit method.

To analyze the iteration and to determine conditions under which it will converge, subtract (6.5.3) from (6.5.2) to obtain

$$y_{n+1} - y_{n+1}^{(j+1)} = \frac{h}{2}\left[ f(x_{n+1}, y_{n+1}) - f(x_{n+1}, y_{n+1}^{(j)}) \right] \tag{6.5.4}$$

Use the Lipschitz condition (6.2.12) to bound this with

$$\left| y_{n+1} - y_{n+1}^{(j+1)} \right| \le \frac{hK}{2} \left| y_{n+1} - y_{n+1}^{(j)} \right| \qquad j \ge 0 \tag{6.5.5}$$

If

$$\frac{hK}{2} < 1 \tag{6.5.6}$$

then the iterates $y_{n+1}^{(j)}$ will converge to $y_{n+1}$ as $j \to \infty$. A more precise estimate of the convergence rate is obtained from applying the mean value theorem to (6.5.4):

$$y_{n+1} - y_{n+1}^{(j+1)} \doteq \frac{h}{2} f_y(x_{n+1}, y_{n+1})[y_{n+1} - y_{n+1}^{(j)}] \tag{6.5.7}$$

Often in practice, the stepsize $h$ and the initial guess $y_{n+1}^{(0)}$ are chosen to ensure that only one iterate need be computed, and then we take $y_{n+1} \doteq y_{n+1}^{(1)}$.

The computation of $y_{n+1}$ from $y_n$ contains a truncation error that is $O(h^3)$ [see (6.5.1)]. To maintain this order of accuracy, the eventual iterate $y_{n+1}^{(i)}$, which is chosen to represent $y_{n+1}$, should satisfy $|y_{n+1} - y_{n+1}^{(i)}| = O(h^3)$. And if we want the iteration error to be less significant (as we do in the next section), then $y_{n+1}^{(i)}$ should be chosen to satisfy

$$|y_{n+1} - y_{n+1}^{(i)}| = O(h^4) \tag{6.5.8}$$

To analyze the error in choosing an initial guess $y_{n+1}$, we must introduce the concept of *local solution*. This will also be important in clarifying exactly what solution is being obtained by most automatic computer programs for solving ordinary differential equations. Let $u_n(x)$ denote the solution of $y' = f(x, y)$ that passes through $(x_n, y_n)$:

$$u_n'(x) = f(x, u_n(x)) \qquad u_n(x_n) = y_n \tag{6.5.9}$$

At step $x_n$, knowing $y_n$, it is $u_n(x_{n+1})$ that we are trying to calculate, rather than $Y(x_{n+1})$.

Applying the derivation that led to (6.5.1), we have

$$u_n(x_{n+1}) = y_n + \frac{h}{2}[f(x_n, y_n) + f(x_{n+1}, u_n(x_{n+1}))] - \frac{h^3}{12} u_n^{(3)}(\xi_n) \tag{6.5.10}$$

for some $x_n \leq \xi_n \leq x_{n+1}$. Let $\tilde{e}_{n+1} = u_n(x_{n+1}) - y_{n+1}$, which we call the *local error* in computing $y_{n+1}$ from $y_n$. Subtract (6.5.2) from the preceding to obtain

$$\tilde{e}_{n+1} = \frac{h}{2}[f(x_{n+1}, u_n(x_{n+1})) - f(x_{n+1}, y_{n+1})] - \frac{h^3}{12} u_n^{(3)}(\xi_n)$$

$$= \frac{h}{2} f_y(x_{n+1}, y_{n+1})\tilde{e}_{n+1} + O(h\tilde{e}_{n+1}^2) - \frac{h^3}{12} u_n^{(3)}(x_n) + O(h^4)$$

where we have twice applied the mean value theorem. It can be shown that for all sufficiently small $h$,

$$\tilde{e}_{n+1} = O(h^3)$$

More precisely,

$$\tilde{e}_{n+1} = \left[1 - \frac{h}{2}f_y(x_{n+1}, y_{n+1})\right]^{-1} \cdot \left[-\frac{h^3}{12}u_n^{(3)}(x_n) + O(h^4)\right]$$

$$u_n(x_{n+1}) - y_{n+1} = -\frac{h^3}{12}u_n^{(3)}(x_n) + O(h^4) \qquad (6.5.11)$$

This shows that the local error is essentially the truncation error.

If Euler's method is used to compute $y_{n+1}^{(0)}$,

$$y_{n+1}^{(0)} = y_n + hf(x_n, y_n) \qquad (6.5.12)$$

then $u_n(x_{n+1})$ can be expanded to show that

$$u_n(x_{n+1}) - y_{n+1}^{(0)} = \frac{h^2}{2}u_n''(\xi_n) \qquad x_n \le \xi_n \le x_{n+1} \qquad (6.5.13)$$

Combined with (6.5.11),

$$y_{n+1} - y_{n+1}^{(0)} = O(h^2) \qquad (6.5.14)$$

To satisfy (6.5.8), the bound (6.5.5) implies two iterates will have to be computed, and then we use $y_{n+1}^{(2)}$ to represent $y_{n+1}$.

Using the midpoint method, we can obtain the more accurate initial guess

$$y_{n+1}^{(0)} = y_{n-1} + 2hf(x_n, y_n) \qquad (6.5.15)$$

To estimate the error, begin by using the derivation that leads to (6.4.1) to obtain

$$u_n(x_{n+1}) = u_n(x_{n-1}) + 2hf(x_n, u_n(x_n)) + \frac{h^3}{3}u_n^{(3)}(\eta_n),$$

for some $x_{n-1} \le \eta_n \le x_{n+1}$. Subtracting (6.5.15),

$$u_n(x_{n+1}) - y_{n+1}^{(0)} = u_n(x_{n-1}) - y_{n-1} + \frac{h^3}{3}u_n^{(3)}(\eta_n)$$

The quantity $u_n(x_{n-1}) - y_{n-1}$ can be computed in a manner similar to that used for (6.5.11) with about the same result:

$$u_n(x_{n-1}) - y_{n-1} = \frac{h^3}{12}u_n^{(3)}(x_n) + O(h^4) \qquad (6.5.16)$$

Then

$$u_n(x_{n+1}) - y_{n+1}^{(0)} = \frac{5h^3}{12}u_n^{(3)}(x_n) + O(h^4)$$

And combining this with (6.5.11),

$$y_{n+1} - y_{n+1}^{(0)} = \frac{h^3}{2}u_n^{(3)}(x_n) + O(h^4) \qquad (6.5.17)$$

With the initial guess (6.5.15), one iterate from (6.5.3) will be sufficient to satisfy (6.5.8), based on the bound in (6.5.5).

The formulas (6.5.12) and (6.5.15) are called *predictor formulas*, and the trapezoidal iteration formula (6.5.3) is called a *corrector formula*. Together they form a *predictor–corrector method*, and they are the basis of a method that can be used to control the size of the local error. This is illustrated in the next section.

**Convergence and stability results**   The convergence of the trapezoidal method is assured by Theorem 6.6. Assuming $hk \leq 1$,

$$\underset{x_0 \leq x_n \leq b}{\text{Max}} |Y(x_n) - y_h(x_n)| \leq e^{2K(b-x_0)}|e_0|$$

$$+ \left[ \frac{e^{2K(b-x_0)} - 1}{K} \right] \left[ \frac{h^2}{12} \|Y^{(3)}\|_\infty \right] \quad (6.5.18)$$

The derivation of an asymptotic error formula is similar to that for Euler's method. Assuming $e_0 = \delta h^2 + O(h^3)$, we can show

$$Y(x_n) - y_h(x_n) = D(x_n)h^2 + O(h^3)$$

$$(6.5.19)$$

$$D'(x) = f_y(x, Y(x))D(x) - \frac{1}{12}Y^{(3)}(x) \quad D(x_0) = \delta_0$$

The standard type of stability result, such as that given in (6.2.28) and (6.2.29) for Euler's method, can also be given for the trapezoidal method. We leave the proof of this to the reader.

As with the midpoint method, we can examine the effect of applying the trapezoidal rule to the model equation

$$y' = \lambda y \quad y(0) = 1 \quad (6.5.20)$$

whose solution is $Y(x) = e^{\lambda x}$. To give further motivation for doing so, consider the trapezoidal method applied to the linear equation

$$y' = \lambda y + g(x) \quad y(0) = Y_0 \quad (6.5.21)$$

namely

$$y_{n+1} = y_n + \frac{h}{2}[\lambda y_n + g(x_n) + \lambda y_{n+1} + g(x_{n+1})] \quad n \geq 0 \quad (6.5.22)$$

with $y_0 = Y_0$. Then consider the perturbed numerical method

$$z_{n+1} = z_n + \frac{h}{2}[\lambda z_n + g(x_n) + \lambda z_{n+1} + g(x_{n+1})] \quad n \geq 0$$

with $z_0 = Y_0 + \epsilon$. To analyze the effect of the perturbation in the initial value, let $w_n = z_n - y_n$. Subtracting,

$$w_{n+1} = w_n + \frac{h}{2}[\lambda w_n + \lambda w_{n+1}] \qquad n \geq 0 \qquad w_0 = \epsilon \qquad (6.5.23)$$

This is simply the trapezoidal method applied to our model problem, except that the initial value is $\epsilon$ rather than 1. The numerical solution in (6.5.23) is simply $\epsilon$ times that obtained in the numerical solution of (6.5.20). Thus the behavior of the numerical solution of the model problem (6.5.20) will give us the stability behavior of the trapezoidal rule applied to (6.5.21).

The model problems in which we are interested are those for which $\lambda$ is real and negative or $\lambda$ is complex with negative real part. The reason for this choice is that then the differential equation problem (6.5.21) is well-conditioned, as noted in (6.1.8), and the major interesting cases excluded are $\lambda = 0$ and $\lambda$ strictly imaginary.

Applying the trapezoidal rule to (6.5.20),

$$y_{n+1} = y_n + \frac{h\lambda}{2}[y_n + y_{n+1}] \qquad y_0 = 1$$

Then

$$y_{n+1} = \left[ \frac{1 + (h\lambda/2)}{1 - (h\lambda/2)} \right] y_n \qquad n \geq 0$$

Inductively

$$y_n = \left[ \frac{1 + (h\lambda/2)}{1 - (h\lambda/2)} \right]^n \qquad n \geq 0 \qquad (6.5.24)$$

provided $h\lambda \neq 2$. For the case of real $\lambda < 0$, write

$$r = \frac{1 + (h\lambda/2)}{1 - (h\lambda/2)} = 1 + \frac{h\lambda}{1 - (h\lambda/2)} = -1 + \frac{2}{1 - (h\lambda/2)}$$

This shows $-1 < r < 1$ for all values of $h > 0$. Thus

$$\underset{n \to \infty}{\text{Limit}} \, y_n = 0 \qquad (6.5.25)$$

There are no limitations on $h$ in order to have boundedness of $\{y_n\}$, and thus stability of the numerical method in (6.5.22) is assured for all $h > 0$ and all $\lambda < 0$. This is a stronger statement than is possible with most numerical methods, where generally $h$ must be sufficiently small to ensure stability. For certain applications, stiff differential equations, this is an important consideration. The property that (6.5.25) holds for all $h > 0$ and all complex $\lambda$ with Real($\lambda$) $< 0$ is called *A-stability*. We explore it further in Section 6.8 and Problem 37.

**Richardson error estimation**   This error estimation was introduced in Section 5.4, and it was used both to predict the error [as in (5.4.42)] and to obtain a more rapidly convergent numerical integration method [as in (5.4.40)]. It can also be used in both of these ways in solving differential equations, although we will use it mainly to predict the error.

Let $y_h(x)$ and $y_{2h}(x)$ denote the numerical solutions to $y' = f(x, y)$ on $[x_0, b]$, obtained using the trapezoidal method (6.5.2). Then using (6.5.19),

$$Y(x_n) - y_h(x_n) = D(x_n)h^2 + O(h^3)$$

$$Y(x_n) - y_{2h}(x_n) = 4D(x_n)h^2 + O(h^3)$$

Multiply the first equation by four, subtract the second, and solve for $Y(x_n)$:

$$Y(x_n) = \frac{1}{3}[4y_h(x_n) - y_{2h}(x_n)] + O(h^3) \tag{6.5.26}$$

The formula on the right side has a higher order of convergence than the trapezoidal method, but note that it requires the computation of $y_h(x_n)$ and $y_{2h}(x_n)$ for all nodes $x_n$ in $[x_0, b]$.

The formula (6.5.26) is of greater use in predicting the global error in $y_h(x)$. Using (6.5.26),

$$Y(x_n) - y_h(x_n) = \frac{1}{3}[y_n(x_n) - y_{2h}(x_n)] + O(h^3)$$

The left side is $O(h^2)$, from (6.5.19), and thus the first term on the right side must also be $O(h^2)$. Thus

$$Y(x_n) - y_h(x_n) \doteq \frac{1}{3}[y_h(x_n) - y_{2h}(x_n)] \tag{6.5.27}$$

is an asymptotic estimate of the error. This is a practical procedure for estimating the global error, although the way we have derived it does not allow for a variable stepsize in the nodes.

***Example***   Consider the problem

$$y' = -y^2 \qquad y(0) = 1$$

**Table 6.8   Trapezoidal method and Richardson error estimation**

| $x$ | $y_{2h}(x)$ | $Y(x) - y_{2h}(x)$ | $y_h(x)$ | $Y(x) - y_h(x)$ | $\frac{1}{3}[y_h(x) - y_{2h}x)]$ |
|-----|-------------|---------------------|----------|------------------|-----------------------------------|
| 1.0 | .483144 | .016856 | .496021 | .003979 | .004292 |
| 2.0 | .323610 | .009723 | .330991 | .002342 | .002460 |
| 3.0 | .243890 | .006110 | .248521 | .001479 | .001543 |
| 4.0 | .194838 | .004162 | .198991 | .001009 | .001051 |
| 5.0 | .163658 | .003008 | .165937 | .000730 | .000759 |

which has the solution $Y(x) = 1/(1 + x)$. The results in Table 6.8 are for stepsizes $h = .25$ and $2h = .5$. The last column is the error estimate (6.5.27), and it is an accurate estimator of the true error $Y(x) - y_h(x)$.

## 6.6  A Low-Order Predictor–Corrector Algorithm

In this section, a fairly simple algorithm is described for solving the initial value problem (6.0.1). It uses the trapezoidal method (6.5.2), and it controls the size of the local error by varying the stepsize $h$. The method is not practical because of its low order of convergence, but it demonstrates some of the ideas and techniques involved in constructing a variable-stepsize predictor–corrector algorithm. It is also simpler to understand than algorithms based on higher order methods.

Each step from $x_n$ to $x_{n+1}$ will consist of constructing $y_{n+1}$ from $y_n$ and $y_{n-1}$, and $y_{n+1}$ will be an approximate solution of (6.5.2) based on using some iterate from (6.5.3). A regular step has $x_{n+1} - x_n = x_n - x_{n-1} = h$, the midpoint predictor (6.5.15) is used, and the local error is predicted using the difference of the predictor and corrector formulas. When the stepsize is being changed, the Euler predictor (6.5.12) is used.

The user of the algorithm will have to specify several parameters in addition to those defining the differential equation problem (6.0.1). The stepsize $h$ will vary, and the user must specify values $h_{min}$ and $h_{max}$ that limit the size of $h$. The user should also specify an initial value for $h$; and the value should be one for which $\frac{1}{2}hf_y(x_0, y_0)$ is sufficiently less than 1 in magnitude, say less than 0.1. This quantity will determine the speed of convergence of the iteration in (6.5.3) and is discussed later in the section, following the numerical example. An error tolerance $\epsilon$ must be given, and the stepsize $h$ is so chosen that the local error *trunc* satisfies

$$\tfrac{1}{4}\epsilon h \le |\text{trunc}| \le \epsilon h \tag{6.6.1}$$

at each step. This is called controlling the *error per unit stepsize*. Its significance is discussed near the end of the section.

The notation of the preceding section is continued. The function $u_n(x)$ is the solution of $y' = f(x, y)$ that passes through $(x_n, y_n)$. The local error to be estimated and controlled is (6.5.11), which is the error in obtaining $u_n(x_{n+1})$ using the trapezoidal method:

$$u_n(x_{n+1}) - y_{n+1} = -\frac{h^3}{12}u_n^{(3)}(x_n) + O(h^4) \qquad h = x_{n+1} - x_n \tag{6.6.2}$$

If $y_n$ is sufficiently close to $Y(x_n)$, then this is a good approximation to the closely related truncation error in (6.5.1):

$$-\frac{h^3}{12}Y^{(3)}(\xi_n)$$

And (6.6.2) is the only quantity for which we have the information needed to control it.

**Choosing the initial stepsize**   The problem is to find an initial value of $h$ and node $x_1 = x_0 + h$, for which $|y_1 - Y(x_1)|$ satisfies the bounds in (6.6.1). With the initial $h$ supplied by the user, the value $y_h(x_1)$ is obtained by using the Euler predictor (6.5.12) and iterating twice in (6.5.3). Using the same procedure, the values $y_{h/2}(x_0 + h/2)$ and $y_{h/2}(x_1)$ are also calculated. The Richardson extrapolation procedure is used to predict the error in $y_h(x_1)$,

$$Y(x_1) - y_h(x_1) \doteq \frac{4}{3} \left[ y_{h/2}(x_1) - y_h(x_1) \right] \tag{6.6.3}$$

If this error satisfies the bounds of (6.6.1), then the value of $h$ is accepted, and the regular trapezoidal step using the midpoint predictor (6.5.15) is begun. But if (6.6.1) is not satisfied by (6.6.3), then a new value of $h$ is chosen.

Using the values

$$f_0 = f(x_0, y_0) \qquad f_1 = f\left(x_0 + \frac{h}{2}, y_{h/2}\left(x_0 + \frac{h}{2}\right)\right) \qquad f_2 = f\left(x_1, y_{h/2}(x_1)\right)$$

obtain the approximation

$$Y^{(3)}(x_0) \doteq D_3 y \equiv \frac{(f_2 - 2f_1 + f_0)}{(h^2/4)} \tag{6.6.4}$$

This is an approximation using the second-order divided difference of $Y' = f(x, Y)$; for example, apply Lemma 2 of Section 3.4. For any small stepsize $h$, the truncation error at $x_0 + h$ is well approximated by

$$-\frac{h^3}{12} Y^{(3)}(\xi_0) \doteq -\frac{h^3}{12} D_3 y$$

The new stepsize $h$ is chosen so that

$$\left| \frac{h^3}{12} D_3 y \right| = \frac{1}{2} \epsilon h$$

$$h = \sqrt{\frac{6\epsilon}{|D_3 y|}} \tag{6.6.5}$$

This should place the initial truncation error in approximately the middle of the range (6.6.1) for the error per unit step criterion. With this new value of $h$, the test using (6.6.3) is again repeated, as a safety check.

By choosing $h$ so that the truncation error will satisfy the bound (6.6.1) when it is doubled or halved, we ensure that the stepsize will not have to be changed for several steps, provided the derivative $Y^{(3)}(x)$ is not changing rapidly. Changing the stepsize will be more expensive than a normal step, and we want to minimize the need for such changes.

**The regular predictor–corrector step**    The stepsize $h$ satisfies $x_n - x_{n-1} = x_{n+1} - x_n = h$. To solve for the value $y_{n+1}$, use the midpoint predictor (6.5.15) and iterate once in (6.5.3). The local error (6.5.11) is estimated using (6.5.17):

$$-\frac{1}{6}\left(y_{n+1} - y_{n+1}^{(0)}\right) = -\frac{h^3}{12}u_n^{(3)}(x_n) + O(h^4)$$

$$= \left[u_n(x_{n+1}) - y_{n+1}\right] + O(h^4) \qquad (6.6.6)$$

Thus we measure the local error using

$$\text{trunc} \equiv -\frac{1}{6}\left(y_{n+1} - y_{n+1}^{(0)}\right) \qquad (6.6.7)$$

If trunc satisfies (6.6.1), then the value of $h$ is not changed and calculation continues with this regular step procedure. But when (6.6.1) is not satisfied, the values $y_{n+1}$ and $x_{n+1}$ are discarded, and a new stepsize is chosen based on the value of trunc.

**Changing the stepsize**    Using (6.6.6),

$$-\frac{u_n^{(3)}(x_n)}{12} \doteq \frac{\text{trunc}}{h_0^3}$$

where $h_0$ denotes the stepsize used in obtaining trunc. For an arbitrary stepsize $h$, the local error in obtaining $y_{n+1}$ is estimated using

$$u_n(x_n + h) - y_h(x_{n+1}) \doteq -\frac{h^3}{12}u_n^{(3)}(x_n) \doteq \left[\frac{h}{h_0}\right]^3 \text{trunc}$$

Choose $h$ so that

$$\left[\frac{h}{h_0}\right]^3 |\text{trunc}| = \frac{1}{2}\epsilon h,$$

$$h = \sqrt{\frac{\epsilon h_0^3}{2 \cdot |\text{trunc}|}} \qquad (6.6.8)$$

Calculate $y_{n+1}$ by using the Euler predictor and iterating twice in (6.5.3). Then return to the regular predictor–corrector step. To avoid rapid changes in $h$ that can lead to significant errors, the new value of $h$ is never allowed to be more than twice the previous value. If the new value of $h$ is less than $h_{\min}$, then calculation is terminated. But if the new value of $h$ is greater than $h_{max}$, we just let $h = h_{max}$ and proceed with the calculation. This has possible problems, which are discussed following the numerical example.

**Algorithm**   $Detrap(f, x_0, y_0, x_{end}, \epsilon, h, h_{min}, h_{max}, ier)$

1. Remark: The problem being solved is $Y' = f(x, Y)$, $Y(x_0) = y_0$, for $x_0 \le x \le x_{end}$, using the method described earlier in the section. The approximate solution values are printed at each node point. The error parameter $\epsilon$ and the stepsize parameters were discussed earlier in the section. The variable ier is an error indicator, output when exiting the algorithm: ier = 0 means a normal return; ier = 1 means that $h = h_{max}$ at some node points; and ier = 2 means that the integration was terminated due to a necessary $h < h_{min}$.

2. Initialize: loop := 1, ier := 0.

3. Remark: Choose an initial value of $h$.

4. Calculate $y_h(x_0 + h)$, $y_{h/2}(x_0 + (h/2))$, $y_{h/2}(x_0 + h)$ using method (6.5.2). In each case, use the Euler predictor (6.5.12) and follow it by two iterations of (6.5.3).

5. For the error in $y_h(x_0 + h)$, use

$$\text{trunc} := \frac{4}{3}\left[ y_{h/2}(x_0 + h) - y_h(x_0 + h) \right]$$

6. If $\frac{1}{4}\epsilon h \le |\text{trunc}| \le \epsilon h$, or if loop = 2, then $x_1 := x_0 + h$, $y_1 := y_h(x_0 + h)$, print $x_1, y_1$, and go to step 10.

7. Calculate $D_3 y \doteq Y^{(3)}(x_0)$ from (6.6.4). If $D_3 y \ne 0$, then

$$h := \left[ \frac{6\epsilon}{|D_3 y|} \right]^{1/2}$$

   If $D_3 y = 0$, then $h := h_{max}$ and loop := 2.

8. If $h < h_{min}$, then ier := 2 and exit. If $h > h_{max}$, then $h := h_{max}$, ier := 1, loop := 2.

9. Go to step 4.

10. Remark: This portion of the algorithm contains the regular predictor–corrector step with error control.

**11.**   Let $x_2 := x_1 + h$, and $y_2^{(0)} := y_0 + 2hf(x_1, y_1)$. Iterate (6.5.3) once to obtain $y_2$.

**12.**   $\text{trunc} := -\dfrac{1}{6}(y_2 - y_2^{(0)})$.

**13.**   If $|\text{trunc}| > \epsilon h$ or $|\text{trunc}| < \frac{1}{4}\epsilon h$, then go to step 16.

**14.**   Print $x_2$, $y_2$.

**15.**   $x_0 := x_1$, $x_1 := x_2$, $y_0 := y_1$, $y_1 := y_2$. If $x_1 < x_{\text{end}}$, then go to step 11. Otherwise exit.

**16.**   Remark: Change the stepsize.

**17.**   $x_0 := x_1$, $y_0 := y_1$, $h_0 := h$, and calculate $h$ using (6.6.8)

**18.**   $h := \text{Min}\{h, 2h_0\}$.

**19.**   If $h < h_{\text{min}}$, then ier $:= 2$ and exit. If $h > h_{\text{max}}$, then ier $:= 1$ and $h := h_{\text{max}}$.

**20.**   $y_1^{(0)} := y_0 + hf(x_0, y_0)$, and iterate twice in (6.5.3) to calculate $y_1$.

**21.**   Print $x_1$, $y_1$.

**22.**   If $x_1 < x_{\text{end}}$, then go to step 10. Otherwise, exit.

The following example uses an implementation of *Detrap* that also prints trunc. A section of code was added to predict the truncation error in $y_1$ of step 20.

***Example***   Consider the problem

$$y' = \frac{1}{1 + x^2} - 2y^2 \qquad y(0) = 0 \qquad\qquad (6.6.9)$$

which has the solution

$$Y(x) = \frac{x}{1 + x^2}$$

This is an interesting problem for testing *Detrap*, and it performs quite well. The equation was solved on $[0, 10]$ with $h_{\text{min}} = .001$, $h_{\text{max}} = 1.0$, $h = .1$, and $\epsilon = .0005$. Table 6.9 contains some of the results, including the true global error and the true local error, labeled True le. The latter was obtained by using another more accurate numerical method. Only selected sections of output are shown because of space.

**Table 6.9    Example of algorithm *Detrap***

| $x_n$ | $h$ | $y_n$ | $Y(x_n) - y_n$ | trunc | True le |
|---|---|---|---|---|---|
| .0227 | .0227 | .022689 | 5.84E − 6 | 5.84E − 6 | 5.84E − 6 |
| .0454 | .0227 | .045308 | 1.17E − 5 | 5.83E − 6 | 5.84E − 6 |
| .0681 | .0227 | .067787 | 1.74E − 5 | 5.76E − 6 | 5.75E − 6 |
| .0908 | .0227 | .090060 | 2.28E − 5 | 5.62E − 6 | 5.61E − 6 |
| .2725 | .0227 | .253594 | 5.16E − 5 | 2.96E − 6 | 2.85E − 6 |
| .3065 | .0340 | .280125 | 5.66E − 5 | 6.74E − 6 | 6.79E − 6 |
| .3405 | .0340 | .305084 | 6.01E − 5 | 6.21E − 6 | 5.73E − 6 |
| .3746 | .0340 | .328411 | 6.11E − 5 | 4.28E − 6 | 3.54E − 6 |
| .4408 | .0662 | .369019 | 5.05E − 5 | −6.56E − 6 | −5.20E − 6 |
| .5070 | .0662 | .403297 | 2.44E − 5 | −1.04E − 5 | −2.12E − 5 |
| .5732 | .0662 | .431469 | −2.03E − 5 | −2.92E − 5 | −4.21E − 5 |
| .6138 | .0406 | .445879 | −2.99E − 5 | −1.12E − 5 | −1.10E − 5 |
| 1.9595 | .135 | .404982 | −1.02E − 4 | −1.64E − 5 | −1.67E − 5 |
| 2.0942 | .135 | .388944 | −1.03E − 4 | −1.79E − 5 | −2.11E − 5 |
| 2.3172 | .223 | .363864 | −6.57E − 5 | 1.27E − 5 | 8.15E − 6 |
| 2.7632 | .446 | .319649 | 3.44E − 4 | 4.41E − 4 | 3.78E − 4 |
| 3.0664 | .303 | .294447 | 3.21E − 4 | 9.39E − 5 | 8.41E − 5 |
| 7.6959 | .672 | .127396 | 3.87E − 4 | 8.77E − 5 | 1.12E − 4 |
| 8.6959 | 1.000 | .113100 | 3.96E − 4 | 1.73E − 4 | 1.57E − 4 |
| 9.6959 | 1.000 | .101625 | 4.27E − 4 | 1.18E − 4 | 1.68E − 4 |
| 10.6959 | 1.000 | .092273 | 4.11E − 4 | 9.45E − 5 | 1.21E − 4 |

We illustrate several points using the example. First, step 18 is necessary in order to avoid stepsizes that are far too large. For the problem (6.6.9), we have

$$Y^{(3)}(x) = \frac{-6(x^4 - 6x^2 + 1)}{(1 + x^2)^4}$$

which is zero at $x \doteq \pm.414, \pm 2.414$. Thus the local error in solving the problem (6.6.9) will be very small near these points, based on (6.6.2) and the close relation of $u_n(x)$ to $Y(x)$. This leads to a prediction of a very large $h$, in (6.6.8), one which will be too large for following points $x_n$. At $x_n = 2.7632$, step 18 was needed to avoid a misleadingly large value of $h$. As can be observed, the local error at $x_n = 2.7632$ increases greatly, due to the larger value of $h$. Shortly thereafter, the stepsize $h$ is decreased to reduce the size of the local error.

In all of the derivations of this and the preceding section, estimates were made that were accurate if $h$ was sufficiently small. In most cases, the crucial quantity is actually $hf_y(x_n, y_n)$, as in (6.5.7) when analyzing the rate of convergence of the iteration (6.5.3). In the case of the trapezoidal iteration of (6.5.3), this rate of convergence is

$$\text{Rate} \doteq \frac{1}{2} hf_y(x_n, y_n) \tag{6.6.10}$$

and for the problem (6.6.9), the rate is

$$\text{Rate} \doteq -2hy_n$$

If this is near 1, then several iterations are necessary to obtain an accurate estimate of $y_{n+1}$. From the table, the rate roughly increases in size as $h$ increases. At $x_n = 2.3172$, this rate is about .162. This still seems small enough, but the local error is more inaccurate than previously, and this may be due to a less accurate iterate being obtained in (6.5.3). The algorithm can be made more sophisticated in order to detect the problems of too large an $h$, but setting a reasonably sized $h_{\max}$ will also help.

**The global error**   We begin by giving an error bound analogous to the bound (6.5.18) for a fixed stepsize. Write

$$Y(x_{n+1}) - y_{n+1} = [Y(x_{n+1}) - u_n(x_{n+1})] + [u_n(x_{n+1}) - y_{n+1}] \quad (6.6.11)$$

For the last term, we assume the error per unit step criterion (6.6.1) is satisfied:

$$|u_n(x_{n+1}) - y_{n+1}| \le \epsilon(x_{n+1} - x_n) \quad (6.6.12)$$

For the other term in (6.6.11), introduce the integral equation reformulations

$$Y(x) = Y(x_n) + \int_{x_n}^{x} f(t, Y(t)) \, dt$$

$$u_n(x) = y_n + \int_{x_n}^{x} f(t, u_n(t)) \, dt \qquad x \ge x_n \quad (6.6.13)$$

Subtract and take bounds using the Lipschitz condition to obtain

$$|Y(x) - u_n(x)| \le |e_n| + K(x - x_n) \underset{x_n \le t \le x}{\text{Max}} |Y(t) - u_n(t)| \qquad x \ge x_n$$

with $e_n = Y(x_n) - y_n$. Using this, we can derive

$$|Y(x_{n+1}) - u_n(x_{n+1})| \le \frac{1}{1 - K(x_{n+1} - x_n)} |e_n| \quad (6.6.14)$$

Introduce

$$H = \underset{x_0 \le x_n \le b}{\text{Max}} (x_{n+1} - x_n) = \text{Max } h$$

and assume that

$$HK < 1$$

Combining (6.6.11), (6.6.12), and (6.6.14), we obtain

$$|e_{n+1}| \le \frac{1}{1 - HK} |e_n| + \epsilon H \qquad x_0 \le x_n \le b$$

This is easily solved, much as in Theorem 6.3, obtaining

$$|Y(x_n) - y_n| \le e^{c(b-x_0)K}|Y(x_0) - y_0| + \left[\frac{e^{c(b-x_0)} - 1}{c}\right]\epsilon \quad (6.6.15)$$

for an appropriate constant $c > 0$. This is the basic error result when using a variable stepsize, and it is a partial justification of the error criterion (6.6.1).

In some situations, we can obtain a more realistic bound. For simplicity, assume $f_y(x, y) \le 0$ for all $(x, y)$. Subtracting in (6.6.13),

$$Y(x) - u_n(x) = e_n + \int_{x_n}^{x}[f(t, Y(t)) - f(t, u_n(t))]\, dt$$

$$= e_n + \int_{x_n}^{x}\frac{\partial f(t, \zeta(t))}{\partial y}[Y(t) - u_n(t)]\, dt$$

The last step uses the mean-value theorem 1.2, and it can be shown that $\partial f(t, \zeta(t))/\partial y$ is a continuous function of $t$. This shows that $v(x) \equiv Y(x) - u_n(x)$ is a solution of the linear problem

$$v'(x) = \frac{\partial f(x, \zeta(x))}{\partial y}v(x) \qquad v(x_n) = e_n$$

The solution of this linear problem, along with the assumption $f_y(x, y) \le 0$, implies

$$|Y(x_{n+1}) - u_n(x_{n+1})| \le |e_n|$$

The condition $f_y(x, y) \le 0$ is associated with well-conditioned initial value problems, as was noted earlier in (6.1.8).

Combining with (6.6.11) and (6.6.12), we obtain

$$|Y(x_{n+1}) - y_{n+1}| \le |Y(x_n) - y_n| + \epsilon(x_{n+1} - x_n)$$

Solving the inequality, we obtain the more realistic bound

$$|Y(x_n) - y_n| \le |Y(x_0) - y_0| + \epsilon(x_n - x_0) \quad (6.6.16)$$

This partially explains the good behavior of the example in Table 6.9; and even better theoretical results are possible. But results (6.6.15) and (6.6.16) are sufficient justification for the use of the test (6.6.1), which controls the error per unit step. For systems of equations $\mathbf{y}' = \mathbf{f}(x, \mathbf{y})$, the condition $f_y(x, y) \le 0$ is replaced by requiring that all eigenvalues of the Jacobian matrix $\mathbf{f_y}(x, \mathbf{Y}(x))$ have real parts that are zero or negative.

The algorithm *Detrap* could be improved in a number of ways. But it illustrates the construction of a predictor–corrector algorithm with variable stepsize. The output is printed at an inconvenient set of node points $x_n$, but a simple interpolation algorithm can take care of this. The predictors can be improved upon, but that too would make the algorithm more complicated. In the next section, we return to a discussion of currently available practical predictor–corrector algorithms, most of which also vary the order.

## 6.7    Derivation of Higher Order Multistep Methods

Recall from (6.3.1) of Section 6.3 the general formula for a $p + 1$ step method for solving the initial value problem (6.0.1):

$$y_{n+1} = \sum_{j=0}^{p} a_j y_{n-j} + h \sum_{j=-1}^{p} b_j f(x_{n-j}, y_{n-j}) \qquad n \geq p \qquad (6.7.1)$$

A theory was given for these methods in Section 6.3. Some specific higher order methods will now be derived. There are two principal means of deriving higher order formulas: (1) The method of undetermined coefficients, and (2) numerical integration. The methods based on numerical integration are currently the most popular, but the perspective of the method of undetermined coefficients is still important in analyzing and developing numerical methods.

The implicit formulas can be solved by iteration, in complete analogy with (6.5.2) and (6.5.3) for the trapezoidal method. If $b_{-1} \neq 0$ in (6.7.1), the iteration is defined by

$$y_{n+1}^{(l+1)} = \sum_{j=0}^{p} \left[ a_j y_{n-j} + h b_j f(x_{n-j}, y_{n-j}) \right] + h b_{-1} f(x_{n+1}, y_{n+1}^{(l)}) \qquad l \geq 0$$

$$(6.7.2)$$

The iteration converges if $h b_{-1} K < 1$, where $K$ is the Lipschitz constant for $f(x, y)$, contained in (6.2.12). The linear rate of convergence will be bounded by $h b_{-1} K$, in analogy with (6.5.5) for the trapezoidal method.

We look for pairs of formulas, a corrector and a predictor. Suppose that the corrector formula has order $m$, that is, the truncation error if $O(h^{m+1})$ at each step. Often only one iterate is computed in (6.7.2), and this means that the predictor must have order at least $m - 1$ in order that the truncation error in $y_{n+1}^{(1)}$ is also $O(h^{m+1})$. See the discussion of the trapezoidal method iteration error in Section 6.5, between (6.5.3) and (6.5.17), using the Euler and midpoint predictors. The essential ideas transfer to (6.7.1) and (6.7.2) without any significant change.

**The method of undetermined coefficients**    If formula (6.7.1) is to have order $m \geq 1$, then from Theorem 6.5 it is necessary and sufficient that

$$\sum_{j=0}^{p} a_j = 1$$

$$\sum_{j=0}^{p} a_j(-j)^i + i \sum_{j=-1}^{p} b_j(-j)^{i-1} = 1 \qquad i = 1, 2, \ldots, m \qquad (6.7.3)$$

For an explicit method, there is the additional condition that $b_{-1} = 0$.

For a general implicit method, there are $2p + 3$ parameters $\{a_j, b_j\}$ to be determined, and there are $m + 1$ equations. It might be thought that we could take $m + 1 = 2p + 3$, but this would be extremely unwise from the viewpoint of the stability and convergence of (6.7.1). This point is illustrated in the next section and in the problems. Generally, it is best to let $m \leq p + 2$ for an implicit method. For an explicit method, stability considerations are often not as important because the method is usually a predictor for an implicit formula.

**Example**    Find all second-order two-step methods. The formula (6.7.1) is

$$y_{n+1} = a_0 y_n + a_1 y_{n-1} + h[b_{-1}f(x_{n+1}, y_{n+1}) + b_0 f(x_n, y_n)$$

$$+ b_1 f(x_{n-1}, y_{n-1})] \qquad n \geq 1 \quad (6.7.4)$$

The coefficients must satisfy (6.7.3) with $m = 2$:

$$a_0 + a_1 = 1 \qquad -a_1 + b_{-1} + b_0 + b_1 = 1 \qquad a_1 + 2b_{-1} - 2b_1 = 1$$

Solving,

$$a_1 = 1 - a_0 \qquad b_{-1} = 1 - \tfrac{1}{4}a_0 - \tfrac{1}{2}b_0 \qquad b_1 = 1 - \tfrac{3}{4}a_0 - \tfrac{1}{2}b_0 \quad (6.7.5)$$

with $a_0, b_0$ variable. The midpoint method is a special case, in which $a_0 = 0$, $b_0 = 2$. The coefficients $a_0, b_0$ can be chosen to improve the stability, give a small truncation error, give an explicit formula, or some combination of these. The conditions to ensure stability and convergence (other than $0 \leq a_0 \leq 1$ and using Theorem 6.6) cannot be given until the general theory for (6.7.1) has been given in the next section.

With (6.7.3) satisfied, Theorem 6.5 implies that the truncation error satisfies

$$T_n(Y) = O(h^{m+1})$$

for all $Y(x)$ that are $m + 1$ times continuously differentiable on $[x_0, b]$. This is sufficient for most practical and theoretical purposes. But sometimes in constructing predictor–corrector algorithms, it is preferable to have an error of the form

$$T_n(Y) = d_m h^{m+1} Y^{(m+1)}(\xi_n) \qquad x_{n-p} \leq \xi_n \leq x_{n+1} \qquad (6.7.6)$$

for some constant $d_m$, independent of $n$. Examples are Euler's method, the midpoint method, and the trapezoidal method.

To obtain a formula (6.7.6), assuming (6.7.3) has been satisfied, we first express the truncation error $T_n(Y)$ as an integral, using the concept of the Peano kernel introduced in the last part of Section 5.1. Expand $Y(x)$ about $x_{n-p}$,

$$Y(x) = \sum_{i=0}^{m} \frac{(x - x_{n-p})^i}{i!} Y^{(i)}(x_{n-p}) + R_{m+1}(x)$$

$$R_{m+1}(x) = \frac{1}{m!} \int_{x_{n-p}}^{x_{n+1}} (x - t)_+^m Y^{(m+1)}(t) \, dt$$

$$(x - t)_+^m = \begin{cases} (x - t)^m & x \geq t \\ 0 & x \leq t \end{cases}$$

Substitute the expansion into the formula (6.3.4) for $T_n(Y)$. Use the assumption that the method is of order $m$ to obtain

$$T_n(Y) = T_n(R_{m+1})$$

$$= R_{m+1}(x_{n+1}) - \left[ \sum_{j=0}^{p} a_j R_{m+1}(x_{n-j}) + h \sum_{j=-1}^{p} b_j R'_{m+1}(x_{n-j}) \right]$$

$$T_n(Y) = \int_{x_{n-p}}^{x_{n+1}} G(t - x_{n-p}) Y^{(m+1)}(t)\, dt \qquad (6.7.7)$$

with the Peano kernel

$$G(s) = \frac{1}{m!} \left\{ (x_{p+1} - s)_+^m - \left[ \sum_{j=0}^{p} a_j (x_{p-j} - s)_+^m + hm \sum_{j=-1}^{p} b_j (x_{p-j} - s)_+^{m-1} \right] \right\}$$

$$= T_p \left( (x - s)_+^m \right) \qquad 0 \le s \le x_{p+1} \qquad (6.7.8)$$

The function $G(s)$ is also often called an *influence function*.

The $m + 1$ conditions (6.7.1) on the coefficients in (6.7.3) mean there are $r = (2p + 3) - (m + 1)$ free parameters [$r = (2p + 3) - (m + 2)$ for an explicit formula]. Thus there are $r$ free parameters in determining $G(s)$. We determine those values of the parameters for which $G(s)$ is of one sign on $[0, x_{p+1}]$. And then by the integral mean value theorem, we will have

$$T_{n+1}(Y) = Y^{(m+1)}(\xi_n) \int_{x_{n-p}}^{x_{n+1}} G(t - x_{n-p})\, dt$$

for some $x_{n-p} \le \xi_n \le x_{n+1}$. By further manipulation, we obtain (6.7.6) with

$$d_m = \frac{1}{m!} \int_0^{p+1} \left[ v^m - \sum_{j=0}^{p} a_j (v - j - 1)_+^m - m \sum_{j=-1}^{p} b_j (v - j - 1)_+^{m-1} \right] dv$$

$$(6.7.9)$$

Again, this is dependent on $G(s)$ being of one sign for $0 \le s \le x_{p+1}$.

***Example*** Consider formula (6.7.4), and assume the formula is explicit [$b_{-1} = 0$]. Then

$$y_{n+1} = a_0 y_n + a_1 y_{n-1} + h \left[ b_0 f(x_n, y_n) + b_1 f(x_{n-1}, y_{n-1}) \right] \qquad n \ge 1 \quad (6.7.10)$$

with

$$a_1 = 1 - a_0 \qquad b_0 = 2 - \tfrac{1}{2} a_0 \qquad b_1 = -\tfrac{1}{2} a_0$$

Using the preceding formulas,

$$G(s) = \tfrac{1}{2}\Big[(x_2 - s)_+^2 - a_0(x_1 - s)_+^2 - a_1(x_0 - s)_+^2$$

$$-2hb_0(x_1 - s)_+ - 2hb_1(x_0 - s)_+\Big]$$

$$= \begin{cases} \tfrac{1}{2}s\big[s(1 - a_0) + a_0h\big] & 0 \le s \le h \\ \tfrac{1}{2}(x_2 - s)^2 & h \le s \le 2h \end{cases}$$

The condition that $G(s)$ be of one sign on $[0, 2h]$ is satisfied if and only if $a_0 \ge 0$. Then

$$T_{n+1}(Y) = \big(\tfrac{1}{12}a_0 + \tfrac{1}{3}\big)h^3 Y^{(3)}(\xi_n) \qquad x_{n-1} \le \xi_n \le x_{n+1} \qquad (6.7.11)$$

Note that the truncation error is a minimum when $a_0 = 0$; thus the midpoint method is optimal in having a small truncation error, among all such explicit two-step second-order methods. For a further discussion of this example, see Problem 31.

**Methods based on numerical integration**    The general idea is the following. Reformulate the differential equation by integrating it over some interval $[x_{n-r}, x_{n+1}]$ to obtain

$$Y(x_{n+1}) = Y(x_{n-r}) + \int_{x_{n-r}}^{x_{n+1}} f(t, Y(t))\, dt \qquad (6.7.12)$$

for some $r \ge 0$, all $n \ge r$. Produce a polynomial $P(t)$ that interpolates the integrand $Y'(t) = f(t, Y(t))$ at some set of node points $\{x_i\}$, and then integrate $P(t)$ over $[x_{n-r}, x_{n+1}]$ to approximate (6.7.12). The three previous methods, Euler, midpoint, and trapezoidal, can all be obtained in this way.

*Example*    Apply Simpson's integration rule, (5.1.13) and (5.1.15), to the equation

$$Y(x_{n+1}) = Y(x_{n-1}) + \int_{x_{n-1}}^{x_{n+1}} f(t, Y(t))\, dt$$

This results in

$$Y_{n+1} = Y_{n-1} + \frac{h}{3}\big[Y'_{n-1} + 4Y'_n + Y'_{n+1}\big] - \frac{h^5}{90} Y^{(5)}(\xi_n)$$

for some $x_{n-1} \le \xi_n \le x_{n+1}$. This approximation is based on integrating the quadratic polynomial that interpolates $Y'(t) = f(t, y(t))$ on the nodes $x_{n-1}, x_n, x_{n+1}$. For simplicity, we use the notation $Y_j = Y(x_j)$, and $Y'_j = Y'(x_j)$.

Dropping the error term results in the implicit fourth-order formula

$$y_{n+1} = y_{n-1} + \frac{h}{3}\left[f(x_{n-1}, y_{n-1}) + 4f(x_n, y_n) + f(x_{n+1}, y_{n+1})\right] \qquad n \geq 1$$

$$(6.7.13)$$

This is a well-known formula, and it is the corrector of a classical predictor–corrector algorithm known as *Milne's method*. From Theorem 6.6, the method converges and

$$\underset{x_0 \leq x_n \leq b}{\text{Max}} |Y(x_n) - y_n| = O(h^4)$$

But the method is only weakly stable, in the same manner as was true of the midpoint method in Section 6.4.

**The Adams methods** These are the most widely used multistep methods. They are used to produce predictor–corrector algorithms in which the error is controlled by varying both the stepsize $h$ and the order of the method. This is discussed in greater detail later in the section, and a numerical example is given.

To derive the methods, use the integral reformulation

$$Y_{n+1} = Y_n + \int_{x_n}^{x_{n+1}} f(t, Y(t)) \, dt \qquad (6.7.14)$$

Polynomials that interpolate $Y'(t) = f(t, Y(t))$ are constructed, and then they are integrated over $[x_n, x_{n+1}]$ to obtain an approximation to $Y_{n+1}$. We begin with the explicit or predictor formulas.

*Case* 1. Adams–Bashforth Methods Let $P_p(t)$ denote the polynomial of degree $\leq p$ that interpolates $Y'(t)$ at $x_{n-p}, \ldots, x_n$. The most convenient form for $P_p(t)$ will be the Newton backward difference formula (3.3.11), expanded about $x_n$:

$$P_p(t) = Y_n' + \frac{(t - x_n)}{h}\nabla Y_n' + \frac{(t - x_n)(t - x_{n-1})}{2!h^2}\nabla^2 Y_n' + \cdots$$

$$+ \frac{(t - x_n)\cdots(t - x_{n-p+1})}{p!h^p}\nabla^p Y_n' \qquad (6.7.15)$$

As an illustration of the notation, $\nabla Y_n' = Y'(x_n) - Y'(x_{n-1})$, and the higher order backward differences are defined accordingly [see (3.3.10)]. The interpolation error is given by

$$E_p(t) = (t - x_{n-p}) \cdots (t - x_n)Y'\left[x_{n-p}, \ldots, x_n, t\right]$$

$$= \frac{(t - x_{n-p}) \cdots (t - x_n)}{(p + 1)!} Y^{(p+2)}(\zeta_t) \qquad x_{n-p} \leq \zeta_t \leq x_{n+1} \quad (6.7.16)$$

**Table 6.10**   **Adams–Bashforth coefficients**

| $\gamma_0$ | $\gamma_1$ | $\gamma_2$ | $\gamma_3$ | $\gamma_4$ | $\gamma_5$ |
|---|---|---|---|---|---|
| 1 | $\dfrac{1}{2}$ | $\dfrac{5}{12}$ | $\dfrac{3}{8}$ | $\dfrac{251}{720}$ | $\dfrac{95}{288}$ |

provided $x_{n-p} \le t \le x_{n+1}$ and $Y(t)$ is $p + 2$ times continuously differentiable. See (3.2.11) and (3.1.8) of Chapter 3 for the justification of (6.7.16).

The integral of $P_p(t)$ is given by

$$\int_{x_n}^{x_{n+1}} P_p(t)\, dt = hY_n' + \sum_{j=1}^{p} \frac{1}{j!h^j} \nabla^j Y_n' \int_{x_n}^{x_{n+1}} (t - x_n) \cdots (t - x_{n+1-j})\, dt$$

$$= h \sum_{j=0}^{p} \gamma_j \nabla^j Y_n' \tag{6.7.17}$$

The coefficients $\gamma_j$ are obtained by introducing the change of variable $s = (t - x_n)/h$, which leads to

$$\gamma_j = \frac{1}{j!} \int_0^1 s(s + 1) \cdots (s + j - 1)\, ds \qquad j \ge 1 \tag{6.7.18}$$

with $\gamma_0 = 1$. Table 6.10 contains the first few values of $\gamma_j$. Gear (1971, pp. 104–111) contains additional information, including a generating function for the coefficients.

The truncation error in using the interpolating polynomial $P_p(t)$ in (6.7.14) is given by

$$T_n(Y) = \int_{x_n}^{x_{n+1}} E_p(t)\, dt = \int_{x_n}^{x_{n+1}} (t - x_n) \cdots (t - x_{n-p}) Y'\big[x_{n-p}, \ldots, x_n, t\big]\, dt$$

Assuming $Y(x)$ is $p + 2$ times continuously differentiable for $x_{n-p} \le x \le x_{n+1}$, Theorem 3.3 of Chapter 3 implies that the divided difference in the last integral is a continuous function of $t$. Since the polynomial $(t - x_n) \cdots (t - x_{n-p})$ is nonnegative on $[x_n, x_{n+1}]$, use the integral mean value theorem (Theorem 1.3) to obtain

$$T_n(Y) = Y'\big[x_{n-p}, \ldots, x_n, \zeta\big] \int_{x_n}^{x_{n+1}} (t - x_n) \cdots (t - x_{n-p})\, dt$$

for some $x_n \le \zeta \le x_{n+1}$. Use (6.7.18) to calculate the integral, and use (3.2.12) to convert the divided difference to a derivative. Then

$$T_n(Y) = \gamma_{p+1} h^{p+2} Y^{(p+2)}(\xi_n) \qquad x_{n-p} \le \xi_n \le x_{n+1} \tag{6.7.19}$$

There is an alternative form that is very useful for estimating the truncation error. Using the mean value theorem (Theorem 1.2) and the analogue for

**Table 6.11**   **Adams–Bashforth formulas**

| | |
|---|---|
| $p = 0$ | $Y_{n+1} = Y_n + hY_n' + \dfrac{1}{2}h^2 Y''(\xi_n)$ |
| $p = 1$ | $Y_{n+1} = Y_n + \dfrac{h}{2}[3Y_n' - Y_{n-1}'] + \dfrac{5}{12}h^3 Y^{(3)}(\xi_n)$ |
| $p = 2$ | $Y_{n+1} = Y_n + \dfrac{h}{12}[23Y_n' - 16Y_{n-1}' + 5Y_{n-2}'] + \dfrac{3}{8}h^4 Y^{(4)}(\xi_n)$ |
| $p = 3$ | $Y_{n+1} = Y_n + \dfrac{h}{24}[55Y_n' - 59Y_{n-1}' + 37Y_{n-2}' - 9Y_{n-3}'] + \dfrac{251}{720}h^5 Y^{(5)}(\xi_n)$ |

backward differences of Lemma 2 of Section 3.4, we have

$$T_n(Y) = h\gamma_{p+1}\nabla^{p+1}Y_n' + O(h^{p+3}) \tag{6.7.20}$$

The principal part of this error, $h\gamma_{p+1}\nabla^{p+1}Y_n'$, is the final term that would be obtained by using the interpolating polynomial $P_{p+1}(t)$ rather than $P_p(t)$. This form of the error is a basic tool used in algorithms that vary the order of the method to help control the truncation error.

Using (6.7.17) and (6.7.19), equation (6.7.14) becomes

$$Y_{n+1} = Y_n + h\sum_{j=0}^{p} \gamma_j \nabla^j Y_n' + \gamma_{p+1}h^{p+2}Y^{(p+2)}(\xi_n) \tag{6.7.21}$$

for some $x_{n-p} \leq \xi_n \leq x_{n+1}$. The corresponding numerical method is

$$y_{n+1} = y_n + h\sum_{j=0}^{p} \gamma_j \nabla^j y_n' \qquad n \geq p \tag{6.7.22}$$

In the formula, $y_j' \equiv f(x_j, y_j)$, and as an example of the backward differences, $\nabla y_j' = y_j' - y_{j-1}'$. Table 6.11 contains the formulas for $p = 0, 1, 2, 3$. They are written in the more usual form of (6.7.1), in which the dependence on the value $f(x_{n-j}, y_{n-j})$ is shown explicitly. Note that the $p = 0$ case is just Euler's method.

The Adams–Bashforth formulas satisfy the hypotheses of Theorem 6.6, and therefore they are convergent and stable methods. In addition, they do not have the instability of the type associated with the midpoint method (6.4.2) and Simpson's method (6.7.13). The proof of this is given in the next section, and further discussion is postponed until near the end of that section, when the concept of relative stability is introduced.

*Case 2.*   **Adams–Moulton Methods**   Again use the integral formula (6.7.14), but interpolate $Y'(t) = f(t, Y(t))$ at the $p + 1$ points $x_{n+1}, \ldots, x_{n-p+1}$ for $p \geq 0$. The derivation is exactly the same as for the Adams–Bashforth methods,

**Table 6.12  Adams–Moulton coefficients**

| $\delta_0$ | $\delta_1$ | $\delta_2$ | $\delta_3$ | $\delta_4$ | $\delta_5$ |
|---|---|---|---|---|---|
| 1 | $-\dfrac{1}{2}$ | $-\dfrac{1}{12}$ | $-\dfrac{1}{24}$ | $-\dfrac{19}{720}$ | $-\dfrac{3}{160}$ |

and we give only the final results. Equation (6.7.14) is transformed to

$$Y_{n+1} = Y_n + h \sum_{j=0}^{p} \delta_j \nabla^j Y'_{n+1} + \delta_{p+1} h^{p+2} Y^{(p+2)}(\zeta_n) \qquad (6.7.23)$$

with $x_{n-p+1} \leq \zeta_n \leq x_{n+1}$. The coefficients $\delta_j$ are defined by

$$\delta_j = \frac{1}{j!} \int_0^1 (s-1)s(s+1) \cdots (s+j-2) \, ds \qquad j \geq 1 \qquad (6.7.24)$$

with $\delta_0 = 1$, and a few values are given in Table 6.12. The truncation error can be put in the form

$$T_{n+1}(Y) = h\delta_{p+1} \nabla^{p+1} Y'_{n+1} + O(h^{p+3}) \qquad (6.7.25)$$

just as with (6.7.20).

The numerical method associated with (6.7.23) is

$$y_{n+1} = y_n + h \sum_{j=0}^{p} \delta_j \nabla^j y'_{n+1} \qquad n \geq p - 1 \qquad (6.7.26)$$

with $y'_j \equiv f(x_j, y_j)$ as before. Table 6.13 contains the low-order formulas for $p = 0, 1, 2, 3$. Note that the $p = 1$ case is the trapezoidal method.

Formula (6.7.26) is an implicit method, and therefore a predictor is necessary for solving it by iteration. The basic ideas involved are exactly the same as those in Section 6.5 for the iterative solution of the trapezoidal method. If a fixed-order predictor–corrector algorithm is desired, and if only one iteration is to be calculated, then an Adams–Moulton formula of order $m \geq 2$ can use a predictor

**Table 6.13  Adams–Moulton formulas**

| | |
|---|---|
| $p = 0$ | $Y_{n+1} = Y_n + hY'_{n+1} - \dfrac{1}{2}h^2 Y''(\zeta_n)$ |
| $p = 1$ | $Y_{n+1} = Y_n + \dfrac{h}{2}[Y'_{n+1} + Y'_n] - \dfrac{1}{12}h^3 Y^{(3)}(\zeta_n)$ |
| $p = 2$ | $Y_{n+1} = Y_n + \dfrac{h}{12}[5Y'_{n+1} + 8Y'_n - Y'_{n-1}] - \dfrac{1}{24}h^4 Y^{(4)}(\zeta_n)$ |
| $p = 3$ | $Y_{n+1} = Y_n + \dfrac{h}{24}[9Y'_{n+1} + 19Y'_n - 5Y'_{n-1} + Y'_{n-2}] - \dfrac{19}{720}h^5 Y^{(5)}(\zeta_n)$ |

of order $m$ or $m - 1$. The advantage of using an order $m - 1$ predictor is that the predictor and corrector would both use derivative values at the same nodes, namely, $x_n, x_{n-1}, \ldots, x_{n-m+2}$. For example, the second-order Adams–Moulton formula with the first-order Adams–Bashforth formula as predictor is just the trapezoidal method with the Euler predictor. This was discussed in Section 6.5 and shown to be adequate; both methods use the single past value of the derivative, $f(x_n, y_n)$.

A less trivial example is the following fourth-order method:

$$y_{n+1}^{(0)} = y_n + \frac{h}{12}\left[23f(x_n, y_n) - 16f(x_{n-1}, y_{n-1}) + 5f(x_{n-2}, y_{n-2})\right]$$

$$y_{n+1}^{(j+1)} = y_n + \frac{h}{24}\left[9f(x_{n+1}, y_{n+1}^{(j)}) + 19f(x_n, y_n) - 5f(x_{n-1}, y_{n-1})\right.$$

$$\left. + f(x_{n-2}, y_{n-2})\right] \quad (6.7.27)$$

Generally only one iterate is calculated, although this will alter the form of the truncation error in (6.7.23). Let $u_n(x)$ denote the solution of $y' = f(x, y)$ passing through $(x_n, y_n)$. Then for the truncation error in using the approximation $y_{n+1}^{(1)}$,

$$u_n(x_{n+1}) - y_{n+1}^{(1)} = \left[u_n(x_{n+1}) - y_{n+1}\right] + \left[y_{n+1} - y_{n+1}^{(1)}\right]$$

Using (6.7.23) and an expansion of the iteration error,

$$u_n(x_{n+1}) - y_{n+1}^{(1)} = \delta_4 h^5 u_n^{(5)}(x_n) + \frac{3h}{8}f_y(x_n, y_n)(y_{n+1} - y_{n+1}^{(0)}) + O(h^6)$$

$$(6.7.28)$$

The first two terms following the equality sign are of order $h^5$. If either (1) more iterates are calculated, or (2) a higher order predictor is used, then the principal part of the truncation error will be simply $\delta_4 h^5 u_n^{(5)}(x_n)$. And this is a more desirable situation from the viewpoint of estimating the error.

The Adams–Moulton formulas have a significantly smaller truncation error than the Adams–Bashforth formulas, for comparable order methods. For example, the fourth-order Adams–Moulton formula has a truncation error .076 times that of the fourth-order Adams–Bashforth formula. This is the principal reason for using the implicit formulas, although there are other considerations. Note also that the fourth-order Adams–Moulton formula has over twice the truncation error of the Simpson method (6.7.13). The reason for using the Adams–Moulton formula is that it has much better stability properties than the Simpson method.

*Example*  Method (6.7.27) was used to solve

$$y' = \frac{1}{1 + x^2} - 2y^2 \qquad y(0) = 0 \qquad (6.7.29)$$

**Table 6.14   Numerical example of the Adams method**

| $x$ | Error for $h = .125$ | Error for $h = .0625$ | Ratio |
|---|---|---|---|
| 2.0 | $2.07E - 5$ | $1.21E - 6$ | 17.1 |
| 4.0 | $2.21E - 6$ | $1.20E - 7$ | 18.3 |
| 6.0 | $3.74E - 7$ | $2.00E - 8$ | 18.7 |
| 8.0 | $1.00E - 7$ | $5.24E - 9$ | 19.1 |
| 10.0 | $3.58E - 8$ | $1.83E - 9$ | 19.6 |

which has the solution $Y(x) = x/(1 + x^2)$. The initial values $y_1$, $y_2$, $y_3$ were taken to be the true values to simplify the example. The solution values were computed with two values of $h$, and the resulting errors at a few node points are given in Table 6.14. The column labeled Ratio is the ratio of the error with $h = .125$ to that with $h = .0625$. Note that the values of Ratio are near 16, which is the theoretical ratio that would be expected since method (6.7.27) is fourth order. ·

**Variable-order methods**   At present, the most popular predictor–corrector algorithms control the truncation error by varying both the stepsize and the order of the method, and all of these algorithms use the Adams family of formulas. The first such computer programs that were widely used were DIFSUB from Gear (1971, pp. 158–166) and codes due to Krogh (1969). Subsequently, other such variable-order Adams codes were written with GEAR from Hindmarsh (1974) and DE/STEP from Shampine and Gordon (1975) among the most popular. The code GEAR has been further improved, to the code LSODE, and is a part of a larger program package called ODEPACK [see Hindmarsh (1983) for a description]. The code DE/STEP has been further improved, to DDEABM, and it is part of another general package, called DEPAC [see Shampine and Watts (1980) for a description]. In all cases, the previously cited codes use error estimation that is based on the formulas (6.7.23) or (6.7.25), although they [and formulas (6.7.22) and (6.7.26)] may need to be modified for a variable stepsize. The codes vary in a number of important but technical aspects, including the amount of iteration of the Adams–Moulton corrector, the form in which past information about derivative $f$ is carried forward, and the form of interpolation of the solution to the current output node point. Because of lack of space, and because of the complexity of the issues involved, we omit any further discussion of the comparative differences in these codes. For some further remarks on these codes, see Gupta et al. (1985, pp. 16–19).

By allowing the order to vary, there is no difficulty in obtaining starting values for the higher order Adams methods. The programs begin with the second-order trapezoidal formula with an Euler predictor; the order is then generally increased as extra starting values become available. If the solution is changing rapidly, then the program will generally choose a low-order formula, while for a smoother and more slowly varying solution, the order will usually be larger.

In the program DE of Shampine and Gordon (1985, 186–209), the truncation error at $x_{n+1}$ [call it *trunc*] is required to satisfy

$$|\text{trunc}_j| \leq \text{ABSERR} + \text{RELERR} * |y_{n,j}| \qquad (6.7.30)$$

This is to hold for each component of the truncation error and for each corresponding component $y_{n,j}$ of the solution $y_n$ of the given system of differential equations. The values ABSERR and RELERR are supplied by the user. The value of trunc is given, roughly speaking, by

$$\text{trunc} \doteq h\delta_{p+1}\nabla^{p+1}y'_{n+1}$$

assuming the spacing is uniform. This is the truncation error for the $p$-step formula

$$y_{n+1}^{(p)} = y_n + h\sum_{j=0}^{p} \delta_j \nabla^j y'_{n+1}$$

Once the test (6.7.30) is satisfied, the value of $y_{n+1}$ is

$$y_{n+1} \equiv y_{n+1}^{(p+1)} = y_{n+1}^{(p)} + \text{trunc} \qquad (6.7.31)$$

Thus the actual truncation error is $O(h^{p+3})$, and combined with (6.7.30), it can be shown that the truncation error in $y_{n+1}$ satisfies an error per unit step criteria, which is similar to that of (6.6.1) for the algorithm *Detrap* of Section 6.6. For a detailed discussion, see Shampine and Gordon (1975, p. 100).

The program DE (and its successor DDEABM) is very sophisticated in its error control, including the choosing of the order and the stepsize. It cannot be discussed adequately in the limited space available in this text, but the best reference is the text of Shampine and Gordon (1975), which is devoted to variable-order Adams algorithms. The programs DE and DDEABM have been well designed from both the viewpoint of error control and user convenience. Each is also written in a portable form, and generally, is a well-recommended program for solving differential equations.

**Example**   Consider the problem

$$y' = \frac{y}{4}\left(1 - \frac{y}{20}\right) \qquad y(0) = 1 \qquad (6.7.32)$$

which has the solution

$$Y(x) = \frac{20}{(1 + 19e^{-(x/4)})}.$$

DDEABM was used to solve this problem with values output at $x = 2, 4, 6, \ldots, 20$. Three values of ABSERR were used, and RELERR $= 0$ in all cases. The true global errors are shown in Table 6.15. The column labeled NFE gives the number of evaluations of $f(x, y)$ necessary to obtain the value $y_h(x)$, beginning from $x_0$.

**Table 6.15   Example of the automatic program DDEABM**

| x | ABSERR = $10^{-3}$ Error | NFE | ABSERR = $10^{-6}$ Error | NFE | ABSERR = $10^{-9}$ Error | NFE |
|---|---|---|---|---|---|---|
| 4.0 | $-3.26E - 5$ | 15 | $1.24E - 6$ | 28 | $2.86E - 10$ | 52 |
| 8.0 | $6.00E - 4$ | 21 | $3.86E - 6$ | 42 | $-1.98E - 9$ | 76 |
| 12.0 | $1.70E - 3$ | 25 | $4.93E - 6$ | 54 | $-2.41E - 9$ | 102 |
| 16.0 | $9.13E - 4$ | 31 | $3.73E - 6$ | 64 | $-1.86E - 9$ | 124 |
| 20.0 | $9.16E - 4$ | 37 | $1.79E - 6$ | 74 | $-9.58E - 10$ | 138 |

**Global error**   The automatic computer codes that are discussed previously control the local error or truncation error. They do *not* control the global error in the solution. Usually the truncation error is kept so small by these codes that the global error is also within acceptable limits, although that is not guaranteed. The reasons for this small global error are much the same as those described in Section 6.6; in particular, recall (6.6.15) and (6.6.16).

The global error can be monitored, and we give an example of this below. But even with an estimate of the global error, we cannot control it for most equations. This is because the global error is composed of the effects of all past truncation errors, and decreasing the present stepsize will not change those past errors. In general, if the global error is too large, then the equation must be solved again, with a smaller stepsize.

There are a number of methods that have been proposed for monitoring the global error. One of these, Richardson extrapolation, is illustrated in Section 6.10 for a Runge–Kutta method. Below, we illustrate another one for the method of Section 6.6. For a general survey of the topic, see Skeel (1986).

For the trapezoidal method, the true solution $Y(x)$ satisfies

$$Y(x_{n+1}) = Y(x_n) + \frac{h}{2}[f(x_n, Y(x_n)) + f(x_{n+1}, Y(x_{n+1}))] - \frac{h^3}{12}Y^{(3)}(\xi_n)$$

with $h = x_{n+1} - x_n$ and $x_n \le \xi_n \le x_{n+1}$. Subtracting the trapezoidal rule

$$y_{n+1} = y_n + \frac{h}{2}[f(x_n, y_n) + f(x_{n+1}, y_{n+1})]$$

we have

$$e_{n+1} = e_n + \frac{h}{2}\{[f(x_n, y_n + e_n) - f(x_n, y_n)]$$

$$+ [f(x_{n+1}, y_{n+1} + e_{n+1}) - f(x_{n+1}, y_{n+1})]\} - \frac{h^3}{12}Y^{(3)}(\xi_n) \quad (6.7.33)$$

with $e_n = Y(x_n) - y_n$, $n \ge 0$. This is the error equation for the trapezoidal method, and we try to solve it approximately in order to calculate $e_{n+1}$.

**Table 6.16   Global error calculation for *Detrap***

| $x_n$ | $h$ | $e_n$ | $\hat{e}_n$ | trunc |
|---|---|---|---|---|
| .0227 | .0227 | 5.84E − 6 | 5.83E − 6 | 5.84E − 6 |
| .0454 | .0227 | 1.17E − 5 | 1.16E − 5 | 5.83E − 6 |
| .0681 | .0227 | 1.74E − 5 | 1.73E − 5 | 5.76E − 6 |
| .0908 | .0227 | 2.28E − 5 | 2.28E − 5 | 5.62E − 6 |
| .2725 | .0227 | 5.16E − 5 | 5.19E − 5 | 2.96E − 6 |
| .3065 | .0340 | 5.66E − 5 | 5.66E − 5 | 6.74E − 6 |
| .3405 | .0340 | 6.01E − 5 | 6.05E − 5 | 6.21E − 6 |
| .3746 | .0340 | 6.11E − 5 | 6.21E − 5 | 4.28E − 6 |
| .4408 | .0662 | 5.05E − 5 | 5.04E − 5 | − 6.56E − 6 |
| .5070 | .0662 | 2.44E − 5 | 3.57E − 5 | − 1.04E − 5 |
| .5732 | .0662 | − 2.03E − 5 | 4.34E − 6 | − 2.92E − 5 |
| .6138 | .0406 | − 2.99E − 5 | − 6.76E − 6 | − 1.12E − 5 |
| 1.9595 | .135 | − 1.02E − 4 | − 8.76E − 5 | − 1.64E − 5 |
| 2.0942 | .135 | − 1.03E − 4 | − 8.68E − 5 | − 1.79E − 5 |
| 2.3172 | .223 | − 6.57E − 5 | − 5.08E − 5 | 1.27E − 5 |
| 2.7632 | .446 | 3.44E − 4 | 3.17E − 4 | 4.41E − 4 |
| 3.0664 | .303 | 3.21E − 4 | 2.96E − 4 | 9.39E − 5 |
| 7.6959 | .672 | 3.87E − 4 | 2.69E − 4 | 8.77E − 5 |
| 8.6959 | 1.000 | 3.96E − 4 | 3.05E − 4 | 1.73E − 4 |
| 9.6959 | 1.000 | 4.27E − 4 | 2.94E − 4 | 1.18E − 4 |
| 10.6959 | 1.000 | 4.11E − 4 | 2.77E − 4 | 9.45E − 5 |

Returning to the algorithm *Detrap* of Section 6.6, we replace the truncation term in (6.7.33) with the variable trunc computed in *Detrap*. Then we solve (6.7.33) for $\hat{e}_{n+1}$, which will be an approximation of the true global error $e_{n+1}$. We can solve for $\hat{e}_{n+1}$ by using various rootfinding methods, but we use simple fixed-point iterations:

$$\hat{e}_{n+1}^{(j+1)} = \hat{e}_n + \frac{h}{2}\left\{\left[f(x_n, y_n + \hat{e}_n) - f(x_n, y_n)\right]\right.$$

$$\left. + \left[f\left(x_{n+1}, y_{n+1} + \hat{e}_{n+1}^{(j)}\right) - f(x_{n+1}, y_{n+1})\right]\right\} + \text{trunc} \quad (6.7.34)$$

for $j \geq 0$. We use $\hat{e}_{n+1}^{(0)} = \hat{e}_n$, and since this is for just illustrative purposes, we iterate several times in (6.7.34). This simple idea is closely related to the difference correction methods of Skeel (1986).

***Example***   We repeat the calculation given in Table 6.9, for *Detrap* applied to Eq. (6.6.9). We use the same parameters for *Detrap*. The results are shown in Table 6.16, for the same values of $x_n$ as in Table 6.9. The results show that $e_n$ and $\hat{e}_n$ are almost always reasonably close, certainly in magnitude. The approximation $e_n \doteq \hat{e}_n$ is poor around $x = .5$, due to the poor estimate of the truncation error in *Detrap*. Even then these poor results damp out for this problem, and for larger values of $x_n$, the approximation $e_n \doteq \hat{e}_n$ is still useful.

## 6.8 Convergence and Stability Theory for Multistep Methods

In this section, a complete theory of convergence and stability is presented for the multistep method

$$y_{n+1} = \sum_{j=0}^{p} a_j y_{n-j} + h \sum_{j=-1}^{p} b_j f(x_{n\,j}, y_{n-j}) \qquad x_{p+1} \le x_{n+1} \le b \quad (6.8.1)$$

This generalizes the work of Section 6.3, and it creates the mathematical tools necessary for analyzing whether method (6.8.1) is only weakly stable, due to instability of the type associated with the midpoint method.

We begin with a few definitions. The concept of stability was introduced with Euler's method [see (6.2.28) and (6.2.20)], and it is now generalized. Let $\{y_n | 0 \le n \le N(h)\}$ be the solution of (6.8.1) for some differential equation $y' = f(x, y)$, for all sufficiently small values of $h$, say $h \le h_0$. Recall that $N(h)$ denotes the largest subscript $N$ for which $x_N \le b$. For each $h \le h_0$, perturb the initial values $y_0, \ldots, y_p$ to new values $z_0, \ldots, z_p$ with

$$\underset{0 \le n \le p}{\text{Max}} |y_n - z_n| \le \epsilon \qquad 0 < h \le h_0 \qquad (6.8.2)$$

Note that these initial values are likely to depend on $h$. We say the family of solutions $\{y_n | 0 \le n \le N(h)\}$ is stable if there is a constant $c$, independent of $h \le h_0$ and valid for all sufficiently small $\epsilon$, for which

$$\underset{0 \le n \le N(h)}{\text{Max}} |y_n - z_n| \le c\epsilon \qquad 0 < h \le h_0 \qquad (6.8.3)$$

Consider all differential equation problems

$$y' = f(x, y) \qquad y(x_0) = Y_0 \qquad (6.8.4)$$

with the derivative $f(x, y)$ continuous and satisfying the Lipschitz condition (6.2.12), and suppose the approximating solutions $\{y_n\}$ are all stable. Then we say that (6.8.1) is a *stable numerical method*.

To define convergence for a given problem (6.8.4), suppose the initial values $y_0, \ldots, y_p$ satisfy

$$\eta(h) \equiv \underset{0 \le n \le p}{\text{Max}} |Y(x_n) - y_n| \to 0 \qquad \text{as} \quad h \to 0 \qquad (6.8.5)$$

Then the solution $\{y_n\}$ is said to converge to $Y(x)$ if

$$\underset{x_0 \le x_n \le b}{\text{Max}} |Y(x_n) - y_n| \to 0 \qquad \text{as} \quad h \to 0 \qquad (6.8.6)$$

If (6.8.1) is convergent for all problems (6.8.4), then it is called a *convergent numerical method*.

Recall the definition of consistency given in Section 6.3. Method (6.8.1) is *consistent* if

$$\frac{1}{h}\, \text{Max}_{x_p \le x_n \le b} |T_n(Y)| \to 0 \qquad \text{as} \quad h \to 0$$

for all functions $Y(x)$ continuously differentiable on $[x_0, b]$. Or equivalently from Theorem 6.5, the coefficients $\{a_j\}$ and $\{b_j\}$ must satisfy

$$\sum_{j=0}^{p} a_j = 1 \qquad -\sum_{j=0}^{p} ja_j + \sum_{j=-1}^{p} b_j = 1 \qquad (6.8.7)$$

Convergence can be shown to imply consistency; consequently, we consider only methods satisfying (6.8.7). As an example of the proof of the necessity of (6.8.7), the assumption of convergence of (6.8.1) for the problem

$$y' \equiv 0 \qquad y(0) = 1$$

will imply the first condition in (6.8.7). Just take $y_0 = \cdots = y_p = 1$, and observe the consequences of the convergence of $y_{p+1}$ to $Y(x) \equiv 1$.

The convergence and stability of (6.8.1) are linked to the roots of the polynomial

$$\rho(r) = r^{p+1} - \sum_{j=0}^{p} a_j r^{p-j} \qquad (6.8.8)$$

Note that $\rho(1) = 0$ from the consistency condition (6.8.7). Let $r_0, \ldots, r_p$ denote the roots of $\rho(r)$, repeated according to their multiplicity, and let $r_0 = 1$. The method (6.8.1) satisfies the *root condition* if

**1.** $$|r_j| \le 1 \qquad j = 0, 1, \ldots, p \qquad (6.8.9)$$

**2.** $$|r_j| = 1 \Rightarrow \rho'(r_j) \ne 0 \qquad (6.8.10)$$

The first condition requires all roots of $\rho(r)$ to lie in the unit circle $\{z: |z| \le 1\}$ in the complex plane. Condition (6.8.10) states that all roots on the boundary of the circle are to be simple roots of $\rho(r)$.

The main results of this section are pictured in Figure 6.6, although some of them will not be proved. The *strong root condition* and the concept of *relative stability* are introduced later in the section.

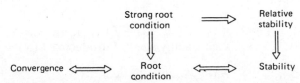

**Figure 6.6**   Schematic of the theory for consistent multistep methods.

**Stability theory**    All of the numerical methods presented in the preceding sections have been stable, but we now give an example of an unstable method. This is to motivate the need to develop a general theory of stability.

***Example***    Recall the general formula (6.7.10) for an explicit two-step second-order method, and choose $a_0 = 3$. Then we obtain the method

$$y_{n+1} = 3y_n - 2y_{n-1} + \frac{h}{2}[f(x_n, y_n) - 3f(x_{n-1}, y_{n-1})] \qquad n \geq 1 \quad (6.8.11)$$

with the truncation error

$$T_n(Y) = \tfrac{7}{12}h^3 Y^{(3)}(\xi_n) \qquad x_{n-1} \leq \xi_n \leq x_{n+1}$$

Consider solving the problem $y' \equiv 0$, $y(0) = 0$, which has the solution $Y(x) \equiv 0$. Using $y_0 = y_1 = 0$, the numerical solution is clearly $y_n = 0$, $n \geq 0$. Perturb the initial data to $z_0 = \epsilon/2$, $z_1 = \epsilon$, for some $\epsilon \neq 0$. Then the corresponding numerical solution can be shown to be

$$z_n = \epsilon \cdot 2^{n-1} \qquad n \geq 0 \qquad (6.8.12)$$

The reasoning used in deriving this solution is given later in a more general context. To see the effect of the perturbation on the original solution,

$$\underset{x_0 \leq x_n \leq b}{\text{Max}} |y_n - z_n| = \underset{0 \leq x_n \leq b}{\text{Max}} |\epsilon|2^{n-1} = |\epsilon|2^{N(h)-1}$$

Since $N(h) \to \infty$ as $h \to 0$, the deviation of $\{z_n\}$ from $\{y_n\}$ becomes increasingly greater as $h \to 0$. The method (6.8.11) is unstable, and it should never be used. Also note that the root condition is violated, since $\rho(r) = r^2 - 3r + 2$ has the roots $r_0 = 1$, $r_1 = 2$.

To investigate the stability of (6.8.1), we consider only the special equation

$$y' = \lambda y \qquad y(0) = 1 \qquad (6.8.13)$$

with the solution $Y(x) = e^{\lambda x}$. The results obtained will transfer to the study of stability for a general differential equation problem. An intuitive reason for this is easily derived. Expand $Y'(x) = f(x, Y(x))$ about $(x_0, Y_0)$ to obtain

$$Y'(x) \doteq f(x_0, Y_0) + f_x(x_0, Y_0)(x - x_0) + f_y(x_0, Y_0)(Y(x) - Y_0)$$

$$= \lambda(Y(x) - Y_0) + g(x) \qquad (6.8.14)$$

with $\lambda = f_y(x_0, Y_0)$ and $g(x) = f(x_0, Y_0) + f_x(x_0, Y_0)(x - x_0)$. This is a valid approximation if $x - x_0$ is sufficiently small. Introducing $V(x) = Y(x) - Y_0$,

$$V'(x) \doteq \lambda V(x) + g(x) \qquad (6.8.15)$$

The inhomogeneous term $g(x)$ will drop out of all derivations concerning numerical stability, because we are concerned with differences of solutions of the equation. Dropping $g(x)$ in (6.8.15), we obtain the model equation (6.8.13). As further motivation, refer back to the stability results (6.1.5)–(6.1.10), and to the trapezoidal stability results in (6.5.20)–(6.5.23).

In the case that $\mathbf{y}' = \mathbf{f}(x, \mathbf{y})$ represents a system of $m$ differentiable equations, as in (6.1.13), the partial derivative $\mathbf{f}_\mathbf{y}(x, \mathbf{y})$ becomes a Jacobian matrix,

$$\left[ \mathbf{f}_\mathbf{y}(x, \mathbf{y}) \right]_{ij} = \frac{\partial f_i}{\partial y_j} \qquad 1 \le i, j \le m$$

as in (6.2.54). Thus the model equation becomes

$$\mathbf{y}' = \Lambda \mathbf{y} + \mathbf{g}(x) \qquad (6.8.16)$$

a system of $m$ linear differential equations with $\Lambda = \mathbf{f}_y(x_0, \mathbf{Y}_0)$. It can be shown that in most cases, this system reduces to an equivalent system

$$z_i' = \lambda_i z_i + \gamma_i(x) \qquad 1 \le i \le m \qquad (6.8.17)$$

with $\lambda_1, \ldots, \lambda_m$ the eigenvalues of $\Lambda$ (see Problem 24). With (6.8.17), we are back to the simple model equation (6.8.13), provided we allow $\lambda$ to be complex in order to include all possible eigenvalues of $\Lambda$.

Applying (6.8.1) to the model equation (6.8.13), we obtain

$$y_{n+1} = \sum_{j=0}^{p} a_j y_{n-j} + h\lambda \sum_{j=-1}^{p} b_j y_{n-j} \qquad (6.8.18)$$

$$(1 - h\lambda b_{-1}) y_{n+1} - \sum_{j=0}^{p} (a_j + h\lambda b_j) y_{n-j} = 0 \qquad n \ge p \qquad (6.8.19)$$

This is a *homogeneous linear difference equation* of order $p + 1$, and the theory for its solvability is completely analogous to that of $(p + 1)$st-order homogeneous linear differential equations. As a general reference, see Henrici (1962, pp. 210–215) or Isaacson and Keller (1966, pp. 405–417).

We attempt to find a general solution by first looking for solutions of the special form

$$y_n = r^n \qquad n \ge 0$$

If we can find $p + 1$ linearly independent solutions, then an arbitrary linear combination will give the general solution of (6.8.19).

Substituting $y_n = r^n$ into (6.8.19) and canceling $r^{n-p}$, we obtain

$$(1 - h\lambda b_{-1}) r^{p+1} - \sum_{j=0}^{p} (a_j + h\lambda b_j) r^{p-j} = 0 \qquad (6.8.20)$$

This is called the *characteristic equation*, and the left-hand side is the *characteris-*

*tic polynomial*. The roots are called *characteristic roots*. Define

$$\sigma(r) = b_{-1}r^{p+1} + \sum_{j=0}^{p} b_j r^{p-j}$$

and recall the definition (6.8.8) of $\rho(r)$. Then (6.8.20) becomes

$$\rho(r) - h\lambda\sigma(r) = 0 \qquad (6.8.21)$$

Denote the characteristic roots by

$$r_0(h\lambda), \ldots, r_p(h\lambda)$$

which can be shown to depend continuously on the value of $h\lambda$. When $h\lambda = 0$, the equation (6.8.21) becomes simply $\rho(r) = 0$, and we have $r_j(0) = r_j$, $j = 0, 1, \ldots, p$, for the earlier roots $r_j$ of $\rho(r) = 0$. Since $r_0 = 1$ is a root of $\rho(r)$, we let $r_0(h\lambda)$ be the root of (6.8.21) for which $r_0(0) = 1$. The root $r_0(h\lambda)$ is called the *principal root*, for reasons that will become apparent later. If the roots $r_j(h\lambda)$ are all distinct, then the general solution of (6.8.19) is

$$y_n = \sum_{j=0}^{p} \gamma_j \left[ r_j(h\lambda) \right]^n \qquad n \geq 0 \qquad (6.8.22)$$

But if $r_j(h\lambda)$ is a root of multiplicity $\nu > 1$, then the following are $\nu$ linearly independent solutions of (6.8.19):

$$\left\{ \left[ r_j(h\lambda) \right]^n \right\}, \left\{ n \left[ r_j(h\lambda) \right]^n \right\}, \ldots, \left\{ n^{\nu-1} \left[ r_j(h\lambda) \right]^n \right\}$$

These can be used with the solution arising from the other roots to generate a general solution for (6.8.19), comparable to (6.8.22).

**Theorem 6.7**  Assume the consistency condition (6.8.7). Then the multistep method (6.8.1) is stable if and only if the root condition (6.8.9), (6.8.10) is satisfied.

**Proof**  1.    We begin by showing the necessity of the root condition for stability. To do so, assume the opposite by letting

$$\left| r_j(0) \right| > 1$$

for some $j$. Consider the differential equation problem $y' \equiv 0$, $y(0) = 0$, with the solution $Y(x) \equiv 0$. Then (6.8.1) becomes

$$y_{n+1} = \sum_{j=0}^{p} a_j y_{n-j} \qquad n \geq p \qquad (6.8.23)$$

If we take $y_0 = y_1 = \cdots = y_p = 0$, then the numerical solution is clearly

$y_n = 0$, with all $n \geq 0$. For the perturbed initial values, take

$$z_0 = \epsilon, \; z_1 = \epsilon r_j(0), \ldots, z_p = \epsilon r_j(0)^p \qquad (6.8.24)$$

For these initial values,

$$\underset{0 \leq n \leq p}{\text{Max}} |y_n - z_n| = \epsilon |r_j(0)|^p$$

which is a uniform bound for all small values of $h$, since the right side is independent of $h$. As $\epsilon \to 0$, the bound also tends to zero.

The solution of (6.8.22) with the initial conditions (6.8.24) is simply

$$z_n = \epsilon \big[ r_j(0) \big]^n \qquad n \geq 0$$

For the deviation from $\{y_n\}$,

$$\underset{x_0 \leq x_n \leq b}{\text{Max}} |y_n - z_n| = \epsilon |r_j(0)|^{N(h)}$$

As $h \to 0$, $N(h) \to \infty$ and the bound becomes infinite. This proves the method is unstable when some $|r_j(0)| > 1$. If the root condition is violated instead by assuming (6.8.10) is false, then a similar proof can be given. This is left as Problem 29.

**2.** Assume the root condition is satisfied. The proof of stability will be restricted to the model equation (6.8.13). A proof can be given for the general equation $y' = f(x, y)$, but it is a fairly involved modification of the following proof. The general proof involves the solution of nonhomogeneous linear difference equations [see Isaacson and Keller (1966), pp. 405–417, for a complete development]. To further simplify the proof, we assume that the roots $r_j(0)$, $j = 0, 1, \ldots, p$ are all distinct. The same will be true of $r_j(h\lambda)$, provided the value of $h$ is kept sufficiently small, say $0 \leq h \leq h_0$.

Let $\{y_n\}$ and $\{z_n\}$ be two solutions of (6.8.19) on $[x_0, b]$, and assume

$$\underset{0 \leq n \leq p}{\text{Max}} |y_n - z_n| \leq \epsilon \qquad 0 < h \leq h_0 \qquad (6.8.25)$$

Introduce the error $e_n = y_n - z_n$. Subtracting using (6.8.19) for each solution,

$$(1 - h\lambda b_{-1})e_{n+1} - \sum_{j=0}^{p} (a_j + h\lambda b_j)e_{n-j} = 0 \qquad (6.8.26)$$

for $x_{p+1} \leq x_{n+1} \leq b$. The general solution is

$$e_n = \sum_{j=0}^{p} \gamma_j \big[ r_j(h\lambda) \big]^n \qquad n \geq 0 \qquad (6.8.27)$$

The coefficients $\gamma_0(h), \ldots, \gamma_p(h)$ must be chosen so that

$$\gamma_0 + \gamma_1 + \cdots + \gamma_p = e_0$$

$$\gamma_0 r_0(h\lambda) + \cdots + \gamma_p r_p(h\lambda) = e_1$$

$$\vdots$$

$$\gamma_0 [r_0(h\lambda)]^p + \cdots + \gamma_p [r_p(h\lambda)]^p = e_p$$

The solution (6.8.27) will then agree with the given initial perturbations $e_0, \ldots, e_p$, and it will satisfy the difference equation (6.8.26). Using the bound (6.8.25) and the theory of systems of linear equations, it is fairly straightforward to show that

$$\underset{0 \le i \le p}{\text{Max}} |\gamma_i| \le c_1 \epsilon \qquad 0 < h \le h_0 \qquad (6.8.28)$$

for some constant $c_1 > 0$. We omit the proof of this, although it can be carried out easily by using concepts introduced in Chapters 7 and 8.

To bound the solution $e_n$ on $[x_0, b]$, we must bound each term $[r_j(h\lambda)]^n$. To do so, consider the expansion

$$r_j(u) = r_j(0) + u r_j'(\zeta) \qquad (6.8.29)$$

for some $\zeta$ between 0 and $u$. To compute $r_j'(u)$, differentiate the characteristic equation

$$\rho(r_j(u)) - u\sigma(r_j(u)) = 0$$

Then

$$r_j'(u) = \frac{\sigma(r_j(u))}{\rho'(r_j(u)) - u\sigma'(r_j(u))} \qquad (6.8.30)$$

By the assumption that $r_j(0)$ is a simple root of $\rho(r) = 0$, $0 \le j \le p$, it follows that $\rho'(r_j(0)) \ne 0$, and by continuity, $\rho'(r_j(u)) \ne 0$ for all sufficiently small values of $u$. The denominator in (6.8.30) is nonzero, and we can bound $r_j'(u)$

$$|r_j'(u)| \le c_2 \qquad \text{all } |u| \le u_0$$

for some $u_0 > 0$.

Using this with (6.8.29) and the root condition (6.8.9), we have

$$|r_j(h\lambda)| \le |r_j(0)| + c_2|h\lambda| \le 1 + c_2|h\lambda|$$

$$\left|[r_j(h\lambda)]^n\right| \le [1 + c_2|h\lambda|]^n \le e^{c_2 n|h\lambda|} \le e^{c_2(b-x_0)|\lambda|} \qquad (6.8.31)$$

for all $0 < h \le h_0$. Combined with (6.8.27) and (6.8.28),

$$\underset{x_0 \le x_n \le b}{\text{Max}} \ |e_n| \le c_3 |\epsilon| e^{c_2 (b - x_0)|\lambda|} \qquad 0 < h \le h_0$$

for an appropriate constant $c_3$. ∎

**Convergence theory**   The following result generalizes Theorem 6.6 of Section 6.3, with necessary and sufficient conditions being given for the convergence of multistep methods.

***Theorem 6.8***   Assume the consistency condition (6.8.7). Then the multistep method (6.8.1) is convergent if and only if the root condition (6.8.9), (6.8.10) is satisfied.

***Proof***   **1.**   We begin by showing the necessity of the root condition for convergence, and again we use the problem $y' \equiv 0$, $y(0) = 0$, with the solution $Y(x) \equiv 0$. The multistep method (6.8.1) becomes

$$y_{n+1} = \sum_{j=0}^{p} a_j y_{n-j} \qquad n \ge p \qquad (6.8.32)$$

with $y_0, \ldots, y_p$ chosen to satisfy

$$\eta(h) \equiv \underset{0 \le n \le p}{\text{Max}} \ |y_n| \to 0 \qquad \text{as} \quad h \to 0 \qquad (6.8.33)$$

Suppose that the root condition is violated. We show that (6.8.32) is not convergent to $Y(x) \equiv 0$.

Assume that some $|r_j(0)| > 1$. Then a satisfactory solution of (6.8.32) is

$$y_n = h\left[r_j(0)\right]^n \qquad x_0 \le x_n \le b \qquad (6.8.34)$$

Condition (6.8.33) is satisfied since

$$\eta(h) = h\left|r_j(0)\right|^p \to 0 \qquad \text{as} \quad h \to 0$$

But the solution $\{y_n\}$ does not converge. First,

$$\underset{x_0 \le x_n \le b}{\text{Max}} \ |Y(x_n) - y_n| = h\left|r_j(0)\right|^{N(h)}$$

Consider those values of $h = b/N(h)$. Then L'Hospital's rule can be used to show that

$$\underset{N \to \infty}{\text{Limit}} \ \frac{b}{N}\left|r_j(0)\right|^N = \infty$$

showing (6.8.32) does not converge to the solution $Y(x) \equiv 0$.

Assume (6.8.9) of the root condition is satisfied, but that some $r_j(0)$ is a multiple root of $\rho(r)$ and $|r_j(0)| = 1$. Then the preceding form of proof is still satisfactory, but we must use the solution

$$y_n = hn\left[r_j(0)\right]^n \qquad 0 \le n \le N(h)$$

This completes the proof of the necessity of the root condition.

**2.**   Assume that the root condition is satisfied. As with the previous theorem, it is too difficult to give a general proof of convergence for an arbitrary differential equation. For that, see the development in Isaacson and Keller (1966, pp. 405–417). The present proof is restricted to the model equation (6.8.13), and again we assume the roots $r_j(0)$ are distinct, in order to simplify the proof.

The multistep method (6.8.1) becomes (6.8.18) for the model equation $y' = \lambda y$, $y(0) = 1$. We show that the term $\gamma_0[r_0(h\lambda)]^n$ in its general solution

$$y_n = \sum_{j=0}^{p} \gamma_j\left[r_j(h\lambda)\right]^n \qquad (6.8.35)$$

will converge to the solution $Y(x) = e^{\lambda x}$ on $[0, b]$. The remaining terms $\gamma_j[r_j(h\lambda)]^n$, $j = 1, 2, \dots, p$, are *parasitic solutions*, and they can be shown to converge to zero as $h \to 0$ (see Problem 30).

Expand $r_0(h\lambda)$ using Taylor's theorem,

$$r_0(h\lambda) = r_0(0) + h\lambda r'(0) + O(h^2)$$

From (6.8.30),

$$r_0'(0) = \frac{\sigma(1)}{\rho'(1)}$$

and using consistency condition (6.8.7), this leads to $r_0'(0) = 1$. Then

$$r_0(h\lambda) = 1 + h\lambda + O(h^2) = e^{\lambda h} + O(h^2)$$

$$\left[r_0(h\lambda)\right]^n = e^{\lambda nh}\left[1 + O(h^2)\right]^n = e^{\lambda x_n}[1 + O(h)] \quad (6.8.36)$$

over every finite interval $0 \le x_n \le b$. Thus

$$\operatorname*{Max}_{0 \le x_n \le b} \left|\left[r_0(h\lambda)\right]^n - e^{\lambda x_n}\right| \to 0 \qquad \text{as} \quad h \to 0 \qquad (6.8.37)$$

We must now show that the coefficient $\gamma_0 \to 1$ as $h \to 0$.

The coefficients $\gamma_0(h), \ldots, \gamma_p(h)$ satisfy the linear system

$$\gamma_0 + \gamma_1 + \cdots + \gamma_p = y_0$$

$$\gamma_0[r_0(h\lambda)] + \cdots + \gamma_p[r_p(h\lambda)] = y_1 \qquad (6.8.38)$$

$$\gamma_0[r_0(h\lambda)]^p + \cdots + \gamma_p[r_p(h\lambda)]^p = y_p$$

The initial values $y_0, \ldots, y_p$ depend on $h$ and are assumed to satisfy

$$\eta(h) \equiv \max_{0 \le n \le p} \left| e^{\lambda x_n} - y_n \right| \to 0 \qquad \text{as} \quad h \to 0$$

But this implies

$$\lim_{h \to 0} y_n = 1 \qquad 0 \le n \le p \qquad (6.8.39)$$

The coefficient $\gamma_0$ can be obtained by using Cramer's rule to solve (6.8.38):

$$\gamma_0 = \frac{\begin{vmatrix} y_0 & 1 & \cdots & 1 \\ y_1 & r_1 & \cdots & r_p \\ \vdots & & & \\ y_p & r_1^p & \cdots & r_p^p \end{vmatrix}}{\begin{vmatrix} 1 & 1 & \cdots & 1 \\ r_0 & r_1 & \cdots & r_p \\ \vdots & & & \\ r_0^p & r_1^p & \cdots & r_p^p \end{vmatrix}} \qquad (6.8.40)$$

The denominator converges to the Vandermonde determinant for $r_0(0) = 1$, $r_1(0), \ldots, r_p(0)$, and this is nonzero since the roots are distinct (see Problem 1 of Chapter 3). By using (6.8.39), the numerator converges to the same quantity as $h \to 0$. Therefore, $\gamma_0 \to 1$ as $h \to 0$. Using this, along with (6.8.37) and Problem 30, the solution $\{y_n\}$ converges to $Y(x) = e^{\lambda x}$ on $[0, b]$.   ∎

The following is a well-known result; it is a trivial consequence of Theorems 6.7 and 6.8.

**Corollary**   Let (6.8.1) be a consistent multistep method. Then it is convergent if and only if it is stable.   ∎

**Relative stability and weak stability**   Consider again the model equation (6.8.13) and its numerical solution (6.8.32). The past theorem stated that the parasitic solutions $\gamma_j[r_j(h\lambda)]^n$ will converge to zero as $h \to 0$. But for a fixed $h$ with increasing $x_n$, we also would like them to remain small relative to the principal part of the solution $\gamma_0[r_0(h\lambda)]^n$. This will be true if the characteristic roots satisfy

$$|r_j(h\lambda)| \leq r_0(h\lambda) \qquad j = 1, 2, \ldots, p \qquad (6.8.41)$$

for all sufficiently small values of $h$. This leads us to the definition of relative stability.

We say that the method (6.8.1) is *relatively stable* if the characteristic roots $r_j(h\lambda)$ satisfy (6.8.41) for all sufficiently small nonzero values of $|h\lambda|$. And the method is said to satisfy the *strong root condition* if

$$|r_j(0)| < 1 \qquad j = 1, 2, \ldots, p \qquad (6.8.42)$$

This is an easy condition to check, and it implies relative stability. Just use the continuity of the roots $r_j(h\lambda)$ with respect to $h\lambda$ to have (6.8.42) imply (6.8.41). Relative stability does not imply the strong root condition, although they are equivalent for most methods [see Problem 36(b)]. If a multistep method is stable but not relatively stable, then it will be called *weakly stable*.

**Example 1.** For the midpoint method, $r_0(h\lambda) = 1 - h|\lambda|$ $\qquad |r_1(h\lambda)| > r_0(h\lambda)$

$r_1(h\lambda) = -1 + h|\lambda|$.

$$r_0(h\lambda) = 1 + h\lambda + O(h^2) \qquad r_1(h\lambda) = -1 + h\lambda + O(h^2) \quad (6.8.43)$$

It is weakly stable according to (6.8.41) when $\lambda < 0$, which agrees with what was shown earlier in Section 6.4.

**2.** The Adams–Bashforth and Adams–Moulton methods, (6.7.22) and (6.7.26), have the same characteristic polynomial when $h = 0$,

$$\rho(r) = r^{p+1} - r^p \qquad (6.8.44)$$

The roots are $r_0 = 1$, $r_j = 0$, $j = 1, 2, \ldots, p$; thus the strong root condition is satisfied and the Adams methods are relatively stable.

**Stability regions** In the preceding discussions of stability, the values of $h$ were required to be sufficiently small in order to carry through the derivations. Little indication was given as to just how small $h$ should be. It is clear that if $h$ is required to be extremely small, then the method is impractical for most problems; thus we need to examine the permissible values of $h$. Since the stability depends on the characteristic roots, and since they in turn depend on $h\lambda$, we are interested in determining the values of $h\lambda$ for which the multistep method (6.8.1) is stable in some sense. To cover situations arising when solving systems of differential equations, it is necessary that the value of $\lambda$ be allowed to be complex, as noted following (6.8.17).

To motivate the later discussion, we consider the stability of Euler's method. Apply Euler's method to the equation

$$y' = \lambda y + g(x) \qquad y(0) = Y_0 \qquad (6.8.45)$$

obtaining

$$y_{n+1} = y_n + h[\lambda y_n + g(x_n)] \qquad n \geq 0 \qquad y_0 = Y_0 \qquad (6.8.46)$$

Then consider the perturbed problem

$$z_{n+1} = z_n + h[\lambda z_n + g(x_n)] \qquad n \geq 0 \quad z_0 = Y_0 + \epsilon \qquad (6.8.47)$$

For the original Eq. (6.8.45), this perturbation of $Y_0$ leads to solutions $Y(x)$ and $Z(x)$ satisfying

$$Y(x) - Z(x) = \epsilon e^{\lambda x} \qquad x \geq 0$$

In this original problem, we would ordinarily be interested in the case with Real $(\lambda) \leq 0$, since then $|Y(x) - Z(x)|$ would remain bounded as $x \to 0$. We further restrict our interest to the case of Real $(\lambda) < 0$, so that $Y(x) - Z(x) \to 0$ as $x \to \infty$. For such $\lambda$, we want to find the values of $h$ so that the numerical solutions of (6.8.46) and (6.8.47) will retain the behavior associated with $Y(x)$ and $Z(x)$.

Let $e_n = z_n - y_n$. Subtracting (6.8.46) from (6.8.47),

$$e_{n+1} = e_n + h\lambda e_n = (1 + h\lambda)e_n \qquad e_0 = \epsilon$$

Inductively,

$$e_n = (1 + h\lambda)^n \epsilon \qquad (6.8.48)$$

Then $e_n \to 0$ as $x_n \to \infty$ if and only if

$$|1 + h\lambda| < 1 \qquad (6.8.49)$$

This yields a set of complex values $h\lambda$ that form a circle of radius 1 about the point $-1$ in the complex plane. If $h\lambda$ belongs to this set, then $y_n - z_n \to 0$ as $x_n \to \infty$, but not otherwise.

To see that this discussion is also important for convergence, realize that the original differential equation can be looked upon as a perturbation of the approximating equation (6.8.46). From (6.2.17), applied to (6.8.45),

$$Y_{n+1} = Y_n + h[\lambda Y_n + g(x_n)] + \frac{h^2}{2} Y''(\xi_n) \qquad (6.8.50)$$

Here we have a perturbation of the equation (6.8.46) at every step, not at just the initial point $x_0 = 0$. Nonetheless, the preceding stability analysis can be shown to apply to this perturbation of (6.8.46). The error formula (6.8.48) will have to be suitably modified, but it will still depend critically on the bound (6.8.49) (see Problem 40).

*Example*  Apply Euler's method to the problem

$$y' = \lambda y + (1 - \lambda)\cos(x) - (1 + \lambda)\sin(x) \qquad y(0) = 1 \qquad (6.8.51)$$

**Table 6.17 Euler's method for (6.8.51)**

| $\lambda$ | $x$ | Error: $h = .5$ | Error: $h = .1$ | Error: $h = .01$ |
|---|---|---|---|---|
| $-1$ | 1 | $-2.46E - 1$ | $-4.32E - 2$ | $-4.22E - 3$ |
| | 2 | $-2.55E - 1$ | $-4.64E - 2$ | $-4.55E - 3$ |
| | 3 | $-2.66E - 2$ | $-6.78E - 3$ | $-7.22E - 4$ |
| | 4 | $2.27E - 1$ | $3.91E - 2$ | $3.78E - 3$ |
| | 5 | $2.72E - 1$ | $4.91E - 2$ | $4.81E - 3$ |
| $-10$ | 1 | $3.98E - 1$ | $-6.99E - 3$ | $-6.99E - 4$ |
| | 2 | $6.90E + 0$ | $-2.90E - 3$ | $-3.08E - 4$ |
| | 3 | $1.11E + 2$ | $3.86E - 3$ | $3.64E - 4$ |
| | 4 | $1.77E + 3$ | $7.07E - 3$ | $7.04E - 4$ |
| | 5 | $2.83E + 4$ | $3.78E - 3$ | $3.97E - 4$ |
| $-50$ | 1 | $3.26E + 0$ | $1.06E + 3$ | $-1.39E - 4$ |
| | 2 | $1.88E + 3$ | $1.11E + 9$ | $-5.16E - 5$ |
| | 3 | $1.08E + 6$ | $1.17E + 15$ | $8.25E - 5$ |
| | 4 | $6.24E + 8$ | $1.23E + 21$ | $1.41E - 4$ |
| | 5 | $3.59E + 11$ | $1.28E + 27$ | $7.00E - 5$ |

whose true solution is $Y(x) = \sin(x) + \cos(x)$. We give results for several values of $\lambda$ and $h$. For $\lambda = -1, -10, -50$, the bound (6.8.49) implies the respective bounds on $h$ of

$$0 < h < 2 \qquad 0 < h < \frac{1}{5} = .2 \qquad 0 < h < \frac{1}{25} = .04$$

The use of larger values of $h$ gives poor numerical results, as seen in Table 6.17.

The preceding derivation with Euler's method motivates our general approach to finding the set of all $h\lambda$ for which the method (6.8.1) is stable. Since we consider only cases with Real $(\lambda) < 0$, we want the numerical solution $\{y_n\}$ of (6.8.1), when applied to the model equation $y' = \lambda y$, to tend to zero as $x_n \rightarrow \infty$, for all choices of initial values $y_0, y_1, \ldots, y_p$. The set of all $h\lambda$ for which this is true is called the *region of absolute stability* of the method (6.8.1). The larger this region, the less the restriction on $h$ in order to have a stable numerical solution.

When (6.8.1) is applied to the model equation, we obtain the earlier equation (6.8.18), and its solution is given by (6.8.22), namely

$$y_n = \sum_{j=0}^{p} \gamma_j \left[ r_j(h\lambda) \right]^n \qquad n \geq 0$$

provided the characteristic roots $r_0(h\lambda), \ldots, r_0(h\lambda)$ are distinct. To have this tend to zero as $n \rightarrow \infty$, for all choices of $y_0, \ldots, y_p$, it is necessary and sufficient to have

$$\left| r_j(h\lambda) \right| < 1 \qquad j = 0, 1, \ldots, p \qquad (6.8.52)$$

The set of all $h\lambda$ satisfying this set of inequalities is also called the region of absolute stability. This region is contained in the set defined in the preceding paragraph, and it is usually equal to that set. We work only with (6.8.52) in finding the region of absolute stability.

***Example***  Consider the second-order Adams–Bashforth method

$$y_{n+1} = y_n + \frac{h}{2}[3y'_n - y'_{n-1}] \qquad n \geq 1 \tag{6.8.53}$$

The characteristic equation is

$$r^2 - \left(1 + \tfrac{3}{2}h\lambda\right)r + \tfrac{1}{2}h\lambda = 0$$

and its roots are

$$r_0 = \tfrac{1}{2}\left\{1 + \tfrac{3}{2}h\lambda + \sqrt{1 + h\lambda + \tfrac{9}{4}h^2\lambda^2}\right\}$$

$$r_1 = \tfrac{1}{2}\left\{1 + \tfrac{3}{2}h\lambda - \sqrt{1 + h\lambda + \tfrac{9}{4}h^2\lambda^2}\right\}$$

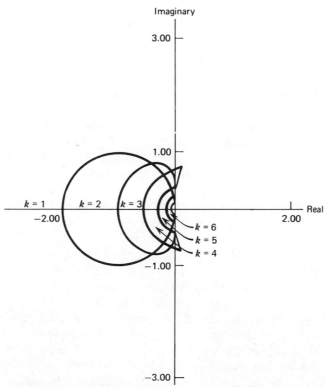

**Figure 6.7**  Stability regions for Adams–Bashforth methods. The method of order $k$ is stable inside the region indicated left of origin. [Taken from Gear (1971), p. 131, with permission.]

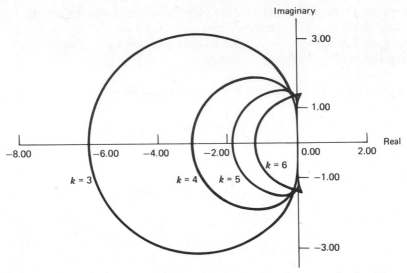

**Figure 6.8**   Stability regions for Adams–Moulton methods. The method of order $k$ is stable inside the region indicated. [Taken from Gear (1971), p. 131, with permission.]

The region of absolute stability is the set of $h\lambda$ for which

$$|r_0(h\lambda)| < 1 \qquad |r_1(h\lambda)| < 1$$

For $\lambda$ real, the acceptable values of $h\lambda$ are $-1 < h\lambda < 0$.

The boundaries of the regions of absolute stability of the Adams–Bashforth and the Adams–Moulton methods are given in Figures 6.7 and 6.8, respectively. For Adams–Moulton formulas with one iteration of an Adams–Bashforth predictor, the regions of absolute stability are given in Shampine and Gordon (1975, pp. 135–140).

From these diagrams, it is clear that the region of absolute stability becomes smaller as the order of the method increases. And for formulas of the same order, the Adams–Moulton formula has a significantly larger region of absolute stability than the Adams–Bashforth formula. The size of these regions is usually quite acceptable from a practical point of view. For example, the real values of $h\lambda$ in the region of absolute stability for the fourth-order Adams–Moulton formula are given by $-3 < h\lambda < 0$. This is not a serious restriction on $h$ in most cases.

The Adams family of formulas is very convenient for creating a variable-order algorithm, and their stability regions are quite acceptable. They will have difficulty with problems for which $\lambda$ is negative and large in magnitude, and these problems are best treated by other methods, which we consider in the next section.

There are special methods for which the region of absolute stability consists of all complex values $h\lambda$ with Real$(h\lambda) < 0$. These methods are called *A-stable*, and with them there is no restriction on $h$ in order to have stability of the type

**Table 6.18    Example of trapezoidal rule: $h = .5$**

| $x$ | Error: $\lambda = -1$ | Error: $\lambda = -10$ | Error: $\lambda = -50$ |
|---|---|---|---|
| 2 | $-1.13E - 2$ | $-2.78E - 3$ | $-7.91E - 4$ |
| 4 | $-1.43E - 2$ | $-8.91E - 5$ | $-8.91E - 5$ |
| 6 | $2.02E - 2$ | $2.77E - 3$ | $4.72E - 4$ |
| 8 | $-2.86E - 3$ | $-2.22E - 3$ | $-5.11E - 4$ |
| 10 | $-1.79E - 2$ | $-9.23E - 4$ | $-1.56E - 4$ |

we have been considering. The trapezoidal rule is an example of such a method [see (6.5.24)–(6.5.25)].

*Example*    Consider the *backward Euler method*:

$$y_{n+1} = y_n + hf(x_{n+1}, y_{n+1}) \qquad n \geq 0 \qquad (6.8.54)$$

Applying this to the model equation $y' = \lambda y$ and solving, we have

$$y_n = \left[\frac{1}{1 - h\lambda}\right]^n y_0 \qquad n \geq 0 \qquad (6.8.55)$$

Then $y_n \to 0$ as $x_n \to \infty$ if and only if

$$\frac{1}{|1 - h\lambda|} < 1$$

This will be true for all $h\lambda$ with $\text{Real}(\lambda) < 0$, and the backward Euler method is an A-stable method.

*Example*    Apply the trapezoidal method to the problem (6.8.51), which was solved earlier with Euler's method. We use a stepsize of $h = .5$ for $\lambda = -1, -10, -50$. The results are given in Table 6.18. They illustrate that the trapezoidal rule does not become unstable as $|\lambda|$ increases, while $\text{Real}(\lambda) < 0$.

It would be useful to have A-stable multistep methods of order greater than 2. But a result of Dahlquist (1963) shows there are no such methods. We examine some higher order methods that have most of the needed stability properties in the following section.

## 6.9    Stiff Differential Equations and the Method of Lines

The numerical solution of stiff differential equations has, within the past ten to fifteen years, become a much studied subject. Such equations (including systems of differential equations) have appeared in an increasing number of applications, in subjects as diverse as chemical kinetics and the numerical solution of partial differential equations. In this section, we sketch some of the main ideas about this subject, and we show its relation to the numerical solution of the simple heat equation.

There are many definitions of the concept of stiff differential equation. The most important common feature of these definitions is that when such equations are being solved with standard numerical methods (e.g., the Adams methods of Section 6.7), the stepsize $h$ is forced to be extremely small in order to maintain stability—far smaller than would appear to be necessary based on a consideration of the truncation error. An indication of this can be seen with Eq. (6.8.51), which was solved in Table 6.17 with Euler's method. In that case, the unknown $Y(x)$ did not change with $\lambda$, and therefore the truncation error was also independent of $\lambda$. But the actual error was strongly affected by the magnitude of $\lambda$, with $h\lambda$ required to satisfy the stability condition $|1 + h\lambda| < 1$ to obtain convergence. As $|\lambda|$ increased, the size of $h$ had to decrease accordingly. This is typical of the behavior of standard numerical methods when applied to stiff differential equations, with the major difference being that the actual values of $|\lambda|$ are far larger in real life examples, for example, $\lambda = -10^6$.

We now look at the most common class of such differential equations, basing our examination on consideration of the linearization of the system $\mathbf{y}' = \mathbf{f}(x, \mathbf{y})$ as developed in (6.8.14)–(6.8.17):

$$\mathbf{y}' = \Lambda \mathbf{y} + \mathbf{g}(x) \tag{6.9.1}$$

with $\Lambda = \mathbf{f_y}(x_0, \mathbf{Y}_0)$ the Jacobian matrix of $\mathbf{f}$. We say the differential equation $\mathbf{y}' = \mathbf{f}(x, \mathbf{y})$ is *stiff* if some of eigenvalues $\lambda_j$ of $\Lambda$, or more generally of $\mathbf{f_y}(x, \mathbf{y})$, have a negative real part of very large magnitude. We study numerical methods for stiff equations by considering their effect on the model equation

$$y' = \lambda y + g(x) \tag{6.9.2}$$

with $\mathrm{Real}(\lambda)$ negative and very large in magnitude. This approach has its limitations, some of which we indicate later, but it does give us a means of rejecting unsatisfactory methods, and it suggests some possibly satisfactory methods.

The concept of a region of absolute stability, introduced in the last section, is the initial tool used in studying the stability of a numerical method for solving stiff differential equations. We seek methods whose stability region contains the entire negative real axis and as much of the left half of the complex plane as possible. There are a number of ways to develop such methods, but we only discuss one of them—obtaining the *backward differentiation formulas* (BDFs).

Let $P_p(x)$ denote the polynomial of degree $\leq p$ that interpolates $Y(x)$ at the points $x_{n+1}, x_n, \ldots, x_{n-p+1}$ for some $p \geq 1$:

$$P_p(x) = \sum_{j=-1}^{p-1} Y(x_{n-j}) l_{j,n}(x) \tag{6.9.3}$$

with $\{l_{j,n}(x)\}$ the Lagrange interpolation basis functions for the nodes

**Table 6.19    Coefficients of BDF method (6.9.6)**

| $p$ | $\beta$ | $\alpha_0$ | $\alpha_1$ | $\alpha_2$ | $\alpha_3$ | $\alpha_4$ | $\alpha_5$ |
|-----|---------|-----------|-----------|-----------|-----------|-----------|-----------|
| 1 | $1$ | $1$ | | | | | |
| 2 | $\dfrac{2}{3}$ | $\dfrac{4}{3}$ | $-\dfrac{1}{3}$ | | | | |
| 3 | $\dfrac{6}{11}$ | $\dfrac{18}{11}$ | $-\dfrac{9}{11}$ | $\dfrac{2}{11}$ | | | |
| 4 | $\dfrac{12}{25}$ | $\dfrac{48}{25}$ | $-\dfrac{36}{25}$ | $\dfrac{16}{25}$ | $-\dfrac{3}{25}$ | | |
| 5 | $\dfrac{60}{137}$ | $\dfrac{300}{137}$ | $-\dfrac{300}{137}$ | $\dfrac{200}{137}$ | $-\dfrac{75}{137}$ | $\dfrac{12}{137}$ | |
| 6 | $\dfrac{60}{147}$ | $\dfrac{360}{147}$ | $-\dfrac{450}{147}$ | $\dfrac{400}{147}$ | $-\dfrac{225}{147}$ | $\dfrac{72}{147}$ | $-\dfrac{10}{147}$ |

$x_{n+1}, \ldots, x_{n-p+1}$ [see (3.1.5)]. Use

$$P_p'(x_{n+1}) \doteq Y'(x_{n+1}) = f(x_{n+1}, Y(x_{n+1})) \qquad (6.9.4)$$

Combining with (6.9.3) and solving for $Y(x_{n+1})$,

$$Y(x_{n+1}) \doteq \sum_{j=0}^{p-1} \alpha_j Y(x_{n-j}) + h\beta f(x_{n+1}, Y(x_{n+1})) \qquad (6.9.5)$$

The $p$-step BDF method is given by

$$y_{n+1} = \sum_{j=0}^{p-1} \alpha_j y_{n-j} + h\beta f(x_{n+1}, y_{n+1}) \qquad (6.9.6)$$

The coefficients for the cases of $p = 1, \ldots, 6$ are given in Table 6.19. The case $p = 1$ is simply the backward Euler method, which was discussed following (6.8.54) in the last section. The truncation error for (6.9.6) can be obtained from the error formula for numerical differentiation, given in (5.7.5):

$$T_n(Y) = -\frac{\beta}{p+1} h^{p+1} Y^{(p+1)}(\xi_n) \qquad (6.9.7)$$

for some $x_{n-p+1} \leq \xi_n \leq x_{n+1}$.

The regions of absolute stability for the formulas of Table 6.19 are given in Gear (1971, pp. 215–216). To create these regions, we must find all values $h\lambda$ for which

$$|r_j(h\lambda)| < 1 \qquad j = 0, 1, \ldots, p \qquad (6.9.8)$$

where the characteristic roots $r_j(h\lambda)$ are the solutions of

$$r^p = \sum_{j=0}^{p-1} \alpha_j r^{p-1-j} + h\beta r^p \tag{6.9.9}$$

It can be shown that for $p = 1$ and $p = 2$, the BDFs are A-stable, and that for $3 \le p \le 6$, the region of absolute stability becomes smaller as $p$ increases, although containing the entire negative real axis in each case. For $p \ge 7$, the regions of absolute stability are not acceptable for the solution of stiff problems. For more discussion of these stability regions, see Gear (1971, chap. 11) and Lambert (1973, chap. 8).

There are still problems with the BDF methods and with other methods that are chosen solely on the basis of their region of absolute stability. First, with the model equation $y' = \lambda y$, if $\text{Real}(\lambda)$ is of large magnitude and negative, then the solution $Y(x)$ goes to the zero very rapidly, and as $|\text{Real}(\lambda)|$ increases, the convergence to zero of $Y(x)$ becomes more rapid. We would like the same behavior to hold for the numerical solution of the model equation $\{y_n\}$. But with the A-stable trapezoidal rule, the solution [from (6.5.24)] is

$$y_n = \left[\frac{1 + \dfrac{h\lambda}{2}}{1 - \dfrac{h\lambda}{2}}\right]^n y_0 \qquad n \ge 0$$

If $|\text{Real}(\lambda)|$ is large, then the fraction inside the brackets is near to $-1$, and $y_n$ decreases to 0 quite slowly. Referring to the type of argument used with the Euler method in (6.8.45)–(6.8.50), the effect of perturbations will not decrease rapidly to zero for larger values of $\lambda$. Thus the trapezoidal method may not be a completely satisfactory choice for stiff problems. In comparison the A-stable backward Euler method has the desired behavior. From (6.8.55), the solution of the model problem is

$$y_n = \left[\frac{1}{1 - h\lambda}\right]^n y_0 \qquad n \ge 0$$

As $|\lambda|$ increases, the sequence $\{y_n\}$ goes to zero more rapidly. Thus the backward Euler solution better reflects the behavior of the true solution of the model equation.

A second problem with the case of stability regions is that it is based on using constant $\lambda$ and linear problems. The linearization (6.9.1) is often valid, but not always. For example, consider the second-order linear problem

$$y'' + ay' + (1 + b \cdot \cos(2\pi x))y = g(x) \qquad x \ge 0 \tag{6.9.10}$$

*in which one coefficient is not constant.* Convert it to the equivalent system

$$y_1' = y_2$$
$$y_2' = -(1 + b \cdot \cos(2\pi x))y_1 - ay_2 + g(x) \tag{6.9.11}$$

We will assume $a > 0$, $|b| < 1$. The eigenvalues of the homogeneous equation $[g(x) \equiv 0]$ are

$$\lambda = \frac{-a \pm \sqrt{a^2 - 4[1 + b \cdot \cos(2\pi x)]}}{2} \tag{6.9.12}$$

These are negative real numbers or are complex numbers with negative real parts. On the basis of the stability theory for the constant coefficient (or constant $\Lambda$) case, we would assume that the effect of all perturbations in the initial data would die away as $x \to \infty$. But in fact, the homogeneous part of (6.9.10) will have unbounded solutions. Thus there will be perturbations of the initial values that will lead to unbounded perturbed solutions in (6.9.10). This calls into question the validity of the use of the model equation $y' = \lambda y + g(x)$. Its use suggests methods that we may want to study further, but by itself, this approach is not sufficient to encompass the vast variety of linear and nonlinear problems. The example (6.9.10) is taken from Aiken (1985, p. 269).

**Solving the finite difference method**   We illustrate the problem by considering the backward Euler method:

$$y_{n+1} = y_n + hf(x_{n+1}, y_{n+1}) \qquad n \ge 0 \tag{6.9.13}$$

If the ordinary iteration formula

$$y_{n+1}^{(j+1)} = y_n + hf\left(x_{n+1}, y_{n+1}^{(j)}\right) \qquad j \ge 0 \tag{6.9.14}$$

is used, then

$$y_{n+1} - y_{n+1}^{(j+1)} \doteq h \frac{\partial f(x_{n+1}, y_{n+1})}{\partial y} \left[y_{n+1} - y_{n+1}^{(j)}\right]$$

For convergence, we would need to have

$$\left| h \frac{\partial f(x_{n+1}, y_{n+1})}{\partial y} \right| < 1 \tag{6.9.15}$$

But with stiff equations, this would again force $h$ to be very small, which we are trying to avoid. Thus another rootfinding method must be used to solve for $y_{n+1}$ in (6.9.13).

The most popular methods for solving (6.9.13) are based on Newton's method. For a single differential equation, Newton's method for finding $y_{n+1}$ is

$$y_{n+1}^{(j+1)} = y_{n+1}^{(j)} - \left[1 - hf_y\left(x_{n+1}, y_{n+1}^{(j)}\right)\right]^{-1} \left[y_{n+1}^{(j)} - y_n - hf\left(x_{n+1}, y_{n+1}^{(j)}\right)\right] \tag{6.9.16}$$

for $j \ge 0$. A crude initial guess is $y_{n+1}^{(0)} = y_n$, although generally this can be improved upon. With systems of differential equations Newton's method becomes very expensive. To decrease the expense, the matrix

$$I - h\mathbf{f_y}(x_{n+1}, \mathbf{z}) \qquad \text{some} \quad \mathbf{z} \doteq \mathbf{y}_n \tag{6.9.17}$$

is used for all $j$ and for a number of successive values of $n$. Thus Newton's method [see Section 2.11] for solving the system version of (6.9.13) is approximated by

$$\left[ I - h\mathbf{f}_\mathbf{y}(x_{n+1}, \mathbf{z}) \right] \boldsymbol{\delta}^{(j)} = \mathbf{y}_{n+1}^{(j)} - \mathbf{y}_n - h\mathbf{f}\left( x_{n+1}, \mathbf{y}_{n+1}^{(j)} \right)$$

$$\mathbf{y}_{n+1}^{(j+1)} = \mathbf{y}_{n+1}^{(j)} + \boldsymbol{\delta}^{(j)}$$

(6.9.18)

for $j \geq 0$. This amounts to solving a number of linear systems with the same coefficient matrix. This can be done much more cheaply than when the matrix is being modified (see the material in Section 8.1). The matrix in (6.9.17) will have to be updated periodically, but the savings will still be very significant when compared to an exact Newton's method. For a further discussion of this topic, see Aiken (1985, p. 7) and Gupta et al. (1985, pp. 22–25). For a survey of computer codes for solving stiff differential equations, see Aiken (1985, chap. 4).

**The method of lines**   Consider the following parabolic partial differential equation problem:

$$U_t = U_{xx} + G(x, t) \qquad 0 < x < 1 \qquad t > 0 \qquad (6.9.19)$$

$$U(0, t) = d_0(t) \qquad U(1, t) = d_1(t) \qquad t \geq 0 \qquad (6.9.20)$$

$$U(x, 0) = f(x) \qquad 0 \leq x \leq 1 \qquad (6.9.21)$$

The notation $U_t$ and $U_{xx}$ refers to partial derivatives with respect to $t$ and $x$, respectively. The unknown function $U(x, t)$ depends on the time $t$ and a spatial variable $x$. The conditions (6.9.20) are called boundary conditions, and (6.9.21) is called an initial condition. The solution $U$ can be interpreted as the temperature of an insulated rod of length 1, with $U(x, t)$ the temperature at position $x$ and time $t$; thus (6.9.19) is often called *the heat equation*. The functions $G$, $d_0$, $d_1$, and $f$ are assumed given and smooth. For a development of the theory of (6.9.19)–(6.9.21), see Widder (1975) or any standard introduction to partial differential equations. We give the *method of lines* for solving for $U$, a numerical method that has become much more popular in the past ten to fifteen years. It will also lead to the solution of a stiff system of ordinary differential equations.

Let $m > 0$ be an integer, and define $\delta = 1/m$,

$$x_j = j\delta \qquad j = 0, 1, \ldots, m$$

We discretize Eq. (6.9.19) by approximating the spatial derivative. Recall the formulas (5.7.17) and (5.7.18) for approximating second derivatives. Using this,

$$U_{xx}(x_j, t) = \frac{U(x_{j+1}, t) - 2U(x_j, t) + U(x_{j-1}, t)}{\delta^2} - \frac{\delta^2}{12} \frac{\partial^4 U(\xi_j, t)}{\partial x^4}$$

for $j = 1, 2, \ldots, m - 1$. Substituting into (6.9.19),

$$U_t(x_j, t) = \frac{U(x_{j+1}, t) - 2U(x_j, t) + U(x_{j-1}, t)}{\delta^2} + G(x_j, t)$$

$$- \frac{\delta^2}{12} \cdot \frac{\partial^4 U(\xi_j, t)}{\partial x^4} \qquad 1 \leq j \leq m - 1 \qquad (6.9.22)$$

Equation (6.9.19) is to be approximated at each interior node point $x_j$. The unknown $\xi_j \in [x_{j-1}, x_{j+1}]$.

Drop the final term in (6.9.22), the truncation error in the numerical differentiation. Forcing equality in the resulting approximate equation, we obtain

$$u'_j(t) = \frac{1}{\delta^2}\left[u_{j+1}(t) - 2u_j(t) + u_{j-1}(t)\right] + G(x_j, t) \qquad (6.9.23)$$

for $j = 1, 2, \ldots, m - 1$. The functions $u_j(t)$ are intended to be approximations of $U(x_j, t)$, $1 \leq j \leq m - 1$. This is the *method of lines* approximation to (6.9.19), and it is a system of $m - 1$ ordinary differential equations. Note that $u_0(t)$ and $u_m(t)$, which are needed in (6.9.23) for $j = 1$ and $j = m - 1$, are given using (6.9.20):

$$u_0(t) = d_0(t) \qquad u_m(t) = d_1(t) \qquad (6.9.24)$$

The initial condition for (6.9.23) is given by (6.9.21):

$$u_j(0) = f(x_j) \qquad 1 \leq j \leq m - 1 \qquad (6.9.25)$$

The name *method of lines* comes from solving for $U(x, t)$ along the lines $(x_j, t)$, $t \geq 0$, $1 \leq j \leq m - 1$, in the $(x, t)$-plane.

Under suitable smoothness assumptions on the functions $d_0$, $d_1$, $G$, and $f$, it can be shown that

$$\underset{\substack{0 \leq j \leq m \\ 0 \leq t \leq T}}{\text{Max}} \left|U(x_j, t) - u_j(t)\right| \leq C_T \delta^2 \qquad (6.9.26)$$

Thus to complete the solution process, we need only solve the system (6.9.23).

It will be convenient to write (6.9.23) in matrix form. Introduce

$$\mathbf{u}(t) = \left[u_1(t), \ldots, u_{m-1}(t)\right]^T \qquad \mathbf{u}_0 = \left[f(x_1), \ldots, f(x_{m-1})\right]^T$$

$$\mathbf{g}(t) = \left[\frac{1}{\delta^2}d_0(t) + G(x_1, t), G(x_2, t), \ldots, G(x_{m-2}, t),\right.$$

$$\left.\frac{1}{\delta^2}d_1(t) + G(x_{m-1}, t)\right]^T$$

$$\Lambda = \frac{1}{\delta^2}\begin{bmatrix} -2 & 1 & 0 & 0 & & \cdots & & 0 \\ 1 & -2 & 1 & 0 & & & & \vdots \\ 0 & 1 & -2 & 1 & & & & \vdots \\ \vdots & & & & \ddots & & & 0 \\ & & & & 1 & -2 & 1 \\ 0 & 0 & \cdots & & & 0 & 1 & -2 \end{bmatrix} \qquad (6.9.27)$$

The matrix $\Lambda$ is of order $m - 1$. In the definitions of $\mathbf{u}$ and $\mathbf{g}$, the superscript $T$ indicates matrix transpose, so that $\mathbf{u}$ and $\mathbf{g}$ are column vectors of length $m - 1$.

Using these matrices, Eqs. (6.9.23)–(6.9.25) can be rewritten as

$$\mathbf{u}'(t) = \Lambda\mathbf{u}(t) + \mathbf{g}(t) \qquad \mathbf{u}(0) = \mathbf{u}_0 \tag{6.9.28}$$

If Euler's method is applied, we have the numerical method

$$\mathbf{V}_{n+1} = \mathbf{V}_n + h(\Lambda\mathbf{V}_n + \mathbf{g}(t_n)] \qquad \mathbf{V}_0 = \mathbf{u}_0 \tag{6.9.29}$$

with $t_n = nh$ and $\mathbf{V}_n \doteq \mathbf{u}(t_n)$. This is a well-known numerical method for the heat equation, called the *simple explicit method*. We analyze the stability of (6.9.29) and some other methods for solving (6.9.28).

Equation (6.9.28) is in the form of the model equation, (6.9.1), and therefore we need the eigenvalues of $\lambda$ to examine the stiffness of the system. It can be shown that these eigenvalues are all real and are given by

$$\lambda_j = -\frac{4}{\delta^2}\sin^2\left(\frac{j\pi}{2m}\right) \qquad 1 \le j \le m - 1 \tag{6.9.30}$$

We leave the proof of this to Problem 6 in Chapter 7. Directly examining this formula,

$$\lambda_{m-1} \le \lambda_j \le \lambda_1 \tag{6.9.31}$$

$$\lambda_{m-1} = \frac{-4}{\delta^2}\sin^2\left(\frac{(m-1)\pi}{2m}\right) \doteq \frac{-4}{\delta^2}$$

$$\lambda_1 = \frac{-4}{\delta^2}\sin^2\left(\frac{\pi}{2m}\right) \doteq -\pi^2$$

with the approximations valid for larger $m$. It can be seen that (6.9.28) is a stiff system if $\delta$ is small.

Applying (6.9.31) and (6.8.49) to the analysis of stability in (6.9.29), we must have

$$|1 + h\lambda_j| < 1 \qquad j = 1, \ldots, m - 1$$

Using (6.9.30), this leads to the equivalent statement

$$0 < \frac{4h}{\delta^2}\sin^2\left(\frac{j\pi}{2m}\right) < 2 \qquad 1 \le j \le m - 1$$

This will be satisfied if $4h/\delta^2 \le 2$, or

$$h \le \tfrac{1}{2}\delta^2 \tag{6.9.32}$$

If $\delta$ is at all small, say $\delta = .01$, then the time step $h$ must be quite small to have stability,

In contrast to the restriction (6.9.32) with Euler's method, the backward Euler method has no such restriction since it is A-stable. The method becomes

$$\mathbf{V}_{n+1} = \mathbf{V}_n + h[\Lambda\mathbf{V}_{n+1} + \mathbf{g}(t_{n+1})] \qquad \mathbf{V}_0 = \mathbf{u}_0 \qquad (6.9.33)$$

To solve this linear problem for $\mathbf{V}_{n+1}$,

$$(I - h\Lambda)\mathbf{V}_{n+1} = \mathbf{V}_n + h\mathbf{g}(t_{n+1}) \qquad (6.9.34)$$

This is a tridiagonal system of linear equations (see Section 8.3). It can be solved very rapidly, with approximately $5m$ arithmetic operations per time step, excluding the cost of computing the right side in (6.9.34). The cost of solving the Euler method (6.9.29) is almost as great, and thus the solution of (6.9.34) is not especially time-consuming.

***Example*** Solve the partial differential equation problem (6.9.19)–(6.9.21) with the functions $G$, $d_0$, $d_1$, and $f$ determined from the known solution

$$U = e^{-.1t} \sin(\pi x) \qquad 0 \le x \le 1 \quad t \ge 0 \qquad (6.9.35)$$

Results for Euler's method (6.9.29) are given in Table 6.20, and results for the backward Euler method (6.9.33) are given in Table 6.21.

For Euler's method, we take $m = 4, 8, 16$, and to maintain stability, we take $h = \delta^2/2$, from (6.9.32). Note this leads to the respective time steps of $h \doteq .031$, .0078, .0020. From (6.9.26) and the error formula for Euler's method, we would expect the error to be proportional to $\delta^2$, since $h = \delta^2/2$. This implies that the

**Table 6.20    The method of lines: Euler's method**

| $t$ | Error $m = 4$ | Ratio | Error $m = 8$ | Ratio | Error $m = 16$ |
|-----|-----|-----|-----|-----|-----|
| 1.0 | 3.89E − 2 | 4.09 | 9.52E − 3 | 4.02 | 2.37E − 3 |
| 2.0 | 3.19E − 2 | 4.09 | 7.79E − 3 | 4.02 | 1.94E − 3 |
| 3.0 | 2.61E − 2 | 4.09 | 6.38E − 3 | 4.01 | 1.59E − 3 |
| 4.0 | 2.14E − 2 | 4.10 | 5.22E − 3 | 4.02 | 1.30E − 3 |
| 5.0 | 1.75E − 2 | 4.09 | 4.28E − 3 | 4.04 | 1.06E − 3 |

**Table 6.21    The method of lines: backward Euler's method**

| $t$ | Error $m = 4$ | Error $m = 8$ | Error $m = 16$ |
|-----|-----|-----|-----|
| 1.0 | 4.45E − 2 | 1.10E − 2 | 2.86E − 3 |
| 2.0 | 3.65E − 2 | 9.01E − 3 | 2.34E − 3 |
| 3.0 | 2.99E − 2 | 7.37E − 3 | 1.92E − 3 |
| 4.0 | 2.45E − 2 | 6.04E − 3 | 1.57E − 3 |
| 5.0 | 2.00E − 2 | 4.94E − 3 | 1.29E − 3 |

error should decrease by a factor of 4 when $m$ is doubled, and the results in Table 6.20 agree. In the table, the column Error denotes the maximum error at the node points $(x_j, t)$, $0 \leq j \leq n$, for the given value of $t$.

For the solution of (6.9.28) by the backward Euler method, there need no longer be any connection between the space step $\delta$ and the time step $h$. By observing the error formula (6.9.26) for the method of lines and the truncation error formula (6.9.7) (use $p = 1$) for the backward Euler method, we see that the error in solving the problem (6.9.19)–(6.9.21) will be proportional to $h + \delta^2$. For the unknown function $U$ of (6.9.34), there is a slow variation with $t$. Thus for the truncation error associated with the time integration, we should be able to use a relatively large time step $h$ as compared to the space step $\delta$, in order to have the two sources of error be relatively equal in size. In Table 6.21, we use $h = .1$ and $m = 4, 8, 16$. Note that this time step is much larger than that used in Table 6.20 for Euler's method, and thus the backward Euler method is much more efficient for this particular example.

For more discussion of the method of lines, see Aiken (1985, pp. 124–148). For some method-of-lines codes to solve systems of nonlinear parabolic partial differential equations, in one and two space variables, see Sincovec and Madsen (1975) and Melgaard and Sincovec (1981).

## 6.10  Single-Step and Runge–Kutta Methods

Single-step methods for solving $y' = f(x, y)$ require only a knowledge of the numerical solution $y_n$ in order to compute the next value $y_{n+1}$. This has obvious advantages over the $p$-step multistep methods that use several past values $\{y_n, \ldots, y_{n-p+1}\}$, since then the additional initial values $\{y_1, \ldots, y_{p-1}\}$ have to be computed by some other numerical method.

The best known one-step methods are the Runge–Kutta methods. They are fairly simple to program, and their truncation error can be controlled in a more straightforward manner than for the multistep methods. For the fixed-order multistep methods that were used more commonly in the past, the Runge–Kutta methods were the usual tool for calculating the needed initial values for the multistep method. The major disadvantage of the Runge–Kutta methods is that they use many more evaluations of the derivative $f(x, y)$ to attain the same accuracy, as compared with the multistep methods. Later we will mention some results on comparisons of variable-order Adams codes and fixed-order Runge–Kutta codes.

The most simple one-step method is based on using the Taylor series. Assume $Y(x)$ is $r + 1$ times continuously differentiable, where $Y(x)$ is the solution of the initial value problem

$$y' = f(x, y) \qquad y(x_0) = Y_0 \qquad (6.10.1)$$

Expand $Y(x_1)$ about $x_0$ using Taylor's theorem:

$$Y(x_1) = Y(x_0) + hY'(x_0) + \cdots + \frac{h^r}{r!} Y^{(r)}(x_0) + \frac{h^{r+1}}{(r+1)!} Y^{(r+1)}(\xi) \quad (6.10.2)$$

for some $x_0 \le \xi_0 \le x_1$. By dropping the remainder term, we have an approximation for $Y(x_1)$, provided we can calculate $Y''(x_0), \ldots, Y^{(r)}(x_0)$. Differentiate $Y'(x) = f(x, Y(x))$ to obtain

$$Y''(x) = f_x(x, Y(x)) + f_y(x, Y(x))Y'(x),$$

$$Y'' = f_x + f_y f$$

and proceed similarly to obtain the higher order derivatives of $Y(x)$.

*Example* Consider the problem

$$y' = -y^2 \qquad y(0) = 1$$

with the solution $Y(x) = 1/(1 + x)$. Then $Y'' = -2YY' = 2Y^3$, and (6.10.2) with $r = 1$ yields

$$Y(x_1) = Y_0 - hY_0^2 + h^2Y_0^3 + \frac{h^3}{6}Y^{(3)}(\xi_0) \qquad x_0 \le \xi_0 \le x_1$$

We drop the remainder to obtain an approximation of $Y(x_1)$. This can then be used in the same manner to obtain an approximation for $Y(x_2)$, and so on. The numerical method is

$$y_{n+1} = y_n - hy_n^2 + h^2y_n^3 \qquad n \ge 0 \qquad (6.10.3)$$

Table 6.22 contains the errors in this numerical solution at a selected set of node points. The grid sizes used are $h = .125$ and $h = .0625$, and the ratio of the resulting errors is also given. Note that when $h$ is halved, the ratio is almost 4. This can be justified theoretically since the rate of convergence can be shown to be $O(h^2)$, with a proof similar to that given in Theorem 6.3 or Theorem 6.9, given later in this section.

The Taylor series method can give excellent results. But it is bothersome to use because of the need to analytically differentiate $f(x, y)$. The derivatives can be very difficult to calculate and very time-consuming to evaluate, especially for systems of equations. These differentiations can be carried out using a symbolic

**Table 6.22    Example of the Taylor series method (6.10.3)**

| | | $h = .0625$ | $h = .125$ | |
|---|---|---|---|---|
| $x$ | $y_h(x)$ | $Y(x) - y_h(x)$ | $Y(x) - y_h(x)$ | Ratio |
| 2.0 | .333649 | $-3.2E - 4$ | $-1.4E - 3$ | 4.4 |
| 4.0 | .200135 | $-1.4E - 4$ | $-5.9E - 4$ | 4.3 |
| 6.0 | .142931 | $-7.4E - 5$ | $-3.2E - 4$ | 4.3 |
| 8.0 | .111157 | $-4.6E - 5$ | $-2.0E - 4$ | 4.3 |
| 10.0 | .090941 | $-3.1E - 5$ | $-1.4E - 4$ | 4.3 |

manipulation language on a computer, and then it is easy to produce a Taylor series method. However, the derivatives are still likely to be quite time-consuming to evaluate, and it appears that methods based on evaluating just $f(x, y)$ will remain more efficient. To imitate the Taylor series method, while evaluating only $f(x, y)$, we turn to the Runge–Kutta formulas.

**Runge–Kutta methods**   The Runge–Kutta methods are closely related to the Taylor series expansion of $Y(x)$ in (6.10.2), but no differentiations of $f$ are necessary in the use of the methods. For notational convenience, we abbreviate Runge–Kutta to RK. All RK methods will be written in the form

$$y_{n+1} = y_n + hF(x_n, y_n, h; f) \qquad n \geq 0 \qquad (6.10.4)$$

We begin with examples of the function $F$, and will later discuss hypotheses for it. But at this point, it should be intuitive that we want

$$F(x, Y(x), h; f) \doteq Y'(x) = f(x, Y(x)) \qquad (6.10.5)$$

for all small values of $h$. Define the truncation error for (6.10.4) by

$$T_n(Y) = Y(x_{n+1}) - Y(x_n) - hF(x_n, Y(x_n), h; f) \qquad n \geq 0 \quad (6.10.6)$$

and define $\tau_n(Y)$ implicitly by

$$T_n(Y) = h\tau_n(Y)$$

Rearranging (6.10.6), we obtain

$$Y(x_{n+1}) = Y(x_n) + hF(x_n, Y(x_n), h; f) + h\tau_n(Y) \qquad n \geq 0 \quad (6.10.7)$$

In Theorem 6.9, this will be compared with (6.10.4) to prove convergence of $\{y_n\}$ to $Y$.

**Example 1.**   Consider the trapezoidal method, solved with one iteration using Euler's method as the predictor:

$$y_{n+1} = y_n + \frac{h}{2}[f(x_n, y_n) + f(x_{n+1}, y_n + hf(x_n, y_n))] \qquad n \geq 0 \quad (6.10.8)$$

In the notation of (6.10.4),

$$F(x, y, h; f) = \tfrac{1}{2}[f(x, y) + f(x + h, y + hf(x, y))]$$

As can be seen in Figure 6.9, $F$ is an *average slope* of $Y(x)$ on $[x, x + h]$.

**2.**   The following method is also based on obtaining an average slope for the solution on $[x_n, x_{n+1}]$:

$$y_{n+1} = y_n + hf(x_n + \tfrac{1}{2}h, y_n + \tfrac{1}{2}hf(x_n, y_n)) \qquad n \geq 0 \qquad (6.10.9)$$

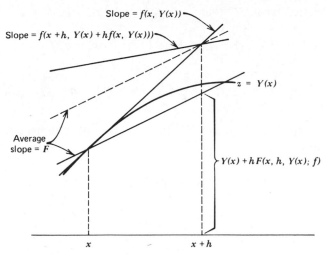

**Figure 6.9**   Illustration of Runge–Kutta method (6.10.8).

For this case,

$$F(x, y, h; f) = f\left(x + \tfrac{1}{2}h, \, y + \tfrac{1}{2}hf(x, y)\right)$$

The derivation of a formula for the truncation error is linked to the derivation of these methods, and this will also be true when considering RK methods of a higher order. The derivation of RK methods will be illustrated by deriving a family of second-order formulas, which will include (6.10.8) and (6.10.9). We suppose $F$ has the general form

$$F(x, y, h; f) = \gamma_1 f(x, y) + \gamma_2 f(x + \alpha h, y + \beta h f(x, y)) \quad (6.10.10)$$

in which the four constants $\gamma_1, \gamma_2, \alpha, \beta$ are to be determined.

We will use Taylor's theorem (Theorem 1.5) for functions of two variables to expand the second term on the right side of (6.10.10), through the second derivative terms. This gives

$$F(x, y, h; f) = \gamma_1 f(x, y) + \gamma_2 \{ f(x, y) + h[\alpha f_x + \beta f f_y]$$

$$+ h^2 \left[ \tfrac{1}{2} \alpha^2 f_{xx} + \alpha \beta f_{xy} f + \tfrac{1}{2} \beta^2 f^2 f_{yy} \right] \} + O(h^3) \quad (6.10.11)$$

Also we will need some derivatives of $Y'(x) = f(x, Y(x))$, namely

$$Y'' = f_x + f_y f$$

$$Y^{(3)} = f_{xx} + 2f_{xy} f + f_{yy} f^2 + f_y f_x + f_y^2 f$$

$$(6.10.12)$$

For the truncation error,

$$T_n(Y) = Y(x_{n+1}) - Y(x_n) - hF(x_n, Y(x_n), h; f)$$

$$= hY_n' + \frac{h^2}{2} Y_n'' + \frac{h^3}{6} Y_n^{(3)} + O(h^4) - hF(x_n, Y_n, h; f)$$

Substituting from (6.10.11) and (6.10.12), and collecting together common powers of $h$, we obtain

$$T_n(Y) = h[1 - \gamma_1 - \gamma_2]f + h^2\left[\left(\tfrac{1}{2} - \gamma_2\alpha\right)f_x + \left(\tfrac{1}{2} - \gamma_2\beta\right)f_yf\right]$$

$$+ h^3\left[\left(\tfrac{1}{6} - \tfrac{1}{2}\gamma_2\alpha^2\right)f_{xx} + \left(\tfrac{1}{3} - \gamma_2\alpha\beta\right)f_{xy}f + \left(\tfrac{1}{6} - \tfrac{1}{2}\gamma_2\beta^2\right)f_{yy}f^2\right.$$

$$\left. + \tfrac{1}{6}f_yf_x + \tfrac{1}{6}f_y^2f\right] + O(h^4) \quad (6.10.13)$$

All derivatives are evaluated at $(x_n, Y_n)$.

We wish to make the truncation error converge to zero as rapidly as possible. The coefficient of $h^3$ cannot be zero in general, if $f$ is allowed to vary arbitrarily. The requirement that the coefficients of $h$ and $h^2$ be zero leads to

$$\gamma_1 + \gamma_2 = 1 \qquad \gamma_2\alpha = \frac{1}{2} \qquad \gamma_2\beta = \frac{1}{2} \qquad (6.10.14)$$

and this yields

$$T_n(Y) = O(h^3)$$

The system (6.10.14) is underdetermined, and its general solution is

$$\gamma_1 = 1 - \gamma_2 \qquad \alpha = \beta = \frac{1}{2\gamma_2} \qquad (6.10.15)$$

with $\gamma_2$ arbitrary. Both (6.10.8) (with $\gamma_2 = \tfrac{1}{2}$) and (6.10.9) (with $\gamma_2 = 1$) are special cases of this solution.

By substituting into (6.10.13), we can obtain the leading term in the truncation error, dependent on only $\gamma_2$. In some cases, the value of $\gamma_2$ has been chosen to make the coefficient of $h^3$ as small as possible, while allowing $f$ to vary arbitrarily. For example, if we write (6.10.13) as

$$T_n(Y) = c(f, \gamma_2)h^3 + O(h^4) \qquad (6.10.16)$$

then the Cauchy–Schwartz inequality [see (7.1.8) in Chapter 7] can be used to show

$$|c(f, \gamma_2)| \le c_1(f)c_2(\gamma_2) \qquad (6.10.17)$$

where

$$c_1(f) = \left[ f_{xx}^2 + f_{xy}^2 f^2 + f_{yy}^2 f^4 + f_y^2 f_x^2 + f_4^4 f^2 \right]^{1/2}$$

$$c_2(\gamma_2) = \left[ \left( \tfrac{1}{6} - \tfrac{1}{2}\gamma_2\alpha^2 \right)^2 + \left( \tfrac{1}{3} - \gamma_2\alpha\beta \right)^2 + \left( \tfrac{1}{6} - \tfrac{1}{2}\gamma_2\beta^2 \right)^2 + \tfrac{1}{18} \right]^{1/2}$$

with $\alpha, \beta$ given by (6.10.15). The minimum value of $c_2(\gamma_2)$ is attained with $\gamma_2 = .75$, and $c_2(.75) = 1/\sqrt{18}$. The resulting second-order numerical method is

$$y_{n+1} = y_n + \frac{h}{4} \left[ f(x_n, y_n) + 3f\left( x_n + \frac{2}{3}h, y_n + \frac{2}{3}hf(x_n, y_n) \right) \right] \qquad n \geq 0$$

(6.10.18)

It is optimal in the sense of minimizing the coefficient $c_2(\gamma_2)$ of the term $c_1(f)h^3$ in the truncation error. For an extensive discussion of this means of analyzing the truncation error in RK methods, see Shampine (1986).

Higher order formulas can be created and analyzed in an analogous manner, although the algebra becomes very complicated. Assume a formula for $F(x, y, h; f)$ of the form

$$F(x, y, h; f) = \sum_{j=1}^{p} \gamma_j V_j \qquad\qquad (6.10.19)$$

$$V_1 = f(x, y)$$

$$V_j = f\left( x + \alpha_j h, y + h \sum_{i=1}^{j-1} \beta_{ji} V_i \right) \qquad j = 2, \ldots, p \quad (6.10.20)$$

These coefficients can be chosen to make the leading terms in the truncation error equal to zero, just as was done with (6.10.10) and (6.10.14). There is obviously a connection between the number of evaluations of $f(x, y)$, call it $p$, and the maximum possible order that can be attained for the truncation error. These are given in Table 6.23, which is due in part to Butcher (1965).

Until about 1970, the most popular RK method has probably been the original classical formula that is a generalization of Simpson's rule. The method is

$$y_{n+1} = y_n + \frac{h}{6} [V_1 + 2V_2 + 2V_3 + V_4] \qquad\qquad (6.10.21)$$

$$V_1 = f(x_n, y_n) \qquad V_2 = f\left( x_n + \frac{1}{2}h, y_n + \frac{1}{2}hV_1 \right)$$

$$V_3 = f\left( x_n + \frac{1}{2}h, y_n + \frac{1}{2}hV_2 \right) \qquad V_4 = f(x_n + h, y_n + hV_3)$$

**Table 6.23    Maximum order of the Runge–Kutta methods**

| Number of function evaluations | 1 | 2 | 3 | 4 | 5 | 6 | 7 | 8 |
|---|---|---|---|---|---|---|---|---|
| Maximum order of method | 1 | 2 | 3 | 4 | 4 | 5 | 6 | 6 |

It can be proved that this is a fourth-order formula with $T_n(Y) = O(h^5)$. If $f(x, y)$ does not depend on $y$, then this formula reduces to Simpson's integration rule.

*Example*    Consider the problem

$$y' = \frac{1}{1 + x^2} - 2y^2 \qquad y(0) = 0 \qquad (6.10.22)$$

with the solution $Y = x/(1 + x^2)$. The method (6.10.21) was used with a fixed stepsize, and the results are shown in Table 6.24. The stepsizes are $h = .25$ and $2h = .5$. The column Ratio gives the ratio of the errors for corresponding node points as $h$ is halved. The last column is an example of formula (6.10.24) from the following material on Richardson extrapolation. Because $T_n(Y) = O(h^5)$ for method (6.10.21), Theorem 6.9 implies that the rate of convergence of $y_h(x)$ to $Y(x)$ is $O(h^4)$. The theoretical value of Ratio is 16, and as $h$ decreases further, this value will be realized more closely.

The RK methods have asymptotic error formulas, the same as the multistep methods. For (6.10.21),

$$Y(x) - y_h(x) = D(x)h^4 + O(h^5) \qquad (6.10.23)$$

where $D(x)$ satisfies a certain initial value problem. The proof of this result is an extension of the proof of Theorem 6.9, and it is similar to the derivation of Theorem 6.4 for Euler's method, in Section 6.2. The result (6.10.23) can be used to produce an error estimate, just as was done with the trapezoidal method in formula (6.5.26) of Section 6.5. For a stepsize of $2h$,

$$Y(x) - y_{2h}(x) = 16D(x)h^4 + O(h^5)$$

**Table 6.24    Example of Runge–Kutta method (6.10.21)**

| $x$ | $y_h(x)$ | $Y(x) - y_h(x)$ | $Y(x) - y_{2h}(x)$ | Ratio | $\frac{1}{15}[y_h(x) - y_{2h}(x)]$ |
|---|---|---|---|---|---|
| 2.0 | .39995699 | 4.3E − 5 | 1.0E − 3 | 24 | 6.7E − 5 |
| 4.0 | .23529159 | 2.5E − 6 | 7.0E − 5 | 28 | 4.5E − 6 |
| 6.0 | .16216179 | 3.7E − 7 | 1.2E − 5 | 32 | 7.7E − 7 |
| 8.0 | .12307683 | 9.2E − 8 | 3.4E − 6 | 36 | 2.2E − 7 |
| 10.0 | .09900987 | 3.1E − 8 | 1.3E − 6 | 41 | 8.2E − 8 |

Proceeding as earlier for (6.5.27), we obtain

$$Y(x) - y_h(x) = \frac{1}{15}[y_h(x) - y_{2h}(x)] + O(h^5) \qquad (6.10.24)$$

and the first term on the right side is an estimate of the left-hand error. This is illustrated in the last column of Table 6.24.

**Convergence analysis**    In order to obtain convergence of the general schema (6.10.4), we need to have $\tau_n(Y) \to 0$ as $h \to 0$. Since

$$\tau_n(Y) = \frac{Y(x_{n+1}) - Y(x_n)}{h} - F(x_n, Y(x_n), h; f)$$

we require that

$$F(x, Y(x), h; f) \to Y'(x) = f(x, Y(x)) \qquad \text{as} \quad h \to 0$$

More precisely, define

$$\delta(h) = \max_{\substack{x_0 \le x \le b \\ -\infty < y < \infty}} |f(x, y) - F(x, y, h; f)| \qquad (6.10.25)$$

and assume

$$\delta(h) \to 0 \qquad \text{as} \quad h \to 0 \qquad (6.10.26)$$

This is occasionally called the *consistency condition* for the RK method (6.10.4). We will also need a Lipschitz condition on $F$:

$$|F(x, y, h; f) - F(x, z, h; f)| \le L|y - z| \qquad (6.10.27)$$

for all $x_0 \le x \le b$, $-\infty < y$, $z < \infty$, and all small $h > 0$. This condition is usually proved by using the Lipschitz condition (6.2.12) on $f(x, y)$. For example, with method (6.10.9),

$$|F(x, y, h; f) - F(x, z, h; f)|$$

$$= \left| f\left(x + \frac{h}{2}, y + \frac{h}{2}f(x, y)\right) - f\left(x + \frac{h}{2}, z + \frac{h}{2}f(x, z)\right) \right|$$

$$\le K\left| y - z + \frac{h}{2}[f(x, y) - f(x, z)] \right|$$

$$\le K\left(1 + \frac{h}{2}K\right)|y - z|$$

Choose $L = K(1 + \frac{1}{2}K)$ for $h \le 1$.

***Theorem 6.9***    Assume that the Runge–Kutta method (6.10.4) satisfies the Lipschitz condition (6.10.27). Then for the initial value problem (6.10.1), the solution $\{y_n\}$ satisfies

$$\underset{x_0 \le x_n \le b}{\text{Max}} |Y(x_n) - y_n| \le e^{(b-x_0)L}|Y_0 - y_0| + \left[\frac{e^{(b-x_0)L} - 1}{L}\right]\tau(h)$$

(6.10.28)

where

$$\tau(h) \equiv \underset{x_0 \le x_n \le b}{\text{Max}} |\tau_n(Y)|$$

(6.10.29)

If the consistency condition (6.10.26) is also satisfied, then the numerical solution $\{y_n\}$ converges to $Y(x)$.

***Proof***    Subtract (6.10.4) from (6.10.7) to obtain

$$e_{n+1} = e_n + h[F(x_n, Y_n, h; f) - F(x_n, y_n, h; f)] + h\tau_{n+1}(Y) \quad (6.10.30)$$

in which $e_n = Y(x_n) - y_n$. Apply the Lipschitz condition (6.10.27) and use (6.10.29) to obtain

$$|e_{n+1}| \le (1 + hL)|e_n| + h\tau(h) \qquad x_0 \le x_N \le b$$

As with the convergence proof for the Euler method, this leads easily to the result (6.10.28) [see (6.2.20)–(6.2.22)].

In most cases, it is known by direct computation that $\tau(h) \to 0$ as $h \to 0$, and in that case, convergence of $\{y_n\}$ to $Y(x)$ is immediately proved. But all that we need to know is that (6.10.26) is satisfied. To see this, write

$$h\tau_n(Y) = Y(x_{n+1}) - Y(x_n) - hF(x_n, Y(x_n), h; f)$$

$$= hY'(x_n) + \frac{h^2}{2}Y''(\xi_n) - hF(x_n, Y(x_n), h; f)$$

$$h|\tau_n(Y)| \le h\delta(h) + \frac{h^2}{2}\|Y''\|_\infty$$

$$\tau(h) \le \delta(h) + \frac{1}{2}h\|Y''\|_\infty$$

Thus $\tau(h) \to 0$ as $h \to 0$, completing the proof.    ∎

The following result is an immediate consequence of (6.10.28).

***Corollary***    If the Runge–Kutta method (6.10.4) has a truncation error $T_n(Y) = O(h^{m+1})$, then the rate of convergence of $\{y_n\}$ to $Y(x)$ is $O(h^m)$. ∎

It is not too difficult to derive an asymptotic error formula for the Runge–Kutta method (6.10.4), provided one is known for the truncation error. Assume

$$T_n(Y) = \varphi(x_n)h^{m+1} + O(h^{m+2}) \tag{6.10.31}$$

with $\varphi(x)$ determined by $Y(x)$ and $f(x, Y(x))$. As an example see the result (6.10.13) to obtain this expansion for second-order RK methods. Strengthened forms of (6.10.26) and (6.10.27) are also necessary. Assume

$$F(x, y, h; f) - F(x, z, h; f) = \frac{\partial F(x, y, h; f)}{\partial y}(y - z) + O\big((y - z)^2\big) \tag{6.10.32}$$

and also that

$$\delta_1(h) \equiv \underset{\substack{x_0 \le x \le b \\ -\infty < y < \infty}}{\text{Max}} \left| \frac{\partial f(x, y)}{\partial y} - \frac{\partial F(x, y, h; f)}{\partial y} \right| \to 0 \quad \text{as} \quad h \to 0 \tag{6.10.33}$$

In practice, both of these results are straightforward to confirm. With these assumptions, we can derive the formula

$$Y(x) - y_h(x) = D(x)h^m + O(h^{m+1}) \tag{6.10.34}$$

with $D(x)$ satisfying the linear initial value problem

$$D' = f_y(x, Y(x))D(x) + \varphi(x) \qquad D(x_0) = 0 \tag{6.10.35}$$

Stability results can be given for RK methods, in analogy with those for multistep methods. The basic type of stability, defined near the beginning of Section 6.8, is easily proved. The proof is a simple modification of that of (6.2.29), the stability of Euler's method, and we leave it to Problem 49. An essential difference with the multistep theory is that there are no parasitic solutions created by the RK methods; thus the concept of relative stability does not apply to RK methods. The regions of absolute stability can be studied as with the multistep theory, but we omit it here, leaving it to the problems.

**Estimation of the truncation error**    In order to control the size of the truncation error, we must first be able to estimate it. Let $u_n(x)$ denote the solution of $y' = f(x, y)$ passing through $(x_n, y_n)$ [see (6.5.9)]. We wish to estimate the error in $y_h(x_n + 2h)$ relative to $u_n(x_n + 2h)$.

Using (6.10.30) for the error in $y_h(x)$ compared to $u_n(x)$, and using the asymptotic formula (6.10.31),

$$u_n(x_{j+1}) - y_h(x_{j+1}) = u_n(x_j) - y_h(x_j)$$

$$+ h\{F(x_j, u_n(x_j), h; f) - F(x_j, y_h(x_j), h; f)\}$$

$$+ \varphi_n(x_j)h^{m+1} + O(h^{m+2}) \qquad j \geq n$$

From this, it is straightforward to prove

$$u_n(x_n + h) - y_h(x_n + h) = \varphi_n(x_n)h^{m+1} + O(h^{m+2})$$

$$u_n(x_n + 2h) - y_h(x_n + 2h) = 2\varphi_n(x_n)h^{m+1} + O(h^{m+2})$$

Applying the same procedure to $y_{2h}(x_j)$,

$$u_n(x_n + 2h) - y_{2h}(x_n + 2h) = 2^{m+1}\varphi_n(x_n)h^{m+1} + O(h^{m+2})$$

From the last two equations,

$$u_n(x_n + 2h) - y_h(x_n + 2h)$$

$$= \frac{1}{2^m - 1}[y_h(x_n + 2h) - y_{2h}(x_n + 2h)] + O(h^{m+2}) \quad (6.10.36)$$

and the first term on the right is an asymptotic estimate of the error on the left side.

Consider the computation of $y_h(x_n + 2h)$ from $y_h(x_n) \equiv y_n$ as a single step in an algorithm. Suppose that a user has given an error tolerance $\epsilon$ and that the value of

$$\text{trunc} \equiv u_n(x_n + 2h) - y_h(x_n + 2h) \doteq \frac{1}{2^m - 1}[y_h(x_n + 2h) - y_{2h}(x_n + 2h)]$$

$$(6.10.37)$$

is to satisfy

$$.5\epsilon h \leq |\text{trunc}| \leq 2\epsilon h \qquad (6.10.38)$$

This controls the error per unit step and it requires that the error be neither too large or too small. Recall the concept of error per unit stepsize in (6.6.1) of Section 6.6.

If the test is satisfied, then computation continues with the same $h$. But if it is not satisfied, then a new value $\hat{h}$ must be chosen. Let it be chosen by

$$2|\varphi_n(x_n)|\hat{h}^{m+1} \doteq \epsilon h$$

where $\varphi_n(x_n)$ is determined from

$$\varphi_n(x_n) \doteq \frac{\text{trunc}}{2h^{m+1}}$$

With the new value of $\hat{h}$, the new truncation error should lie near the midpoint of (6.10.38). This form of algorithm has been implemented with a number of methods. For example, see Gear (1971, pp. 83–84) for a similar algorithm for a fourth-order RK method.

With many programs, the error estimation (6.10.37) is added to the current calculated value of $y_h(x_n + 2h)$, giving a more accurate result. This is called *local extrapolation*. When it is used, the *error per unit step* criterion of (6.10.38) is replaced by an *error per step* criterion:

$$.5\epsilon \leq |\text{trunc}| \leq 2\epsilon \qquad (6.10.39)$$

In such cases, it can be shown that the local error in the extrapolated value of $y_h(x_n + 2h)$ satisfies a modified error per unit step criteria [see Shampine and Gordon (1975), p. 100]. For implementations of the same method, programs that use local extrapolation and the error per step criterion appear to be more efficient than those using (6.10.38) and not using local extrapolation.

To better understand the expense of RK methods with the error estimation previously given, consider only fourth-order RK methods with four evaluations of $f(x, y)$ per RK step. In going from $x_n$ to $x_n + 2h$, eight evaluations will be required to obtain $y_h(x_n + 2h)$, and three additional evaluations to obtain $y_{2h}(x_n + 2h)$. Thus a single step of the variable-step algorithms will require eleven evaluations of $f$. Although fairly expensive to use when compared with a multistep method, a variable-stepsize RK method is very stable, reliable, and is comparatively easy to program for a computer.

**Runge–Kutta–Fehlberg methods**    The Runge–Kutta–Fehlberg methods are RK methods in which the truncation error is computed by comparing the computed answer $y_{n+1}$ with the result of an associated higher order RK formula. The most popular of such methods are due to E. Fehlberg, [e.g., see Fehlberg (1970)]; these are currently the most popular RK methods. To clarify the presentation, we consider only the most popular pair of Runge–Kutta–Fehlberg (RKF) formulas of order 4 and 5. These formulas are computed simultaneously, and their difference is taken as an estimate of the truncation error in the fourth-order method.

Note from Table 6.23 that a fifth-order RK method requires six evaluations of $f$ per step. Consequently, Fehlberg chose to use five evaluations of $f$ for the fourth-order formula, rather than the usual four. This extra degree of freedom in choosing the fourth-order formula allowed it to be chosen with a smaller truncation error, and this is illustrated later.

As before, define

$$V_1 = f(x_n, y_n)$$

$$V_j = f\left(x_n + \alpha_j h, \, y_n + h \sum_{i=1}^{j-1} \beta_{ji} V_i\right) \qquad j = 2, \ldots, 6 \qquad (6.10.40)$$

**Table 6.25** Coefficients $\alpha_j$ and $\beta_{ji}$ for the RKF method

| | | | | $\beta_{ji}$ | | |
|---|---|---|---|---|---|---|
| $j$ | $\alpha_j$ | $i = 1$ | 2 | 3 | 4 | 5 |
| 2 | $\dfrac{1}{4}$ | $\dfrac{1}{4}$ | | | | |
| 3 | $\dfrac{3}{8}$ | $\dfrac{3}{32}$ | $\dfrac{9}{32}$ | | | |
| 4 | $\dfrac{12}{13}$ | $\dfrac{1932}{2197}$ | $-\dfrac{7200}{2197}$ | $\dfrac{7296}{2197}$ | | |
| 5 | $1$ | $\dfrac{439}{216}$ | $-8$ | $\dfrac{3680}{513}$ | $-\dfrac{845}{4104}$ | |
| 6 | $\dfrac{1}{2}$ | $-\dfrac{8}{27}$ | $2$ | $-\dfrac{3544}{2565}$ | $\dfrac{1859}{4104}$ | $-\dfrac{11}{40}$ |

**Table 6.26** Coefficients $\gamma_j, \hat{\gamma}_j, c_j$ for the RKF method

| $j$ | 1 | 2 | 3 | 4 | 5 | 6 |
|---|---|---|---|---|---|---|
| $\gamma_j$ | $\dfrac{25}{216}$ | $0$ | $\dfrac{1408}{2565}$ | $\dfrac{2197}{4104}$ | $-\dfrac{1}{5}$ | |
| $\hat{\gamma}_j$ | $\dfrac{16}{135}$ | $0$ | $\dfrac{6656}{12825}$ | $\dfrac{28561}{56430}$ | $-\dfrac{9}{50}$ | $\dfrac{2}{55}$ |
| $c_j$ | $\dfrac{1}{360}$ | $0$ | $-\dfrac{128}{4275}$ | $-\dfrac{2197}{75240}$ | $\dfrac{1}{50}$ | $\dfrac{2}{55}$ |

The fourth- and fifth-order formulas are, respectively,

$$y_{n+1} = y_n + h \sum_{i=1}^{5} \gamma_j V_j \qquad (6.10.41)$$

$$\hat{y}_{n+1} = y_n + h \sum_{i=1}^{6} \hat{\gamma}_j V_j \qquad (6.10.42)$$

The truncation error in $y_{n+1}$ is approximately

$$\text{trunc} = \hat{y}_{n+1} - y_{n+1} = h \sum_{j=1}^{6} c_j V_j \qquad (6.10.43)$$

The coefficients are given in Tables 6.25 and 6.26.

To compare the classical RK method (6.10.21) and the preceding RKF method, consider the truncation errors in solving the simple problem

$$y' = x^4 \qquad y(0) = 0$$

**Table 6.27   Example of the RKF method**

| $x$ | $y_h(x)$ | $Y(x) - y_h(x)$ | $Y(x) - y_{2h}(x)$ | Ratio | $\frac{1}{15}[y_h(x) - y_{2h}(x)]$ |
|-----|----------|-----------------|--------------------|-------|-----------------------------------|
| 2.0 | .40000881 | $-8.8E - 6$ | $-5.0E - 4$ | 57 | $-3.3E - 5$ |
| 4.0 | .23529469 | $-5.8E - 7$ | $-4.0E - 5$ | 69 | $-2.6E - 6$ |
| 6.0 | .16216226 | $-9.5E - 8$ | $-7.9E - 6$ | 83 | $-5.2E - 7$ |
| 8.0 | .12307695 | $-2.6E - 8$ | $-2.5E - 6$ | 95 | $-1.6E - 7$ |
| 10.0 | .09900991 | $-9.4E - 9$ | $-1.0E - 6$ | 106 | $-6.6E - 8$ |

The truncation errors for (6.10.41) and (6.10.21) are, respectively,

$$\text{RKF}\,(6.10.41): T_{n+1}(Y) \doteq .00048h^5$$

$$\text{RK}\,(6.10.21): T_{n+1}(Y) \doteq -.0083h^5 \qquad n \geq 0$$

This suggests that the RKF method should generally have a smaller truncation error, although in practice, the difference is generally not this great. Note that the classical method (6.10.21) with a stepsize $h$ and using the error estimate (6.10.37) will require eleven evaluations of $f$ to go from $y_h(x_n)$ to $y_h(x_n + 2h)$. And the RFK method (6.10.41) will require twelve evaluations to go from $y_h(x_n)$ to $y_h(x_n + 2h)$. Consequently, the computational effort in going from $x_n$ to $x_n + 2h$ is comparable, and it is fair to compare their errors by using the same value of $h$.

*Example*   Use the RKF method (6.10.41) to solve the problem (6.10.22). It was previously an example for the classical RK method (6.10.21). As before, $h = .25$; and the results are given in Table 6.27. The theoretical value for Ratio is again 16, and clearly it has not yet settled down to that value. As $h$ decreases, it approaches 16 more closely. The use of the Richardson extrapolation formula (6.10.24) is given in the last column, and it clearly overestimates the error. Nonetheless, this is still a useful error estimate in that it gives some idea of the magnitude of the global error.

The method (6.10.41) and (6.10.42) is generally used with local extrapolation, as is illustrated later. The method has been much studied, to see whether improvements were possible. Recently, Shampine (1986) has given an analysis that suggests some improved RKF formulas, based on several criteria for comparing Runge–Kutta formulas. To date, these have not been made a part of a high-quality production computer code, although it is expected they will be.

**Automatic Runge–Kutta–Fehlberg programs**   A variable-stepsize RKF program can be written by using (6.10.43) to estimate and control the truncation error in the fourth-order formula (6.10.41). Such a method has been written by L. Shampine and H. Watts, and it is described in Shampine and Watts (1976a). Its general features are as follows. The program is named RKF45, and a user of the program must specify two error parameters ABSERR and RELERR. The truncation error in (6.10.43) for $y_{n+1}$ is forced to satisfy the *error per step* criterion

$$|\text{trunc}_j| \leq \text{ABSERR} + \text{RELERR} * |y_{n,j}| \tag{6.10.44}$$

for each component $y_{n,j}$ of the computed solution $y_n$ of the system of differential equations being solved. But then the final result of the computation, to be used in further calculations, is taken to be the fifth-order formula $\hat{y}_{n+1}$ from (6.10.42) rather than the fourth-order formula $y_{n+1}$. The terms $y_n$ on the right side in (6.10.41) and (6.10.42) should therefore be replaced by $\hat{y}_n$. Thus RKF45 is really a fifth-order method, which chooses the stepsize $h$ by controlling the truncation error in the fourth-order formula (6.10.41). As before, this is called local extrapolation, and it can be shown that the bound (6.10.44) for $y_{n+1}$ will imply that $\hat{y}_{n+1}$ satisfies a modified error per unit stepsize bound on its truncation error. The argument is similar to that given in Shampine and Gordon (1975, p. 100) for variable-order Adams methods.

The tests of Shampine et al. (1976) show that RKF45 is a superior RK program, one that is an excellent candidate for inclusion in a library of programs for solving ordinary differential equations. It has become widely used, and is given in several texts [e.g., Forsythe et al. (1977), p. 129].

In general, the comparisons given in Enright and Hull (1976) have shown RKF methods to be superior to other RK methods. Comparisons with multistep methods are more difficult. Multistep methods require fewer evaluations of the derivative $f(x, y)$ than RK methods, but the overhead costs per step are much greater with multistep than with RK methods. A judgment as to which kind of method to use depends on how costly it is to evaluate $f(x, y)$ as compared with the overhead costs in the multistep methods. There are other considerations, for example, the size of the system of differential equations being solved. A general discussion of these factors and their influence is given in Enright and Hall (1976) and Shampine et al. (1976).

**Example** Solve the problem

$$y' = \frac{y}{4}\left(1 - \frac{y}{20}\right) \qquad y(0) = 1$$

which has the solution

$$Y(x) = \frac{20}{(1 + 19e^{-x/4})}.$$

The problem was solved using RKF45, and values were output at $x = $

**Table 6.28    Example of Runge–Kutta–Fehlberg program RKF45**

| $x$ | ABSERR $= 10^{-3}$ | | ABSERR $= 10^{-6}$ | | ABSERR $= 10^{-9}$ | |
|---|---|---|---|---|---|---|
| | Error | NFE | Error | NFE | Error | NFE |
| 4.0 | $-1.4E - 5$ | 19 | $-2.2E - 7$ | 43 | $-5.5E - 10$ | 121 |
| 8.0 | $-3.4E - 5$ | 31 | $-5.6E - 7$ | 79 | $-1.3E - 9$ | 229 |
| 12.0 | $-3.7E - 5$ | 43 | $-5.5E - 7$ | 103 | $-8.6E - 10$ | 312 |
| 16.0 | $-3.3E - 5$ | 55 | $-5.4E - 7$ | 127 | $-1.2E - 9$ | 395 |
| 20.0 | $-1.8E - 6$ | 67 | $-1.6E - 7$ | 163 | $-4.3E - 10$ | 503 |

$2, 4, 6, \ldots, 20$. Three values of ABSERR were used, and RELERR $= 10^{-12}$ in all cases. The resulting global errors are given in Table 6.28. The column headed by NFE gives the number of evaluations of $f(x, y)$ needed to obtain the given answer. Compare the results with those given in Table 6.15, obtained using the variable-order multistep program DDEABM.

A version of RKF45 is available that estimates the global error in the computed solution, using Richardson error estimation as in (6.10.24). For a discussion of this code, called GERK, see Shampine and Watts (1976b), which also contains a discussion of the general problem of global error estimation.

**Implicit Runge–Kutta methods**   We have considered only the explicit RK methods, since these are the main methods that are in current use. But there are implicit RK formulas. Generalizing the explicit formula (6.10.19) and (6.10.20), we consider

$$y_{n+1} = y_n + hF(x_n, y_n, h; f) \tag{6.10.45}$$

$$F(x, y, h; f) = \sum_{j=1}^{p} \gamma_1 V_j$$

$$V_j = f\left(x + \alpha_j h, y + h\sum_{i=1}^{p} \beta_{ji}V_i\right) \qquad j = 1, \ldots, p \tag{6.10.46}$$

The coefficients $\{\alpha_j, \beta_{ji}, \gamma_j\}$ specify the method. For an explicit method, we have $\beta_{ji} = 0$ for $i \geq j$.

The implicit methods have been studied extensively in recent years, since some of them possess stability properties favorable for solving stiff differential equations. For any order of convergence, there are A-stable methods of that order. In this way, the implicit RK methods are superior to the multistep methods, for which there are no A-stable methods of order greater than 2. The most widely used codes for solving stiff differential equations, at present, are based on the backward differentiation formulas introduced in Section 6.9. But there is much work on developing similar codes based on implicit RK methods. For an introductory survey of this topic, see Aiken (1985, Sec. 3.1).

## 6.11   Boundary Value Problems

Up to this point, we have considered numerical methods for solving only *initial value problems* for differential equations. For such problems, conditions on the solution of the differential equation are specified at a single point, called the initial point. We now consider problems in which the unknown solution has conditions imposed on it at more than one point. Such problems for differential equations are called *boundary value problems*, or BVPs for short.

A typical problem that we study is the two-point BVP:

$$y'' = f(x, y, y') \qquad a < x < b$$

$$A \begin{bmatrix} u(a) \\ u'(a) \end{bmatrix} + B \begin{bmatrix} u(b) \\ u'(b) \end{bmatrix} = \begin{bmatrix} \gamma_1 \\ \gamma_2 \end{bmatrix} \qquad (6.11.1)$$

The terms $A$ and $B$ denote given square matrices of order $2 \times 2$, and $\gamma_1$ and $\gamma_2$ are given constants. The theory for BVPs such as this one is much more complicated than for the initial value problem (see Theorems 6.1 and 6.2 in Section 6.1). To illustrate the possible difficulties, we give the following examples.

**Example 1.**    Consider the two-point linear BVP

$$y'' = -\lambda y \qquad 0 < x < 1$$

$$y(0) = y(1) = 0 \qquad (6.11.2)$$

If $\lambda$ is not one of the numbers

$$\pi^2, 4\pi^2, 9\pi^2, \ldots, n^2\pi^2, \ldots \qquad (6.11.3)$$

then this BVP has the unique solution $Y(x) \equiv 0$. Otherwise, there are an infinite number of solutions,

$$Y(x) = C \sin(\sqrt{\lambda}\, x) \qquad (6.11.4)$$

with $\lambda$ chosen from (6.11.3) and $C$ an arbitrary constant.

**2.**    Consider the related problem

$$y'' = -\lambda y + g(x) \qquad 0 < x < 1$$

$$y(0) = y(1) = 0 \qquad (6.11.5)$$

If $\lambda$ is not chosen from (6.11.3), and if $g(x)$ is continuous for $a \le x \le b$, then this problem has a unique solution $Y(x)$ that is twice continuously differentiable on $[0, 1]$. In contrast, if $\lambda = \pi^2$, then the problem (6.11.5) has a solution if and only if $g(x)$ satisfies

$$\int_0^1 g(x) \sin(\pi x)\, dx = 0$$

In the case that this is satisfied, the solution is given by

$$Y(x) = C \sin(\pi x) + \frac{1}{\pi} \int_0^x g(t) \sin(\pi(x - t))\, dt \qquad (6.11.6)$$

with $C$ an arbitrary constant. A similar result holds for other $\lambda$ chosen from (6.11.3).

A few results from the theory of two-point BVPs is now given, to help in further understanding them and to aid in the presentation of numerical methods for their solution. We begin with the two-point problem for the second-order linear equation:

$$y'' = p(x)y' + q(x)y + g(x) \qquad a < x < b$$

$$A \begin{bmatrix} u(a) \\ u'(a) \end{bmatrix} + B \begin{bmatrix} u(b) \\ u'(b) \end{bmatrix} = \begin{bmatrix} \gamma_1 \\ \gamma_2 \end{bmatrix} \qquad (6.11.7)$$

The *homogeneous problem* is the case in which $g(x) \equiv 0$ and $\gamma_1, \gamma_2 = 0$.

**Theorem 6.10**   The nonhomogeneous problem (6.11.7) has a unique solution $Y(x)$ on $[a, b]$, for each set of given data $\{g(x), \gamma_1, \gamma_2\}$, if and only if the homogeneous problem has only the trivial solution $Y(x) \equiv 0$.

**Proof**   See Stakgold (1979, p. 197). The preceding examples are illustrations of the theorem. ∎

For conditions under which the homogeneous problem for (6.11.7) has only the zero solution, we consider the following more special linear problem:

$$y'' = p(x)y' + q(x)y + g(x) \qquad a < x < b$$

$$a_0 y(a) - a_1 y'(a) = \gamma_1 \qquad b_0 y(b) + b_1 y'(b) = \gamma_2 \qquad (6.11.8)$$

In this problem, we say the boundary conditions are *separated*. Assume the following conditions are satisfied:

$$q(x) > 0 \qquad a \le x \le b$$

$$a_0 a_1 \ge 0 \qquad b_0 b_1 \ge 0 \qquad (6.11.9)$$

$$|a_0| + |a_1| \ne 0 \qquad |b_0| + |b_1| \ne 0 \qquad |a_0| + |b_0| \ne 0$$

Then the homogeneous problem for (6.11.8) has only the zero solution, Theorem 6.10 is applicable, and the nonhomogeneous problem has a unique solution, for each set of data $\{g(x), \gamma_1, \gamma_2\}$. For a proof of this result, see Keller (1968, p. 11). Example (6.11.5) illustrates this result. It also shows that the conditions (6.11.9) are not necessary; the problem (6.11.5) is uniquely solvable for most negative choices of $q(x) = \lambda$.

The theory for the nonlinear problem (6.11.1) is far more complicated than that for the linear problem (6.11.7). We give an introduction to that theory for the following more limited problem:

$$y'' = f(x, y, y') \qquad a < x < b$$

$$a_0 y(a) - a_1 y'(a) = \gamma_1 \qquad b_0 y(b) + b_1 y'(b) = \gamma_2 \qquad (6.11.10)$$

The function $f$ is assumed to satisfy the following Lipschitz condition:

$$|f(x, u_1, v) - f(x, u_2, v)| \leq K|u_1 - u_2|$$

$$|f(x, u, v_1) - f(x, u, v_2)| \leq K|v_1 - v_2|$$

(6.11.11)

for all points $(x, u_i, v)$, $(x, u, v_i)$ in the region

$$R = \{(x, u, v) | a \leq x \leq b, \quad -\infty < u, v < \infty\}$$

This is far stronger than needed, but it simplifies the statement of the following theorem and the analysis of the numerical methods given later.

**Theorem 6.11** For the problem (6.11.10), assume $f(x, u, v)$ is continuous on the region $R$ and that it satisfies the Lipschitz condition (6.11.11). In addition, assume that on $R$, $f$ satisfies

$$\frac{\partial f(x, u, v)}{\partial u} > 0 \qquad \left| \frac{\partial f(x, u, v)}{\partial v} \right| \leq M \quad (6.11.12)$$

for some constant $M > 0$. For the boundary conditions of (6.11.10), assume

$$a_0 a_1 \geq 0 \qquad b_0 b_1 \geq 0$$

$$|a_0| + |a_1| \neq 0 \qquad |b_0| + |b_1| \neq 0 \qquad |a_0| + |b_0| \neq 0 \quad (6.11.13)$$

Then the BVP (6.11.10) has a unique solution.

**Proof** See Keller (1968, p. 9). The earlier uniqueness result for the linear problem (6.11.8) is a special case of this theorem. ∎

Nonlinear BVPs may be nonuniquely solvable, with only a finite number of solutions. This is in contrast to the situation for linear problems, in which nonuniqueness always means an infinity of solutions, as illustrated in (6.11.2)–(6.11.6). An example of such nonunique solvability is the second-order problem

$$\frac{d}{dx}\left[ I(x) \frac{dy}{dx} \right] + \lambda \sin(y) = 0 \qquad 0 < x < 1$$

$$y'(0) = y'(1) = 0, \qquad |y(x)| < \pi \quad (6.11.14)$$

which arises in studying the buckling of a column. The parameter $\lambda$ is proportional to the load on the column; when $\lambda$ exceeds a certain size, there is a solution to the problem (6.11.14) other than the zero solution. For further detail on this problem, see Keller and Antman (1969, p. 43).

As with the earlier material on initial value problems [see (6.1.11)–(6.1.15) in Section 6.1], all boundary value problems for higher order equations can be reformulated as problems for systems of first-order equations. The general form of a two-point BVP for a system of first-order equations is

$$\mathbf{y}' = \mathbf{f}(x, \mathbf{y}) \qquad a < x < b$$

$$A\mathbf{y}(a) + B\mathbf{y}(b) = \gamma \tag{6.11.15}$$

This represents a system of $n$ first-order equations. The quantities $\mathbf{y}(x)$, $\mathbf{f}(x, \mathbf{y})$, and $\gamma$ are vectors with $n$ components, and $A$ and $B$ are matrices of order $n \times n$. There is a theory for such BVPs, analogous to that for the two-point problem (6.11.1), but for reasons of space, we omit it here.

In the remainder of this section, we describe briefly the principal numerical methods for solving the two-point BVP (6.11.1). These methods generalize to first-order systems such as (6.11.15), but for reasons of space, we omit those results. Much of our presentation follows Keller (1968), and a theory for first-order systems is given there. Unlike the situation with initial value problems, it is often advantageous to directly treat higher order BVPs rather than to numerically solve their reformulation as a first-order system. The numerical methods for the two-point problem are also less complicated to present, and therefore we have opted to discuss the two-point problem rather than the system (6.11.15).

**Shooting methods**   One of the very popular approaches to solving a two-point BVP is to reduce it to a problem in which a program for solving initial value problems can be used. We now develop such a method for the BVP (6.11.10).

Consider the initial value problem

$$y'' = f(x, y, y') \qquad a < x < b$$

$$y(a) = a_1 s - c_1 \gamma_1 \qquad y'(a) = a_0 s - c_0 \gamma_1 \tag{6.11.16}$$

depending on the parameter $s$, where $c_0, c_1$ are arbitrary constants satisfying

$$a_1 c_0 - a_0 c_1 = 1$$

Denote the solution of (6.11.16) by $Y(x; s)$. Then it is straightforward to see that

$$a_0 Y(a; s) - a_1 Y'(a; s) = \gamma_1$$

for all $s$ for which $Y$ exists.

Since $Y$ is a solution of (6.11.1), all that is needed for it to be a solution of (6.11.10) is to have it satisfy the remaining boundary condition at $b$. This means that $Y(x; s)$ must satisfy

$$\varphi(s) \equiv b_0 Y(b; s) + b_1 Y'(b; s) - \gamma_2 = 0 \tag{6.11.17}$$

This is a nonlinear equation for $s$. If $s^*$ is a root of $\varphi(s)$, then $Y(x; s)$ will satisfy the BVP (6.11.10). It can be shown that under suitable assumptions on $f$ and its boundary conditions, the equation (6.11.17) will have a unique solution $s^*$ [see Keller (1968), p. 9]. We can use a rootfinding method for nonlinear equations to solve for $s^*$. This way of finding a solution to a BVP is called a *shooting method*. The name comes from ballistics, in which one attempts to determine the needed initial conditions at $x = a$ in order to obtain a certain value at $x = b$.

Any of the rootfinding methods of Chapter 2 can be applied to solving $\varphi(s) = 0$. Each evaluation of $\varphi(s)$ involves the solution of the initial value problem (6.11.16) over $[a, b]$, and consequently, we want to minimize the number of such evaluations. As a specific example of an important and rapidly convergent method, we look at Newton's method:

$$s_{m+1} = s_m + \frac{\varphi(s_m)}{\varphi'(s_m)} \qquad m = 0, 1, \ldots \qquad (6.11.18)$$

To calculate $\varphi'(s)$, differentiate the definition (6.11.17) to obtain

$$\varphi'(s) = b_0 \xi_s(b) + b_1 \xi_s'(b) \qquad (6.11.19)$$

where

$$\xi_s(x) = \frac{\partial Y(x; s)}{\partial s} \qquad (6.11.20)$$

To find $\xi_s(x)$, differentiate the equation

$$Y''(x; s) = f(x, Y(x; s), Y'(x; s))$$

with respect to $s$. Then $\xi_s$ satisfies the initial value problem

$$\xi_s''(x) = f_2(x, Y(x; s), Y'(x; s))\xi_s(x)$$

$$+ f_3(x, Y(x; s), Y'(x; s))\xi_s'(x) \qquad (6.11.21)$$

$$\xi_s(a) = a_1 \qquad \xi_s'(a) = a_0$$

The functions $f_2$ and $f_3$ denote the partial derivatives of $f(x, u, v)$ with respect to $u$ and $v$, respectively. The initial values obtained from those in (6.11.16) and from the definition of $\xi_s$.

In practice we convert the problems (6.11.16) and (6.11.21) to a system of four first-order equations with the unknowns $Y$, $Y'$, $\xi_s$, and $\xi_s'$. This system is solved numerically, say with a method of order $p$ and stepsize $h$. Let $y_h(x; s)$ denote the approximation to $Y(x; s)$, with a similar notation for the remaining unknowns. From earlier results for solving initial value problems, it can be shown that these approximate solutions will be in error by $O(h^p)$. With suitable assumptions on the original problem (6.11.10), it can then be shown that the root $s_h^*$ obtained will also be in error by $O(h^p)$, and similarly for the approximate solution $y_h(x; s_h^*)$ when compared to the solution $Y(x; s^*)$ of the boundary value problem. For the details of this analysis, see Keller (1968, pp. 47–54).

***Example*** Consider the two-point BVP

$$y'' = -y + \frac{2(y')^2}{y} \qquad -1 < x < 1$$

$$\qquad\qquad (6.11.22)$$

$$y(-1) = y(1) = (e + e^{-1})^{-1} \doteq .324027137$$

The true solution is $Y(x) = (e^x + e^{-x})^{-1}$. The initial value problem (6.11.15) for the shooting method is

$$y'' = -y + \frac{(2y')^2}{y} \qquad -1 < x \le 1$$

$$\qquad\qquad (6.11.23)$$

$$y(-1) = (e + e^{-1})^{-1} \qquad y'(-1) = s$$

The associated problem (6.11.21) for $\xi_s(x)$ is

$$\xi_s'' = \left[ -1 - 2\left(\frac{y'}{y}\right)^2 \right] \xi_s + 4\frac{y'}{y}\xi_s'$$

$$\qquad\qquad (6.11.24)$$

$$\xi_s(-1) = 0 \qquad \xi_s'(-1) = 1$$

The equation for $\xi_s''$ uses the solution $Y(x; s)$ of (6.11.23). The function $\varphi(s)$ for computing $s^*$ is given by

$$\varphi(s) \equiv Y(1; s) - (e + e^{-1})^{-1}$$

For use in defining Newton's method, we have

$$\varphi'(s) = \xi_s(1)$$

From the true solution $Y$ of (6.11.22) and the condition $y'(-1) = s$ in (6.11.23), the desired root $s^*$ of $\varphi(s)$ is simply

$$s^* = Y'(-1) = \frac{e - e^{-1}}{(e + e^{-1})^2} \doteq .245777174$$

To solve the initial value problem (6.11.23)–(6.11.24), we used the second-order Runge–Kutta method (6.10.9) with a stepsize of $h = 2/n$. The results for several values of $n$ are given in Table 6.29. The solution of (6.11.24) is denoted by

**Table 6.29   Shooting method for solving (6.11.22)**

| $n = \dfrac{2}{h}$ | $s^* - s_h^*$ | Ratio | $E_h$ | Ratio |
|---|---|---|---|---|
| 4 | 4.01E − 3 | | 2.83E − 2 | |
| 8 | 1.52E − 3 | 2.64 | 7.30E − 3 | 3.88 |
| 16 | 4.64E − 4 | 3.28 | 1.82E − 3 | 4.01 |
| 32 | 1.27E − 4 | 3.64 | 4.54E − 4 | 4.01 |
| 64 | 3.34E − 5 | 3.82 | 1.14E − 4 | 4.00 |

$y_h(x; s)$, and the resulting root for

$$\varphi_h(s) \equiv y_h(1; s) - (e + e^{-1})^{-1} = 0$$

is denoted by $s_h^*$. For the error in $y_h(x; s_h^*)$, let

$$E_h = \operatorname*{Max}_{0 \le i \le n} |Y(x_i) - y_h(x_i; s_h^*)|$$

where $\{x_i\}$ are the node points used in solving the initial value problem. The columns labeled Ratio give the factors by which the errors decreased when $n$ was doubled (or $h$ was halved). Theoretically these factors should approach 4 since the Runge–Kutta method has an error of $O(h^2)$. Empirically, the factors approach 4.0, as expected. For the Newton iteration (6.11.18), $s_0 = .2$ was used in each case. The iteration was terminated when the test

$$|s_{m+1} - s_m| \le 10^{-10}$$

was satisfied. With these choices, the Newton method needed six iterations in each case, except that of $n = 4$ (when seven iterations were needed). However, if $s_0 = 0$ was used, then 25 iterations were needed for the $n = 4$ case, showing the importance of a good choice of the initial guess $s_0$.

There are a number of problems that can arise with the shooting method. First, there is no general guess $s_0$ for the Newton iteration, and with a poor choice, the iteration may diverge. For this reason, the modified Newton method of (2.11.11)–(2.11.12) in Section 2.11 may be needed to force convergence. A second problem is that the choice of $y_h(x; s)$ may be very sensitive to $h$, $s$, and other characteristics of the boundary value problem. For example, if the linearization of the initial value problem (6.11.16) has large positive eigenvalues, then the choice of $Y(x; s)$ is likely to be sensitive to variations in $s$. For a thorough discussion of these and other problems, see Keller (1968, chap. 2), Stoer and Burlirsch (1980, Sec. 7.3), and Fox (1980, p. 184–186). Some of these problems are more easily examined for linear BVPs, such as (6.11.8), as is done in Keller (1968, Chap. 2).

**Finite difference methods**    We consider the two-point BVP

$$y'' = f(x, y, y') \qquad a < x < b$$

$$y(a) = \gamma_1 \qquad y(b) = \gamma_2 \tag{6.11.25}$$

with the true solution denoted by $Y(x)$. Let $n > 1$, $h = (b - a)/n$, $x_j = a + jh$ for $j = 0, 1, \ldots, n$. At each interior node point $x_i$, $0 < i < n$, approximate $Y''(x_i)$ and $Y'(x_i)$:

$$Y''(x_i) = \frac{Y_{i+1} - 2Y_i + Y_{i-1}}{h^2} - \frac{h^2}{12} Y^{(4)}(\xi_i)$$

$$\tag{6.11.26}$$

$$Y'(x_i) = \frac{Y_{i+1} - Y_{i-1}}{2h} - \frac{h^2}{6} Y^{(3)}(\eta_i)$$

for some $x_{i-1} \le \xi_i$, $\eta_i \le x_{i+1}$, $i = 1, \ldots, n - 1$. Dropping the final error terms and using these approximations in the differential equation, we have the approximating nonlinear system:

$$\frac{y_{i+1} - 2y_i + y_{i-1}}{h^2} = f\left(x_i, y_i, \frac{y_{i+1} - y_{i-1}}{2h}\right) \qquad i = 1, \ldots, n - 1 \tag{6.11.27}$$

This is a system of $n - 1$ nonlinear equations in the $n - 1$ unknowns $y_1, \ldots, y_{n-1}$. The values $y_0 = \gamma_1$ and $y_n = \gamma_2$ are known from the boundary conditions.

The analysis of the error in $\{y_i\}$ compared to $\{Y(x_i)\}$ is too complicated to be given here, as it requires the methods of analyzing the solvability of systems of nonlinear equations. In essence, if $Y(x)$ is four times differentiable, if the problem (6.11.25) is uniquely solvable for some region about the graph on $[a, b]$ of $Y(x)$, and if $f(x, u, v)$ is sufficiently differentiable, then there is a solution to (6.11.27) and it satisfies

$$\underset{0 \le i \le n}{\text{Max}} |Y(x_i) - y_i| = O(h^2) \tag{6.11.28}$$

For an analysis, see Keller (1976, Sec. 3.2) or Keller (1968, Sec. 3.2). Moreover, with additional assumptions on $f$ and the smoothness of $Y$, it can be shown that

$$Y(x_i) - y_i = \tau(x_i)h^2 + O(h^4) \tag{6.11.29}$$

with $\tau(x)$ independent of $h$. This can be used to justify Richardson extrapolation, to obtain results that converge more rapidly. [Other methods to improve convergence are based on correcting for the error in the central difference approximations of (6.11.27).]

The system (6.11.27) can be solved in a variety of ways, some of which are simple modifications of the methods described in Sections 2.10 and 2.11. In matrix form, we have

$$
\frac{1}{h^2}
\begin{bmatrix}
-2 & 1 & 0 & & \cdots & & 0 \\
1 & -2 & 1 & & & & \vdots \\
\vdots & & & \ddots & & & \\
& & & & 1 & -2 & 1 \\
0 & \cdots & & & 0 & 1 & -2
\end{bmatrix}
\begin{bmatrix}
y_1 \\
y_2 \\
\vdots \\
\\
y_{n-1}
\end{bmatrix}
$$

$$
=
\begin{bmatrix}
f\left(x_1, y_1, \dfrac{y_2 - y_0}{2h}\right) \\[2mm]
f\left(x_2, y_2, \dfrac{y_3 - y_1}{2h}\right) \\[2mm]
\vdots \\[2mm]
f\left(x_{n-1}, y_{n-1}, \dfrac{y_n - y_{n-2}}{2h}\right)
\end{bmatrix}
-
\begin{bmatrix}
\dfrac{\gamma_1}{h^2} \\[2mm]
0 \\[2mm]
\vdots \\[2mm]
\dfrac{\gamma_2}{h^2}
\end{bmatrix}
$$

which we denote by

$$
\frac{1}{h^2} A\mathbf{y} = \hat{\mathbf{f}}(\mathbf{y}) + \mathbf{g} \tag{6.11.30}
$$

The matrix $A$ is nonsingular [see Theorem 8.2, Chapter 8]; and linear systems $A\mathbf{z} = \mathbf{b}$ are easily solvable, as described preceding Theorem 8.2 in Chapter 8. Newton's method (see Section 2.11) for solving (6.11.30) is given by

$$
\mathbf{y}^{(m+1)} = \mathbf{y}^{(m)} - \left[\frac{1}{h^2} A - F(\mathbf{y}^{(m)})\right]^{-1}\left[\frac{1}{h^2} A\mathbf{y}^{(m)} - \hat{\mathbf{f}}(\mathbf{y}^{(m)}) - \mathbf{g}\right] \tag{6.11.31}
$$

$$
F(\mathbf{y}) = \left[\frac{\partial \hat{f}_i}{\partial y_j}\right] \qquad 1 \le i, j \le n - 1
$$

The Jacobian matrix simplifies considerably because of the special form of $\hat{\mathbf{f}}(\mathbf{y})$:

$$
[F(\mathbf{y})]_{ij} = \frac{\partial f\left(x_i, y_i, \dfrac{y_{i+1} - y_{i-1}}{2h}\right)}{\partial y_j}
$$

This is zero unless $j = i - 1$, $i$, or $i + 1$:

$$
[F(y)]_{ii} = f_2\left(x_i, y_i, \frac{y_{i+1} - y_{i-1}}{2h}\right) \qquad 1 \le i \le n - 1
$$

$$
[F(y)]_{i, i-1} = \frac{-1}{2h} f_3\left(x_i, y_i, \frac{y_{i+1} - y_{i-1}}{2h}\right) \qquad 2 \le i \le n - 1
$$

$$
[F(y)]_{i, i+1} = \frac{1}{2h} f_3\left(x_i, y_i, \frac{y_{i+1} - y_{i-1}}{2h}\right) \qquad 1 \le i \le n - 2
$$

with $f_2(x, u, v)$ and $f_3(x, u, v)$ denoting partial derivatives with respect to $u$ and $v$, respectively. Thus the matrix being inverted in (6.11.31) is of the special form we call *tridiagonal*. Letting

$$B_m = \frac{1}{h^2} A - F(y^{(m)})  \qquad (6.11.32)$$

we can rewrite (6.11.31) as

$$y^{(m+1)} = y^{(m)} - \delta^{(m)}$$

$$B_m \delta^{(m)} = \frac{1}{h^2} A y^{(m)} - f(y^{(m)}) - g  \qquad (6.11.33)$$

This linear system is easily and rapidly solvable, as shown in Section 8.2 of Chapter 8. The number of multiplications and divisions can be shown to about $5n$, a relatively small number of operations for solving a linear system of $n - 1$ equations. Additional savings can be made by not varying $B_m$ or by only changing it after several iterations of (6.11.33). For an extensive survey and discussion of the solution of nonlinear systems that arise in connection with solving BVPs, see Deuflhard (1979).

***Example***  We applied the preceding finite difference procedure (6.11.27) to the solution of the BVP (6.11.22), used earlier to illustrate the shooting method. The results are given in Table 6.30 for successive doublings of $n = 2/h$. The nonlinear system in (6.11.27) was solved using Newton's method, as described in (6.11.33). The initial guess was

$$y_h^{(0)}(x_i) = (e + e^{-1})^{-1} \qquad i = 0, 1, \dots, n$$

based on connecting the boundary values by a straight line. The quantity

$$d_h = \underset{0 \le i \le n}{\mathrm{Max}} \left| y_i^{(m+1)} - y_i^{(m)} \right|$$

was computed for each iterate, and when the condition

$$d_h \le 10^{-10}$$

**Table 6.30    The finite difference method for solving (6.11.22)**

| $n = \dfrac{2}{h}$ | $E_h$ | Ratio |
|---|---|---|
| 4 | 2.63E − 2 | |
| 8 | 5.87E − 3 | 4.48 |
| 16 | 1.43E − 3 | 4.11 |
| 32 | 3.55E − 4 | 4.03 |
| 64 | 8.86E − 5 | 4.01 |

was satisfied, the iteration was terminated. In all cases, the number of iterates computed was 5 or 6. For the error, let

$$E_h = \operatorname*{Max}_{0 \le i \le n} |Y(x_i) - y_h(x_i)|$$

with $y_h$ the solution of (6.11.27) obtained with Newton's method. According to (6.11.28) and (6.11.29), we should expect the values $E_h$ to decrease by about a factor of 4 when $h$ is halved, and that is what we observe in the table.

Higher order methods can be obtained in several ways. (1) Use higher order approximations to the derivatives, improving (6.11.26); (2) use Richardson extrapolation, based on (6.11.29); as with Romberg integration, it can be repeated to obtain methods of arbitrarily high order; (3) the truncation errors in (6.11.26) can be approximated with higher order differences using the calculated values of $y_h$. Using these values as corrections to (6.11.27), we can obtain a new more accurate approximation to the differential equation in (6.11.25), leading to a more accurate solution. All of these techniques have been used, and some have been implemented as quite sophisticated computer codes. For a further discussion and for examples of computer codes, see Fox (1980, p. 191), Jain (1984, Chap. 4), and Pereyra (1979).

**Other methods and problems**   There are a number of other methods used for solving boundary value problems. The most important of these is probably the *collocation method*. For discussions referring to collocation methods, see Reddien (1979), Deuflhard (1979), and Ascher and Russell (1985). For an important collocation computer code, see Ascher et al. (1981a) and (1981b).

Another approach to solving a boundary value problem is to solve an equivalent reformulation as an integral equation. There is much less development of such numerical methods, although they can be very effective in some situations. For an introduction to this approach, see Keller (1968, Chap. 4).

There are also many other types of boundary value problems, some containing some type of these singular behavior, that we have not discussed here. For all of these, see the papers in the proceedings of Ascher and Russell (1985), Aziz (1975), Childs et al. (1979), and Gladwell and Sayers (1980); also see Keller (1976, Chap. 4) for singular problems. For discussions of software, see Childs et al. (1979), Gladwell and Sayers (1980), and Enright (1985).

# Discussion of the Literature

Ordinary and partial differential equations are the principal form of mathematical model occurring in the sciences and engineering, and consequently, the numerical solution of differential equations is a very large area of study. Two classical books that reflect the state of knowledge before the widespread use of digital computers are Collatz (1966) and Milne (1953). Some important and general books, since 1960, in the numerical solution of ordinary differential equations are Henrici (1962), Gear (1971), Lapidus and Seinfeld (1971), Lambert (1973), Stetter (1973), Hall and Watt (1976), Shampine and Gordon (1975), Van der Houwen (1977), Ortega and Poole (1981), and Butcher (1987). A useful survey is given in Gupta et al. (1985).

The modern theory of convergence and stability of multistep methods, introduced in Section 6.8, dates from Dahlquist (1956). An historical account is given in Dahlquist (1985). The text by Henrici (1962) has become a classic account of that theory, including extensions and applications of it. Gear (1971) is a more modern account of all methods, especially variable order methods. Stetter (1973) gives a very general and complete abstract analysis of the numerical theory for solving initial value problems. A complete account up to 1970 of Runge–Kutta methods, their development and error analysis, is given in Lapidus and Seinfeld (1971). Hall and Watt (1976) gives a survey of all aspects of the solution of ordinary differential equations, including the many special topics that have become of greater interest in the past ten years.

The first significant use of the concept of a variable order method is due to Gear (1971) and Krogh (1969). Such methods are superior to a fixed-order multistep method in efficiency, and they do not require any additional method for starting the integration or for changing the stepsize. A very good account of the variable-order Adams method is given in Shampine and Gordon (1975) and the excellent code DE/STEP is included. Other important early codes based on the Adams family of formulas were those in Krogh (1969), DIFSUB from Gear (1971), and GEAR from Hindmarsh (1974). The latter program GEAR has been further developed into a large multifunction package, called ODEPACK, and it is described in Hindmarsh (1983). Variants of these codes and other differential equation solvers are available in the IMSL and NAG libraries.

Runge–Kutta methods are a continuing active area of theoretical research and program development, and a very general development is given in Butcher (1987). New methods are being developed for nonstiff problems; for example, see Shampine (1986) and Shampine and Baca (1986). There is also great interest in implicit Runge–Kutta methods, for use in solving stiff differential equations. For a survey of the latter, see Aiken (1985, pp. 70–92). An important competitor to the code RKF45 is the code DVERK described in Hull et al. (1976). It is based on a Fehlberg-type scheme, with a pair of formulas of orders 5 and 6.

A third class of methods has been ignored in our presentation, those based on extrapolation. Current work in this area began with Gragg (1965) and Bulirsch and Stoer (1966). The main idea is to perform repeated extrapolation on some simple method, to obtain methods of increasingly higher order. In effect, this gives another way to produce variable-order methods. These methods have performed fairly well in the tests of Enright and Hull (1976) and Shampine et al. (1976), but they were judged to not be as advanced in their practical and theoretical development as are the multistep and Runge–Kutta methods. For a recent survey of the area, see Deuflhard (1985). Also, see Shampine and Baca (1986), in which extrapolation methods are discussed as one example of variable order Runge–Kutta methods.

Global error estimation is an area in which comparatively little has been published. For a general survey, see Skeel (1986). To our knowledge, the only running computer code is GERK from Shampine and Watts (1976a). Additional work is needed in this area. Many users of automatic packages are under the mistaken impression that the automatic codes they are using are controlling the global error, but such global error control is not possible in a practical sense. In many cases, it would seem to be important to have some idea of the actual magnitude of the global error present in the numerical solution.

Boundary value problems for ordinary differential equations are another important topic; but both their general theory and their numerical analysis are much more sophisticated than for the initial value problem. Important texts are Keller (1968) and (1976), and the proceedings of Child et al. (1979) gives many important papers on producing computer codes. For additional papers, see the collections of Aziz (1975), Hall and Watt (1976), and Ascher and Russell (1985). This area is still comparatively young relative to the initial value problem for nonstiff equations. The development of computer codes is proceeding in a number of directions, and some quite good codes have been produced in recent years. More work has been done on codes using the shooting method, but there are also excellent codes being produced for collocation and finite difference methods. For some discussion of such codes, see Enright (1985) and Gladwell and Sayers (1980, pp. 273–303). Some boundary value codes are given in the IMSL and NAG libraries.

Stiff differential equations is one of several special areas that have become much more important in the past ten years. The best general survey of this area is given in Aiken (1985). It gives examples of how such problems arise, the theory of numerical methods for solving stiff problems, and a survey of computer codes that exist for their solution. Many of the other texts in our bibliography also address the problem of stiff differential equations. We also recommend the paper of Shampine and Gear (1979).

Equations with a highly oscillatory solution occur in a number of applications. For some discussion of this, see Aiken (1985, pp. 111–123). The method of lines for solving time-dependent partial differential equations is a classical procedure that has become more popular in recent years. It is discussed in Aiken (1985, pp. 124–138), Sincovec and Madsen (1975), and Melgaard and Sincovec (1981).

Yet another area of interest is the solution of mixed systems of differential and algebraic equations (DAEs). This refers to systems in which there are $n$ unknowns, $m < n$ differential equations, and $n - m$ algebraic equations, involving the $n$ unknown functions. Such problems occur in many areas of applications. One such area of much interest in recent years is that of computer aided design (CAD). For papers applicable to such problems and to other DAEs, see Rheinboldt (1984) and (1986).

Because of the creation of a number of automatic programs for solving differential equations, several empirical studies have been made to assess their performance and to make comparisons between programs. Some of the major comparisons are given in Enright and Hull (1976), Enright et al. (1975), and Shampine et al. (1976). It is clear from their work that programs must be compared, as well as methods. Different program implementations of the same method can vary widely in their performance. No similar results are known for comparisons of boundary value codes.

# Bibliography

Aiken, R., Ed. (1985). *Stiff Computation*. Oxford Univ. Press, Oxford, England.

Ascher, U. (1986). Collocation for two-point boundary value problems revisited, *SIAM J. Numer. Anal.* **23**, 596–609.

Ascher, U., J. Christiansen, and R. Russell (1981a). Collocation software for boundary-value ODEs, *ACM Trans. Math. Softw.* **7**, 209–222.

Ascher, U., J. Christiansen, and R. Russell (1981b). COLSYS: collocation software for boundary-value ODEs, *ACM Trans. Math. Softw.* **7**, 223–229.

Ascher, U., and R. Russell, Eds. (1985). *Numerical Boundary Value ODEs*. Birkhäuser, Basel.

Aziz, A. K., Ed. (1975). *Numerical Solutions of Boundary Value Problems for Ordinary Differential Equations*. Academic Press, New York.

Boyce, W., and R. Diprima (1986). *Elementary Differential Equations and Boundary Value Problems*, 4th ed. Wiley, New York.

Bulirsch, R., and J. Stoer (1966). Numerical treatment of ordinary differential equations by extrapolation, *Numer. Math.* **8**, 1–13.

Butcher, J. (1965). On the attainable order of Runge–Kutta methods, *Math. Comput.* **19**, 408–417.

Butcher, J. (1987). *The Numerical Analysis of Ordinary Differential Equations*. Wiley, New York.

Childs, B., M. Scott, J. Daniel, E. Denman, and P. Nelson, Eds. (1979). *Codes for Boundary-Value Problems in Ordinary Differential Equations*. Lecture Notes in Computer Science 76, Springer-Verlag, New York.

Coddington, E., and N. Levinson (1955). *Theory of Ordinary Differential Equations*. McGraw-Hill, New York.

Collatz, L. (1966). *The Numerical Treatment of Differential Equations*, 3rd ed. Springer-Verlag, New York.

Dahlquist, G. (1956). Numerical integration of ordinary differential equations, *Math. Scandinavica* **4**, 33–50.

Dahlquist, G. (1963). A special stability property for linear multistep methods, *BIT* **3**, 27–43.

Dahlquist, G. (1985). 33 years of numerical instability, part 1, *BIT* **25**, 188–204.

Deuflhard, P. (1979). Nonlinear equation solvers in boundary value problem codes. In B. Childs, M. Scott, J. Daniel, E. Denman, and P. Nelson (Eds.), *Codes for Boundary-Value Problems in Ordinary Differential Equations*, pp. 40–66. Lecture Notes in Computer Science 76, Springer-Verlag, New York.

Deuflhard, P. (1985). Recent progress in extrapolation methods for ordinary differential equations, *SIAM Rev.* **27**, 505–536.

Enright, W. (1985). Improving the performance of numerical methods for two-point boundary value problems. In U. Ascher and R. Russell (Eds.), *Numerical Boundary Value ODEs*, pp. 107–120. Birkhäuser, Basel.

Enright, W., and T. Hull (1976). Test results on initial value methods for non-stiff ordinary differential equations, *SIAM J. Numer. Anal.* **13**, 944–961.

Enright, W., T. Hull, and B. Lindberg (1975). Comparing numerical methods for stiff systems of O.D.E.'s, *BIT* **15**, 10–48.

Fehlberg, E. (1970). Klassische Runge–Kutta–Formeln vierter und niedrigerer Ordnumg mit Schrittweiten–Kontrolle und ihre Anwendung auf Wärmeleitungsprobleme, *Computing* **6**, 61–71.

Forsythe, G., M. Malcolm, and C. Moler (1977). *Computer Methods for Mathematical Computations*. Prentice-Hall, Englewood Cliffs, N.J.

Fox, L. (1980). Numerical methods for boundary-value problems. In I. Gladwell and D. Sayers (Eds.), *Computational Techniques for Ordinary Differential Equations*, pp. 175–217. Academic Press, New York.

Gear, C. W. (1971). *Numerical Initial Value Problems in Ordinary Differential Equations*. Prentice-Hall, Englewood Cliffs, N.J.

Gladwell, I., and D. Sayers, Eds. (1980). *Computational Techniques for Ordinary Differential Equations*. Academic Press, New York.

Gragg, W. (1965). On extrapolation algorithms for ordinary initial value problems, *SIAM J. Numer. Anal.* **2**, 384–403.

Gupta, G., R. Sacks-Davis, and P. Tischer (1985). A review of recent developments in solving ODEs, *Comput. Surv.* **17**, 5–47.

Hall, G., and J. Watt, Eds. (1976). *Modern Numerical Methods for Ordinary Differential Equations*. Oxford Univ. Press, Oxford, England.

Henrici, P. (1962). *Discrete Variable Methods in Ordinary Differential Equations*. Wiley, New York.

Hindmarsh, A. (1974). GEAR: Ordinary differential equation solver. Lawrence Livermore Rep. UCID-30001, Rev. 3, Livermore, Calif.

Hindmarsh, A. (1983). ODEPACK: A systematized collection of ODE solvers. In *Numerical Methods for Scientific Computation*, R. Stepleman, Ed. North-Holland, Amsterdam.

Hull, T., W. Enright, and K. Jackson (1976). User's guide for DVERK: A subroutine for solving non-stiff ODEs. Dept. Computer Sci. Tech. Rep. 100, Univ. of Toronto, Toronto, Ont., Canada.

Isaacson, E., and H. Keller (1966). *Analysis of Numerical Methods*. Wiley, New York.

Jain, M. K. (1984). *Numerical Solution of Differential Equations*, 2nd ed. Halstead Press, New York.

Keller, H. (1968). *Numerical Methods for Two-Point Boundary Value Problems*. Ginn (Blaisdell), Boston.

Keller, H. (1976). *Numerical Solution of Two-Point Boundary Value Problems*. Regional Conference Series in Applied Mathematics 24, Society for Industrial and Applied Mathematics, Philadelphia.

Keller, J., and S. Antman, Eds. (1969). *Bifurcation Theory and Nonlinear Eigenvalue Problems*. Benjamin, New York.

Krogh, F. (1969), VODQ/SVDQ/DVDQ—Variable order integrators for the numerical solution of ordinary differential equations. Section 314 Subroutine Writeup, Jet Propulsion Lab., Pasadena, Calif.

Lambert, J. (1973). *Computational Methods in Ordinary Differential Equations.* Wiley, New York.

Lapidus, L., and W. Schiesser, Eds. (1976). *Numerical Methods for Differential Equations: Recent Developments in Algorithms, Software.* New Applications, Academic Press, New York.

Lapidus, L., and J. Seinfeld (1971). *Numerical Solution of Ordinary Differential Equations.* Academic Press, New York.

Lentini, M., M. Osborne, and R. Russell (1985). The close relationship between methods for solving two-point boundary value problems, *SIAM J. Numer. Anal.* **22**, 280–309.

Melgaard, D., and R. Sincovec (1981). Algorithm 565: PDETWO/ PSETM/GEARB: Solution of systems of two-dimensional nonlinear partial differential equations, *ACM Trans. Math. Softw.* **7**, 126–135.

Miller, R. K. and A. Michel (1982). *Ordinary Differential Equations.* Academic Press, New York.

Milne, W. (1953). *Numerical Solution of Differential Equations.* Wiley, New York.

Ortega, J., and W. Poole (1981). *An Introduction to Numerical Methods for Differential Equations*, Pitman, New York.

Pereyra, V. (1979). PASVA3: An adaptive finite difference Fortran program for first order nonlinear, ordinary differential equation problems. In B. Childs, M. Scott, J. Daniel, E. Denman, and P. Nelson (Eds.), *Codes for Boundary-Value Problems in Ordinary Differential Equations*, pp. 67–88. Lecture Notes in Computer Science 76, Springer-Verlag, New York.

Reddien, G. (1979). Projection methods. In B. Childs, M. Scott, J. Daniel, E. Denman, and P. Nelson (Eds.), *Codes for Boundary-Value Problems in Ordinary Differential Equations*, pp. 206–227. Lecture Notes in Computer Science 76, Springer-Verlag, New York.

Rheinboldt, W. (1984). Differential-algebraic systems as differential equations on manifolds, *Math. Comput.* **43**, 473–482.

Rheinboldt, W. (1986). *Numerical Analysis of Parametrized Nonlinear Equations.* Wiley, New York.

Shampine, L. (1985). Local error estimation by doubling, *Computing* **34**, 179–190.

Shampine, L. (1986). Some practical Runge–Kutta formulas, *Math. Comput.* **46**, 135–150.

Shampine, L., and L. Baca (1986). Fixed versus variable order Runge–Kutta, *ACM Trans. Math. Softw.* **12**, 1–23.

Shampine, L., and C. W. Gear (1979). A user's view of solving stiff ordinary differential equations, *SIAM Rev.* **21**, 1–17.

Shampine, L., and M. Gordon (1975). *Computer Solution of Ordinary Differential Equations.* Freeman, San Francisco.

Shampine, L., and H. Watts (1976a). Global error estimation for ordinary differential equations, *ACM Trans. Math. Softw.* **2**, 172–186.

Shampine, L., and H. Watts (1976b). Practical solution of ordinary differential equations by Runge–Kutta methods. Sandia Labs. Tech. Rep. SAND 76-0585, Albuquerque, N.Mex.

Shampine, L., and H. Watts (1980). DEPAC—Design of a user oriented package of ODE solvers. Sandia National Labs. Rep. SAND79-2374, Albuquerque, N.Mex.

Shampine, L., H. Watts, and S. Davenport (1976). Solving nonstiff ordinary differential equations—The state of the art, *SIAM Rev.* **18**, 376–411.

Sincovec, R., and N. Madsen (1975). Software for nonlinear partial differential equations, *ACM Trans. Math. Softw.* **1**, 232–260.

Skeel, R. (1986). Thirteen ways to estimate global error, *Numer. Math.* **48**, 1–20.

Stakgold, I. (1979). *Green's Functions and Boundary Value Problems*. Wiley, New York.

Stetter, H. (1973). *Analysis of Discretization Methods for Ordinary Differential Equations*. Springer-Verlag, New York.

Stoer, J., and R. Bulirsch (1980). *Introduction to Numerical Analysis*. Springer-Verlag, New York.

Van der Houwen, P. J. (1977). *Construction of Integration Formulas for Initial Value Problems*. North-Holland, Amsterdam.

Widder, D. (1975). *The Heat Equation*. Academic Press, New York.

## Problems

1. Draw the direction field for $y' = x - y^2$, and then draw in some sample solution curves. Attempt to guess the behavior of the solutions $Y(x)$ as $x \to \infty$.

2. Determine Lipschitz constants for the following functions, as in (6.1.2).

   (a) $f(x, y) = 2y/x$, $x \geq 1$

   (b) $f(x, y) = \tan^{-1}(y)$

   (c) $f(x, y) = (x^3 - 2)^{27}/(17x^2 + 4)$

   (d) $f(x, y) = x - y^2$, $|y| \leq 10$

3. Convert the following problems to first-order systems.

   (a) $y'' - 3y' + 2y = 0$, $y(0) = 1$, $y'(0) = 1$

**(b)** $y'' - .1(1 - y^2)y' + y = 0$, $y(0) = 1$, $y'(0) = 0$
(Van der Pol's equation)

**(c)** $x''(t) = -x/r^3$, $y''(t) = -y/r^3$, $r = \sqrt{x^2 + y^2}$
(orbital equation)
$x(0) = .4$, $x'(0) = 0$, $y(0) = 0$, $y'(0) = 2$

4. Let $Y(x)$ be the solution, if it exists, to the initial value problem (6.0.1). By integrating, show that $Y$ satisfies

$$Y(x) = Y_0 + \int_{x_0}^{x} f(t, Y(t))\, dt$$

Conversely, show that if this equation has a continuous solution on the interval $x_0 \leq x \leq b$, then the initial value problem (6.0.1) has the same solution.

5. The integral equation of Problem 4 is solved, at least in theory, by using the iteration

$$Y_{m+1}(x) = Y_0 + \int_{x_0}^{x} f(t, Y_m(t))\, dt, \qquad x_0 \leq x \leq b$$

for $m \geq 0$, with $Y_0(x) \equiv Y_0$. This is called *Picard iteration*, and under suitable assumptions, the iterates $\{Y_m(x)\}$ can be shown to converge uniformly to $Y(x)$. Illustrate this by computing the Picard iterates $Y_1$, $Y_2$, and $Y_3$ for the following problems; compare them to the true solution $Y$.

**(a)** $y' = -y$, $y(0) = 1$

**(b)** $y' = -xy$, $y(1) = 2$

**(c)** $y' = y + 2\cos(x)$, $y(0) = 1$

6. Write a computer program to solve $y' = f(x, y)$, $y(x_0) = y_0$, using Euler's method. Write it to be used with an arbitrary $f$, stepsize $h$, and interval $[x_0, b]$. Using the program, solve $y' = x^2 - y$, $y(0) = 1$, for $0 \leq x \leq 4$, with stepsizes of $h = .25, .125, .0625$, in succession. For each value of $h$, print the true solution, approximate solution, error, and relative error at the nodes $x = 0, .25, .50, .75, \ldots, 4.00$. The true solution is $Y(x) = x^2 - 2x + 2 - e^{-x}$. Analyze your output and supply written comments on it. Analysis of output is as important as obtaining it.

7. For the problem $y' = y$, $y(0) = 1$, give the explicit solution $\{y_n\}$ for Euler's approximation to the equation. Use this to show that for $x_n = 1$, $Y(1) - y_n \doteq (h/2)e$ as $h \to 0$.

8.  Consider solving the problem

    $$y' = -y^2 \qquad x \geq 0 \qquad y(0) = 1$$

    by Euler's method. Compare (1) the bound (6.2.13), using $K = 2$ as the Lipschitz constant, and (2) the asymptotic estimate from (6.2.36). The true solution is $Y(x) = 1/(1 + x)$.

9.  Show that Euler's method fails to approximate the solution $Y(x) = (\frac{2}{3}x)^{3/2}$, $x \geq 0$, of the problem $y' = y^{1/3}$, $y(0) = 0$. Explain why.

10. For the equations in Problem 3, write out the approximating difference equations obtained by using Euler's method.

11. Recall the rounding error example for Euler's method, with the results shown in Table 6.3. Attempt to produce a similar behavior on your computer by letting $h$ become smaller, until eventually the error begins to increase.

12. Convert the problem

    $$y^{(3)} + 4y'' + 5y' + 2y = -4\sin(x) - 2\cos(x)$$

    $$y(0) = 1 \qquad y'(0) = 0 \qquad y''(0) = -1$$

    to a system of first-order equations. Using Euler's method, solve this system and empirically study the error. The true solution is $Y(x) = \cos(x)$.

13. (a) Derive the Lipschitz condition (6.2.52) for a system of two differential equations.

    (b) Prove that the method (6.2.51) will converge to the solution of (6.2.50).

14. Consider the two-step method

    $$y_{n+1} = \frac{1}{2}(y_n + y_{n-1}) + \frac{h}{4}[4y'_{n+1} - y'_n + 3y'_{n-1}] \qquad n \geq 1$$

    with $y'_n \equiv f(x_n, y_n)$. Show it is a second-order method, and find the leading term in the truncation error, written as in (6.3.15).

15. Assume that the multistep method (6.3.1) is consistent and that it satisfies $a_j \geq 0$, $j = 0, 1, \ldots, p$, the same as in Theorem 6.6. Prove stability of (6.3.1), in analogy with (6.2.28), for Euler's method, but letting $\delta \equiv 0$.

16. Write a program to solve $y' = f(x, y)$, $y(x_0) = y_0$, using the midpoint rule (6.4.2). Use a fixed stepsize $h$. For the initial value $y_1$, use the Euler

method:

$$y_1 = y_0 + hf(x_0, y_0)$$

With the program, solve the following problems:

(a)  $y' = -y^2$, $y(0) = 1$; $Y(x) = 1/(1 + x)$

(b)  $y' = \dfrac{y}{4}\left[1 - \dfrac{y}{20}\right]$, $y(0) = 1$; $Y(x) = \dfrac{20}{[1 + 19e^{-x/4}]}$

(c)  $y' = -y + 2\cos(x)$, $y(0) = 1$; $Y(x) = \cos(x) + \sin(x)$

(d)  $y' = y - 2\sin(x)$, $y(0) = 1$; $Y(x) = \cos(x) + \sin(x)$

Solve on the interval $[x_0, b] = [0, 5]$, with $h = .5, .25$. Print the numerical solution $y_n = y_h(x_n)$ at each node, along with the true error. Discuss your results.

17. Write a program to solve $y' = f(x, y)$, $y(x_0) = y_0$, using the trapezoidal rule (6.5.2) with a fixed stepsize $h$. In the iteration (6.5.3), use the midpoint method as the predictor of $y_{n+1}^{(0)}$. Allow the number of iterates $J$ to be an input variable, $J \geq 1$. Solve the equations in Problem 16 with $h = .5$ and .25, for $[x_0, b] = [0, 10]$. Solve over the entire interval with $J = 1$, then repeat the process with $J = 2$, and then $J = 3$. Discuss the results. How does the total error vary with $J$?

18. Derive the asymptotic error formula (6.5.19) for the trapezoidal method.

19. Write a program to implement the algorithm *Detrap* of Section 6.6. Using it, solve the equations given in Problem 16, with $\epsilon = .001$.

20. Use the quadratic interpolant to $Y'(x) = f(x, Y(x))$ at $x_n, x_{n-1}, x_{n-2}$ to obtain the formula

$$Y(x_{n+1}) = Y_{n-3} + \frac{4h}{3}[2Y_n' - Y_{n-1}' + 2Y_{n-2}'] + \frac{28}{90}h^5 Y^{(5)}(\xi_n)$$

When the truncation error is dropped, we obtain the method

$$y_{n+1} = y_{n-3} + \frac{4h}{3}[2y_n' - y_{n-1}' + 2y_{n-2}'] \qquad n \geq 3$$

This is the predictor formula for the Milne method (6.7.13).

21. Show that the Simpson method (6.7.13) is only weakly stable, in the same sense as was true of the midpoint method in Section 6.4.

22.    Write a program to solve $y' = f(x, y)$, $y(x_0) = y_0$, $x_0 \le x \le b$, using the fourth-order Adams–Moulton formula and a fixed stepsize $h$. Use the fourth-order Adams–Bashforth formula as the predictor. Generate the initial values $y_1$, $y_2$, $y_3$ using true solution $Y(x)$. Solve the equations of Problem 16 with $h = .5$ and $h = .25$. Print the calculated answers and the true errors. For comparison with method (6.7.27), also solve the example used in Table 6.14. Check the accuracy of the Richardson extrapolation error estimate

$$Y(x) - y_h(x) \doteq \tfrac{1}{15}\left[ y_h(x) - y_{2h}(x) \right]$$

that is based on the global error being of the fourth order. Discuss all your results.

23.    (a)    For the coefficients $\gamma_i$ and $\delta_i$ of the Adams–Bashforth and Adams–Moulton formulas, show that $\delta_i = \gamma_i - \gamma_{i-1}$, $i \ge 1$.

(b)    For the $p$-step Adams–Moulton formula (6.7.26), prove

$$y_{n+1} = y_{n+1}^{(0)} + h\gamma_p \nabla^{p+1} y'_{n+1} \tag{1}$$

with $y_{n+1}^{(0)}$ the $p + 1$ step Adams–Bashforth formula from (6.7.22),

$$y_{n+1}^{(0)} = y_n + h \sum_{j=0}^{p} \gamma_j \nabla^j y'_n \tag{2}$$

These formulas are both of order $p + 1$. The result (1) is of use in calculating the corrector from the predictor, and is based on carrying certain backward differences from one step to the next. There is a closely related result when a $p$-step predictor (order $p$) is used to solve the $p$-step corrector (order $p + 1$) [see Shampine and Gordon (1975), p. 51].

24.    Consider the model equation (6.8.16) with $\Lambda$ a square matrix of order $m$. Assume $\Lambda = P^{-1}DP$, with $D$ a diagonal matrix with entries $\lambda_1, \ldots, \lambda_m$. Introduce the new unknown vector function $\mathbf{z} = P^{-1}\mathbf{y}(x)$. Show that (6.8.16) converts to the form given in (6.8.17), demonstrating the reduction to the one-dimensional model equation.

25.    Following the ideas given in Sections 6.4 and 6.8, give the general procedure for solving the linear difference equation

$$y_{n+1} = a_0 y_n + a_1 y_{n-1}$$

Apply this to find the general solution of the following equations.

(a)    $y_{n+1} = -\tfrac{1}{2}y_n + \tfrac{1}{2}y_{n-1}$

(b)    $y_{n+1} = y_n - \tfrac{1}{4}y_{n-1}$    *Hint:* See the formula following (6.8.22).

**26.**   Solve the third-order linear difference equation

$$u_{n+1} = u_n + c(u_{n-1} - u_{n-2}) \qquad n \geq 2 \qquad 0 < c < 1$$

with $u_0, u_1, u_2$ given. What can be said about

$$\underset{n \to \infty}{\text{Limit}} \, u_n$$

**27.**   Consider the numerical method

$$y_{n+1} = 4y_n - 3y_{n-1} - 2hf(x_{n-1}, y_{n-1}) \qquad n \geq 1$$

Determine its order. Illustrate with an example that the method is unstable.

**28.**   Show that the two-step method

$$y_{n+1} = 2y_{n-1} - y_n + h\left[\tfrac{5}{2}y'_n + \tfrac{1}{2}y'_{n-1}\right]$$

is of order 2 and unstable. Also, show directly that it need not converge when solving $y' = f(x, y)$.

**29.**   Complete part (1) of the proof of Theorem 6.7, in which the root condition is violated by assuming $|r_j| = 1$, $\rho'(r_j) = 0$, for some $j$.

**30.**   For part (2) of the proof of Theorem 6.8, show that $\gamma_j[r_j(h\lambda)]^n \to 0$ as $h \to 0$, $1 \leq j \leq p$.

**31.**   **(a)**   Determine the values of $a_0$ in the explicit second-order method (6.7.10) for which the method is stable.

   **(b)**   If only the truncation error (6.7.11) is considered, subject to the stability restriction in part (a), how should $a_0$ be chosen?

   **(c)**   To ensure a large region of stability, subject to part (a), how should $a_0$ be chosen?

**32.**   **(a)**   Find the general formula for all two-step third-order methods. These will be a one-parameter family of methods, say, depending on the coefficient $a_1$.

   **(b)**   What are the restrictions on $a_1$ for the method in part (a) to be stable?

   **(c)**   If the truncation error is written as

$$T_{n+1}(Y) = \beta h^4 Y^{(4)}(x_n) + O(h^5)$$

give a formula for $\beta$ in terms of $a_1$. (It is not necessary to construct a Peano kernel or influence function for the method.) How should $a_1$ be chosen if the truncation error is to be minimized, subject to the stability restriction from part (b)?

(d)   Consider the region of absolute stability for the methods of part (a). What is this region for the method of part (c) that minimizes the truncation error coefficient $\beta$? Give another value of $a_1$ that gives a stable method and that has a larger region of absolute stability. Discuss finding an optimal region by choosing $a_1$ approximately.

33.   (a)   Find all explicit fourth-order formulas of the form

$$y_{n+1} = a_0 y_n + a_1 y_{n-1} + a_2 y_{n-2} + h\left[b_0 y_n' + b_1 y_{n-1}' + b_2 y_{n-2}'\right] \quad n \geq 2$$

(b)   Show that every such method is unstable.

34.   Derive an implicit fourth-order multistep method, other than those given in the text. Make it be relatively stable.

35.   For the polynomial $\rho(r) = r^{p+1} - \sum_0^p a_j r^{p-j}$, assume $a_j \geq 0$, $0 \leq j \leq p$, and $\sum_0^p a_j = 1$. Show that the roots of $\rho(r)$ will satisfy the root condition (6.8.9) and (6.8.10). This shows directly that Theorem 6.6 is a corollary of Theorem 6.8.

36.   (a)   Consider methods of the form

$$y_{n+1} = y_{n-q} + h \sum_{j=-1}^{p} b_j f\left(x_{n-j}, y_{n-j}\right)$$

with $q \geq 1$. Show that such methods do not satisfy the strong root condition. As a consequence, most such methods are only weakly stable.

(b)   Find an example with $q = 1$ that is relatively stable.

37.   Show that the region of absolute stability for the trapezoidal method is the set of all complex $h\lambda$ with $\text{Real}(\lambda) < 0$.

38.   Use the backward Euler method to solve the problem (6.8.51). Because the equation is linear, the implicit equation for $y_{n+1}$ can be solved exactly. Compare your results with those given in Table 6.17 for Euler's method.

39.   Repeat Problem 38 using the second-order BDF formula (6.9.6). To find $y_1$ for use in (6.8.6), use the backward Euler method.

40.   Recall the model equation (6.8.50) where it is regarded as a perturbation of Euler's method (6.8.46). For the special case $Y''(x) \equiv$ constant, analyze the

behavior of the error $Y(x_n) - y_n$ as it depends on $h\lambda$. Show that again the condition (6.8.49) is needed in order that the error be well-behaved.

41.  Derive the truncation error formula (6.9.7) for backward differentiation formulas.

42.  For solving $y' = f(x, y)$, consider the numerical method

$$y_{n+1} = y_n + \frac{h}{2}[y_n' + y_{n+1}'] + \frac{h^2}{12}[y_n'' - y_{n+1}''] \qquad n \geq 0$$

Here $y_n' = f(x_n, y_n)$

$$y_n'' = \frac{\partial f(x_n, y_n)}{\partial x} + f(x_n, y_n)\frac{\partial f(x_n, y_n)}{\partial y}$$

with this formula based on differentiating $Y'(x) = f(x, Y(x))$.

(a)  Show that this is a fourth-order method: $T_n(Y) = O(h^5)$.

(b)  Show that the region of absolute stability contains the entire negative real axis of the complex $h\lambda$-plane.

43.  Generalize the method of lines, given in (6.9.23)–(6.9.25), to the problem

$$U_t = a(x, t)U_{xx} + G(x, t, U(x, t)) \qquad 0 < x < 1 \qquad t > 0$$

$$U(0, t) = d_0(t) \qquad U(1, t) = d_1(t) \qquad t \geq 0$$

$$U(x, 0) = f(x) \qquad 0 \leq x \leq 1$$

For it to be well-defined, we assume $a(x, t) > 0$, $0 \leq x \leq 1$, $t \geq 0$.

44.  (a)  If you have a solver of tridiagonal linear algebraic systems available to you, then write a program to implement the method of lines for the problem (6.9.19)–(6.9.21). The example in the text, with the unknown (6.9.35), was solved using the backward Euler method. Now implement the method of lines using the trapezoidal rule. Compare your results with those in Table 6.20 for the backward Euler method.

(b)  Repeat with the second-order BDF method.

45.  Derive a third-order Taylor series method to solve $y' = -y^2$. Compare the numerical results to those in Table 6.22.

46.  Using the Taylor series method of Section 6.10, produce a fourth-order method to solve $y' = x - y^2$, $y(0) = 0$. Use fixed stepsizes, $h = .5$, .25, .125 in succession, and solve for $0 \leq x \leq 10$. Estimate the global error using the error estimate (6.10.24) based on Richardson extrapolation.

47. Write a program to solve $y' = f(x, y)$, $y(x_0) = y_0$, using the classical Runge–Kutta method (6.10.21), and let the stepsize $h$ be fixed.

   (a) Using the program, solve the equations of Problem 16.

   (b) Solve $y' = x - y^2$, $y(0) = 0$, for $h = .5, .25, .125$. Compare the results with those of Problem 46.

48. Consider the three stage Runge–Kutta formula

$$y_{n+1} = y_n + h[\gamma_1 V_1 + \gamma_2 V_2 + \gamma_3 V_3]$$

$$V_1 = f(x_n, y_n), \quad V_2 = f(x_n + \alpha_2 h, y_n + h\beta_{21} V_1)$$

$$V_3 = f(x_n + \alpha_3 h, y_n + h(\beta_{31} V_1 + \beta_{32} V_2))$$

   Determine the set of equations that the coefficients $\{\gamma_j, \alpha_j, \beta_{ji}\}$ must satisfy if the formula is to be of order 3. Find a particular solution of these equations.

49. Prove that if the Runge–Kutta method (6.10.4) satisfies (6.10.27), then it is stable.

50. Apply the classical Runge–Kutta method (6.10.21) to the test problem (6.8.51), for various values of $\lambda$ and $h$. For example, try $\lambda = -1, -10, -50$ and $h = .5, .1, .01$, as in Table 6.17.

51. Calculate the real part of the region of absolute stability for the Runge–Kutta method of (a) (6.10.8), (b) (6.10.9), (c) (6.10.21). We are interested in the behavior of the numerical solution for the differential equation $y' = \lambda y$ with $\text{Real}(\lambda) < 0$. In particular, we are interested in those values of $h\lambda$ for which the numerical solution tends to zero as $x_n \to \infty$.

52. (a) Using the Runge–Kutta method (6.10.8), solve

$$y' = -y + x^{-1}[1.1 + x] \qquad y(0) = 0$$

   whose solution is $Y(x) = x^{1.1}$. Solve the equation on $[0, 5]$, printing the errors at $x = 1, 2, 3, 4, 5$. Use stepsizes $h = .1, .05, .025, .0125, .00625$. Calculate the errors by which the errors decrease when $h$ is halved. How does this compare with the usual theoretical rate of convergence of $O(h^2)$? Explain your results.

   (b) What difficulty arises when trying to use a Taylor method of order $\geq 2$ to solve the equation of part (a)? What does it tell us about the solution?

53.   Convert the boundary value problem (6.11.1) to an equivalent boundary value problem for a system of first-order equations, as in (6.11.15).

54.   **(a)**   Consider the two-point boundary value problem (6.11.25). To convert this to an equivalent problem with zero boundary conditions, write $y(x) = z(x) + w(x)$, with $w(x)$ a straight line satisfying the following boundary conditions: $w(a) = \gamma_1$, $w(b) = \gamma_2$. Derive a new boundary value problem for $z(x)$.

      **(b)**   Generalize this procedure to problem (6.11.10). Obtain a new problem with zero boundary conditions. What assumptions, if any, are needed for the coefficients $a_0, a_1, b_0, b_1$?

55.   Using the shooting method of Section 6.11, solve the following boundary value problems. Study the convergence rate as $h$ is varied.

      **(a)**   $y'' = \dfrac{-2}{x} yy'$, $1 < x < 2$;   $y(1) = \dfrac{1}{2}$, $y(2) = \dfrac{2}{3}$

      True solution: $Y(x) = x/(1 + x)$.

      **(b)**   $y'' = 2yy'$, $0 < x < \dfrac{\pi}{4}$;   $y(0) = 0$, $y\left(\dfrac{\pi}{4}\right) = 1$

      True solution: $Y(x) = \tan(x)$.

56.   Investigate the differential equation programs provided by your computer center. Note those that automatically control the truncation error by varying the stepsize, and possibly the order. Classify the programs as multistep (fixed-order or variable-order), Runge–Kutta, or extrapolation. Compare one of these with the programs DDEABM [of Section 6.7 and Shampine and Gordon (1975)] and RKF45 [of Section 6.9 and Shampine and Watts (1976b)] by solving the problem

$$y' = \frac{y}{4}\left[1 - \frac{y}{20}\right] \qquad y(0) = 1$$

with desired absolute errors of $10^{-3}$, $10^{-6}$, and $10^{-9}$. Compare the results with those given in Tables 6.15 and 6.28.

57.   Consider the problem

$$y' = \frac{1}{t + 1} + c \cdot \tan^{-1}(y(t)) - \frac{1}{2} \qquad y(0) = 0$$

with $c$ a given constant. Since $y'(0) = \frac{1}{2}$, the solution $y(t)$ is initially increasing as $t$ increases, regardless of the value of $c$. As best you can, show that there is a value of $c$, call it $c^*$, for which (1) if $c > c^*$, the solution $y(t)$

increases indefinitely, and (2) if $c < c^*$, then $y(t)$ increases initially, but then peaks and decreases. Determine $c^*$ to within .00005, and then calculate the associated solution $y(t)$ for $0 \leq t \leq 50$.

58.    Consider the system

$$x'(t) = Ax - Bxy \qquad y'(t) = Cxy - Dy$$

This is known as the Lotka–Volterra predator–prey model for two populations, with $x(t)$ being the number of prey and $y(t)$ the number of predators, at time $t$.

(a)    Let $A = 4$, $B = 2$, $C = 1$, $D = 3$, and solve the model to at least three significant digits for $0 \leq t \leq 5$. The initial values are $x(0) = 3$, $y(0) = 5$. Plot $x$ and $y$ as functions of $t$, and plot $x$ versus $y$.

(b)    Solve the same model with $x(0) = 3$ and, in succession, $y(0) = 1, 1.5$, 2. Plot $x$ versus $y$ in each case. What do you observe? Why would the point $(3, 2)$ be called an equilibrium point?

# SEVEN

# LINEAR ALGEBRA

The solution of systems of simultaneous linear equations and the calculation of the eigenvalues and eigenvectors of a matrix are two very important problems that arise in a wide variety of contexts. As a preliminary to the discussion of these problems in the following chapters, we present some results from linear algebra. The first section contains a review of material on vector spaces, matrices, and linear systems, which is taught in most undergraduate linear algebra courses. These results are summarized only, and no derivations are included. The remaining sections discuss eigenvalues, canonical forms for matrices, vector and matrix norms, and perturbation theorems for matrix inverses. If necessary, this chapter can be skipped, and the results can be referred back to as they are needed in Chapters 8 and 9. For notation, Section 7.1 and the norm notation of Section 7.3 should be skimmed.

## 7.1  Vector Spaces, Matrices, and Linear Systems

Roughly speaking a *vector space* $V$ is a set of objects, called *vectors*, for which operations of *vector addition* and *scalar multiplication* have been defined. A vector space $V$ has a set of scalars associated with it, and in this text, this set can be either the real numbers **R** or complex numbers **C**. The vector operations must satisfy certain standard associative, commutative, and distributive rules, which we will not list. A subset $W$ of a vector space $V$ is called a *subspace* of $V$ if $W$ is a vector space using the vector operations inherited from $V$. For a complete development of the theory of vector spaces, see any undergraduate text on linear algebra [for example, Anton (1984), chap. 3; Halmos (1958), chap. 1; Noble (1969), chaps. 4 and 14; Strang (1980), chap. 2].

*Example 1.*  $V = \mathbf{R}^n$, the set of all $n$-tuples $(x_1, \ldots, x_n)$ with real entries $x_i$, and **R** is the associated set of scalars.

**2.**  $V = \mathbf{C}^n$, the set of all $n$-tuples with complex entries, and **C** is the set of scalars.

**3.**  $V =$ the set of all polynomials of degree $\leq n$, for some given $n$, is a vector space. The scalars can be **R** or **C**, as desired for the application.

**4.** $V = C[a, b]$, the set of all continuous real valued [or complex valued] functions on the interval $[a, b]$, is a vector space with scalar set equal to **R** [or **C**]. The example in (3) is a subspace of $C[a, b]$.

***Definition***   Let $V$ be a vector space and let $v_1, v_2, \ldots, v_m \in V$.

1. We say that $v_1, \ldots, v_m$ are *linearly dependent* if there is a set of scalars $\alpha_1, \ldots, \alpha_m$, with at least one nonzero scalar, for which

$$\alpha_1 v_1 + \cdots + \alpha_m v_m = 0$$

   Since at least one scalar is nonzero, say $\alpha_i \neq 0$, we can solve for

$$v_i = -\frac{\alpha_1}{\alpha_i} v_1 \cdots - \frac{\alpha_{i-1}}{\alpha_i} v_{i-1} - \frac{\alpha_{i+1}}{\alpha_i} v_{i+1} - \cdots - \frac{\alpha_m}{\alpha_i} v_m$$

   We say that $v_i$ is a *linear combination* of the vectors $v_1, \ldots, v_{i-1}, v_{i+1}, \ldots, v_m$. For a set of vectors to be linearly dependent, one of them must be a linear combination of the remaining ones.

2. We say $v_1, \ldots, v_m$ are *linearly independent* if they are not dependent. Equivalently, the only choice of scalars $\alpha_1, \ldots, \alpha_m$ for which

$$\alpha_1 v_1 + \cdots + \alpha_m v_m = 0$$

   is the trivial choice $\alpha_1 = \cdots = \alpha_m = 0$. No $v_i$ can be written as a combination of the remaining ones.

3. $\{v_1, \ldots, v_m\}$ is a *basis* for $V$ if for every $v \in V$, there is a unique choice of scalars $\alpha_1, \ldots, \alpha_m$ for which

$$v = \alpha_1 v_1 + \cdots + \alpha_m v_m$$

   Note that this implies $v_1, \ldots, v_m$ are independent. If such a finite basis exists, we say $V$ is *finite dimensional*. Otherwise, it is called *infinite dimensional*.

***Theorem 7.1***   If $V$ is a vector space with a basis $\{v_1, \ldots, v_m\}$, then every basis for $V$ will contain exactly $m$ vectors. The number $m$ is called the dimension of $V$.

***Example 1.***   $\{1, x, x^2, \ldots, x^n\}$ is a basis for the space $V$ of polynomials of degree $\leq n$. Thus dimension $V = n + 1$.

**2.**   **R**$^n$ and **C**$^n$ have the basis $\{e_1, \ldots, e_n\}$, in which

$$e_i = (0, 0, \ldots, 0, 1, 0, \ldots, 0) \tag{7.1.1}$$

with the 1 in position $i$. Dimension $\mathbf{R}^n, \mathbf{C}^n = n$. This is called the standard basis for $\mathbf{R}^n$ and $\mathbf{C}^n$, and the vectors in it are called *unit vectors*.

**3.**    $C[a, b]$ is infinite dimensional.

**Matrices and linear systems**    Matrices are rectangular arrays of real or complex numbers, and the general matrix of order $m \times n$ has the form

$$A = \begin{bmatrix} a_{11} & a_{12} \cdots & a_{1n} \\ \vdots & & \vdots \\ a_{m1} & a_{m2} \cdots & a_{mn} \end{bmatrix} \tag{7.1.2}$$

A matrix of order $n$ is shorthand for a square matrix of order $n \times n$. Matrices will be denoted by capital letters, and their entries will normally be denoted by lowercase letters, usually corresponding to the name of the matrix, as just given. The following definitions give the common operations on matrices.

**Definition 1.**    Let $A$ and $B$ have order $m \times n$. The sum of $A$ and $B$ is the matrix $C = A + B$, of order $m \times n$, given by

$$c_{ij} = a_{ij} + b_{ij}$$

**2.**    Let $A$ have order $m \times n$, and let $\alpha$ be a scalar. Then the scalar multiple $C = \alpha A$ is of order $m \times n$ and is given by

$$c_{ij} = \alpha a_{ij}$$

**3.**    Let $A$ have order $m \times n$ and $B$ have order $n \times p$. Then the product $C = AB$ is of order $m \times p$, and it is given by

$$c_{ij} = \sum_{k=1}^{n} a_{ik} b_{kj}$$

**4.**    Let $A$ have order $m \times n$. The *transpose* $C = A^T$ has order $n \times m$, and is given by

$$c_{ij} = a_{ji}$$

The *conjugate transpose* $C = A^*$ also has order $n \times m$, and

$$c_{ij} = \bar{a}_{ji}$$

The notation $\bar{z}$ denotes the complex conjugate of the complex number $z$, and $z$ is real if and only if $\bar{z} = z$. The conjugate transpose $A^*$ is also called the *adjoint* of $A$.

The following arithmetic properties of matrices can be shown without much difficulty, and they are left to the reader.

(a) $A + B = B + A$

(b) $(A + B) + C = A + (B + C)$

(c) $A(B + C) = AB + AC$

(d) $A(BC) = (AB)C$      (7.1.3)

(e) $(A + B)^T = A^T + B^T$

(f) $(AB)^T = B^T A^T$

It is important for many applications to note that the matrices need not be square for the preceding properties to hold.

The vector spaces $\mathbf{R}^n$ and $\mathbf{C}^n$ will usually be identified with the set of column vectors of order $n \times 1$, with real and complex entries, respectively. The linear system

$$a_{11}x_1 + \cdots + a_{1n}x_n = b_1$$

$$\vdots$$

$$a_{m1}x_1 + \cdots + a_{mn}x_n = b_m$$

(7.1.4)

can be written as $Ax = b$, with $A$ as in (7.1.2), and

$$x = [x_1, \ldots, x_n]^T \qquad b = [b_1, \ldots, b_m]^T$$

The vector $b$ is a given vector in $\mathbf{R}^m$, and the solution $x$ is an unknown vector in $\mathbf{R}^n$. The use of matrix multiplication reduces the linear system (7.1.4) to the simpler and more intuitive form $Ax = b$.

We now introduce a few additional definitions for matrices, including some special matrices.

**Definition 1.** The *zero matrix* of order $m \times n$ has all entries equal to zero. It is denoted by $0_{m \times n}$, or more simply, by 0. For any matrix $A$ of order $m \times n$,

$$A + 0 = 0 + A = A$$

**2.** The *identity matrix* of order $n$ is defined by $I = [\delta_{ij}]$,

$$\delta_{ij} = \begin{cases} 1 & i = j \\ 0 & i \neq j \end{cases}$$

(7.1.5)

for all $1 \leq i, j \leq n$. For all matrices $A$ of order $m \times n$ and $B$ of order $n \times p$,

$$AI = A \qquad IB = B$$

The notation $\delta_{ij}$ denotes the *Kronecker delta function*.

**3.** Let $A$ be a square matrix of order $n$. If there is a square matrix $B$ of order $n$ for which $AB = BA = I$, then we say $A$ is invertible, with *inverse B*. The matrix $B$ can be shown to be unique, and we denote the inverse of $A$ by $A^{-1}$.

**4.** A matrix $A$ is called *symmetric* if $A^T = A$, and it is called *Hermitian* if $A^* = A$. The term symmetric is generally used only with real matrices. The matrix $A$ is *skew-symmetric* if $A^T = -A$. Of necessity, all matrices that are symmetric, Hermitian, or skew-symmetric must also be square.

**5.** Let $A$ be an $m \times n$ matrix. The row rank of $A$ is the number of linearly independent rows in $A$, regarded as elements of $\mathbf{R}^n$ or $\mathbf{C}^n$, and the column rank is the number of linearly independent columns. It can be shown (Problem 4) that these two numbers are always equal, and this is called the *rank* of $A$.

For the definition and properties of the determinant of a square matrix $A$, see any linear algebra text [for example, Anton (1984), chap. 2; Noble (1969), chap. 7; and Strang (1980), chap. 4]. We summarize many of the results on matrix inverses and the solvability of linear systems in the following theorem.

**Theorem 7.2**    Let $A$ be a square matrix with elements from $\mathbf{R}$ (or $\mathbf{C}$), and let the vector space be $V = \mathbf{R}^n$ (or $\mathbf{C}^n$). Then the following are equivalent statements.

**1.** $Ax = b$ has a unique solution $x \in V$ for every $b \in V$.

**2.** $Ax = b$ has a solution $x \in V$ for every $b \in V$.

**3.** $Ax = 0$ implies $x = 0$.

**4.** $A^{-1}$ exists.

**5.** Determinant $(A) \neq 0$.

**6.** Rank $(A) = n$.

Although no proof is given here, it is an excellent exercise to prove the equivalence of some of these statements. Use the concepts of linear independence and basis, along with Theorem 7.1. Also, use the decomposition

$$Ax = x_1 A_{*1} + \cdots + x_n A_{*n} \qquad x \in \mathbf{R}^n \quad \text{or} \quad \mathbf{C}^n \qquad (7.1.6)$$

with $A_{*j}$ denoting column $j$ in $A$. This says that the space of all vectors of the form $Ax$ is spanned by the columns of $A$, although they may be linearly dependent.

**Inner product vector spaces**    One of the important reasons for reformulating problems as equivalent linear algebra problems is to introduce some geometric insight. Important to this process are the concepts of inner product and orthogonality.

***Definition* 1.**   The *inner product* of two vectors $x, y \in \mathbf{R}^n$ is defined by

$$(x, y) = \sum_{i=1}^{n} x_i y_i = x^T y = y^T x$$

and for vectors $x, y \in \mathbf{C}^n$, define the inner product by

$$(x, y) = \sum_{i=1}^{n} x_i \bar{y}_i = y^* x$$

**2.**   The *Euclidean norm* of $x$ in $\mathbf{C}^n$ or $\mathbf{R}^n$ is defined by

$$\|x\|_2 = \sqrt{(x, x)} = \sqrt{|x_1|^2 + \cdots + |x_n|^2} \qquad (7.1.7)$$

The following results are fairly straightforward to prove, and they are left to the reader. Let $V$ denote $\mathbf{C}^n$ or $\mathbf{R}^n$.

**1.**   For all $x, y, z \in V$,

$$(x, y + z) = (x, y) + (x, z), \qquad (x + y, z) = (x, z) + (y, z)$$

**2.**   For all $x, y \in V$,

$$(\alpha x, y) = \alpha(x, y)$$

and for $V = \mathbf{C}^n$, $\alpha \in \mathbf{C}$,

$$(x, \alpha y) = \bar{\alpha}(x, y)$$

**3.**   In $\mathbf{C}^n$, $(x, y) = \overline{(y, x)}$; and in $\mathbf{R}^n$, $(x, y) = (y, x)$.

**4.**   For all $x \in V$,

$$(x, x) \geq 0$$

and $(x, x) = 0$ if and only if $x = 0$.

**5.**   For all $x, y \in V$,

$$|(x, y)|^2 \leq (x, x)(y, y) \qquad (7.1.8)$$

This is called the *Cauchy–Schwartz inequality*, and it is proved in exactly the same manner as (4.4.3) in Chapter 4. Using the Euclidean norm, we can write it as

$$|(x, y)| \leq \|x\|_2 \|y\|_2 \qquad (7.1.9)$$

**6.**   For all $x, y \in V$,

$$\|x + y\|_2 \leq \|x\|_2 + \|y\|_2 \qquad (7.1.10)$$

This is the *triangle inequality*. For a geometric interpretation, see the earlier comments in Section 4.1 of Chapter 4 for the norm $\|f\|_\infty$ on $C[a, b]$. For a proof of (7.1.10), see the derivation of (4.4.4) in Chapter 4.

**7.** For any square matrix $A$ of order $n$, and for any $x, y \in \mathbf{C}^n$,

$$(Ax, y) = (x, A^*y) \qquad (7.1.11)$$

The inner product was used to introduce the Euclidean length, but it is also used to define a sense of angle, at least in spaces in which the scalar set is $\mathbf{R}$.

**Definition 1.**  For $x, y$ in $\mathbf{R}^n$, the *angle between x and y* is defined by

$$\mathscr{A}(x, y) = \cos^{-1}\left[\frac{(x, y)}{\|x\|_2\|y\|_2}\right]$$

Note that the argument is between $-1$ and $1$, due to the Cauchy–Schwartz inequality (7.1.9). The preceding definition can be written implicitly as

$$(x, y) = \|x\|_2\|y\|_2 \cos(\mathscr{A}) \qquad (7.1.12)$$

a familiar formula from the use of the dot product in $\mathbf{R}^2$ and $\mathbf{R}^3$.

**2.** Two vectors $x$ and $y$ are *orthogonal* if and only if $(x, y) = 0$. This is motivated by (7.1.12). If $\{x^{(1)}, \ldots, x^{(n)}\}$ is a basis for $\mathbf{C}^n$ or $\mathbf{R}^n$, and if $(x^{(i)}, x^{(j)}) = 0$ for all $i \ne j$, $1 \le i, j \le n$, then we say $\{x^{(1)}, \ldots, x^{(n)}\}$ is an *orthogonal basis*. If all basis vectors have Euclidean length 1, the basis is called *orthonormal*.

**3.** A square matrix $U$ is called *unitary* if

$$U^*U = UU^* = I$$

If the matrix $U$ is real, it is usually called *orthogonal*, rather than unitary. The rows [or columns] of an order $n$ unitary matrix form an orthonormal basis for $\mathbf{C}^n$, and similarly for orthogonal matrices and $\mathbf{R}^n$.

**Example 1.**  The angle between the vectors

$$x = (1, 2, 3) \qquad y = (3, 2, 1)$$

is given by

$$\mathscr{A} = \cos^{-1}\left[\frac{10}{14}\right] \doteq .775 \text{ radians}$$

**2.** The matrices

$$U_1 = \begin{bmatrix} \cos\theta & \sin\theta \\ -\sin\theta & \cos\theta \end{bmatrix} \qquad U_2 = \begin{bmatrix} \dfrac{1}{\sqrt{2}} & \dfrac{i}{\sqrt{2}} \\ \dfrac{i}{\sqrt{2}} & \dfrac{1}{\sqrt{2}} \end{bmatrix}$$

are unitary, with the first being orthogonal.

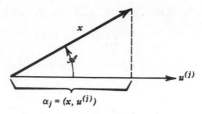

**Figure 7.1**  Illustration of (7.1.15).

An orthonormal basis for a vector space $V = \mathbf{R}^n$ or $\mathbf{C}^n$ is desirable, since it is then easy to decompose an arbitrary vector into its components in the direction of the basis vectors. More precisely, let $\{u^{(1)}, \ldots, u^{(n)}\}$ be an orthonormal basis for $V$, and let $x \in V$. Using the basis,

$$x = \alpha_1 u^{(1)} + \cdots + \alpha_n u^{(n)}$$

for some unique choice of coefficients $\alpha_1, \ldots, \alpha_n$. To find $\alpha_j$, form the inner product of $x$ with $u^{(j)}$, and then

$$(x, u^{(j)}) = \alpha_1(u^{(1)}, u^{(j)}) + \cdots + \alpha_n(u^{(n)}, u^{(j)})$$

$$= \alpha_j \tag{7.1.13}$$

using the orthonormality properties of the basis. Thus

$$x = \sum_{j=1}^{n} (x, u^{(j)}) u^{(j)} \tag{7.1.14}$$

This can be given a geometric interpretation, which is shown in Figure 7.1. Using (7.1.13)

$$\alpha_j = (x, u^{(j)}) = \|x\|_2 \|u^{(j)}\|_2 \cos\left(\mathscr{A}(x, u^{(j)})\right)$$

$$= \|x\|_2 \cos\left(\mathscr{A}(x, u^{(j)})\right) \tag{7.1.15}$$

Thus the coefficient $\alpha_j$ is just the length of the orthogonal projection of $x$ onto the axis determined by $u^{(j)}$. The formula (7.1.14) is a generalization of the decomposition of a vector $x$ using the standard basis $\{e^{(1)}, \ldots, e^{(n)}\}$, defined earlier.

**Example**  Let $V = \mathbf{R}^2$, and consider the orthonormal basis

$$u^{(1)} = \left(\frac{1}{2}, \frac{\sqrt{3}}{2}\right) \qquad u^{(2)} = \left(-\frac{\sqrt{3}}{2}, \frac{1}{2}\right)$$

Then for a given vector $x = (x_1, x_2)$, it can be written as

$$x = \alpha_1 u^{(1)} + \alpha_2 u^{(2)}$$

$$\alpha_1 = \left(x, u^{(1)}\right) = \frac{x_1 + x_2\sqrt{3}}{2} \qquad \alpha_2 = \left(x, u^{(2)}\right) = \frac{x_2 - x_1\sqrt{3}}{2}$$

For example,

$$(1,0) = \frac{1}{2}u^{(1)} - \frac{\sqrt{3}}{2}u^{(2)}$$

## 7.2   Eigenvalues and Canonical Forms for Matrices ·

The number $\lambda$, complex or real, is an *eigenvalue* of the square matrix $A$ if there is a vector $x \in \mathbf{C}^n$, $x \neq 0$, such that

$$Ax = \lambda x \tag{7.2.1}$$

The vector $x$ is called an *eigenvector* corresponding to the eigenvalue $\lambda$. From Theorem 7.2, statements (3) and (5), $\lambda$ is an eigenvalue of $A$ if and only if

$$\det\left(A - \lambda I\right) = 0 \tag{7.2.2}$$

This is called the *characteristic equation* for $A$, and to analyze it we introduce the function

$$f_A(\lambda) \equiv \det\left(A - \lambda I\right)$$

If $A$ has order $n$, then $f_A(\lambda)$ will be a polynomial of degree exactly $n$, called the *characteristic polynomial* of $A$. To prove it is a polynomial, expand the determinant by minors repeatedly to get

$$f_A(\lambda) = \det\left(A - \lambda I\right)$$

$$= \det \begin{bmatrix} a_{11} - \lambda & a_{12} & \cdots & & a_{1n} \\ a_{21} & a_{22} - \lambda & a_{23} & \cdots & a_{2n} \\ \vdots & & & \ddots & \\ a_{n1} & & \cdots & & a_{nn} - \lambda \end{bmatrix}$$

$$= (a_{11} - \lambda)(a_{22} - \lambda) \cdots (a_{nn} - \lambda)$$

$$+ \text{ terms of degree } \leq n - 2$$

$$f_A(\lambda) = (-1)^n \lambda^n + (-1)^{n-1}(a_{11} + \cdots + a_{nn})\lambda^{n-1}$$

$$+ \text{ terms of degree } \leq n - 2 \tag{7.2.3}$$

Also note that the constant term is

$$f_A(0) = \det(A) \tag{7.2.4}$$

From the coefficient of $\lambda^{n-1}$, define

$$\text{trace}(A) = a_{11} + a_{22} + \cdots + a_{nn} \tag{7.2.5}$$

which is often a quantity of interest in the study of $A$.

Since $f_A(\lambda)$ is of degree $n$, there are exactly $n$ eigenvalues for $A$, if we count multiple roots according to their multiplicity. Every matrix has at least one eigenvalue–eigenvector pair, and the $n \times n$ matrix $A$ has at most $n$ distinct eigenvalues.

**Example 1.**    The characteristic polynomial for

$$A = \begin{bmatrix} 2 & 1 & 0 \\ 1 & 3 & 1 \\ 0 & 1 & 2 \end{bmatrix}$$

is

$$f_A(\lambda) = -\lambda^3 + 7\lambda^2 - 14\lambda + 8$$

The eigenvalues are $\lambda_1 = 1$, $\lambda_2 = 2$, $\lambda_3 = 4$, and the corresponding eigenvectors are

$$u^{(1)} = \begin{bmatrix} 1 \\ -1 \\ 1 \end{bmatrix} \quad u^{(2)} = \begin{bmatrix} 1 \\ 0 \\ -1 \end{bmatrix} \quad u^{(3)} = \begin{bmatrix} 1 \\ 2 \\ 1 \end{bmatrix}$$

Note that these eigenvectors are orthogonal to each other, and therefore they are linearly independent. Since the dimension of $\mathbf{R}^3$ (and $\mathbf{C}^3$) is three, these eigenvectors form an orthogonal basis for $\mathbf{R}^3$ (and $\mathbf{C}^3$). This illustrates Theorem 7.4, which is presented later in the section.

**2.**    For the matrix

$$A = \begin{bmatrix} 1 & 0 & 0 \\ 0 & 1 & 0 \\ 0 & 0 & 1 \end{bmatrix} \quad f_A(\lambda) = (1 - \lambda)^3$$

and there are three linearly independent eigenvectors for the eigenvalue $\lambda = 1$, for example,

$$[1,0,0]^T \quad [0,1,0]^T \quad [0,0,1]^T$$

All other eigenvectors are linear combinations of these three vectors.

3.    For the matrix

$$A = \begin{bmatrix} 1 & 1 & 0 \\ 0 & 1 & 1 \\ 0 & 0 & 1 \end{bmatrix} \qquad f_A(\lambda) = (1 - \lambda)^3$$

The matrix $A$ has only one linearly independent eigenvector for the eigenvalue $\lambda = 1$, namely

$$x = [1, 0, 0]^T$$

and multiples of it.

The *algebraic multiplicity* of an eigenvalue of a matrix $A$ is its multiplicity as a root of $f_A(\lambda)$, and its *geometric multiplicity* is the maximum number of linearly independent eigenvectors associated with the eigenvalue. The sum of the algebraic multiplicities of the eigenvalues of an $n \times n$ matrix $A$ is constant with respect to small perturbations in $A$, namely $n$. But the sum of the geometric multiplicities can vary greatly with small perturbations, and this causes the numerical calculation of eigenvectors to often be a very difficult problem. Also, the algebraic and geometric multiplicities need not be equal, as the preceding examples show.

**Definition**    Let $A$ and $B$ be square matrices of the same order. Then $A$ is *similar* to $B$ if there is a nonsingular matrix $P$ for which

$$B = P^{-1}AP \tag{7.2.6}$$

Note that this is a symmetric relation since

$$A = Q^{-1}BQ \qquad Q = P^{-1}$$

The relation (7.2.6) can be interpreted to say that $A$ and $B$ are matrix representations of the same linear transformation $T$ from $V$ to $V$ [$V = \mathbf{R}^n$ or $\mathbf{C}^n$], but with respect to different bases for $V$. The matrix $P$ is called the *change of basis matrix*, and it relates the two representations of a vector $x \in V$ with respect to the two bases being used [see Anton (1984), sec. 5.5 or Noble (1969), sec. 14.5 for greater detail].

We now present a few simple properties about similar matrices and their eigenvalues.

1.    If $A$ and $B$ are similar, then $f_A(\lambda) = f_B(\lambda)$. To prove this, use (7.2.6) to show

$$f_B(\lambda) = \det(B - \lambda I) = \det\left[P^{-1}(A - \lambda I)P\right]$$

$$= \det(P^{-1}) \det(A - \lambda I) \det(P) = f_A(\lambda)$$

since

$$\det(P)\det(P^{-1}) = \det(PP^{-1}) = \det(I) = 1$$

2.  The eigenvalues of similar matrices $A$ and $B$ are exactly the same, and there is a one-to-one correspondence of the eigenvectors. If $Ax = \lambda x$, then using

$$P^{-1}AP(P^{-1}x) = \lambda P^{-1}x$$

$$Bz = \lambda z \qquad z = P^{-1}x \qquad (7.2.7)$$

Trivially, $z \neq 0$, since otherwise $x$ would be zero. Also, given any eigenvector $z$ of $B$, this argument can be reversed to produce a corresponding eigenvector $x = Pz$ for $A$.

3.  Since $f_A(\lambda)$ is invariant under similarity transformations of $A$, the coefficients of $f_A(\lambda)$ are also invariant under such similarity transformations. In particular, for $A$ similar to $B$,

$$\text{trace}(A) = \text{trace}(B) \qquad \det(A) = \det(B) \qquad (7.2.8)$$

**Canonical forms**   We now present several important canonical forms for matrices. These forms relate the structure of a matrix to its eigenvalues and eigenvectors, and they are used in a variety of applications in other areas of mathematics and science.

***Theorem 7.3***   (Schur Normal Form)   Let $A$ have order $n$ with elements from $\mathbf{C}$. Then there exists a unitary matrix $U$ such that

$$T \equiv U^*AU \qquad (7.2.9)$$

is upper triangular.
   Since $T$ is triangular, and since $U^* = U^{-1}$,

$$f_A(\lambda) = f_T(\lambda) = (\lambda - t_{11}) \cdots (\lambda - t_{nn}) \qquad (7.2.10)$$

and thus the eigenvalues of $A$ are the diagonal elements of $T$.

***Proof***   The proof is by induction on the order $n$ of $A$. The result is trivially true for $n = 1$, using $U = [1]$. We assume the result is true for all matrices of order $n \leq k - 1$, and we will then prove it has to be true for all matrices of order $n = k$.
   Let $\lambda_1$ be an eigenvalue of $A$, and let $u^{(1)}$ be an associated eigenvector with $\|u^{(1)}\|_2 = 1$. Beginning with $u^{(1)}$, pick an orthonormal basis for $\mathbf{C}^k$, calling it $\{u^{(1)}, \ldots, u^{(k)}\}$. Define the matrix $P_1$ by

$$P_1 = \left[ u^{(1)}, u^{(2)}, \ldots, u^{(k)} \right]$$

which is written in partitioned form, with columns $u^{(1)}, \ldots, u^{(k)}$ that are orthogonal. Then $P_1^*P_1 = I$, and thus $P_1^{-1} = P_1^*$. Define

$$B_1 = P_1^*AP_1$$

Claim:

$$B_1 = \begin{bmatrix} \lambda_1 & \alpha_2 & \cdots & \alpha_k \\ 0 & & & \\ \vdots & & A_2 & \\ 0 & & & \end{bmatrix}$$

with $A_2$ of order $k-1$ and $\alpha_2, \ldots, \alpha_k$ some numbers. To prove this, multiply using partitioned matrices:

$$AP_1 = A\left[u^{(1)}, \ldots, u^{(k)}\right] = \left[Au^{(1)}, \ldots, Au^{(k)}\right]$$

$$= \left[\lambda_1 u^{(1)}, v^{(2)}, \ldots, v^{(k)}\right] \qquad v^{(j)} = Au^{(j)}$$

$$B_1 = P_1^* AP_1 = \left[\lambda_1 P_1^* u^{(1)}, P_1^* v^{(2)}, \ldots, P_1^* v^{(k)}\right]$$

Since $P_1^* P_1 = I$, it follows that $P_1^* u^{(1)} = e^{(1)} = [1, 0, \ldots, 0]^T$. Thus

$$B_1 = \left[\lambda_1 e^{(1)}, w^{(2)}, \ldots, w^{(k)}\right] \qquad w^{(j)} = P_1^* v^{(j)}$$

which has the desired form.

By the induction hypothesis, there exists a unitary matrix $\hat{P}_2$ of order $k-1$ for which

$$\hat{T} = \hat{P}_2^* A_2 \hat{P}_2$$

is an upper triangular matrix of order $k-1$. Define

$$P_2 = \begin{bmatrix} 1 & 0 & \cdots & 0 \\ 0 & & & \\ \vdots & & \hat{P}_2 & \\ 0 & & & \end{bmatrix}$$

Then $P_2$ is unitary, and

$$P_2^* B_1 P_2 = \begin{bmatrix} \lambda_1 & \gamma_2 & & \gamma_k \\ 0 & & & \\ \vdots & & \hat{P}_2^* A_2 \hat{P}_2 & \\ 0 & & & \end{bmatrix}$$

$$= \begin{bmatrix} \lambda_1 & \gamma_2 & \cdots & \gamma_k \\ 0 & & & \\ \vdots & & \hat{T} & \\ 0 & & & \end{bmatrix} \equiv T$$

an upper triangular matrix. Thus

$$T = P_2^* B_1 P_2 = P_2^* P_1^* A P_1 P_2 = (P_1 P_2)^* A (P_1 P_2)$$

$$T = U^* A U \qquad U = P_1 P_2$$

and $U$ is easily unitary. This completes the induction and the proof. ∎

**Example**   For the matrix

$$A = \begin{bmatrix} .2 & .6 & 0 \\ 1.6 & -.2 & 0 \\ -1.6 & 1.2 & 3.0 \end{bmatrix}$$

the matrices of the theorem and (7.2.9) are

$$T = \begin{bmatrix} 1 & 0 & -1 \\ 0 & 3 & 2 \\ 0 & 0 & -1 \end{bmatrix} \qquad U = \begin{bmatrix} .6 & 0 & -.8 \\ .8 & 0 & .6 \\ 0 & 1.0 & 0 \end{bmatrix}$$

This is not the usual way in which eigenvalues are calculated, but should be considered only as an illustration of the theorem. The theorem is used generally as a theoretical tool, rather than as a computational tool.

Using (7.2.8) and (7.2.9),

$$\text{trace}\,(A) = \lambda_1 + \lambda_2 + \cdots + \lambda_n \qquad \det\,(A) = \lambda_1 \lambda_2 \ldots \lambda_n \qquad (7.2.11)$$

where $\lambda_1, \ldots, \lambda_n$ are the eigenvalues of $A$, which must form the diagonal elements of $T$. As a much more important application, we have the following well-known theorem.

**Theorem 7.4** (Principal Axes Theorem)   Let $A$ be a Hermitian matrix of order $n$, that is, $A^* = A$. Then $A$ has $n$ real eigenvalues $\lambda_1, \ldots, \lambda_n$, not necessarily distinct, and $n$ corresponding eigenvectors $u^{(1)}, \ldots, u^{(n)}$ that form an orthonormal basis for $\mathbf{C}^n$. If $A$ is real, the eigenvectors $u^{(1)}, \ldots, u^{(n)}$ can be taken as real, and they form an orthonormal basis of $\mathbf{R}^n$. Finally there is a unitary matrix $U$ for which

$$U^* A U = D \equiv \text{diag}\,[\lambda_1, \ldots, \lambda_n] \qquad (7.2.12)$$

is a diagonal matrix with diagonal elements $\lambda_1, \ldots, \lambda_n$. If $A$ is also real, then $U$ can be taken as orthogonal.

**Proof**   From Theorem 7.3, there is a unitary matrix $U$ with

$$U^* A U = T$$

with $T$ upper triangular. Form the conjugate transpose of both sides to obtain

$$T^* = (U^*AU)^* = U^*A^*(U^*)^* = U^*AU = T$$

Since $T^*$ is lower triangular, we must have

$$T = \text{diag}[\lambda_1, \ldots, \lambda_n]$$

Also, $T^* = T$ involves complex conjugation of all elements of $T$, and thus all diagonal elements of $T$ must be real.

Write $U$ as

$$U = [u^{(1)}, \ldots, u^{(n)}]$$

Then $T = U^*AU$ implies $AU = UT$,

$$A[u^{(1)}, \ldots, u^{(n)}] = [u^{(1)}, \ldots, u^{(n)}]\begin{bmatrix} \lambda_1 & & 0 \\ & \ddots & \\ 0 & & \lambda_n \end{bmatrix}$$

$$[Au^{(1)}, \ldots, Au^{(n)}] = [\lambda_1 u^{(1)}, \ldots, \lambda_n u^{(n)}]$$

and

$$Au^{(j)} = \lambda_j u^{(j)} \qquad j = 1, \ldots, n \qquad (7.2.13)$$

Since the columns of $U$ are orthonormal, and since the dimension of $\mathbf{C}^n$ is $n$, these must form an orthonormal basis for $\mathbf{C}^n$. We omit the proof of the results that follow from $A$ being real. This completes the proof. ∎

*Example*   From an earlier example in this section, the matrix

$$A = \begin{bmatrix} 2 & 1 & 0 \\ 1 & 3 & 1 \\ 0 & 1 & 2 \end{bmatrix}$$

has the eigenvalues $\lambda_1 = 1$, $\lambda_2 = 2$, $\lambda_3 = 4$ and corresponding orthonormal eigenvectors

$$u^{(1)} = \frac{1}{\sqrt{3}}\begin{bmatrix} 1 \\ -1 \\ 1 \end{bmatrix} \qquad u^{(2)} = \frac{1}{\sqrt{2}}\begin{bmatrix} 1 \\ 0 \\ -1 \end{bmatrix} \qquad u^{(3)} = \frac{1}{\sqrt{6}}\begin{bmatrix} 1 \\ 2 \\ 1 \end{bmatrix}$$

These form an orthonormal basis for $\mathbf{R}^3$ or $\mathbf{C}^3$.

There is a second canonical form that has recently become more important for problems in numerical linear algebra, especially for solving overdetermined systems of linear equations. These systems arise from the fitting of empirical data using the linear least squares procedures [see Golub and Van Loan (1983), chap. 6, and Lawson and Hanson (1974)].

**Theorem 7.5** (Singular Value Decomposition)    Let $A$ be order $n \times m$. Then there are unitary matrices $U$ and $V$, of orders $m$ and $n$, respectively, such that

$$V^*AU = F \qquad (7.2.14)$$

is a "diagonal" rectangular matrix of order $n \times m$,

$$F = \begin{bmatrix} \mu_1 & & & & 0 & \\ & \mu_2 & & & & \\ & & \ddots & & & \\ & 0 & & \mu_r & & \\ & & & & 0 & \\ & & & & & \ddots \end{bmatrix} \qquad (7.2.15)$$

The numbers $\mu_1, \ldots, \mu_r$ are called the *singular values* of $A$. They are all real and positive, and they can be arranged so that

$$\mu_1 \geq \mu_2 \geq \cdots \geq \mu_r > 0 \qquad (7.2.16)$$

where $r$ is the rank of the matrix $A$.

**Proof**    Consider the square matrix $A^*A$ of order $m$. It is a Hermitian matrix, and consequently Theorem 7.4 can be applied to it. The eigenvalues of $A^*A$ are all real; moreover, they are all nonnegative. To see this, assume

$$A^*Ax = \lambda x \qquad x \neq 0.$$

Then

$$(x, A^*Ax) = (x, \lambda x) = \lambda \|x\|_2^2$$

$$(x, A^*Ax) = (Ax, Ax) = \|Ax\|_2^2$$

$$\lambda = \left( \frac{\|Ax\|_2}{\|x\|_2} \right)^2 \geq 0$$

This result also proves that

$$Ax = 0 \qquad \text{if and only if} \qquad A^*Ax = 0 \qquad x \in \mathbf{C}^n \quad (7.2.17)$$

From Theorem 7.3, there is an $m \times m$ unitary matrix $U$ such that

$$U^*A^*AU = \text{diag}[\lambda_1, \ldots, \lambda_r, 0, \ldots, 0] \qquad (7.2.18)$$

where all $\lambda_i \neq 0$, $1 \leq i \leq r$, and all are positive. Because $A^*A$ has order

$m$, the index $r \leq m$. Introduce the *singular values*

$$\mu_i = \sqrt{\lambda_i} \qquad i = 1, \ldots, r \qquad (7.2.19)$$

The $U$ can be chosen so that the ordering (7.2.16) is obtained. Using the diagonal matrix

$$D = \text{diag}\, [\mu_1, \ldots, \mu_r, 0, \ldots, 0]$$

of order $m$, we can write (7.2.18) as

$$(AU)^*(AU) = D^2 \qquad (7.2.20)$$

Let $W = AU$. Then (7.2.20) says $W^*W = D^2$. Writing $W$ as

$$W = \left[ W^{(1)}, \ldots, W^{(m)} \right] \qquad W^{(j)} \in \mathbb{C}^n$$

we have

$$\left( W^{(j)}, W^{(j)} \right) = \begin{cases} \mu_j^2 & 1 \leq j \leq r \\ 0 & j > r \end{cases} \qquad (7.2.21)$$

and

$$\left( W^{(i)}, W^{(j)} \right) = 0 \qquad \text{if } i \neq j \qquad (7.2.22)$$

From (7.2.21), $W^{(j)} = 0$ if $j > r$. And from (7.2.22), the first $r$ columns of $W$ are orthogonal elements in $\mathbb{C}^n$. Thus the first $r$ columns are linearly independent, and this implies $r \leq n$.

Define

$$V^{(j)} = \frac{1}{\mu_j} W^{(j)} \qquad j = 1, \ldots, r \qquad (7.2.23)$$

This is an orthonormal set in $\mathbb{C}^n$. If $r < n$, then choose $V^{(r+1)}, \ldots, V^{(n)}$ so that $\{ V^{(1)}, \ldots, V^{(n)} \}$ is an orthonormal basis for $\mathbb{C}^n$. Define

$$V = \left[ V^{(1)}, \ldots, V^{(n)} \right] \qquad (7.2.24)$$

Easily $V$ is an $n \times n$ unitary matrix, and it can be verified directly that $VF = W$, with $F$ as in (7.2.15). Thus

$$VF = AU$$

which proves (7.2.14). The proof that $r = \text{rank}\,(A)$ and the derivation of other properties of the singular value decomposition are left to Problem 19. The singular value decomposition is used in Chapter 9, in the least squares solution of overdetermined linear systems. ∎

To give the most basic canonical form, introduce the following notation. Define the $n \times n$ matrix

$$J_n(\lambda) = \begin{bmatrix} \lambda & 1 & 0 & \cdots & 0 \\ 0 & \lambda & 1 & & \\ \vdots & & \ddots & & \\ & & & \ddots & 1 \\ 0 & \cdots & & & \lambda \end{bmatrix} \qquad n \geq 1 \qquad (7.2.25)$$

where $J_n(\lambda)$ has the single eigenvalue $\lambda$, of algebraic multiplicity $n$ and geometric multiplicity 1. It is called a *Jordan block*.

**Theorem 7.6** (Jordan Canonical Form)   Let $A$ have order $n$. Then there is a nonsingular matrix $P$ for which

$$P^{-1}AP = \begin{bmatrix} J_{n_1}(\lambda_1) & & & 0 \\ & J_{n_2}(\lambda_2) & & \\ & & \ddots & \\ 0 & & & J_{n_r}(\lambda_r) \end{bmatrix} \qquad (7.2.26)$$

The eigenvalues $\lambda_1, \lambda_2, \ldots, \lambda_r$ need not be distinct. For $A$ Hermitian, Theorem 7.4 implies we must have $n_1 = n_2 = \cdots = n_r = 1$, for in that case the sum of the geometric multiplicities must be $n$, the order of the matrix $A$.

It is often convenient to write (7.2.26) as

$$P^{-1}AP = D + N$$
$$D = \text{diag}[\lambda_1, \ldots, \lambda_r] \qquad (7.2.27)$$

with each $\lambda_i$ appearing $n_i$ times on the diagonal of $D$. The matrix $N$ has all zero entries, except for possible 1s on the superdiagonal. It is a *nilpotent* matrix, and more precisely, it satisfies

$$N^n = 0 \qquad (7.2.28)$$

The Jordan form is not an easy theorem to prove, and the reader is referred to any of the large number of linear algebra texts for a development of this rich topic [e.g., see Franklin (1968), chap. 5; Halmos (1958), sec. 58; or Noble (1969), chap. 11].

## 7.3   Vector and Matrix Norms

The Euclidean norm $\|x\|_2$ has already been introduced, and it is the way in which most people are used to measuring the size of a vector. But there are many situations in which it is more convenient to measure the size of a vector in other ways. Thus we introduce a general concept of the *norm* of a vector.

***Definition***    Let $V$ be a vector space, and let $N(x)$ be a real valued function defined on $V$. Then $N(x)$ is a *norm* if:

**(N1)**    $N(x) \geq 0$ for all $x \in V$, and $N(x) = 0$ if and only if $x = 0$.

**(N2)**    $N(\alpha x) = |\alpha| N(x)$, for all $x \in V$ and all scalars $\alpha$.

**(N3)**    $N(x + y) \leq N(x) + N(y)$, for all $x, y \in V$.

The usual notation is $\|x\| = N(x)$. The notation $N(x)$ is used to emphasize that the norm is a function, with domain $V$ and range the nonnegative real numbers. Define the distance from $x$ to $y$ as $\|x - y\|$. Simple consequences are the *triangular inequality* in its alternative form

$$\|x - z\| \leq \|x - y\| + \|y - z\|$$

and the *reverse triangle inequality*,

$$\big|\,\|x\| - \|y\|\,\big| \leq \|x - y\| \qquad x, y \in V \tag{7.3.1}$$

***Example 1.***    For $1 \leq p < \infty$, define the *p-norm*,

$$\|x\|_p = \left[ \sum_1^n |x_j|^p \right]^{1/p} \qquad x \in \mathbf{C}^n \tag{7.3.2}$$

**2.**    The *maximum norm* is

$$\|x\|_\infty = \operatorname*{Max}_{1 \leq j \leq n} |x_j| \qquad x \in \mathbf{C}^n \tag{7.3.3}$$

The use of the subscript $\infty$ on the norm is motivated by the result in Problem 23.

**3.**    For the vector space $V = C[a, b]$, the function norms $\|f\|_2$ and $\|f\|_\infty$ were introduced in Chapters 4 and 1, respectively.

***Example***    Consider the vector $x = (1, 0, -1, 2)$. Then

$$\|x\|_1 = 4 \qquad \|x\|_2 = \sqrt{6} \qquad \|x\|_\infty = 2$$

To show that $\|\cdot\|_p$ is a norm for a general $p$ is nontrivial. The cases $p = 1$ and $\infty$ are straightforward, and $\|\cdot\|_2$ has been treated in Section 4.1. But for $1 < p < \infty$, $p \neq 2$, it is difficult to show that $\|\cdot\|_p$ satisfies the triangle inequality. This is not a significant problem for us since the main cases of interest are $p = 1, 2, \infty$. To give some geometrical intuition for these norms, the *unit circles*

$$S_p = \left\{ x \in \mathbf{R}^2 \mid \|x\|_p = 1 \right\} \qquad p = 1, 2, \infty \tag{7.3.4}$$

are sketched in Figure 7.2.

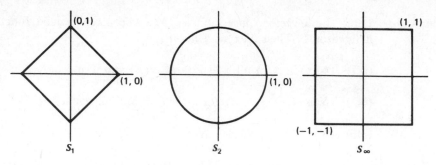

**Figure 7.2** The unit sphere $S_p$ using vector norm $\|\cdot\|_p$.

We now prove some results relating different norms. We begin with the following result on the continuity of $N(x) \equiv \|x\|$, as a function of $x$.

**Lemma** Let $N(x)$ be a norm on $\mathbf{C}^n$ (or $\mathbf{R}^n$). Then $N(x)$ is a continuous function of the components $x_1, x_2, \ldots, x_n$ of $x$.

**Proof** We want to show that

$$x_i \doteq y_i \qquad i = 1, 2, \ldots, n$$

implies

$$N(x) \doteq N(y)$$

Using the reverse triangle inequality (7.3.1),

$$|N(x) - N(y)| \leq N(x - y) \qquad x, y \in \mathbf{C}^n$$

Recall from (7.1.1) the definition of the standard basis $\{e^{(1)}, \ldots, e^{(n)}\}$ for $\mathbf{C}^n$. Then

$$x - y = \sum_{j=1}^{n} (x_j - y_j) e^{(j)}$$

$$N(x - y) \leq \sum_{j=1}^{n} |x_j - y_j| N(e^{(j)}) \leq \|x - y\|_\infty \sum_{j=1}^{n} N(e^{(j)})$$

$$|N(x) - N(y)| \leq c\|x - y\|_\infty \qquad c = \sum_{j=1}^{n} N(e^{(j)}) \qquad (7.3.5)$$

This completes the proof. ∎

Note that it also proves that for every vector norm $N$ on $\mathbf{C}^n$, there is a $c > 0$ with

$$N(x) \leq c\|x\|_\infty \qquad \text{all } x \in \mathbf{C}^n \qquad (7.3.6)$$

Just let $y = 0$ in (7.3.5). The following theorem proves the converse of this result.

***Theorem 7.7*** (Equivalence of Norms)   Let $N$ and $M$ be norms on $V = \mathbf{C}^n$ or $\mathbf{R}^n$. Then there are constants $c_1, c_2 > 0$ for which

$$c_1 M(x) \leq N(x) \leq c_2 M(x) \qquad \text{all } x \in V \qquad (7.3.7)$$

***Proof***   It is sufficient to consider the case in which $N$ is arbitrary and $M(x) = \|x\|_\infty$. Combining two such statements then leads to the general result. Thus we wish to show there are constants $c_1, c_2$ for which

$$c_1 \|x\|_\infty \leq N(x) \leq c_2 \|x\|_\infty \qquad (7.3.8)$$

or equivalently,

$$c_1 \leq N(z) \leq c_2 \qquad \text{all } z \in S \qquad (7.3.9)$$

in which $S$ is the set of all points $z$ in $\mathbf{C}^n$ for which $\|z\|_\infty = 1$. The upper inequality of (7.3.9) follows immediately from (7.3.6).

Note that $S$ is a closed and bounded set in $\mathbf{C}^n$, and $N$ is a continuous function on $S$. It is then a standard result of advanced calculus that $N$ attains its maximum and minimum on $S$ at points of $S$, that is, there are constants $c_1, c_2$ and points $z_1, z_2$ in $S$ for which

$$c_1 = N(z_1) \leq N(z) \leq N(z_2) = c_2 \quad \text{all } z \in S$$

Clearly, $c_1, c_2 \geq 0$. And if $c_1 = 0$, then $N(z_1) = 0$. But then $z_1 = 0$, contrary to the construction of $S$ that requires $\|z_1\|_\infty = 1$. This proves (7.3.9), completing the proof of the theorem. *Note:* This theorem does not generalize to infinite dimensional spaces.  ∎

Many numerical methods for problems involving linear systems produce a sequence of vectors $\{x^{(m)} | m \geq 0\}$, and we want to speak of convergence of this sequence to a vector $x$.

***Definition***   A sequence of vectors $\{x^{(1)}, x^{(2)}, \ldots, x^{(m)}, \cdots\}$, in $\mathbf{C}^n$ or $\mathbf{R}^n$ is said to *converge* to a vector $x$ if and only if

$$\|x - x^{(m)}\| \to 0 \qquad \text{as} \quad m \to \infty$$

Note that the choice of norm is left unspecified. For finite dimensional spaces, it doesn't matter which norm is used. Let $M$ and $N$ be two norms on $\mathbf{C}^n$. Then from (7.3.7),

$$c_1 M(x - x^{(m)}) \leq N(x - x^{(m)}) \leq c_2 M(x - x^{(m)}) \qquad m \geq 0$$

and $M(x - x^{(m)})$ converges to zero if and only if $N(x - x^{(m)})$ does the same. Thus $x^{(m)} \to x$ with the $M$ norm if and only if it converges with the $N$ norm. This is an important result, and it is not true for infinite dimensional spaces.

**Matrix norms**   The set of all $n \times n$ matrices with complex entries can be considered as equivalent to the vector space $\mathbf{C}^{n^2}$, with a special multiplicative operation added onto the vector space. Thus a matrix norm should satisfy the usual three requirements N1–N3 of a vector norm. In addition, we also require two other conditions.

**Definition**  A *matrix norm* satisfies N1–N3 and the following:

**(N4)**  $\|AB\| \leq \|A\|\,\|B\|$.

**(N5)**  Usually the vector space we will be working with, $V = \mathbf{C}^n$ or $\mathbf{R}^n$, will have some vector norm, call it $\|x\|_v$, $x \in V$. We require that the matrix and vector norms be *compatible*:

$$\|Ax\|_v \leq \|A\|\,\|x\|_v \qquad \text{all } x \in V \qquad \text{all } A$$

**Example**  Let $A$ be $n \times n$, $\|\cdot\|_v = \|\cdot\|_2$. Then for $x \in \mathbf{C}^n$,

$$\|Ax\|_2 = \left[ \sum_{i=1}^{n} \left| \sum_{j=1}^{n} a_{ij}x_j \right|^2 \right]^{1/2}$$

$$\leq \left[ \sum_{i=1}^{n} \left\{ \sum_{j=1}^{n} |a_{ij}|^2 \right\} \left\{ \sum_{j=1}^{n} |x_j|^2 \right\} \right]^{1/2}$$

by using the Cauchy–Schwartz inequality (7.1.8). Then

$$\|Ax\|_2 \leq F(A)\|x\|_2 \qquad F(A) = \left[ \sum_{i,\,j=1}^{n} |a_{ij}|^2 \right]^{1/2} \qquad (7.3.10)$$

$F(A)$ is called the *Frobenius norm* of $A$. Property N5 is shown using (7.3.10) directly. Properties N1–N3 are satisfied since $F(A)$ is just the Euclidean norm on $\mathbf{C}^{n^2}$. It remains to show N4. Using the Cauchy–Schwartz inequality,

$$F(AB) = \left[ \sum_{i,\,j=1}^{n} \left| \sum_{k=1}^{n} a_{ik}b_{kj} \right|^2 \right]^{1/2}$$

$$\leq \left[ \sum_{i,\,j=1}^{n} \left\{ \sum_{k=1}^{n} |a_{ik}|^2 \right\} \left\{ \sum_{k=1}^{n} |b_{kj}|^2 \right\} \right]^{1/2}$$

$$= F(A)F(B)$$

Thus $F(A)$ is a matrix norm, compatible with the Euclidean norm.

Usually when given a vector space with a norm $\|\cdot\|_v$, an associated matrix norm is defined by

$$\|A\| = \operatorname*{Supremum}_{x \neq 0} \frac{\|Ax\|_v}{\|x\|_v} \qquad (7.3.11)$$

**Table 7.1    Vector norms and associated operator matrix norms**

| Vector Norm | Matrix Norm |
|:---:|:---:|
| $\|x\|_1 = \displaystyle\sum_{i=1}^{n} \|x_i\|$ | $\|A\|_1 = \displaystyle\operatorname*{Max}_{1 \leq j \leq n} \sum_{i=1}^{n} \|a_{ij}\|$ |
| $\|x\|_2 = \left[ \displaystyle\sum_{j=1}^{n} \|x_i\|^2 \right]^{1/2}$ | $\|A\|_2 = \sqrt{r_\sigma(A^*A)}$ |
| $\|x\|_\infty = \displaystyle\operatorname*{Max}_{1 \leq i \leq n} \|x_i\|$ | $\|A\|_\infty = \displaystyle\operatorname*{Max}_{1 \leq i \leq n} \sum_{j=1}^{n} \|a_{ij}\|$ |

It is often called the *operator norm*. By its definition, it satisfies N5:

$$\|Ax\|_v \leq \|A\| \, \|x\|_v \qquad x \in \mathbf{C}^n \qquad (7.3.12)$$

For a matrix $A$, the operator norm induced by the vector norm $\|x\|_p$ will be denoted by $\|A\|_p$. The most important cases are given in Table 7.1, and the derivations are given later. We need the following definition in order to define $\|A\|_2$.

**Definition**    Let $A$ be an arbitrary matrix. The *spectrum* of $A$ is the set of all eigenvalues of $A$, and it is denoted by $\sigma(A)$. The *spectral radius* is the maximum size of these eigenvalues, and it is denoted by

$$r_\sigma(A) = \operatorname*{Max}_{\lambda \in \sigma(A)} |\lambda| \qquad (7.3.13)$$

To show (7.3.11) is a norm in general, we begin by showing it is finite. Recall from Theorem 7.7 that there are constants $c_1, c_2 > 0$ with

$$c_1\|x\|_2 \leq \|x\|_v \leq c_2\|x\|_2 \qquad x \in \mathbf{C}^n$$

Thus,

$$\frac{\|Ax\|_v}{\|x\|_v} \leq \frac{c_2\|Ax\|_2}{c_1\|x\|_2} \leq \frac{c_2}{c_1} F(A)$$

which proves $\|A\|$ is finite.

At this point it is interesting to note the geometric significance of $\|A\|$:

$$\|A\| = \operatorname*{Supremum}_{x \neq 0} \frac{\|Ax\|_v}{\|x\|_v} = \operatorname*{Supremum}_{x \neq 0} \left\| A\left( \frac{x}{\|x\|_v} \right) \right\|_v = \operatorname*{Supremum}_{\|z\|_v = 1} \|Az\|_v$$

By noting that the supremum doesn't change if we let $\|z\|_v \leq 1$,

$$\|A\| = \operatorname*{Supremum}_{\|z\|_v \leq 1} \|Az\| \qquad (7.3.14)$$

Let

$$B = \left\{ z \in \mathbf{C}^n \middle| \|z\|_v \leq 1 \right\}$$

the unit ball with respect to $\| \cdot \|_v$. Then

$$\|A\| = \operatorname*{Supremum}_{z \in B} \|Az\|_v = \operatorname*{Supremum}_{w \in A(B)} \|w\|_v$$

with $A(B)$ the image of $B$ when $A$ is applied to it. Thus $\|A\|$ measures the effect of $A$ on the unit ball, and if $\|A\| > 1$, then $\|A\|$ denotes the maximum stretching of the ball $B$ under the action of $A$.

**Proof**  Following is a proof that the operator norm $\|A\|$ is a matrix norm.

1. Clearly $\|A\| \geq 0$, and if $A = 0$, then $\|A\| = 0$. Conversely, if $\|A\| = 0$, then $\|Ax\|_v = 0$ for all $x$. Thus $Ax = 0$ for all $x$, and this implies $A = 0$.

2. Let $\alpha$ be any scalar. Then

$$\|\alpha A\| = \operatorname*{Supremum}_{\|x\|_v \leq 1} \|\alpha Ax\|_v = \operatorname*{Supremum}_{\|x\|_v \leq 1} |\alpha| \|Ax\|_v = |\alpha| \operatorname*{Supremum}_{\|x\|_v \leq 1} \|Ax\|_v$$

$$= |\alpha| \|A\|$$

3. For any $x \in \mathbf{C}^n$,

$$\|(A + B)x\|_v = \|Ax + Bx\|_v \leq \|Ax\|_v + \|Bx\|_v$$

since $\| \cdot \|_v$ is a norm. Using the property (7.3.12),

$$\|(A + B)x\|_v \leq \|A\| \|x\|_v + \|B\| \|x\|_v$$

$$\frac{\|(A + B)x\|_v}{\|x\|_v} \leq \|A\| + \|B\|$$

This implies

$$\|A + B\| \leq \|A\| + \|B\|$$

4. For any $x \in \mathbf{C}^n$, use (7.3.12) to get

$$\|(AB)x\|_v = \|A(Bx)\|_v \leq \|A\| \|Bx\|_v \leq \|A\| \|B\| \|x\|_v$$

$$\frac{\|ABx\|_v}{\|x\|_v} \leq \|A\| \|B\|$$

This implies

$$\|AB\| \leq \|A\| \|B\| \qquad \blacksquare$$

We now comment more extensively on the results given in Table 7.1.

***Example* 1.**    Use the vector norm

$$\|x\|_1 = \sum_{j=1}^{n} |x_j| \qquad x \in \mathbf{C}^n$$

Then

$$\|Ax\|_1 = \sum_{i=1}^{n} \left| \sum_{j=1}^{n} a_{ij} x_j \right| \leq \sum_{i=1}^{n} \sum_{j=1}^{n} |a_{ij}| \, |x_j|$$

Changing the order of summation, we can separate the summands,

$$\|Ax\|_1 \leq \sum_{j=1}^{n} |x_j| \sum_{i=1}^{n} |a_{ij}|$$

Let

$$c = \operatorname*{Max}_{1 \leq j \leq n} \sum_{i=1}^{n} |a_{ij}| \qquad (7.3.15)$$

Then

$$\|Ax\|_1 \leq c\|x\|_1$$

and thus

$$\|A\|_1 \leq c$$

To show this is an equality, we demonstrate an $x$ for which

$$\frac{\|Ax\|_1}{\|x\|_1} = c$$

Let $k$ be the column index for which the maximum in (7.3.15) is attained. Let $x = e^{(k)}$, the $k$th unit vector. Then $\|x\|_1 = 1$ and

$$\|Ax\|_1 = \sum_{i=1}^{n} \left| \sum_{j=1}^{n} a_{ij} x_j \right| = \sum_{i=1}^{n} |a_{ik}| = c$$

This proves that for the vector norm $\| \cdot \|_1$, the operator norm is

$$\|A\|_1 = \operatorname*{Max}_{1 \leq j \leq n} \sum_{i=1}^{n} |a_{ij}| \qquad (7.3.16)$$

This is often called the *column norm*.

2.    For $\mathbf{C}^n$ with the norm $\|x\|_\infty$, the operator norm is

$$\|A\|_\infty = \operatorname*{Max}_{1 \leq i \leq n} \sum_{j=1}^{n} |a_{ij}| \qquad (7.3.17)$$

This is called the *row norm* of $A$. The proof of the formula is left as Problem 25, although it is similar to that for $\|A\|_1$.

3.  Use the norm $\|x\|_2$ on $\mathbf{C}^n$. From (7.3.10), we conclude that

$$\|A\|_2 \le F(A) \tag{7.3.18}$$

In general, these are not equal. For example, with $A = I$, the identity matrix, use (7.3.10) and (7.3.11) to obtain

$$F(I) = \sqrt{n} \qquad \|I\|_2 = 1$$

We prove

$$\|A\|_2 = \sqrt{r_\sigma(A^*A)} \tag{7.3.19}$$

as stated earlier in Table 7.1. The matrix $A^*A$ is Hermitian and all of its eigenvalues are nonnegative, as shown in the proof of Theorem 7.5. Let it have the eigenvalues

$$\lambda_1 \ge \lambda_2 \ge \cdots \ge \lambda_n \ge 0$$

counted according to their multiplicity, and let $u^{(1)}, \ldots, u^{(n)}$ be the corresponding eigenvectors, arranged as an orthonormal basis for $\mathbf{C}^n$.

For a general $x \in \mathbf{C}^n$,

$$\|Ax\|_2^2 = (Ax, Ax) = (x, A^*Ax)$$

Write $x$ as

$$x = \sum_{j=1}^{n} \alpha_j u^{(j)} \qquad \alpha_j \equiv (x, u^{(j)}) \tag{7.3.20}$$

Then

$$A^*Ax = \sum_{j=1}^{n} \alpha_j A^* A u^{(j)} = \sum_{j=1}^{n} \alpha_j \lambda_j u^{(j)}$$

and

$$\|Ax\|_2^2 = \left( \sum_{i=1}^{n} \alpha_i u^{(i)}, \sum_{j=1}^{n} \alpha_j \lambda_j u^{(j)} \right) = \sum_{j=1}^{n} \lambda_j |\alpha_j|^2$$

$$\le \lambda_1 \sum_{j=1}^{n} |\alpha_j|^2 = \lambda_1 \|x\|_2^2$$

using (7.3.20) to calculate $\|x\|_2$. Thus

$$\|A\|_2^2 \le \lambda_1$$

Equality follows by noting that if $x = u^{(1)}$, then $\|x\|_2 = 1$ and

$$\|Ax\|_2^2 = (x, A^*Ax) = \left(u^{(1)}, \lambda_1 u^{(1)}\right) = \lambda_1$$

This proves (7.3.19), since $\lambda_1 = r_\sigma(A^*A)$. It can be shown that $AA^*$ and $A^*A$ have the same nonzero eigenvalues (see Problem 19); thus, $r_\sigma(AA^*) = r_\sigma(A^*A)$, an alternative formula for (7.3.19). It also proves

$$\|A\|_2 = \|A^*\|_2 . \tag{7.3.21}$$

This is not true for the previous matrix norms.

It can be shown fairly easily that if $A$ is Hermitian, then

$$\|A\|_2 = r_\sigma(A) \tag{7.3.22}$$

This is left as Problem 27.

**Example**   Consider the matrix

$$A = \begin{bmatrix} 1 & -2 \\ -3 & 4 \end{bmatrix}$$

Then

$$\|A\|_1 = 6 \qquad \|A\|_2 = \sqrt{15 + \sqrt{221}} \doteq 5.46 \qquad \|A\|_\infty = 7$$

As an illustration of the inequality (7.3.23) of the following theorem,

$$r_\sigma(A) = \frac{5 + \sqrt{33}}{2} \doteq 5.37 < \|A\|_2$$

**Theorem 7.8**   Let $A$ be an arbitrary square matrix. Then for any operator matrix norm,

$$r_\sigma(A) \le \|A\| \tag{7.3.23}$$

Moreover, if $\epsilon > 0$ is given, then there is an operator matrix norm, denoted here by $\|\cdot\|_\epsilon$, for which

$$\|A\|_\epsilon \le r_\sigma(A) + \epsilon \tag{7.3.24}$$

**Proof**   To prove (7.3.23), let $\|\cdot\|$ be any matrix norm with an associated compatible vector norm $\|\cdot\|_v$. Let $\lambda$ be the eigenvalue in $\sigma(A)$ for which

$$|\lambda| = r_\sigma(A)$$

and let $x$ be an associated eigenvector, $\|x\|_v = 1$. Then

$$r_\sigma(A) = |\lambda| = \|\lambda x\|_v \le \|Ax\|_v \le \|A\| \, \|x\|_v = \|A\|$$

which proves (7.3.23).

The proof of (7.3.24) is a nontrivial construction, and a proof is given in Isaacson and Keller (1966, p. 12).   ∎

The following corollary is an easy, but important, consequence of Theorem 7.8.

*Corollary*   For a square matrix $A$, $r_\sigma(A) < 1$ if and only if $\|A\| < 1$ for some operator matrix norm.   ∎

This result can be used to prove Theorem 7.9 in the next section, but we prefer to use the Jordan canonical form, given in Theorem 7.6. The results (7.3.22) and Theorem 7.8 show that $r_\sigma(A)$ is almost a matrix norm, and this result is used in analyzing the rates of convergence for some of the iteration methods given in Chapter 8 for solving linear systems of equations.

## 7.4   Convergence and Perturbation Theorems

The following results are the theoretical framework from which we later construct error analyses for numerical methods for linear systems of equations.

*Theorem 7.9*   Let $A$ be a square matrix of order $n$. Then $A^m$ converges to the zero matrix as $m \to \infty$ if and only if $r_\sigma(A) < 1$.

*Proof*   We use Theorem 7.6 as a fundamental tool. Let $J$ be the Jordan canonical form for $A$,

$$P^{-1}AP = J$$

Then

$$A^m = \left(PJP^{-1}\right)^m = PJ^mP^{-1} \tag{7.4.1}$$

and $A^m \to 0$ if and only if $J^m \to 0$. Recall from (7.2.27) and (7.2.28) that $J$ can be written as

$$J = D + N$$

in which

$$D = \operatorname{diag}[\lambda_1, \ldots, \lambda_n]$$

contains the eigenvalues of $J$ (and $A$), and $N$ is a matrix for which

$$N^n = 0$$

Thus

$$J^m = (D + N)^m = \sum_{j=0}^m \binom{m}{j} D^{m-j} N^j$$

and using $N^j = 0$ for $j \geq n$,

$$J^m = \sum_{j=0}^{n} \binom{m}{j} D^{m-j} N^j \qquad (7.4.2)$$

Notice that the powers of $D$ satisfy

$$m - j \geq m - n \to \infty \qquad \text{as} \quad m \to \infty \qquad (7.4.3)$$

We need the following limits: For any positive $c < 1$ and any $r \geq 0$,

$$\operatorname*{Limit}_{m \to \infty} m^r c^m = 0 \qquad (7.4.4)$$

This can be proved using L'Hospital's rule from elementary calculus.

In (7.4.2), there are a fixed number of terms, $n + 1$, regardless of the size of $m$, and we can consider the convergence of $J^m$ by considering each of the individual terms. Assuming $r_\sigma(A) < 1$, we know that all $|\lambda_i| < 1$, $i = 1, \ldots, n$. And for any matrix norm

$$\left\| \binom{m}{j} D^{m-j} N^j \right\| \leq \frac{m^j}{j!} \|N\|^j \|D^{m-j}\|$$

Using the row norm, we have that the preceding is bounded by

$$\frac{1}{j!} \|N\|_\infty^j m^j [r_\sigma(A)]^{m-j}$$

which converges to zero as $m \to \infty$, using (7.4.3) and (7.4.4), for $0 \leq j \leq n$. This proves half of the theorem, namely that if $r_\sigma(A) < 1$, then $J^m$ and $A^m$, from (7.4.1), converge to zero as $m \to \infty$.

Suppose that $r_\sigma(A) \geq 1$. Then let $\lambda$ be an eigenvalue of $A$ for which $|\lambda| \geq 1$, and let $x$ be an associated eigenvector, $x \neq 0$. Then

$$A^m x = \lambda^m x$$

and clearly this does not converge to zero as $m \to \infty$. Thus it is not possible that $A^m \to 0$, as that would imply $A^m x \to 0$. This completes the proof.    ∎

**Theorem 7.10** (Geometric Series)    Let $A$ be a square matrix. If $r_\sigma(A) < 1$, then $(I - A)^{-1}$ exists, and it can be expressed as a convergent series,

$$(I - A)^{-1} = I + A + A^2 + \cdots + A^m + \cdots \qquad (7.4.5)$$

Conversely, if the series in (7.4.5) is convergent, then $r_\sigma(A) < 1$.

**Proof**    Assume $r_\sigma(A) < 1$. We show the existence of $(I - A)^{-1}$ by proving the equivalent statement (3) of Theorem 7.2. Assume

$$(I - A)x = 0$$

Then $Ax = x$, and this implies that 1 is an eigenvalue of $A$ if $x \neq 0$. But we assumed $r_\sigma(A) < 1$, and thus we must have $x = 0$, concluding the proof of the existence of $(I - A)^{-1}$.

We need the following identity:

$$(I - A)(I + A + A^2 + \cdots + A^m) = I - A^{m+1} \qquad (7.4.6)$$

which is true for any matrix $A$. Multiplying by $(I - A)^{-1}$,

$$I + A + A^2 + \cdots + A^m = (I - A)^{-1}(I - A^{m+1})$$

The left-hand side has a limit if the right-hand side does. By Theorem 7.9, $r_\sigma(A) < 1$ implies that $A^{m+1} \to 0$ as $m \to \infty$. Thus we have the result (7.4.5).

Conversely, assume the series converges and denote it by

$$B = I + A + A^2 + \cdots + A^m + \cdots$$

Then $B - AB = B - BA = I$, and thus $I - A$ has an inverse, namely $B$. Taking limits on both sides of (7.4.6), the left-hand side has the limit $(I - A)B = I$, and thus the same must be true of the right-hand limit. But that implies

$$A^{m+1} \to 0 \quad \text{as} \quad m \to \infty$$

By Theorem 7.9, we must have $r_\sigma(A) < 1$. ∎

**Theorem 7.11**  Let $A$ be a square matrix. If for some operator matrix norm, $\|A\| < 1$, then $(I - A)^{-1}$ exists and has the geometric series expansion (7.4.5). Moreover,

$$\left\| (I - A)^{-1} \right\| \le \frac{1}{1 - \|A\|} \qquad (7.4.7)$$

**Proof**  Since $\|A\| < 1$, it follows from (7.3.23) of Theorem 7.8 that $r_\sigma(A) < 1$. Except for (7.4.7), the other conclusions follow from Theorem 7.10. For (7.4.7), let

$$B_m = I + A + \cdots + A^m$$

From (7.4.6),

$$B_m = (I - A)^{-1}(I - A^{m+1})$$

$$(I - A)^{-1} - B_m = (I - A)^{-1}\left[ I - (I - A^{m+1}) \right]$$

$$= (I - A)^{-1} A^{m+1} \qquad (7.4.8)$$

Using the reverse triangle inequality,

$$\left| \|(I - A)^{-1}\| - \|B_m\| \right| \le \|(I - A)^{-1} - B_m\|$$

$$< \|(I - A)^{-1}\| \, \|A\|^{m+1}$$

Since this converges to zero as $m \to \infty$, we have

$$\|B_m\| \to \|(I - A)^{-1}\| \qquad \text{as} \quad m \to \infty \qquad (7.4.9)$$

From the definition of $B_m$ and the properties of a matrix norm,

$$\|B_m\| \le \|I\| + \|A\| + \|A\|^2 + \cdots + \|A\|^m$$

$$= \frac{1 - \|A\|^{m+1}}{1 - \|A\|} \le \frac{1}{1 - \|A\|}$$

Combined with (7.4.9), this concludes the proof of (7.4.7).  ∎

**Theorem 7.12**  Let $A$ and $B$ be square matrices of the same order. Assume $A$ is nonsingular and suppose that

$$\|A - B\| < \frac{1}{\|A^{-1}\|} \qquad (7.4.10)$$

Then $B$ is also nonsingular,

$$\|B^{-1}\| \le \frac{\|A^{-1}\|}{1 - \|A^{-1}\| \, \|A - B\|} \qquad (7.4.11)$$

and

$$\|A^{-1} - B^{-1}\| \le \frac{\|A^{-1}\|^2 \|A - B\|}{1 - \|A^{-1}\| \, \|A - B\|} \qquad (7.4.12)$$

**Proof**  Note the identity

$$B = A - (A - B) = A[I - A^{-1}(A - B)] \qquad (7.4.13)$$

The matrix $[I - A^{-1}(A - B)]$ is nonsingular using Theorem 7.11, based on the inequality (7.4.10), which implies

$$\|A^{-1}(A - B)\| \le \|A^{-1}\| \, \|A - B\| < 1$$

Since $B$ is the product of nonsingular matrices, it too is nonsingular,

$$B^{-1} = [I - A^{-1}(A - B)]^{-1} A^{-1}$$

The bound (7.4.11) follows by taking norms and applying Theorem 7.11. To prove (7.4.12), use

$$A^{-1} - B^{-1} = A^{-1}(B - A)B^{-1}$$

Take norms again and apply (7.4.11). ■

This theorem is important in a number of ways. But for the moment, it says that all sufficiently close perturbations of a nonsingular matrix are nonsingular.

***Example*** We illustrate Theorem 7.11 by considering the invertibility of the matrix

$$A = \begin{bmatrix} 4 & 1 & 0 & 0 & \cdots & & 0 \\ 1 & 4 & 1 & 0 & & & \\ 0 & 1 & 4 & 1 & & & \vdots \\ \vdots & & & \ddots & & & \\ & & & 1 & 4 & 1 \\ 0 & \cdots & & & 0 & 1 & 4 \end{bmatrix}$$

Rewrite $A$ as

$$A = 4(I + B),$$

$$B = \begin{bmatrix} 0 & \dfrac{1}{4} & 0 & 0 & \cdots & & 0 \\ \dfrac{1}{4} & 0 & \dfrac{1}{4} & 0 & & & 0 \\ 0 & & & \ddots & & & \vdots \\ \vdots & & & & & & \dfrac{1}{4} \\ 0 & \cdots & & & & \dfrac{1}{4} & 0 \end{bmatrix}$$

Using the row norm (7.3.17), $\|B\|_\infty = \frac{1}{2}$. Thus $(I + B)^{-1}$ exists from Theorem 7.11, and from (7.4.7),

$$\left\|(I + B)^{-1}\right\|_\infty \le \frac{1}{1 - \frac{1}{2}} = 2$$

Thus $A^{-1}$ exists, $A^{-1} = \frac{1}{4}(I + B)^{-1}$, and

$$\|A^{-1}\|_\infty \le \frac{1}{2}$$

By use of the row norm and inequality (7.3.23),

$$r_\sigma(A) \le 6 \qquad r_\sigma(A^{-1}) \le \frac{1}{2}$$

Since the eigenvalues of $A^{-1}$ are the reciprocals of those of $A$ (see problem 27), and since all eigenvalues of $A$ are real because $A$ is Hermitian, we have the bound

$$2 \le |\lambda| \le 6 \qquad \text{all } \lambda \in \sigma(A)$$

For better bounds in this case, see the Gerschgorin Circle Theorem of Chapter 9.

## Discussion of the Literature

The subject of this chapter is linear algebra, especially selected for use in deriving and analyzing methods of numerical linear algebra. The books by Anton (1984) and Strang (1980) are introductory-level texts for undergraduate linear algebra. Franklin's (1968) is a higher level introduction to matrix theory, and Halmos's (1958) is a well-known text on abstract linear algebra. Noble's (1969) is a wide-ranging applied linear algebra text. Introductions to the foundations are also contained in Fadeeva (1959), Golub and Van Loan (1982), Parlett (1980), Stewart (1973), and Wilkinson (1965), all of which are devoted entirely to numerical linear algebra. For additional theory at a more detailed and higher level, see the classical accounts of Gantmacher (1960) and Householder (1965). Additional references are given in the bibliographies of Chapters 8 and 9.

## Bibliography

Anton, H. (1984). *Elementary Linear Algebra*, 4th ed. Wiley, New York.

Fadeeva, V. (1959). *Computational Methods of Linear Algebra*. Dover, New York.

Franklin, J. (1968). *Matrix Theory*. Prentice-Hall, Englewood Cliffs, N.J.

Gantmacher, F. (1960). *The Theory of Matrices*, vols. I and II. Chelsea, New York.

Golub, G., and C. Van Loan (1983). *Matrix Computations*. Johns Hopkins Press, Baltimore.

Halmos, P. (1958). *Finite-Dimensional Vector Spaces*. Van Nostrand, Princeton, N.J.

Householder, A. (1965). *The Theory of Matrices in Numerical Analysis*. Ginn (Blaisdell), Boston.

Isaacson, E., and H. Keller (1966). *Analysis of Numerical Methods*. Wiley, New York.

Lawson, C., and R. Hanson (1974). *Solving Least Squares Problems*. Prentice-Hall, Englewood Cliffs, N.J.

Noble, B. (1969). *Applied Linear Algebra*. Prentice-Hall, Englewood Cliffs, N.J.

Parlett, B. (1980). *The Symmetric Eigenvalue Problem*. Prentice-Hall, Englewood Cliffs, N.J.

Stewart, G. (1973). *Introduction to Matrix Computations*. Academic Press, New York.

Strang, G. (1980). *Linear Algebra and Its Applications*, 2nd ed. Academic Press, New York.

Wilkinson, J. (1965). *The Algebraic Eigenvalue Problem*. Oxford Univ. Press, Oxford, England.

## Problems

1.  Determine whether the following sets of vectors are dependent or independent.

    **(a)**   $(1, 2, -1, 3), (3, -1, 1, 1), (1, 9, -5, 11)$

    **(b)**   $(1, 1, 0), (0, 1, 1), (1, 0, 1)$

2.  Let $A$, $B$, and $C$ be matrices of order $m \times n$, $n \times p$, and $p \times q$, respectively.

    **(a)**   Prove the associative law $(AB)C = A(BC)$.

    **(b)**   Prove $(AB)^T = B^T A^T$.

3.  **(a)**   Produce square matrices $A$ and $B$ for which $AB \neq BA$.

    **(b)**   Produce square matrices $A$ and $B$, with no zero entries, for which $AB = 0$, $BA \neq 0$.

4.  Let $A$ be a matrix of order $m \times n$, and let $r$ and $c$ denote the row and column rank of $A$, respectively. Prove that $r = c$. *Hint:* For convenience, assume that the first $r$ rows of $A$ are independent, with the remaining rows dependent on these first $r$ rows, and assume the same for the first $c$ columns of $A$. Let $\hat{A}$ denote the $r \times n$ matrix obtained by deleting the last $m - r$ rows of $A$, and let $\hat{r}$ and $\hat{c}$ denote the row and column rank of $\hat{A}$, respectively. Clearly $\hat{r} = r$. Also, the columns of $\hat{A}$ are elements of $\mathbf{C}^r$, which has dimension $r$, and thus we must have $\hat{c} \leq r$. Show that $\hat{c} = c$, thus proving that $c \leq r$. The reverse inequality will follow by applying the same argument to $A^T$, and taken together, these two inequalities imply $r = c$.

5.  Prove the equivalence of statements (1)–(4) and (6) of Theorem 7.2. *Hint:* Use Theorem 7.1, the result in Problem 4, and the decomposition (7.1.6).

**6.**   Let

$$f_n(x) = \det \begin{bmatrix} x & 1 & 0 & \cdots & & 0 \\ 1 & x & 1 & 0 & & 0 \\ 0 & 1 & x & 1 & & \vdots \\ \vdots & & & \ddots & & 0 \\ & & & & & 1 \\ 0 & \cdots & & 0 & 1 & x \end{bmatrix}$$

with the matrix of order $n$. Also define $f_0(x) \equiv 1$.

**(a)**   Show

$$f_{n+1}(x) = xf_n(x) - f_{n-1}(x) \qquad n \geq 1$$

**(b)**   Show

$$f_n(x) = S_n\left(\frac{x}{2}\right) \qquad n \geq 0$$

with $S_n(x)$ the Chebyshev polynomial of the second kind of degree $n$ (see Problem 24 in Chapter 4).

**7.**   Let $A$ be a square matrix of order $n$ with real entries. The function

$$q(x_1,\ldots, x_n) = (Ax, x) = \sum_{i=1}^{n} \sum_{j=1}^{n} a_{ij}x_i x_j \qquad x \in \mathbf{R}^n,$$

is called the *quadratic form* determined by $A$. It is a quadratic polynomial in the $n$ variables $x_1,\ldots, x_n$, and it occurs when considering the maximization or minimization of a function of $n$ variables.

**(a)**   Prove that if $A$ is skew-symmetric, then $q(x) \equiv 0$.

**(b)**   For a general square matrix $A$, define $A_1 = \frac{1}{2}(A + A^T)$, $A_2 = \frac{1}{2}(A - A^T)$. Then $A = A_1 + A_2$. Show that $A_1$ is symmetric, and $(Ax, x) = (A_1 x, x)$, all $x \in \mathbf{R}^n$. This shows that the coefficient matrix $A$ for a quadratic form can always be assumed to be symmetric, without any loss of generality.

**8.**   Given the orthogonal vectors

$$u^{(1)} = (1, 2, -1) \qquad u^{(1)} = (1, 1, 3)$$

produce a third vector $u^{(3)}$ such that $\{u^{(1)}, u^{(2)}, u^{(3)}\}$ is an orthogonal basis for $\mathbf{R}^3$. Normalize these vectors to obtain an orthonormal basis.

9.   For the column vector $w \in \mathbf{C}^n$ with $\|w\|_2 = \sqrt{w^*w} = 1$, define the $n \times n$ matrix

$$A = I - 2ww^*$$

(a)   For the special case $w = [\frac{1}{3}, \frac{2}{3}, \frac{2}{3}]^T$, produce the matrix $A$. Verify that it is symmetric and orthogonal.

(b)   Show that, in general, all such matrices $A$ are Hermitian and unitary.

10.   Let $W$ be a subspace of $\mathbf{R}^n$. For $x \in \mathbf{R}^n$, define

$$\rho(x) = \underset{y \in W}{\text{Infimum}} \|x - y\|_2$$

Let $\{u_1, \ldots, u_m\}$ be an orthonormal basis of $W$, where $m$ is the dimension of $W$. Extend this to an orthonormal basis $\{u_1, \ldots, u_m, \ldots, u_n\}$ of all of $\mathbf{R}^n$.

(a)   Show that

$$\rho(x) = \left[ \sum_{j=m+1}^{n} |(x, u_j)|^2 \right]^{1/2}$$

and that it is uniquely attained at

$$y = Px \qquad P = \sum_{j=1}^{m} u_j u_j^T$$

This is called the *orthogonal projection* of $x$ onto $W$.

(b)   Show $P^2 = P$. Such a matrix is called a *projection matrix*.

(c)   Show $P^T = P$.

(d)   Show $(Px, z - Pz) = 0$ for all $x, z \in \mathbf{R}^n$.

(e)   Show $\|x\|_2^2 = \|Px\|_2^2 + \|x - Px\|_2^2$, for all $x \in \mathbf{R}^n$. This is a version of the *theorem of Pythagoras*.

11.   Calculate the eigenvalues and eigenvectors of the following matrices.

(a)   $\begin{bmatrix} 1 & 4 \\ 1 & 1 \end{bmatrix}$ 
 
(b)   $\begin{bmatrix} 1 & 1 \\ -1 & 3 \end{bmatrix}$

12.   Let $y \neq 0$ in $\mathbf{R}^n$, and define $A = yy^T$, an $n \times n$ matrix. Show that $\lambda = 0$ is an eigenvalue of multiplicity exactly $n - 1$. What is the single nonzero eigenvalue?

13. Let $U$ be an $n \times n$ unitary matrix.

    **(a)** Show $\|Ux\|_2 = \|x\|_2$, all $x \in \mathbf{C}^n$. Use this to prove that the distance between points $x$ and $y$ is the same as the distance between $Ux$ and $Uy$, showing that unitary transformations of $\mathbf{C}^n$ preserve distances between all points.

    **(b)** Let $U$ be orthogonal, and show that

$$(Ux, Uy) = (x, y) \qquad x, y \in \mathbf{R}^n$$

    This shows that orthogonal transformations of $\mathbf{R}^n$ also preserve angles between lines, as defined in (7.1.12).

    **(c)** Show that all eigenvalues of a unitary matrix have magnitude one.

14. Let $A$ be a Hermitian matrix of order $n$. It is called *positive definite* if and only if $(Ax, x) > 0$ for all $x \neq 0$ in $\mathbf{C}^n$. Show that $A$ is positive definite if and only if all of its eigenvalues are real and positive. *Hint:* Use Theorem 7.4, and expand $(Ax, x)$ by using an eigenvector basis to express an arbitrary $x \in \mathbf{C}^n$.

15. Let $A$ be real and symmetric, and denote its eigenvalues by $\lambda_1, \ldots, \lambda_n$, repeated according to their multiplicity. Using a basis of orthonormal eigenvectors, show that the quadratic form of Problem 7, $q(x) = (Ax, x)$, $x \in \mathbf{R}^n$, can be reduced to the simpler form

$$q(x) = \sum_{j=1}^{n} \alpha_j^2 \lambda_j$$

    with the $\{\alpha_j\}$ determined from $x$. Using this, explore the possible graphs for

$$(Ax, x) = \text{constant}$$

when $A$ is of order 3.

16. Assume $A$ is real, symmetric, positive definite, and of order $n$. Define

$$f(x) = \frac{1}{2}x^T A x - b^T x \qquad x, b \in \mathbf{R}^n$$

    Show that the unique minimum of $f(x)$ is given by solving $Ax = b$ for $\alpha = A^{-1}b$.

17. Let $f(x)$ be a real valued function of $x \in \mathbf{R}^n$, and assume $f(x)$ is three times continuously differentiable with respect to the components of $x$.

Apply Taylor's theorem 1.5 of Chapter 1, generalized to $n$ variables, to obtain

$$f(x) = f(\alpha) + (x - \alpha)^T \nabla f(\alpha)$$

$$+ \frac{1}{2}(x - \alpha)^T H(\alpha)(x - \alpha) + O(\|x - \alpha\|^3)$$

Here

$$\nabla f(x) = \left[ \frac{\partial f}{\partial x_1}, \ldots, \frac{\partial f}{\partial x_n} \right]^T$$

is the *gradient* of $f$, and

$$H(x) = \left[ \frac{\partial^2 f(x)}{\partial x_i \partial x_j} \right] \qquad 1 \leq i, j \leq n$$

is the *Hessian matrix* for $f(x)$. The final term indicates that the remaining terms are smaller than some multiple of $\|x - \alpha\|^3$ for $x$ close to $\alpha$.

If $\alpha$ is to be a local maximum or minimum, then a necessary condition is that $\nabla f(x) = 0$. Assuming $\nabla f(\alpha) = 0$, show that $\alpha$ is a strict (or unique) local minimum of $f(x)$ if and only if $H(\alpha)$ is positive definite. [Note that $H(x)$ is always symmetric.]

**18.**   Demonstrate the relation (7.2.28).

**19.**   Recall the notation used in Theorem 7.5 on the singular value decomposition of a matrix $A$.

**(a)**   Show that $\mu_1^2, \ldots, \mu_r^2$ are the nonzero eigenvalues of $A^*A$ and $AA^*$, with corresponding eigenvectors $U^{(1)}, \ldots, U^{(r)}$ and $V^{(1)}, \ldots, V^{(r)}$, respectively. The vector $U^{(j)}$ denotes column $j$ of $U$, and similarly for $V^{(j)}$ and $V$.

**(b)**   Show that $AU^{(j)} = \mu_j V^{(j)}$, $A^*V^{(j)} = \mu_j U^{(j)}$, $1 \leq j \leq r$.

**(c)**   Prove $r = \text{rank}(A)$.

**20.**   For any polynomial $p(x) = b_0 + b_1 x + \cdots + b_m x^m$, and for $A$ any square matrix, define

$$p(A) = b_0 I + b_1 A + \cdots + b_m A^m$$

Let $A$ be a matrix for which the Jordan canonical form is a diagonal matrix,

$$P^{-1}AP = D = \text{diag}[\lambda_1, \ldots, \lambda_n]$$

For the characteristic polynomial $f_A(\lambda)$ of $A$, prove $f_A(A) = 0$. (This result is the *Cayley–Hamilton theorem*. It is true for any square matrix, not just those that have a diagonal Jordan canonical form.) *Hint:* Use the result $A = PDP^{-1}$ to simplify $f_A(A)$.

21. Prove the following: for $x \in \mathbf{C}^n$

    **(a)**   $\|x\|_\infty \le \|x\|_1 \le n\|x\|_\infty$

    **(b)**   $\|x\|_\infty \le \|x\|_2 \le \sqrt{n}\,\|x\|_\infty$

    **(c)**   $\|x\|_2 \le \|x\|_1 \le \sqrt{n}\,\|x\|_2$

22. Let $A$ be a real nonsingular matrix of order $n$, and let $\|\cdot\|_v$, denote a vector norm on $\mathbf{R}^n$. Define

    $$\|x\|_* = \|Ax\|_v \qquad x \in \mathbf{R}^n$$

    Show that $\|\cdot\|_*$ is a vector norm on $\mathbf{R}^n$.

23. Show

    $$\underset{p \to \infty}{\text{Limit}} \left[ \sum_{j=1}^{n} |x_i|^p \right]^{1/p} = \underset{1 \le i \le n}{\text{Max}} |x_i| \qquad x \in \mathbf{C}^n$$

    This justifies the use of the notation $\|x\|_\infty$ for the right side.

24. For any matrix norm, show that (a) $\|I\| \ge 1$, and (b) $\|A^{-1}\| \ge (1/\|A\|)$. For an operator norm, it is immediate from (7.3.11) that $\|I\| = 1$.

25. Derive formula (7.3.17) for the operator matrix norm $\|A\|_\infty$.

26. Define a vector norm on $\mathbf{R}^n$ by

    $$\|x\| = \frac{1}{n} \sum_{j=1}^{n} |x_j| \qquad x \in \mathbf{R}^n$$

    What is the operator matrix norm associated with this vector norm?

27. Let $A$ be a square matrix of order $n \times n$.

    **(a)** Given the eigenvalues and eigenvectors of $A$, determine those of (1) $A^m$ for $m \ge 2$, (2) $A^{-1}$, assuming $A$ is nonsingular, and (3) $A + cI$, $c = $ constant.

**(b)**   Prove $\|A\|_2 = r_\sigma(A)$ when $A$ is Hermitian.

**(c)**   For $A$ arbitrary and $U$ unitary of the same order, show $\|AU\|_2 = \|UA\|_2 = \|A\|_2$.

**28.**   Let $A$ be square of order $n \times n$.

**(a)**   Show that $F(AU) = F(UA) = F(A)$, for any unitary matrix $U$.

**(b)**   If $A$ is Hermitian, then show that

$$F(A) = \sqrt{\lambda_1^2 + \cdots + \lambda_n^2}$$

where $\lambda_1, \ldots, \lambda_n$ are the eigenvalues of $A$, repeated according to their multiplicity. Furthermore,

$$\frac{1}{\sqrt{n}} F(A) \le \|A\|_2 \le F(A)$$

**29.**   Recalling the notation of Theorem 7.5, show

$$\|A\|_2 = \mu_1 \qquad F(A) = \sqrt{\mu_1^2 + \cdots + \mu_r^2}$$

**30.**   Let $A$ be of order $n \times n$. Show

$$|\operatorname{trace}(A)| \le n r_\sigma(A)$$

If $A$ is symmetric and positive definite, show

$$\operatorname{trace}(A) \ge r_\sigma(A)$$

**31.**   Show that the infinite series

$$I + A + \frac{A^2}{2!} + \frac{A^3}{3!} + \cdots + \frac{A^n}{n!} + \cdots$$

converges for any square matrix $A$, and denote the sum of the series by $e^A$.

**(a)**   If $A = P^{-1}BP$, show that $e^A = P^{-1}e^B P$.

**(b)**   Let $\lambda_1, \ldots, \lambda_n$ denote the eigenvalues of $A$, repeated according to their multiplicity, and show that the eigenvalues of $e^A$ are $e^{\lambda_1}, \ldots, e^{\lambda_n}$.

**32.** Consider the matrix

$$
A = \begin{bmatrix}
6 & 1 & 1 & 0 & & \cdots & & & & 0 \\
1 & 6 & 1 & 1 & 0 & & & & & \\
1 & 1 & 6 & 1 & 1 & 0 & & & & \vdots \\
0 & 1 & 1 & 6 & 1 & 1 & 0 & & & \\
\vdots & & & & & & & & & 0 \\
 & & & & & & & & & 1 \\
 & & & & & & 1 & 1 & 6 & 1 \\
0 & & \cdots & & & & 0 & 1 & 1 & 6
\end{bmatrix}
$$

Show $A$ is nonsingular. Find a bound for $\|A^{-1}\|_\infty$ and $\|A^{-1}\|_2$.

**33.** In producing cubic interpolating splines in Section 3.7 of Chapter 3, it was necessary to solve the linear system $AM = D$ of (3.7.21) with

$$
A = \begin{bmatrix}
\dfrac{h_1}{3} & \dfrac{h_1}{6} & 0 & \cdots & & 0 \\
\dfrac{h_1}{6} & \dfrac{h_1 + h_2}{3} & \dfrac{h_2}{6} & & & \vdots \\
\vdots & & \ddots & & & \\
 & & \dfrac{h_{m-1}}{6} & \dfrac{h_{m-1} + h_m}{3} & \dfrac{h_m}{6} \\
0 & \cdots & 0 & \dfrac{h_m}{6} & \dfrac{h_m}{3}
\end{bmatrix}
$$

All $h_i > 0$, $i = 1, \ldots, m$. Using one or more of the results of Section 7.4, show that $A$ is nonsingular. In addition, derive the bounds

$$
\frac{1}{6} \operatorname{Min}(h_i) \le |\lambda| \le \operatorname{Max} h_i \qquad \lambda \in \sigma(A)
$$

for the eigenvalues of $A$.

**34.** Let $A$ be a square matrix, with $A^m = 0$ for some $m \ge 2$. Show that $I - A$ is nonsingular. Such a matrix $A$ is called *nilpotent*.

# EIGHT

# NUMERICAL SOLUTION
# OF SYSTEMS OF LINEAR EQUATIONS

Systems of linear equations arise in a large number of areas, both directly in modeling physical situations and indirectly in the numerical solution of other mathematical models. These applications occur in virtually all areas of the physical, biological, and social sciences. In addition, linear systems are involved in the following: optimization theory; solving systems of nonlinear equations; the approximation of functions; the numerical solution of boundary value problems for ordinary differential equations, partial differential equations, and integral equations; statistical inference; and numerous other problems. Because of the widespread importance of linear systems, much research has been devoted to their numerical solution. Excellent algorithms have been developed for the most common types of problems for linear systems, and some of these are defined, analyzed, and illustrated in this chapter.

The most common type of problem is to solve a square linear system

$$Ax = b$$

of moderate order, with coefficients that are mostly nonzero. Such linear systems, of any order, are called *dense*. For such systems, the coefficient matrix $A$ must generally be stored in the main memory of the computer in order to efficiently solve the linear system, and thus memory storage limitations in most computers will limit the order of the system. With the rapid decrease in the cost of computer memory, quite large linear systems can be accommodated on some machines, but it is expected that for most smaller machines, the practical upper limits on the order will be of size 100 to 500. Most algorithms for solving such dense systems are based on *Gaussian elimination*, which is defined in Section 8.1. It is a direct method in the theoretical sense that if rounding errors are ignored, then the exact answer is found in a finite number of steps. Modifications for improved error behavior with Gaussian elimination, variants for special classes of matrices, and error analyses are given in Section 8.2 through Section 8.5.

A second important type of problem is to solve $Ax = b$ when $A$ is square, sparse, and of large order. A *sparse* matrix is one in which most coefficients are zero. Such systems arise in a variety of ways, but we restrict our development to those for which there is a simple, known pattern for the nonzero coefficients. These systems arise commonly in the numerical solution of partial differential equations, and an example is given in Section 8.8. Because of the large order of most sparse systems of linear equations, sometimes as large as $10^5$ or more, the linear system cannot usually be solved by a direct method such as Gaussian elimination. Iteration methods are the preferred method of solution, and these are introduced in Section 8.6 through Section 8.9.

For solving dense square systems of moderate order, most computer centers have a set of programs that can be used for a variety of problems. Students should become acquainted with those at their university computer center and use them to further illustrate the material of this chapter. An excellent package is called LINPACK. and it is described in Dongarra et al. (1979). It is widely available, and we will make further reference to it later in this chapter.

## 8.1   Gaussian Elimination

This is the formal name given to the method of solving systems of linear equations by successively eliminating unknowns and reducing to systems of lower order. It is the method most people learn in high school algebra or in an undergraduate linear algebra course (in which it is often associated with producing the row-echelon form of a matrix). A precise definition is given of Gaussian elimination, which is necessary when implementing it on a computer and when analyzing the effects of rounding errors that occur when computing with it.

To solve $Ax = b$, we reduce it to an equivalent system $Ux = g$, in which $U$ is upper triangular. This system can be easily solved by a process of *back-substitution*. Denote the original linear system by $A^{(1)}x = b^{(1)}$,

$$A^{(1)} = \left[ a_{ij}^{(1)} \right] \qquad b^{(1)} = \left[ b_1^{(1)}, \ldots, b_n^{(1)} \right]^T \qquad 1 \le i, j \le n$$

in which $n$ is the order of the system. We reduce the system to the triangular form $Ux = g$ by adding multiples of one equation to another equation, eliminating some unknown from the second equation. Additional row operations are used in the modifications given in succeeding sections. To keep the presentation simple, we make some technical assumptions in defining the algorithm; they are removed in the next section.

**Gaussian elimination algorithm**

**STEP 1:**   Assume $a_{11}^{(1)} \ne 0$. Define the row multipliers by

$$m_{i1} = \frac{a_{i1}^{(1)}}{a_{11}^{(1)}} \qquad i = 2, 3, \ldots, n$$

These are used in eliminating the $x_1$ term from equations 2 through $n$. Define

$$a_{ij}^{(2)} = a_{ij}^{(1)} - m_{i1}a_{1j}^{(1)} \qquad i, j = 2, \ldots, n$$

$$b_i^{(2)} = b_i^{(1)} - m_{i1}b_1^{(1)} \qquad i = 2, \ldots, n$$

Also, the first rows of $A$ and $b$ are left undisturbed, and the first column of $A^{(1)}$, below the diagonal, is set to zero. The system $A^{(2)}x = b^{(2)}$ looks like

$$
\begin{bmatrix}
a_{11}^{(1)} & a_{12}^{(1)} & \cdots & a_{1n}^{(1)} \\
0 & a_{22}^{(2)} & \cdots & a_{2n}^{(2)} \\
\vdots & \vdots & & \vdots \\
0 & a_{n2}^{(2)} & \cdots & a_{nn}^{(2)}
\end{bmatrix}
\begin{bmatrix}
x_1 \\
x_2 \\
\vdots \\
x_n
\end{bmatrix}
=
\begin{bmatrix}
b_1^{(1)} \\
b_2^{(2)} \\
\vdots \\
b_n^{(2)}
\end{bmatrix}
$$

We continue to eliminate unknowns, going onto columns 2, 3, etc., and this is expressed generally in the following.

**STEP k:**  Let $1 \le k \le n - 1$. Assume that $A^{(k)}x = b^{(k)}$ has been constructed, with $x_1, x_2, \ldots, x_{k-1}$ eliminated at successive stages, and $A^{(k)}$ has the form

$$
A^{(k)} =
\begin{bmatrix}
a_{11}^{(1)} & a_{12}^{(1)} & & \cdots & & a_{1n}^{(1)} \\
0 & a_{22}^{(2)} & & & & a_{2n}^{(2)} \\
\vdots & & \ddots & & & \vdots \\
0 & \cdots & 0 & a_{kk}^{(k)} & \cdots & a_{kn}^{(k)} \\
\vdots & & & \vdots & & \vdots \\
0 & \cdots & 0 & a_{nk}^{(k)} & \cdots & a_{nn}^{(k)}
\end{bmatrix}
$$

Assume $a_{kk}^{(k)} \ne 0$. Define the *multipliers*

$$
m_{ik} = \frac{a_{ik}^{(k)}}{a_{kk}^{(k)}} \qquad i = k + 1, \ldots, n \tag{8.1.1}
$$

Use these to remove the unknown $x_k$ from equations $k + 1$ through $n$. Define

$$
a_{ij}^{(k+1)} = a_{ij}^{(k)} - m_{ik}a_{kj}^{(k)}
$$

$$
b_i^{(k+1)} = b_i^{(k)} - m_{ik}b_k^{(k)} \qquad i, j = k + 1, \ldots, n \tag{8.1.2}
$$

The earlier rows 1 through $k$ are left undisturbed, and zeros are introduced into column $k$ below the diagonal element.

By continuing in this manner, after $n - 1$ steps we obtain $A^{(n)}x = b^{(n)}$:

$$
\begin{bmatrix}
a_{11}^{(1)} & \cdots & a_{1n}^{(1)} \\
0 & & \\
\vdots & \ddots & \vdots \\
0 & \cdots & a_{nn}^{(n)}
\end{bmatrix}
\begin{bmatrix}
x_1 \\
\vdots \\
x_n
\end{bmatrix}
=
\begin{bmatrix}
b_1^{(1)} \\
\vdots \\
b_n^{(n)}
\end{bmatrix}
$$

For notational convenience, let $U = A^{(n)}$ and $g = b^{(n)}$. The system $Ux = g$ is upper triangular, and it is quite easy to solve. First

$$x_n = \frac{g_n}{u_{nn}}$$

and then

$$x_k = \frac{1}{u_{kk}} \left[ g_k - \sum_{j=k+1}^{n} u_{kj} x_j \right] \qquad k = n-1, n-2, \ldots, 1 \quad (8.1.3)$$

This completes the Gaussian elimination algorithm.

**Example**    Solve the linear system

$$x_1 + 2x_2 + x_3 = 0$$

$$2x_1 + 2x_2 + 3x_3 = 3 \qquad (8.1.4)$$

$$-x_1 - 3x_2 \qquad = 2$$

To simplify the notation, we note that the unknowns $x_1, x_2, x_3$ never enter into the algorithm until the final step. Thus we represent the preceding linear system with the *augmented matrix*

$$[A|b] = \begin{bmatrix} 1 & 2 & 1 & | & 0 \\ 2 & 2 & 3 & | & 3 \\ -1 & -3 & 0 & | & 2 \end{bmatrix}$$

The row operations are performed on this augmented matrix, and the unknowns are given in the final step. In the following diagram, the multipliers are given next to an arrow, corresponding to the changes they cause:

$$\begin{bmatrix} 1 & 2 & 1 & | & 0 \\ 2 & 2 & 3 & | & 3 \\ -1 & -3 & 0 & | & 2 \end{bmatrix} \xrightarrow[\substack{m_{21}=2 \\ m_{31}=-1}]{} \begin{bmatrix} 1 & 2 & 1 & | & 0 \\ 0 & -2 & 1 & | & 3 \\ 0 & -1 & 1 & | & 2 \end{bmatrix}$$

$$\Big\downarrow m_{32}=\tfrac{1}{2}$$

$$[U|g] \equiv \begin{bmatrix} 1 & 2 & 1 & | & 0 \\ 0 & -2 & 1 & | & 3 \\ 0 & 0 & \tfrac{1}{2} & | & \tfrac{1}{2} \end{bmatrix}$$

Solving $Ux = g$,

$$x_3 = 1 \qquad x_2 = -1 \qquad x_1 = 1$$

**Triangular factorization of a matrix**   It is convenient to keep the multipliers $m_{ij}$, since we often want to solve $Ax = b$ with the same $A$ but a different vector $b$. In the computer the elements $a_{ij}^{(k+1)}$, $j \geq i$, always are stored into the storage for $a_{ij}^{(k)}$. The elements below the diagonal are being zeroed, and this provides a convenient storage for the elements $m_{ij}$. Store $m_{ij}$ into the space originally used to store $a_{ij}$, $i > j$.

There is yet another reason for looking at the multipliers $m_{ij}$ as the elements of a matrix. First, introduce the lower triangular matrix

$$
L = \begin{bmatrix}
1 & 0 & 0 & \cdots & 0 \\
m_{21} & 1 & 0 & \cdots & 0 \\
\vdots & & & \ddots & \vdots \\
m_{n1} & m_{n2} & \cdots & & 1
\end{bmatrix}
$$

**Theorem 8.1**   If $L$ and $U$ are the lower and upper triangular matrices defined previously using Gaussian elimination, then

$$ A = LU \tag{8.1.5} $$

**Proof**   This proof is basically an algebraic manipulation, making use of definitions (8.1.1) and (8.1.2). To visualize the matrix element $(LU)_{ij}$, use the vector formula

$$
(LU)_{ij} = \begin{bmatrix} m_{i1}, \ldots, m_{i,i-1}, 1, 0, \ldots, 0 \end{bmatrix}
\begin{bmatrix}
u_{1j} \\
\vdots \\
u_{jj} \\
0 \\
\vdots \\
0
\end{bmatrix}
$$

For $i \leq j$,

$$ (LU)_{ij} = m_{i1}u_{1j} + m_{i2}u_{2j} + \cdots + m_{i,i-1}u_{i-1,j} + u_{i,j} $$

$$ = \sum_{k=1}^{i-1} m_{ik}a_{kj}^{(k)} + a_{ij}^{(i)} $$

$$ = \sum_{k=1}^{i-1} \left[ a_{ij}^{(k)} - a_{ij}^{(k+1)} \right] + a_{ij}^{(i)} $$

$$ = a_{ij}^{(1)} = a_{ij} $$

For $i > j$,

$$(LU)_{ij} = m_{i1}u_{1j} + \cdots + m_{ij}u_{jj}$$

$$= \sum_{k=1}^{j-1} m_{ik}a_{kj}^{(k)} + m_{ij}a_{jj}^{(j)}$$

$$= \sum_{k=1}^{j-1} \left[ a_{ij}^{(k)} - a_{ij}^{(k+1)} \right] + a_{ij}^{(j)}$$

$$= a_{ij}^{(1)} = a_{ij}$$

This completes the proof.    ∎

The decomposition (8.1.5) is an important result, and extensive use is made of it in developing variants of Gaussian elimination for special classes of matrices. But for the moment we give only the following corollary.

**Corollary**    With the matrices $A$, $L$, and $U$ as in Theorem 8.1,

$$\det(A) = u_{11}u_{22} \cdots u_{nn}$$

$$= a_{11}^{(1)}a_{22}^{(2)} \cdots a_{nn}^{(n)}$$

**Proof**    By the product rule for determinants,

$$\det(A) = \det(L)\det(U)$$

Since $L$ and $U$ are triangular, their determinants are the product of their diagonal elements. The desired result follows easily, since $\det(L) = 1$.    ∎

**Example**    For the system (8.1.4) of the previous example,

$$L = \begin{bmatrix} 1 & 0 & 0 \\ 2 & 1 & 0 \\ -1 & \frac{1}{2} & 1 \end{bmatrix} \qquad U = \begin{bmatrix} 1 & 2 & 1 \\ 0 & -2 & 1 \\ 0 & 0 & \frac{1}{2} \end{bmatrix}$$

It is easily verified that $A = LU$. Also $\det(A) = \det(U) = -1$.

**Operation count**    To analyze the number of operations necessary to solve $Ax = b$ using Gaussian elimination, we will consider separately the creation of $L$ and $U$ from $A$, the modification of $b$ to $g$, and finally the solution of $x$.

1.    Calculation of $L$ and $U$. At step 1, $n - 1$ divisions were used to calculate the multipliers $m_{i1}$, $2 \le i \le n$. Then $(n - 1)^2$ multiplications and $(n - 1)^2$ additions were used to create the new elements $a_{ij}^{(2)}$. We can continue in this

**Table 8.1    Operation count for *LU* decomposition of a matrix**

| Step $k$ | Additions | Multiplications | Divisions |
|---|---|---|---|
| 1 | $(n-1)^2$ | $(n-1)^2$ | $n-1$ |
| 2 | $(n-2)^2$ | $(n-2)^2$ | $n-2$ |
| $\vdots$ | $\vdots$ | $\vdots$ | $\vdots$ |
| $n-1$ | 1 | 1 | 1 |
| Total | $\dfrac{n(n-1)(2n-1)}{6}$ | $\dfrac{n(n-1)(2n-1)}{6}$ | $\dfrac{n(n-1)}{2}$ |

way for each step. The results are summarized in Table 8.1. The total value for each column was obtained using the identities

$$\sum_{j=1}^{p} j = \frac{p(p+1)}{2} \qquad \sum_{j=1}^{p} j^2 = \frac{p(p+1)(2p+1)}{6} \qquad p \geq 1$$

Traditionally, it is the number of multiplications and divisions, counted together, that is used as the operation count for Gaussian elimination. On earlier computers, additions were much faster than multiplications and divisions, and thus additions were ignored in calculating the cost of many algorithms. However, on modern computers, the time of additions, multiplications, and divisions are quite close in size. For a convenient notation, let $MD(\cdot)$ and $AS(\cdot)$ denote the number of multiplications and divisions, and the number of additions and subtractions, respectively, for the computation of the quantity in the parentheses.

For the *LU* decomposition of $A$, we have

$$MD(LU) = \frac{n(n^2-1)}{3} \doteq \frac{n^3}{3}$$

$$AS(LU) = \frac{n(n-1)(2n-1)}{6} \doteq \frac{n^3}{3} \tag{8.1.6}$$

The final estimates are valid for larger values of $n$.

2.  Modification of $b$ to $g = b^{(n)}$:

$$MD(g) = (n-1) + (n-2) + \cdots + 1 = \frac{n(n-1)}{2}$$

$$AS(g) = \frac{n(n-1)}{2} \tag{8.1.7}$$

3.  Solution of $Ux = g$

$$MD(x) = \frac{n(n+1)}{2} \qquad AS(x) = \frac{n(n-1)}{2} \tag{8.1.8}$$

**4.**    Solution of $Ax = b$. Combine (1) through (3) to get

$$MD(LU, x) = \frac{n^3}{3} + n^2 - \frac{n}{3} \doteq \frac{1}{3}n^3$$

$$(8.1.9)$$

$$AS(LU, x) = \frac{n(n-1)(2n+5)}{6} \doteq \frac{1}{3}n^3$$

The number of additions is always about the same as the number of multiplications and divisions, and thus from here on, we consider only the latter. The first thing to note is that solving $Ax = b$ is comparatively cheap when compared to such a supposedly simple operation as multiplying two $n \times n$ matrices. The matrix multiplication requires $n^3$ operations, and the solution of $Ax = b$ requires only about $\frac{1}{3}n^3$ operations.

Second, the main cost of solving $Ax = b$ is in producing the decomposition $A = LU$. Once it has been found, only $n^2$ additional operations are necessary to solve $Ax = b$. After once solving $Ax = b$, it is comparatively cheap to solve additional systems with the same coefficient matrix, provided the $LU$ decomposition has been saved.

Finally, Gaussian elimination is much cheaper than *Cramer's rule*, which uses determinants and is often taught in linear algebra courses [for example, see Anton (1984), sec. 2.4]. If the determinants in Cramer's rule are computed using *expansion by minors*, then the operation count is $(n + 1)!$. For $n = 10$, Gaussian elimination uses 430 operations, and Cramer's rule uses 39,916,800 operations. This should emphasize the point that Cramer's rule is not a practical computational tool, and that it should be considered as just a theoretical mathematics tool.

**5.**    Inversion of $A$. The inverse $A^{-1}$ is generally not needed, but it can be produced by using Gaussian elimination. Finding $A^{-1}$ is equivalent to solving the equation $AX = I$, with $X$ an $n \times n$ unknown matrix. If we write $X$ and $I$ in terms of their columns,

$$X = \left[ x^{(1)}, \ldots, x^{(n)} \right] \qquad I = \left[ e^{(1)}, \ldots, e^{(n)} \right]$$

then solving $AX = I$ is equivalent to solving the $n$ systems

$$Ax^{(1)} = e^{(1)}, \ldots, Ax^{(n)} = e^{(n)} \qquad (8.1.10)$$

all having the same coefficient matrix $A$. Using (1)–(3)

$$MD(A^{-1}) = \frac{4}{3}n^3 - \frac{n}{3} \doteq \frac{4}{3}n^3$$

Calculating $A^{-1}$ is four times the expense of solving $Ax = b$ for a single vector $b$, not $n$ times the work as one might first imagine. By careful attention to the details of the inversion process, taking advantage of the special form of the right-hand vectors $e^{(1)}, \ldots, e^{(n)}$, it is possible to further reduce the operation count to exactly

$$MD(A^{-1}) = n^3 \tag{8.1.11}$$

However, it is still wasteful in most situations to produce $A^{-1}$ to solve $Ax = b$. And there is no advantage in saving $A^{-1}$ rather than the $LU$ decomposition to solve future systems $Ax = b$. In both cases, the number of multiplications and divisions necessary to solve $Ax = b$ is exactly $n^2$.

## 8.2 Pivoting and Scaling in Gaussian Elimination

At each stage of the elimination process in the last section, we assumed the appropriate pivot element $a_{kk}^{(k)} \neq 0$. To remove this assumption, begin each step of the elimination process by switching rows to put a nonzero element in the pivot position. If none such exists, then the matrix must be singular, contrary to assumption.

It is not enough, however, to just ask that the pivot element be nonzero. Often an element would be zero except for rounding errors that have occurred in calculating it. Using such an element as the pivot element will result in gross errors in the further calculations in the matrix. To guard against this, and for other reasons involving the propagation of rounding errors, we introduce *partial pivoting* and *complete pivoting*.

**Definition 1.** Partial Pivoting. For $1 \leq k \leq n - 1$, in the Gaussian elimination process at stage $k$, let

$$c_k = \underset{k \leq i \leq n}{\text{Max}} |a_{ik}^{(k)}| \tag{8.2.1}$$

Let $i$ be the smallest row index, $i \geq k$, for which the maximum $c_k$ is attained. If $i > k$, then switch rows $k$ and $i$ in $A$ and $b$, and proceed with step $k$ of the elimination process. All of the multipliers will now satisfy

$$|m_{ik}| \leq 1 \qquad i = k + 1, \dots, n \tag{8.2.2}$$

This aids in preventing the growth of elements in $A^{(k)}$ of greatly varying size, and thus lessens the possibility for large loss of significance errors.

**2.** Complete Pivoting. Define

$$c_k = \underset{k \leq i, \, j \leq n}{\text{Max}} |a_{ij}^{(k)}|$$

Switch rows of $A$ and $b$ and columns of $A$ to bring to the pivot position an element giving the maximum $c_k$. Note that with a column switch, the order of the unknowns is changed. At the completion of the elimination and back substitution process, this must be reversed.

Complete pivoting has been proved to cause the roundoff error in Gaussian elimination to propagate at a reasonably slow speed, compared with what can happen when no pivoting is used. The theoretical results on the use of partial pivoting are not quite as good, but in virtually all practical problems, the error behavior is essentially the same as that for complete pivoting. Comparing operation times, complete pivoting is the more expensive strategy, and thus, partial pivoting is used in most practical algorithms. Henceforth we always mean *partial pivoting* when we used the word *pivoting*. The entire question of roundoff error propagation in Gaussian elimination has been analyzed very thoroughly by J. H. Wilkinson [e.g., see Wilkinson (1965), pp. 209–220], and some of his results are presented in Section 8.4.

*Example*    We illustrate the effect of using pivoting by solving the system

$$.729x + .81y + .9z = .6867$$

$$x + y + z = .8338$$

$$1.331x + 1.21y + 1.1z = 1.000 \qquad (8.2.3)$$

The exact solution, rounded to four significant digits, is

$$x = .2245 \qquad y = .2814 \qquad z = .3279 \qquad (8.2.4)$$

Floating-point decimal arithmetic, with four digits in the mantissa, will be used to solve the linear system. The reason for using this arithmetic is to show the effect of working with only a finite number of digits, while keeping the presentation manageable in size. The augmented matrix notation will be used to represent the system (8.2.3), just as was done with the earlier example (8.1.4).

1.    Solution without pivoting.

$$\begin{bmatrix} .7290 & .8100 & .9000 & | & .6867 \\ 1.000 & 1.000 & 1.000 & | & .8338 \\ 1.331 & 1.210 & 1.100 & | & 1.000 \end{bmatrix}$$

$$\downarrow \begin{array}{l} m_{21} = 1.372 \\ m_{31} = 1.826 \end{array}$$

$$\begin{bmatrix} .7290 & .8100 & .9000 & | & .6867 \\ 0.0 & -.1110 & -.2350 & | & -.1084 \\ 0.0 & -.2690 & -.5430 & | & -.2540 \end{bmatrix}$$

$$\downarrow m_{32} = 2.423$$

$$\begin{bmatrix} .7290 & .8100 & .9000 & | & .6867 \\ 0.0 & -.1110 & -.2350 & | & -.1084 \\ 0.0 & 0.0 & .02640 & | & .008700 \end{bmatrix}$$

The solution is

$$x = .2251 \qquad y = .2790 \qquad z = .3295 \qquad (8.2.5)$$

**2.** Solution with pivoting. To indicate the interchange of rows $i$ and $j$, we will use the notation $r_i \leftrightarrow r_j$.

$$\begin{bmatrix} .7290 & .8100 & .9000 & .6867 \\ 1.0000 & 1.0000 & 1.0000 & .8338 \\ 1.331 & 1.210 & 1.100 & 1.000 \end{bmatrix}$$

$$r_1 \leftrightarrow r_3 \begin{vmatrix} m_{21} = .7513 \\ m_{31} = .5477 \end{vmatrix} \downarrow$$

$$\begin{bmatrix} 1.331 & 1.210 & 1.100 & 1.000 \\ 0.0 & .09090 & .1736 & .08250 \\ 0.0 & .1473 & .2975 & .1390 \end{bmatrix}$$

$$r_2 \leftrightarrow r_3 \begin{vmatrix} m_{32} = .6171 \end{vmatrix} \downarrow$$

$$\begin{bmatrix} 1.331 & 1.210 & 1.100 & 1.000 \\ 0.0 & .1473 & .2975 & .1390 \\ 0.0 & 0.0 & -.01000 & -.003280 \end{bmatrix}$$

The solution is

$$x = .2246 \qquad y = .2812 \qquad z = .3280 \qquad (8.2.6)$$

The error in (8.2.5) is from seven to sixteen times larger than it is for (8.2.6), depending on the component of the solution being considered. The results in (8.2.6) have one more significant digit than do those of (8.2.5). This illustrates the positive effect that the use of pivoting can have on the error behavior for Gaussian elimination.

Pivoting changes the factorization result (8.1.5) given in Theorem 8.1. The result is still true, but in a modified form. If the row interchanges induced by pivoting were carried out on $A$ before beginning elimination, then pivoting would be unnecessary. Row interchanges on $A$ can be represented by premultiplication of $A$ by an approximate *permutation matrix* $P$, to get $PA$. Then Gaussian elimination on $PA$ leads to

$$LU = PA \qquad (8.2.7)$$

where $U$ is the upper triangular matrix obtained in the elimination process with pivoting. The lower triangular matrix $L$ can be constructed using the multipliers from Gaussian elimination with pivoting. We omit the details, as the actual construction is unimportant.

***Example***   From the preceding example with pivoting, we form

$$
L = \begin{bmatrix} 1.000 & 0.0 & 0.0 \\ .5477 & 1.000 & 0.0 \\ .7513 & .6171 & 1.000 \end{bmatrix}
\qquad
U = \begin{bmatrix} 1.331 & 1.210 & 1.100 \\ 0.0 & .1473 & .2975 \\ 0.0 & 0.0 & -.01000 \end{bmatrix}
$$

When multiplied,

$$
LU = \begin{bmatrix} 1.331 & 1.210 & 1.100 \\ .7289 & .8100 & .9000 \\ 1.000 & 1.000 & 1.000 \end{bmatrix} = PA
\qquad
P = \begin{bmatrix} 0 & 0 & 1 \\ 1 & 0 & 0 \\ 0 & 1 & 0 \end{bmatrix}
$$

The result $PA$ is the matrix $A$ with first, rows 1 and 3 interchanged, and then rows 2 and 3 interchanged. This illustrates (8.2.7).

**Scaling**   It has been observed empirically that if the elements of the coefficient matrix $A$ vary greatly in size, then it is likely that large loss of significance errors will be introduced and the propagation of rounding errors will be worse. To avoid this problem, we usually *scale* the matrix $A$ so that the elements vary less. This is usually done by multiplying the rows and columns by suitable constants. The subject of scaling is not well understood currently, especially how to guarantee that the effect of rounding errors in Gaussian elimination will be made smaller by such scaling. Computational experience suggests that often all rows should be scaled to make them approximately equal in magnitude. In addition, all columns can be scaled to make them of about equal magnitude. The latter is equivalent to scaling the unknown components $x_i$ of $x$, and it can often be interpreted to say that the $x_i$ should be measured in units of comparable size.

There is no known a priori strategy for picking the scaling factors so as to always decrease the effect of rounding error propagation, based solely on a knowledge of $A$ and $b$. Stewart (1977) is somewhat critical of the general use of scaling as described in the preceding paragraph. He suggests choosing scaling factors so as to obtain a rescaled matrix in which the errors in the coefficients are of about equal magnitude. When rounding is the only source of error, this leads to the strategy of scaling to make all elements of about equal size. The LINPACK programs do not include scaling, but they recommend a strategy along the lines indicated by Stewart [see Dongarra et al. (1979), pp. I7–I12 for a more extensive discussion; for other discussions of scaling, see Forsythe and Moler (1967), chap. 11 and Golub and Van Loan (1983), pp. 72–74].

If we let $B$ denote the result of row and column scaling in $A$, then

$$ B = D_1 A D_2 $$

where $D_1$ and $D_2$ are diagonal matrices, with entries the scaling constants. To solve $Ax = b$, observe that

$$ D_1 A D_2 \left( D_2^{-1} x \right) = D_1 b $$

Thus we solve for $x$ by solving

$$ Bz = D_1 b \qquad x = D_2 z \qquad\qquad (8.2.8) $$

The remaining discussion is restricted to row scaling, since some form of it is fairly widely agreed upon.

Usually we attempt to choose the coefficients so as to have

$$\underset{1 \leq j \leq n}{\text{Max}} |b_{ij}| \doteq 1 \qquad i = 1, \ldots, n \qquad (8.2.9)$$

where $B = [b_{ij}]$ is the result of scaling $A$. The most straightforward approach is to define

$$s_i = \underset{1 \leq j \leq n}{\text{Max}} |a_{ij}| \qquad i = 1, \ldots, n$$

$$b_{ij} = \frac{a_{ij}}{s_i}, \qquad j = 1, \ldots, n \qquad (8.2.10)$$

But because this introduces an additional rounding error into each element of the coefficient matrix, two other techniques are more widely used.

**1.**   Scaling using computer number base. Let $\beta$ denote the base used in the computer arithmetic, for example, $\beta = 2$ on binary machines. Let $r_i$ be the smallest integer for which $\beta^{r_i} \geq s_i$. Define the scaled matrix $B$ by

$$b_{ij} = \frac{a_{ij}}{\beta^{r_i}} \qquad i, j = 1, \ldots, n \qquad (8.2.11)$$

No rounding is involved in defining $b_{ij}$, only a change in the exponent in the floating-point form for $a_{ij}$. The values of $B$ satisfy

$$\beta^{-1} < \underset{1 \leq j \leq n}{\text{Max}} |b_{ij}| \leq 1$$

and thus (8.2.9) is satisfied fairly well.

**2.**   Implicit scaling. The use of scaling will generally change the choice of pivot elements when pivoting is used with Gaussian elimination. And it is only with such a change of pivot elements that the results in Gaussian elimination will be changed. This is due to a result of F. Bauer [in Forsythe and Moler (1967), p. 38], which states that if the scaling (8.2.11) is used, and if the choice of pivot elements is forced to remain the same as when solving $Ax = b$, then the solution of (8.2.8) will yield exactly the same computed value for $x$. Thus the only significance of scaling is in the choice of the pivot elements.

For implicit scaling, we continue to use the matrix $A$. But we choose the pivot element in step $k$ of the Gaussian elimination algorithm by defining

$$c_k = \underset{k \leq i \leq n}{\text{Max}} \frac{|a_{ik}^{(k)}|}{s_i} \qquad (8.2.12)$$

replacing the definition (8.2.1) used in defining partial pivoting. Choose the smallest index $i \geq k$ that yields $c_k$ in (8.2.12), and if $i \neq k$, then interchange rows $i$ and $k$. Then proceed with the elimination algorithm of Section 8.1, as before. This form of scaling seems to be the form most commonly used in current published algorithms, if scaling is being used.

**An algorithm for Gaussian elimination**   We first give an algorithm, called *Factor*, for the triangular factorization of a matrix $A$. It uses Gaussian elimination with partial pivoting, combined with the implicit scaling of (8.2.10) and (8.2.12). We then give a second algorithm, called *Solve*, for using the results of *Factor* to solve a linear system $Ax = b$. The reason for separating the elimination procedure into these two steps is that we will often want to solve several systems $Ax = b$, with the same $A$, but different values for $b$.

*Algorithm*   *Factor* $(A, n, \text{Pivot}, \det, \text{ier})$

    **1**   Remarks: $A$ is an $n \times x$ matrix, to be factored using the *LU* decomposition. Gaussian elimination is used, with partial pivoting and implicit scaling in the rows. Upon completion of the algorithm, the upper triangular matrix $U$ will be stored in the upper triangular part of $A$; and the multipliers of (8.1.1), which make up $L$ below its diagonal, will be stored in the corresponding positions of $A$. The vector Pivot will contain a record of all row interchanges. If Pivot$(k) = k$, then no interchange was used in step $k$ of the elimination process. But if Pivot$(k) = i \neq k$, then rows $i$ and $k$ were interchanged in step $k$ of the elimination process. The variable det will contain $\det(A)$ on exit.

        The variable ier is an error indicator. If ier $= 0$, then the routine was completed satisfactorily. But for ier $= 1$, the matrix $A$ was singular, in the sense that all possible pivot elements were zero at some step of the elimination process. In this case, all computation ceased, and the routine was exited. No attempt is made to check on the accuracy of the computed decomposition of $A$, and it can be nearly singular without being detected.

    **2**   $\det := 1$

    **3**   $s_i := \underset{1 \leq j \leq n}{\text{Max}} |a_{ij}|, \ i = 1, \ldots, n.$

    **4**   Do through step 16 for $k = 1, \ldots, n - 1$.

    **5**   $c_k := \underset{k \leq i \leq n}{\text{Max}} \left| \dfrac{a_{ik}}{s_i} \right|$

    **6**   Let $i_0$ be the smallest index $i \geq k$ for which the maximum in step 5 is attained. Pivot$(k) := i_0$.

    **7**   If $c_k = 0$, then ier $:= 1$, det $:= 0$, and exit from the algorithm.

**8**  If $i_0 = k$, then go to step 11.

**9**  det $:= -\det$

**10**  Interchange $a_{kj}$ and $a_{i_0j}$, $j = k, \ldots, n$. Interchange $s_k$ and $s_{i_0}$.

**11**  Do through step 14 for $i = k + 1, \ldots, n$.

**12**  $a_{ik} := m_i := \dfrac{a_{ik}}{a_{kk}}$

**13**  $a_{ij} := a_{ij} - m_i a_{kj}$, $j = k + 1, \ldots, n$

**14**  End loop on $i$.

**15**  det $:= a_{kk} \cdot \det$

**16**  End loop on $k$.

**17**  det $= a_{nn} \cdot \det$; ier $= 0$ and exit the algorithm.

***Algorithm***   *Solve* $(A, n, b, \text{Pivot})$

1.  Remarks: This algorithm will solve the linear system $Ax = b$. It is assumed that the original matrix $A$ has been factored using the algorithm *Factor*, with the row interchanges recorded in Pivot. The solution will be stored in $b$ on exit. The matrix $A$ and vector Pivot are left unchanged.

2.  Do through step 5 for $k = 1, 2, \ldots, n - 1$.

3.  If $i := \text{Pivot}(k) \neq k$, then interchange $b_i$ and $b_k$.

4.  $b_i := b_i - a_{ik}b_k$, $i = k + 1, \ldots, n$

5.  End loop on $k$.

6.  $b_n := \dfrac{b_n}{a_{nn}}$

7.  Do through step 9 for $i = n - 1, \ldots, 1$.

8.  $b_i := \dfrac{1}{a_{ii}}\left\{ b_i - \sum_{j=i+1}^{n} a_{ij}b_j \right\}$

9.  End loop on $i$.

10.  Exit from algorithm.

The earlier example (8.2.3) will serve as an illustration. The use of implicit scaling in this case will not require a change in the choice of pivot elements with partial pivoting. Algorithms similar to *Factor* and *Solve* are given in a number of references [see Forsythe and Moler (1967), chaps. 16 and 17 and Dongarra et al. (1979), chap. 1, for improved versions of these algorithms]. The programs in LINPACK will also compute information concerning the condition or stability of the problem $Ax = b$ and the accuracy of the computed solution.

An important aspect of the LINPACK programs is the use of *Basic Linear Algebra Subroutines* (BLAS). These perform simple operations on vectors, such as forming the dot product of two vectors or adding a scalar multiple of one vector to another vector. The programs in LINPACK use these BLAS to replace many of the inner loops in a method. The BLAS can be optimized, if desired, for each computer; thus, the performance of the main LINPACK programs can also be easily improved while keeping the main source code machine-independent. For a more complete discussion of BLAS, see Lawson et al. (1979).

## 8.3 Variants of Gaussian Elimination

There are many variants of Gaussian elimination. Some are modifications or simplifications, based on the special properties of some class of matrices, for example, symmetric, positive definite matrices. Other variants are ways to rewrite Gaussian elimination in a more compact form, sometimes in order to use special techniques to reduce the error. We consider only a few such variants, and later make reference to others.

**Gauss–Jordon method**  This procedure is much the same as regular elimination including the possible use of pivoting and scaling. It differs in eliminating the unknown in equations above the diagonal as well as below it. In step $k$ of the elimination algorithm choose the pivot element as before. Then define

$$a_{kj}^{(k+1)} = \frac{a_{kj}^{(k)}}{a_{kk}^{(k)}} \qquad j = k, \ldots, n$$

$$b_k^{(k+1)} = \frac{b_k^{(k)}}{a_{kk}^{(k)}}$$

Eliminate the unknown $x_k$ in equations both above and below equation $k$. Define

$$a_{ij}^{(k+1)} = a_{ij}^{(k)} - a_{ik}^{(k)} a_{kj}^{(k+1)}$$

$$b_i^{(k+1)} = b_i^{(k)} - a_{ik}^{(k)} b_k^{(k+1)}$$

(8.3.1)

for $j = k, \ldots, n$, $i = 1, \ldots, n$, $i \neq k$. The Gauss–Jordan method is equivalent to the use of the *reduced row-echelon form* of linear algebra texts [for example, see Anton (1984), pp. 8–9].

This procedure will convert the augmented matrix $[A|b]$ to $[I|b^{(n)}]$, so that at the completion of the preceding elimination, $x = b^{(n)}$. To solve $Ax = b$ by this technique requires

$$\frac{n(n-1)^2}{2} \doteq \frac{n^3}{2} \qquad (8.3.2)$$

multiplications and divisions. This is 50 percent more than the regular elimination method; consequently, the Gauss–Jordan method should usually not be used for solving linear systems. However, it can be used to produce a matrix inversion program that uses a minimum of storage. By taking special advantage of the special structure of the right side in $AX = I$, the Gauss–Jordan method can produce the solution $X = A^{-1}$ using only $n$ extra storage locations, rather than the normal $n^2$ extra storage locations. Partial pivoting and implicit scaling can still be used.

**Compact methods**    It is possible to move directly from a matrix $A$ to its $LU$ decomposition, and this can be combined with partial pivoting and scaling. If we disregard the possibility of pivoting for the moment, then the result

$$A = LU \qquad (8.3.3)$$

leads directly to a set of recursive formulas for the elements of $L$ and $U$.

There is some nonuniqueness in the choice of $L$ and $U$, if we insist only that $L$ and $U$ be lower and upper triangular, respectively. If $A$ is nonsingular, and if we have two decompositions

$$A = L_1 U_1 = L_2 U_2 \qquad (8.3.4)$$

then

$$L_2^{-1} L_1 = U_2 U_1^{-1} \qquad (8.3.5)$$

The inverse and the products of lower triangular matrices are again lower triangular, and similarly for upper triangular matrices. The left and right sides of (8.3.5) are lower and upper triangular, respectively. Thus they must equal a diagonal matrix, call it $D$, and

$$L_1 = L_2 D \qquad U_1 = D^{-1} U_2 \qquad (8.3.6)$$

The choice of $D$ is tied directly to the choice of the diagonal elements of either $L$ or $U$, and once they have been chosen, $D$ is uniquely determined.

If the diagonal elements of $L$ are all required to equal 1, then the resulting decomposition $A = LU$ is that given by Gaussian elimination, as in Section 8.1. The associated compact method gives explicit formulas for $l_{ij}$ and $u_{ij}$, and it is known as *Doolittle's method*. If we choose to have the diagonal elements of $U$ all equal 1, the associated compact method for calculating $A = LU$ is called *Crout's method*. There is only a multiplying diagonal matrix to distinguish it from Doolittle's method. For an algorithm using Crout's algorithm for the factorization (8.3.3), with partial pivoting and implicit scaling, see the program *unsymdet* in Wilkinson and Reinsch (1971, pp. 93–110). In some situations, Crout's method has advantages over the usual Doolittle method.

The principal advantage of the compact formula is that the elements $l_{ij}$ and $u_{ij}$ all involve inner products, as illustrated below in formula (8.3.14)–(8.3.15) for the factorization of a symmetric positive definite matrix. These inner products can be accumulated using double precision arithmetic, possibly including a concluding division, and then be rounded to single precision. This way of computing inner products was discussed in Chapter 1 preceding the error formula (1.5.19). This limited use of double precision can greatly increase the accuracy of the factors $L$ and $U$, and it is not possible to do this with the regular elimination method unless all operations and storage are done in double precision [for a complete discussion of these compact methods, see Wilkinson (1965), pp. 221–228, and Golub and Van Loan (1983), sec. 5.1].

**The Cholesky method**   Let $A$ be a symmetric and positive definite matrix of order $n$. The matrix $A$ is positive definite if

$$(Ax, x) = \sum_{i=1}^{n} \sum_{j=1}^{n} a_{ij} x_i x_j > 0 \tag{8.3.7}$$

for all $x \in \mathbf{R}^n$, $x \neq 0$. Some of the properties of positive definite matrices are given in Problem 14 of Chapter 7 and Problems 9, 11, and 12 of this chapter. Symmetric positive definite matrices occur in a wide variety of applications.

For such a matrix $A$, there is a very convenient factorization, and it can be carried through without any need for pivoting or scaling. This is called *Choleski's method*, and it states that we can find a lower triangular real matrix $L$ such that

$$A = LL^T \tag{8.3.8}$$

The method requires only $\frac{1}{2}n(n + 1)$ storage locations for $L$, rather than the usual $n^2$ locations, and the number of operations is about $\frac{1}{6}n^3$, rather than the number $\frac{1}{3}n^3$ required for the usual decomposition.

To prove that (8.3.8) is possible, we give a derivation of $L$ based on induction. Assume the result is true for all positive definite symmetric matrices of order $\leq n - 1$. We show it is true for all such matrices $A$ of order $n$. Write the desired $L$, of order $n$, in the form

$$L = \begin{bmatrix} \hat{L} & 0 \\ \gamma^T & x \end{bmatrix}$$

with $\hat{L}$ a square matrix of order $n - 1$, $\gamma \in \mathbf{R}^{n-1}$, and $x$ a scalar. The $L$ is to be chosen to satisfy $A = LL^T$:

$$\begin{bmatrix} \hat{L} & 0 \\ \gamma^T & x \end{bmatrix} \begin{bmatrix} \hat{L}^T & \gamma \\ 0 & x \end{bmatrix} = A = \begin{bmatrix} \hat{A} & c \\ c^T & d \end{bmatrix} \tag{8.3.9}$$

with $\hat{A}$ of order $n - 1$, $c \in \mathbf{R}^{n-1}$, and $d = a_{nn}$ real. Since (8.3.7) is true for $A$, let $x_n = 0$ in it to obtain the analogous statement for $\hat{A}$, showing $\hat{A}$ is also positive definite and symmetric. In addition, $d > 0$, by letting $x_1 = \cdots = x_{n-1} = 0$, $x_n = 1$ in (8.3.7). Multiplying in (8.3.9), choose $\hat{L}$, by the induction hypothesis to satisfy

$$\hat{L}\hat{L}^T = \hat{A} \tag{8.3.10}$$

Then choose $\gamma$ by solving

$$\hat{L}\gamma = c \tag{8.3.11}$$

since $\hat{L}$ is nonsingular, because $\det(\hat{A}) = [\det(\hat{L})]^2$. Finally, $x$ must satisfy

$$\gamma^T\gamma + x^2 = d \tag{8.3.12}$$

To see that $x^2$ must be positive, form the determinant of both sides in (8.3.9), obtaining

$$[\det(\hat{L})]^2 x^2 = \det(A) \tag{8.3.13}$$

Since $\det(A)$ is the product of the eigenvalues of $A$, and since all eigenvalues of positive definite symmetric are positive (see Problem 14 of Chapter 7), $\det(A)$ is positive. Also, by the induction hypothesis, $\hat{L}$ is real. Thus $x^2$ is positive in (8.3.13), and we let $x$ be its positive square root. Since the result (8.2.8) is trivially true for matrices of order $n = 1$, this completes the proof of the factorization (8.3.8). For another approach, see Golub and Van Loan (1983, sec. 5.2).

A practical construction of $L$ can be based on (8.3.9)–(8.3.12), but we give one based on directly finding the elements of $L$. Let $L = [l_{ij}]$, with $l_{ij} = 0$ for $j > i$. Begin the construction of $L$ by multiplying the first row of $A$ times the first column of $L^T$ to get

$$l_{11}^2 = a_{11}$$

Because $A$ is positive definite, $a_{11} > 0$, and $l_{11} = \sqrt{a_{11}}$. Multiply the second row of $L$ times the first two columns of $L^T$ to get

$$l_{21}l_{11} = a_{21} \qquad l_{21}^2 + l_{22}^2 = a_{22}$$

Again, we can solve for the unknowns $l_{21}$ and $l_{22}$. In general for $i = 1, 2, \ldots, n$,

$$l_{ij} = \frac{a_{ij} - \sum_{k=1}^{j-1} l_{ik}l_{jk}}{l_{jj}} \qquad j = 1, \ldots, i-1 \tag{8.3.14}$$

$$l_{ii} = \left[ a_{ii} - \sum_{k=1}^{i-1} l_{ik}^2 \right]^{1/2} \tag{8.3.15}$$

The argument in this square root is the term $x^2$ in the earlier derivation (8.3.12), and $l_{ii}$ is real and positive. For programs implementing Cholesky's method, see Dongarra et al. (1979, chap. 3) and Wilkinson and Reinsch (1971, pp. 10–30).

Note the inner products in (8.3.14) and (8.3.15). These can be accumulated in double precision, minimizing the number of rounding errors, and the elements $l_{ij}$ will be in error by much less than if they had been calculated using only single precision arithmetic. Also note that the elements of $L$ remain bounded relative to $A$, since (8.3.15) yields a bound for the elements of row $i$, using

$$l_{i1}^2 + \cdots + l_{ii}^2 = a_{ii} \tag{8.3.16}$$

The square roots in (8.3.15) of Choleski's method can be avoided by using a slight modification of (8.3.8). Find a diagonal matrix $D$ and a lower triangular matrix $\tilde{L}$, with 1s on the diagonal, such that

$$A = \tilde{L} D \tilde{L}^T \tag{8.3.17}$$

This factorization can be done with about the same number of operations as Choleski's method, about $\frac{1}{6}n^3$, with no square roots. For further discussion and a program, see Wilkinson and Reinsch (1971, pp. 10–30).

***Example*** Consider the Hilbert matrix of order three,

$$A = \begin{bmatrix} 1 & \dfrac{1}{2} & \dfrac{1}{3} \\[2mm] \dfrac{1}{2} & \dfrac{1}{3} & \dfrac{1}{4} \\[2mm] \dfrac{1}{3} & \dfrac{1}{4} & \dfrac{1}{5} \end{bmatrix} \tag{8.3.18}$$

For the Choleski decomposition,

$$L = \begin{bmatrix} 1 & 0 & 0 \\[2mm] \dfrac{1}{2} & \dfrac{1}{2\sqrt{3}} & 0 \\[2mm] \dfrac{1}{3} & \dfrac{1}{2\sqrt{3}} & \dfrac{1}{6\sqrt{5}} \end{bmatrix}$$

and for (8.3.17),

$$\tilde{L} = \begin{bmatrix} 1 & 0 & 0 \\[2mm] \dfrac{1}{2} & 1 & 0 \\[2mm] \dfrac{1}{3} & 1 & 1 \end{bmatrix} \qquad D = \begin{bmatrix} 1 & 0 & 0 \\[2mm] 0 & \dfrac{1}{12} & 0 \\[2mm] 0 & 0 & \dfrac{1}{180} \end{bmatrix}$$

For many linear systems in applications, the coefficient matrix $A$ is *banded*, which means

$$a_{ij} = 0 \quad \text{if} \quad |i - j| > m \tag{8.3.19}$$

for some small $m > 0$. The preceding algorithms simplify in this case, with a considerable savings in computation time. For such algorithms when $A$ is symmetric and positive definite, see the LINPACK programs in Dongarra et al. (1979, chap. 4). We next describe an algorithm in the case $m = 1$ in (8.3.19).

**Tridiagonal systems** The matrix $A = [a_{ij}]$ is *tridiagonal* if

$$a_{ij} = 0 \quad \text{for} \quad |i - j| > 1 \tag{8.3.20}$$

This gives the form

$$
A = \begin{bmatrix}
a_1 & c_1 & 0 & 0 & \cdots & & 0 \\
b_2 & a_2 & c_2 & 0 & & & \\
0 & b_3 & a_3 & c_3 & & & \vdots \\
\vdots & & & \ddots & & & \\
& & & & b_{n-1} & a_{n-1} & c_{n-1} \\
0 & \cdots & & & 0 & b_n & a_n
\end{bmatrix} \tag{8.3.21}
$$

Tridiagonal matrices occur in a variety of applications. Recall the linear system (3.7.22) for spline functions in Section 3.7 of Chapter 3. In addition, many numerical methods for solving boundary value problems for ordinary and partial differential equations involve the solution of tridiagonal systems. Virtually all of these applications yield tridiagonal matrices for which the $LU$ factorization can be formed without pivoting, and for which there is no large increase in error as a consequence. The precise assumptions on $A$ are given below in Theorem 8.2.

By considering the factorization $A = LU$ without pivoting, we find that most elements of $L$ and $U$ will be zero. And we are lead to the following general formula for the decomposition:

$$A = LU$$

$$
= \begin{bmatrix}
\alpha_1 & 0 & \cdots & & 0 \\
b_2 & \alpha_2 & 0 & & \vdots \\
0 & b_3 & \alpha_3 & & \\
\vdots & & & \ddots & \\
0 & & \cdots & b_n & \alpha_n
\end{bmatrix}
\begin{bmatrix}
1 & \gamma_1 & 0 & \cdots & & 0 \\
0 & 1 & \gamma_2 & 0 & & \\
\vdots & & & \ddots & & \vdots \\
& & & & 1 & \gamma_{n-1} \\
0 & & \cdots & & 0 & 1
\end{bmatrix}.
$$

We can multiply to obtain a way to recursively compute $\{\alpha_i\}$ and $\{\gamma_i\}$:

$$a_1 = \alpha_1 \qquad \alpha_1\gamma_1 = c_1$$

$$a_i = \alpha_i + b_i\gamma_{i-1} \qquad i = 2, \ldots, n \tag{8.3.22}$$

$$\alpha_i\gamma_i = c_i \qquad i = 2, 3, \ldots, n - 1$$

These can be solved to give

$$\alpha_1 = a_1 \qquad \gamma_1 = \frac{c_1}{\alpha_1}$$

$$\alpha_i = a_i - b_i\gamma_{i-1} \qquad \gamma_i = \frac{c_i}{\alpha_i} \qquad i = 2, 3, \ldots, n-1 \qquad (8.3.23)$$

$$\alpha_n = a_n - b_n\gamma_{n-1}$$

To solve $LUx = f$, let $Ux = z$ and $Lz = f$. Then

$$z_1 = \frac{f_1}{\alpha_1} \qquad z_i = \frac{f_i - b_i z_{i-1}}{\alpha_i} \qquad i = 2, 3, \ldots, n$$

$$x_n = z_n \qquad x_i = z_i - \gamma_i x_{i+1} \qquad i = n-1, n-2, \ldots, 1 \qquad (8.3.24)$$

The constants in (8.3.23) can be stored for later use in solving the linear system $Ax = f$, for as many right sides $f$ as desired.

Counting only multiplications and divisions, the number of operations to calculate $L$ and $U$ is $2n - 2$; to solve $Ax = f$ takes an additional $3n - 2$ operations. Thus we need only $5n - 4$ operations to solve $Ax - f$ the first time, and for each additional right side, with the same $A$, we need only $3n - 2$ operations. This is extremely rapid. To illustrate this, note that $A^{-1}$ generally is dense and has mostly nonzero entries; thus, the calculation of $x = A^{-1}f$ will require $n^2$ operations. In many applications $n$ may be larger than 1000, and thus there is a significant savings in using (8.3.23)–(8.3.24) as compared with other methods of solution.

To justify the preceding decomposition of $A$, especially to show that all the coefficients $\alpha_i \neq 0$, we have the following theorem.

**Theorem 8.2**   Assume the coefficients $\{a_i, b_i, c_i\}$ of (8.3.21) satisfy the following conditions:

1.   $|a_1| > |c_1| > 0$

2.   $|a_i| \geq |b_i| + |c_i| \qquad b_i, c_i \neq 0 \qquad i = 2, \ldots, n-2$

3.   $|a_n| > |b_n| > 0$

Then $A$ is nonsingular,

$$|\gamma_i| < 1 \qquad i = 1, \ldots, n-1$$

$$|a_i| - |b_i| < |\alpha_i| < |a_i| + |b_i| \qquad i = 2, \ldots, n$$

**Proof**   For the proof, see Isaacson and Keller (1966, p. 57). Note that the last bound shows $|\alpha_i| > |c_i|$, $i = 2, \ldots, n-2$. Thus the coefficients of $L$

and $U$ remain bounded, and no divisors are used that are almost zero except for rounding error.

The condition that $b_i, c_i \neq 0$ is not essential. For example, if some $b_i = 0$, then the linear system can be broken into two new systems, one of order $i - 1$ and the other of order $n - i + 1$. For example, if

$$A = \begin{bmatrix} a_1 & c_1 & 0 & 0 \\ b_2 & a_2 & c_2 & 0 \\ 0 & 0 & a_3 & c_3 \\ 0 & 0 & b_4 & a_4 \end{bmatrix}$$

then solve $Ax = f$ by reducing it to the following two linear systems,

$$\begin{bmatrix} a_3 & c_3 \\ b_4 & a_4 \end{bmatrix} \begin{bmatrix} x_3 \\ x_4 \end{bmatrix} = \begin{bmatrix} f_3 \\ f_4 \end{bmatrix} \qquad \begin{bmatrix} a_1 & c_1 \\ b_2 & a_2 \end{bmatrix} \begin{bmatrix} x_1 \\ x_2 \end{bmatrix} = \begin{bmatrix} f_1 \\ f_2 - c_2 x_3 \end{bmatrix}$$

This completes the proof. ■

***Example*** Consider the coefficient matrix for spline interpolation, in (3.7.22) of Chapter 3. Consider $h_i = $ constant in that matrix, and then factor $h/6$ from every row. Restricting our interest to the matrix of order four, the resulting matrix is

$$A \equiv \begin{bmatrix} 2 & 1 & 0 & 0 \\ 1 & 4 & 1 & 0 \\ 0 & 1 & 4 & 1 \\ 0 & 0 & 1 & 2 \end{bmatrix}$$

Using the method (8.3.23), this has the $LU$ factorization

$$L = \begin{bmatrix} 2 & 0 & 0 & 0 \\ 1 & \dfrac{7}{2} & 0 & 0 \\ 0 & 1 & \dfrac{26}{7} & 0 \\ 0 & 0 & 1 & \dfrac{45}{26} \end{bmatrix} \qquad U = \begin{bmatrix} 1 & \dfrac{1}{2} & 0 & 0 \\ 0 & 1 & \dfrac{2}{7} & 0 \\ 0 & 0 & 1 & \dfrac{7}{26} \\ 0 & 0 & 0 & 1 \end{bmatrix}$$

This completes the example. And it should indicate that the solution of the cubic spline interpolation problem, described in Section 3.7 of Chapter 3, is not difficult to compute.

## 8.4   Error Analysis

We begin the error analysis of methods for solving $Ax = b$ by examining the stability of the solution $x$ relative to small perturbations in the right side $b$. We will follow the general schemata of Section 1.6 of Chapter 1, and in particular, we will study the condition number of (1.6.6).

Let $Ax = b$, of order $n$, be uniquely solvable, and consider the solution of the perturbed problem

$$A\tilde{x} = b + r \tag{8.4.1}$$

Let $e = \tilde{x} - x$, and subtract $Ax = b$ to get

$$Ae = r \qquad e = A^{-1}r \tag{8.4.2}$$

To examine the stability of $Ax = b$ as in (1.6.6), we want to bound the quantity

$$\frac{\|e\|}{\|x\|} \div \frac{\|r\|}{\|b\|} \tag{8.4.3}$$

as $r$ ranges over all elements of $\mathbf{R}^n$, which are small relative to $b$.

From (8.4.2), take norms to obtain

$$\|r\| \le \|A\| \, \|e\| \qquad \|e\| \le \|A^{-1}\| \, \|r\|$$

Divide by $\|A\| \, \|x\|$ in the first inequality and by $\|x\|$ in the second one to obtain

$$\frac{\|r\|}{\|A\| \, \|x\|} \le \frac{\|e\|}{\|x\|} < \frac{\|A^{-1}\| \, \|r\|}{\|x\|}$$

The matrix norm is the operator matrix norm induced by the vector norm. Using the bounds

$$\|b\| \le \|A\| \, \|x\| \qquad \|x\| \le \|A^{-1}\| \, \|b\|$$

we obtain

$$\frac{1}{\|A\| \, \|A^{-1}\|} \cdot \frac{\|r\|}{\|b\|} \le \frac{\|e\|}{\|x\|} \le \|A\| \, \|A^{-1}\| \cdot \frac{\|r\|}{\|b\|} \tag{8.4.4}$$

Recalling (8.4.3), this result is justification for introducing the *condition number* of $A$:

$$\text{cond}(A) = \|A\| \, \|A^{-1}\| \tag{8.4.5}$$

For each given $A$, there are choices of $b$ and $r$ for which either of the inequalities in (8.4.4) can be made an equality. This is a further reason for introducing cond($A$) when considering (8.4.3). We leave the proof to Problem 20.

The quantity cond($A$) will vary with the norm being used, but it is always bounded below by one, since

$$1 = \|I\| = \|AA^{-1}\| \le \|A\| \, \|A^{-1}\| = \text{cond}(A)$$

If the condition number is nearly 1, then we see from (8.4.4) that small relative perturbations in $b$ will lead to similarly small relative perturbations in the solution $x$. But if cond$(A)$ is large, then (8.4.4) suggests that there may be small relative perturbations of $b$ that will lead to large relative perturbations in $x$.

Because (8.4.5) will vary with the choice of norm, we sometimes use another definition of condition number, one independent of the norm. From Theorem 7.8 of Chapter 7,

$$\text{cond}(A) \geq r_\sigma(A) r_\sigma(A^{-1})$$

Since the eigenvalues of $A^{-1}$ are the reciprocals of those of $A$, we have the result

$$\text{cond}(A) \geq \frac{\underset{\lambda \in \sigma(A)}{\text{Max }} |\lambda|}{\underset{\lambda \in \sigma(A)}{\text{Min }} |\lambda|} \equiv \text{cond}(A)_* \tag{8.4.6}$$

in which $\sigma(A)$ denotes the set of all eigenvalues of $A$.

**Example**   Consider the linear system

$$7x_1 + 10x_2 = b_1$$

$$5x_1 + 7x_2 = b_2 \tag{8.4.7}$$

For the coefficient matrix,

$$A = \begin{bmatrix} 7 & 10 \\ 5 & 7 \end{bmatrix} \qquad A^{-1} = \begin{bmatrix} -7 & 10 \\ 5 & -7 \end{bmatrix}$$

Let the condition number in (8.4.5) be denoted by cond$(A)_p$ when it is generated using the matrix norm $\|\cdot\|_p$. For this example,

$$\text{cond}(A)_1 = \text{cond}(A)_\infty = (17)(17) = 289,$$

$$\text{cond}(A)_2 \doteq 223 \qquad \text{cond}(A)_* \doteq 198$$

These condition numbers all suggest that (8.4.7) may be sensitive to changes in the right side $b$. To illustrate this possibility, consider the particular case

$$7x_1 + 10x_2 = 1$$

$$5x_1 + 7x_2 = .7$$

which has the solution

$$x_1 = 0 \qquad x_2 = .1$$

For the perturbed system, solve

$$7\tilde{x}_1 + 10\tilde{x}_2 = 1.01$$

$$5\tilde{x}_1 + 7\tilde{x}_2 = .69$$

It has the solution

$$\tilde{x}_1 = -.17 \qquad \tilde{x}_2 = .22$$

The relative changes in $x$ are quite large when compared with the size of the relative changes in the right side $b$.

A linear system whose solution $x$ is unstable with respect to small relative changes in the right side $b$ is called *ill-conditioned*. The preceding system (8.4.7) is somewhat ill-conditioned, especially if only three or four decimal digit floating-point arithmetic is used in solving it. The condition numbers $\text{cond}(A)$ and $\text{cond}(A)_*$ are fairly good indicators of ill-conditioning. As they increase by a factor of 10, it is likely that one less digit of accuracy will be obtained in the solution.

In general, if $\text{cond}(A)_*$ is large, then there will be values of $b$ for which the system $Ax = b$ is quite sensitive to changes $r$ in $b$. Let $\lambda_l$ and $\lambda_u$ denote eigenvalues of $A$ for which

$$|\lambda_l| = \underset{\lambda \in \sigma(A)}{\text{Min}} |\lambda| \qquad |\lambda_u| = \underset{\lambda \in \sigma(A)}{\text{Max}} |\lambda|$$

and thus

$$\text{cond}(A)_* = \left| \frac{\lambda_u}{\lambda_l} \right| \tag{8.4.8}$$

Let $x_l$ and $x_u$ be corresponding eigenvectors, with $\|x_l\|_\infty = \|x_u\|_\infty = 1$. Then

$$Ax = \lambda_u x_u$$

has the solution $x = x_u$. And the system

$$A\tilde{x} = \lambda_u x_u + \lambda_l x_l = \lambda_u \left[ x_u + \frac{1}{\text{cond}(A)_*} x_l \right]$$

has the solution

$$\tilde{x} = x_u + x_l$$

If $\text{cond}(A)_*$ is large, then the right-hand side has only a small relative perturbation,

$$\frac{\|r\|_\infty}{\|b\|_\infty} = \frac{1}{\text{cond}(A)_*} \tag{8.4.9}$$

But for the solution, we have the much larger relative perturbation

$$\frac{\|\tilde{x} - x\|_\infty}{\|x\|_\infty} = \frac{\|x_l\|_\infty}{\|x_u\|_\infty} = 1 \tag{8.4.10}$$

There are systems that are not ill-conditioned in actual practice, but for which the preceding condition numbers are quite large. For example,

$$A = \begin{bmatrix} 1 & 0 \\ 0 & 10^{-10} \end{bmatrix}$$

has all condition numbers $\operatorname{cond}(A)_p$ and $\operatorname{cond}(A)_*$ equal to $10^{10}$. But usually the matrix is not considered ill-conditioned. The difficulty is in using norms to measure changes in a vector, rather than looking at each component separately. If scaling has been carried out on the coefficient matrix and unknown vector, then this problem does not usually arise, and then the condition numbers are usually an accurate predictor of ill-conditioning.

As a final justification for the use of $\operatorname{cond}(A)$ as a condition number, we give the following result.

**Theorem 8.3** (Gastinel)    Let $A$ be any nonsingular matrix of order $n$, and let $\|\cdot\|$ denote an operator matrix norm. Then

$$\frac{1}{\operatorname{cond}(A)} = \operatorname{Min}\left\{ \frac{\|A - B\|}{\|A\|} \,\middle|\, B \text{ a singular matrix} \right\} \quad (8.4.11)$$

with $\operatorname{cond}(A)$ defined in (8.4.5).

**Proof**    See Kahan (1966, p. 775).    ∎

The theorem states that $A$ can be well approximated in a relative error sense by a singular matrix $B$ if and only if $\operatorname{cond}(A)$ is quite large. And from our view, a singular matrix $B$ is the ultimate in ill-conditioning. There are nonzero perturbations of the solution, by the eigenvector for the eigenvalue $\lambda = 0$, which correspond to a zero perturbation in the right side $b$. More importantly, there are values of $b$ for which $Bx = b$ is no longer solvable.

**The Hilbert matrix**    The Hilbert matrix of order $n$ is defined by

$$H_n = \begin{bmatrix} 1 & \dfrac{1}{2} & \dfrac{1}{3} & \cdots & \dfrac{1}{n} \\[2mm] \dfrac{1}{2} & \dfrac{1}{3} & \dfrac{1}{4} & & \dfrac{1}{n+1} \\[2mm] \vdots & & & & \vdots \\[2mm] \dfrac{1}{n} & \dfrac{1}{n+1} & & \cdots & \dfrac{1}{2n-1} \end{bmatrix} \quad (8.4.12)$$

This matrix occurs naturally in solving the continuous least squares approximation problem. Its derivation is given near the end of Section 4.3 of Chapter 4, with the resulting linear system given in (4.3.14). As was indicated in Section 4.3 and illustrated following (1.6.9) in Section 1.4 of Chapter 1, the Hilbert matrix is

**Table 8.2    Condition numbers of Hilbert matrix**

| $n$ | cond $(H_n)_*$ | $n$ | cond $(H_n)_*$ |
|---|---|---|---|
| 3 | 5.24E + 2 | 7 | 4.75E + 8 |
| 4 | 1.55E + 4 | 8 | 1.53E + 10 |
| 5 | 4.77E + 5 | 9 | 4.93E + 11 |
| 6 | 1.50E + 7 | 10 | 1.60E + 13 |

very ill-conditioned, and increasingly so as $n$ increases. As such, it has been a favorite numerical example for checking programs for solving linear systems of equations, to determine the limits of effectiveness of the program when dealing with ill-conditioned problems. Table 8.2 gives the condition number cond $(H_n)_*$ for a few values of $n$. The inverse matrix $H_n^{-1} = [\alpha_{ij}^{(n)}]$ is known explicitly:

$$\alpha_{ij}^{(n)} = \frac{(-1)^{i+j}(n+i-1)!(n+j-1)!}{(i+j-1)[(i-1)!(j-1)!]^2(n-i)!(n-j)!} \qquad 1 \le i, j \le n$$

$$(8.4.13)$$

For additional information on $H_n$, including an asymptotic formula for cond $(H_n)_*$, see Gregory and Karney (1969, pp. 33–38, 66–73).

Although widely used as an example, some care must be taken as to what is the true answer. Let $\overline{H}_n$ denote the version of $H_n$ after it is entered into the finite arithmetic of a computer. For a matrix inversion program, the results of the program should be compared with $\overline{H}_n^{-1}$, not with $H_n^{-1}$; these two inverse matrices can be quite different. For example, if we use four decimal digit floating-point arithmetic with rounding, then

$$\overline{H}_3 = \begin{bmatrix} 1.000 & .5000 & .3333 \\ .5000 & .3333 & .2500 \\ .3333 & .2500 & .2000 \end{bmatrix} \qquad (8.4.14)$$

Rounding has occurred only in expanding $\frac{1}{3}$ in decimal fraction form. Then

$$H_3^{-1} = \begin{bmatrix} 9.000 & -36.00 & 30.00 \\ -36.00 & 192.0 & -180.0 \\ 30.00 & -180.0 & 180.0 \end{bmatrix}$$

$$\overline{H}_3^{-1} = \begin{bmatrix} 9.062 & -36.32 & 30.30 \\ -36.32 & 193.7 & -181.6 \\ 30.30 & -181.6 & 181.5 \end{bmatrix} \qquad (8.4.15)$$

Any program for matrix inversion, when applied to $\overline{H}_3$, should have its resulting solution compared with $\overline{H}_3^{-1}$, not $H_3^{-1}$. We return to this example later in Section 8.5.

**Error bounds**    We consider the effects of rounding error on the solution $\hat{x}$ to $Ax = b$, obtained using Gaussian elimination. We begin by giving a result bounding the error when $b$ and $A$ are changed by small amounts. This is a useful result by itself, and it is necessary for the error analysis of Gaussian elimination that follows later.

**Theorem 8.4**    Consider the system $Ax = b$, with $A$ nonsingular. Let $\delta A$ and $\delta b$ be perturbations of $A$ and $b$, and assume

$$\|\delta A\| < \frac{1}{\|A^{-1}\|} \qquad (8.4.16)$$

Then $A + \delta A$ is nonsingular. And if we define $\delta x$ implicitly by

$$(A + \delta A)(x + \delta x) = b + \delta b \qquad (8.4.17)$$

then

$$\frac{\|\delta x\|}{\|x\|} \leq \frac{\operatorname{cond}(A)}{1 - \operatorname{cond}(A)\dfrac{\|\delta A\|}{\|A\|}} \cdot \left\{ \frac{\|\delta A\|}{\|A\|} + \frac{\|\delta b\|}{\|b\|} \right\} \qquad (8.4.18)$$

**Proof**    First note that $\delta A$ represents any matrix satisfying (8.4.16), not a constant $\delta$ times the matrix $A$, and similarly for $\delta b$ and $\delta x$. Using (8.4.16), the nonsingularity of $A + \delta A$ follows immediately from Theorem 7.12 of Chapter 7. From (7.4.11),

$$\|(A + \delta A)^{-1}\| \leq \frac{\|A^{-1}\|}{1 - \|A^{-1}\|\,\|\delta A\|} \qquad (8.4.19)$$

Solving for $\delta x$ in (8.4.17), and using $Ax = b$,

$$(A + \delta A)\,\delta x + Ax + (\delta A)x = b + \delta b$$

$$\delta x = (A + \delta A)^{-1}[\delta b - (\delta A)x]$$

Using (8.4.19) and the definition (8.4.5) of $\operatorname{cond}(A)$,

$$\|\delta x\| \leq \frac{\operatorname{cond}(A)}{1 - \operatorname{cond}(A)\dfrac{\|\delta A\|}{\|A\|}} \cdot \left\{ \frac{\|\delta b\|}{\|A\|} + \|x\|\frac{\|\delta A\|}{\|A\|} \right\}$$

Divide by $\|x\|$ on both sides, and use $\|b\| \leq \|A\|\,\|x\|$ to obtain (8.4.18).    ∎

The analysis of the effect of rounding errors on Gaussian elimination is due to J. H. Wilkinson, and it can be found in Wilkinson (1963, pp. 94–99), (1965, pp.

209–216), Forsythe and Moler (1967, chap. 21), and Golub and Van Loan (1983, chap. 4). Let $\hat{x}$ denote the computed solution of $Ax = b$. It is very difficult to compute directly the effects on $x$ of rounding at each step, as a means of obtaining a bound on $\|x - \hat{x}\|$. Rather, it is easier, although nontrivial, to take $\hat{x}$ and the elimination algorithm and to work backwards to show that $\hat{x}$ is the exact solution of a system

$$(A + \delta A)\hat{x} = b$$

in which bounds can be given for $\delta A$. This approach is known as *backward error analysis*. We can then use the preceding Theorem 8.4 to bound $\|x - \hat{x}\|$. In the following result, the matrix norm will be $\|A\|_\infty$, the row norm (7.3.17) induced by the vector norm $\|x\|_\infty$.

**Theorem 8.5**   Let $A$ be of order $n$ and nonsingular, and assume partial or complete pivoting is used in the Gaussian elimination process. Define

$$\rho = \frac{1}{\|A\|_\infty} \underset{1 \le i,\, j,\, k \le n}{\text{Max}} |a_{ij}^{(k)}| \tag{8.4.20}$$

Let $u$ denote the unit round on the computer being used. [See (1.2.11)–(1.2.12) for the definition of $u$.]

1. The matrices $L$ and $U$ computed using Gaussian elimination satisfy

$$LU = A + E,$$

$$\|E\|_\infty \le n^2 \rho \|A\|_\infty u \tag{8.4.21}$$

2. The approximate solution $\hat{x}$ of $Ax = b$, computed using Gaussian elimination, satisfies

$$(A + \delta A)\hat{x} = b \tag{8.4.22}$$

with

$$\frac{\|\delta A\|_\infty}{\|A\|_\infty} \le \{1.01(n^3 + 3n^2)\rho u\} \tag{8.4.23}$$

3. Using Theorem 8.4,

$$\frac{\|x - \hat{x}\|_\infty}{\|x\|_\infty} \le \frac{\text{cond}(A)_\infty}{1 - \text{cond}(A)_\infty \dfrac{\|\delta A\|_\infty}{\|A\|_\infty}} \left[1.01(n^3 + 3n^2)\rho u\right] \tag{8.4.24}$$

**Proof**   The proofs of (1) and (2) are given in Forsythe and Moler (1967, chap. 21). Variations on these results are given in Golub and Van Loan (1983, chap. 4).   ∎

Empirically, the bound (8.4.24) is too large, due to cancellation of rounding errors of varying magnitude and sign. According to Wilkinson (1963, p. 108), a better empirical bound for most cases is

$$\frac{\|\delta A\|_\infty}{\|A\|_\infty} \le nu \qquad (8.4.25)$$

The result (8.4.24) shows the importance of the size of cond$(A)$.

The quantity $\rho$ in the bounds can be computed during the elimination process, and it can also be bounded a priori. For complete pivoting, an a priori bound is

$$\rho \le 1.8 n^{(\ln n)/4} \qquad n \ge 1$$

and it is conjectured that $\rho \le cn$ for some $c$. For partial pivoting, an a priori bound is $2^{n-1}$, and pathological examples are known for which this is possible. Nonetheless, in all empirical studies to date, $\rho$ has been bounded by a relatively small number, independent of $n$. Because of the differing theoretical bounds for $\rho$, complete pivoting is sometimes considered superior. In actual practice, however, the error behavior with partial pivoting is as good as with complete pivoting. Moreover, complete pivoting requires many more comparisons at each step of the elimination process. Consequently, partial pivoting is the approach used in all modern Gaussian elimination codes.

One of the most important consequences of the preceding analysis is to show that Gaussian elimination is a very stable process, provided only that the matrix $A$ is not badly ill-conditioned. Historically, researchers in the early 1950s were uncertain as to the stability of Gaussian elimination for larger systems, for example, $n \ge 10$, but that question has now been settled.

The size of the residual in the computed solution $\hat{x}$, namely

$$r = b - A\hat{x} \qquad (8.4.26)$$

is sometimes linked, mistakenly, to the size of the error $x - \hat{x}$. In fact, the error in $\hat{x}$ can be large even though $r$ is small, and this is usually the case with ill-conditioned problems. From (8.4.26) and $Ax = b$,

$$r = A(x - \hat{x})$$

$$x - \hat{x} = A^{-1}r \qquad (8.4.27)$$

and thus $x - \hat{x}$ can be much larger than $r$ if $A^{-1}$ has large elements.

In practice, the residual $r$ is quite small, even for ill-conditioned problems. To suggest why this should happen, use (8.4.22) to obtain

$$r = (\delta A)\hat{x}$$

$$\|r\|_\infty \le \|\delta A\|_\infty \|\hat{x}\|$$

$$\frac{\|r\|_\infty}{\|A\|_\infty \|\hat{x}\|_\infty} \le \frac{\|\delta A\|_\infty}{\|A\|_\infty} \qquad (8.4.28)$$

The bounds for $\|\delta A\|/\|A\|$, in (8.4.23) or (8.4.25), are independent of the conditioning of the problem. Thus $\|r\|_\infty$ will generally be small relative to $\|A\| \|\hat{x}\|$. The latter is often close to $\|b\|$ or is of the same magnitude, since $b = Ax$, and then $\|r\|$ will be small relative to $\|b\|$. As a final note on the size of the residual, there are some problems in which it is important only to have $r$ be small, without $x - \hat{x}$ needing to be small. In such cases, ill-conditioning will not have the same meaning.

The bounds (8.4.18) and (8.4.24) indicate the importance of $\text{cond}(A)$ in determining the error. Generally if $\text{cond}(A) \doteq 10^m$, some $m \geq 0$, then about $m$ digits of accuracy will be lost in computing $\hat{x}$, relative to the number of digits in the arithmetic being used. Thus measuring $\text{cond}(A) = \|A\| \|A^{-1}\|$ is desirable. The term $\|A\|$ is easy and inexpensive to evaluate, and $\|A^{-1}\|$ is the main problem in computing $\text{cond}(A)$. Calculating $A^{-1}$ requires $n^3$ operations, and this is too expensive a way to compute $\|A^{-1}\|$. A less expensive approach, using $O(n^2)$ operations, was developed for the LINPACK package.

For any system $Ay = d$,

$$y = A^{-1}d$$

$$\|y\| \leq \|A^{-1}\| \|d\|$$

$$\|A^{-1}\| \geq \frac{\|y\|}{\|d\|} \tag{8.4.29}$$

We want to choose $d$ to make this ratio as large as possible. Write $A = LU$, with $LU$ obtained in the Gaussian elimination. Then solving $Ay = d$ is equivalent to solving

$$Lw = d \qquad Uy = w$$

While solving $Lw = d$, develop $d$ to make $w$ as large as possible, while retaining $\|d\|_\infty = 1$. Then solve $Uy = w$ for $y$. This will give a better bound in (8.4.29) than a randomly chosen $d$. An algorithm for choosing $d$ is given in Golub and Van Loan (1983, p. 77). The algorithm in LINPACK is a more complicated extension of the preceding. For a description see Golub and Van Loan (1983, p. 78) or Dongarra et al. (1979, pp. 1.12–1.13).

**A posteriori error bounds**    We begin with error bounds for a computed inverse $C$ of a given matrix $A$. Define the residual matrix by

$$R = I - CA$$

**Theorem 8.6**    If $\|R\| < 1$, then $A$ and $C$ are nonsingular, and

$$\frac{\|R\|}{\|A\| \|C\|} \leq \frac{\|A^{-1} - C\|}{\|C\|} \leq \frac{\|R\|}{1 - \|R\|} \tag{8.4.30}$$

***Proof***   Since $\|R\| < 1$, $I - R$ is nonsingular by Theorem 7.11 of Chapter 7, and

$$\|(I - R)^{-1}\| \leq \frac{1}{1 - \|R\|}$$

But

$$I - R = CA \qquad (8.4.31)$$

$$0 \neq \det(I - R) = \det(CA) = \det(C)\det(A)$$

and thus both $\det(C)$ and $\det(A)$ are nonzero. This shows that both $A$ and $C$ are nonsingular.

For the lower bound in (8.4.30),

$$R = I - CA = (A^{-1} - C)A,$$

$$\|R\| \leq \|A^{-1} - C\|\,\|A\|$$

and dividing by $\|A\|\,\|C\|$ proves the result. For the upper bound, (8.4.31) implies

$$(I - R)^{-1} = A^{-1}C^{-1}$$

$$A^{-1} = (I - R)^{-1}C \qquad (8.4.32)$$

For the error in $C$,

$$A^{-1} - C = (I - CA)A^{-1} = RA^{-1} = R(I - R)^{-1}C$$

$$\|A^{-1} - C\| \leq \frac{\|R\|\,\|C\|}{1 - \|R\|}$$

This completes the proof.   ∎

This result is generally of more theoretical than practical interest. Inverse matrices should not be produced for solving a linear system, as was pointed out earlier in Section 8.1. And as a consequence, there is seldom any real need for the preceding type of error bound. The main exception is when $C$ has been produced as an approximation by means other than Gaussian elimination, often by some theoretical derivation. Such approximate inverses are then used to solve $Ax = b$ by the residual correction procedure (8.5.3) described in the next section. In this case, the bound (8.4.30) can furnish some useful information on $C$.

***Corollary***   Let $A$, $C$, and $R$ be as given in Theorem 8.6. Let $\hat{x}$ be an approximate solution to $Ax = b$, and define $r = b - A\hat{x}$. Then

$$\|x - \hat{x}\| \leq \frac{\|Cr\|}{1 - \|R\|} \qquad (8.4.33)$$

∎

***Proof***   From

$$r = b - A\hat{x} = Ax - A\hat{x} = A(x - \hat{x})$$

$$x - \hat{x} = A^{-1}r = (I - R)^{-1}Cr \qquad (8.4.34)$$

with (8.4.32) used in the last equality. Taking norms, we obtain (8.4.33).
    ∎

This bound (8.4.33) has been found to be quite accurate, especially when compared with a number of other bounds that are commonly used. For a complete discussion of computable error bounds, including a number of examples, see Aird and Lynch (1975).

The error bound (8.4.33) is relatively expensive to produce. If we suppose that $\hat{x}$ was obtained by Gaussian elimination, then about $n^3/3$ operations were used to calculate $\hat{x}$ and the $LU$ decomposition of $A$. To produce $C \doteq A^{-1}$ by elimination will take at least $\frac{2}{3}n^3$ additional operations, producing $CA$ requires $n^3$ multiplications, and producing $Cr$ requires $n^2$. Thus the error bound requires at least a fivefold increase in the number of operations. It is generally preferable to estimate the error by solving approximately the error equation

$$A(x - \hat{x}) = r$$

using the $LU$ decomposition stored earlier. This requires $n^2$ operations to evaluate $r$, and an additional $n^2$ to solve the linear system. Unless, the residual matrix $R = I - CA$ has norm nearly one, this approach will give a very reasonable error estimate. This is pursued and illustrated in the next section.

## 8.5   The Residual Correction Method

We assume that $Ax = b$ has been solved for an approximate solution $\hat{x} \equiv x^{(0)}$. Also the $LU$ decomposition along with a record of all row or column interchanges should have been stored. Calculate

$$r^{(0)} = b - Ax^{(0)} \qquad (8.5.1)$$

Define $e^{(0)} = x - x^{(0)}$. Then as before in (8.4.34),

$$Ae^{(0)} = r^{(0)}$$

Solve this system using the stored $LU$ decomposition, and call the resulting approximate solution $\hat{e}^{(0)}$. Define a new approximate solution to $Ax = b$ by

$$x^{(1)} = x^{(0)} + \hat{e}^{(0)} \qquad (8.5.2)$$

The process can be repeated, calculating $x^{(2)}, \ldots$, to continually decrease the error. To calculate $r^{(0)}$ takes $n^2$ operations, and the calculation of $\hat{e}^{(0)}$ takes an additional $n^2$ operations. Thus the calculation of the improved values $x^{(1)}, x^{(2)}, \ldots,$ is inexpensive compared with the calculation of the original value $x^{(0)}$. This method is also known as *iterative improvement* or the *residual correction method*.

It is extremely important to obtain accurate values for $r^{(0)}$. Since $x^{(0)}$ approximately solves $Ax = b$, $r^{(0)}$ will generally involve loss-of-significance errors in its calculation, with $Ax^{(0)}$ and $b$ agreeing to almost the full precision of the machine arithmetic. Thus to obtain accurate values for $r^{(0)}$, we must usually go to higher precision arithmetic. If only regular arithmetic is used to calculate $r^{(0)}$, the same arithmetic as used in calculating $x^{(0)}$ and $LU$, then the resulting inaccuracy in $r^{(0)}$ will usually leads to $\hat{e}^{(0)}$ being a poor approximation to $e^{(0)}$. In single precision arithmetic, we calculate $r^{(0)}$ in double precision. But if the calculations are already in double precision, it is often hard to go to a higher precision arithmetic.

**Example**   Solve the system $Ax = b$, with $A = \overline{H}_3$ from (8.4.14). The arithmetic will be four decimal digit floating-point with rounding. For the right side, use

$$b = [1, 0, 0]^T$$

The true solution is the first column of $\overline{H}_3^{-1}$, which from (8.4.15) is

$$x = [9.062, -36.32, 30.30]^T$$

to four significant digits.

Using elimination with partial pivoting,

$$x^{(0)} = [8.968, -35.77, 29.77]^T$$

The residual $r^{(0)}$ is calculated with double precision arithmetic, and is then rounded to four significant digits. The value obtained is

$$r^{(0)} = [-.005341, -.004359, -.005344]^T$$

Solving $Ae^{(0)} = r^{(0)}$ with the stored $LU$ decomposition,

$$\hat{e}^{(0)} = [.09216, -.5442, .5239]^T$$

$$x^{(1)} = [9.060, -36.31, 30.29]^T$$

Repeating these operations,

$$r^{(1)} = [-.0006570, -.0003770, -.0001980]^T$$

$$\hat{e}^{(1)} = [.001707, -.01300, .01241]^T$$

$$x^{(2)} = [9.062, -36.32, 30.30]^T$$

The vector $x^{(2)}$ is accurate to four significant digits. Also, note that $x^{(1)} - x^{(0)} = \hat{e}^{(0)}$ is an accurate predictor of the error $e^{(0)}$ in $x^{(0)}$.

Formulas can be developed to estimate how many iterates should be calculated in order to get essentially full accuracy in the solution $x$. For a discussion of what is involved and for some algorithms implementing this method, see Dongarra et al. (1979, p. 1.9), Forsythe and Moler (1967, chaps. 13, 16, 17), Golub and Van Loan (1983, p. 74), and Wilkinson and Reinsch (1971, pp. 93–110).

**Another residual correction method**  There are situations in which we can calculate an approximate inverse $C$ to the given matrix $A$. This is generally done by carefully considering the structure of $A$, and then using a variety of approximation techniques to estimate $A^{-1}$. Without considering the origin of $C$, we show how to use it to iteratively solve $Ax = b$.

Let $x^{(0)}$ be an initial guess, and define $r^{(0)} = b - Ax^{(0)}$. As before, $A(x - x^{(0)}) = r^{(0)}$. Define $x^{(1)}$ implicitly by

$$x^{(1)} - x^{(0)} = Cr^{(0)}$$

In general, define

$$r^{(m)} = b - Ax^{(m)} \qquad x^{(m+1)} = x^{(m)} + Cr^{(m)} \qquad m = 0,1,2,\dots \quad (8.5.3)$$

If $C$ is a good approximation to $A^{-1}$, the iteration will converge rapidly, as shown in the following analysis.

We first obtain a recursion formula for the error:

$$x - x^{(m+1)} = x - x^{(m)} - Cr^{(m)} = x - x^{(m)} - C[b - Ax^{(m)}]$$

$$= x - x^{(m)} - C[Ax - Ax^{(m)}]$$

$$x - x^{(m+1)} = (I - CA)(x - x^{(m)}) \qquad (8.5.4)$$

By induction

$$x - x^{(m)} = (I - CA)^m (x - x^{(0)}) \qquad m \geq 0 \qquad (8.5.5)$$

If

$$\|I - CA\| < 1 \qquad (8.5.6)$$

for some matrix norm, then using the associated vector norm,

$$\|x - x^{(m)}\| \leq \|I - CA\|^m \|x - x^{(0)}\| \qquad (8.5.7)$$

And this converges to zero as $m \to \infty$, for any choice of initial guess $x^{(0)}$. More generally, (8.5.5) implies that $x^{(m)}$ converges to $x$, for any choice of $x^{(0)}$, if and only if

$$(I - CA)^m \to 0 \qquad \text{as} \quad m \to \infty$$

And by Theorem 7.9 of Chapter 7, this is equivalent to

$$r_\sigma(I - CA) < 1 \qquad (8.5.8)$$

for the special radius of $I - CA$. This may be possible to show, even when (8.5.6) fails for the common matrix norms. Also note that

$$I - AC = A(I - CA)A^{-1}$$

and thus $I - AC$ and $I - CA$ are similar matrices and have the same eigenvalues. If

$$\|I - AC\| < 1 \qquad (8.5.9)$$

then (8.5.8) is true, even if (8.5.6) is not true, and convergence will still occur.

Statement (8.5.4) shows that the rate of convergence of $x^{(m)}$ to $x$ is linear:

$$\|x - x^{(m+1)}\| \le c\|x - x^{(m)}\| \qquad m \ge 0 \qquad (8.5.10)$$

with $c < 1$ unknown. The constant $c$ is often estimated computationally with

$$c = \text{Max} \frac{\|x^{(m+2)} - x^{(m+1)}\|}{\|x^{(m+1)} - x^{(m)}\|} \qquad (8.5.11)$$

with the maximum performed over some or all of the iterates that have been computed. This is not rigorous, but is motivated by the formula

$$x^{(m+2)} - x^{(m+1)} = (I - CA)(x^{(m+1)} - x^{(m)}) \qquad (8.5.12)$$

To prove this, simply use (8.5.4), subtracting formulas for successive values of $m$.

If we assume (8.5.10) is valid for the iterates that we are calculating, and if we have an estimate for $c$, then we can produce an error bound.

$$\|x^{(m+1)} - x^{(m)}\| = \|[x - x^{(m)}] - [x - x^{(m+1)}]\|$$

$$\ge \|x - x^{(m)}\| - \|x - x^{(m+1)}\|$$

$$\ge \|x - x^{(m)}\| - c\|x - x^{(m)}\|$$

$$\|x - x^{(m)}\| \le \frac{1}{1 - c}\|x^{(m+1)} - x^{(m)}\|$$

$$\|x - x^{(m+1)}\| \le \frac{c}{1 - c}\|x^{(m+1)} - x^{(m)}\| \qquad (8.5.13)$$

For slowly convergent iterates [with $c \doteq 1$], this bound is important, since $\|x^{(m+1)} - x^{(m)}\|$ can then be much smaller than $\|x - x^{(m)}\|$. Also, recall the earlier derivation in Section 2.5 of Chapter 2. A similar bound, (2.5.5), was derived for the error in a linearly convergent method.

***Example***   Define $A(\epsilon) = A_0 + \epsilon B$, with

$$A_0 = \begin{bmatrix} 2 & 1 & 0 \\ 1 & 2 & 1 \\ 0 & 1 & 2 \end{bmatrix} \qquad B = \begin{bmatrix} 0 & 1 & 1 \\ -1 & 0 & 1 \\ -1 & -1 & 0 \end{bmatrix}$$

As an approximate inverse to $A(\epsilon)$, use

$$A(\epsilon)^{-1} \doteq C = A_0^{-1} = \begin{bmatrix} \frac{3}{4} & -\frac{1}{2} & \frac{1}{4} \\ -\frac{1}{2} & 1 & -\frac{1}{2} \\ \frac{1}{4} & -\frac{1}{2} & \frac{3}{4} \end{bmatrix}$$

We can solve the system $A(\epsilon)x = b$ using the residual correction method (8.5.3). For the convergence analysis,

$$I - CA(\epsilon) = I - A_0^{-1}[A_0 + \epsilon B] = -\epsilon A_0^{-1} B$$

$$= -\epsilon \begin{bmatrix} \frac{1}{4} & \frac{1}{2} & \frac{1}{4} \\ -\frac{1}{2} & 0 & \frac{1}{2} \\ -\frac{1}{4} & -\frac{1}{2} & -\frac{1}{4} \end{bmatrix}$$

Convergence is assured if

$$\|I - CA(\epsilon)\|_\infty = |\epsilon| < 1$$

and from (8.5.4),

$$\|x - x^{(m+1)}\|_\infty \le |\epsilon| \, \|x - x^{(m)}\|_\infty \qquad m \ge 0$$

There are many situations of the kind in this example. We may have to solve linear systems of a general form $A(\epsilon)x = b$ for any $\epsilon$ near zero. To save time, we obtain either $A(0)^{-1}$ or the $LU$ decomposition of $A(0)$. This is then used as an approximate inverse to $A(\epsilon)$, and we solve $A(\epsilon)x = b$ using the residual correction method.

## 8.6   Iteration Methods

As was mentioned in the introduction to this chapter, many linear systems are too large to be solved by direct methods based on Gaussian elimination. For these systems, iteration methods are often the only possible method of solution, as well as being faster than elimination in many cases. The largest area for the application of iteration methods is to the linear systems arising in the numerical solution of partial differential equations. Systems of orders $10^3$ to $10^5$ are not unusual, although almost all of the coefficients of the system will be zero. As an example of such problems, the numerical solution of Poisson's equation is studied in Section 8.8. The reader may want to combine reading that section with the present one.

Besides being large, the linear systems to be solved, $Ax = b$, often have several other important properties. They are usually *sparse*, which means that only a small percentage of the coefficients are nonzero. The nonzero coefficients generally have a special pattern in the way they occur in $A$, and there is usually a simple formula that can be used to generate the coefficients $a_{ij}$ as they are needed, rather than having to store them. As one consequence of these properties, the storage space for the vectors $x$ and $b$ may be a more important consideration than is storage for $A$. The matrices $A$ will often have special properties, which are discussed in this and the next two sections.

We begin by defining and analyzing two classical iteration methods; following that, a general abstract framework is presented for studying iteration methods. The special properties of the linear system $Ax = b$ are very important when setting up an iteration method for its solution. The results of this section are just a beginning to the design of a method for any particular area of applications.

**The Gauss–Jacobi method**  (Simultaneous displacements)  Rewrite $Ax = b$ as

$$x_i = \frac{1}{a_{ii}} \left\{ b_i - \sum_{\substack{j=1 \\ j \neq i}}^{n} a_{ij}x_j \right\} \qquad i = 1, 2, \ldots, n \qquad (8.6.1)$$

assuming all $a_{ii} \neq 0$. Define the iteration as

$$x_i^{(m+1)} = \frac{1}{a_{ii}} \left\{ b_i - \sum_{\substack{j=1 \\ j \neq i}}^{n} a_{ij}x_j^{(m)} \right\} \qquad i = 1, \ldots, n \qquad m \geq 0 \qquad (8.6.2)$$

and assume initial guesses $x_i^{(0)}$, $i = 1, \ldots, n$, are given. There are other forms to the method. For example, many problems are given naturally in the form

$$(I - B)x = b$$

and then we would usually first consider the iteration

$$x^{(m+1)} = b + Bx^{(m)} \qquad m \geq 0 \qquad (8.6.3)$$

Our initial error analysis is restricted to (8.6.2), but the same ideas can be used for (8.6.3).

To analyze the convergence, let $e^{(m)} = x - x^{(m)}$, $m \geq 0$. Subtracting (8.6.2) from (8.6.1),

$$e_i^{(m+1)} = -\sum_{\substack{j=1 \\ j \neq i}}^{n} \frac{a_{ij}}{a_{ii}} e_j^{(m)} \qquad i = 1, \ldots, n \qquad m \geq 0 \qquad (8.6.4)$$

$$|e_i^{(m+1)}| \leq \sum_{\substack{j=1 \\ j \neq i}}^{n} \left| \frac{a_{ij}}{a_{ii}} \right| \|e^{(m)}\|_\infty$$

Define

$$\mu = \underset{1 \le i \le n}{\text{Max}} \sum_{\substack{j=1 \\ j \ne i}}^{n} \left| \frac{a_{ij}}{a_{ii}} \right| \tag{8.6.5}$$

Then

$$|e_i^{(m+1)}| \le \mu \|e^{(m)}\|_\infty$$

and since the right side is independent of $i$,

$$\|e^{(m+1)}\|_\infty \le \mu \|e^{(m)}\|_\infty \tag{8.6.6}$$

If $\mu < 1$, then $e^{(m)} \to 0$ as $m \to \infty$ with a linear rate bounded by $\mu$, and

$$\|e^{(m)}\|_\infty \le \mu^m \|e^{(0)}\|_\infty \tag{8.6.7}$$

In order for $\mu < 1$ to be true, the matrix $A$ must be *diagonally dominant*, that is, it must satisfy

$$\sum_{\substack{j=1 \\ j \ne i}}^{n} |a_{ij}| < |a_{ii}| \qquad i = 1, 2, \ldots, n \tag{8.6.8}$$

Such matrices occur in a number of applications, and often the associated matrix is sparse.

To have a more general result, write (8.6.4) as

$$e^{(m+1)} = Me^{(m)} \qquad m \ge 0 \tag{8.6.9}$$

$$M = - \begin{bmatrix} 0 & \dfrac{a_{12}}{a_{11}} & \cdots & \dfrac{a_{1n}}{a_{11}} \\ \dfrac{a_{21}}{a_{22}} & 0 & \dfrac{a_{23}}{a_{22}} & \cdots & \dfrac{a_{2n}}{a_{22}} \\ \vdots & & \ddots & \vdots \\ \dfrac{a_{n1}}{a_{nn}} & & \cdots & 0 \end{bmatrix}$$

Inductively,

$$e^{(m)} = M^m e^{(0)} \tag{8.6.10}$$

If we want $e^{(m)} \to 0$ as $m \to \infty$, independent of the choice of $x^{(0)}$ (and thus of $e^{(0)}$), it is necessary and sufficient that

$$M^m \to 0 \qquad \text{as} \quad m \to \infty$$

Or equivalently from Theorem 7.9 of Chapter 7,

$$r_\sigma(M) < 1 \tag{8.6.11}$$

The condition $\mu < 1$ is merely the requirement that the row norm of $M$ be less than 1, $\|M\|_\infty < 1$, and this implies (8.6.11). But now we see that $e^{(m)} \to 0$ if $\|M\| < 1$ for any operator matrix norm.

***Example***   Consider solving $Ax = b$ by the Gauss–Jacobi method, with

$$A = \begin{bmatrix} 10 & 3 & 1 \\ 2 & -10 & 3 \\ 1 & 3 & 10 \end{bmatrix} \quad b = \begin{bmatrix} 14 \\ -5 \\ 14 \end{bmatrix} \quad x^{(0)} = \begin{bmatrix} 0 \\ 0 \\ 0 \end{bmatrix} \tag{8.6.12}$$

Solving for unknown $i$ in equation $i$, we have $x = g + Mx$,

$$M = \begin{bmatrix} 0 & -.3 & -.1 \\ .2 & 0 & .3 \\ -.1 & -.3 & 0 \end{bmatrix} \quad g = \begin{bmatrix} 1.4 \\ .5 \\ 1.4 \end{bmatrix}$$

The true solution is $x = [1, 1, 1]^T$. To check for convergence, note that $\|M\|_\infty = .5$, $\|M\|_1 = .6$. Hence

$$\|e^{(m+1)}\|_\infty \le .5\|e^{(m)}\|_\infty \qquad m \ge 0 \tag{8.6.13}$$

A similar statement holds for $\|e^{(m+1)}\|_1$. Thus convergence is guaranteed, and the errors will decrease by at least one-half with each iteration. Actual numerical results are given in Table 8.3, and they confirm the result (8.6.13). The final column is

$$\text{Ratio} \equiv \frac{\|e^{(m)}\|_\infty}{\|e^{(m-1)}\|_\infty} \tag{8.6.14}$$

It demonstrates that the convergence can vary from one step to the next, while satisfying (8.6.13), or more generally (8.6.6).

**Table 8.3   Numerical results for the Gauss–Jacobi method**

| $m$ | $x_1^{(m)}$ | $x_2^{(m)}$ | $x_3^{(m)}$ | $\|e^{(m)}\|_\infty$ | Ratio |
|---|---|---|---|---|---|
| 0 | 0 | 0 | 0 | 1.0 | |
| 1 | 1.4 | .5 | 1.4 | .5 | .5 |
| 2 | 1.11 | 1.20 | 1.11 | .2 | .4 |
| 3 | .929 | 1.055 | .929 | .071 | .36 |
| 4 | .9906 | .9645 | .9906 | .0355 | .50 |
| 5 | 1.01159 | .9953 | 1.01159 | .01159 | .33 |
| 6 | 1.000251 | 1.005795 | 1.000251 | .005795 | .50 |

**The Gauss–Seidel method**   (Successive displacements)   Using (8.6.1), define

$$x_i^{(m+1)} = \frac{1}{a_{ii}}\left\{ b_i - \sum_{j=1}^{i-1} a_{ij}x_j^{(m+1)} - \sum_{j=i+1}^{n} a_{ij}x_j^{(m)} \right\} \qquad i = 1, 2, \ldots, n \quad (8.6.15)$$

Each new component $x_i^{(m+1)}$ is immediately used in the computation of the next component. This is convenient for computer calculations, since the new value can be immediately stored in the location that held the old value, and this minimizes the number of necessary storage locations. The storage requirements for $x$ with the Gauss–Seidel method is only half what it would be with the Gauss–Jacobi method.

To analyze the error, subtract (8.6.15) from (8.6.1):

$$e_i^{(m+1)} = -\sum_{j=1}^{i-1} \frac{a_{ij}}{a_{ii}}e_j^{(m+1)} - \sum_{j=i+1}^{n} \frac{a_{ij}}{a_{ii}}e_j^{(m)} \qquad i = 1, 2, \ldots, n \quad (8.6.16)$$

Define

$$\alpha_i = \sum_{1}^{i-1} \left|\frac{a_{ij}}{a_{ii}}\right| \qquad \beta_i = \sum_{j=i+1}^{n} \left|\frac{a_{ij}}{a_{ii}}\right| \qquad i = 1, \ldots, n$$

with $\alpha_1 = \beta_n = 0$. Using the same definition (8.6.5) for $\mu$ as with the Jacobi method,

$$\mu = \underset{1 \le i \le n}{\text{Max}} (\alpha_i + \beta_i)$$

We assume $\mu < 1$. Then define

$$\eta = \underset{1 \le i \le n}{\text{Max}} \frac{\beta_i}{1 - \alpha_i} \qquad\qquad (8.6.17)$$

From (8.6.16),

$$|e_i^{(m+1)}| \le \alpha_i \|e^{(m+1)}\|_\infty + \beta_i \|e^{(m)}\|_\infty \qquad i = 1, \ldots, n \qquad (8.6.18)$$

Let $k$ be a subscript for which

$$\|e^{(m+1)}\|_\infty = |e_k^{(m+1)}|$$

Then with $i = k$ in (8.6.18),

$$\|e^{(m+1)}\|_\infty \le \alpha_k \|e^{(m+1)}\|_\infty + \beta_k \|e^{(m)}\|_\infty$$

$$\|e^{(m+1)}\|_\infty \le \frac{\beta_k}{1 - \alpha_k}\|e^{(m)}\|_\infty$$

and thus

$$\|e^{(m+1)}\|_\infty \le \eta \|e^{(m)}\|_\infty \qquad\qquad (8.6.19)$$

**Table 8.4   Numerical results for the Gauss–Seidel method**

| $m$ | $x_1^{(m)}$ | $x_2^{(m)}$ | $x_3^{(m)}$ | $\|e^{(m)}\|_\infty$ | Ratio |
|---|---|---|---|---|---|
| 0 | 0 | 0 | 0 | 1 | |
| 1 | 1.4 | .78 | 1.026 | .4 | .4 |
| 2 | 1.063400 | 1.020480 | .987516 | 6.34E − 2 | .16 |
| 3 | .995104 | .995276 | 1.001907 | 4.90E − 3 | .077 |
| 4 | 1.001227 | 1.000817 | .999632 | 1.23E − 3 | .25 |
| 5 | .999792 | .999848 | 1.000066 | 2.08E − 4 | .17 |
| 6 | 1.000039 | 1.000028 | .999988 | 3.90E − 5 | .19 |

Since for each $i$,

$$(\alpha_i + \beta_i) - \frac{\beta_i}{1 - \alpha_i} = \frac{\alpha_i[1 - (\alpha_i + \beta_i)]}{1 - \alpha_i} \geq \frac{\alpha_i}{1 - \alpha_i}[1 - \mu] \geq 0$$

we have

$$\eta \leq \mu < 1 \qquad (8.6.20)$$

Combined with (8.6.19), this shows the convergence of $e^{(m)} \to 0$ as $m \to \infty$. Also, the rate of convergence will be linear, but with a faster rate than with the Jacobi method.

***Example***   Use the system (8.6.12) of the previous example, and solve it with the Gauss–Seidel method. By a simple calculation from (8.6.17) and (8.6.12),

$$\eta = .4$$

The numerical results are given in Table 8.4. The speed of convergence is significantly better than for the previous example of the Gauss–Jacobi method, given in Table 8.3. The values of Ratio appear to converge to about .18.

**General framework for iteration methods**   To solve $Ax = b$, form a split of $A$:

$$A = N - P \qquad (8.6.21)$$

and write $Ax = b$ as

$$Nx = b + Px \qquad (8.6.22)$$

The matrix $N$ is chosen in such a way that the linear system $Nz = f$ is "easily solvable" for any $f$. For example, $N$ might be diagonal, triangular, or tridiagonal. Define the iteration method by

$$Nx^{(m+1)} = b + Px^{(m)} \qquad m \geq 0 \qquad (8.6.23)$$

with $x^{(0)}$ given.

***Example* 1.**    The Jacobi method.

$$N = \text{diag}\,[a_{11}, a_{22}, \ldots, a_{nn}] \qquad P = N - A$$

**2.    Gauss–Seidel method.**

$$
N =
\begin{bmatrix}
a_{11} & 0 & \cdots & & 0 \\
a_{21} & a_{22} & 0 & \cdots & 0 \\
\vdots & & & \ddots & \vdots \\
a_{n1} & & \cdots & & a_{nn}
\end{bmatrix}
\tag{8.6.24}
$$

To analyze the error, subtract (8.6.23) from (8.6.22) to get

$$Ne^{(m+1)} = Pe^{(m)},$$

$$e^{(m+1)} = Me^{(m)} \qquad M = N^{-1}P \tag{8.6.25}$$

By induction,

$$e^{(m)} = M^m e^{(0)} \qquad m \geq 0 \tag{8.6.26}$$

In order that $e^{(m)} \to \infty$ as $n \to \infty$, for arbitrary initial guesses $x^{(0)}$ (and thus arbitrary $e^{(0)}$), it is necessary and sufficient that

$$M^m \to 0 \qquad \text{as} \quad m \to \infty$$

Or equivalently from Theorem 7.9,

$$r_\sigma(M) < 1 \tag{8.6.27}$$

This general framework for iteration methods is adapted from Isaacson and Keller (1966, pp. 61–81).

The condition (8.6.27) was derived earlier in (8.6.11) for the Gauss–Jacobi method. For the Gauss–Seidel method the matrix $N^{-1}P$, with $N$ given by (8.6.24), is more difficult to work with. We must examine the values of $\lambda$ for which

$$\det\,(\lambda I - N^{-1}P) = 0$$

or equivalently,

$$\det\,(\lambda N - P) = 0 \tag{8.6.28}$$

For applications to the numerical solution of the partial differential equations in Section 8.8, the preceding convergence analysis of the Gauss–Seidel method is not adequate. The constants $\mu$ and $\eta$ of (8.6.5) and (8.6.17) will both equal 1, although empirically the method still converges. To deal with many of these systems, the following important theorem is often used.

***Theorem 8.7***    Let $A$ be Hermitian with positive diagonal elements. Then the Gauss–Seidel method (8.6.15) for solving $Ax = b$ will converge, for any choice of $x^{(0)}$, if and only if $A$ is positive definite.

***Proof***    The proof is given in Isaacson and Keller (1966, pp. 70–71). For the definition of a positive definite matrix, recall Problem 14 of Chapter 7. The theorem is illustrated in Section 8.8.    ∎

**Other iteration methods**    The best iteration methods are based on a thorough knowledge of the problem being solved, taking into account its special features in the design of the iteration scheme. This usually includes looking at the form of the matrix and the source of the linear system.

The matrix $A$ may have a special form that leads to a simple iteration method. For example, suppose $A$ is of block tridiagonal form:

$$
A = \begin{bmatrix} B_1 & C_1 & & \cdots & & 0 \\ A_2 & B_2 & C_2 & & & \\ \vdots & & & \ddots & & \vdots \\ 0 & & & \cdots & A_r & B_r \end{bmatrix}
\tag{8.6.29}
$$

The matrices $A_i, B_i, C_i$ are square, of order $m$, and $A$ is of order $n = rm$. For $x, b \in \mathbf{R}^n$, write $x$ and $b$ in partitioned form as

$$
x = \begin{bmatrix} x_{(1)} \\ \vdots \\ x_{(r)} \end{bmatrix} \qquad b = \begin{bmatrix} b_{(1)} \\ \vdots \\ b_{(r)} \end{bmatrix} \qquad x_{(i)}, b_{(i)} \in \mathbf{R}^m
$$

Then $Ax = b$ can be rewritten as

$$
B_1 x_{(1)} + C_1 x_{(2)} = b_{(1)}
$$

$$
A_i x_{(i-1)} + B_i x_{(i)} + C_i x_{(i+1)} = b_{(i)} \qquad 2 \le i \le r - 1
\tag{8.6.30}
$$

$$
A_r x_{(r-1)} + B_r x_{(r)} = b_{(r)}
$$

We assume the linear systems

$$
B_j x_{(j)} = d_{(j)} \qquad 1 \le j \le r
\tag{8.6.31}
$$

are easily solvable, probably directly for all right sides $d_{(j)}$. For example, we often have all $B_j = T$, a constant tridiagonal matrix to which the procedure in (8.3.20)–(8.3.24) can be applied.

A Jacobi-type method can be applied to (8.6.30):

$$
B_1 x_{(1)}^{(\nu+1)} = b_{(1)} - C_1 x_{(2)}^{(\nu)}
$$

$$
B_i x_{(i)}^{(\nu+1)} = b_{(i)} - A_i x_{(i-1)}^{(\nu)} - C_i x_{(i+1)}^{(\nu)} \qquad 2 \le i \le r - 1
\tag{8.6.32}
$$

$$
B_r x_{(r)}^{(\nu+1)} = b_{(r)} - A_r x_{(r-1)}^{(\nu)}
$$

for $v \geq 0$. The analysis of convergence is more complicated than for the Gauss–Jacobi and Gauss–Seidel methods; some results are suggested in Problem 29. Similar methods are used with the linear systems arising from solving some partial differential equations, and these are indicated in Section 8.8.

Another important aspect of solving linear systems $Ax = b$ is to look at their origin. In many cases we have a differential or integral equation, say

$$\mathscr{A}x = y \tag{8.6.33}$$

where $x$ and $y$ are functions. This is discretized to give a family of problems

$$A_n x_n = y_n \qquad x_n, y_n \in \mathbf{R}^n \tag{8.6.34}$$

with $A_n$ of order $n$. As $n \to \infty$, the solutions $x_n$ of (8.6.34) approach (in some sense) the solution $x$ of (8.6.33). Thus the linear systems in (8.6.34) are closely related. For example, in some sense $A_m^{-1} \doteq A_n^{-1}$ for $m$ and $n$ sufficiently large, even though they are matrices of different orders. This can be given a more precise meaning, leading to ways of iteratively solving large systems by using the solvability of lower order systems. Recently, many such methods have been developed under the name of *multigrid methods*, with applications particularly to partial differential equations [see Hackbusch and Trottenberg (1982)]. For iterative methods for integral equations, see the related but different development in Atkinson (1976, part II, chap. 4). Multigrid methods are very effective and efficient iterative methods for differential and integral equations.

## 8.7  Error Prediction and Acceleration

From (8.6.25), we have the error relation

$$x - x^{(m+1)} = M(x - x^{(m)}) \qquad m \geq 0 \tag{8.7.1}$$

The manner of convergence of $x^{(m)}$ to $x$ can be quite complicated, depending on the eigenvalues and eigenvectors of $M$. But in most practical cases, the behavior of the errors is quite simple: The size of $\|x - x^{(m)}\|_\infty$ decreases by approximately a constant factor at each step, and

$$\|x - x^{(m+1)}\|_\infty \leq c\|x - x^{(m)}\|_\infty \tag{8.7.2}$$

for some $c < 1$, closely related to $r_\sigma(M)$. To measure this constant $c$, note from (8.7.1) that

$$x^{(m+1)} - x^{(m)} = e^{(m)} - e^{(m+1)} = Me^{(m-1)} - Me^{(m)}$$

$$x^{(m+1)} - x^{(m)} = M(x^{(m)} - x^{(m-1)}) \qquad m \geq 0 \tag{8.7.3}$$

This motivates the use of

$$c \doteq \frac{\|x^{(m+1)} - x^{(m)}\|_\infty}{\|x^{(m)} - x^{(m-1)}\|_\infty} \tag{8.7.4}$$

**Table 8.5   Example of Gauss–Seidel iteration**

| $m$ | $\|u^{(m)} - u^{(m-1)}\|_\infty$ | Ratio | Est. Error | Error |
|---|---|---|---|---|
| 20 | 1.20E − 3 | .966 | 3.42E − 2 | 3.09E − 2 |
| 21 | 1.16E − 3 | .966 | 3.24E − 2 | 2.98E − 2 |
| 22 | 1.12E − 3 | .965 | 3.08E − 2 | 2.86E − 2 |
| 23 | 1.08E − 3 | .965 | 2.93E − 2 | 2.76E − 2 |
| 24 | 1.04E − 3 | .964 | 2.80E − 2 | 2.65E − 2 |
| 60 | 2.60E − 4 | .962 | 6.58E − 3 | 6.58E − 3 |
| 61 | 2.50E − 4 | .962 | 6.33E − 3 | 6.33E − 3 |
| 62 | 2.41E − 4 | .962 | 6.09E − 3 | 6.09E − 3 |

or for greater safety, the maximum of several successive such ratios. In many applications, this ratio is about constant for large values of $m$.

Once this constant $c$ has been obtained, and assuming (8.7.2), we can bound the error in $x^{(m+1)}$ by using (8.5.13):

$$\|x - x^{(m+1)}\|_\infty \le \frac{c}{1 - c}\|x^{(m+1)} - x^{(m)}\|_\infty \qquad (8.7.5)$$

This bound is important when $c \doteq 1$ and the convergence is slow. In that case, the difference $\|x^{(m+1)} - x^{(m)}\|_\infty$ can be much smaller than the actual error $\|x - x^{(m+1)}\|_\infty$.

*Example*  The linear system (8.8.5) of Section 8.8 was solved using the Gauss–Seidel method. In keeping with (8.8.5), we denote our unknown vector by $u$. In (8.8.4), the function $f = x^2 y^2$, and in (8.8.5), the function $g = 2(x^2 + y^2)$. The region was $0 \le x,\ y \le 1$, and the mesh size in each direction was $h = \frac{1}{16}$. This gave an order of 225 for the linear system (8.8.5). The initial guess $u^{(0)}$ in the iteration was based on a "bilinear" interpolant of $f = x^2 y^2$ over the region $0 \le x,\ y \le 1$ [see (8.8.17)]. A selection of numerical results is given in Table 8.5. The column Ratio is calculated from (8.7.4), the column Est. Error uses (8.7.5), and the column Error is the true error $\|u - u^{(m)}\|_\infty$.

As can be seen in the table, the convergence was quite slow, justifying the need for (8.7.5) rather than the much smaller $\|u^{(m)} - u^{(m-1)}\|_\infty$. As $m \to \infty$, the value of Ratio converges to .962, and the error estimate (8.7.5) is an accurate estimator of the true iteration error.

**Speed of convergence**  We now discuss how many iterates to calculate in order to obtain a desired error. And when is iteration preferable to Gaussian elimination in solving $Ax = b$? We find the value of $m$ for which

$$\|x - x^{(m)}\|_\infty \le \epsilon\|x - x^{(0)}\|_\infty \qquad (8.7.6)$$

with $\epsilon$ a given factor by which the initial error is to be reduced. We base the analysis on the assumption (8.7.2). Generally the constant $c$ is almost equal to $r_\sigma(M)$, with $M$ as in (8.7.1).

The relation (8.7.2) implies

$$\|x - x^{(m)}\|_\infty \le c^m \|x - x^{(0)}\|_\infty \qquad m \ge 0 \qquad (8.7.7)$$

Thus we find the smallest value of $m$ for which

$$c^m \le \epsilon$$

Solving this, we must have

$$m \ge \frac{-\ln \epsilon}{R(c)} \equiv m^* \qquad R(c) = -\ln c \qquad (8.7.8)$$

Doubling $R(c)$ leads to halving the number of iterates that must be calculated.

To make this result more meaningful, we apply it to the solution of a dense linear system by iteration. Assume that the Gauss–Jacobi or Gauss–Seidel method is being used to solve $Ax = b$ to single precision accuracy on an IBM mainframe computer, that is, to about six significant digits. Assume $x^{(0)} = 0$, and that we want to find $m$ such that

$$\frac{\|x - x^{(m)}\|_\infty}{\|x\|_\infty} \le 10^{-6} = \epsilon \qquad (8.7.9)$$

Assuming $A$ has order $n$, the number of operations (multiplications and divisions) per iteration is $n^2$. To obtain the result (8.7.9), the necessary number of iterates is

$$m^* = \frac{6 \ln_e 10}{R(c)}$$

and the number of operations is

$$m^* n^2 = (6 \ln_e 10) \frac{n^2}{R(c)}$$

If Gaussian elimination is used to solve $Ax = b$ with the same accuracy, the number of operations is about $n^3/3$. The iteration method will be more efficient than the Gaussian elimination method if

$$m^* n^2 < \frac{n^3}{3},$$

$$m^* < \frac{n}{3} \qquad (8.7.10)$$

**Example**   Consider a matrix $A$ of order $n = 51$. Then iteration is more efficient if $m^* < 17$. Table 8.6 gives the values of $m^*$ for various values of $c$. For $c \le .44$, the iteration method will be more efficient than Gaussian elimination. And if less

**Table 8.6    Example of iteration count**

| $c$ | $R(c)$ | $m^*$ |
|-----|--------|-------|
| .9  | .105   | 131   |
| .8  | .223   | 62    |
| .6  | .511   | 27    |
| .4  | .916   | 15    |
| .2  | 1.61   | 9     |

than full precision accuracy in (8.7.9) is desired, then iteration will be more efficient with even larger values of $c$. In practice, we also will usually know an initial guess $x^{(0)}$ that is better than $x^{(0)} = 0$, further decreasing the number of needed iterates.

The main use of iteration methods is for the solution of large sparse systems, in which case Gaussian elimination is often not possible. And even when elimination is possible, iteration may still be preferable. Some examples of such systems are discussed in Section 8.8.

**Acceleration methods**    Most iteration methods have a regular pattern in which the error decreases. This can often be used to accelerate the convergence, just as was done in earlier chapters with other numerical methods. Rather than giving a general theory for the acceleration of iteration methods for solving $Ax = b$, we just describe an acceleration of the Gauss–Seidel method. This is one of the main cases of interest in applications.

Recall the definition (8.6.15) of the Gauss–Seidel method. Introduce an *acceleration parameter* $\omega$, and consider the following modification of (8.6.15):

$$z_i^{(m+1)} = \frac{1}{a_{ii}}\left\{ b_i - \sum_{j=1}^{i-1} a_{ij}x_j^{(m+1)} - \sum_{j=i+1}^{n} a_{ij}x_j^{(m)} \right\}$$

$$x_i^{(m+1)} = \omega z_i^{(m+1)} + \left(1 - \omega\right)x_i^{(m)} \qquad i = 1,\dots,n \qquad (8.7.11)$$

for $m \geq 0$. The case $\omega = 1$ is the regular Gauss–Seidel method. The acceleration is to optimally choose some linear combination of the preceding iterate and the regular Gauss–Seidel iterate. The method (8.7.11), with an optimal choice of $\omega$, is called the *SOR method*, which is an abbreviation for *successive overrelaxation*, an historical term.

To understand how $\omega$ should be chosen, we rewrite (8.7.11) in matrix form. Decompose $A$ as

$$A = D + L + U$$

with $D = \text{diag}[a_{11},\dots, a_{nn}]$, $L$ lower triangular, and $U$ upper triangular, with both $L$ and $U$ having zeros on the diagonal. Then (8.7.11) becomes

$$z^{(m+1)} = D^{-1}[b - Lx^{(m+1)} - Ux^{(m)}]$$

$$x^{(m+1)} = \omega z^{(m+1)} + \left(1 - \omega\right)x^{(m)} \qquad m \geq 0$$

Eliminating $z^{(m+1)}$ and solving for $x^{(m+1)}$,

$$[I + \omega D^{-1}L]x^{(m+1)} = \omega D^{-1}b + [(1 - \omega)I - \omega D^{-1}U]x^{(m)}$$

For the error,

$$e^{(m+1)} = M(\omega)e^{(m)} \qquad m \geq 0 \tag{8.7.12}$$

$$M(\omega) = [I + \omega D^{-1}L]^{-1}[(1 - \omega)I - \omega D^{-1}U] \tag{8.7.13}$$

The parameter $\omega$ is to be chosen to minimize $r_\sigma(M(\omega))$, in order to make $x^{(m)}$ converge to $x$ as rapidly as possible. Call the optimal value $\omega^*$.

The calculation of $\omega^*$ is difficult except in the simplest of cases. And usually it is obtained only approximately, based on trying several values of $\omega$ and observing the effect on the speed of convergence. In spite of the problem of calculating $\omega^*$, the resulting increase in the speed of convergence of $x^{(m)}$ to $x$ is very dramatic, and the calculation of $\omega^*$ is well worth the effort. This is illustrated in the next section.

*Example*  We apply the acceleration (8.7.11) to the preceding example of the Gauss–Seidel method, given following (8.7.5). The optimal acceleration parameter is $\omega^* \doteq 1.6735$. A more extensive discussion of the SOR method for solving the linear systems arising in solving partial differential equations is given in the following section. The initial guess was the same as before. The results are given in Table 8.7.

The results show a much faster rate of convergence than for the Gauss–Seidel method. For example, with the Gauss–Seidel method, we have $\|u - u^{(228)}\|_\infty = 9.70E - 6$. In comparison, the SOR method leads to $\|u - u^{(32)}\|_\infty = 8.71E - 6$. But we have lost the regular behavior in the convergence of the iterates, as can be seen from the values of Ratio. The value of $c$ used in the error test (8.7.5) needs to be chosen more carefully than our choice of $c = $ Ratio in the table. You may want to use an average or the maximum of several successive preceding values of Ratio.

**Table 8.7    Example of SOR method (8.7.11)**

| $m$ | $\|u^{(m)} - u^{(m-1)}\|_\infty$ | Ratio | Est. Error | Error |
|---|---|---|---|---|
| 21 | $2.06E - 4$ | .693 | $4.65E - 4$ | $3.64E - 4$ |
| 22 | $1.35E - 4$ | .657 | $2.59E - 4$ | $2.65E - 4$ |
| 23 | $8.76E - 5$ | .648 | $1.61E - 4$ | $1.87E - 4$ |
| 24 | $5.11E - 5$ | .584 | $7.17E - 5$ | $1.39E - 4$ |
| 25 | $3.48E - 5$ | .680 | $7.40E - 5$ | $1.06E - 4$ |
| 26 | $2.78E - 5$ | .800 | $1.11E - 4$ | $8.04E - 5$ |
| 27 | $2.46E - 5$ | .884 | $1.87E - 4$ | $6.15E - 5$ |
| 28 | $2.07E - 5$ | .842 | $1.11E - 4$ | $4.16E - 5$ |

## 8.8   The Numerical Solution of Poisson's Equation

The most important application of linear iteration methods is to the large linear systems arising from the numerical solution of partial differential equations by finite difference methods. To illustrate this, we solve the Dirichlet problem for Poisson's equation on the unit square in the $xy$-plane:

$$\frac{\partial^2 u}{\partial x^2} + \frac{\partial^2 u}{\partial y^2} = g(x, y) \qquad 0 < x, y < 1$$

$$u(x, y) = f(x, y) \qquad (x, y) \text{ a boundary point}$$

(8.8.1)

The functions $g(x, y)$ and $f(x, y)$ are given, and we must find $u(x, y)$.

For $N > 1$, define $h = 1/N$, and

$$(x_j, y_k) = (jh, kh) \qquad 0 \le j, k \le N$$

These are called the grid points or mesh points (see Figure 8.1). To approximate (8.8.1), we use approximations to the second derivatives. For a four times continuously differentiable function $G(x)$ on $[x - h, x + h]$, the results (5.7.17)–(5.7.18) of Section 5.7 give

$$G''(x) = \frac{G(x + h) - 2G(x) + G(x - h)}{h^2}$$

$$- \frac{h^2}{12} G^{(4)}(\xi) \qquad x - h \le \xi \le x + h$$

(8.8.2)

When applied to (8.8.1) at each interior grid point, we obtain

$$\frac{u(x_{j+1}, y_k) - 2u(x_j, y_k) + u(x_{j-1}, y_k)}{h^2} + \frac{u(x_j, y_{k+1}) - 2u(x_j, y_k) + u(x_j, y_{k-1})}{h^2}$$

$$= g(x_j, y_k) + \frac{h^2}{12} \left\{ \frac{\partial^4 u(\xi, y_k)}{\partial x^4} + \frac{\partial^4 u(x_j, \eta)}{\partial y^4} \right\}$$

(8.8.3)

for some $x_{j-1} \le \xi \le x_{j+1}$, $y_{k-1} \le \eta \le y_{k+1}$, $1 \le j, k \le N - 1$.

For the numerical approximation $u_h(x, y)$ of (8.8.1), let

$$u_h(x_j, y_k) = f(x_j, y_k) \qquad (x_j, y_k) \text{ a boundary grid point} \qquad (8.8.4)$$

At all interior mesh points, drop the right-hand truncation errors in (8.8.3) and solve for the approximating solution $u_h(x_j, y_k)$:

$$u_h(x_j, y_k) = \frac{1}{4} \left\{ u_h(x_{j+1}, y_k) + u_h(x_j, y_{k+1}) + u_h(x_{j-1}, y_k) + u_h(x_j, y_{k-1}) \right\}$$

$$- \frac{h^2}{4} g(x_j, y_k) \qquad 1 \le j, k \le N - 1$$

(8.8.5)

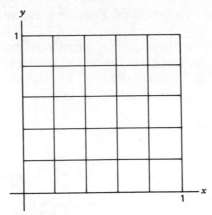

**Figure 8.1** Finite difference mesh.

The number of equations in (8.8.4)–(8.8.5) is equal to the number of unknowns, $(N + 1)^2$.

**Theorem 8.8** For each $N \geq 2$, the linear system (8.8.4)–(8.8.5) has a unique solution $\{u_h(x_j, y_k) | 0 \leq j, k \leq N\}$. If the solution $u(x, y)$ of (8.8.1) is four times continuously differentiable, then

$$\underset{0 \leq j, k \leq N}{\text{Max}} |u(x_j, y_k) - u_h(x_j, y_k)| \leq ch^2 \qquad (8.8.6)$$

$$c = \frac{1}{24} \left\{ \underset{0 \leq x, y \leq 1}{\text{Max}} \left| \frac{\partial^4 u(x, y)}{\partial x^4} \right| + \underset{0 \leq x, y \leq 1}{\text{Max}} \left| \frac{\partial^4 u(x, y)}{\partial y^4} \right| \right\}$$

**Proof** 1. We prove the unique solvability of (8.8.4)–(8.8.5) by using Theorem 7.2 of Chapter 7. We consider the homogeneous system

$$v_h(x_j, y_k) = \tfrac{1}{4} \left[ v_h(x_{j+1}, y_k) + v_h(x_j, y_{k+1}) + v_h(x_{j-1}, y_k) \right.$$

$$\left. + v_h(x_j, y_{k-1}) \right] \qquad 1 \leq j, k \leq N - 1 \quad (8.8.7)$$

$$v_h(x_j, y_k) = 0 \qquad (x_j, y_k) \text{ a boundary point} \qquad (8.8.8)$$

By showing that this system has only the trivial solution $v_h(x_j, y_k) \equiv 0$, it will follow from Theorem 7.2 that the nonhomogeneous system (8.8.4)–(8.8.5) will have a unique solution.

Let

$$\alpha = \underset{0 \leq j, k \leq N}{\text{Max}} v_h(x_j, y_k)$$

From (8.8.8), $\alpha \geq 0$. Assume $\alpha > 0$. Then there must be an interior grid point $(\bar{x}_j, \bar{y}_k)$ for which this maximum is attained. But using (8.8.7),

$v_h(\bar{x}_j, \bar{y}_k)$ is the average of the values of $v_h$ at the four points neighboring $(\bar{x}_j, \bar{y}_k)$. The only way that this can be compatible with $(\bar{x}_j, \bar{y}_k)$ being a maximum point is if $v_h$ also equals $\alpha$ at the four neighboring grid points. Continue the same argument to these neighboring points. Since there are only a finite number of grid points, we eventually have $v_h(x_j, y_k) = \alpha$ for a boundary point $(x_j, y_k)$. But then $\alpha > 0$ will contradict (8.8.8). Thus the maximum of $v_h(x_j, y_k)$ is zero. A similar argument will show that the minimum of $v_h(x_j, y_k)$ is also zero. Taken together, these results show that the only solution of (8.8.7)–(8.8.8) is $v_h(x_j, y_k) \equiv 0$.

**2.** To consider the convergence of $u_h(x_j, y_k)$ to $u(x_j, y_k)$, define

$$e_h(x_j, y_k) = u(x_j, y_k) - u_h(x_j, y_k)$$

Subtracting (8.8.5) from (8.8.3), we obtain

$$e_h(x_j, y_k) = \frac{1}{4}\left[e_h(x_{j+1}, y_k) + e_h(x_j, y_{k+1}) + e_h(x_{j-1}, y_k) + e_h(x_j, y_{k-1})\right]$$
$$-\frac{h^2}{12}\left[\frac{\partial^4 u(\xi_j, y_k)}{\partial x^4} + \frac{\partial^4 u(x_j, \eta_k)}{\partial y^4}\right] \tag{8.8.9}$$

and from (8.8.4),

$$e_h(x_j, y_k) = 0 \qquad (x_j, y_k) \text{ a boundary grid point} \tag{8.8.10}$$

This system can be treated in a manner similar to that used in part (1), and the result (8.8.6) will follow. Because it is not central to the discussion of the linear systems, the argument is omitted [see Isaacson and Keller (1966), pp. 447–450]. ∎

*Example*   Solve

$$\frac{\partial^2 u}{\partial x^2} + \frac{\partial^2 u}{\partial y^2} = 0 \qquad 0 \le x, y \le 1 \tag{8.8.11}$$

$$u(0, y) = \cos(\pi y) \qquad u(1, y) = e^\pi \cos(\pi y)$$
$$u(x, 0) = e^{\pi x} \qquad u(x, 1) = -e^{\pi x}$$

The true solution is

$$u(x, y) = e^{\pi x} \cos(\pi y)$$

Numerical results for several values of $N$ are given in Table 8.8. The error is the maximum over all grid points, and the column Ratio gives the factor by which the maximum error decreases when the grid size $h$ is halved. Theoretically from (8.8.6), it should be 4.0. The numerical results confirm this.

**Table 8.8   Numerical solution of (8.8.11)**

| N | $\|u - u_h\|_\infty$ | Ratio |
|---|---|---|
| 4 | .144 | |
| 8 | .0390 | 3.7 |
| 16 | .0102 | 3.8 |
| 32 | .00260 | 3.9 |
| 64 | .000654 | 4.0 |

**Iterative Solution**   Because the Gauss–Seidel method is generally faster than the Gauss–Jacobi method, we only consider the former. For $k = 1, 2, \ldots, N - 1$, define

$$u_h^{(m+1)}(x_j, y_k) = \frac{1}{4}\left[ u_h^{(m)}(x_{j+1}, y_k) + u_h^{(m)}(x_j, y_{k+1}) + u_h^{(m+1)}(x_{j-1}, y_k) \right.$$

$$\left. + u_h^{(m+1)}(x_j, y_{k-1}) \right] - \frac{h^2}{4} g(x_j, y_k) \qquad j = 1, 2, \ldots, N - 1 \quad (8.8.12)$$

For boundary points, use

$$u_h^{(m)}(x_j, y_k) = f(x_j, y_k) \qquad \text{all } m \geq 0$$

The values of $u_h^{(m+1)}(x_j, y_k)$ are computed row by row, from the bottom row of grid points first to the top row of points last. And within each row, we solve from left to right.

For the iteration (8.8.12), the convergence analysis must be based on Theorem 8.7. It can be shown easily that the matrix is symmetric, and thus all eigenvalues are real. Moreover, the eigenvalues can all be shown to lie in the interval $0 < \lambda < 2$. From this and Problem 14 of Chapter 7, the matrix is positive definite. Since all diagonal coefficients of the matrix are positive, it then follows from Theorem 8.7 that the Gauss–Seidel method will converge. To show that all eigenvalues lie in $0 < \lambda < 2$, see Isaacson and Keller (1966, pp. 458–459) or Problem 3 of Chapter 9.

The calculation of the speed of convergence $r_\sigma(M)$ from (8.6.28) is nontrivial. The argument is quite sophisticated, and we only refer to the very complete development in Isaacson and Keller (1966, pp. 463–470), including the material on the acceleration of the Gauss–Seidel method. It can be shown that

$$r_\sigma(M) = 1 - \pi^2 h^2 + O(h^4) \qquad (8.8.13)$$

The Gauss–Seidel method converges, but the speed of convergence is quite slow for even moderately small values of $h$. This is illustrated in Table 8.5 of Section 8.7.

To accelerate the Gauss–Seidel method, use

$$v_h^{(m+1)}(x_j, y_k) = \frac{1}{4}\left[u_h^{(m)}(x_{j+1}, y_k) + u_h^{(m)}(x_j, y_{k+1})\right.$$

$$\left. + u_h^{(m+1)}(x_{j-1}, y_k) + u_h^{(m+1)}(x_j, y_{k-1})\right] - \frac{h^2}{4}g(x_j, y_k)$$

$$u_h^{(m+1)}(x_j, y_k) = \omega v_h^{(m+1)}(x_j, y_k)$$

$$+ (1 - \omega)u_h^{(m)}(x_j, y_k) \qquad j = 1, \ldots, N - 1 \qquad (8.8.14)$$

for $k = 1, \ldots, N - 1$. The optimal acceleration parameter is

$$\omega^* = \frac{2}{1 + \sqrt{1 - \xi^2}}$$

$$\xi = 1 - 2\sin^2\left(\frac{\pi}{2N}\right) \qquad (8.8.15)$$

The correspondence rate of convergence is

$$r_\sigma(M(\omega^*)) = \omega^* - 1 = 1 - 2h\pi + O(h^2) \qquad (8.8.16)$$

This is a much better rate than that given by (8.8.13). The accelerated Gauss–Seidel method (8.8.14) with the optimal value $\omega^*$ of (8.8.15) is known as the *SOR method*. The name SOR is an abbreviation for *successive overrelaxation*, a name that is based on a physical interpretation of the method, first used in deriving it.

***Example***    Recall the previous example (8.8.11). This was solved with both the Gauss–Seidel method and the SOR method. The initial guess for the iteration was taken to be the "bilinear" interpolation formula for the boundary data $f$:

$$u_h^{(0)}(x, y) = (1 - x)f(0, y) + xf(1, y) + (1 - y)f(x, 0) + yf(x, 1)$$

$$- [(1 - y)(1 - x)f(0, 0) + (1 - y)xf(1, 0)$$

$$+ y(1 - x)f(0, 1) + xyf(1, 1)] \qquad (8.8.17)$$

at all interior grid points. The error test to stop the iteration was

$$\max_{1 \le j, k \le N-1} \left|u_h(x_j, y_k) - u_h^{(m)}(x_j, y_k)\right| \le \epsilon$$

with $\epsilon > 0$ given and the right-hand side of (8.7.5) used to predict the error in the iterate. The numerical results for the necessary number of iterates are given in Table 8.9. The SOR method requires far fewer iterates for the smaller values of $h$ than does the Gauss–Seidel method.

**Table 8.9    Number of iterates necessary to solve (8.8.5)**

| $N$ | $\epsilon$ | Gauss–Seidel | SOR |
|-----|------------|--------------|-----|
| 8   | .01        | 25           | 12  |
| 8   | .001       | 40           | 16  |
| 16  | .001       | 142          | 32  |
| 32  | .001       | 495          | 65  |
| 8   | .0001      | 54           | 18  |
| 16  | .0001      | 201          | 35  |
| 32  | .0001      | 733          | 71  |

Recall from the previous section that the number of iterates, called $m^*$, necessary to reduce the iteration error by a factor of $\epsilon$ is proportional to $1/\ln(c)$, where $c$ is the ratio by which the iteration error decreases at each step. For the methods of this section, we take $c = r_\sigma(M)$. If we write $r_\sigma(M) = 1 - \delta$, then

$$\frac{1}{-\ln r_\sigma(M)} = \frac{1}{-\ln(1-\delta)} \doteq \frac{1}{\delta}$$

When $\delta$ is halved, the number of iterates to be computed is doubled. For the Gauss–Seidel method,

$$\frac{1}{\ln r_\sigma(M)} \doteq \frac{1}{\pi^2 h^2}$$

When $h$ is halved, the number of iterates to be computed increases by a factor of 4. For the SOR method,

$$\frac{1}{\ln r_\sigma(M(\omega^*))} \doteq \frac{1}{2\pi h}$$

and when $h$ is halved, the number of iterates is doubled. These two results are illustrated in Table 8.9 by the entries for $\epsilon = 10^{-3}$ and $\epsilon = 10^{-4}$.

With either method, note that doubling $N$ will increase the number of equations to be solved by a factor of 4, and thus the work per iteration will increase by the same amount. The use of SOR greatly reduces the resulting work, although it still is large when $N$ is large.

## 8.9    The Conjugate Gradient Method

The iteration method presented in this section was developed in the 1950s, but has gained its main popularity in more recent years, especially in solving linear systems associated with the numerical solution of linear partial differential equations. The literature on the conjugate gradient method (CG method) has become quite large and sophisticated, and there are numerous connections to other topics in linear algebra. Thus, for reasons of space, we are able to give only a brief introduction, defining the CG method and stating some of the principal theoretical results concerning it.

The CG method differs from earlier methods of this chapter, in that it is based on solving a nonlinear problem; in fact, the CG method is also a commonly used method for minimizing nonlinear functions. The linear system to be solved,

$$Ax = b \qquad (8.9.1)$$

is assumed to have a coefficient matrix $A$ that is real, symmetric, and positive definite. The solution of this system is equivalent to the minimization of the function

$$f(x) = \tfrac{1}{2}x^TAx - b^Tx \qquad x \in \mathbf{R}^n \qquad (8.9.2)$$

The unique solution $x^*$ of $Ax = b$ is also the unique minimizer of $f(x)$ as $x$ varies over $\mathbf{R}^n$. To see this, first show

$$f(x) = E(x) - \tfrac{1}{2}b^Tx^*$$

$$E(x) \equiv \tfrac{1}{2}(x^* - x)^TA(x^* - x) \qquad (8.9.3)$$

Using $Ax^* = b$, the proof is straightforward. The functions $E(x)$ and $f(x)$ differ by a constant, and thus they will have the same minimizers. By the positive definiteness of $A$, $E(x)$ is minimized uniquely by $x = x^*$, and thus the same is true of $f(x)$.

A well-known iteration method for finding a minimum for a nonlinear function is the *method of steepest descent*, which was introduced briefly in Section 2.12. For minimizing $f(x)$ by this method, assume that an initial guess $x_0$ is given. Choose a path in which to search for a new minimum by looking along the direction in which $f(x)$ decreases most rapidly at $x_0$. This is given by $g_0 = -\nabla f(x_0)$, the negative of the *gradient* of $f(x)$ at $x_0$:

$$g(x_0) \equiv g_0 = b - Ax_0 \qquad (8.9.4)$$

Then solve the one-dimensional minimization problem

$$\underset{0 \le \alpha < \infty}{\text{Min}} \; f(x_0 + \alpha g_0)$$

calling the solution $\alpha_1$. Using it, define the new iterate

$$x_1 = x_0 + \alpha_1 g_0 \qquad (8.9.5)$$

Continue this process inductively. The method of steepest descent will converge, but the convergence is generally quite slow. The optimal local strategy of using a direction of fastest descent is not a good strategy for finding an optimal direction for finding the global minimum. In comparison, the CG method will be more rapid, and it will take no more that $n$ iterates, assuming there are no rounding errors.

For the remainder of this section, we assume the given initial guess is $x_0 = 0$. If it is not, then we can always solve the modified problem

$$Az = b - Ax_0$$

Denoting its solution by $z^*$, we have $x^* = x_0 + z^*$. An initial guess of $z^* \doteq z_0 = 0$ corresponds to $x^* \doteq x_0$ in the original problem. Henceforth, assume $x_0 = 0$.

**Conjugate direction methods**   Assuming $A$ is $n \times n$, we say a set of nonzero vectors $p_1, \ldots, p_n$ in $\mathbf{R}^n$ is *A-conjugate* if

$$p_i^T A p_j = 0 \qquad 1 \le i, j \le n \quad i \ne j \tag{8.9.6}$$

The vectors $p_j$ are often called *conjugate directions*. An equivalent geometric definition can be given by introducing a new inner product and norm for $\mathbf{R}^n$:

$$(x, y)_A = x^T A y$$

$$\|x\|_A = \sqrt{(x, x)_A} = \sqrt{x^T A x} \qquad x \in \mathbf{R}^n \tag{8.9.7}$$

The condition (8.9.6) is equivalent to requiring $p_1, \ldots, p_n$ to be an orthogonal basis for $\mathbf{R}^n$ with respect to the inner product $(\cdot, \cdot)_A$. Thus we also say that $\{p_1, \ldots, p_n\}$ are *A-orthogonal* if they satisfy (8.9.6). With the norm $\|\cdot\|_A$, the function $E(x)$ of (8.9.3) is seen to be

$$E(x) = \tfrac{1}{2}\|x^* - x\|_A^2 \tag{8.9.8}$$

which is more clearly a measure of the error $x - x^*$. The relationship of $\|x\|_2$ and $\|x\|_A$ is further explored in Problem 36.

Given a set of conjugate directions $\{p_1, \ldots, p_n\}$, it is straightforward to solve $Ax = b$. Let

$$x^* = \alpha_1 p_1 + \cdots + \alpha_n p_n$$

Using (8.9.6),

$$\alpha_k = \frac{p_k^T A x^*}{p_k^T A p_k} = \frac{p_k^T b}{p_k^T A p_k} \qquad k = 1, \ldots, n \tag{8.9.9}$$

We use this formula for $x^*$ to introduce the *conjugate direction iteration method*. Let $x_0 = 0$,

$$x_k = \alpha_1 p_1 + \cdots + \alpha_k p_k \qquad 1 \le k \le n \tag{8.9.10}$$

Introduce

$$r_k = b - Ax_k = -\nabla f(x_k)$$

the residual of $x_k$ in $Ax = b$. Easily, $r_0 = b$ and

$$x_k = x_{k-1} + \alpha_k p_k \qquad r_k = r_{k-1} - \alpha_k Ap_k \qquad k = 1,\ldots, n \quad (8.9.11)$$

For $k = n$, we have $x_n = x^*$ and $r_n = 0$, and $x_k$ may equal $x^*$ with a smaller value of $k$.

**Lemma 1**    The term $r_k$ is orthogonal to $p_1,\ldots, p_k$, that is, $r_k^T p_i = 0$, $1 \leq i \leq k$.

We leave the proof to Problem 37.

**Lemma 2**    **(a)**    The minimization problem

$$\underset{-\infty < \alpha < \infty}{\text{Min}} f(x_{k-1} + \alpha p_k)$$

is solved uniquely by $\alpha = \alpha_k$, yielding $f(x_k)$ as the minimum.

**(b)**    Let $\mathscr{S}_k = \text{Span}\{p_1,\ldots, p_k\}$, the $k$-dimensional subspace generated by $\{p_1,\ldots, p_k\}$. Then the problem

$$\underset{x \in \mathscr{S}_k}{\text{Min}} f(x)$$

is uniquely solved by $x = x_k$, yielding the minimum $f(x_k)$.

**Proof**    **(a)**    Expand $\varphi(\alpha) \equiv f(x_{k-1} + \alpha p_k)$:

$$\varphi(\alpha) = f(x_{k-1}) + \alpha p_k^T Ax_{k-1} + \tfrac{1}{2}\alpha^2 p_k^T Ap_k - \alpha b^T p_k$$

The term $p_k^T Ax_{k-1}$ equals 0, because $x_{k-1} \in \mathscr{S}_{k-1}$ and $p_k$ is $A$-orthogonal to $\mathscr{S}_{k-1}$. Solve $\varphi'(\alpha) = 0$, obtaining $\alpha = \alpha_k$, to complete the proof.

**(b)**    Expand $f(x_k + h)$, for any $h \in \mathscr{S}_k$:

$$f(x_k + h) = f(x_k) + h^T Ax_k + \tfrac{1}{2}h^T Ah - h^T b$$

$$= f(x_k) + \tfrac{1}{2}h^T Ah - h^T r_k$$

By Lemma 1 and the assumption $h \in \mathscr{S}_k$, it follows that $h^T r_k = 0$. Thus

$$f(x_k + h) = f(x_k) + \tfrac{1}{2}h^T Ah \geq f(x_k)$$

since $A$ is positive definite. The minimum is attained uniquely in $\mathscr{S}_k$ by letting $h = 0$, proving (b). ∎

Lemma 2 gives an optimality property for conjugate direction methods, defined by (8.9.11) and (8.9.9). The problem is knowing how to choose the conjugate directions $\{p_j\}$. There are many possible choices, and some of them lead to well-known methods for directly solving $Ax = b$ [see Stewart (1973b)].

**The conjugate gradient method**   We give a way to simultaneously generate the directions $\{p_k\}$ and the iterates $\{x_k\}$. For the first direction $p_1$, we use the steepest descent direction:

$$p_1 = -\nabla f(x_0) = r_0 = b \tag{8.9.12}$$

since $x_0 = 0$. An inductive construction is given for the remaining directions. Assume $x_1, \ldots, x_k$ have been generated, along with the conjugate directions $p_1, \ldots, p_k$. A new direction $p_{k+1}$ must be chosen, one that is $A$-conjugate to $p_1, \ldots, p_k$. Also, assume $x_k \neq x^*$, and thus $r_k \neq 0$; otherwise, we would have the solution $x^*$ and there would be no point to proceeding.

By Lemma 1, $r_k$ is orthogonal to $\mathscr{S}_k$, and thus $r_k$ does not belong to $\mathscr{S}_k$. We use $r_k$ to generate $p_{k+1}$, choosing a component of $r_k$. For reasons too complicated to consider here, it suffices to consider

$$p_{k+1} = r_k + \beta_{k+1} p_k \tag{8.9.13}$$

Then the condition $p_k^T A p_{k+1} = 0$ implies

$$\beta_{k+1} = -\frac{p_k^T A r_k}{p_k^T A p_k} \tag{8.9.14}$$

The denominator is nonzero since $A$ is positive definite and $p_k \neq 0$. It can be shown [Luenberger (1984), p. 245] that this definition of $p_{k+1}$ also satisfies

$$p_j^T A p_{k+1} = 0 \qquad j = 1, 2, \ldots, k - 1 \tag{8.9.15}$$

thus showing $\{p_1, \ldots, p_{k+1}\}$ is $A$-conjugate.

The conjugate gradient method consists of choosing $\{p_k\}$ from (8.9.12)–(8.9.14) and $\{x_k, r_k\}$ from (8.9.11) and (8.9.9). Ignoring rounding errors, the method converges in $n$ or fewer iterations. The actual speed of convergence varies a great deal with the eigenvalues of $A$. The error analysis of the CG method is based on the following optimality result.

***Theorem 8.9***   The iterates $\{x_k\}$ of the CG method satisfy

$$\|x^* - x_k\|_A = \underset{\deg(q) < k}{\text{Min}} \|x^* - q(A)b\|_A \tag{8.9.16}$$

***Proof***   For $q(\lambda)$ a polynomial, the notation $q(A)$ denotes the matrix expression with each power $\lambda^j$ replaced by $A^j$. For example,

$$q(\lambda) = a_0 + a_1\lambda + a_2\lambda^2 \quad \Rightarrow \quad q(A) = a_0 I + a_1 A + a_2 A^2$$

The proof of (8.9.16) is given in Luenberger (1984, p. 246).   ∎

Using this theorem, a number of error results can be given, varying with the properties of the matrix $A$. For example, let the eigenvalues of $A$ be denoted by

$$0 < \lambda_1 \leq \cdots \leq \lambda_n \tag{8.9.17}$$

repeated according to their multiplicity, and let $v_1, \ldots, v_n$ denote a corresponding orthonormal basis of eigenvectors. Using this basis, write

$$x^* = \sum_{j=1}^{n} c_j v_j \qquad b = Ax^* = \sum_{j=1}^{n} c_j \lambda_j v_j \qquad (8.9.18)$$

Then

$$q(A)b = \sum_{j=1}^{n} c_j \lambda_j q(\lambda_j) v_j$$

$$\|x^* - q(A)b\|_A = \left[ \sum_{j=1}^{n} c_j^2 \lambda_j \left[1 - \lambda_j q(\lambda_j)\right]^2 \right]^{1/2} \qquad (8.9.19)$$

Any choice of a polynomial $q(\lambda)$ of degree $< k$ will give a bound for $\|x^* - x_k\|_A$. One of the better known bounds is

$$\|x^* - x_k\|_A \leq 2 \left[ \frac{1 - \sqrt{c}}{1 + \sqrt{c}} \right]^k \|x^*\|_A \qquad (8.9.20)$$

with $c = \lambda_1 / \lambda_n$, the reciprocal of the condition number $\text{cond}(A)_2$. This is a conservative bound, implying poor convergence for ill-conditioned problems. Its proof is sketched in Luenberger (1984, p. 258, prob. 10). Other bounds can be derived, based on the behavior of the eigenvalues $\{\lambda_i\}$ and the coefficients $c_i$ of (8.9.18). In many applications in which the $\lambda_i$ vary greatly, it often happens that the $c_j$ for the smaller $\lambda_j$ are quite close to zero. Then the formula in (8.9.19) can be manipulated to give an improved bound over that in (8.9.20). In other cases, the eigenvalues may coalesce around a small number of values, and then (8.9.19) can be used to show convergence with a small $k$. For other results, see Luenberger (1984, p. 250), Jennings (1977), and van der Sluis and van der Horst (1986).

The formulas for $\{\alpha_j, \beta_j\}$ in defining the CG method can be further modified, to give simpler and more efficient formulas. We incorporate those into the following.

*Algorithm*   CG $(A, b, x, n)$

1. Remark: This algorithm calculates the solution of $Ax = b$ using the conjugate gradient method.

2. $x_0 := 0$, $r_0 := b$, $p_0 := 0$

3. For $k = 0, \ldots, n - 1$, do through step 7.

4. If $r_k = 0$, then set $x = x_k$ and exit.

5. For $k = 0$, $\beta_1 := 0$; and
   for $k > 0$, $\beta_{k+1} := r_k^T r_k / r_{k-1}^T r_{k-1}$.
   $p_{k+1} := r_k + \beta_{k+1} p_k$.

**6.**  $\alpha_{k+1} := r_k^T r_k / p_{k+1}^T A p_{k+1}$
$x_{k+1} := x_k + \alpha_{k+1} p_{k+1}$
$r_{k+1} := b - A x_{k+1}$

**7.**  End loop on $k$.

**8.**  $x := x_n$ and exit.

This algorithm does not consider the problems of using finite precision arithmetic. For a more complete algorithm, see Wilkinson and Reinsch (1971, pp. 57–69).

Our discussion of the CG method has followed closely that in Luenberger (1984, chap 8). For another approach, with more geometric motivation, see Golub and Van Loan (1983, sec. 10.2). They also have extensive references to the literature.

*Example*   As a simpleminded test case, we use the order five matrix

$$A = \begin{bmatrix} 5 & 4 & 3 & 2 & 1 \\ 4 & 5 & 4 & 3 & 2 \\ 3 & 4 & 5 & 4 & 3 \\ 2 & 3 & 4 & 5 & 4 \\ 1 & 2 & 3 & 4 & 5 \end{bmatrix} \tag{8.9.21}$$

The smallest and largest eigenvalues are $\lambda_1 \doteq .5484$ and $\lambda_5 \doteq 17.1778$, respectively. For the error bound (8.9.20), $c \doteq .031925$, and

$$\frac{1 - \sqrt{c}}{1 + \sqrt{c}} \doteq .697$$

For the linear system, we chose

$$b \doteq [7.9380, 12.9763, 17.3057, 19.4332, 18.4196]^T$$

which leads to the true solution

$$x^* \doteq [-0.3227, 0.3544, 1.1010, 1.5705, 1.6897]^T$$

**Table 8.10   Example of the conjugate gradient method**

| $k$ | $\|r_k\|_\infty$ | $\|x - x_k\|_\infty$ | $\|x - x_k\|_A$ | Bound (8.9.20) |
|---|---|---|---|---|
| 1 | 4.27 | 8.05E − 1 | 2.62 | 12.7 |
| 2 | 8.98E − 2 | 7.09E − 2 | 1.31E − 1 | 8.83 |
| 3 | 2.75E − 3 | 3.69E − 3 | 4.78E − 3 | 6.15 |
| 4 | 7.59E − 5 | 1.38E − 4 | 1.66E − 4 | 4.29 |
| 5 | $\doteq 0$ | $\doteq 0$ | $\doteq 0$ | 2.99 |

The results from using CG are shown in Table 8.10, along with the error bound in (8.9.20). As stated earlier, the bound (8.9.20) is very conservative.

The residuals decrease, as expected. But from the way the directions $\{p_k\}$ are constructed, this implies that obtaining accurate directions $p_k$ for larger $k$ will likely be difficult because of the smaller number of digits of accuracy in the residuals $r_k$. For some discussion of this, see Golub and Van Loan (1983, p. 373), which also contains additional references to the literature for this problem.

**The preconditioned conjugate gradient method**    The bound (8.9.20) indicates or seems to imply that the CG iterates can converge quite slowly, even for methods with a moderate condition number such as $\text{cond}(A)_2 = 1/c = 100$. To increase the rate of convergence, or at least to guarantee a rapid rate of convergence, the problem $Ax = b$ is transformed to an equivalent problem with a smaller condition number. The bound in (8.9.20) will be smaller, and one expects that the iterates will converge more rapidly.

For a nonsingular matrix $Q$, transform $Ax = b$ by

$$(Q^{-1}AQ^{-T})(Q^Tx) = Q^{-1}b \tag{8.9.22}$$

with $Q^{-T} \equiv (Q^T)^{-1}$. Write

$$\tilde{A} = Q^{-1}AQ^{-T} \qquad \tilde{x} = Q^Tx \qquad \tilde{b} = Q^{-1}b \tag{8.9.23}$$

Then (8.9.22) is simply $\tilde{A}\tilde{x} = \tilde{b}$. The matrix $Q$ is to be chosen so that $\text{cond}(\tilde{A})_2$ is significantly smaller than $\text{cond}(\tilde{A})_2$. The actual CG method is not applied explicitly to solving $\tilde{A}\tilde{x} = \tilde{b}$, but rather the algorithm CG is modified slightly. For the resulting algorithm when $Q$ is symmetric, see Golub and Van Loan (1983, p. 374).

Finding $Q$ requires a careful analysis of the original problem $Ax = b$, understanding the structure of $A$ in order to pick $Q$. From (8.9.23),

$$A = Q\tilde{A}Q^T$$

with $\tilde{A}$ to be chosen with eigenvalues near 1 in magnitude. For example, if $\tilde{A}$ is about the identity $I$, then $A \doteq QQ^T$. This decomposition could be accomplished with a Cholesky triangular factorization. Approximate Cholesky factors are used in defining preconditioners in some cases. For an introduction to the problem of selecting preconditioners, see Golub and Van Loan (1983, sec. 10.3) and Axelson (1985).

## Discussion of the Literature

The references that have most influenced the presentation of Gaussian elimination and other topics in this chapter are the texts of Forsythe and Moler (1967),

Golub and Van Loan (1983), Isaacson and Keller (1966), and Wilkinson (1963), (1965), along with the paper of Kahan (1966). Other very good general treatments are given in Conte and de Boor (1980), Noble (1969), Rice (1981), and Stewart (1973a). More elementary introductions are given in Anton (1984) and Strang (1980).

The best codes for the direct solution of both general and special forms of linear systems, of small to moderate size, are based on those given in the package LINPACK, described in Dongarra et al. (1979). These are completely portable programs, and they are available in single and double precision, in both real and complex arithmetic. Along with the solution of the systems, they also can estimate the condition number of the matrix under consideration. The linear equation programs in IMSL and NAG are variants and improvements of the programs in LINPACK.

Another feature of the LINPACK is the use of the *Basic Linear Algebra Subroutines* (BLAS). These are low-level subprograms that carry out basic vector operations, such as the dot product of two vectors and the sum of two vectors. These are available in Fortran, as part of LINPACK; but by giving assembly language implementations of them, it is often possible to significantly improve the efficiency of the main LINPACK programs. For a more general discussion of the BLAS, see Lawson et al. (1979). The LINPACK programs are widely available, and they have greatly influenced the development of linear equation programs in other packages.

There is a very large literature on solving the linear systems arising from the numerical solution of partial differential equations (PDEs). For some general texts on the numerical solution of PDEs, see Birkhoff and Lynch (1984), Forsythe and Wasow (1960), Gladwell and Wait (1979), Lapidus and Pinder (1982), and Richtmyer and Morton (1967). For texts devoted to classical iterative methods for solving the linear systems arising from the numerical solution of PDEs, see Hageman and Young (1981) and Varga (1962). For other approaches of more recent interest, see Swarztrauber (1984), Swarztrauber and Sweet (1979), George and Liu (1981), and Hackbusch and Trottenberg (1982).

The numerical solution of PDEs is the source of a large percentage of the sparse linear systems that are solved in practice. However, sparse systems of large order also occur with other applications [e.g., see Duff (1981)]. There is a large variety of approaches to solving large sparse systems, some of which we discussed in Sections 8.6–8.8. Other direct and iteration methods are available, depending on the structure of the matrix. For a sample of the current research in this very active area, see the survey of Duff (1977), the proceedings of Björck et al. (1981), Duff (1981), Duff and Stewart (1979), and Evans (1985), and the texts of George and Liu (1981) and Pissanetzky (1984). There are several software packages for the solution of various types of sparse systems, some associated with the preceding books. For a general index of many of the packages that are available, see the compilation of Heath (1982). For iteration methods for the systems associated with solving partial differential equations, the books of Varga (1962) and Hageman and Young (1981) discuss many of the classical approaches.

Integral equations lead to dense linear systems, and other types of iteration methods have been used for their solution. For some quite successful methods, see Atkinson (1976, part II, chap. 4).

The conjugate gradient method dates to Hestenes and Stiefel (1952), and its use in solving integral and partial differential equations is still under development. For more extensive discussions relating the conjugate direction method to other numerical methods, see Hestenes (1980) and Stewart (1973b). For references to the recent literature, including discussions of the preconditioned conjugate gradient method, see Axelsson (1985), Axelsson and Lindskog (1986), and Golub and Van Loan (1983, secs. 10.2 and 10.3). A generalization for nonsymmetric systems is proposed in Eisenstat et al. (1983).

One of the most important forces that will be determining the direction of future research in numerical linear algebra is the growing use of vector and parallel processor computers. The vector machines, such as the CRAY-2, work best when doing basic operations on vector quantities, such as those specified in the BLAS used in LINPACK. In recent years, there has been a vast increase in the availability of time on these machines, on newly developed nationwide computer networks. This has changed the scale of many physical problems that can be attempted, and it has led to a demand for ever more efficient computer programs for solving a wide variety of linear systems. The use of parallel computers is even more recent, and only in the middle to late 1980s have they become widespread. There is a wide variety of architectures for such machines. Some have the multiple processors share a common memory, with a variety of possible designs; others are based on each processor having its own memory and being linked in various ways to other processors. Parallel computers often lead to quite different types of numerical algorithms than those we have been studying for sequential computers, algorithms that can take advantage of several concurrent processors working on a problem. There is little literature available, although that is changing quite rapidly. As a survey of the solution of the linear systems associated with partial differential equations, on both vector and parallel computers, see Ortega and Voigt (1985). For a proposed text for the solution of linear systems, see Ortega (1987).

# Bibliography

Aird, T., and R. Lynch (1975). Computable accurate upper and lower error bounds for approximate solutions of linear algebraic systems, *ACM Trans. Math. Softw.* **1**, 217–231.

Anton, H. (1984). *Elementary Linear Algebra*, 4th ed. Wiley, New York.

Atkinson, K. (1976). *A Survey of Numerical Methods for the Solution of Fredholm Integral Equations of the Second Kind*. SIAM Pub., Philadelphia.

Axelsson, O. (1985). A survey of preconditioned iterative methods for linear systems of algebraic equations, *BIT* **25**, 166–187.

Axelsson, O., and G. Lindskop (1986). On the rate of convergence of the preconditioned conjugate gradient method, *Numer. Math.* **48**, 499–523.

Birkhoff, G., and R. Lynch (1984). *Numerical Solution of Elliptic Problems*. SIAM Pub., Philadelphia.

Björck, A., R. Plemmons, and H. Schneider, Eds. (1981). *Large Scale Matrix Problems*. North-Holland, Amsterdam.

Concus, P., G. Golub, and D. O'Leary (1984). A generalized conjugate gradient method for the numerical solution of elliptic partial differential equations. In *Studies in Numerical Analysis*, G. Golub, Ed. Mathematical Association of America, pp. 178–198.

Conte, S., and C. de Boor (1980). *Elementary Numerical Analysis*, 3rd ed. McGraw-Hill, New York.

Dongarra, J., J. Bunch, C. Moler, and G. Stewart (1979). *Linpack User's Guide*. SIAM Pub., Philadelphia.

Dorr, F. (1970). The direct solution of the discrete Poisson equations on a rectangle, *SIAM Rev.* **12**, 248–263.

Duff, I. (1977). A survey of sparse matrix research, *Proc. IEEE* **65**, 500–535.

Duff, I., Ed. (1981). *Sparse Matrices and Their Uses*. Academic Press, New York.

Duff, I., and G. Stewart, Eds. (1979). *Sparse Matrix Proceedings 1978*. SIAM Pub., Philadelphia.

Eisenstat, S., H. Elman, and M. Schultz (1983). Variational iterative methods for nonsymmetric systems of linear equations, *SIAM J. Numer. Anal.* **20**, 345–357.

Evans, D., Ed. (1985). *Sparsity and Its Applications*. Cambridge Univ. Press, Cambridge, England.

Forsythe, G., and C. Moler (1967). *Computer Solution of Linear Algebraic Systems*. Prentice-Hall, Englewood Cliffs, N.J.

Forsythe, G., and W. Wasow (1960). *Finite Difference Methods for Partial Differential Equations*. Wiley, New York.

George, A., and J. Liu (1981). *Computer Solution of Large Sparse Positive Definite Systems*. Prentice-Hall, Englewood Cliffs, N.J.

Gladwell, I., and R. Wait, Eds. (1979). *A Survey of Numerical Methods for Partial Differential Equations*. Oxford Univ. Press, Oxford, England.

Golub, G., and C. Van Loan (1983). *Matrix Computations*. Johns Hopkins Press, Baltimore.

Gregory, R., and D. Karney (1969). *A Collection of Matrices for Testing Computational Algorithms*. Wiley, New York.

Hackbusch, W., and U. Trottenberg, Eds. (1982). *Multigrid Methods*, Lecture Notes Math. *960*. Springer-Verlag, New York.

Hageman, L., and D. Young (1981). *Applied Iterative Methods*. Academic Press, New York.

Heath, M., Ed. (1982). *Sparse Matrix Software Catalog*. Oak Ridge National Laboratory, Oak Ridge, Tenn. (Published in connection with the Sparse Matrix Symposium 1982.)

Hestenes, M. (1980). *Conjugate Direction Methods in Optimization*. Springer-Verlag, New York.

Hestenes, M., and E. Stiefel (1952). Methods of conjugate gradients for solving linear systems, *J. Res. Nat. Bur. Stand.* **49**, 409–439.

Isaacson, E., and H. Keller (1966). *Analysis of Numerical Methods*. Wiley, New York.

Jennings, A., (1977). Influence of the eigenvalue spectrum on the convergence rate of the conjugate gradient method, *J. Inst. Math. Its Appl.* **20**, 61–72.

Kahan, W. (1966). Numerical linear algebra, *Can. Math. Bull.* **9**, 756–801.

Lapidus, L., and G. Pinder (1982). *Numerical Solution of Partial Differential Equations in Science and Engineering*. Wiley, New York.

Lawson, C., and R. Hanson (1974). *Solving Least Squares Problems*. Prentice-Hall, Englewood Cliffs, N.J.

Lawson, C., R. Hanson, D. Kincaid, and F. Krogh (1979). Basic linear algebra subprograms for Fortran usage, *ACM Trans. Math. Softw.* **5**, 308–323.

Luenberger, D. (1984). *Linear and Nonlinear Programming*, 2nd ed. Addison-Wesley, Reading, Mass.

Noble, B. (1969). *Applied Linear Algebra*. Prentice-Hall, Englewood Cliffs, N.J.

Ortega, J. (1987). *Parallel and Vector Solution of Linear Systems*. Preprint, Univ. of Virginia, Charlottesville.

Ortega, J., and R. Voigt (1985). Solution of partial differential equations on vector and parallel computers, *SIAM Rev.* **27**, 149–240.

Pissanetzky, S. (1984). *Sparse Matrix Technology*. Academic Press, New York.

Rice, J. (1981). *Matrix Computations and Mathematical Software*. McGraw-Hill, New York.

Richtmyer, R., and K. Morton (1967). *Difference Methods for Initial Value Problems*, 2nd ed. Wiley, New York.

Stewart, G. (1973a). *Introduction to Matrix Computations*. Academic Press, New York.

Stewart, G. (1973b). Conjugate direction methods for solving systems of linear equations, *Numer. Math.* **21**, 284–297.

Stewart, G. (1977). Research, development, and LINPACK. In *Mathematical Software III*, John Rice (Ed.). Academic Press, New York.

Stone, H. (1968). Iterative solution of implicit approximations of multidimensional partial differential equations, *SIAM J. Numer. Anal.* **5**, 530–558.

Strang, G. (1980). *Linear Algebra and Its Applications*, 2nd ed. Academic Press, New York.

Swarztrauber, P. (1984). Fast Poisson solvers. In *Studies in Numerical Analysis*, G. Golub, Ed. Mathematical Association of America, pp. 319–370.

Swarztrauber, P., and R. Sweet (1979). Algorithm 541: Efficient Fortran subprograms for the solution of separable elliptic partial differential equations, *ACM Trans. Math. Softw.* **5**, 352–364.

Van der Sluis, A., and H. van der Horst (1986). The rate of convergence of conjugate gradients, *Numer. Math.* **48**, 543–560.

Varga, R. (1962). *Matrix Analysis*. Prentice-Hall, Englewood Cliffs, N.J.

Wilkinson, J. (1963). *Rounding Errors in Algebraic Processes*. Prentice-Hall, Englewood Cliffs, N.J.

Wilkinson, J. (1965). *The Algebraic Eigenvalue Problem*. Oxford Univ. Press, Oxford, England.

Wilkinson, J., and C. Reinsch (1971). *Linear Algebra, Handbook for Automatic Computation*, Vol. 2. Springer-Verlag, New York.

## Problems

1. Solve the following systems $Ax = b$ by Gaussian elimination without pivoting. Check that $A = LU$, as in (8.1.5).

   (a) $A = \begin{bmatrix} 1 & 1 & -1 \\ 1 & 2 & -2 \\ -2 & 1 & 1 \end{bmatrix}$ $\quad b = \begin{bmatrix} 1 \\ 0 \\ 1 \end{bmatrix}$

   (b) $A = \begin{bmatrix} 4 & 3 & 2 & 1 \\ 3 & 4 & 3 & 2 \\ 2 & 3 & 4 & 3 \\ 1 & 2 & 3 & 4 \end{bmatrix}$ $\quad b = \begin{bmatrix} 1 \\ 1 \\ -1 \\ -1 \end{bmatrix}$

   (c) $A = \begin{bmatrix} 1 & -1 & 1 & -1 \\ -1 & 3 & -3 & 3 \\ 2 & -4 & 7 & -7 \\ -3 & 7 & -10 & 14 \end{bmatrix}$ $\quad b = \begin{bmatrix} 0 \\ 2 \\ -2 \\ 8 \end{bmatrix}$

2. Consider the linear system

$$6x_1 + 2x_2 + 2x_3 = -2$$

$$2x_1 + \tfrac{2}{3}x_2 + \tfrac{1}{3}x_3 = 1$$

$$x_1 + 2x_2 - x_3 = 0,$$

and verify its solution is

$$x_1 = 2.6 \qquad x_2 = -3.8 \qquad x_3 = -5.0$$

   (a) Using four-digit floating-point decimal arithmetic with rounding, solve the preceding system by Gaussian elimination without pivoting.

**(b)** Repeat part (a), using partial pivoting. In performing the arithmetic operations, remember to round to four significant digits after each operation, just as would be done on a computer.

3. **(a)** Implement the algorithms *Factor* and *Solve* of Section 8.2, or implement the analogous programs given in Forsythe and Moler (1967, chaps. 16 and 17).

   **(b)** To test the program, solve the system $Ax = b$ of order $n$, with $A = [a_{ij}]$ defined by

   $$a_{ij} = \text{Max}\,(i,\,j)$$

   Also define $b = [1, 1, \ldots, 1]^T$. The true solution is $x = [0, 0, \ldots, 0, (1/n)]^T$. This matrix is taken from Gregory and Karney (1969, p. 42).

4. Consider solving the integral equation

   $$\lambda x(s) - \int_0^1 \cos(\pi st) x(t) dt = 1 \qquad 0 \le s \le 1$$

   by discretizing the integral with the midpoint numerical integration rule (5.2.18). More precisely, let $n > 0$, $h = 1/n$, $t_i = (i - \frac{1}{2})h$ for $i = 1, \ldots, n$. We solve for approximate values of $x(t_1), \ldots, x(t_n)$ by solving the linear system

   $$\lambda z_i - \sum_{j=1}^n h \cos\left(\pi t_i t_j\right) z_j = 1 \qquad i = 1, \ldots, n$$

   Denote this linear system by $(\lambda I - K_n)z = b$, with $K_n$ of order $n \times n$,

   $$(K_n)_{ij} = h \cdot \cos\left(\pi t_i t_j\right) \qquad b_i = 1 \qquad 1 \le i, j \le n$$

   For sufficiently large $n$, $z_i \doteq x(t_i)$, $1 \le i \le n$. The value of $\lambda$ is nonzero, and it is assumed here to not be an eigenvalue of $K_n$.

   Solve $(\lambda I - K_n)z = b$ for several values of $n$, say $n = 2, 4, 8, 16, 32, 64$, and print the vector solutions $z$. If possible, also graph these solutions, to gain some idea of the solution function $x(s)$ of the original integral equation. Use $\lambda = 4, 2, 1, .5$.

5. **(a)** Consider solving $Ax = b$, with $A$ and $b$ complex and order$(A) = n$. Convert this problem to that of solving a real square system of order $2n$. *Hint:* Write $A = A_1 + iA_2$, $b = b_1 + ib_2$, $x = x_1 + ix_2$, with $A_1, A_2, b_1, b_2, x_1, x_2$ all real. Determine equations to be satisfied by $x_1$ and $x_2$.

   **(b)** Determine the storage requirements and the number of operations for the method in (a) of solving the complex system $Ax = b$. Compare these results with those based on directly solving $Ax = b$ using Gaussian elimination and complex arithmetic. Note the greater expense of complex arithmetic operations.

6. Let $A, B, C$ be matrices of orders $m \times n$, $n \times p$, $p \times q$, respectively. Do an operations count for computing $A(BC)$ and $(AB)C$. Give examples of when one order of computation is preferable over the other.

7. (a) Show that the number of multiplications and divisions for the Gauss–Jordan method of Section 8.3 is about $\frac{1}{2}n^3$.

   (b) Show how the Gauss–Jordan method, with partial pivoting, can be used to invert an $n \times n$ matrix within only $n(n + 1)$ storage locations. Can complete pivoting be used?

8. Use either the programs of Problem 3(a) or the Gauss–Jordan method to invert the matrices in Problems 1 and 3(b).

9. Prove that if $A = LL^T$ with $L$ real and nonsingular, then $A$ is symmetric and positive definite.

10. Using the Choleski method, calculate the decomposition $A = LL^T$ for

   (a) $\begin{bmatrix} 2.25 & -3.0 & 4.5 \\ -3.0 & 5.0 & -10.0 \\ 4.5 & -10.0 & 34.0 \end{bmatrix}$   (b) $\begin{bmatrix} 15 & -18 & 15 & -3 \\ -18 & 24 & -18 & 4 \\ 15 & -18 & 18 & -3 \\ -3 & 4 & -3 & 1 \end{bmatrix}$

11. Let $A = LU = LDM$, with all $l_{ii}$, $m_{ii} = 1$, and $D$ diagonal. Further assume $A$ is symmetric. Show that $M = L^T$, and thus $A = LDL^T$. Show $A$ is positive definite if and only if all $d_{ii} > 0$.

12. Let $A$ be real, symmetric, positive definite, and of order $n$. Consider solving $Ax = b$ using Gaussian elimination without pivoting. The purpose of this problem is to justify that the pivots will be nonzero.

   (a) Show that all of the diagonal elements satisfy $a_{ii} > 0$. This shows that $a_{11}$ can be used as a pivot element.

   (b) After elimination of $x_1$ from equations 2 through $n$, let the resulting matrix $A^{(2)}$ be written as

$$A^{(2)} = \begin{bmatrix} a_{11} & a_{12} & \cdots & a_{1n} \\ 0 & & & \\ \vdots & & \hat{A}^{(2)} & \\ 0 & & & \end{bmatrix}$$

   Show that $\hat{A}^{(2)}$ is symmetric and positive definite.

This procedure can be continued inductively to each stage of the elimination process, thus justifying the existence of nonzero pivots at every step. *Hint:* To prove $\hat{A}^{(2)}$ is positive definite, first prove the identity

$$\sum_{i, j=2}^{n} a_{ij}^{(2)} x_i x_j = \sum_{i, j=1}^{n} a_{ij} x_i x_j - a_{11} \left[ x_1 + \sum_{j=2}^{n} \frac{a_{j1}}{a_{11}} x_j \right]^2$$

for any choice of $x_1, x_2, \ldots, x_n$. Then choose $x_1$ suitably.

13. As another approach to developing a compact method for producing the *LU* factorization of *A*, consider the following matrix-oriented approach. Write

$$A = \begin{bmatrix} \hat{A} & d \\ c^T & \alpha \end{bmatrix} \qquad c, d \in \mathbf{R}^{n-1} \qquad \alpha \in \mathbf{R}$$

and $\hat{A}$ square of order $n - 1$. Assume $A$ is nonsingular. As a step in an induction process, assume $\hat{A} = \hat{L}\hat{U}$ is known. Look for $A = LU$ in the form

$$A = \begin{bmatrix} \hat{L} & 0 \\ m^T & 1 \end{bmatrix} \begin{bmatrix} \hat{U} & q \\ 0 & \gamma \end{bmatrix} \qquad m, q \in \mathbf{R}^{n-1} \qquad \gamma \in \mathbf{R}$$

Show that $m$, $q$, and $\gamma$ can be found, and describe how to do so. (This method is applied to an original $A$, factoring each principal submatrix in the upper left corner, in increasing order.)

14. Using the algorithm (8.3.23)–(8.3.24) for solving tridiagonal systems, solve $Ax = b$ with

$$A = \begin{bmatrix} 2 & -1 & 0 & 0 & 0 \\ 1 & 2 & -1 & 0 & 0 \\ 0 & 1 & 2 & -1 & 0 \\ 0 & 0 & 1 & 2 & -1 \\ 0 & 0 & 0 & 1 & 2 \end{bmatrix} \qquad b = \begin{bmatrix} 3 \\ -2 \\ 2 \\ -2 \\ 1 \end{bmatrix}$$

Check that the hypotheses and conclusions of Theorem 8.2 are satisfied by this example.

15. Define the order $n$ tridiagonal matrix

$$A_n = \begin{bmatrix} 2 & -1 & 0 & & \cdots & 0 \\ -1 & 2 & -1 & 0 & & \\ 0 & -1 & 2 & -1 & & \vdots \\ \vdots & & & \ddots & & \\ 0 & & \cdots & & -1 & 2 \end{bmatrix}$$

Find a general formula for $A_n = LU$. *Hint:* Consider the cases $n = 3, 4, 5$, and then guess the general pattern and verify it.

**16.** Write a subroutine to solve tridiagonal systems using (8.3.23)–(8.3.24). Check it using the examples in Problems 14 and 15. There are also a number of tridiagonal systems in Gregory and Karney (1969, chap. 2) for which the true inverses are known.

**17.** There are families of linear systems $A_k x = b$ in which $A_k$ changes in some simple way into a matrix $A_{k+1}$, and it may then be simpler to find the *LU* factorization of $A_{k+1}$ by modifying that of $A_k$. As an example that arises in the simplex method for linear programming, let $A_1 = [a_1, \ldots, a_n]$ and $A_2 = [a_2, \ldots, a_{n+1}]$, with all $a_j \in \mathbf{R}^n$. Suppose $A_1 = L_1 U_1$ is known, with $L_1$ lower triangular and $U_1$ upper triangular. Find a simple way to obtain the *LU* factorization $A_2 = L_2 U_2$ from that for $A_1$, assuming pivoting is not needed. *Hint:* Using $L_1 u_i = a_i$, $1 \le i \le n$, write

$$A_2 = L_1 \left[ u_2, u_3, \ldots u_n, L_1^{-1} a_{n+1} \right] \equiv L_1 \tilde{U}$$

Show that $\tilde{U}$ can be simply modified into an upper triangular form $U_2$, and that this corresponds to the conversion of $L_1$ into the desired $L_2$. More precisely, $U_2 = M\tilde{U}$, $L_2 = L_1 M^{-1}$. Show that the operation cost for obtaining $A_2 = L_2 U_2$ is $O(n^2)$.

**18.** **(a)** Calculate the condition numbers $\text{cond}(A)_p$, $p = 1, 2, \infty$, for

$$A = \begin{bmatrix} 100 & 99 \\ 99 & 98 \end{bmatrix}$$

**(b)** Find the eigenvalues and eigenvectors of $A$, and use them to illustrate the remarks following (8.4.8) in Section 8.4.

**19.** Prove that if $A$ is unitary, then $\text{cond}(A)_* = 1$.

**20.** Show that for every $A$, the upper bound in (8.4.4) can be attained for suitable choices of $b$ and $r$. *Hint:* From the definitions of $\|A\|$ and $\|A^{-1}\|$ in Section 7.3, there are vectors $\hat{x}$ and $\hat{r}$ for which $\|A\hat{x}\| = \|A\| \|\hat{x}\|$ and $\|A^{-1}\hat{r}\| = \|A^{-1}\| \|\hat{r}\|$. Use this to complete the construction of equality in the upper bound of (8.4.4).

**21.** The condition number $\text{cond}(A)_*$ of (8.4.6) can be quite small for matrices $A$ that are ill-conditioned. To see this, define the $n \times n$ matrix

$$A_R = \begin{bmatrix} 1 & -1 & -1 & \cdots & -1 \\ 0 & 1 & -1 & \cdots & -1 \\ \vdots & & \ddots & & \vdots \\ & & & 1 & -1 \\ 0 & & \cdots & 0 & 1 \end{bmatrix}$$

Easily $\text{cond}(A)_* = 1$. Verify that $A_n^{-1}$ is given by the upper triangular matrix $B = [b_{ij}]$, with $b_{ii} = 1$,

$$b_{ij} = 2^{j-i-1} \qquad i < j \le n$$

Compute $\text{cond}(A)_\infty$.

22. As in Section 8.4, let $H_n = [1/(i + j - 1)]$ denote the Hilbert matrix of order $n$, and let $\overline{H}_n$ denote the matrix obtained when $H_n$ is entered into your computer in single precision arithmetic. To compare $H_n^{-1}$ and $\overline{H}_n^{-1}$, convert $\overline{H}_n$ to a double precision matrix by appending additional zeros to the mantissa of each entry. Then use a double precision matrix inversion computer program to calculate $\overline{H}_n^{-1}$ numerically. This will give an accurate value of $\overline{H}_n^{-1}$ to single precision accuracy, for lower values of $n$. After obtaining $\overline{H}_n^{-1}$, compare it with $H_n^{-1}$, given in (8.4.13) or Gregory and Karney (1969, pp. 34–37).

23. Using the programs of Problem 3 or the LINPACK programs SGECO and SGESL, solve $\overline{H}_n x = b$ for several values of $n$. Use $b = [1, -1, 1, -1, \ldots]^T$, and calculate the true answer by using $\overline{H}_n^{-1}$ from Problem 22. Comment on your results.

24. Using the residual correction method, described at the beginning of Section 8.5, calculate accurate single precision answers to the linear systems of Problem 23. Print the residuals and corrections, and examine the rate of decrease in the correction terms as the order $n$ is increased. Attempt to explain your results.

25. Consider the linear system of Problem 4, for solving approximately an integral equation. Occasionally we want to solve such a system for several values of $\lambda$ that are close together. Write a program to first solve the system for $\lambda_0 = 4.0$, and then save the $LU$ decomposition of $\lambda_0 I - K$. To solve $(\lambda I - K)x = b$ with other values of $\lambda$ nearby $\lambda_0$, use the residual correction method (8.5.3) with $C = [LU]^{-1}$. For example, solve the system when $\lambda = 4.1, 4.5, 5,$ and $10$. In each case, print the iterates and calculate the ratio in (8.5.11). Comment on the behavior of the iterates as $\lambda$ increases.

26. The system $Ax = b$,

$$A = \begin{bmatrix} 4 & -1 & 0 & -1 & 0 & 0 \\ -1 & 4 & -1 & 0 & -1 & 0 \\ 0 & -1 & 4 & 0 & 0 & -1 \\ -1 & 0 & 0 & 4 & -1 & 0 \\ 0 & -1 & 0 & -1 & 4 & -1 \\ 0 & 0 & -1 & 0 & -1 & 4 \end{bmatrix} \qquad b = \begin{bmatrix} 2 \\ 1 \\ 2 \\ 2 \\ 1 \\ 2 \end{bmatrix}$$

has the solution $x = [1, 1, 1, 1, 1, 1]^T$. Solve the system using the Gauss–Jacobi iteration method, and then solve it again using the

Gauss–Seidel method. Use the initial guess $x^{(0)} = 0$. Note the rate at which the iteration error decreases. Find the answers with an accuracy of $\epsilon = .0001$.

27. Let $A$ and $B$ have order $n$, with $A$ nonsingular. Consider solving the linear system

$$Az_1 + Bz_2 = b_1 \qquad Bz_1 + Az_2 = b_2$$

with $z_1, z_2, b_1, b_2 \in \mathbf{R}^n$.

(a) Find necessary and sufficient conditions for convergence of the iteration method

$$Az_1^{(m+1)} = b_1 - Bz_2^{(m)} \qquad Az_2^{(m+1)} = b_2 - Bz_1^{(m)} \qquad m \geq 0$$

(b) Repeat part (a) for the iteration method

$$Az_1^{(m+1)} = b_1 - Bz_2^{(m)} \qquad Az_2^{(m+1)} = b_2 - Bz_1^{(m+1)} \qquad m \geq 0$$

Compare the convergence rates of the two methods.

28. For the error equation (8.6.25), show that $r_\sigma(M) < 1$ if

$$\|P\| < \frac{1}{2\|A^{-1}\|}$$

for some matrix norm.

29. For the iteration of a block tridiagonal systems, given in (8.6.30), show convergence under the assumptions that

$$\|C_1\| < \frac{1}{\|B_1^{-1}\|}; \ \|A_i\| + \|C_i\| < \frac{1}{\|B_i^{-1}\|}, \quad 2 \leq i \leq r-1; \quad \|A_r\| < \frac{1}{\|B_r^{-1}\|}$$

Bound the rate of convergence.

30. Recall the matrix $A_n$ of Problem 15, and consider the linear system $A_n x = b$. This system is important as it arises in the standard finite difference approximation (6.11.30) to the two-point boundary value problem

$$y''(x) = f(x, y(x)) \qquad \alpha < x < \beta \qquad y(\alpha) = a_0 \qquad y(\beta) = a_1$$

It is also important because it arises in the analysis of iterative methods for solving discretizations of Poissons equation, as in (8.8.5). In line with this, consider using Jacobi's method to solve $A_n x = b$ iteratively. Show that Jacobi's method converges by showing $r_\sigma(M) < 1$ for the appropriate matrix $M$. *Hint:* Use the results of Problem 6 of Chapter 7.

**31.** As an example that convergent iteration methods can behave in unusual ways, consider

$$x^{(k+1)} = b + Ax^{(k)} \quad k \geq 0$$

with

$$A = \begin{bmatrix} \lambda & c \\ 0 & -\lambda \end{bmatrix} \quad b \in \mathbf{R}^2$$

Assuming $|\lambda| < 1$, we have $(I - A)^{-1}$ exists and $x^{(k)} \to x^* = (I - A)^{-1}b$ for all initial guesses $x^0$. Find explicit formulas for $A^k$, $x^* - x^{(k)}$, and $x^{(k+1)} - x^{(k)}$. By suitably adjusting $c$ relative to $\lambda$, show that it is possible for $\|x^* - x^{(k)}\|_\infty$ to alternately increase and decrease as it converges to zero. Look at the corresponding values for $\|x^{(k+1)} - x^{(k)}\|_\infty$. For simplicity, use $x^{(0)} = 0$ in all calculations.

**32. (a)** Let $C_0$ be an approximate inverse to $A$. Define $R_0 = I - AC_0$, and assume $\|R_0\| < 1$ for some matrix norm. Define the iteration method

$$C_{m+1} = C_m(I + R_m) \qquad R_{m+1} = I - AC_{m+1} \qquad m \geq 0$$

This is a well-known iteration method for calculating the inverse $A^{-1}$. Show the convergence of $C_m$ to $A^{-1}$ by first relating the error $A^{-1} - C_m$ to the residual $R_m$. And then examine the behavior of the residual $R_m$ by showing that $R_{m+1} = R_m^2$, $m \geq 0$.

**(b)** Relate $C_m$ to the expansion

$$A^{-1} = C_0(I - R_0)^{-1} = C_0 \sum_{j=0}^{\infty} R_0^j$$

Observe the relation of this method for inverting $A$ to the iteration method (2.0.6) of Chapter 2 for calculating $1/a$ for nonzero numbers $a$. Also, see Problem 1 of Chapter 2.

**33.** Implement programs for iteratively solving the discretization (8.8.5) of Poisson's equation on the unit square. To have a situation for which you have a true solution of the linear system, choose Poisson equations in which there is no discretization error in going to (8.8.5). This will be true if the truncation errors in (8.8.3) are identically zero, as, for example, with $u(x, y) = x^2 y^2$.

**(a)** Solve (8.8.5) with Jacobi's method. Observe the actual error $\|x - x^{(\nu)}\|_\infty$ in each iterate, as well as $\|x^{(\nu+1)} - x^{(\nu)}\|_\infty$. Estimate the constant $c$ of (8.7.2), and compute the estimated error bound of (8.7.5). Compare with the true iteration error.

**(b)**    Repeat with the Gauss–Seidel method. Also compare the iteration rate $c$ with that predicted by (8.8.13).

**(c)**    Implement the SOR method, using the optimal acceleration parameter $\omega^*$ from (8.8.15).

**34.**   **(a)**    Generalize the discretization of the Poisson equation in (8.8.1) to the equation

$$\frac{\partial^2 u}{\partial x^2} + \frac{\partial^2 u}{\partial y^2} - c(x, y)u = g(x, y) \qquad 0 < x, y < 1$$

with $u = f(x, y)$ on the boundary as before.

**(b)**    Assume $c(x, y) \geq 0$ for $0 \leq x, y \leq 1$. Generalize part (1) of the proof of Theorem 8.8 to show that the linear system of part (a) will have a unique solution.

**35.**   Implement the conjugate gradient algorithm CG of Section 8.9. Test it with the systems of Problems 1, 3, and 4. Whenever possible, for testing purposes, use systems with a known true solution. Using it, compute the true errors in each iterate and see how rapidly they decrease. For the linear system in Problem 4, that is based on solving an integral equation, solve the system for several values of $n$. Comment on the results.

**36.**   Recall the vector norm $\|x\|_A$ of (8.9.7), with $A$ symmetric and positive definite. Let the eigenvalues of $A$ be denoted by

$$0 < \lambda_1 \leq \lambda_2 \leq \cdots \leq \lambda_n$$

Show that

$$\sqrt{\lambda_1}\|x\|_2 \leq \|x\|_A \leq \sqrt{\lambda_n}\|x\|_2$$

with both equalities attainable for suitable choices of $x$. *Hint:* Use an orthonormal basis of eigenvectors of $A$.

**37.**   Prove Lemma 1, following (8.9.11). *Hint:* Use mathematical induction on $k$. Prove it for $k = 1$. Then assume it is true for $k \leq l$, and prove it for $k = l + 1$. Break the proof into two parts: (1) $p_i^T r_{l+1} = 0$ for $i \leq l$, and (2) $p_{l+1}^T r_{l+1} = 0$.

**38.**   Let $A$ be symmetric, positive definite, and order $n \times n$. Let $U = \{u_1, \ldots, u_n\}$ be a set of nonzero vectors in $\mathbf{R}^n$. Then if $U$ is both an orthogonal set and an $A$-orthogonal set, then $Au_i = \lambda_i u_i$, $i = 1, \ldots, n$ for suitable $\lambda_i > 0$. Conversely, one can always choose a set of eigenvectors $\{u_1, \ldots, u_n\}$ of $A$ to have them be both orthogonal and $A$-orthogonal.

**39.** Let $A$ be symmetric, positive definite, and of order $n$. Let $\{v_1, \ldots, v_n\}$ be an $A$-orthogonal set in $\mathbf{R}^n$, with all $v_i \neq 0$. Define

$$Q_j = \frac{v_j v_j^T A}{v_j^T A v_j} \qquad j = 1, \ldots, n$$

Showing the following properties for $Q_j$.

1. $Q_j v_i = 0$ if $i \neq j$; and $Q_j v_j = v_j$.
2. $Q_j^2 = Q_j$.
3. $(x, Q_j y)_A = (Q_j x, y)_A$, for all $x, y \in \mathbf{R}^n$.
4. $(Q_j x, (I - Q_j)y)_A = 0$, for all $x, y \in \mathbf{R}^n$.
5. $\|x\|_A^2 = \|Q_j x\|_A^2 + \|(I - Q_j)x\|_A^2$, for all $x \in \mathbf{R}^n$.

Properties (2)–(5) say that $Q_j$ is an *orthogonal projection* on the vector space $\mathbf{R}^n$ with the inner product $(\cdot, \cdot)_A$. Define

$$S_k = \text{Span}\{v_1, \ldots, v_k\}$$

Show that the solution to the minimization problem

$$\text{Min}_{y \in S_k} \|x - y\|_A$$

is given by

$$y = \left[\sum_{j=1}^{k} Q_j\right] x \equiv P_k x$$

The matrix $P_k$ also satisfies properties (2)–(5).

# NINE

# THE MATRIX EIGENVALUE PROBLEM

We study the problem of calculating the eigenvalues and eigenvectors of a square matrix. This problem occurs in a number of contexts and the resulting matrices may take a variety of forms. These matrices may be sparse or dense, may have greatly varying order and structure, and often are symmetric. In addition, what is to be calculated can vary enough as to affect the choice of method to be used. If only a few eigenvalues are to be calculated, then the numerical method will be different than if all eigenvalues are required.

The general problem of finding all eigenvalues and eigenvectors of a nonsymmetric matrix $A$ can be quite unstable with respect to perturbations in the coefficients of $A$, and this makes more difficult the design of general methods and computer programs. The eigenvalues of a symmetric matrix $A$ are quite stable with respect to perturbations in $A$. This is investigated in Section 9.1, along with the possible instability for nonsymmetric matrices. Because of the greater stability of the eigenvalue problem for symmetric matrices and because of its common occurrence, many methods have been developed especially for it. This will be a major emphasis of the development of this chapter, although methods for the nonsymmetric matrix eigenvalue problem are also discussed.

The eigenvalues of a matrix are usually calculated first, and they are used in calculating the eigenvectors, if these are desired. The main exception to this rule is the *power method* described in Section 9.2, which is useful in calculating a single dominant eigenvalue of a matrix. The usual procedure for calculating the eigenvalues of a matrix $A$ is two-stage. First, similarity transformations are used to reduce $A$ to a simpler form, which is usually tridiagonal for symmetric matrices. And second, this simpler matrix is used to calculate the eigenvalues, and also the eigenvectors if they are required. The main form of similarity transformations used are certain special unitary or orthogonal matrices, which are discussed in Section 9.3. For the calculation of the eigenvalues of a symmetric tridiagonal matrix, the theory of Sturm sequences is introduced in Section 9.4 and the $QR$ algorithm is discussed in Section 9.5. Once the eigenvalues have been calculated, the most powerful technique for calculating the eigenvectors is the method of *inverse iteration*. It is discussed and illustrated in Section 9.6. It should be noted that we will be using the words *symmetric* and *nonsymmetric* quite generally, where they ordinarily should be used only in connection with real matrices. For complex matrices, always substitute *Hermitian* and *non-Hermitian*, respectively.

Most numerical methods used at present have been developed since 1950. They are nontrivial to implement as computer programs, especially those that are to be used for nonsymmetric matrices. Beginning in the mid-1960s, algorithms for a variety of matrix eigenvalue problems were published, in ALGOL, in the journal *Numerische Mathematik*. These were tested extensively, and were subsequently revised based on the tests and on new theoretical results. These algorithms have been collected together in Wilkinson and Reinsch (1971, part II). A project within the Applied Mathematics Division of the Argonne National Laboratory translated these programs into Fortran, and further testing and improvement was carried out. This package of programs is called EISPACK and it is available from the Argonne National Laboratory and other sources (see the appendix). A complete description of the package, including all programs, is given in Smith et al. (1976) and Garbow et al. (1977).

## 9.1   Eigenvalue Location, Error, and Stability Results

We begin by giving some results for locating and bounding the eigenvalues of a matrix $A$. As a crude upper bound, recall from Theorem 7.8 of Chapter 7 that

$$\underset{\lambda \in \sigma(A)}{\text{Max}} |\lambda| \leq \|A\| \tag{9.1.1}$$

for any matrix norm. The notation $\sigma(A)$ denotes the set of all eigenvalues of $A$. The next result is a simple computational technique for giving better estimates for the location of the eigenvalues of $A$.

For $A = [a_{ij}]$ of order $n$, define

$$r_i = \sum_{\substack{j=1 \\ j \neq i}}^{n} |a_{ij}| \qquad i = 1, 2, \ldots, n \tag{9.1.2}$$

and let $Z_i$ denote the circle in the complex plane with center $a_{ii}$ and radius $r_i$:

$$Z_i = \{ z \in \mathbf{C} \mid |z - a_{ii}| \leq r_i \} \tag{9.1.3}$$

***Theorem 9.1***   (Gerschgorin)   Let $A$ have order $n$ and let $\lambda$ be an eigenvalue of $A$. Then $\lambda$ belongs to one of the circles $Z_i$. Moreover if $m$ of the circles form a connected set $S$, disjoint from the remaining $n - m$ circles, then $S$ contains exactly $m$ of the eigenvalues of $A$, counted according to their multiplicity as roots of the characteristic polynomial of $A$.

Since $A$ and $A^T$ have the same eigenvalues and characteristic polynomial, these results are also valid if summation within the column, rather than in the row, is used in defining the radii in (9.1.2).

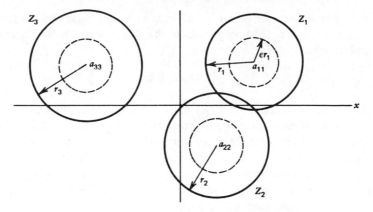

**Figure 9.1**   Example of Gerschgorin circle theorem.

***Proof***   Figure 9.1 gives a picture in the complex plane of what the circles might look like for a complex matrix of order three. The solid circles are the ones given by (9.1.3), and the dotted ones occur later in the proof. According to the theorem, there should be one eigenvalue in $Z_3$, and two eigenvalues in the union of $Z_1$ and $Z_2$.

Let $\lambda$ be an eigenvalue of $A$, and let $x$ be a corresponding eigenvector. Let $k$ be the subscript of a component of $x$ for which

$$|x_k| = \operatorname*{Max}_{1 \le i \le n} |x_i| = \|x\|_\infty$$

Then from $Ax = \lambda x$, the $k$th component yields

$$\sum_{j=1}^{n} a_{kj} x_j = \lambda x_k$$

$$(\lambda - a_{kk}) x_k = \sum_{\substack{j=1 \\ j \ne k}}^{n} a_{kj} x_j$$

$$|\lambda - a_{kk}| \, |x_k| \le \sum_{\substack{j=1 \\ j \ne k}}^{n} |a_{kj}| \, |x_j| \le r_k \|x\|_\infty$$

Canceling $\|x\|_\infty$ proves the first part of the theorem.
Define

$$D = \operatorname{diag}[a_{11}, a_{22}, \ldots, a_{nn}] \qquad E = A - D$$

For $0 \le \epsilon \le 1$, define

$$A(\epsilon) = D + \epsilon E \qquad\qquad (9.1.4)$$

and denote its eigenvalues by $\lambda_1(\epsilon), \ldots, \lambda_n(\epsilon)$. Note that $A(1) = A$, the original matrix. The eigenvalues are the roots of the characteristic polynomial

$$f_\epsilon(\lambda) \equiv \det[A(\epsilon) - \lambda I]$$

Since the coefficients of $f_\epsilon(\lambda)$ are continuous functions of $\epsilon$, and since the roots of any polynomial are continuous functions of its coefficients [see Henrici (1974) p. 281], we have that $\lambda_1(\epsilon), \ldots, \lambda_n(\epsilon)$ are continuous functions of $\epsilon$. As the parameter $\epsilon$ changes, each eigenvalue $\lambda_i(\epsilon)$ will vary in the complex plane, marking out a path from $\lambda_i(0)$ to $\lambda_i(1)$.

From the first part of the theorem, we know the eigenvalues $\lambda_i(\epsilon)$ are contained in the circles

$$Z_i(\epsilon) = \{z \in \mathbf{C} | \ |z - a_{ii}| \leq \epsilon r_i\} \qquad i = 1, \ldots, n \qquad (9.1.5)$$

with $r_i$ defined as before in (9.1.2). Examples of these circles are given in Figure 9.1 by the dotted circles. These circles decrease as $\epsilon$ goes from 1 to 0, and the eigenvalue $\lambda_i(\epsilon)$ must remain within them. When $\epsilon = 0$, the eigenvalues are simply

$$\lambda_i(0) = a_{ii}$$

By considering the path $\lambda_i(\epsilon)$, $0 \leq \epsilon \leq 1$, it must remain in the circle $Z_i(1)$ in which it begins at $\epsilon = 0$. Thus if $m$ of the circles $Z_i(1)$ form a connected set $S$, disjoint from the remaining $n - m$ circles, then $S$ must contain exactly $m$ eigenvalues $\lambda_i(1)$, as it contains the $m$ eigenvalues $\lambda_i(0)$. This proves the second result.

Since

$$\det[A - \lambda I] = \det[A - \lambda I]^T = \det[A^T - \lambda I]$$

we have $\sigma(A) = \sigma(A^T)$. Thus apply the theorem to the rows of $A^T$ in order to prove it for the columns of $A$. This completes the proof. ∎

This theorem can be used in a number of ways, but we just provide two simple numerical examples.

***Example*** Consider the matrix

$$A = \begin{bmatrix} 4 & 1 & 0 \\ 1 & 0 & -1 \\ 1 & 1 & -4 \end{bmatrix}.$$

From the preceding theorem, the eigenvalues must be contained in the circles

$$|\lambda - 4| \leq 1 \qquad |\lambda| \leq 2 \qquad |\lambda + 4| \leq 2 \qquad (9.1.6)$$

Since the first circle is disjoint from the remaining ones, there must be a single root in the circle. Since the coefficients of

$$f(\lambda) = \det[A - \lambda I]$$

are real, the complex eigenvalues must occur in conjugate pairs, if they occur at all. This will easily imply, with (9.1.6), that there is a real eigenvalue in the interval [3, 5]. The last two circles touch at the single point $(-2, 0)$. Using the same reasoning as before, the eigenvalues in these two circles must be real. And by using the construction (9.1.4) of $A(\epsilon)$, $\epsilon < 1$, there is one eigenvalue in $[-6, -2]$ and one in $[-2, 2]$. Since it is easily checked that $\lambda = -2$ is not an eigenvalue, we can conclude that $A$ has one real eigenvalue in each of the intervals

$$[-6, -2), (-2, 2], [3, 5]$$

The true eigenvalues are

$$-3.76010, -.442931, 4.20303$$

***Example*** Consider

$$A = \begin{bmatrix} 4 & 1 & 0 & & \cdots & 0 \\ 1 & 4 & 1 & 0 & & \\ 0 & 1 & 4 & 1 & & \vdots \\ \vdots & & & \ddots & & \\ & & & 1 & 4 & 1 \\ 0 & \cdots & & 0 & 1 & 4 \end{bmatrix} \qquad (9.1.7)$$

a matrix of order $n$. Since $A$ is symmetric, all eigenvalues of $A$ are real. The radii $r_i$ of (9.1.2) are all 1 or 2, and all the centers of the circles are $a_{ii} = 4$. Thus from the preceding theorem, the eigenvalues must all lie in the interval [2, 6]. Since the eigenvalues of $A^{-1}$ are the reciprocals of those of $A$, we must have

$$\frac{1}{6} \le \mu \le \frac{1}{2}$$

for all eigenvalues $\mu$ of $A^{-1}$. Using the matrix norm (7.3.22) induced by the Euclidean vector norm, we have

$$\|A^{-1}\|_2 = r_\sigma(A^{-1}) \le \frac{1}{2}$$

independent of the size of $n$.

**Bounds for perturbed eigenvalues**    Given a matrix $A$, we wish to perturb it and to then observe the effect on the eigenvalues of $A$. Analytical bounds are derived for the perturbations in the eigenvalues based on the perturbations in the matrix $A$. These bounds also suggest a definition of *condition number* that can be used to indicate the degree of stability or instability present in the eigenvalues.

To simplify the arguments considerably we assume that the Jordan canonical form of $A$ is diagonal (see Theorem 7.6):

$$P^{-1}AP = \text{diag}[\lambda_1, \ldots, \lambda_n] \equiv D \qquad (9.1.8)$$

for some nonsingular matrix $P$. The columns of $P$ will be the eigenvectors of $A$, corresponding to the eigenvalues $\lambda_1, \ldots, \lambda_n$. Matrices for which (9.1.8) holds are the most important case in practice. For a brief discussion of the case in which the Jordan canonical form is not diagonal, see the last topic of this section.

We also need to assume a special property for the matrix norms to be used. For any diagonal matrix

$$G = \text{diag}[g_1, \ldots, g_n]$$

we must have that

$$\|G\| = \underset{1 \leq i \leq n}{\text{Max}} |g_i| \qquad (9.1.9)$$

All of the operator matrix norms induced by the vector norms $\|x\|_p$, $1 \leq p \leq \infty$, have this property. We can now state the following result.

**Theorem 9.2** (Bauer–Fike)    Let $A$ be a matrix with a diagonal Jordan canonical form, as in (9.1.8). And assume the matrix norm satisfies (9.1.9). Let $A + E$ be a perturbation of $A$, and let $\lambda$ be an eigenvalue of $A + E$. Then

$$\underset{1 \leq i \leq n}{\text{Min}} |\lambda - \lambda_i| \leq \|P\|\,\|P^{-1}\|\,\|E\| \qquad (9.1.10)$$

**Proof**    If $\lambda$ is also an eigenvalue of $A$, then (9.1.10) is trivially true. Thus assume $\lambda \neq \lambda_1, \lambda_2, \ldots, \lambda_n$, and let $x$ be an eigenvector for $A + E$ corresponding to $\lambda$. Then

$$(A + E)x = \lambda x$$

$$(\lambda I - A)x = Ex$$

Substitute from (9.1.8) and multiply by $P^{-1}$ to obtain

$$(\lambda I - PDP^{-1})x = Ex$$

$$(\lambda I - D)(P^{-1}x) = (P^{-1}EP)(P^{-1}x)$$

Since $\lambda \neq \lambda_1, \ldots, \lambda_n$, $\lambda I - D$ is nonsingular,

$$(\lambda I - D)^{-1} = \text{diag}\left[(\lambda - \lambda_1)^{-1}, \ldots, (\lambda - \lambda_n)^{-1}\right]$$

Then

$$P^{-1}x = (\lambda I - D)^{-1}(P^{-1}EP)(P^{-1}x)$$

$$\|P^{-1}x\| \leq \left\|(\lambda I - D)^{-1}\right\|\,\|P^{-1}EP\|\,\|P^{-1}x\|$$

Canceling $\|P^{-1}x\|$ and using (9.1.9),

$$1 \leq \left[\text{Max } |\lambda - \lambda_i|^{-1}\right] \|P^{-1}\| \|P\| \|E\|$$

This is equivalent to (9.1.10), completing the proof.    ∎

**Corollary**    If $A$ is Hermitian, and if $A + E$ is any perturbation of $A$, then

$$\underset{1 \leq i \leq n}{\text{Min}} |\lambda - \lambda_i| \leq \|E\|_2 \tag{9.1.11}$$

for any eigenvalue $\lambda$ of $A + E$.    ∎

**Proof**    Since $A$ is Hermitian, the matrix $P$ can be chosen to be unitary. And using the operator norm (7.3.19) induced by the Euclidean vector norm, $\|P\|_2 = \|P^{-1}\|_2 = 1$ (see Problem 13 of Chapter 7). This completes the proof.    ∎

The statement (9.1.11) proves that small perturbations of a Hermitian matrix lead to equally small perturbations in the eigenvalues, as was asserted in the introduction to this chapter. Note that the relative error in some or all of the eigenvalues may still be large, and that this occurs commonly when the eigenvalues of a matrix vary greatly in magnitude.

**Example**    Consider the Hilbert matrix of order three,

$$H_3 = \begin{bmatrix} 1 & \dfrac{1}{2} & \dfrac{1}{3} \\[2mm] \dfrac{1}{2} & \dfrac{1}{3} & \dfrac{1}{4} \\[2mm] \dfrac{1}{3} & \dfrac{1}{4} & \dfrac{1}{5} \end{bmatrix} \tag{9.1.12}$$

Its eigenvalues to seven significant digits are

$$\lambda_1 = 1.408319 \qquad \lambda_2 = .1223271 \qquad \lambda_3 = .002687340 \tag{9.1.13}$$

Now consider the perturbed matrix $\hat{H}_3$, representing $H_3$ to four significant digits:

$$\hat{H}_3 = \begin{bmatrix} 1.000 & .5000 & .3333 \\ .5000 & .3333 & .2500 \\ .3333 & .2500 & .2000 \end{bmatrix} \tag{9.1.14}$$

Its eigenvalues to seven significant digits are

$$\hat{\lambda}_1 = 1.408294 \qquad \hat{\lambda}_2 = .1223415 \qquad \hat{\lambda}_3 = .002664489 \tag{9.1.15}$$

To verify the validity of (9.1.11) for this case, it is straightforward to calculate

$$\|E\| = r_\sigma(E) = \tfrac{1}{3} \times 10^{-4} \doteq .000033$$

For the errors and relative errors in (9.1.15),

$$\lambda_1 - \hat{\lambda}_1 = .0000249 \qquad \text{Rel}(\hat{\lambda}_1) = .0000177$$

$$\lambda_2 - \hat{\lambda}_2 = -.0000144 \qquad \text{Rel}(\hat{\lambda}_2) = -.000118$$

$$\lambda_3 - \hat{\lambda}_3 = .0000229 \qquad \text{Rel}(\hat{\lambda}_3) = .0085$$

All of the errors satisfy (9.1.11). But the relative error in $\hat{\lambda}_3$ is quite significant compared to the relative perturbations in $\hat{H}_3$.

For a nonsymmetric matrix $A$ with $P$ as in (9.1.8), the number

$$K(A) = \|P\| \, \|P^{-1}\|$$

will be called the condition number for the eigenvalue problem for $A$. This is based on the bound (9.1.10) for the perturbations in the eigenvalues of the matrix $A$ when it is perturbed. Another choice, even more difficult to compute, would be to use

$$K(A) = \text{Infimum} \, \|P\| \, \|P^{-1}\| \tag{9.1.16}$$

with the infimum taken over all matrices $P$ for which (9.1.8) holds and over all matrix norms satisfying (9.1.9). The reason for having condition numbers is that for nonsymmetric matrices $A$, small perturbations $E$ can lead to relatively large perturbations in the eigenvalues of $A$.

***Example*** To illustrate the pathological problems that can occur with nonsymmetric matrices, consider

$$A = \begin{bmatrix} 101 & -90 \\ 110 & -98 \end{bmatrix} \qquad A + E = \begin{bmatrix} 101 - \epsilon & -90 - \epsilon \\ 110 & -98 \end{bmatrix} \tag{9.1.17}$$

The eigenvalues of $A$ are $\lambda = 1, 2$, and the eigenvalues of $A + E$ are

$$\lambda = \frac{3 - \epsilon \pm \sqrt{1 - 828\epsilon + \epsilon^2}}{2}$$

As a specific example to give better intuition, take $\epsilon = .001$. Then

$$A + E = \begin{bmatrix} 100.999 & -90.001 \\ 110 & -98 \end{bmatrix}$$

and its eigenvalues are

$$\lambda \doteq 1.298, 1.701 \tag{9.1.18}$$

This problem should not be taken to imply that nonsymmetric matrices are ill-conditioned. Most cases in practice are fairly well-conditioned. But in writing a general algorithm, we always seek to cover as many cases as possible, and this example shows that this is likely to be difficult for the class of all nonsymmetric matrices.

For symmetric matrices, the result (9.1.11) can be improved upon in several ways. There is a *minimax characterization* for the eigenvalues of symmetric matrices. For a discussion of this theory and the resultant error bounds, see Parlett (1980, sec. 10.2) or Wilkinson (1965, p. 101). Instead, we give the following result, which will be more useful for error analyses of methods presented later.

**Theorem 9.3** (Wielandt–Hoffman)    Let $A$ and $E$ be real, symmetric matrices of order $n$, and define $\hat{A} = A + E$. Let $\lambda_i$ and $\hat{\lambda}_i$, $i = 1, \ldots, n$, be the eigenvalues of $A$ and $\hat{A}$, respectively, arranged in increasing order. Then

$$\left[ \sum_{j=1}^{n} \left( \lambda_j - \hat{\lambda}_j \right)^2 \right]^{1/2} \leq F(E) \qquad (9.1.19)$$

where $F(E)$ is the Frobenius norm of $E$, defined in (7.3.10).

**Proof**    See Wilkinson (1965, pp. 104–108).    ∎

This result will be used later in bounding the effect of the rounding errors that occur in reducing a symmetric matrix to tridiagonal form.

**A computable error bound for symmetric matrices**    Let $A$ be a symmetric matrix for which an approximate eigenvalue $\lambda$ and approximate eigenvector $x$ have been computed. Define the residual

$$\eta = Ax - \lambda x \qquad (9.1.20)$$

Since $A$ is symmetric, there is a unitary matrix $U$ for which

$$U^*AU = \text{diag}\, [\lambda_1, \ldots, \lambda_n] \equiv D \qquad (9.1.21)$$

Then we will show that

$$\underset{1 \leq i \leq n}{\text{Min}}\, |\lambda - \lambda_i| \leq \frac{\|\eta\|_2}{\|x\|_2} \qquad (9.1.22)$$

Using (9.1.21)

$$\eta = UDU^*x - \lambda x$$

$$U^*\eta = DU^*x - \lambda U^*x = (D - \lambda I)U^*x$$

If $\lambda$ is an eigenvalue of $A$, then (9.1.22) is trivially true. Thus there is no loss of generality in assuming $\lambda \neq \lambda_1, \ldots, \lambda_n$. Thus $D - \lambda I$ is nonsingular, and

$$U^*x = (D - \lambda I)^{-1}U^*\eta$$

$$\|U^*x\|_2 \le \|(D - \lambda I)^{-1}\|_2\|U^*\eta\|_2$$

Recall Problem 13 of Chapter 7, which implies

$$\|U^*x\|_2 = \|x\|_2 \qquad \|U^*\eta\|_2 = \|\eta\|_2$$

Then using the definition of the matrix norm, we have

$$\|x\|_2 \le \left[ \underset{1 \le i \le n}{\text{Max}} \left|(\lambda - \lambda_i)^{-1}\right| \right]\|\eta\|_2$$

which is equivalent to (9.1.22).

The use of (9.1.22) is illustrated later in (9.2.15) of Section 9.2, using an approximate eigenvalue–eigenvector pair produced by the power method.

**Stability of eigenvalues for nonsymmetric matrices**    In order to deal effectively with the potential for instability in the nonsymmetric matrix eigenvalue problem, it is necessary to have a better understanding of the nature of that instability. For example, one consequence of the analysis of instability will be that unitary similarity transformations will not make worse the conditioning of the problem.

As before, assume that $A$ has a diagonal Jordan canonical form:

$$P^{-1}AP = \text{diag}\,[\lambda_1, \ldots, \lambda_n] \equiv D \qquad (9.1.23)$$

Then $\lambda_1, \ldots, \lambda_n$ are the eigenvalues of $A$, and the columns of $P$ are the corresponding eigenvectors, call them $u_1, \ldots, u_n$. The matrix $P$ is not unique. For example, if $F$ is any nonsingular diagonal matrix, then

$$(PF)^{-1}A(PF) = F^{-1}DF = D$$

By choosing $F$ appropriately, the columns of $PF$ will have length one. Thus without loss of generality, assume the columns of $P$ will have length one:

$$Au_i = \lambda_i u_i \qquad u_i^*u_i = 1 \qquad i = 1, \ldots, n \qquad (9.1.24)$$

with

$$P = [u_1, \ldots, u_n]$$

By taking the conjugate transpose in (9.1.23),

$$P^*A^*(P^*)^{-1} = D^* = \text{diag}\,[\bar{\lambda}_1, \ldots, \bar{\lambda}_n]$$

which shows that the eigenvalues of $A^*$ are the complex conjugates of those of $A$.

Writing

$$(P^*)^{-1} = [w_1, \ldots, w_n] \tag{9.1.25}$$

we have

$$A^* w_i = \overline{\lambda}_i w_i \qquad i = 1, \ldots, n \tag{9.1.26}$$

Equivalently, by forming the conjugate transpose,

$$w_i^* A = \lambda_i w_i^* \tag{9.1.27}$$

This says $w_i^*$ is a *left eigenvector* of $A$, for the eigenvalue $\lambda_i$. Since $P^{-1}P = I$, and since

$$P^{-1} = \begin{bmatrix} w_1^* \\ \vdots \\ w_n^* \end{bmatrix}$$

we have

$$w_i^* u_j = \begin{cases} 1 & i = j \\ 0 & i \neq j \end{cases} \tag{9.1.28}$$

This says the eigenvectors $\{u_i\}$ of $A$ and the eigenvectors $\{w_i\}$ of $A^*$ form a *biorthogonal* set.

Normalize the eigenvectors $w_i$ by

$$v_i = \frac{w_i}{\|w_i\|_2} \qquad i = 1, \ldots, n$$

Define

$$s_i = v_i^* u_i = \frac{1}{\|w_i\|_2} \tag{9.1.29}$$

a positive real number. The matrix $(P^*)^{-1}$ can now be written

$$(P^*)^{-1} = \left[ \frac{v_1}{s_1}, \ldots, \frac{v_n}{s_n} \right]$$

And

$$A^* v_i = \overline{\lambda}_i v_i \qquad \|v_i\|_2 = 1 \qquad i = 1, \ldots, n \tag{9.1.30}$$

We now examine the stability of a simple eigenvalue $\lambda_k$ of $A$. Being simple means that $\lambda_k$ has multiplicity one as a root of the characteristic polynomial of $A$. The results can be extended to eigenvalues of multiplicity greater than one, but we omit that case. Consider the perturbed matrix

$$A(\epsilon) = A + \epsilon B \qquad \epsilon > 0$$

for some matrix $B$, independent of $\epsilon$. Denote the eigenvalues of $A(\epsilon)$ by $\lambda_1(\epsilon), \dots, \lambda_n(\epsilon)$. Then

$$P^{-1}A(\epsilon)P = D + \epsilon C \qquad C = P^{-1}BP$$

$$c_{ij} = \frac{1}{s_i} v_i^* B u_j \qquad 1 \le i, j \le n \qquad (9.1.31)$$

We will prove that

$$\lambda_k(\epsilon) = \lambda_k + \frac{\epsilon}{s_k} v_k^* B u_k + O(\epsilon^2) \qquad (9.1.32)$$

The derivation of this result uses the Gerschgorin Theorem 9.1. We also need to note that for any nonsingular diagonal matrix $F$,

$$FP^{-1}A(\epsilon)PF^{-1} = D + \epsilon FCF^{-1} \qquad (9.1.33)$$

and this leaves the eigenvalues of $A(\epsilon)$ unchanged. Pick $F$ as follows:

$$f_{ii} = \begin{cases} \epsilon\alpha & i = k \\ 1 & i \ne k \end{cases}$$

with $\alpha$ a positive constant to be determined later. Most of the coefficients of the matrix (9.1.33) are not changed, and only those in row $k$ and column $k$ need to be considered. They are

$$[D + \epsilon FCF^{-1}]_{kj} = \begin{cases} \epsilon^2 \alpha c_{kj} & j \ne k \\ \lambda_k + \epsilon c_{kk} & j = k \end{cases}$$

$$[D + \epsilon FCF^{-1}]_{ik} = \frac{1}{\alpha} c_{ik} \qquad i \ne k$$

Apply Theorem 9.1 to the matrix (9.1.33). The circle centers and radii are

$$\text{center} = \lambda_k + \epsilon c_{kk} \qquad r_k = \epsilon^2 \alpha \sum_{j \ne k} |c_{kj}|$$

$$\text{center} = \lambda_i + \epsilon c_{ii} \qquad r_i = \epsilon \sum_{j \ne i, k} |c_{ij}| + \frac{1}{\alpha} |c_{ik}| \qquad i \ne k$$

(9.1.34)

We wish to pick $\alpha$ so large and $\epsilon$ sufficiently small so as to isolate the circle about $\lambda_k + \epsilon c_{kk}$ from the remaining circles, and in that way, know there is exactly one eigenvalue of (9.1.33) within circle $k$. The distance between the centers of circles $k$ and $i \ne k$ is bounded from below by

$$|\lambda_i - \lambda_k| - \epsilon |c_{ii} - c_{kk}|$$

which is about $|\lambda_i - \lambda_k|$ for small values of $\epsilon$. Pick $\alpha$ such that

$$\frac{1}{\alpha}|c_{ik}| \leq \tfrac{1}{2}|\lambda_i - \lambda_k| \qquad \text{all } i \neq k \tag{9.1.35}$$

Then choose $\epsilon_0$ such that for all $0 < \epsilon \leq \epsilon_0$, circle $k$ does not intersect any of the remaining circles. This can be done because $\lambda_k$ is distinct from the remaining eigenvalues $\lambda_i$, and because of the inequality (9.1.35).

Using this construction, Theorem 9.1 implies that circle $k$ contains exactly one eigenvalue of $A(\epsilon)$, call it $\lambda_k(\epsilon)$. From (9.1.34),

$$|\lambda_k(\epsilon) - \lambda_k - \epsilon c_{kk}| \leq r_k = O(\epsilon^2)$$

and using the formula for $c_{kk}$ in (9.1.31), this proves the desired result (9.1.32). Taking bounds in (9.1.32), we obtain

$$|\lambda_k(\epsilon) - \lambda_k| \leq \frac{\epsilon}{s_k}\|v_k\|_2\|B\|_2\|u_k\|_2 + O(\epsilon^2)$$

and using (9.1.24) and (9.1.30),

$$|\lambda_k(\epsilon) - \lambda_k| \leq \frac{\epsilon}{s_k}\|B\|_2 + O(\epsilon^2) \tag{9.1.36}$$

The number $s_k$ is intimately related to the stability of the eigenvalue of $\lambda_k$, when the matrix $A$ is perturbed by small amounts $E = \epsilon B$. If $A$ were symmetric, we would have $u_k = v_k$, and thus $s_k = 1$, giving the same qualitative result for symmetric matrices as derived previously. For nonsymmetric matrices, if $s_k$ is quite small, then small perturbations $E = \epsilon B$ can lead to a large perturbation in the eigenvalue $\lambda_k$. Such problems are called ill-conditioned.

***Example*** Recall the example (9.1.17). Then $\epsilon = .001$,

$$P^{-1}AP = \begin{bmatrix} 1 & 0 \\ 0 & 2 \end{bmatrix} \qquad B = \begin{bmatrix} -1 & -1 \\ 0 & 0 \end{bmatrix}$$

$$P = \begin{bmatrix} \dfrac{9}{\sqrt{181}} & \dfrac{-10}{\sqrt{221}} \\ \dfrac{10}{\sqrt{181}} & \dfrac{-11}{\sqrt{221}} \end{bmatrix} \qquad P^{-1} = \begin{bmatrix} -11\sqrt{181} & 10\sqrt{181} \\ -10\sqrt{221} & 9\sqrt{221} \end{bmatrix} \tag{9.1.37}$$

If we use the row norm to estimate the condition number of (9.1.16), then

$$K(A) \leq \|P\|_\infty\|P^{-1}\|_\infty \doteq 419$$

The columns of $P$ give $u_1, u_2$, and the columns of $(P^{-1})^T$ give the vectors $w_1, w_2$ [see (9.1.25)].

To calculate (9.1.36) for $\lambda_1 = 1$,

$$s_1 = v_1^T u_1 = \frac{1}{\|w_1\|_2} = \frac{1}{\sqrt{(221)(181)}} \doteq .005$$

and $\|B\|_2 = \sqrt{2}$. Formula (9.1.36) yields

$$|\lambda_1(\epsilon) - \lambda_1| \le \frac{\sqrt{2}\,\epsilon}{.005} + O(\epsilon^2) \doteq 283\epsilon + O(\epsilon^2) \qquad (9.1.38)$$

The actual error is $\lambda_1(.001) - \lambda_1 \doteq 1.298 - 1 = .298$, and the preceding gives the error estimate of $283\epsilon \doteq .283$. Thus (9.1.36) is a reasonable estimated bound of the error.

In the following sections, some of the numerical methods first convert the matrix $A$ to a simpler form using similarity transformations. We wish to use transformations that will not make the numbers $s_k$ even smaller, which would make an ill-conditioned problem even worse. From this viewpoint, unitary or orthogonal transformations are the best ones to use.

Let $U$ be unitary, and let $\hat{A} = U^*AU$. For a simple eigenvalue $\lambda_k$, let $s_k$ and $\hat{s}_k$ denote the numbers (9.1.29) for the two matrices $A$ and $\hat{A}$. If $\{u_i\}$ and $\{v_i\}$ are the eigenvectors of $A$ and $A^*$, then $\{U^*u_i\}$ and $\{U^*v_i\}$ are the corresponding eigenvectors for $\hat{A}$ and $\hat{A}^*$. For $\hat{s}_k$,

$$\hat{s}_k = (U^*v_k)^*(U^*u_k) = v_k^*UU^*u_k = v_k^*u_k$$

$$= s_k \qquad (9.1.39)$$

Thus the stability of the eigenvalue $\lambda_k$ is made neither better nor worse. Unitary transformations also preserve vector length and the angles between vectors (see Problem 13 of Chapter 7). In general, unitary matrix operations on a given matrix $A$ will not cause any deterioration in the conditioning of the eigenvalue problem, and that is one of the major reasons that they are the preferred form of similarity transformation in solving the matrix eigenvalue problem.

Techniques similar to the preceding, in (9.1.23)–(9.1.36), can be used to give a stability result for eigenvectors of isolated eigenvalues. Using the same assumptions on $\{\lambda_i\}$, $\{u_i \equiv u_i(0)\}$, and $\{v_i\}$, consider the eigenvector problem

$$(A + \epsilon B)u_k(\epsilon) = \lambda_k(\epsilon)u_k(\epsilon) \qquad \|u_k(\epsilon)\| = 1 \qquad (9.1.40)$$

with $\lambda_k$ a simple eigenvalue of $A$. Then

$$u_k(\epsilon) = u_k + \epsilon \sum_{\substack{j=1 \\ j \ne k}}^{n} \left[ \frac{v_j^*Bu_k}{(\lambda_k - \lambda_j)s_j} \right] u_j + O(\epsilon^2) \qquad (9.1.41)$$

The proof of this is left to Problem 6.

The result (9.1.41) shows that stability of $u_k(\epsilon)$ depends on the condition numbers $s_j$ and on the nearness of the eigenvalues $\lambda_j$ to $\lambda_k$. This indicates probable instability, and further examples of this are given in Problem 7. A deeper examination of the behavior of eigenvectors when $A$ is subjected to perturbations requires an examination of the eigenvector subspaces and their relation to each other, especially when the eigenvalue $\lambda_k$ is not simple. For more on this, see Golub and Van Loan (1983, pp. 203–207, 271–275).

**Matrices with nondiagonal Jordan canonical form**    We have avoided discussing the eigenvalue problem for those matrices for which the Jordan canonical form is not diagonal. There are problems of instability in the eigenvalue problem, worse than that given in Theorem 9.2. And there are significant problems in the determination of a correct basis for the eigenvectors.

Rather than giving a general development, we examine the difficulties for this class of matrices by examining one simple case in detail. Let

$$A = \begin{bmatrix} 1 & 1 & 0 & & \cdots & 0 \\ 0 & 1 & 1 & 0 & & \vdots \\ \vdots & & \ddots & & & \vdots \\ & & & & 1 & 1 \\ 0 & & \cdots & & 0 & 1 \end{bmatrix} \tag{9.1.42}$$

a matrix of order $n$. The characteristic polynomial is

$$f(\lambda) = (1 - \lambda)^n$$

and $\lambda = 1$ is a root of multiplicity $n$. There is only a one-dimensional set of eigenvectors, spanned by

$$x = [1, 0, \ldots, 0]^T \tag{9.1.43}$$

For $\epsilon > 0$, perturb $A$ to

$$A(\epsilon) = \begin{bmatrix} 1 & 1 & 0 & & \cdots & 0 \\ 0 & 1 & 1 & 0 & & \vdots \\ \vdots & & & \ddots & & \vdots \\ 0 & & & & 1 & 1 \\ \epsilon & 0 & \cdots & & 0 & 1 \end{bmatrix}$$

Its characteristic polynomial is

$$f_\epsilon(\lambda) = (1 - \lambda)^n - (-1)^n \epsilon$$

There are $n$ distinct roots,

$$\lambda_k(\epsilon) = 1 + \omega_k \epsilon^{1/n} \qquad k = 1, \ldots, n \tag{9.1.44}$$

with $\{\omega_k\}$ the $n$th roots of unity,

$$\omega_k = e^{2\pi ki/n} \qquad k = 1, \ldots, n$$

For the perturbations in the eigenvalue of $A$,

$$|\lambda_k(\epsilon) - \lambda_k| = \epsilon^{1/n} \tag{9.1.45}$$

For example, if $n = 10$ and $\epsilon = 10^{-10}$, then

$$|\lambda_k(\epsilon) - \lambda_k| = .1 \tag{9.1.46}$$

The earlier result (9.1.10) gave a bound that was linear in $\epsilon$, and (9.1.45) is much worse, as shown by (9.1.46).

Since $A(\epsilon)$ has $n$ distinct eigenvalues, it also has a complete set of $n$ linearly independent eigenvectors, call them $x_1(\epsilon), \ldots, x_n(\epsilon)$. The first thing that must be done is to give the relationship of these eigenvectors to the single eigenvector $x$ in (9.1.43). This is a difficult problem to deal with, and it always must be dealt with for matrices whose Jordan form is not diagonal.

The matrices $A$ and $A(\epsilon)$ are in extremely simple form, and they merely hint at the difficulties that can occur when a matrix is not similar to a diagonal matrix. In actual practice, rounding errors will always ensure that such a matrix will have distinct eigenvalues. And this example is correct from the qualitative point of view in showing the difficulties that will arise.

## 9.2 The Power Method

This is a classical method, of use mainly in finding the dominant eigenvalue and associated eigenvector of a matrix. It is not a general method, but is useful in a number of situations. For example, it is sometimes a satisfactory method with large sparse matrices, where the methods of later sections cannot be used because of computer memory size limitations. In addition the method of inverse iteration, described in Section 9.6, is the power method applied to an appropriate inverse matrix. And the considerations of this section are an introduction to that later material. It is extremely difficult to implement the power method as a general-purpose computer program, treating a large and quite varied class of matrices. But it is easy to implement for more special classes.

We assume that $A$ is a real $n \times n$ matrix for which the Jordan canonical form is diagonal. Let $\lambda_1, \ldots, \lambda_n$ denote the eigenvalues of $A$, and let $x_1, \ldots, x_n$ be the corresponding eigenvectors, which form a basis for $\mathbf{C}^n$. We further assume that

$$|\lambda_1| > |\lambda_2| \geq |\lambda_3| \geq \cdots \geq |\lambda_n| \geq 0 \tag{9.2.1}$$

Although quite special, this is the main case of interest for the application of the power method. And the development can be extended fairly easily to the case of a single dominant eigenvalue of geometric multiplicity $r > 1$ (see Problem 10). Note that these assumptions imply $\lambda_1$ and $x_1$ are real.

Let $z^{(0)}$ be a real initial guess of some multiple of $x_1$. If there is no rational method for choosing $z^{(0)}$, then use a random number generator to choose each component. For the power method, define

$$w^{(m)} = Az^{(m-1)} \tag{9.2.2}$$

Let $\alpha_m$ be a component of $w^{(m)}$ that is maximum in size. Define

$$z^{(m)} = \frac{w^{(m)}}{\alpha_m} \qquad m \geq 1 \tag{9.2.3}$$

We show that the vectors $\{z^{(m)}\}$ will approximate $\hat{\sigma}_m x^{(1)} / \|x^{(1)}\|_\infty$, with each $\hat{\sigma}_m = \pm 1$, as $m \to \infty$.

We begin by showing that

$$z^{(m)} = \sigma_m \frac{A^m z^{(0)}}{\|A^m z^{(0)}\|_\infty} \qquad \sigma_m = \pm 1 \qquad m \geq 1 \tag{9.2.4}$$

First, $w^{(1)} = Az^{(0)}$.

$$\alpha_1 = \sigma_1 \|w^{(1)}\|_\infty = \sigma_1 \|Az^{(0)}\|_\infty \qquad \sigma_1 = \pm 1$$

Then

$$z^{(1)} = \frac{w^{(1)}}{\alpha_1} = \sigma_1 \frac{Az^{(0)}}{\|Az^{(0)}\|_\infty}$$

The proof of (9.2.4) for general $m > 1$ uses mathematical induction. For the case $m = 2$, as an example,

$$w^{(2)} = Az^{(1)} = \sigma_1 \frac{A^2 z^{(0)}}{\|Az^{(0)}\|_\infty}$$

$$\alpha_2 = \mu \frac{\|A^2 z^{(0)}\|_\infty}{\|Az^{(0)}\|_\infty} \qquad \mu = \pm 1$$

$$z^{(2)} = \frac{w^{(2)}}{\alpha_2} = \sigma_1 \frac{A^2 z^{(0)}}{\|Az^{(0)}\|_\infty} \div \frac{\mu \|A^2 z^{(0)}\|_\infty}{\|Az^{(0)}\|_\infty} = \sigma_2 \frac{A^2 z^{(0)}}{\|A^2 z^{(0)}\|_\infty}$$

with $\sigma_2 = \sigma_1 \mu$. The case of general $m$ is essentially the same.

To examine the convergence of $\{z^{(m)}\}$, first expand $z^{(0)}$ using the eigenvector basis $\{x_j\}$:

$$z^{(0)} = \sum_{j=1}^{n} \alpha_j x_j$$

We will assume $\alpha_1 \neq 0$, which a random choice of $z^{(0)}$ will generally ensure. Also $\alpha_1$ can be shown to be real. Then

$$A^m z^{(0)} = \sum_{j=1}^{n} \alpha_j A^m x_j = \sum_{j=1}^{n} \alpha_j \lambda_j^m x_j$$

$$= \lambda_1^m \left[ \alpha_1 x_1 + \sum_{j=2}^{n} \alpha_j \left( \frac{\lambda_j}{\lambda_1} \right)^m x_j \right] \qquad (9.2.5)$$

From (9.2.1),

$$\left( \frac{\lambda_j}{\lambda_1} \right)^m \to 0 \qquad \text{as} \quad m \to \infty \qquad 2 \le j \le n \qquad (9.2.6)$$

Using this in (9.2.5) and (9.2.4), we have

$$z^{(m)} \doteq \left( \frac{\lambda_1}{|\lambda_1|} \right)^m \frac{\sigma_m \alpha_1 x_1}{|\alpha_1| \, \|x_1\|_\infty} \equiv \hat{\sigma}_m \cdot \frac{x_1}{\|x_1\|_\infty} \qquad (9.2.7)$$

Then $\sigma_m = \pm 1$, and generally it is independent of $m$. This will be the case if $x_1$ has a unique maximal component. In cases with $x_1$ having more than one maximal component, it is possible that $\hat{\sigma}_m$ will vary with $m$ (see Problem 9). The rate of convergence in (9.2.7) depends on $|\lambda_2/\lambda_1|$:

$$\left\| z^{(m)} - \hat{\sigma}_m \cdot \frac{x_1}{\|x_1\|_\infty} \right\|_\infty \le C \left| \frac{\lambda_2}{\lambda_1} \right|^m \qquad (9.2.8)$$

since all of the remaining ratios $|\lambda_j/\lambda_1|$ are bounded by $|\lambda_2/\lambda_1|$.

To obtain a sequence of approximate eigenvalues, let $k$ be the index of a nonzero component of $x_1$. Generally, we pick $k$ as the subscript of the component $\alpha_m$, and thus it will possibly vary with $m$. Define

$$\lambda_1^{(m)} = \frac{w_k^{(m)}}{z_k^{(m-1)}} \qquad m \ge 1 \qquad (9.2.9)$$

To examine the rate of convergence, use (9.2.4) and (9.2.5):

$$\lambda_1^{(m)} = \left[ \sigma_m \cdot \frac{A^m z^{(0)}}{\|A^{m-1} z^{(0)}\|_\infty} \right]_k \div \left[ \sigma_m \cdot \frac{A^{m-1} z^{(0)}}{\|A^{m-1} z^{(0)}\|_\infty} \right]_k$$

$$= \frac{\lambda_1^m \left[ \alpha_1 x_1 + \sum_{j=2}^{n} \alpha_j \left( \frac{\lambda_j}{\lambda_1} \right)^m x_j \right]_k}{\lambda_1^{m-1} \left[ \alpha_1 x_1 + \sum_{j=2}^{n} \alpha_j \left( \frac{\lambda_j}{\lambda_1} \right)^{m-1} x_j \right]_k} \qquad (9.2.10)$$

$$\lambda_1^{(m)} = \lambda_1 \left[ 1 + O\left( \left| \frac{\lambda_2}{\lambda_1} \right|^m \right) \right] \qquad (9.2.11)$$

The rate of convergence is linear, and it depends on $|\lambda_2/\lambda_1|$. The index $k$ is usually chosen fixed. If we choose $k$ as the index of the maximal component $\alpha_m$ of $w^{(m)}$, and if $x_1$ has a single maximum component, then $k$ will become constant as $m \rightarrow \infty$. With more than one maximal component, it can move about, as is shown in Problem 9. An alternative method of defining $\lambda_1^{(m)}$ is given in Conte and de Boor (1980, p. 192), avoiding the need to select a particular component index:

$$\lambda_1^{(m)} = \frac{u^T w^{(m)}}{u^T z^{(m-1)}}$$

The vector $u$ is to satisfy $u^T x_1 \neq 0$, and a random choice of $u$ will generally suffice. The error in $\lambda_1^{(m)}$ will again satisfy (9.2.11).

***Example***   Let

$$A = \begin{bmatrix} 1 & 2 & 3 \\ 2 & 3 & 4 \\ 3 & 4 & 5 \end{bmatrix} \qquad (9.2.12)$$

The true eigenvalues are

$$\lambda_1 = 9.623475383 \qquad \lambda_2 = -.6234753830 \qquad \lambda_3 = 0 \qquad (9.2.13)$$

An initial guess $z^{(0)}$ was generated with a random number generator. The first five iterates are shown in Table 9.1, along with the ratios

$$R_m = \frac{\lambda_1^{(m)} - \lambda_1^{(m-1)}}{\lambda_1^{(m-1)} - \lambda_1^{(m-2)}} \qquad (9.2.14)$$

The iterates $\lambda_1^{(m)}$ were defined using (9.2.9), with $k = 3$. According to a later discussion, these ratios should approximate the ratio $\lambda_2/\lambda_1$ as $m \rightarrow \infty$.

We use the computable error bound (9.1.22) that was derived earlier in Section 9.1. Calculate

$$\eta = Ax - \lambda x$$

with

$$x = x^{(5)} \qquad \lambda = \lambda_1^{(5)}$$

**Table 9.1   Example of the power method**

| $m$ | $z_1^{(m)}$ | $z_2^{(m)}$ | $z_3^{(m)}$ | $\lambda_1^{(m)}$ | $R_m$ |
|---|---|---|---|---|---|
| 1 | .50077 | .75038 | 1.0000 | 11.7628133 | |
| 2 | .52626 | .76313 | 1.0000 | 9.5038496 | |
| 3 | .52459 | .76230 | 1.0000 | 9.6313231 | −.05643 |
| 4 | .52470 | .76235 | 1.0000 | 9.6229674 | −.06555 |
| 5 | .52469 | .76235 | 1.0000 | 9.6235083 | −.06474 |

Then from (9.1.22),

$$\text{Min}_i |\lambda_i - \lambda_1^{(5)}| < \frac{\|\eta\|_2}{\|x\|_2} = 3.30 \times 10^{-5} \qquad (9.2.15)$$

A direct comparison with the true answer $\lambda_1$ gives

$$\lambda_1 - \lambda_1^{(5)} = -.0000329$$

which shows that (9.2.15) is a very accurate estimate in this case.

**Acceleration methods**    Since there is a known regular pattern with which the error decreases for both $\lambda_1^{(m)}$ and $z^{(m)}$, this can be used to obtain more rapidly convergent methods. We give three different approaches for accelerating the convergence.

*Case* **1.**    Translation of the Eigenvalues.    Choose a constant $b$, and replace the calculation of the eigenvalues of $A$ by those of

$$B = A - bI \qquad (9.2.16)$$

The eigenvalues of $B$ are $\lambda_i - b$, $i = 1, \ldots, n$. Pick $b$ so that $\lambda_1 - b$ is the dominant eigenvalue of $B$, and choose $b$ to minimize the ratio of convergence. As a particular case in order to be more explicit, suppose that all eigenvalues of $A$ are real and that they have been so arranged that

$$\lambda_1 > \lambda_2 \geq \lambda_3 \geq \cdots \geq \lambda_n \qquad \lambda_1 > |\lambda_n|$$

Then the dominant eigenvalue of $B$ could be either $\lambda_1 - b$ or $\lambda_n - b$, depending on the size of $b$. We first require that $b$ satisfy

$$|\lambda_1 - b| > |\lambda_n - b|$$

The rate of convergence will be

$$\text{Max}\left\{\frac{|\lambda_2 - b|}{|\lambda_1 - b|}, \frac{|\lambda_n - b|}{|\lambda_1 - b|}\right\} \qquad (9.2.17)$$

If we look carefully at the behavior of these two ratios as $b$ varies, we see that the minimum of (9.2.17) occurs when

$$(\lambda_1 - b) - (\lambda_2 - b) = (\lambda_n - b) - [-(\lambda_1 - b)]$$

and

$$b^* = \tfrac{1}{2}(\lambda_2 + \lambda_n) \qquad (9.2.18)$$

is the optimal choice of $b$. The resulting ratio of convergence is

$$\frac{\lambda_2 - b^*}{\lambda_1 - b^*} = -\frac{\lambda_n - b^*}{\lambda_1 - b^*} = \frac{\lambda_2 - \lambda_n}{2\lambda_1 - \lambda_2 - \lambda_n} \qquad (9.2.19)$$

Experimental methods based on this formula and on (9.2.11) can be used to determine approximate values of $b^*$.

Transformations other than (9.2.16) can be used to transform the set of eigenvalues in such a way as to obtain even more rapid convergence. For a further discussion of these ideas, see Wilkinson (1965, pp. 570–584).

**Example**    In the previous example (9.2.12), the theoretical ratio of convergence was

$$\frac{\lambda_2}{\lambda_1} \doteq -.0648$$

Using the optimal value $b$ given by (9.2.18), and using (9.2.13) in a rearranged order,

$$b^* = \tfrac{1}{2}(\lambda_2 + \lambda_n) \doteq -.31174 \qquad (9.2.20)$$

The eigenvalues of $A - bI$ are

$$9.93522 \qquad .31174 \qquad -.31174 \qquad (9.2.21)$$

The ratio of convergence for the power method applied to $A - bI$ will be

$$\pm \frac{.31174}{9.93522} \doteq \pm .0314$$

which is less than half the magnitude of the original ratio.

**Case 2.**    Aitken Extrapolation.    The form of convergence in (9.2.11) is completely analogous to the linearly convergent rootfinding methods of Section 2.5 of Chapter 2. Following the same development as in Section 2.6, we consider the use of Aitken extrapolation to accelerate the convergence of $\{\lambda_1^{(m)}\}$ and $\{z^{(m)}\}$. To use the following development, we must assume in (9.2.1) that

$$|\lambda_2| \neq |\lambda_3| \qquad (9.2.22)$$

This can be weakened to

$$|\lambda_2| = |\lambda_j| \qquad \text{implies} \qquad \lambda_2 = \lambda_j$$

But we do not allow two ratios of convergence of equal magnitude and opposite sign (see the preceding example of (9.2.21) for such a case). The Aitken procedure can also be modified so as to remove the restriction (9.2.22).

With (9.2.22), and using (9.2.10),

$$\lambda_1 - \lambda_1^{(m)} \doteq cr^m \qquad (9.2.23)$$

where $c$ is some constant, $r$ is the unknown rate of convergence, and theoretically $r = \lambda_2/\lambda_1$. Proceeding exactly as in Section 2.6, implies that

$$R_m = \frac{\lambda_1^{(m)} - \lambda_1^{(m-1)}}{\lambda_1^{(m-1)} - \lambda_1^{(m-2)}} \to r \quad \text{as} \quad m \to \infty \tag{9.2.24}$$

And Aitken extrapolation gives the improved value $\hat{\lambda}_1$:

$$\hat{\lambda}_1 = \lambda_1^{(m)} - \frac{\left[\lambda_1^{(m)} - \lambda_1^{(m-1)}\right]^2}{\left[\lambda_1^{(m)} - \lambda_1^{(m-1)}\right] - \left[\lambda_1^{(m-1)} - \lambda_1^{(m-2)}\right]} \qquad m \geq 3 \tag{9.2.25}$$

A similar derivation can be applied to the eigenvector approximants, to accelerate each component of the sequence $\{z^{(m)}\}$, although some care must be used.

**Example** Consider again the example (9.2.12). In Table 9.1,

$$R_5 = -.06474 \doteq \frac{\lambda_2}{\lambda_1} = -.06479$$

As an example of (9.2.25), extrapolate with the values $\lambda_1^{(2)}$, $\lambda_1^{(3)}$, and $\lambda_1^{(4)}$ from that table. Then

$$\hat{\lambda}_1 = 9.6234814 \qquad \lambda_1 - \hat{\lambda}_1 = -6.03 \times 10^{-6}$$

In comparison using the more accurate table value $\lambda_1^{(5)}$, the error is $\lambda_1 - \lambda_1^{(5)} = -3.29 \times 10^{-5}$. This again shows the value of using extrapolation whenever the theory justifies its use.

**Case 3.** The Rayleigh–Ritz Quotient. Whenever $A$ is symmetric, it is better to use the following eigenvalue approximations:

$$\lambda_1^{(m+1)} = \frac{\left(Az^{(m)}, z^{(m)}\right)}{\left(z^{(m)}, z^{(m)}\right)} = \frac{\left(w^{(m+1)}, z^{(m)}\right)}{\left(z^{(m)}, z^{(m)}\right)} \qquad m \geq 0 \tag{9.2.26}$$

We are using standard inner product notation:

$$(w, z) = \sum_{1}^{n} w_i z_i \qquad w, z \in \mathbf{R}^n$$

To analyze this sequence (9.2.26), note that all eigenvalues of $A$ are real and that the eigenvectors $x_1, \ldots, x_n$ can be chosen to be orthonormal. Then (9.2.2), (9.2.4), (9.2.5), together with (9.2.26), imply

$$\lambda_1^{(m+1)} = \frac{\sum\limits_{j=1}^{n} |\alpha_j|^2 \lambda_j^{2m+1}}{\sum\limits_{j=1}^{n} |\alpha_j|^2 \lambda_j^{2m}}$$

$$\lambda_1^{(m+1)} = \lambda_1 \left[1 + O\left(\left(\frac{\lambda_2}{\lambda_1}\right)^{2m}\right)\right] \tag{9.2.27}$$

The ratio of convergence of $\lambda_1^{(m)}$ to $\lambda_1$ is $(\lambda_2/\lambda_1)^2$, an improvement on the original ratio in (9.47) of $\lambda_2/\lambda_1$.

This is a well-known classical procedure, and it has many additional aspects that are of use in some problems. For additional discussion, see Wilkinson (1965, pp. 172–178).

***Example*** In the example (9.2.12), use the approximate eigenvector $z^{(2)}$ in (9.2.26). Then

$$\lambda_1^{(3)} = \frac{\left(Az^{(2)}, z^{(2)}\right)}{\left(z^{(2)}, z^{(2)}\right)} \doteq 9.623464$$

which is as accurate as the value $\lambda_1^{(5)}$ obtained earlier.

The power method can be used when there is not a single dominant eigenvalue, but the algorithm is more complicated. The power method can also be used to determine eigenvalues other than the dominant one. This involves a process called *deflation* of $A$ to remove $\lambda$ as an eigenvalue. For a complete discussion of all aspects of the power method, see Golub and Van Loan (1983, 208–218) and Wilkinson (1965, chap. 9). Although it is a useful method in some circumstances, it should be stressed that the methods of the following sections are usually more efficient. For a rapidly convergent variation on the power method and the Rayleigh–Ritz quotient, see the *Rayleigh quotient iteration* for symmetric matrices in Parlett (1980, p. 70).

## 9.3  Orthogonal Transformations Using Householder Matrices

As one step in finding the eigenvalues of a matrix, it is often reduced to a simpler form using similarity transformations. Orthogonal matrices will be the class of matrices we use for these transformations. It was shown in (9.1.39) that orthogonal transformations will not worsen the condition or stability of the eigenvalues of a nonsymmetric matrix. Also, orthogonal matrices have other desirable error propagation properties, an example of which is given later in the section. For these reasons, we restrict our transformations to those using orthogonal matrices.

We begin the section by looking at a special class of orthogonal matrices known as *Householder matrices*. Then we show how to construct a Householder matrix that will transform a given vector to a simpler form. With this construction as a tool, we look at two transformations of a given matrix $A$: (1) obtain its *QR* factorization, and (2) construct a similar tridiagonal matrix when $A$ is a symmetric matrix. These forms are used in the next two sections in the calculation of the eigenvalues of $A$. As a matter of notation, note that we should be restricting the use of the term orthogonal to real matrices. But it has become common usage in this area to use *orthogonal* rather than *unitary* for the general complex case, and we will adopt the same convention. The reader should understand *unitary* when *orthogonal* is used for a complex matrix.

Let $w \in \mathbf{C}^n$ with $\|w\|_2 = \sqrt{w^*w} = 1$. Define

$$U = I - 2ww^* \tag{9.3.1}$$

This is the general form of a *Householder matrix*.

***Example***    For $n = 3$, we require

$$w = [w_1, w_2, w_3]^T \qquad |w_1|^2 + |w_2|^2 + |w_3|^2 = 1$$

The matrix $U$ is given by

$$U = \begin{bmatrix} 1 - 2|w_1|^2 & -2w_1\bar{w}_2 & -2w_1\bar{w}_3 \\ -2\bar{w}_1w_2 & 1 - 2|w_2|^2 & -2w_2\bar{w}_3 \\ -2\bar{w}_1w_3 & -2\bar{w}_2w_3 & 1 - 2|w_3|^2 \end{bmatrix}$$

For the particular case

$$w = \left[ \frac{1}{3}, \frac{2}{3}, \frac{2}{3} \right]^T$$

we have

$$U = \frac{1}{9} \begin{bmatrix} 7 & -4 & -4 \\ -4 & 1 & -8 \\ -4 & -8 & 1 \end{bmatrix}$$

We first prove $U$ is Hermitian and orthogonal. To show it is Hermitian,

$$U^* = (I - 2ww^*)^* = I^* - 2(ww^*)^*$$

$$= I - 2(w^*)^*w^* = I - 2ww^* = U$$

To show it is orthogonal,

$$U^*U = U^2 = (I - 2ww^*)^2$$

$$= I - 4ww^* + 4(ww^*)(ww^*)$$

$$= I$$

since using the associative law and $w^*w = 1$ implies

$$(ww^*)(ww^*) = w(w^*w)w^* = ww^*$$

The matrix $U$ of the preceding example illustrates these properties. In Problem 12, we give a geometric meaning to the linear function $T(x) = Ux$ for $U$ a Householder matrix.

We will usually use vectors $w$ with leading zero components:

$$w = [0,\dots, 0, w_r,\dots, w_n]^T = [0_{r-1}, \hat{w}^T]^T \qquad (9.3.2)$$

with $\hat{w} \in \mathbf{C}^{n-r+1}$. Then

$$U = \begin{bmatrix} I_{r-1} & 0 \\ 0 & I_{n-r+1} - 2\hat{w}\hat{w}^* \end{bmatrix} \qquad (9.3.3)$$

Premultiplication of a matrix $A$ by this $U$ will leave the first $r - 1$ rows of $A$ unchanged, and postmultiplication of $A$ will leave its first $r - 1$ columns unchanged. For the remainder of this section we assume all matrices and vectors are real, in order to avoid having to deal with possible complex values for $w$.

The Householder matrices are used to transform a nonzero vector into a new vector containing mainly zeros. Let $b \neq 0$ be given, $b \in \mathbf{R}^n$, and suppose we want to produce $U$ of form (9.3.1) such that $Ub$ contains zeros in positions $r + 1$ through $n$, for some given $r \geq 1$. Choose $w$ as in (9.3.2). Then the first $r - 1$ elements of $b$ and $Ub$ are the same.

To simplify the later work, write $m = n - r + 1$,

$$w = \begin{bmatrix} 0_{r-1} \\ v \end{bmatrix} \qquad b = \begin{bmatrix} c \\ d \end{bmatrix}$$

with $c \in \mathbf{R}^{r-1}$, $v, d \in \mathbf{R}^m$. Then our restriction on the form of $Ub$ requires the first $r - 1$ components of $Ub$ to be $c$, and

$$(I - 2vv^T)d = [\alpha, 0, \dots, 0]^T \qquad \|v\|_2 = 1 \tag{9.3.4}$$

for some $\alpha$. Since $I - 2vv^T$ is orthogonal, the length of $d$ is preserved (Problem 13 of Chapter 7); and thus

$$|\alpha| = \|d\|_2 \equiv S$$

$$\alpha = \pm S = \pm \sqrt{d_1^2 + \cdots + d_m^2} \tag{9.3.5}$$

Define

$$p = v^T d$$

From (9.3.4),

$$d - 2pv = [\alpha, 0, \dots, 0]^T \tag{9.3.6}$$

Multiplication by $v^T$ and use of $\|v\|_2 = 1$ implies

$$p = -\alpha v_1 \tag{9.3.7}$$

Substituting this into the first component of (9.3.6) gives

$$d_1 + 2\alpha v_1^2 = \alpha$$

$$v_1^2 = \frac{1}{2}\left[1 - \frac{d_1}{\alpha}\right] \tag{9.3.8}$$

Choose the sign of $\alpha$ in (9.3.5) by

$$\text{sign}(\alpha) = -\text{sign}(d_1) \tag{9.3.9}$$

This choice maximizes $v_1^2$, and it avoids any possible loss of significance errors in the calculation of $v_1$. The sign for $v_1$ is irrelevant. Having $v_1$, obtain $p$ from (9.3.7). Return to (9.3.6), and then using components 2 through $m$,

$$v_j = \frac{d_j}{2p} \qquad j = 2, 3, \ldots, m \qquad (9.3.10)$$

The statements (9.3.5), (9.3.7)–(9.3.9) completely define $v$, and thus $w$ and $U$. The operation count is $2m + 2$ multiplications and divisions, and two square roots. The square root defining $v_1$ can be avoided in practice, because it will disappear when the matrix $ww^T$ is formed. A sequence of such transformations of vectors $b$ will be used to systematically reduce matrices to simpler forms.

*Example*   Consider the given vector

$$b = [2, 2, 1]^T$$

We calculate a matrix $U$ for which $Ub$ will have zeros in its last two positions. To help in following the construction, some of the intermediate calculations are listed. Note that $w = v$ and $b = d$ for this case. Then

$$\alpha = -3 \qquad v_1 = \sqrt{\frac{5}{6}} \qquad p = \sqrt{\frac{15}{2}}$$

$$v_2 = \frac{2}{\sqrt{30}} \qquad v_3 = \frac{1}{\sqrt{30}}$$

The matrix $U$ is given by

$$U = \begin{bmatrix} -\dfrac{2}{3} & -\dfrac{2}{3} & -\dfrac{1}{3} \\[2mm] -\dfrac{2}{3} & \dfrac{11}{15} & -\dfrac{2}{15} \\[2mm] -\dfrac{1}{3} & -\dfrac{2}{15} & \dfrac{14}{15} \end{bmatrix}$$

and

$$Ub = [-3, 0, 0]^T$$

**The QR factorization of a matrix**   Given a real matrix $A$, we show there is an orthogonal matrix $Q$ and an upper triangular matrix $R$ for which

$$A = QR \qquad (9.3.11)$$

Let

$$P_r = I - 2w^{(r)}w^{(r)T} \qquad r = 1, \ldots, n - 1 \qquad (9.3.12)$$

with $w^{(r)}$ as in (9.3.2) with $r - 1$ leading zeros. Writing $A$ in terms of its columns $A_{*1}, \ldots, A_{*n}$, we have

$$P_1 A = [P_1 A_{*1}, \ldots, P_1 A_{*n}]$$

Pick $P_1$ and $w^{(1)}$ using the preceding construction (9.3.5)–(9.3.10) with $b = A_{*1}$. Then $P_1 A$ contains zeros below the diagonal in its first column.

Choose $P_2$ similarly, so that $P_2 P_1 A$ will contain zeros in its second column below the diagonal. First note that because $w^{(2)}$ contains a zero in position 1, and because $P_1 A$ is zero in the first column below position 1, the products $P_2 P_1 A$ and $P_1 A$ contain the same elements in row one and column one. Now choose $P_2$ and $w^{(2)}$ as before in (9.3.5)–(9.3.10), with $b$ equal to the second column of $P_1 A$.

By carrying this out with each column of $A$, we obtain an upper triangular matrix

$$R \equiv P_{n-1} \cdots P_1 A \tag{9.3.13}$$

If at step $r$ of the construction, all elements below the diagonal of column $r$ are zero, then just choose $P_r = I$ and go onto the next step. To complete the construction, define

$$Q^T = P_{n-1} \cdots P_1 \tag{9.3.14}$$

which is orthogonal. Then $A = QR$, as desired.

***Example*** Consider

$$A = \begin{bmatrix} 4 & 1 & 1 \\ 1 & 4 & 1 \\ 1 & 1 & 4 \end{bmatrix}$$

Then

$$w^{(1)} = [.985599, .119573, .119573]^T$$

$$A_2 = P_1 A = \begin{bmatrix} -4.24264 & -2.12132 & -2.12132 \\ 0 & 3.62132 & .621321 \\ 0 & .621321 & 3.62132 \end{bmatrix}$$

$$w^{(2)} = [0, .996393, .0848572]^T$$

$$R = P_2 A_2 = \begin{bmatrix} -4.24264 & -2.12132 & -2.12132 \\ 0 & -3.67423 & -1.22475 \\ 0 & 0 & 3.46410 \end{bmatrix}.$$

For the factorization $A = QR$, evaluate $Q = P_1 P_2$. But in most applications, it would be inefficient to explicitly produce $Q$. We comment further on this shortly.

Since $Q$ orthogonal implies $\det(Q) = \pm 1$, we have

$$|\det(A)| = |\det(Q) \det(R)| = |\det(R)| = 53.9999$$

This number is consistent with the fact that the eigenvalues of $A$ are $\lambda = 3, 3, 6$, and their product is $\det(A) = 54$.

**Discussion of the QR factorization**    It is useful to know to what extent the factorization $A = QR$ is unique. For $A$ nonsingular, suppose

$$A = Q_1 R_1 = Q_2 R_2 \qquad (9.3.15)$$

Then $R_1$ and $R_2$ must also be nonsingular, and

$$Q_2^T Q_1 = R_2 R_1^{-1}$$

The inverse of an upper triangular matrix is upper triangular, and the product of two upper triangular matrices is upper triangular. Thus $R_2 R_1^{-1}$ is upper triangular. Also, the product of two orthogonal matrices is orthogonal; thus, the product $Q_2^T Q_1$ is orthogonal. This says $R_2 R_1^{-1}$ is orthogonal. But it is not hard to show that the only upper triangular orthogonal matrices are the diagonal matrices. For some diagonal matrix $D$,

$$R_2 R_1^{-1} = D$$

Since $R_2 R_1^{-1}$ is orthogonal,

$$D^2 = I$$

Since we are only dealing with real matrices, $D$ has diagonal elements equal to $+1$ or $-1$. Combining these results,

$$Q_2 = Q_1 D \qquad R_2 = D R_1 \qquad (9.3.16)$$

This says the signs of the diagonal elements of $R$ in $A = QR$ can be chosen arbitrarily, but then the rest of the decomposition is uniquely determined.

Another practical matter is deciding how to evaluate the matrix $R$ of (9.3.13). Let

$$A_r = P_r A_{r-1} = \left[ I - 2w^{(r)} w^{(r)T} \right] A_{r-1} \qquad r = 1, 2, \ldots, n-1 \quad (9.3.17)$$

with $A_0 = A$, $A_{n-1} = R$. If we calculate $P_r$ and then multiply it times $A_{r-1}$ to form $A_r$, the number of multiplications will be

$$(n - r + 1)^3 + \frac{1}{2}(n - r + 2)(n - r + 1)$$

There is a much more efficient method for calculating $A_r$. Rewrite (9.3.17) as

$$A_r = A_{r-1} - 2w^{(r)} \left[ w^{(r)T} A_{r-1} \right] \qquad (9.3.18)$$

First calculate $w^{(r)T} A_{r-1}$, and then calculate $w^{(r)}[w^{(r)T} A_{r-1}]$ and $A_r$. This requires about

$$2(n - r)(n - r + 1) + (n - r + 1) \qquad (9.3.19)$$

multiplications, which shows (9.3.18) is a preferable way to evaluate each $A_r$ and finally $R = A_{n-1}$. This does not include the cost of obtaining $w^{(r)}$, which was discussed earlier, following (9.3.10).

If it is necessary to store the matrices $P_1, \ldots, P_{n-1}$ for later use, just store each column $w^{(r)}$, $r = 1, \ldots, n - 1$. Save the first nonzero element of $w^{(r)}$, the one in position $r$, in a special storage location, and save the remaining nonzero elements of $w^{(r)}$, those in positions $r + 1$ through $n$, in column $r$ of the matrix $A_r$ and $R$, below the diagonal. The matrix $Q$ of (9.3.14) could be produced explicitly. But as the construction (9.3.18) shows, we do not need $Q$ explicitly in order to multiply it times some other matrix.

The main use of the $QR$ factorization of $A$ will be in defining the $QR$ method for calculating the eigenvalues of $A$, which is presented in Section 9.5. The factorization can also be used to solve a linear system of equations $Ax = b$. The factorization leads directly to the equivalent system $Rx = Q^Tb$, and very little error is introduced because $Q$ is orthogonal. The system $Rx = Q^Tb$ is upper triangular, and it can be solved in a stable manner using back substitution. For $A$ an ill-conditioned matrix, this may be a superior way to solve the linear system $Ax = b$. For a discussion of the errors involved in obtaining and using the $QR$ factorization and for a comparison of it and Gaussian elimination for solving $Ax = b$, see Wilkinson (1965, pp. 236, 244–249). We pursue this topic further in Section 9.7, when we discuss the least squares solution of overdetermined linear systems.

**The transformation of a symmetric matrix to tridiagonal form**    Let $A$ be a real symmetric matrix. To find the eigenvalues of $A$, it is usually first reduced to tridiagonal form by orthogonal similarity transformations. The eigenvalues of the tridiagonal matrix are then calculated using the theory of Sturm sequences, presented in Section 9.4, or the $QR$ method, presented in Section 9.5. For the orthogonal matrices, we use the Householder matrices of (9.3.3).

Let

$$P_r = I - 2w^{(r+1)}w^{(r+1)T} \qquad r = 1, \ldots, n - 2 \qquad (9.3.20)$$

with $w^{(r+1)}$ defined as in (9.3.2):

$$w^{(r+1)} = [0, \ldots, 0, w_{r+1}, \ldots, w_n]^T$$

[Note the change in notation from that of the $P_r$ of (9.3.12) used in defining the $QR$ factorization.] The matrix

$$A_2 = P_1^T A P_1 = P_1 A P_1$$

is similar to $A$, the element $a_{11}$ is unchanged, and $A_2$ will be symmetric. Produce $w^{(2)}$ and $P_1$ to obtain the form

$$P_1 A_{*1} = [a_{11}, \hat{a}_{21}, 0, \ldots, 0]^T$$

for some $\hat{a}_{21}$. The vector $A_{*1}$ is the first column of $A$. Use (9.3.5)–(9.3.10) with

$m = n - 1$ and

$$d = [a_{21}, a_{31}, \ldots, a_{n1}]^T$$

For example, from (9.3.8),

$$w_2^2 = v_1^2 = \frac{1}{2}\left[1 - \frac{a_{21}}{\alpha}\right]$$

$$\alpha = -\operatorname{sign}(a_{21})\sqrt{a_{21}^2 + \cdots + a_{n1}^2}$$

Having obtained $P_1$ and $P_1 A$, postmultiplication by $P_1$ will not change the first column of $P_1 A$. (This should be checked by the reader.) The symmetry of $A_2$ follows from

$$A_2^T = (P_1 A P_1)^T = P_1^T A^T P_1^T = P_1 A P_1 = A_2$$

Since $A_2$ is symmetric, the construction on the first column of $A$ will imply that $A_2$ has zeros in positions 3 through $n$ of both the first row and column.

Continue this process, letting

$$A_{r+1} = P_r^T A_r P_r \qquad r = 1, 2, \ldots, n - 2 \tag{9.3.21}$$

with $A_1 = A$. Pick $P_r$ to introduce zeros into positions $r + 2$ through $n$ of column $r$. Columns 1 through $r - 1$ will remain undisturbed in calculating $P_r A_{r-1}$, due to the special form of $P_r$. Pick the vector $w^{(r+1)}$ in analogy with the preceding description for $w^{(2)}$.

The final matrix $T \equiv A_{n-1}$ is tridiagonal and symmetric.

$$T = \begin{bmatrix} \alpha_1 & \beta_1 & 0 & & \cdots & & 0 \\ \beta_1 & \alpha_2 & \beta_2 & & & & \vdots \\ 0 & \beta_2 & \alpha_3 & & & & \\ \vdots & & & \ddots & & & \\ & & & & \alpha_{n-1} & \beta_{n-1} \\ 0 & & \cdots & & \beta_{n-1} & \alpha_n \end{bmatrix} \tag{9.3.22}$$

This will be a much more convenient form for the calculation of the eigenvalues of $A$, and the eigenvectors of $A$ can easily be obtained from those of $T$.

$A$ is related to $A$ by

$$T = Q^T A Q \qquad Q = P_1 \cdots P_{n-2} \tag{9.3.23}$$

As before with the $QR$ factorization, we seldom produce $Q$ explicitly, preferring to work with the individual matrices $P_r$ in analogy with (9.3.18). For an eigenvector $x$ of $A$, say $Ax = \lambda x$, we have

$$Tz = \lambda z \qquad x = Qz \tag{9.3.24}$$

If we produce an orthonormal set of eigenvectors $\{z_i\}$ for $T$, then $\{Qz_i\}$ will be an orthonormal set of eigenvectors for $A$, since $Q$ preserves length and angles (see Problem 13 of Chapter 7).

**Example**  Let

$$A = \begin{bmatrix} 1 & 3 & 4 \\ 3 & 1 & 2 \\ 4 & 2 & 1 \end{bmatrix}$$

Then

$$w^{(2)} = \left[ 0, \frac{2}{\sqrt{5}}, \frac{1}{\sqrt{5}} \right]^T$$

$$P_1 = \begin{bmatrix} 1 & 0 & 0 \\ 0 & -\dfrac{3}{5} & -\dfrac{4}{5} \\ 0 & -\dfrac{4}{5} & \dfrac{3}{5} \end{bmatrix}$$

$$T = P_1^T A P_1 = \begin{bmatrix} 1 & -5 & 0 \\ -5 & \dfrac{73}{25} & \dfrac{14}{25} \\ 0 & \dfrac{14}{25} & -\dfrac{23}{25} \end{bmatrix}$$

For an error analysis of this reduction to tridiagonal form, we give some results from Wilkinson (1965). Let the computer arithmetic be binary floating-point with rounding, with $t$ binary digits in the mantissa. Furthermore, assume that all inner products

$$\sum_{i=1}^{m} a_i b_i$$

that occur in the calculation are accumulated in double precision, with rounding to single precision at the completion of the summation. These inner products occur in a variety of places in the computation of $T$ from $A$. Let $\hat{T}$ denote the actual symmetric tridiagonal matrix that is computed from $A$ using the preceding computer arithmetic. Let $\hat{P}_r$ denote the actual matrix produced in converting $A_{r-1}$ to $A_r$, let $P_r$ be the theoretically exact version of this matrix if no rounding errors occurred, and let $Q = P_1 \cdots P_{r-1}$ be the exact product of these $P_r$, an orthogonal matrix.

**Theorem 9.4**  Let $A$ be a real symmetric matrix of order $n$. Let $\hat{T}$ be the real symmetric tridiagonal matrix resulting from applying the Householder similarity transformations (9.3.20) to $A$, as in (9.3.21).

Assume the floating-point arithmetic used has the characteristics described in the preceding paragraph. Let $\{\lambda_i\}$ and $\{\tau_i\}$ be the eigenvalues of $A$ and $\hat{T}$, respectively, arranged in increasing order. Then

$$\left[\frac{\sum_{i=1}^{n} (\tau_i - \lambda_i)^2}{\sum_{i=1}^{n} \lambda_i^2}\right]^{1/2} \leq c_n 2^{-t} \qquad (9.3.25)$$

with

$$c_n = 25(n-1)\left[1 + (12.36)2^{-t}\right]^{2n-4}$$

For small and moderate values of $n$, $c_n \doteq 25(n-1)$.

**Proof**    From Wilkinson (1965, p. 161) using the Frobenius matrix norm $F$,

$$F(\hat{T} - Q^T A Q) \leq 2x(n-1)(1+x)^{2n-4}F(A) \qquad (9.3.26)$$

with $x = (12.36)2^{-t}$. From the Wielandt–Hoffman result (9.1.19) of Theorem 9.3, we have

$$\left[\sum_{i=1}^{n} (\tau_i - \lambda_i)^2\right]^{1/2} \leq F(\hat{T} - Q^T A Q) \qquad (9.3.27)$$

since $A$ and $Q^T A Q$ have the same eigenvalues. And from Problem 28(b) of Chapter 7,

$$F(A) = \left[\sum_{i=1}^{n} \lambda_i^2\right]^{1/2}$$

Combining these results yields (9.3.25).                                         ■

For a further discussion of the error, including the case in which inner products are not accumulated in double precision, see Wilkinson (1965, pp. 297–299). The result (9.3.25) shows that the reduction to tridiagonal form is an extremely stable operation, with little new error introduced for the eigenvalues.

**Planar rotation orthogonal matrices**    There are other classes of orthogonal matrices that can be used in place of the Householder matrices. The principal class is the set of plane rotations, which can be given the geometric interpretation of rotating a pair of coordinate axes through a given angle $\theta$ in the plane of the axes. For integers $k, l,\ 1 \leq k < l \leq n$, define the $n \times n$ orthogonal matrix $R^{(k,l)}$ by altering four elements of the identity matrix $I_n$. For any real number $\theta$, define the elements of $R^{(k,l)}$ by

$$R_{i,j}^{(k,l)} = \begin{cases} \cos\theta & (i,j) = (k,k) \quad \text{or} \quad (l,l) \\ \sin\theta & (i,j) = (k,l) \\ -\sin\theta & (i,j) = (l,k) \\ (I_n)_{ij} & \text{all other } (i,j) \end{cases} \tag{9.3.28}$$

for $1 \le i,\ j \le n$.

***Example***    For $n = 3$,

$$R^{(1,3)} = \begin{bmatrix} \cos\theta & 0 & \sin\theta \\ 0 & 1 & 0 \\ -\sin\theta & 0 & \cos\theta \end{bmatrix}$$

As a particular case, take $\theta = \pi/4$. Then

$$R^{(1,3)} = \begin{bmatrix} \dfrac{1}{\sqrt{2}} & 0 & \dfrac{1}{\sqrt{2}} \\ 0 & 1 & 0 \\ -\dfrac{1}{\sqrt{2}} & 0 & \dfrac{1}{\sqrt{2}} \end{bmatrix}$$

The plane rotations $R^{(k,l)}$ can be used to accomplish the same reductions for which the Householder matrices are used. In most situations, the Householder matrices are more efficient, but the plane rotations are more efficient for part of the $QR$ method of Section 9.5. The idea of solving the symmetric matrix eigenvalue problem by first reducing it to tridiagonal form is due to W. Givens in 1954. He also proposed the use of the techniques of the next section for the calculation of the eigenvalues of the tridiagonal matrix. Givens used the plane rotations $R^{(k,l)}$, and the Householder matrices were introduced in 1958 by A. Householder. For additional discussion of rotation matrices and their properties and uses, see Golub and Van Loan (1983, sec. 3.4), Parlett (1980, sec. 6.4), Wilkinson (1965, p. 131), and Problems 15 and 17(b).

## 9.4   The Eigenvalues of a Symmetric Tridiagonal Matrix

Let $T$ be a real symmetric tridiagonal matrix of order $n$, as in (9.3.22). We compute the characteristic polynomial of $T$ and use it to calculate the eigenvalues of $T$.

To compute

$$f_n(\lambda) \equiv \det(T - \lambda I) \tag{9.4.1}$$

introduce the sequence

$$f_k(\lambda) = \det \begin{bmatrix} \alpha_1 - \lambda & \beta_1 & 0 & \cdots & & 0 \\ \beta_1 & \alpha_2 - \lambda & \beta_2 & & & \vdots \\ 0 & & \ddots & & & \\ \vdots & & & & \beta_{k-1} \\ 0 & & \cdots & & \beta_{k-1} & \alpha_k - \lambda \end{bmatrix} \qquad (9.4.2)$$

for $1 \le k \le n$, and $f_0(\lambda) \equiv 1$. By direct evaluation,

$$f_1(\lambda) = \alpha_1 - \lambda$$

$$f_2(\lambda) = (\alpha_2 - \lambda)(\alpha_1 - \lambda) - \beta_1^2$$

$$= (\alpha_2 - \lambda)f_1(\lambda) - \beta_1^2 f_0(\lambda)$$

The formula for $f_2(\lambda)$ illustrates the general triple recursion relation that the sequence $\{f_k(\lambda)\}$ satisfies:

$$f_k(\lambda) = (\alpha_k - \lambda)f_{k-1}(\lambda) - \beta_{k-1}^2 f_{k-2}(\lambda) \qquad 2 \le k \le n \qquad (9.4.3)$$

To prove this, expand the determinant (9.4.2) in its last row using minors and the result will follow easily. This method for evaluating $f_n(\lambda)$ will require $2n - 3$ multiplications, once the coefficients $\{\beta_k^2\}$ have been evaluated.

**Example**    Let

$$T = \begin{bmatrix} 2 & 1 & 0 & 0 & 0 & 0 \\ 1 & 2 & 1 & 0 & 0 & 0 \\ 0 & 1 & 2 & 1 & 0 & 0 \\ 0 & 0 & 1 & 2 & 1 & 0 \\ 0 & 0 & 0 & 1 & 2 & 1 \\ 0 & 0 & 0 & 0 & 1 & 2 \end{bmatrix} \qquad (9.4.4)$$

Then

$$f_0(\lambda) = 1 \qquad f_1(\lambda) = 2 - \lambda$$

$$f_j(\lambda) = (2 - \lambda)f_{j-1}(\lambda) - f_{j-2}(\lambda) \qquad j = 2, 3, 4, 5, 6 \qquad (9.4.5)$$

Without the triple recursion relation (9.4.5), the evaluation of $f_6(\lambda)$ would be much more complicated.

At this point, we might consider the problem as solved since $f_n(\lambda)$ is a polynomial and there are many polynomial rootfinding methods. Or we might use a more general method, such as the secant method or Brent's method, both described in Chapter 2. But the sequence $\{f_k(\lambda) | 0 \le k \le n\}$ has special properties that make it a *Sturm sequence*, and these properties make it comparatively easy to isolate the eigenvalues of $T$. Once the eigenvalues have been isolated, a method such as Brent's method [see Section 2.8] can be used to rapidly calculate the roots. The theory of Sturm sequences is discussed in Henrici (1974, p. 444), but we only consider the special case of $\{ f_k(\lambda) \}$.

Before stating the consequences of the Sturm theory for $\{ f_k(\lambda) \}$, we consider what happens when some $\beta_l = 0$. Then the eigenvalue problem can be broken apart into two smaller eigenvalue problems of orders $l$ and $n - l$. As an example, consider

$$
T = \begin{bmatrix}
\alpha_1 & \beta_1 & 0 & 0 & 0 \\
\beta_1 & \alpha_2 & 0 & 0 & 0 \\
0 & 0 & \alpha_3 & \beta_3 & 0 \\
0 & 0 & \beta_3 & \alpha_4 & \beta_4 \\
0 & 0 & 0 & \beta_4 & \alpha_5
\end{bmatrix}
$$

Define $T_1$ and $T_2$ as the two blocks along the diagonal, of orders 2 and 3, respectively, and then

$$
T = \begin{bmatrix} T_1 & 0 \\ 0 & T_2 \end{bmatrix}
$$

From this,

$$
\det [T - \lambda I_5] = \det [T_1 - \lambda I_2] \det [T_2 - \lambda I_3]
$$

and we can find the eigenvalues of $T$ by finding those of $T_1$ and $T_2$. The eigenvector problem also can be solved in the same way. For example, if $T_1 \hat{x} = \lambda x$, with $\hat{x} \neq 0$ in $\mathbf{R}^2$, define

$$
x = [\hat{x}^T, 0, 0, 0]^T
$$

Then $Tx = \lambda x$. This construction can be used to calculate a complete set of eigenvectors for $T$ from those for $T_1$ and $T_2$. For the remainder of the section, we assume that all $\beta_l \neq 0$ in the matrix $T$. Under this assumption, all eigenvalues of $T$ will be simple roots of $f_n(\lambda)$.

**The Sturm sequence property of $\{ f_k(\lambda) \}$**   The sequences $\{ f_k(a) \}$ and $\{ f_k(b) \}$ can be used to determine the number of roots of $f_n(\lambda)$ that are contained in $[a, b]$. To do this, introduce the following integer-valued function $s(\lambda)$. Define $s(\lambda)$ to be the number of agreements in sign of consecutive members of the sequence $\{ f_k(\lambda) \}$, and if the value of some member $f_j(\lambda) = 0$, let its sign be chosen opposite to that of $f_{j-1}(\lambda)$. It can be shown that $f_j(\lambda) = 0$ implies $f_{j-1}(\lambda) \neq 0$.

**Example**   Consider the sequence $f_0(\lambda), \ldots, f_6(\lambda)$ given in (9.4.5) of the last example. For $\lambda = 3$,

$$
(f_0(\lambda), \ldots, f_6(\lambda)) = (1, -1, 0, -1, 0, 1)
$$

The corresponding sequence of signs is

$$(+, -, +, +, -, +, +)$$

and $s(3) = 2$.

We now state the basic result used in computing the roots of $f_n(\lambda)$ and thus the eigenvalues of $T$. The proof follows from the general theory given in Henrici (1974).

**Theorem 9.5**   Let $T$ be a real symmetric tridiagonal matrix of order $n$, as given in (9.3.22). Let the sequence $\{ f_k(\lambda) | 0 \leq k \leq n \}$ be defined as in (9.4.2), and assume all $\beta_l \neq 0$, $l = 1, \ldots, n - 1$. Then the number of roots of $f_n(\lambda)$ that are greater than $\lambda = a$ is given by $s(a)$, which is defined in the preceding paragraph. For $a < b$, the number of roots in the interval $a < \lambda \leq b$ is given by $s(a) - s(b)$.

**Calculation of the eigenvalues**   Theorem 9.5 will be the basic tool in locating and separating the roots of $f_n(\lambda)$. To begin, calculate an interval that contains the roots. Using the Gerschgorin circle Theorem 9.1, all eigenvalues are contained in the interval $[a, b]$, with

$$a = \operatorname*{Min}_{1 \leq i \leq n} \{ \alpha_i - |\beta_i| - |\beta_{i-1}| \}$$

$$b = \operatorname*{Max}_{1 \leq i \leq n} \{ \alpha_i + |\beta_i| + |\beta_{i-1}| \}$$

where $\beta_0 = \beta_n = 0$.

We use the bisection method on $[a, b]$ to divide it into smaller subintervals. Theorem 9.5 is used to determine how many roots are contained in a subinterval, and we seek to obtain subintervals that will each contain one root. If some eigenvalues are nearly equal, then we continue subdividing until the root is found with sufficient accuracy. Once a subinterval is known to contain a single root, we can switch to a more rapidly convergent method.

**Example**   Consider further the example (9.4.4). By the Gerschgorin Theorem 9.1, all eigenvalues lie in $[0, 4]$. And it is easily checked that neither $\lambda = 0$ nor $\lambda = 4$ is an eigenvalue. A systematic bisection process was carried out on $[0, 4]$ to separate the six roots of $f_6(\lambda)$ into six separate subintervals. The results are shown in Table 9.2 in the order they were calculated. The roots are labeled as follows:

$$0 \leq \lambda_6 \leq \lambda_5 \leq \cdots \leq \lambda_1 \leq 4$$

The roots can be found by continuing with the bisection method, although Theorem 9.5 is no longer needed. But it would be better to use some other rootfinding method.

Although all roots of a tridiagonal matrix may be found by this technique, it is generally faster in that case to use the $QR$ algorithm of the next section. With

**Table 9.2    Example of use of Theorem 9.5**

| $\lambda$ | $f_6(\lambda)$ | $s(\lambda)$ | Comment |
|---|---|---|---|
| 0.0 | 7.0 | 6 | $\lambda_6 > 0$ |
| 4.0 | 7.0 | 0 | $\lambda_1 < 4$ |
| 2.0 | $-1.0$ | 3 | $\lambda_4 < 2 < \lambda_3$ |
| 1.0 | 1.0 | 4 | $\lambda_5 < 1 < \lambda_4 < 2$ |
| .5 | $-1.421875$ | 5 | $0 < \lambda_6 < 0.5 < \lambda_5 < 1$ |
| 3.0 | 1.0 | 2 | $2 < \lambda_3 < 3 < \lambda_2$ |
| 3.5 | $-1.421875$ | 1 | $3 < \lambda_2 < 3.5 < \lambda_1 < 4$ |

large matrices, we usually do not want all of the roots, in which case the methods of this section are preferable. If we want only certain specific roots, for example, the five largest or all roots in a given interval or all roots in $[1, 3]$, then it is easy to locate them using Theorem 9.5.

## 9.5   The $QR$ Method

At the present time, the $QR$ method is the most efficient and widely used general method for the calculation of all of the eigenvalues of a matrix. The method was first published in 1961 by J. G. F. Francis and it has since been the subject of intense investigation. The $QR$ method is quite complex in both its theory and application, and we are able to give only an introduction to the theory of the method. For actual algorithms for both symmetric and nonsymmetric matrices, refer to those in EISPACK and Wilkinson and Reinsch (1971).

Given a matrix $A$, there is a factorization

$$A = QR$$

with $R$ upper triangular and $Q$ orthogonal. With $A$ real, both $Q$ and $R$ can be chosen real; their construction is given in Section 9.3. We assume $A$ is real throughout this section. Let $A_1 = A$, and define a sequence of matrices $A_m$, $Q_m$, and $R_m$ by

$$A_m = Q_m R_m \qquad A_{m+1} = R_m Q_m \qquad m = 1, 2, \ldots \qquad (9.5.1)$$

Since $R_m = Q_m^T A_m$, we have

$$A_{m+1} = Q_m^T A_m Q_m \qquad (9.5.2)$$

The matrix $A_{m+1}$ is orthogonally similar to $A_m$, and thus by induction, to $A_1$.

The sequence $\{A_m\}$ will converge to either a triangular matrix with the eigenvalues of $A$ on its diagonal or to a near-triangular matrix from which the eigenvalues can be easily calculated. In this form the convergence is usually slow, and a technique known as *shifting* is used to accelerate the convergence. The technique of shifting will be introduced and illustrated later in the section.

***Example***   Let

$$A_1 = \begin{bmatrix} 2 & 1 & 0 \\ 1 & 3 & 1 \\ 0 & 1 & 4 \end{bmatrix} \tag{9.5.3}$$

The eigenvalues are

$$\lambda_1 = 3 + \sqrt{3} \doteq 4.7321 \qquad \lambda_2 = 3.0 \qquad \lambda_3 = 3 - \sqrt{3} \doteq 1.2679$$

The iterates $A_m$ do not converge rapidly, and only a few are given to indicate the qualitative behavior of the convergence:

$$A_2 = \begin{bmatrix} 3.0000 & 1.0954 & 0 \\ 1.0954 & 3.0000 & -1.3416 \\ 0 & -1.3416 & 3.0000. \end{bmatrix} \qquad A_3 = \begin{bmatrix} 3.7059 & .9558 & 0 \\ .9558 & 3.5214 & .9738 \\ 0 & .9738 & 1.7727 \end{bmatrix}$$

$$A_7 = \begin{bmatrix} 4.6792 & .2979 & 0 \\ .2979 & 3.0524 & .0274 \\ 0 & .0274 & 1.2684 \end{bmatrix} \qquad A_8 = \begin{bmatrix} 4.7104 & .1924 & 0 \\ .1924 & 3.0216 & -.0115 \\ 0 & -.0115 & 1.2680 \end{bmatrix}$$

$$A_9 = \begin{bmatrix} 4.7233 & .1229 & 0 \\ .1229 & 3.0087 & .0048 \\ 0 & .0048 & 1.2680 \end{bmatrix} \qquad A_{10} = \begin{bmatrix} 4.7285 & .0781 & 0 \\ .0781 & 3.0035 & -.0020 \\ 0 & -.0020 & 1.2680 \end{bmatrix}$$

The elements in the $(1, 2)$ position decrease geometrically with a ratio of about .64 per iterate, and those in the $(2, 3)$ position decrease with a ratio of about .42 per iterate. The value in the $(3, 3)$ position of $A_{15}$ will be 1.2679, which is correct to five places.

**The preliminary reduction of $A$ to simpler form**   The $QR$ method can be relatively expensive because the $QR$ factorization is time-consuming when repeated many times. To decrease the expense the matrix is prepared for the $QR$ method by reducing it to a simpler form, one for which the $QR$ factorization is much less expensive.

If $A$ is symmetric, it is reduced to a similar symmetric tridiagonal matrix exactly as described in Section 9.3. If $A$ is nonsymmetric, it is reduced to a similar *Hessenberg matrix*. A matrix $B$ is Hessenberg if

$$b_{ij} = 0 \qquad \text{for all } i > j + 1 \tag{9.5.4}$$

It is upper triangular except for a single nonzero subdiagonal. The matrix $A$ is reduced to Hessenberg form using the same algorithm as was used for reducing symmetric matrices to tridiagonal form.

With $A$ tridiagonal or Hessenberg, the Householder matrices of Section 9.3 take a simple form when calculating the $QR$ factorization. But generally the plane rotations (9.3.28) are used in place of the Householder matrices because they are more efficient to compute and apply in this situation. Having produced $A_1 = Q_1R_1$ and $A_2 = R_1Q_1$, we need to know that the form of $A_2$ is the same as that of $A_1$ in order to continue using the less expensive form of $QR$ factorization.

Suppose $A_1$ is in the Hessenberg form. From Section 9.3 the factorization $A_1 = Q_1R_1$ has the following value of $Q_1$:

$$Q_1 = H_1 \ldots H_{n-1} \tag{9.5.5}$$

with each $H_k$ a Householder matrix (9.3.12):

$$H_k = I - 2w^{(k)}w^{(k)T} \qquad 1 \le k \le n-1 \tag{9.5.6}$$

Because the matrix $A_1$ is of Hessenberg form, the vectors $w^{(k)}$ can be shown to have the special form

$$w_i^{(k)} = 0 \qquad \text{for } i < k \text{ and } i > k+1 \tag{9.5.7}$$

This can be shown from the equations for the components of $w^{(k)}$, and in particular (9.3.10). From (9.5.7), the matrix $H_k$ will differ from the identity in only the four elements in positions $(k, k)$, $(k, k+1)$, $(k+1, k)$, and $(k+1, k+1)$. And from this it is a fairly straightforward computation to show that $Q_1$ must be Hessenberg in form. Another necessary lemma is that the product of an upper triangular matrix and a Hessenberg matrix is again Hessenberg. Just multiply the two forms of matrices, observing the respective patterns of zeros, in order to prove this lemma. Combining these results, observing that $R_1$ is upper triangular, we have that $A_2 = R_1Q_1$ must be in Hessenberg form.

If $A_1$ is symmetric and tridiagonal, then it is trivially Hessenberg. From the preceding result, $A_2$ must also be Hessenberg. But $A_2$ is symmetric, since

$$A_2^T = \left(Q_1^TA_1Q_1\right)^T = Q_1^TA_1^TQ_1 = Q_1^TA_1Q_1 = A_2$$

Since any symmetric Hessenberg matrix is tridiagonal, we have shown that $A_2$ is tridiagonal. Note that the iterates in the example (9.5.3) illustrate this result.

**Convergence of the $QR$ method**    Convergence results for the $QR$ method can be found in Golub and Van Loan (1983, secs. 7.5 and 8.2), Parlett (1968), (1980, chap. 8), and Wilkinson (1965, chap. 8). The following theorem is taken from the latter reference.

**Theorem 9.6**    Let $A$ be a real matrix of order $n$, and let its eigenvalues $\{\lambda_i\}$ satisfy

$$|\lambda_1| > |\lambda_2| \cdots > |\lambda_n| > 0 \tag{9.5.8}$$

Then the iterates $R_m$ of the $QR$ method, defined in (9.5.1), will converge to an upper triangular matrix $D$, which contains the eigenvalues $\{\lambda_i\}$ in the diagonal positions. If $A$ is symmetric, the sequence $\{A_m\}$ converges to a diagonal matrix. For the speed of convergence,

$$\|D - A_m\| \le c \, \underset{i}{\text{Max}} \left| \frac{\lambda_{i+1}}{\lambda_i} \right| \tag{9.5.9}$$

As an example of this error bound, consider the example (9.5.3). In it, the ratios of the successive eigenvalues are

$$\frac{\lambda_2}{\lambda_1} \doteq .63 \qquad \frac{\lambda_3}{\lambda_2} \doteq .42 \tag{9.5.10}$$

If any one of the off-diagonal elements of $A_m$ in the example is examined, it will be seen to decrease by one of the factors in (9.5.10).

For matrices whose eigenvalues do not satisfy (9.5.8), the iterates $A_m$ may not converge to a triangular matrix. For $A$ symmetric, the sequence $\{A_m\}$ will converge to a block diagonal matrix

$$A_m \to D = \begin{bmatrix} B_1 & & & 0 \\ & B_2 & & \\ & & \ddots & \\ 0 & & & B_r \end{bmatrix} \tag{9.5.11}$$

in which all blocks $B_i$ have order 1 or 2. Thus the eigenvalues of $A$ can be easily computed from those of $D$. If $A$ is real and nonsymmetric, the situation is more complicated, but acceptable. For a discussion, see Wilkinson (1965, chap. 8) and Parlett (1968).

To see that $\{A_m\}$ does not always converge to a diagonal matrix, consider the simple symmetric example

$$A = \begin{bmatrix} 0 & 1 \\ 1 & 0 \end{bmatrix}$$

Its eigenvalues are $\lambda = \pm 1$. Since $A$ is orthogonal, we have

$$A = Q_1 R_1 \quad \text{with} \quad Q_1 = A \quad R_1 = I$$

And thus

$$A_2 = R_1 Q_1 = A$$

and all iterates $A_m = A$. The sequence $\{A_m\}$ does not converge to a diagonal matrix.

**The QR method with shift**   The QR algorithm is generally applied with a shift of origin for the eigenvalues in order to increase the speed of convergence. For a sequence of constants $\{c_m\}$, define $A_1 = A$ and

$$A_m - c_m I = Q_m R_m$$

$$A_{m+1} = c_m I + R_m Q_m \qquad m = 1, 2, \dots \tag{9.5.12}$$

The matrices $A_m$ are similar to $A_1$, since

$$R_m = Q_m^T(A_m - c_m I)$$

$$A_{m+1} = c_m I + Q_m^T(A_m - c_m I)Q_m$$

$$= c_m I + Q_m^T A_m Q_m - c_m I$$

$$A_{m+1} = Q_m^T A_m Q_m \qquad m \geq 1 \tag{9.5.13}$$

The eigenvalues of $A_{m+1}$ are the same as those of $A_m$, and thence the same as those of $A$.

To be more specific on the choice of shifts $\{c_m\}$, we consider only a symmetric tridiagonal matrix $A$. For $A_m$, let

$$A_m = \begin{bmatrix} \alpha_1^{(m)} & \beta_1^{(m)} & 0 & \cdots & & 0 \\ \beta_1^{(m)} & \alpha_2^{(m)} & \beta_2^{(m)} & & & \vdots \\ 0 & & \ddots & & & \\ \vdots & & & & \beta_{n-1}^{(m)} \\ 0 & & \cdots & & \beta_{n-1}^{(m)} & \alpha_n^{(m)} \end{bmatrix} \tag{9.5.14}$$

There are two methods by which $\{c_m\}$ is chosen: (1) Let $c_m = \alpha_n^{(m)}$, and (2) let $c_m$ be the eigenvalue of

$$\begin{bmatrix} \alpha_{n-1}^{(m)} & \beta_{n-1}^{(m)} \\ \beta_{n-1}^{(m)} & \alpha_n^{(m)} \end{bmatrix} \tag{9.5.15}$$

which is closest to $\alpha_n^{(m)}$. The second strategy is preferred, but in either case the matrices $A_m$ converge to a block diagonal matrix in which the blocks have order 1 or 2, as in (9.5.11). It can be shown that either choice of $\{c_m\}$ ensures

$$\beta_{n-1}^{(m)}\beta_{n-2}^{(m)} \to 0 \qquad \text{as} \quad n \to \infty \tag{9.5.16}$$

generally at a much more rapid rate than with the original $QR$ method (9.5.1). From (9.5.29),

$$\|A_{m+1}\|_2 = \|Q_m^T A_m Q_m\|_2 = \|A_m\|_2$$

using the operator matrix norm (7.3.19) and Problem 27(c) of Chapter 7. The matrices $\{A_m\}$ are uniformly bounded, and consequently the same is true of their elements. From (9.5.16) and the uniform boundedness of $\{\beta_{n-1}^{(m)}\}$ and $\{\beta_{n-2}^{(m)}\}$, we have either $\beta_{n-1}^{(m)} \to 0$ or $\beta_{n-2}^{(m)} \to 0$ as $m \to \infty$. In the former case, $\alpha_n^{(m)}$ converges to an eigenvalue of $A$. And in the latter case, two eigenvalues can easily be extracted from the limit of the submatrix (9.5.15).

Once one or two eigenvalues have been obtained due to $\beta_{n-1}^{(m)}$ or $\beta_{n-2}^{(m)}$ being essentially zero, the matrix $A_m$ can be reduced in order by one or two rows, respectively. Following this, the $QR$ method with shift can be applied to the reduced matrix. The choice of shifts is designed to make the convergence to zero be more rapid for $\beta_{n-1}^{(m)}\beta_{n-2}^{(m)}$ than for the remaining off-diagonal elements of the matrix. In this way, the $QR$ method becomes a rapid general-purpose method, faster than any other method at the present time. For a proof of convergence of the $QR$ method with shift, see Wilkinson (1968). For a much more complete discussion of the $QR$ method, including the choice of a shift, see Parlett (1980, chap. 8).

**Example**   Use the previous example (9.5.3), and use the first method of choosing the shift, $c_m = \alpha_n^{(m)}$. The iterates are

$$A_1 = \begin{bmatrix} 2 & 1 & 0 \\ 1 & 3 & 1 \\ 0 & 1 & 4 \end{bmatrix} \qquad A_2 = \begin{bmatrix} 1.4000 & .4899 & 0 \\ .4899 & 3.2667 & .7454 \\ 0 & .7454 & 4.3333 \end{bmatrix}$$

$$A_3 = \begin{bmatrix} 1.2915 & .2017 & 0 \\ .2017 & 3.0202 & .2724 \\ 0 & .2724 & 4.6884 \end{bmatrix} \qquad A_4 = \begin{bmatrix} 1.2737 & .0993 & 0 \\ .0993 & 2.9943 & .0072 \\ 0 & .0072 & 4.7320 \end{bmatrix}$$

$$A_5 = \begin{bmatrix} 1.2694 & .0498 & 0 \\ .0498 & 2.9986 & 0 \\ 0 & 0 & 4.7321 \end{bmatrix}$$

The element $\beta_2^{(m)}$ converges to zero extremely rapidly, but the element $\beta_1^{(m)}$ converges to zero geometrically with a ratio of only about .5.

Mention should be made of the antecedent to the $QR$ method, motivating much of it. In 1958, H. Rutishauser introduced an $LR$ method based on the Gaussian elimination decomposition of a matrix into a lower triangular matrix times an upper triangular matrix. Define

$$A_m = L_m R_m \qquad A_{m+1} = R_m L_m = L_m^{-1} A_m L_m$$

with $L_m$ lower triangular, $R_m$ upper triangular. When applicable, this method will generally be more efficient than the $QR$ method. But the nonorthogonal similarity transformations can cause a deterioration of the conditioning of the eigenvalues of some nonsymmetric matrices. And generally it is a more complicated algorithm to implement in an automatic program. A complete discussion is given in Wilkinson (1965, chap. 8).

## 9.6   The Calculation of Eigenvectors and Inverse Iteration

The most powerful tool for the calculation of the eigenvectors of a matrix is inverse iteration, a method attributed to H. Wielandt in 1944. We first define and illustrate inverse iteration, and then comment more generally on the calculation of eigenvectors.

To simplify the analysis, let $A$ be a matrix whose Jordan canonical form is diagonal,

$$P^{-1}AP = \text{diag}[\lambda_1, \ldots, \lambda_n] \qquad (9.6.1)$$

Let the columns of $P$ be denoted by $x_1, \ldots, x_n$. Then

$$Ax_i = \lambda_i x_i \qquad i = 1, \ldots, n \qquad (9.6.2)$$

Without loss of generality, it can also be assumed that $\|x_i\|_\infty = 1$, for all $i$.

Let $\lambda$ be an approximation to a simple eigenvalue $\lambda_k$ of $A$. Given an initial $z^{(0)}$, define $\{w^{(m)}\}$ and $\{z^{(m)}\}$ by

$$(A - \lambda I)w^{(m+1)} = z^{(m)}, \qquad z^{(m+1)} = \frac{w^{(m+1)}}{\|w^{(m+1)}\|_\infty} \qquad m \geq 0. \quad (9.6.3)$$

This is essentially the power method, with $(A - \lambda I)^{-1}$ replacing $A$ in (9.2.2)–(9.2.3). The matrix $A - \lambda I$ is ill-conditioned from the viewpoint of the material in Section 8.4 of Chapter 8. But any resulting large perturbations in the solution will be rich in the eigenvector $x_k$ of the eigenvalue $\lambda_k - \lambda$ for $A - \lambda I$, and this is the vector we desire. For a further discussion of this source of instability in solving the linear system, see the material following formula (8.4.8) in Section 8.4. For the method (9.6.3) to work, we do not want $A - \lambda I$ to be singular. Thus $\lambda$ shouldn't be exactly $\lambda_k$, although it can be quite close, as a later example demonstrates.

For a more precise analysis, let $z^{(0)}$ be expanded in terms of the eigenvector basis of (9.6.2):

$$z^{(0)} = \sum_{i=1}^{n} \alpha_i x_i \qquad (9.6.4)$$

And assume $\alpha_k \neq 0$. In analogy with formula (9.2.4) for the power method, we can show

$$z^{(m)} = \frac{\sigma_m (A - \lambda I)^{-m} z^{(0)}}{\|(A - \lambda I)^{-m} z^{(0)}\|_\infty}, \qquad |\sigma_m| = 1 \qquad (9.6.5)$$

Using (9.6.4),

$$(A - \lambda I)^{-m} z^{(0)} = \sum_{i=1}^{n} \alpha_i \left[\frac{1}{\lambda_i - \lambda}\right]^m x_i \qquad (9.6.6)$$

Let $\lambda_k - \lambda = \epsilon$, and assume

$$|\lambda_i - \lambda| \geq c > 0 \qquad i = 1, \ldots, n \qquad i \neq k \qquad (9.6.7)$$

From (9.6.6) and (9.6.5),

$$z^{(m)} = \sigma_m \frac{x_k + \epsilon^m \sum_{i \neq k} \frac{\alpha_i}{\alpha_k} \left[ \frac{1}{\lambda_i - \lambda} \right]^m x_i}{\left\| x_k + \epsilon^m \sum_{i \neq k} \frac{\alpha_i}{\alpha_k} \left[ \frac{1}{\lambda_i - \lambda} \right]^m x_i \right\|_\infty} \tag{9.6.8}$$

with $|\sigma_m| = 1$. If $|\epsilon| < c$, then

$$\left\| \epsilon^m \sum_{i \neq k} \frac{\alpha_i}{\alpha_k} \left[ \frac{1}{\lambda_i - \lambda} \right]^m x_i \right\|_\infty \leq \left[ \frac{\epsilon}{c} \right]^m \sum_{i \neq k} \left| \frac{\alpha_i}{\alpha_k} \right| \tag{9.6.9}$$

This quantity goes to zero as $m \to \infty$. Combined with (9.6.8), this shows $z^{(m)}$ converges to a multiple of $x_k$ as $m \to \infty$. This convergence is linear, with a ratio of $|\epsilon/c|$ decrease in the error in each iterate. In practice $|\epsilon|$ is quite small and this will usually mean $|\epsilon/c|$ is also quite small, ensuring rapid convergence.

In implementing (9.6.3), begin by factoring $A - \lambda I$ using the $LU$ decomposition of Section 8.1 of Chapter 8. To simplify the notation, write

$$A - \lambda I = LU$$

in which pivoting is not involved. In practice, pivoting would be used. Solve for each iterate $z^{(m+1)}$ as follows:

$$Ly^{(m+1)} = z^{(m)} \qquad Uw^{(m+1)} = y^{(m+1)}$$

$$z^{(m+1)} = \frac{w^{(m+1)}}{\|w^{(m+1)}\|_\infty} \tag{9.6.10}$$

Since $A - \lambda I$ is nearly singular, the last diagonal element of $U$ will be nearly zero. If it is exactly zero, then change it to some small number or else change $\lambda$ very slightly and recalculate $L$ and $U$.

For the initial guess $z^{(0)}$, Wilkinson (1963, p. 147) suggests using

$$z^{(0)} = Le \qquad e = [1, 1, \ldots, 1]^T$$

Thus in (9.6.10),

$$y^{(1)} = e \qquad Uw^{(m+1)} = e \tag{9.6.11}$$

This choice is intended to ensure that $\alpha_k$ is neither nonzero nor small in (9.6.4). But even if it were small, the method would usually converge rapidly. For example, suppose that some or all of the values $\alpha_i/\alpha_k$ in (9.6.9) are about $10^4$. And suppose $|\epsilon/c| = 10^{-5}$, a realistic value for many cases. Then the bound in (9.6.9) becomes

$$(10^{-5})^m \cdot n \cdot 10^4$$

and this will decrease very rapidly as $m$ increases.

***Example***   Use the earlier matrix (9.5.8).

$$A = \begin{bmatrix} 2 & 1 & 0 \\ 1 & 3 & 1 \\ 0 & 1 & 4 \end{bmatrix} \qquad (9.6.12)$$

Let $\lambda = 1.2679 \doteq \lambda_3 = 3 - \sqrt{3}$, which is accurate to five places. This leads to

$$L = \begin{bmatrix} 1.0 & 0 & 0 \\ 1.3659 & 1.0 & 0 \\ 0 & 2.7310 & 1.0 \end{bmatrix} \qquad U = \begin{bmatrix} .7321 & 1.0 & 0 \\ 0 & .3662 & 1.0 \\ 0 & 0 & .0011 \end{bmatrix}$$

subject to the effects of rounding errors. Using $y^{(1)} = [1, 1, 1]^T$,

$$w^{(1)} = [3385.2, -2477.3, 908.20]^T$$

$$z^{(1)} = [1.0000, -.73180, .26828]^T$$

$$w^{(2)} = [20345, -14894, 5451.9]^T$$

$$z^{(2)} = [1.000, -.73207, .26797]^T \qquad (9.6.13)$$

and the vector of $z^{(3)} = z^{(2)}$. The true answer is

$$x_3 = \left[ 1, 1 - \sqrt{3}, 2 - \sqrt{3} \right]^T$$

$$\doteq [1.0000, -.73205, .26795]^T \qquad (9.6.14)$$

and $z^{(2)}$ equals $x_3$ to within the limits of rounding error accumulations.

**Eigenvectors for symmetric tridiagonal matrices**   Let $A$ be a real symmetric tridiagonal matrix of order $n$. As previously, we assume that some or all of its eigenvalues have been computed accurately. Inverse iteration is the preferred method of calculation for the eigenvectors, and it is quite easy to implement. For $\lambda$ an approximate eigenvalue of $A$, the calculation of the $LU$ decomposition is inexpensive in time and storage, even with pivoting. For example, see the material on tridiagonal systems in Section 8.3 of Chapter 8. The previous numerical example also illustrates the method for tridiagonal matrices.

Some error results are given as further justification for the use of inverse iteration. Suppose that the computer arithmetic is binary floating point with rounding and with $t$ digits in the mantissa. In Wilkinson (1963, pp. 143–147) it is shown that the computed solution $\hat{w}$ of

$$(A - \lambda I)w^{(m+1)} = z^{(m)}$$

is the exact solution of

$$(A - \lambda I + E)\hat{w} = z^{(m)} \qquad (9.6.15)$$

with

$$\|E\|_2 \le K\sqrt{n} \cdot 2^{-t} \qquad (9.6.16)$$

for some constant $K$ of order unity. This bound is of the size that would be expected from errors of the order of the rounding error.

If the solution $\hat{w}$ of (9.6.15) is quite large, then it will be a good approximation to an eigenvector of $A$. To prove this, we begin by introducing

$$\hat{z} = \frac{\hat{w}}{\|\hat{w}\|_2}$$

Then

$$(A - \lambda I + E)\hat{z} = \frac{z^{(m)}}{\|\hat{w}\|_2}$$

$$\eta \equiv (A - \lambda I)\hat{z} = -E\hat{z} + \frac{\hat{z}}{\|\hat{w}\|_2}$$

For the residual $\eta$,

$$\|\eta\|_2 \leq \|E\|_2 + \frac{1}{\|\hat{w}\|_2}$$

$$\leq K\sqrt{n} \cdot 2^{-t} + \frac{1}{\|\hat{w}\|_2} \qquad (9.6.17)$$

which is small if $\|\hat{w}\|_2$ is large. To prove that this implies $\hat{z}$ is close to an eigenvector of $A$, we let $\{x_i | i = 1, \ldots, n\}$ be an orthonormal set of eigenvectors. And assume

$$\lambda \doteq \lambda_k \qquad \hat{z} = \sum_{i=1}^{n} \alpha_i x_i \qquad \|\hat{z}\|_2^2 = \sum_{1}^{n} \alpha_i^2 = 1$$

with $\lambda_k$ an isolated eigenvalue of $A$. Also, suppose

$$|\lambda_i - \lambda| \geq c > 0 \qquad \text{all} \quad i \neq k$$

with

$$c \gg |\lambda_k - \lambda|$$

With these assumptions, we can now derive a bound for the error in $\hat{z}$. Expanding $\eta$ using the eigenvector basis:

$$\eta = Az - \lambda z = \sum_{1}^{n} \alpha_i (\lambda_i - \lambda) x_i$$

$$\|\eta\|_2^2 = \sum_{1}^{n} \alpha_i^2 (\lambda_i - \lambda)^2$$

$$\geq \sum_{i \neq k} \alpha_i^2 (\lambda_i - \lambda)^2 \geq c^2 \sum_{i \neq k} \alpha_i^2$$

$$\sum_{i \neq k} \alpha_i^2 \leq \frac{1}{c^2} \|\eta\|_2^2$$

which is quite small using (9.6.17). From $\|\hat{z}\| = 1$, this implies $\alpha_k \doteq 1$ and

$$\|\hat{z} - \alpha_k x_k\|_2 = \sqrt{\sum_{i \neq k} \alpha_i^2} \leq \frac{1}{c}\|\eta\|_2 \qquad (9.6.18)$$

showing the desired result. For a further discussion of the error, see Wilkinson (1963, pp. 142–146) and (1965, pp. 321–330).

Another method for calculating eigenvectors would appear to be the direct solution of

$$(A - \lambda I)x = 0$$

after deleting one equation and setting one of the unknown components to a nonzero constant, for example $x_1 = 1$. This is often the procedure used in undergraduate linear algebra courses. But as a general numerical method, it can be disastrous. A complete discussion of this problem is given in Wilkinson (1965, pp. 315–321), including an excellent example. We just use the previous example to show that the results need not be as good as those obtained with inverse iteration.

***Example*** Consider the preceding example (9.6.12) with $\lambda = 1.2679$. We consider $(A - \lambda I)x = 0$ and delete the last equation to obtain

$$.7321x_1 + x_2 = 0$$

$$x_1 + 1.7321x_2 + x_3 = 0$$

Taking $x_1 = 1.0$, we have the approximate eigenvector

$$x = [1.0000, -.73210, .26807]$$

Compared with the true answer (9.6.14), this is a slightly poorer result than (9.6.13) obtained by inverse iteration. In general, the results of using this approach can be very poor, and great care must be taken when using it.

The inverse iteration method requires a great deal of care in its implementation. For dealing with a particular matrix, any difficulties can be dealt with on an ad hoc basis. But for a general computer program we have to deal with eigenvalues that are multiple or close together, which can cause some difficulty if not dealt with carefully. For nonsymmetric matrices whose Jordan canonical form is not diagonal, there are additional difficulties in selecting a correct basis of eigenvectors. The best reference for this topic is Wilkinson (1965). Also see Golub and Van Loan (1983, pp. 238–240) and Parlett (1980, pp. 62–69). For several excellent programs, see Wilkinson and Reinsch (1971, pp. 418–439) and Garbow et al. (1977).

## 9.7   Least Squares Solution of Linear Systems

We now consider the solution of overdetermined systems of linear equations

$$\sum_{j=1}^{n} a_{ij}x_j = b_i \qquad i = 1, \ldots, m \qquad (9.7.1)$$

with $m > n$. These systems arise in a variety of applications, with the best known being the fitting of functions to a set of data $\{(t_i, b_i) | i = 1, \ldots, m\}$, about which we say more later. It might appear that the logical place for considering such systems would be in Chapter 8, but some of the tools used in the solution of (9.7.1) involve the orthogonal transformations introduced in this chapter. The numerical solution of (9.7.1) can be quite involved, both theoretically and practically, and we give only some major highlights of the subject.

An overdetermined system (9.7.1) will generally not have a solution. For that reason, we seek a vector $x = (x_1, \ldots, x_n)$ that solves (9.7.1) approximately in some sense. Introduce

$$A = [a_{ij}] \qquad x = [x_1, \ldots, x_n]^T \qquad b = [b_1, \ldots, b_m]^T$$

with $A$ $m \times n$. Then (9.7.1) can be written as

$$Ax = b \tag{9.7.2}$$

For simplicity, assume $A$ and $b$ are real. Among the possible ways of finding an approximate solution, we can seek a vector $x$ that minimizes

$$\|Ax - b\|_p \tag{9.7.3}$$

for some $p$, $1 \le p \le \infty$. In this section, only the classical case of $p = 2$ is considered, although in recent years, much work has also been done for the cases $p = 1$ and $p = \infty$.

The solution $x^*$ of

$$\underset{x \in \mathbf{R}^n}{\text{Minimize}} \|Ax - b\|_2 \tag{9.7.4}$$

is called the *least squares solution* of the linear system $Ax = b$. There are a number of reasons for this approach to solving $Ax = b$. First, it is easier to develop the theory and the practical solution techniques for minimizing $\|Ax - b\|_2$, partly because it is a continuously differentiable function of $x_1, \ldots, x_n$. Second, the curve fitting problems that lead to systems (9.7.1) often have a statistical framework that leads to (9.7.4), in preference to minimizing $\|Ax - b\|_p$ with some $p \ne 2$.

To better understand the nature of the solution of (9.7.4), we give the following theoretical construction. It also can be used as a practical numerical approach, although there are usually other more efficient constructions. Crucial to the theory is the singular value decomposition (SVD)

$$V^T A U = F = \begin{bmatrix} \mu_1 & & \cdots & & & 0 \\ 0 & \ddots & & & & \\ & & \mu_r & & & \\ & & & 0 & & \\ & & & & \ddots & \\ \vdots & & & & & 0 \\ & & & & & \vdots \\ 0 & & \cdots & & & 0 \end{bmatrix} \tag{9.7.5}$$

The matrices $U$ and $V$ are orthogonal, and the singular values $\mu_i$ satisfy

$$\mu_1 \geq \mu_2 \geq \cdots \geq \mu_r > 0$$

See Theorem 7.5 in Chapter 7 for more information; and later in this section, we describe a way to construct the SVD of $A$.

**Theorem 9.7**  Let $A$ be real and $m \times n$, $m \geq n$. Define $z = U^T x$, $c = V^T b$. Then the solution $x^* = Uz^*$ of (9.7.4) is given by

$$z_i^* = \frac{c_i}{\mu_i} \qquad i = 1, \ldots, r \tag{9.7.6}$$

with $z_{r+1}, \ldots, z_n$ arbitrary. When $r = n$, $x^*$ is unique. When $r < n$, the solution of (9.7.4) of minimal Euclidean norm is obtained by setting

$$z_i^* = 0 \qquad i = r + 1, \ldots, n \tag{9.7.7}$$

[This is also called the least squares solution of (9.7.4), even though it is not the unique minimizer of $\|Ax - b\|_2$.] The minimum in (9.7.4) is given by

$$\|Ax^* - b\|_2 = \left[ \sum_{j=r+1}^{m} c_i^2 \right]^{1/2} \tag{9.7.8}$$

**Proof**  Recall Problem 13(a) of Chapter 7. For any $y \in \mathbf{R}^n$ and any orthogonal matrix $P$,

$$\|Px\|_2 = \|x\|_2$$

Applying this to $\|Ax - b\|_2$ and using (9.7.5),

$$\|Ax - b\|_2 = \|V^T Ax - V^T b\|_2 = \|V^T A U U^T x - c\|_2$$

$$= \|Fz - c\|_2$$

$$= \left[ \sum_{j=1}^{r} (\mu_j z_j - c_j)^2 + \sum_{j=r+1}^{m} c_j^2 \right]^{1/2} \tag{9.7.9}$$

Then (9.7.6) and (9.7.8) follow immediately. For (9.7.7), use

$$\|x^*\|_2 = \|z\|_2 = \left[ \sum_{j=1}^{r} \left(z_j^*\right)^2 + \sum_{j=r+1}^{n} z_j^2 \right]^{1/2}$$

with $z_{r+1}, \ldots, z_n$ arbitrary according to (9.7.9). Choosing (9.7.7) leads to a unique minimum value for $\|x^*\|_2$.  ∎

Define the $n \times m$ matrix

$$F^+ = \begin{bmatrix} \mu_1^{-1} & 0 & & \cdots & & & 0 \\ & \ddots & & & & & \\ \vdots & & \mu_r^{-1} & & & & \vdots \\ & & & 0 & & & \\ & & & & \ddots & & \\ 0 & & \cdots & & 0 & \cdots & 0 \end{bmatrix} \qquad (9.7.10)$$

and

$$A^+ = UF^+V^T \qquad (9.7.11)$$

Looking at (9.7.6)–(9.7.8),

$$x^* = Uz^* = UF^+c = UF^+V^Tb$$

$$x^* = A^+b \qquad (9.7.12)$$

The matrix $A^+$ is called the *generalized inverse* of $A$, and it yields the least squares solution of $Ax = b$. The formula (9.7.12) shows that $x^*$ depends linearly on $b$. This representation of $x^*$ is an important tool in studying the numerical solution of $Ax = b$. Some further properties of $A^+$ are left to Problems 27 and 28.

To simplify the remaining development of methods for finding $x^*$ and analyzing its stability, we restrict $A$ to having full rank, $r = n$. This is the most important case for applications. For the singular values of $A$,

$$\mu_1 \geq \mu_2 \geq \cdots \geq \mu_n > 0 \qquad (9.7.13)$$

The concept of matrix norm can be generalized to $A$, from that given for square matrices in Section 7.3. Define

$$\|A\| = \underset{\substack{x \in \mathbf{R}^n \\ x \neq 0}}{\text{Supremum}} \frac{\|Ax\|_2}{\|x\|_2} \qquad (9.7.14)$$

It can be shown, using the SVD of $A$, that

$$\|A\| = \sqrt{r_\sigma(A^TA)} = \mu_1 \qquad (9.7.15)$$

In analogy with the error analysis in Section 8.4, define a condition number for $Ax = b$ by

$$\text{cond}\,(A)_2 = \|A\|\,\|A^+\| = \frac{\mu_1}{\mu_n} \qquad (9.7.16)$$

Using this notation, we give a stability result from Golub and Van Loan (1983,

p. 141). It is the analogue of Theorem 8.4, for the perturbation analysis of square nonsingular linear systems.

Let $b + \delta b$ and $A + \delta A$ be perturbations of $b$ and $A$, respectively. Define

$$x^* = A^+ b \qquad \hat{x}^* = (A + \delta A)^+ (b + \delta b)$$

$$r = b - Ax^* \qquad \hat{r} = (b + \delta b) - (A + \delta A)\hat{x}^* \qquad (9.7.17)$$

Assume

$$\epsilon \equiv \mathrm{Max}\left[\frac{\|\delta A\|}{\|A\|}, \frac{\|\delta b\|_2}{\|b\|_2}\right] < \frac{1}{\mathrm{cond}\,(A)_2} \qquad (9.7.18)$$

and

$$\sin(\theta) \equiv \frac{\|r\|_2}{\|b\|_2} < 1 \qquad (9.7.19)$$

implicitly defining $\theta$, $0 \le \theta < \pi/2$. Then

$$\frac{\|\hat{x}^* - x^*\|_2}{\|x^*\|_2} \le \epsilon\left[\frac{2\,\mathrm{cond}\,(A)_2}{\cos\theta} + \tan\theta\,[\mathrm{cond}\,(A)_2]^2\right] + O(\epsilon^2) \quad (9.7.20)$$

$$\frac{\|\hat{r} - r\|_2}{\|b\|_2} \le \epsilon[1 + 2\,\mathrm{cond}\,(A)_2]\,\mathrm{Min}\,\{1, m - n\} + O(\epsilon^2) \quad (9.7.21)$$

For the case $m = n$ with rank$(A) = n$, the residual $r$ will be zero, and then (8.7.20) will reduce to the earlier Theorem 8.4.

The preceding results say that the change in $r$ can be quite small, while the change in $x^*$ can be quite large. Note that the bound in (9.7.20) depends on the square of cond$(A)_2$, as compared to the dependence on cond$(A)$ for the nonsingular case with $m = n$ [see (8.4.18)]. If the columns of $A$ are nearly dependent, then cond$(A)_2$ can be very large, resulting in a larger bound in (9.7.20) than in (9.7.21) [see Problem 34(a)]. Whether this is acceptable or not will depend on the problem, on whether one wants small values of $r$ or accurate values of $x^*$.

**The least squares data-fitting problem**    The origin of most overdetermined linear systems is that of fitting data by a function from a prescribed family of functions. Let $\{(t_i, b_i)\,|\,i = 1, \ldots, m\}$ be a given set of data, presumably representing some function $b = g(t)$. Let $\varphi_1(t), \ldots, \varphi_n(t)$ be given functions, and let $\mathscr{F}$ be the family of all linear combinations of $\varphi_1, \ldots, \varphi_n$:

$$\mathscr{F} = \left\{\sum_{j=1}^n x_j \varphi_j(t)\,\bigg|\,x_j \in \mathbf{R}\right\} \qquad (9.7.22)$$

We want to choose an element of $\mathcal{F}$ to approximately fit the given data:

$$\sum_{j=1}^{n} x_j \varphi_j(t_i) = b_i \qquad i = 1, \ldots, m \tag{9.7.23}$$

This is the system (9.7.1), with $a_{ij} = \varphi_j(t_i)$.

For statistical modeling reasons, we seek to minimize

$$E(x) = \left[ \frac{1}{m} \sum_{i=1}^{m} \left[ b_i - \sum_{j=1}^{n} x_j \varphi_j(t_i) \right]^2 \right]^{1/2} \tag{9.7.24}$$

hence the description of *fitting data in the sense of least squares*. The quantity $E(x^*)$, for which $E(x)$ is a minimum, is called the *root-mean-square error* in the approximation of the data by the function

$$g^*(t) = \sum_{j=1}^{n} x_j^* \varphi_j(t) \tag{9.7.25}$$

Using earlier notation,

$$E(x) = \frac{1}{\sqrt{m}} \|b - Ax\|_2$$

and minimizing $E(x)$ is equivalent to finding the least squares solution of (9.7.23).

Forming the partial derivatives of (9.7.24) with respect to each $x_i$, and setting these equal to zero, we obtain the system of equations

$$A^T A x = A^T b \tag{9.7.26}$$

This system is a necessary condition for any minimizer of $E(x)$, and it can also be shown to be sufficient. The system (9.7.26) is called the *normal equation* for the least squares problem. If $A$ has rank $n$, then $A^T A$ will be $n \times n$ and nonsingular, and (9.7.26) has a unique solution.

To establish the equivalency of (9.7.26) with the earlier solution of the least squares problem, we use the SVD of $A$ to convert (9.7.26) to a simpler form. Substituting $A = VFU^T$ into (9.7.26),

$$UF^T F U^T x = UF^T V^T b$$

Multiply by $U$, and use the earlier notation $z = U^T x$, $c = V^T b$. Then

$$F^T F z = F^T c$$

This gives a complete mathematical equivalence of the normal equation to the earlier minimization of $\|Ax - b\|_2$ given in Theorem 9.7.

Assuming that rank$(A) = n$, the solution $x^*$ can be found by solving the normal equation. Since $A^T A$ is symmetric and positive definite, the Cholesky decomposition can be used for the solution [see (8.3.8)–(8.3.17)]. The effect on $x^*$ of rounding errors will be proportional to both the unit round of the computer and to the condition number of $A^T A$. From the SVD of $A$, this is easily seen to be

$$\text{cond}\,(A^T A)_2 = \frac{\mu_1^2}{\mu_n^2} = [\text{cond}\,(A)_2]^2 \qquad (9.7.27)$$

Thus the sensitivity of $x^*$ to errors will be proportional to $[\text{cond}\,(A)_2]^2$, which is consistent with the earlier perturbation error bound (9.7.20).

The result (9.7.27) used to be cited as the main reason for avoiding the use of the normal equation for solving the least squares problem. This is still good advice, but the reasons are more subtle. From (9.7.20), if $\|r\|_2$ is nearly zero, then $\sin \theta \doteq 0$, and the bound will be proportional to $\text{cond}\,(A)_2$. In contrast, the error bound for Cholesky's method will feature $[\text{cond}\,(A)_2]^2$, which is larger when $\text{cond}\,(A)_2$ is large. A second reason occurs when $A$ has columns that are nearly dependent. The use of finite computer arithmetic can then lead to an approximate normal equation that has lost vital information present in $A$. In such case, $A^T A$ will be nearly singular, and solution of the normal equation will yield much less accuracy in $x^*$ than will some other methods that work directly with $Ax = b$. For a more extensive discussion of this, see Lawson and Hanson (1974, pp. 126–129).

***Example*** Consider the data in Table 9.3 and its plot in Figure 9.2. We use a cubic polynomial to fit these data, and thus are led to minimizing the expression

$$E(x) = \left[ \frac{1}{m} \sum_{i=1}^{m} \left[ b_i - \sum_{j=1}^{4} x_j t_i^{j-1} \right]^2 \right]^{1/2}$$

This yields the overdetermined linear system

$$\sum_{j=1}^{4} x_j t_i^{j-1} = b_i \qquad i = 1, \ldots, m \qquad (9.7.28)$$

**Table 9.3    Data for a cubic least squares fit**

| $t_i$ | $b_i$ | $t_i$ | $b_i$ |
|-------|-------|-------|-------|
| 0.00 | .486 | .55 | 1.102 |
| .05 | .866 | .60 | 1.099 |
| .10 | .944 | .65 | 1.017 |
| .15 | 1.144 | .70 | 1.111 |
| .20 | 1.103 | .75 | 1.117 |
| .25 | 1.202 | .80 | 1.152 |
| .30 | 1.166 | .85 | 1.265 |
| .35 | 1.191 | .90 | 1.380 |
| .40 | 1.124 | .95 | 1.575 |
| .45 | 1.095 | 1.00 | 1.857 |
| .50 | 1.122 | | |

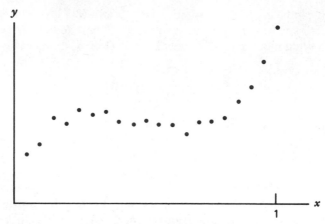

**Figure 9.2**   Plot of data of Table 9.3.

and the normal equations

$$\sum_{j=1}^{n} x_j \left[ \sum_{i=1}^{m} t_i^{j+k-2} \right] = \sum_{i=1}^{m} t_i^{k-1} b_i \qquad k = 1, 2, 3, 4$$

Writing this in the form (9.7.26),

$$A^T A = \begin{bmatrix} 21 & 10.5 & 7.175 & 5.5125 \\ 10.5 & 7.175 & 5.5125 & 4.51666 \\ 7.175 & 5.5125 & 4.51666 & 3.85416 \\ 5.5125 & 4.51666 & 3.85416 & 3.38212 \end{bmatrix}$$

$$A^T b = [24.1180, 13.2345, 9.468365, 7.5594405]^T \qquad (9.7.29)$$

The solution is

$$x^* = [.5747, 4.7259, -11.1282, 7.6687]^T \qquad (9.7.30)$$

This solution is very sensitive to changes in $b$. This can be inferred from the condition number

$$\text{cond}(A^T A) = 12105 \qquad (9.7.31)$$

As a further indication, perturb the right-hand vector $A^T b$ by adding it to the vector

$$[.01, -.01, .01, -.01]^T$$

This is consistent with the size of the errors present in the data values $b_i$. With this new right side, the normal equation has the perturbed solution

$$\hat{x}^* = [.7408, 2.6825, -6.1538, 4.4550]^T$$

which differs significantly from $x^*$.

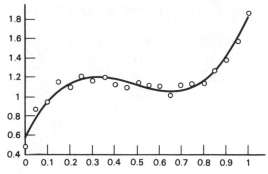

**Figure 9.3**   The least squares fit $g^*(t)$.

The plot of the least squares fit

$$g^*(t) = x_1^* + x_2^* t + x_3^* t^2 + x_4^* t^3$$

is shown in Figure 9.3, together with the data. Its root mean square error is

$$E(x^*) = .0421$$

The columns in the example matrix $A$ for the preceding example has columns that are almost linearly dependent, and $A^T A$ has a large condition number. To improve on this, we can choose a better set of basis functions $\{\varphi_i(t)\}$ for the family $\mathscr{F}$, of polynomials of degree $\leq 3$. Examining the coefficients of $A^T A$,

$$[A^T A]_{jk} = \sum_{i=1}^{m} \varphi_j(t_i) \varphi_k(t_i) \qquad 1 \leq j, k \leq n \qquad (9.7.32)$$

If the points $\{t_i\}$ are well distributed throughout the interval $[a, b]$, then the previous sum, when multiplied by $(b - a)/m$, is an approximation to

$$\int_a^b \varphi_k(t) \varphi_j(t) \, dt$$

To obtain a matrix $A^T A$ that has a smaller condition number, choose functions $\varphi_j(t)$ that are orthonormal. Then $A^T A$ will approximate the identity, and $A$ will have approximately orthonormal columns, leading to condition numbers close to 1. In fact, all that is really important is that the family $\{\varphi_j(t)\}$ be orthogonal, since then the matrix $A^T A$ will be nearly diagonal, a well-conditioned matrix.

**Example**   We repeat the preceding example, using the Legendre polynomials that are orthonormal over $[0, 1]$. The first four orthonormal Legendre polynomials on $[0, 1]$ are

$$\varphi_0(t) = 1 \quad \varphi_1(t) = \sqrt{3}\, s \quad \varphi_2(t) = \frac{\sqrt{5}}{2}(3s^2 - 1) \quad \varphi_3(t) = \frac{\sqrt{7}}{2}(5s^3 - 3s)$$

$$(9.7.33)$$

with $s = 2t - 1, 0 \le t \le 1$. For the normal equation (9.7.26),

$$A^T A = \begin{bmatrix} 21.0000 & 0 & 2.3479 & 0 \\ 0 & 23.1000 & 0 & 5.1164 \\ 2.3479 & 0 & 25.4993 & 0 \\ 0 & 5.1164 & 0 & 28.3889 \end{bmatrix}$$

$$A^T b = [24.1180, 4.0721, 3.4015, 4.8519]^T$$

$$x^* = [1.1454, .1442, .0279, .1449]^T \qquad (9.7.34)$$

The condition number of $A^T A$ is now

$$\text{cond}(A^T A) = 1.58 \qquad (9.7.35)$$

much less than earlier in (9.7.31).

**The QR method of solution**    Recall the $QR$ factorization of Section 9.3, following (9.3.11). As there, we consider Householder matrices of order $m \times m$

$$P_j = I - 2w^{(j)} w^{(j)T} \qquad j = 1, \dots, n$$

to reduce to zero the elements below the diagonal in $A$. The orthogonal matrices $P_j$ are applied in succession, to reduce to zero the elements below the diagonal in columns 1 through $n$ of $A$. The vector $w^{(j)}$ has nonzero elements in positions $j$ through $m$. This process leads to a matrix

$$R = P_n \cdots P_1 A = Q^T A \qquad (9.7.36)$$

If these are also applied to the right side in the system $Ax = b$, we obtain the equivalent system

$$Rx = Q^T b \qquad (9.7.37)$$

The matrix $R$ has the form

$$R = \begin{bmatrix} R_1 \\ 0 \end{bmatrix} \qquad (9.7.38)$$

with $R_1$ an upper triangular square matrix of order $n \times n$. The matrix $R_1$ must be nonsingular, since $A$ and $R = Q^T A$ have the same rank, namely $n$. In line with (9.7.38), write

$$Q^T b = \begin{bmatrix} g_1 \\ g_2 \end{bmatrix} \qquad g_1 \in \mathbf{R}^n \quad g_2 \in \mathbf{R}^{m-n}$$

Then

$$\|Ax - b\|_2 = \|Q^T Ax - Q^T b\|_2 = \|Rx - Q^T b\|_2$$

$$= \left[ \|R_1 x - g_1\|_2^2 + \|g_2\|_2^2 \right]^{1/2}$$

The least squares solution of $Ax = b$ is obtained by solving the nonsingular upper triangular system

$$R_1 x = g_1 \tag{9.7.39}$$

Then the minimum is

$$\|Ax^* - b\|_2 = \|g_2\|_2 \tag{9.7.40}$$

The $QR$ method for calculating $x^*$ is slightly more expensive in operations than the Cholesky method. The Cholesky method, including the formation of $A^T A$, has an operation count (multiplications and divisions) of about

$$\frac{1}{2}mn^2 + \frac{n^3}{6}$$

and the Householder $QR$ method has an operation count of about

$$mn^2 - \frac{n^3}{3}$$

Nonetheless, the $QR$ method is generally the recommended method for calculating the least squares solution. It works directly on the matrix $A$, and because of that and the use of orthogonal transformations, the effect of rounding errors is better than with the use of the Cholesky factorization to solve the normal equation. For a thorough discussion, see Golub and Van Loan (1983, pp. 147–149) and Lawson and Hanson (1974, chap. 16).

**Example**  We consider the earlier example of the linear system (9.7.28) with

$$A_{ij} = \left[t_i^{j-1}\right] \qquad 1 \le i \le 21, \quad 1 \le j \le 4$$

for the data in Table 9.3. Then $\text{cond}(A) = 110.01$,

$$R_1 = \begin{bmatrix} -4.5826 & -2.2913 & -1.5657 & -1.2029 \\ 0 & 1.3874 & 1.3874 & 1.2688 \\ 0 & 0 & -.3744 & -.5617 \\ 0 & 0 & 0 & -.0987 \end{bmatrix}$$

$$g_1 = \left[-5.2630, .8472, -.1403, -.7566\right]^T$$

The solution $x^*$ is the same as in (9.7.30), as is the root-mean-square error.

**The singular value decomposition**  The SVD is a very valuable tool for analyzing and solving least squares problems and other problems of linear algebra. For least squares problems of less than full rank, the $QR$ method just described will probably lead to a triangular matrix $R_1$ that is nonsingular, but has some very small diagonal elements. The SVD of $A$ can then be quite useful in making clearer the structure of $A$. If some singular values $\mu_i$ are nearly zero, then the effect of setting them to zero can be determined more easily than with some other methods for solving for $x^*$. Thus there is ample justification for finding efficient ways to calculate the SVD of $A$.

One of the best known ways to calculate the SVD of $A$ is due to G. Golub, C. Reinsch, and W. Kahan, and a complete discussion of it is given in Golub and Van Loan (1983, sec. 6.5). We instead merely show how the singular value decomposition in (9.7.5) can be obtained from the solution of a symmetric matrix eigenvalue problem together with a $QR$ factorization.

From $A$ real and $m \times n$, $m \geq n$, we have that $A^TA$ is $n \times n$ and real. In addition, it is straightforward to show that $A^TA$ is symmetric and positive semidefinite [$x^TAx \geq 0$ for all $x$]. Using a program to solve the symmetric eigenvalue problem, find a diagonal matrix $D$ and an orthogonal matrix $U$ for which

$$U^T(A^TA)U = D \qquad (9.7.41)$$

Let $D = \text{diag}[\lambda_1, \ldots, \lambda_n]$ with the eigenvalues arranged in descending order. If any $\lambda_i$ is a small negative number, then set it to zero, since all eigenvalues of $A^TA$ should be nonnegative except for possible perturbations due to rounding errors.

From (9.7.41), define $B = AU$, of order $m \times n$. Then (9.7.41) implies

$$B^TB = D$$

Then the columns of $B$ are orthogonal. Moreover, if some $\lambda_i = 0$, then the corresponding column of $B$ must be identically zero, because its norm is zero. Using the $QR$ method, calculate an orthogonal matrix $V$ for which

$$V^TB = R \qquad (9.7.42)$$

is zero below the diagonal in all columns. The matrix $R$ satisfies

$$R^TR = B^TV^TVB = B^TB = D$$

Again, the columns of $R$ must be orthogonal, and if some $\lambda_i = 0$, then the corresponding column of $R$ must be zero. Since $R$ is upper triangular, we can use the orthogonality to show that the columns of $R$ must be zero in all positions above the diagonal. Thus $R$ has the form of the matrix $F$ of (9.7.5). We will then have $R = F$ with $\mu_i = \sqrt{\lambda_i}$. Letting $B = AU$ in (9.7.42), we have the desired SVD:

$$V^TAU = R$$

One of the possible disadvantages of this procedure is $A^TA$ must be formed, and this may lead to a loss of information due to the use of finite-length computer arithmetic. But the method is simple to implement, if the symmetric eigenvalue problem is solvable.

**Example** Consider again the matrix $A$ of (9.7.28), based on the data of Table 9.3. The matrix $A^TA$ is given in (9.7.29). Using EISPACK and LINPACK programs,

$$U = \begin{bmatrix} .7827 & .5963 & -.1764 & .0256 \\ .4533 & -.3596 & .7489 & -.3231 \\ .3326 & -.4998 & -.0989 & .7936 \\ .2670 & -.5150 & -.6311 & -.5150 \end{bmatrix}$$

The singular values are

$$32.0102, \ 3.8935, \ .1674, \ .0026$$

The matrix $V$ is orthogonal and of order $21 \times 21$, and we omit it for obvious reasons. In practice it would not be computed, since it is a product of four Householder matrices, which can be stored in a simpler form.

For a much more extensive discussion of the solution of least squares problems, see Golub and Van Loan (1983, chap. 6) and the book by Lawson and Hanson (1974). There are many additional practical problems that must be discussed, including that of determining the rank of a matrix when rounding error causes it to falsely have full rank. For programs, see the appendix to Lawson and Hanson (1974) and LINPACK. For the SVD, see LINPACK or EISPACK.

## Discussion of the Literature

The main source of information for this chapter was the well-known and encyclopedic book of Wilkinson (1965). Other sources were Golub and Van Loan (1983), Gourlay and Watson (1976), Householder (1964), Noble (1969, chaps. 9–12), Parlett (1980), Stewart (1973), and Wilkinson (1963). For matrices of moderate size, the numerical solution of the eigenvalue problem is fairly well understood. For another perspective on the $QR$ method, see Watkins (1982), and for an in-depth look at inverse iteration, see Peters and Wilkinson (1979). Excellent algorithms for most eigenvalue problems are given in Wilkinson and Reinsch (1971) and the EISPACK guides by Smith et al. (1976), and Garbow et al. (1977). For a history of the EISPACK project, see Dongarra and Moler (1984). An excellent general account of the problems of developing mathematical software for eigenvalue problems and other matrix problems is given in Rice (1981). The EISPACK package is the basis for most of the eigenvalue programs in the IMSL and NAG libraries.

A number of problems and numerical methods have not been discussed in this chapter, often for reasons of space. For the symmetric eigenvalue problem, the Jacobi method has been omitted. It is an elegant and rapidly convergent method for computing all of the eigenvalues of a symmetric matrix, and it is relatively easy to program. For a description of the Jacobi method, see Golub and Van Loan (1983, sec. 8.4), Parlett (1980, chap. 9), and Wilkinson (1965, pp. 266–282).

An ALGOL program is given in Wilkinson and Reinsch (1971, pp. 202–211). The generalized eigenvalue problem, $Ax = \lambda Bx$, has also been omitted. This has become an important problem in recent years. The most popular method for its solution is due to Moler and Stewart (1973), and other descriptions of the problem and its solution are given in Golub and Van Loan (1983, secs. 7.7 and 8.6) and Parlett (1980, chap. 15). EISPACK programs for the generalized eigenvalue problem are given in Garbow et al. (1977).

The problem of finding the eigenvalues and eigenvectors of large sparse matrices is an active area of research. When the matrices have large order (e.g., $n \geq 300$), most of the methods of this chapter are more difficult to apply because of storage considerations. In addition, the methods often do not take special account of the sparseness of most large matrices that occur in practice. One common form of problem involves a symmetric banded matrix. Programs for this problem are given in Wilkinson and Reinsch (1971, pp. 266–283) and Garbow et al. (1977). For more general discussions of the eigenvalue problem for sparse matrices, see Jennings (1985) and Pissanetzky (1984, chap. 6). For a discussion of software for the eigenvalue problem for sparse matrices, see Duff (1984, pp. 179–182) and Heath (1982). An important method for the solution of the eigenvalue problem for sparse symmetric matrices is the Lanczos method. For a discussion of it, see Scott (1981) and the very extensive books and programs of Cullum and Willoughby (1984, 1985).

The least squares solution of overdetermined linear systems is a very important tool, one that is very widely used in the physical, biological, and social sciences. We have just introduced some aspects of the subject, showing the crucial role of the singular value decomposition. A very comprehensive introduction to the least squares solution of linear systems is given in Lawson and Hanson (1974). It gives a complete treatment of the theory, the practical implementation of methods, and ways for handling large data sets efficiently. In addition, the book contains a complete set of programs for solving a variety of least squares problems. For other references to the least squares solutions of linear systems, see Golub and Van Loan (1983, chap. 6) and Rice (1981, chap. 11). Programs for some least squares problems are also given in LINPACK.

In discussing the least squares solution of overdetermined systems of linear equations, we have avoided any discussion of the statistical aspect of the subject. Partly this was for reasons of space, and partly it was a mistrust of using the statistical justification, since it often depends on assumptions about the distribution of the error that are difficult to validate. We refer the reader to any of the many statistics textbooks for a development of the statistical framework for the least squares method for curve fitting of data.

# Bibliography

Chatelin, F. (1987). *Eigenvalues of Matrices*. Wiley, London.

Conte, S., and C. de Boor (1980). *Elementary Numerical Analysis*, 3rd ed. McGraw-Hill, New York.

Cullum, J., and R. Willoughby (1984, 1985). *Lanczos Algorithms for Large Symmetric Eigenvalue Computations*, Vol. 1, *Theory*; Vol. 2, *Programs*. Birkhäuser, Basel.

Dongarra, J., and C. Moler (1984). EISPACK—A package for solving matrix eigenvalue problems. In *Sources and Development of Mathematical Software*, W. Cowell (Ed.), pp. 68–87. Prentice-Hall, Englewood Cliffs, N.J.

Dongarra, J., J. Bunch, C. Moler, and G. Stewart (1979). *LINPACK User's Guide*. SIAM Pub., Philadelphia.

Duff, I. (1984). A survey of sparse matrix software, In *Sources and Development of Mathematical Software*, W. Cowell (Ed.). Prentice-Hall, Englewood Cliffs, N.J.

Garbow, B., J. Boyle, J. Dongarra, and C. Moler (1977). *Matrix Eigensystems Routines—EISPACK Guide Extension*, Lecture Notes in Computer Science, Vol. 51. Springer-Verlag, New York.

Golub, G., and C. Van Loan (1983). *Matrix Computations*. Johns Hopkins Press, Baltimore.

Gourlay, A., and G. Watson (1976). *Computational Methods for Matrix Eigenproblems*. Wiley, New York.

Gregory, R., and D. Karney (1969). *A Collection of Matrices for Testing Computational Algorithms*. Wiley, New York.

Heath, M., Ed. (1982). *Sparse Matrix Software Catalog*. Oak Ridge National Laboratory, Mathematics and Statistics Dept., Tech. Rep. Oak Ridge, Tenn.

Henrici, P. (1974). *Applied and Computational Complex Analysis*, Vol. I. Wiley, New York.

Householder, A. (1964). *The Theory of Matrices in Numerical Analysis*. Ginn (Blaisdell), Boston.

Jennings, A. (1985). Solutions of sparse eigenvalue problems. In *Sparsity and Its Applications*, D. Evans (Ed.), pp. 153–184. Cambridge Univ. Press, Cambridge, England.

Lawson, C., and R. Hanson (1974). *Solving Least Squares Problems*. Prentice-Hall, Englewood Cliffs, N.J.

Moler, C., and G. Stewart (1973). An algorithm for generalized matrix eigenvalue problems, *SIAM J. Numer. Anal.* **10**, 241–256.

Noble, B. (1969). *Applied Linear Algebra*. Prentice-Hall, Englewood Cliffs, N.J.

Parlett, B. (1968). Global convergence of the basic $QR$ algorithm on Hessenberg matrices, *Math. Comput.* **22**, 803–817.

Parlett, B. (1980). *The Symmetric Eigenvalue Problem*. Prentice-Hall, Englewood Cliffs, N.J.

Peters, G., and J. Wilkinson (1979). Inverse iteration, ill-conditioned equations and Newton's method, *SIAM Rev.* **21**, 339–360.

Pissanetzky, S. (1984). *Sparse Matrix Technology*. Academic Press, New York.

Rice, J. (1981). *Matrix Computations and Mathematical Software*. McGraw-Hill, New York.

Scott, D. (1981). The Lanczos algorithm. In *Sparse Matrices and Their Uses*, I. Duff (Ed.), pp. 139–160. Academic Press, London.

Smith, B. T., J. Boyle, B. Garbow, Y. Ikebe, V. Klema, and C. Moler (1976). *Matrix Eigensystem Routines—EISPACK Guide*, 2nd ed., *Lecture Notes in Computer Science*, Vol. 6. Springer-Verlag, New York.

Stewart, G. (1973). *Introduction to Matrix Computations*. Academic Press, New York.

Watkins, D. (1982). Understanding the *QR* algorithm, *SIAM Rev.* **24**, 427–440.

Wilkinson, J. (1963). *Rounding Errors in Algebraic Processes*. Prentice-Hall, Englewood Cliffs, N.J.

Wilkinson, J. (1965). *The Algebraic Eigenvalue Problem*. Oxford Univ. Press, Oxford, England.

Wilkinson, J. (1968). Global convergence of the tridiagonal *QR* algorithm with origin shifts. *Linear Algebra Its Appl.* **1**, 409–420.

Wilkinson, J., and C. Reinsch, Eds. (1971). *Linear Algebra*. Springer-Verlag, New York.

## Problems

1. Use the Gerschgorin theorem 9.1 to determine the approximate location of the eigenvalues of

   (a) $\begin{bmatrix} 1 & -1 & 0 \\ 1 & 5 & 1 \\ -2 & -1 & 9 \end{bmatrix}$    (b) $\begin{bmatrix} -2 & 1 & 1 \\ 1 & 3 & 1 \\ 1 & -1 & 3 \end{bmatrix}$

   Where possible, use these results to infer whether the eigenvalues are real or complex. To check these results, compute the eigenvalues directly by finding the roots of the characteristic polynomial.

2. (a) Given a polynomial

   $$p(\lambda) = \lambda^n + a_{n-1}\lambda^{n-1} + \cdots + a_0$$

   show $p(\lambda) = \det[\lambda I - A]$ for the matrix

   $$A = \begin{bmatrix} 0 & 1 & 0 & & \cdots & & 0 \\ 0 & 0 & 1 & 0 & & & \vdots \\ \vdots & & & \ddots & & & \\ 0 & & & & & 0 & 1 \\ -a_0 & -a_1 & & \cdots & & & -a_{n-1} \end{bmatrix}$$

The roots of $p(\lambda)$ are the eigenvalues of $A$. The matrix $A$ is called the *companion matrix* for the polynomial $p(\lambda)$.

**(b)** Apply the Gerschgorin theorem 9.1 to obtain the following bounds for the roots $r$ of $p(\lambda)$: $|r| \le 1$, or $|r + a_{n-1}| \le |a_0| + \cdots + |a_{n-2}|$. If these bounds give disjoint regions in the complex plane, what can be said about the number of roots within each region.

**(c)** Use the Gerschgorin theorem on the columns of $A$ to obtain additional bounds for the roots of $p(\lambda)$.

**(d)** Use the results of parts (b) and (c) to bound the roots of the following polynomial equations:

(i)  $\lambda^{10} + 8\lambda^9 + 1 = 0$

(ii)  $\lambda^6 - 4\lambda^5 + \lambda^4 - \lambda^3 + \lambda^2 - \lambda + 1 = 0$

3. Recall the linear system (8.8.5) of Chapter 8, which arises when numerically solving Poisson's equation. If the equations are ordered in the manner described in (8.8.12) and following, then the linear system is symmetric with positive diagonal elements. For the Gauss–Seidel iteration method in (8.8.12) to converge, it is necessary and sufficient that $A$ be positive definite, according to Theorem 8.7. Use the Gerschgorin theorem 9.1 to prove $A$ is positive definite. It will also be necessary to quote Theorem 8.8, that $\lambda = 0$ is not an eigenvalue of $A$.

4. The values $\lambda = -8.02861$ and

$$x = [1.0, 2.50146, -.75773, -2.56421]$$

are an approximate eigenvalue and eigenvector for the matrix

$$A = \begin{bmatrix} 2 & 1 & 3 & 4 \\ 1 & -3 & 1 & 5 \\ 3 & 1 & 6 & -2 \\ 4 & 5 & -2 & -1 \end{bmatrix}$$

Use the result (9.1.22) to compute an error bound for $\lambda$.

5. For the matrix example (9.1.17) with $\epsilon = .001$ and $\lambda = 2$, compute the perturbation error bound (9.1.36). The same bound was given in (9.1.38) for the other eigenvalue $\lambda = 1$.

6. Prove the eigenvector perturbation result (9.1.41). *Hint:* Assume $\lambda_k(\epsilon)$ and $u_k(\epsilon)$ are continuously differentiable functions of $\epsilon$. From (9.1.32), $\lambda'_k(0) = v_k^* B u_k / s_k$. Write

$$u_k(\epsilon) = u_k(0) + \epsilon u'_k(0) + O(\epsilon^2)$$

and solve for $u'_k(0)$. Since $\{u_1, \ldots, u_n\}$ is a basis, write

$$u'_k(0) = \sum_{j=1}^{n} a_j u_j$$

To find $a_j$, first differentiate (9.1.40) with respect to $\epsilon$, and then let $\epsilon = 0$. Substitute the previous representation for $u'_k(0)$. Use (9.1.29) and the biorthogonality relation

$$v_i^* u_j = 0 \qquad i \neq j$$

from (9.1.28).

7.  For the following matrices $A(\epsilon)$, determine the eigenvalues and eigenvectors for both $\epsilon = 0$ and $\epsilon > 0$. Observe the behavior as $\epsilon \to 0$.

(a) $\begin{bmatrix} 1 & 1 \\ \epsilon & 1 \end{bmatrix}$     (b) $\begin{bmatrix} 1 & 1 \\ 0 & 1+\epsilon \end{bmatrix}$     (c) $\begin{bmatrix} 1 & \epsilon \\ 0 & 1 \end{bmatrix}$     (d) $\begin{bmatrix} 1 & 1 & 0 \\ 0 & 1 & \epsilon \\ 0 & \epsilon & 1 \end{bmatrix}$

What do these examples say about the stability of eigenvector subspaces?

8.  Use the power method to calculate the dominant eigenvalue and associated eigenvector for the following matrices.

(a) $\begin{bmatrix} 6 & 4 & 4 & 1 \\ 4 & 6 & 1 & 4 \\ 4 & 1 & 6 & 4 \\ 1 & 4 & 4 & 6 \end{bmatrix}$     (b) $\begin{bmatrix} 2 & 1 & 3 & 4 \\ 1 & -3 & 1 & 5 \\ 3 & 1 & 6 & -2 \\ 4 & 5 & -2 & -1 \end{bmatrix}$

(c) $\begin{bmatrix} 1 & 3 & -2 \\ -1 & -2 & 3 \\ 1 & 1 & 2 \end{bmatrix}$

Check the speed of convergence, calculating the ratios $R_m$ of (9.2.14). When the ratios $R_m$ are fairly constant, use Aitken extrapolation to improve the speed of convergence of both the eigenvalue and eigenvector, using the eigenvalue ratios $R_m$ to accelerate the eigenvectors $\{z^{(m)}\}$.

9.  Use the power method to find the dominant eigenvalue of

$$A = \begin{bmatrix} 7 & 13 & -16 \\ 13 & -10 & 13 \\ -16 & 13 & 7 \end{bmatrix}$$

Use the initial guess $z^{(0)} = [1, 0, 1]^T$. Print each iterate $z^{(m)}$ and $\lambda_1^{(m)}$. Comment on the results. What would happen if $\lambda_1^{(m)}$ were defined by $\lambda_1^{(m)} = \alpha_m$?

10. For a matrix $A$ of order $n$, assume its Jordan canonical form is diagonal and denote the eigenvalues by $\lambda_1, \ldots, \lambda_n$. Assume that $\lambda_1 = \lambda_2 = \cdots = \lambda_r$ for some $r > 1$, and

$$|\lambda_r| > |\lambda_{r+1}| \geq \cdots \geq |\lambda_n| \geq 0$$

Show that the power method (9.2.2)–(9.2.3) will still converge to $\lambda_1$ and some associated eigenvector, for most choices of initial vector $z^{(0)}$.

11. Let $A$ be a symmetric matrix of order $n$, with the eigenvalues ordered by

$$\lambda_1 \geq \lambda_2 \geq \cdots \geq \lambda_n$$

Define

$$\mathscr{R}(x) = \frac{(Ax, x)}{(x, x)} \qquad x \neq 0 \qquad x \in \mathbf{R}^n,$$

using the standard inner product. Show

$$\operatorname{Max} \mathscr{R}(x) = \lambda_1 \qquad \operatorname{Min} \mathscr{R}(x) = \lambda_n.$$

as $x \neq 0$ ranges over $\mathbf{R}^n$. The function $\mathscr{R}(x)$ is called the *Rayleigh quotient*, and it can be used to characterize the remaining eigenvalues of $A$, in addition to $\lambda_1$ and $\lambda_n$. Using these maximizations and minimizations for $\mathscr{R}(x)$ forms the basis of some classical numerical methods for calculating the eigenvalues of $A$.

12. To give a geometric meaning to the $n \times n$ Householder matrix $P = I - 2ww^T$, let $u^{(2)}, \ldots, u^{(n)}$ be an orthonormal basis of the $(n - 1)$ dimensional subspace that is perpendicular to $w$. Define

$$T(x) = (I - 2ww^T)x \qquad x \in \mathbf{R}^n.$$

Use the basis $\{w, u^{(2)}, \ldots, u^{(n)}\}$ for $\mathbf{R}^n$ to write

$$x = a_1 w + a_2 u^{(2)} + \cdots + a_n u^{(n)}$$

Apply $T$ to this representation and interpret the results.

13. (a) Let $A$ be a symmetric matrix, and let $\lambda$ and $x$ be an eigenvalue–eigenvector pair for $A$ with $\|x\|_2 = 1$. Let $P$ be an orthogonal matrix for which

$$Px = e_1 \equiv [1, 0, \ldots, 0]^T$$

Consider the similar matrix $B = PAP^T$, and show that the first row and column are zero except for the diagonal element, which equals $\lambda$. *Hint:* Calculate and use $Be_1$.

**(b)**    For the matrix

$$A = \begin{bmatrix} 2 & 10 & 2 \\ 10 & 5 & -8 \\ 2 & -8 & 11 \end{bmatrix}$$

$\lambda = 9$ is an eigenvalue with associated eigenvector $x = [\frac{2}{3}, \frac{1}{3}, \frac{2}{3}]^T$. Produce a Householder matrix $P$ for which $Px = e_1$, and then produce $B = PAP^T$. The matrix eigenvalue problem for $B$ can then be reduced easily to a problem for a $2 \times 2$ matrix. Use this procedure to calculate the remaining eigenvalues and eigenvectors of $A$. The process of changing $A$ to $B$ and of then solving a matrix eigenvalue problem of order one less than for $A$, is known as *deflation*. It can be used to extend the applicability of the power method to other than the dominant eigenvalue. For an extensive discussion, see Wilkinson (1965, pp. 584–598) and Parlett (1980, chap. 5).

**14.**    Use Householder matrices to produce the $QR$ factorization of

**(a)**    $A = \begin{bmatrix} 1 & 1 & 1 \\ 2 & -1 & -1 \\ 2 & -4 & 5 \end{bmatrix}$    **(b)**    $\begin{bmatrix} 1 & 3 & -2 \\ -1 & -2 & 3 \\ 1 & 1 & 2 \end{bmatrix}$

**15.**    Consider the rotation matrix of order $n$,

$$R^{(k,l)} = \begin{bmatrix} 1 & 0 & 0 & & & & & \cdots & & & 0 \\ 0 & 1 & 0 & & & & & & & & \vdots \\ \vdots & & \ddots & & & & & & & & \\ 0 & & & \alpha & 0 & \cdots & \beta & 0 & \cdots & 0 & \text{row } k \\ & & & 0 & 1 & & 0 & & & & \\ \vdots & & & \vdots & & \ddots & & & & & \\ & & & -\beta & 0 & & \alpha & 0 & & 0 & \text{row } l \\ & & & & & & & 1 & & & \\ & & & & & & & & \ddots & \vdots & \\ 0 & & & & \cdots & & & & & 1 \end{bmatrix}$$

with $\alpha^2 + \beta^2 = 1$. If we compute $Rb$ for a given $b \in \mathbf{R}^n$, then the only elements that will be changed are in positions $k$ and $l$. By choosing $\alpha$ and $\beta$ suitably, we can force $Rb$ to have a zero in position $l$. Choose $\alpha, \beta$ so that

$$\begin{bmatrix} \alpha & \beta \\ -\beta & \alpha \end{bmatrix} \begin{bmatrix} b_k \\ b_l \end{bmatrix} = \begin{bmatrix} \gamma \\ 0 \end{bmatrix}$$

for some $\gamma$.

**(a)**    Derive formulas for $\alpha, \beta$, and show $\gamma = \sqrt{b_k^2 + b_l^2}$.

**(b)**   Reduce $b = [1, 1, 1, 1]^T$ to a form $\hat{b} = [c, 0, 0, 0]$ by a sequence of multiplications by rotation matrices:

$$\hat{b} = R^{(1,2)}R^{(1,3)}R^{(1,4)}b$$

**16.**   Show how the rotation matrices $R^{(k,l)}$ can be used to produce the $QR$ factorization of a matrix.

**17.   (a)**   Do an operations count for producing the $QR$ factorization of a matrix using Householder matrices, as in Section 9.3. As usual, combine multiplications and divisions, and keep a separate count for the number of square roots.

**(b)**   Repeat part (a), but use the rotation matrices $R^{(k,l)}$ for the reduction.

**18.**   Give the explicit formulas for the calculation of the $QR$ factorization of a symmetric tridiagonal matrix. Do an operations count, and compare the result with those of Problem 17.

**19.**   Use Theorem 9.5 to separate the roots of

**(a)**
$$\begin{bmatrix} 0 & 1 & 0 & 0 & 0 \\ 1 & 1 & 1 & 0 & 0 \\ 0 & 1 & 1 & 1 & 0 \\ 0 & 0 & 1 & 1 & 1 \\ 0 & 0 & 0 & 1 & 2 \end{bmatrix}$$
**(b)**
$$\begin{bmatrix} 1 & 2 & 0 & 0 & 0 \\ 2 & 2 & 3 & 0 & 0 \\ 0 & 3 & 3 & 4 & 0 \\ 0 & 0 & 4 & 4 & 5 \\ 0 & 0 & 0 & 5 & 5 \end{bmatrix}$$

Then obtain accurate approximations using the bisection method or some other rootfinding technique.

**20.   (a)**   Write a program to reduce a symmetric matrix to tridiagonal form using Householder matrices for the similarity transformations. For efficiency in the matrix multiplications, use the analogue of the form of multiplication shown in (9.3.18).

**(b)**   Use the program to reduce the following matrices to tridiagonal form:

**(i)**
$$\begin{bmatrix} 1 & 2 & 3 \\ 2 & 3 & 5 \\ 3 & 5 & 8 \end{bmatrix}$$
**(ii)**
$$\begin{bmatrix} 5 & 4 & 1 & 1 \\ 4 & 5 & 1 & 1 \\ 1 & 1 & 4 & 2 \\ 1 & 1 & 2 & 4 \end{bmatrix}$$

**(iii)**
$$\begin{bmatrix} 4 & 6 & 242 & 12 \\ 6 & 225 & 3 & 18 \\ 242 & 3 & 25 & 6 \\ 12 & 18 & 6 & 0 \end{bmatrix}$$

**(c)**   Calculate the eigenvalues of your reduced tridiagonal matrix as accurately as possible.

**21.** Let $\{ p_n(x)|n \geq 0 \}$ denote a family of orthogonal polynomials with respect to a weight function $w(x)$ on an interval $a < x < b$. Further, assume that the polynomials have leading coefficient 1:

$$p_n(x) = x^n + \sum_{j=0}^{n-1} a_{n,j} x^j$$

Find a symmetric tridiagonal matrix $R_n$ for which $p_n(\lambda)$ is the characteristic polynomial. Thus, calculating the roots of an orthogonal polynomial (and the nodes of a Gaussian quadrature formula) is reduced to the solution of an eigenvalue problem for a symmetric tridiagonal matrix.
*Hint:* Recall the formula for the triple recursion relation for $\{ p_n(x) \}$, and compare it to the formula (9.4.3).

**22.** Use the $QR$ method (a) without shift, and (b) with shift, to calculate the eigenvalues of

**(a)** $\begin{bmatrix} 3 & 1 & 0 \\ 1 & 2 & 1 \\ 0 & 1 & 1 \end{bmatrix}$    **(b)** $\begin{bmatrix} 2 & 1 & 0 & 0 \\ 1 & 2 & 1 & 0 \\ 0 & 1 & 2 & 1 \\ 0 & 0 & 1 & 2 \end{bmatrix}$

**(c)** $\begin{bmatrix} 0 & 1 & 0 & 0 & 0 \\ 1 & 1 & 1 & 0 & 0 \\ 0 & 1 & 1 & 1 & 0 \\ 0 & 0 & 1 & 1 & 1 \\ 0 & 0 & 0 & 1 & 2 \end{bmatrix}$

**23.** Let $A$ be a Hessenberg matrix and consider the factorization $A = QR$, with $Q$ orthogonal and $R$ upper triangular.

**(a)** Recalling the discussion following (9.5.5), show that (9.5.7) is true.

**(b)** Show that the result (9.5.7) implies a form for $H_k$ in (9.5.6) such that $Q$ will be a Hessenberg matrix.

**(c)** Show the product of a Hessenberg matrix and an upper triangular matrix, in either order, is again a Hessenberg matrix.

When combined, these results show that $RQ$ is again a Hessenberg matrix, as claimed in the paragraph following (9.5.7).

**24.** For the matrix $A$ of Problem 4, two additional approximate eigenvalues are $\lambda = 7.9329$ and $\lambda = 5.6689$. Use inverse iteration to calculate the associated eigenvectors.

**25.** Investigate the programs available at your computer center for the calculation of the eigenvalues of a real symmetric matrix. Using such a program, compute the eigenvalues of the Hilbert matrices $H_n$ for $n = 3, 4, 5, 6, 7$. To check your answers, see the very accurate values given in Gregory and Karney (1969, pp. 66–73).

**26.** Consider calculating the eigenvalues and associated eigenfunctions $x(t)$ for which

$$\int_0^1 \frac{x(t)\, dt}{1 + (s - t)^2} = \lambda x(s) \qquad 0 \le s \le 1$$

One way to obtain approximate eigenvalues is to discretize the equation using numerical integration. Let $h = 1/n$ for some $n \ge 1$ and define $t_j = (j - \frac{1}{2})h$, $j = 1, \ldots, n$. Substitute $t_i$ for $s$ in the equation, and approximate the integral using the midpoint numerical integration method. This leads to the system

$$h \sum_{j=1}^n \frac{\hat{x}(t_j)}{1 + (t_i - t_j)^2} = \lambda \hat{x}(t_i) \qquad i = 1, \ldots, n$$

in which $\hat{x}(s)$ denotes a function that we expect approximates $x(s)$. This system is the eigenvalue problem for a symmetric matrix of order $n$. Find the two largest eigenvalues of this matrix for $n = 2, 4, 8, 16, 32$. Examine the convergence of these eigenvalues as $n$ increases, and attempt to predict the error in the most accurate case, $n = 32$, as compared with the unknown true eigenvalues for the integral equation.

**27.** Show that the generalized inverse $A^+$ of (9.7.11) satisfies the following *Moore–Penrose conditions*.

1. $AA^+A = A$            3. $(AA^+)^T = AA^+$

2. $A^+AA^+ = A^+$          4. $(A^+A)^T = A^+A$

Also show

5. $(A^+A)^2 = A^+A$        6. $(AA^+)^2 = AA^+$

Conditions (3)–(6) show that $A^+A$ and $AA^+$ represent *orthogonal projections* on $\mathbf{R}^n$ and $\mathbf{R}^m$, respectively.

**28.** For an arbitrary $m \times n$ matrix $A$, show that

$$\underset{\alpha \to 0+}{\text{Limit}} \left( \alpha I + A^T A \right)^{-1} A = A^+$$

where $\alpha > 0$. *Hint:* Use the SVD of $A$.

**29.** Unlike the situation with nonsingular square matrices, the generalized inverse $A^+$ need not vary continuously with changes in $A$. To support this, find a family of matrices $\{A(\epsilon)\}$ where $A(\epsilon)$ converges to $A(0)$, but $A(\epsilon)^+$ does not converge to $A(0)^+$.

**30.** Calculate the linear polynomial least squares fit for the following data. Graph the data and the least squares fit. Also, find the root-mean-square error in the least squares fit.

| $t_i$ | $b_i$ | $t_i$ | $b_i$ | $t_i$ | $b_i$ |
|---|---|---|---|---|---|
| $-1.0$ | 1.032 | $-.3$ | 1.139 | .4 | $-.415$ |
| $-.9$ | 1.563 | $-.2$ | .646 | .5 | $-.112$ |
| $-.8$ | 1.614 | $-.1$ | .474 | .6 | $-.817$ |
| $-.7$ | 1.377 | 0.0 | .418 | .7 | $-.234$ |
| $-.6$ | 1.179 | .1 | .067 | .8 | $-.623$ |
| $-.5$ | 1.189 | .2 | .371 | .9 | $-.536$ |
| $-.4$ | .910 | .3 | .183 | 1.0 | $-1.173$ |

**31.** Do a quadratic least squares fit to the following data. Use the standard form

$$g(t) = x_1 + x_2 t + x_3 t^2$$

and use the normal equation (9.7.26). What is the condition number of $A^T A$?

| $t_i$ | $b_i$ | $t_i$ | $b_i$ | $t_i$ | $b_i$ |
|---|---|---|---|---|---|
| $-1.0$ | 7.904 | $-.3$ | .335 | .4 | $-.711$ |
| $-.9$ | 7.452 | $-.2$ | $-.271$ | .5 | .224 |
| $-.8$ | 5.827 | $-.1$ | $-.963$ | .6 | .689 |
| $-.7$ | 4.400 | 0.0 | $-.847$ | .7 | .861 |
| $-.6$ | 2.908 | .1 | $-1.278$ | .8 | 1.358 |
| $-.5$ | 2.144 | .2 | $-1.335$ | .9 | 2.613 |
| $-.4$ | .581 | .3 | $-.656$ | 1.0 | 4.599 |

**32.** For the matrix $A$ arising in the least squares curve fitting of Problem 31, calculate its $QR$ factorization, its SVD, and its generalized inverse. Use these to again solve the least squares problem.

**33.** Find the $QR$ factorization, singular value decomposition, and generalized inverse of the following matrices. Also give $\text{cond}(A)_2$.

(a) $\quad A = \begin{bmatrix} .9 & 1.1 \\ -1.0 & -1.0 \\ 1.1 & .9 \end{bmatrix}$     (b) $\quad A = \begin{bmatrix} 1 & 2 & 3 \\ 2 & 3 & 4 \\ 3 & 4 & 5 \\ 4 & 5 & 6 \end{bmatrix}$

**34.**  **(a)**  Let $A$ be $m \times n$, $m \geq n$, and suppose that the columns of $A$ are nearly dependent. More precisely, let $A = [u_1, \ldots, u_n]$, $u_j \in \mathbf{R}^m$, and suppose the vector

$$v = \alpha_1 u_1 + \cdots + \alpha_n u_n$$

is quite small compared to $\|\alpha\|_2$, $\alpha = [\alpha_1, \ldots, \alpha_n]^T$. Show that $A$ will have a large condition number.

   **(b)**  In contrast to part (a), suppose the columns of $A$ are orthonormal. Show $\text{cond}(A)_2 = 1$.

# APPENDIX

# MATHEMATICAL SOFTWARE

Beginning with the early 1970s, some excellent computer programs have been written for the basic areas of numerical analysis. These are efficient, reliable, and portable to all brands of computers. This area of numerical analysis and computer science is now called *mathematical software*, a term dating from Rice (1971). There are now many high-quality program packages for most of the basic areas of numerical analysis. These are generally available as public domain software, and they have often been incorporated into the major commercial numerical analysis libraries. For a survey and historical account of many of these numerical analysis packages, see Cowell (1984), and for other texts that discuss such packages in a significant way, see Rice (1977), (1981), and (1983). The text Forsythe et al. (1977) supplies a few carefully chosen programs that are efficient and relatively easy to understand, for various numerical analysis problems.

There are two major commercial numerical analysis libraries for mainframe computers. These are the IMSL libraries and the NAG library, and both are available on all major lines of computers, including microcomputers. Both libraries have incorporated many of the specific packages that will be referred to in later paragraphs. Both companies are also developing special versions for use on vector pipeline computers. There are other commercial numerical analysis libraries, but generally they do not serve as wide a variety of computers.

We strongly urge the student and researcher to look first to these commercial libraries for their program needs. The programs are well written, with much care given to ease of use, accuracy and error control, and efficiency. They are also being improved continually, to reflect current research on numerical methods and the changing hardware of computers. When compared to the specific software packages that we list below, these large commercial libraries are much easier to use, and their implementation has been done by your local computer center staff, avoiding the need for the user to become involved with it. The specialized packages to be described later should be used only in those cases where the commercial packages do not fulfill the user's needs. Most of these specialized packages are static, having been written at a particular time and not having been updated to reflect changes in algorithms or computers.

Whenever possible, the individual should avoid writing his or her own codes for problems where there are standard packages available. It is very difficult and time consuming to write good computer programs, and it is unlikely that the typical user can come close to matching the performance of the codes in the standard commercial program libraries.

**661**

The IMSL library is divided into three major components: (1) numerical analysis library, (2) statistics library, and (3) special functions library. Each library is composed of subprograms written in Fortran 77, and these are meant to be called from a user-supplied program. In addition, IMSL has additional packages that make it easier to use their main library in a high-level interactive manner. IMSL also acts as a distribution agent for several of the public domain programs listed later. The address of IMSL is:

> IMSL, Inc.
> NBC Building
> 7500 Bellaire Boulevard
> Houston, Texas 77036-5085

As their libraries are being upgraded continually, write to IMSL for a description of the contents of current libraries.

The NAG library covers much the same subject material as the IMSL library, and it too has auxiliary packages to make easier use of the main library. The library is available in both Fortran and ALGOL. The programs in the library have been developed by academics at various United Kingdom universities, and many of the major special packages described later have also been absorbed into the NAG library. For more information, write to:

> Numerical Algorithms Group, Ltd.
> NAG Central Office
> Mayfield House
> 256 Banbury Road
> Oxford OX2 7DE, United Kingdom

A large number of high-quality computer packages for specific problem areas have been written since 1972. The first of these, and probably the best known, is the EISPACK package for solving matrix eigenvalue problems. Since then, there have been packages developed, in a variety of settings, for most of the basic areas of numerical analysis, and many of these have already been alluded to in the text. An excellent introduction to much of this software is provided by the book by Cowell (1984). We now list some of the packages of which we have knowledge, and apologize for omitting many others. In some cases, we also give a source for the codes. In addition, almost all of these codes (especially those produced at U.S. national laboratories) can be obtained, with some restrictions, from:

> National Energy Software Center
> Argonne National Laboratory
> 9700 South Cass Avenue
> Argonne, Illinois 60439

1.  *EISPACK*.   This package solves the matrix eigenvalue problem for a variety of types of matrices. It also contains programs for the generalized eigenvalue problem $Ax = \lambda Bx$ and for calculating the singular value decomposition of a matrix. The users guide for the package is given in Smith et al.

(1976) and Garbow et al. (1977). For a description of the development of the package, see Cowell (1984, chap. 4). IMSL is one of the distributors of this package.

2. **LINPACK.** This package is for the solution of systems of linear equations. It has four versions: real and complex arithmetic in both single and double precision arithmetic. In addition to the usual programs for various kinds of square nonsingular systems, LINPACK also contains programs for the least squares solution of linear systems. For the users guide, see Dongarra et al. (1979), and for a description of the development of LINPACK, see Cowell (1984, chap. 2). IMSL is one of the distributors of this package.

3. **MINPACK.** This package is for solving systems of nonlinear equations and optimization problems. At present, only the first version of the package is available, treating (a) the solution of $n$ nonlinear equations in $n$ unknowns, and (b) the least squares solution of overdetermined systems of nonlinear equations. A future version is intended to cover various other unconstrained and constrained optimization problems. For a users guide, see Moré et al. (1980), and for a description of the project, see Cowell (1984, chap. 5). IMSL is one of the distributors of the package.

4. **LLSQ.** This is a package for the least squares solution of systems of linear equations. It accompanies the book by Lawson and Hanson (1974), and it is available from IMSL.

5. **QUADPACK.** This package is for numerical integration of functions of a single variable. There are both general programs and programs for integrals with a special form. For a users guide, see Piessens et al. (1983).

6. **DEPAC.** This is a package of programs for solving systems of ordinary differential equations. At present it contains three main codes: a variable-order Adams code, a Runge–Kutta–Fehlberg fixed-order code, and a variable-order code for stiff problems based on backward differentiation formulas. These codes are available from Sandia National Laboratories in Albuquerque, New Mexico. For a discussion of software development for ordinary differential equations, see Cowell (1984, chap. 6).

7. **GRD1.** This is a package of programs for solving ordinary and partial differential equations, with the latter based on the method of lines. The codes for ordinary differential equations are based on those developed by Hindmarsh and his colleagues at Lawrence Livermore Laboratory. For information on obtaining the codes, write

> Dr. W. E. Schiesser
> Whitaker Laboratory No. 5
> Lehigh University
> Bethlehem, Pennsylvania 18015

8. **FISHPAK.** This is a package for solving the finite difference approximations to certain classic separable partial differential equations (e.g., the Poisson equation and the Helmholtz equation) on special regions, such as rectangles and spheres. It was developed for fluid mechanics calculations at

NCAR (National Center for Atmospheric Research). For a discussion of the package, see Swarztrauber and Sweet (1979). For a more general introduction to software for partial differential equations, see Cowell (1984, chap. 9). This includes discussions of ELLPACK [for elliptic partial differential equations, see Rice (1977), pp. 319–341] and ITPACK [for iterative methods of solving finite difference equations, see Hageman and Young (1981)].

In addition to the preceding packages, many programs are available as published algorithms in the *ACM Transactions on Mathematical Software*. See an issue of this journal for a list of some of the available algorithms. The actual programs are available from IMSL.

**An electronic software exchange**   A means of obtaining much of the preceding software by electronic mail has been set up at the Argonne National Laboratory in Chicago, Illinois. The system is called *NETLIB*, and it can be accessed through the electronic mail address:

NETLIB@MCS.ANL.GOV

Send the following message to this address to obtain directions for using *NETLIB*:

send index

This system is available at the time of writing this text, but there is no guarantee of its continued existence. It is a very useful means for obtaining high-quality mathematical software.

**Software for microcomputers**   The IMSL and NAG libraries have special subsets of their full libraries available on a number of micrcomputers, including IBM PC compatible machines. In addition, there are other packages continually being developed. Of special note are those packages that provide a friendlier access to sophisticated numerical analysis tools. Two current such packages are PC-MATLAB and GAUSS, both of which provide access to a number of the programs available in the EISPACK and LINPACK packages, along with other numerical analysis and graphics facilities. They are available on most microcomputers and scientific workstations. For information on these packages, write to these addresses:

GAUSS:

APTECH Systems, Inc.
P.O. Box 6487
Kent, Washington 98064

PC-MATLAB:

The MathWorks, Inc.
21 Eliot Street
South Natick, Massachusetts 01760
Telex: 9102405521
FAX: (508) 653-2997

# Bibliography

Cowell, W., Ed. (1984). *Sources and Development of Mathematical Software.* Prentice-Hall, Englewood Cliffs, N.J.

Dongarra, J., J. Bunch, C. Moler, and G. Stewart (1979). *LINPACK User's Guide.* SIAM Pub., Philadelphia.

Forsythe, G., M. Malcolm, and C. Moler (1977). *Computer Methods for Mathematical Computations.* Prentice-Hall, Englewood Cliffs, N.J.

Garbow, B., J. Boyle, J. Dongarra, and C. Moler (1977). *Matrix Eigensystem Routines—EISPACK Guide Extension. Lecture Notes in Computer Science 51.* Springer-Verlag, New York.

Hageman, L., and D. Young (1981). *Applied Iterative Methods.* Academic Press, New York.

Lawson, C., and R. Hanson (1974). *Solving Least Squares Problems.* Prentice-Hall, Englewood Cliffs, N.J.

Moré, J., B. Garbow, and K. Hillstrom (1980). *User Guide for MINPACK-1.* Argonne National Laboratory Rep. ANL-80-74, Chicago, Ill.

Piessens, R., E. deDoncker-Kapenga, C. Überhuber, and D. Kahaner (1983). *QUADPACK: A Subroutine Package for Automatic Integration.* Springer-Verlag, New York.

Rice, J., Ed. (1971). *Mathematical Software.* Academic Press, New York.

Rice, J., Ed. (1977). *Mathematical Software III.* Academic Press, New York.

Rice, J. (1981). *Matrix Computations and Mathematical Software.* McGraw-Hill, New York.

Rice, J. (1983). *Numerical Methods, Software, and Analysis.* McGraw-Hill, New York.

Smith, B., J. Boyle, B. Garbow, Y. Ikebe, V. Klema, and C. Moler (1976). *Matrix Eigensystem Routines—EISPACK Guide,* 2nd ed., *Lecture Notes in Computer Science 6.* Springer-Verlag, New York.

Swarztrauber, P., and R. Sweet (1979). Algorithm 541: Efficient Fortran subprograms for the solution of separable elliptic partial differential equations, *ACM Trans. Math. Softw.* **5**, 352–364.

# ANSWERS TO SELECTED EXERCISES

## Chapter 1

5. **(a)** $p_{2n-2}(x) = \sum_{j=0}^{n-1} (-1)^j \dfrac{x^{2j}}{(2j+1)(j!)}$

$|\text{Error}| \leq \dfrac{|x|^{2n}}{(2n+1)(n!)}$

7. **(a)** $p_1(x, y) = 1 + x - \frac{1}{2}y$, $p_2(x, y) = p_1(x+y) - \frac{1}{2}x^2 + \frac{1}{2}xy - \frac{1}{8}y^2$

10. **(a)** 21.625 **(c)** $\frac{2}{3}$ **(d)** $\frac{2}{3}$

11. **(b)** 1111111

12. **(a)** .1101 **(c)** .00011001100...

21. **(a)** 4 **(b)** 2 **(c)** 3

22. **(a)** [2.05265, 2.05375]

**(d)** $\left[ \dfrac{8.4725}{.0645}, \dfrac{8.4735}{.0635} \right] \doteq [131.3566, 133.4409]$

26. **(b)** For $x \doteq 0$, the quadratic Taylor series approximation is

$$f(x) \doteq \tfrac{-5}{24} - \tfrac{11}{48}x - \tfrac{379}{1920}x^2$$

Also, $f(0) = \frac{-5}{24}$ exactly.

28. **(a)** $-.0030$ **(c)** $.00068$

30. **(a)** Error $\doteq -1.46(x_T - x_A) - 1.71(y_T - y_A)$, with $x_A = 3.14$, $y_A = 2.685$, and $x_T, y_T$ denoting the true unrounded numbers associated with $x_A, y_A$. Also, $|\text{Relative error}| \leq .0098$

## Chapter 2

1. $\prod_{j=0}^{\infty}[1 + r^{2^j}] = 1/(1-r)$, and the infinite product converges if and only if $|r| < 1$.

3. **(a)** root = 4.493409458

**4.** **(b)** 1.8392868 **(d)** 1.1284251

**5.** **(a)** 4.493409458 **(b)** 98.95006282

**6.**

| B | Root |
|---|---|
| 1 | − .5884017765 |
| 5 | − .4049115482 |
| 10 | − .3265020101 |
| 25 | − .2374362439 |
| 50 | − .1832913333 |

**12.** **(e)** $|\text{Rel}(x_4)| \le 2^{-15}10^{-16} \doteq 3.05 \times 10^{-21}$

**13.** The iterate $x_4$ will be sufficiently accurate.

**15.** **(a)** does not converge **(b)** converges

**19.** $\alpha = 2.1322677$, $[a, b] = [1, 1 + \pi/2]$

**24.** **(a)** diverges **(c)** converges

**25.** $\displaystyle \text{Limit}_{n \to \infty} \frac{\sqrt{a} - x_{n+1}}{\left(\sqrt{a} - x_n\right)^3} = \frac{1}{4a}$

**29.** **(a)** rate $\doteq$ .625

**39.** **(b)** 2.470638970 + 4.640533162i

**40.** **(a)** This polynomial is the degree 12 Legendre polynomial on the interval $-1 \le x \le 1$. Its roots are

| | |
|---|---|
| ± .1252334085 | ± .7699026742 |
| ± .3678314990 | ± .9041172564 |
| ± .5873179543 | ± .9815606342 |

**51.**

| $(x_0, y_0)$ | Method Converges To |
|---|---|
| (1.2, 2.5) | (1.336355377, 1.754235198) |
| (−2, 2.5) | (−.9012661908, −2.086587595) |
| (−1.2, −2.5) | (−.9012661908, −2.086587595) |
| (2.0, −2.5) | (−3.001624887, .1481079950) |

In the last case, the method jumps around in a seemingly random fashion, and after 16 iterations it begins to converge to the root previously given. The equations do have another root.

**52.** **(a)** There are two roots, $(a, b)$ and $(b, a)$, with $a = .673007170$, $b = 1.94502682$

# Chapter 3

**4.** |Interpolation error| $\le (h^2/8)e^2 \doteq 9.2 \times 10^{-5}$
|Error due to rounding| $\le 5 \times 10^{-5}$
|Total error| $\le 1.42 \times 10^{-4}$

**7.** Choose $h = .001$, and let the table entries have seven significant digits. Then the total error will be bounded by $8.4 \times 10^{-7}$.

**10.** Choose the table entries to have six significant digits. One possible partition of $1 \leq x \leq 10$ with the suggested grid size and resulting total interpolation error is given below.

| Interval | $h$ | Total error |
|---|---|---|
| $1 \leq x \leq 2$ | .01 | $6.81 \times 10^{-7}$ |
| $2 \leq x \leq 3$ | .025 | $7.35 \times 10^{-7}$ |
| $3 \leq x \leq 6$ | .05 | $8.85 \times 10^{-7}$ |
| $6 \leq x \leq 10$ | .1 | $8.85 \times 10^{-7}$ |

**20.** degree 3

**22.** The sixth entry 419327 should be changed to 419237 or 419238. There are two other errors, and their higher order differences have overlapping effects.

**29.** In order that there be a unique interpolating polynomial of degree $\leq 2$, it is necessary and sufficient that $x_1 \neq \frac{1}{2}(x_0 + x_2)$.

**34.** $s(x)$ on $[x_0, x_2]$ is given by

$$s(x) = \frac{(x_2 - x)^3 M_0 + (x - x_0)^3 M_2}{6(h_0 + h_1)} + \frac{(x_2 - x)y_0 + (x - x_0)y_2}{h_0 + h_1}$$

$$- \frac{h_0 + h_1}{6}\left[(x_2 - x)M_0 + (x - x_0)M_2\right]$$

Form $s(x_1) = y_1$ to set up an interpolating condition at $x_1$, giving an equation for $M_0$ and $M_2$. A similar derivation applies for $s(x)$ on $[x_{n-2}, x_n]$, with $s(x_{n-1}) = y_{n-1}$.

**36.** **(i)** For $n = 96$ subintervals, the error in $s(x)$ for the various boundary conditions are

**(a)**  1.35E − 10   **(b)**  8.32E − 11   **(c)**  8.25E − 10

**39.** $B_i^{(m)}(x) = \dfrac{B_{i+1}^{(m-1)}(x)}{x_{i+m} - x_{i+1}} - \dfrac{B_i^{(m-1)}(x)}{x_{i+m-1} - x_i}$

**43.** **(a)** $d_0 = 1$; $d_k = 0$ for $1 \leq k \leq m - 1$

## Chapter 4

**1.** The Bernstein polynomial of degree 4 is

$$p_4(x) = 2\sqrt{2}\,x(1 - x)^3 + 6x^2(1 - x)^2 + 2\sqrt{2}\,x^3(1 - x)$$

The Taylor polynomial of degree 4 is

$$f_4(x) = 1 - \frac{\pi^2}{2}\left(x - \frac{1}{2}\right)^2 + \frac{\pi^4}{24}\left(x - \frac{1}{2}\right)^4$$

The following table gives the values of $p_4(x)$ and $f_4(x)$ at several points, along with the errors when compared to $\sin(\pi x)$.

| $x$ | $p_4(x)$ | Error | $f_4(x)$ | Error |
|-----|----------|-------|----------|-------|
| 0.0 | 0.0 | 0.0 | 0.200 | $-2.0E-2$ |
| .1 | .2573 | .0517 | .3143 | $-5.3E-3$ |
| .2 | .4613 | .1265 | .5887 | $-9.6E-4$ |
| .3 | .6091 | .1999 | .8091 | $-8.5E-5$ |
| .4 | .6986 | .2525 | .9511 | $-1.3E-6$ |
| .5 | .7286 | .2714 | 1.0000 | 0.0 |

3. **(b)** Let

$$S = \sum_{1}^{\infty} \frac{(-1)^j x^{2j}}{j^2}$$

The series has the interval of convergence $|x| \le 1$. For the error,

$$\left| S - \sum_{j=1}^{n} (-1)^j \cdot \frac{x^{2j}}{j^2} \right| \le \frac{|x|^{2n+2}}{(n+1)^2}$$

If $n \ge 316$, the error is less than $10^{-5}$ for all $|x| \le 1$. But this is likely to be much too large for $|x| < 1$, since it ignores the factor of $|x|^{2n+2}$ in the error.

6. The Pade approximation is $R(x) = (1 + \frac{1}{2}x)/(1 - \frac{1}{2}x)$. For the error,

$$e^x - R(x) = -\left[\frac{1}{12}x^3 + \frac{1}{12}x^4 + \frac{13}{240}x^5 + \cdots\right]$$

and for the Taylor series error, $e^x - p_2(x) = \frac{1}{6}x^3 + \frac{1}{24}x^4 + \cdots$. The error in $R(x)$ is about half that of $p_2(x)$ for small $|x|$. But for $x = 1$, the error is larger with $R(x)$.

10. **(a)** $p_1(x) = \ln(\frac{3}{2}) - 1 + \frac{2}{3}x$; $\|\ln(x) - p_1\|_\infty = .072$

 **(b)** $q_1^*(x) = -.6633 + .6931x$; $\|\ln(x) - q_1^*\|_\infty = .030$

13. The minimizing value of $\alpha$ is $\alpha = \sqrt{e}$. The minimum value is $e + 1 - 2\sqrt{e} \doteq .42$.

20. **(a)** $\psi_0(x) \equiv 1$, $\psi_1(x) = x - \frac{1}{4}$, $\psi_2(x) = x^2 - \frac{5}{7}x + \frac{17}{252}$

28. **(a)** $\rho_n(\sin x) \le \dfrac{\pi^{n+1}}{(n+1)!2^{3n+2}}$

31. **(c)** $I_3(x) = .968706x - .187130x^3$; $\|\tan^{-1}x - I_3\|_\infty = .00590$

 **(d).**

| $n$ | $\|f - I_n\|_\infty$ | $n$ | $\|f - I_n\|_\infty$ |
|-----|----------------------|-----|----------------------|
| 1 | 3.72E $-$ 2 | 5 | 1.14E $-$ 5 |
| 2 | 4.37E $-$ 3 | 6 | 1.69E $-$ 6 |
| 3 | 5.72E $-$ 4 | 7 | 2.55E $-$ 7 |
| 4 | 7.94E $-$ 5 | 8 | 3.91E $-$ 8 |

**32.** **(c)** $F_3(x) = .972420x - .191898x^3$; $\|\tan^{-1} x - F_3\|_\infty = .00500$

**36.** $q_{n-1}^*(x) = p(x) - \dfrac{a_n}{2^{n-1}} T_n(x)$, $\rho_{n-1}(p) = \dfrac{|a_n|}{2^{n-1}}$

**40.** **(d)**

| $n$ | $\rho_n(f)$ | $n$ | $\rho_n(f)$ |
|---|---|---|---|
| 1 | 2.98E − 2 | 5 | 8.69E − 6 |
| 2 | 3.42E − 3 | 6 | 1.28E − 6 |
| 3 | 4.42E − 4 | 7 | 1.92E − 7 |
| 4 | 6.07E − 5 | 8 | 2.93E − 8 |

# Chapter 5

**1.**

(a) $\displaystyle\int_0^1 e^{-x^2}\,dx$   (c) $\displaystyle\int_{-4}^4 \frac{dx}{1+x^2}$

| $n$ | $I_n$ | $R_n$ | $n$ | $I_n$ | $R_n$ |
|---|---|---|---|---|---|
| 2 | .73137025 | | 2 | 4.235294 | |
| 4 | .74298410 | | 4 | 2.917647 | |
| 8 | .74586561 | 4.03 | 8 | 2.658824 | 5.09 |
| 16 | .74658460 | 4.01 | 16 | 2.650507 | 31.1 |
| 32 | .74676425 | 4.00 | 32 | 2.651347 | −9.90 |
| 64 | .74680916 | 4.00 | 64 | 2.651563 | 3.89 |
| 128 | .74682039 | 4.00 | 128 | 2.651617 | 4.00 |
| 256 | .74682320 | 4.00 | 256 | 2.651631 | 4.00 |

**2.**

(a) $\displaystyle\int_0^1 e^{-x^2}\,dx$   (c) $\displaystyle\int_{-4}^4 \frac{dx}{1+x^2}$

| $n$ | $I_n$ | $R_n$ | $n$ | $I_n$ | $R_n$ |
|---|---|---|---|---|---|
| 2 | .7471804289095 | | 2 | 5.4901960784 | |
| 4 | .7468553797910 | | 4 | 2.4784313725 | |
| 8 | .7468261205275 | 11.1 | 8 | 2.5725490196 | −32.0 |
| 16 | .7468242574357 | 15.7 | 16 | 2.6477345635 | 1.25 |
| 32 | .7468241406070 | 15.9 | 32 | 2.6516272830 | 19.3 |
| 64 | .7468241332997 | 16.0 | 64 | 2.6516352807 | 487 |
| 128 | .7468241328429 | 16.0 | 128 | 2.6516353244 | 183 |
| 256 | .7468241328143 | 16.0 | 256 | 2.6516353272 | 16.0 |

**6.** The ratios $R_n$ approached constants with increasing $n$. The values for $n = 16$ are given in the following:

| $\alpha$ | .25 | .5 | .75 | 1.0 |
|---|---|---|---|---|
| $R_n$ | 2.11 | 2.52 | 3.03 | 3.64 |

**9.**  $I_h(f) = \frac{3}{4}h[f(0) + 3f(2h)]$

**15.**

| $n$ | $I_n$ | Error |
|---|---|---|
| 2 | 1.26316 | 1.388 |
| 4 | 2.04729 | .6044 |
| 6 | 2.41169 | .2399 |
| 8 | 2.56008 | .09156 |
| 10 | 2.61725 | .03438 |

**17.**  $\int_0^1 f(x) \ln\left(\frac{1}{x}\right) dx \doteq w_1 f(x_1) + w_2 f(x_2)$

$x_1 = \dfrac{15 - \sqrt{106}}{42} \doteq .1120088062$   $\qquad$   $x_2 = \dfrac{15 + \sqrt{106}}{42} \doteq .6022769081$

$w_1 = \dfrac{21}{\sqrt{106}}\left[x_2 - \dfrac{1}{4}\right] \doteq .718539319$   $\qquad$   $w_2 = 1 - w_1 = .281460681$

**20.**  For Gaussian quadrature, $|I - I_n| \le (.080/n^4)\|f^{(4)}\|_\infty$. For Simpson's rule on $[-1, 1]$, $|I - I_n| \le (.18/n^4)\|f^{(4)}\|_\infty$.

**24.**  $\displaystyle\sum_1^\infty \frac{1}{n^{5/4}} = 4.59511254 + E$   $\qquad$   $0 < -E < 1.34 \times 10^{-6}$

**26.**  Assuming that $I - I_n \doteq c/n^p$, and using $I_{16}$, $I_{32}$, and $I_{64}$, we obtain $p \doteq 3.44$, $c \doteq 0.0154$. For the Aitken extrapolate, $\tilde{I}_{64} = .28571428586$, and $I - \tilde{I}_{64} \doteq \tilde{I}_{64} - I_{64} = 9.43 \times 10^{-9}$. To have $|I - I_n| \le 10^{-11}$ will require $n \ge 469$. The logical choice to use would be $n = 512$.

**28.**  $\tilde{I}_n = \frac{1}{3}[I_n^{(T)} + 2I_n^{(M)}]$, which will be Simpson's rule.

**38.**  **(a)**  Split integral, $I = \left(\displaystyle\int_0^1 + \int_1^{4\pi}\right) \cos(x) \ln(x)\, dx$. Using a Taylor series,

$$\int_0^1 \cos(x) \ln(x)\, dx \doteq -\left(1 - \frac{1}{2!9} + \frac{1}{4!25} - \frac{1}{6!49} + \frac{1}{8!81}\right)$$

$$\doteq -.9460830727$$

which is in error by less than $2.28 \times 10^{-9}$. For the remaining integral, a standard method can be used to compute it. For example, Romberg integration with $n = 129$ nodes leads to the value $-.5460781580$, which is in error by less than $2.8 \times 10^{-9}$. Thus the integral is $I = 1.4921612307$, in error by less than $5.1 \times 10^{-9}$.

**(b)**  Since $|x^2 \sin(1/x)| \le x^2$, pick $\epsilon > 0$ so that

$$\left|\int_0^\epsilon x^2 \sin\left(\frac{1}{x}\right) dx\right| \le \int_0^\epsilon x^2\, dx \le .0005$$

A convenient choice is $\epsilon = .1$. Then compute $\displaystyle\int_{.1}^{2/\pi}(x^2)\sin(1/x)\, dx$ with an accuracy of at least .0005.

**41. (a)** $D_h^{(2)}f(x) = \dfrac{f(x) - 2f(x + h) + f(x + 2h)}{h^2}$

$$f''(x) - D_h^{(2)}f(x) \doteq hf^{(3)}(x)$$

# Chapter 6

**2. (a)** 2 **(b)** 1 **(c)** 0 **(d)** 20

**3. (b)** $\begin{aligned} y_1' &= y_2 \\ y_2' &= .1(1 - y_1^2)y_2 - y_1 \end{aligned}$   $\begin{aligned} y_1(0) &= 1 \\ y_2(0) &= 0 \end{aligned}$

**5. (c)** True solution $Y(x) = 2e^x + \sin(x) - \cos(x)$. The Picard iterates are

$$Y_0(x) \equiv 1, \quad Y_1(x) = 1 + x + 2\sin(x)$$

$$Y_2(x) = 3 + x + \frac{x^2}{2} + 2\sin(x) - 2\cos(x)$$

$$Y_3(x) = 3 + 3x + \tfrac{1}{2}x^2 + \tfrac{1}{6}x^3 - 2\cos(x)$$

Compare them for $x \doteq 0$ by using Taylor series expansions of them and $Y(x)$.

**8.** Assume the initial error $e_0 = 0$. For the bound from (6.2.13),

$$\underset{0 \le x \le b}{\text{Max}} |Y(x_n) - y_n| \le \frac{h}{2}(e^{2b} - 1)$$

The asymptotic error is given by

$$Y(x) - y_h(x) = h\frac{\ln(1 + x)}{(1 + x)^2} + O(h^2) \qquad x \ge 0$$

This shows the error decreases with increasing $x$, whereas the previous bound predicts an increasing error as the interval $[0, b]$ increases.

**10. (b)** $y_{1, n+1} = y_{1, n} + hy_{2, n}$

$$y_{2, n+1} = y_{2, n} + h[.1(1 - y_{1, n}^2)y_{2, n} - y_{1, n}] \qquad n \ge 0$$

**14.** $T_{n+1}(Y) = -\tfrac{5}{8}h^3Y^{(3)}(x_n) + O(h^4)$

**26.** For $c = \tfrac{1}{4}$, $u_n = c_1 + c_2(\tfrac{1}{2})^n + c_3(-\tfrac{1}{2})^n$, with $c_1$, $c_2$, and $c_3$ arbitrary. As $n \to \infty$, $u_n \to c_1$, a constant.

**31. (a)** $0 \le a_0 < 2$   **(b)** $a_0 = 0$

**(c)** At $a_0 = 0$, there is no region of absolute stability. As $a_0$ goes from 0 toward 2, the region of absolute stability increases to $-2 < h\lambda < 0$ in the limit. This is only for the real values in the absolute stability region.

**33. (a)** $a_0 = -9 - 3b_2$     $a_1 = 9$     $a_2 = 1 + 3b_2$
        $b_0 = 6 + b_2$       $b_1 = 6 + 4b_2$

**(b)** The characteristic equation for $h = 0$ is

$$p(r) \equiv r^3 - a_0 r^2 - 9r + a_0 + 8 = 0$$

Factoring out $r - 1$, the remaining two roots satisfy

$$r^2 + (1 - a_0)r - (a_0 + 8) = 0 \qquad\qquad (\#)$$

The discriminant is

$$(1 - a_0)^2 + 4(a_0 + 8) = (1 + a_0)^2 + 32 \geq 32 > 0$$

for all real values of $a_0$. Thus both of the roots $r_1$ and $r_2$ of equation $(\#)$ must be real, for any value of $a_0$. A direct examination of these roots will show that they cannot both satisfy $-1 < r < 1$ simultaneously for the same value of $a_0$.

**36.** $y_{n+1} = y_{n-1} + 2hf(x_{n-1}, y_{n-1})$ is first order and relatively stable, but it doesn't satisfy the strong root condition.

**40.** Use $Y(x) = \frac{1}{2}x^2$, so that $Y''(x) \equiv 1$. Then the error $e_n = Y(x_n) - y_n$ satisfies

$$e_n = \frac{h}{2\lambda}\left[(1 + h\lambda)^n - 1\right] \qquad n \geq 0$$

**48.**

$$\gamma_1 + \gamma_2 + \gamma_3 = 1$$

$$\gamma_2\alpha_2 + \gamma_3\alpha_3 = \tfrac{1}{2}$$

$$\gamma_2\beta_{21} + \gamma_3(\beta_{31} + \beta_{32}) = \tfrac{1}{2}$$

$$\gamma_2\alpha_2^2 + \gamma_3\alpha_3^2 = \tfrac{1}{3}$$

$$\gamma_2\alpha_2\beta_{21} + \gamma_3\alpha_3(\beta_{31} + \beta_{32}) = \tfrac{1}{3}$$

$$\gamma_2\beta_{21}^2 + \gamma_3(\beta_{31} + \beta_{32})^2 = \tfrac{1}{3}$$

$$\gamma_3\alpha_3\beta_{32} = \tfrac{1}{6}$$

$$\gamma_3\beta_{32}\beta_{21} = \tfrac{1}{6}$$

These equations are dependent and can be reduced to six independent equations. One particular solution is

$$Y_{n+1} = Y_n + \frac{h}{6}(V_1 + 4V_2 + V_3)$$

$$V_1 = f(x_n, y_n) \qquad V_2 = f(x_n + \tfrac{1}{2}h, y_n + \tfrac{1}{2}hV_1)$$

$$V_3 = f[x_n + h, y_n + h(2V_2 - V_1)]$$

This is Simpson's rule if $f(x, y)$ does not depend on $y$.

51.  **(a)**  The roots $h\lambda$ of $|1 + h\lambda + \tfrac{1}{2}(h\lambda)^2| < 1$. For $\lambda$ real, this is the interval $-2 < h\lambda < 0$.

# Chapter 7

1.  **(a)** dependent    **(b)** independent
7.  **(a)**  $(Ax, x) = x^T(Ax)$. But also,

$$(Ax, x) = (x, Ax) = (Ax)^T x = x^T A^T x = -x^T A x$$

Since $x^T A x = -x^T A x$, we must have $x^T A x = 0$, the desired result.

8.  $u^{(3)} = (-7, 4, 1)$. To normalize, divide the respective vectors by their lengths:

$$\|u^{(1)}\|_2 = \sqrt{6} \qquad \|u^{(2)}\|_2 = \sqrt{11} \qquad \|u^{(3)}\|_2 = \sqrt{66}$$

9.  **(a)**  $A = \frac{1}{9}\begin{bmatrix} 7 & -4 & -4 \\ -4 & 1 & -8 \\ -4 & -8 & 1 \end{bmatrix}$

11.  **(a)**  $\lambda_1 = -1, \qquad x = [2, -1]^T$
     $\lambda_2 = 3, \qquad x = [2, 1]^T$

13.  **(a)**  $\|Ux\|_2^2 = (Ux, Ux) = (x, U^*Ux) = (x, Ix) = (x, x) = \|x\|_2^2$, showing the desired result: $\|Ux\|_2 = \|x\|_2$ for all $x$. For the distance between $Ux$ and $Uy$,

$$\|Ux - Uy\|_2 = \|U(x - y)\|_2 = \|x - y\|_2$$

from the earlier result.

14.  Since $A$ is Hermitian, let $x^{(1)}, \dots, x^{(n)}$ be an orthonormal basis for $\mathbf{C}^n$ corresponding to the real eigenvalues $\lambda_1, \dots, \lambda_n$. For any $x$ in $\mathbf{C}^n$, we can write

$$x = \sum_1^n \alpha_i x^{(i)} \qquad \text{with } \alpha_i = (x, x^{(i)}) \quad \text{for} \quad i = 1, \dots, n$$

From the orthonormality, $\|x\|_2^2 = \sum_1^n |\alpha_i|^2$. Using this form for $x$, show

$$(Ax, x) = \sum_1^n \lambda_i |\alpha_i|^2$$

Since $\alpha_1, \ldots, \alpha_n$ can be varied arbitrarily to obtain various elements $x$ of $\mathbf{C}^n$, this formula can be used to prove that $A$ is positive definite if and only if all $\lambda_i$ are positive.

23.    First prove

$$\|x\|_\infty \le \|x\|_p \le n^{1/p}\|x\|_\infty \qquad \text{all } x \in \mathbf{C}^n$$

It then follows easily that $\|x\|_p \to \|x\|_\infty$ as $p \to \infty$.

27.    (a)    Let $\lambda$ and $x$ be an eigenvalue–eigenvector pair for the matrix $A$. Then the associated eigen pairs for various modifications of $A$ are: (i) $\lambda^m$ and $x$ for $A^m$; (ii) $1/\lambda$ and $x$ for $A^{-1}$; and (iii) $\lambda + c$ and $x$ for $A + cI$.

28.    Write $A = [A_{*1}, \ldots, A_{*n}]$ using the columns $A_{*j}$ of $A$. Then

$$F(A) = \sqrt{\|A_{*1}\|_2^2 + \cdots + \|A_{*n}\|_2^2}$$

(a)    $F(UA) = \sqrt{\|UA_{*1}\|_2^2 + \cdots + \|UA_{*n}\|_2^2}$. From Problem 13(a), $\|UA_{*j}\|_2 = \|A_{*j}\|_2$ for all $j$. Thus $F(UA) = F(A)$.

(b)    Let $U^*AU = D = \text{diag}[\lambda_1, \ldots, \lambda_n]$. Apply part (a) to obtain

$$F(A) = F(U^*AU) = F(D) = \sqrt{\lambda_1^2 + \cdots + \lambda_n^2}$$

32.    $\|A\|_\infty \le 10$, $\|A^{-1}\|_\infty < \frac{1}{2}$

# Chapter 8

1.    (a)

$$L = \begin{bmatrix} 1 & 0 & 0 \\ 1 & 1 & 0 \\ -2 & 3 & 1 \end{bmatrix} \qquad U = \begin{bmatrix} 1 & 1 & -1 \\ 0 & 1 & -1 \\ 0 & 0 & 2 \end{bmatrix} \qquad x = \begin{bmatrix} 2 \\ 2 \\ 3 \end{bmatrix}$$

2.    (a)    Without pivoting, the augmented matrix $[A|b]$ is reduced to

$$\begin{bmatrix} 6.000 & 2.000 & 2.000 & -2.000 \\ 0 & .0001000 & -.3333 & 1.667 \\ 0 & 0 & 5555 & -27790 \end{bmatrix} \begin{matrix} m_{21} = .3333 \\ m_{31} = .1667 \\ m_{32} = 16670 \end{matrix}$$

The solution by back substitution is

$$x_1 = 1.335 \qquad x_2 = 0 \qquad x_3 = -5.003$$

All arithmetic operations were carried out using four-digit decimal floating-point arithmetic with rounding, and the same is true in the following part (b).

**(b)** $[A|b]$ is reduced, with pivoting, to

$$\begin{bmatrix} 6.000 & 2.000 & 2.000 & -2.000 \\ 0 & 1.667 & -1.333 & .3334 \\ 0 & 0 & .3332 & 1.667 \end{bmatrix} \begin{matrix} m_{21} = .3333 \\ m_{31} = 1.667 \\ m_{32} = .00005999 \end{matrix}$$

The solution by back substitution is

$$x_1 = 2.602 \qquad x_2 = -3.801 \qquad x_3 = -5.003$$

**5. (a)** In partitioned form,

$$\begin{bmatrix} A_1 & -A_2 \\ A_2 & A_1 \end{bmatrix} \begin{bmatrix} x_1 \\ x_2 \end{bmatrix} = \begin{bmatrix} b_1 \\ b_2 \end{bmatrix}$$

**(b)** Let system 1 denote the real system of part (a), and let system 2 denote the original complex system $Ax = b$. For the matrix storage requirements, system 1 requires $4n^2$ locations, and system 2 requires $2n^2$ locations (each complex number requires two storage locations). To solve system 1 requires about $\frac{1}{3}(2n)^3 = \frac{8}{3}n^3$ multiplications and divisions. System 2 requires $\frac{1}{3}n^3$ complex multiplications and divisions. Since each complex multiplication requires four real multiplications, the actual operation count is $\frac{4}{3}n^3$. Thus system 2 requires half the storage requirements and about half the operation time of system 1.

**9.** $(Ax, x) = (LL^T x, x) = (L^T x, L^T x) = \|L^T x\|_2^2 > 0$ for all $x \neq 0$, since $L^T$ in nonsingular. Also $\det(A) = \det(L)^2$.

**10. (a)**

$$L = \begin{bmatrix} 1.5 & 0 & 0 \\ -2.0 & 1.0 & 0 \\ 3.0 & -4.0 & 3.0 \end{bmatrix}$$

**14.**

$$L = \begin{bmatrix} 2 & 0 & 0 & 0 & 0 \\ 1 & \frac{5}{2} & 0 & 0 & 0 \\ 0 & 1 & \frac{12}{5} & 0 & 0 \\ 0 & 0 & 1 & \frac{29}{12} & 0 \\ 0 & 0 & 0 & 1 & \frac{70}{29} \end{bmatrix} \qquad U = \begin{bmatrix} 1 & -\frac{1}{2} & 0 & 0 & 0 \\ 0 & 1 & -\frac{2}{5} & 0 & 0 \\ 0 & 0 & 1 & -\frac{5}{12} & 0 \\ 0 & 0 & 0 & 1 & -\frac{12}{29} \\ 0 & 0 & 0 & 0 & 1 \end{bmatrix}$$

$Lz = b$ has solution $z = [\frac{3}{2}, -\frac{7}{5}, \frac{17}{12}, -\frac{41}{29}, 1]^T$
$Ux = z$ has solution $x = [1, -1, 1, -1, 1]^T$

**18. (a)** $\mathrm{cond}(A)_1 = \mathrm{cond}(A)_\infty = 39601$, $\mathrm{cond}(A)_2 = 39206$

**26.**    For the rates of convergence $\mu$ of (8.6.5) and $\eta$ of (8.6.17) in this case, $\mu = \frac{3}{4}$ and $\eta = \frac{2}{3}$. The actual observed rates of decrease are .61 for the Gauss–Jacobi and .37 for the Gauss–Seidel methods.

**27.**    **(a)**    Converges if and only if $r_\sigma(A^{-1}B) < 1$, and the rate of convergence is essentially $r_\sigma(A^{-1}B)$ or $\|A^{-1}B\|$, depending on the norm used.

   **(b)**    Again it converges if and only if $r_\sigma(A^{-1}B) < 1$. But the rate of convergence is about $r_\sigma(A^{-1}B)^2$ or $\|A^{-1}B\|^2$, which is faster than part (a).

**32.**    **(a)**    $A^{-1} - C_m = A^{-1}R_m = A^{-1}R_0^{2^m}$, $m \geq 0$. The term $C_m$ converges to $A^{-1}$ if and only if $r_\sigma(R_0) < 1$.

   **(b)**    Writing $(I - R_0)^{-1}$ as an infinite product,

$$A^{-1} = C_0(I + R_0)(I + R_0^2)(I + R_0^4)(I + R_0^8)\ldots$$

$$C_m = C_0(I + R_0)(I + R_0^2)(I + R_0^4)\cdots(I + R_0^{2^{m-1}})$$

# Chapter 9

**1.**    **(a)**    Applying the Gerschgorin theorem to the rows of $A$, the circles are $|\lambda - 1| \leq 1$, $|\lambda - 5| \leq 2$, and $|\lambda - 9| \leq 3$. The second and third circles intersect, and thus they must contain two eigenvalues in their union. The first circle is distinct from the others. Since the characteristic polynomial of $A$ has real coefficients, the eigenvalues occur in conjugate pairs if they are complex. Thus $|\lambda - 1| \leq 1$ contains one real eigenvalue, in the interval $[0, 2]$.

   Applying the theorem to the columns of $A$, the circles are $|\lambda - 1| \leq 3$, $|\lambda - 5| \leq 2$, and $|\lambda - 9| \leq 1$. By the same type of argument as before, and using the previous results, there are real eigenvalues in each of the intervals $[0, 2]$, $[3, 7]$, and $[8, 10]$. The true eigenvalues are

$$1.331927689 \qquad 4.856933692 \qquad 8.811138616$$

**2.**    **(c)**    The $n$ circles are

$$|r| \leq |a_0| \qquad |r| \leq 1 + |a_j| \qquad j = 1, \ldots, n - 2 \quad \text{and} \quad |r + a_{n-1}| \leq 1$$

   **(d)**    **(i)**    There are nine roots satisfying $|r - 1| \leq 1$ and one real root satisfying $-9 \leq r \leq -7$.

**4.**    $\displaystyle \min_{1 \leq i \leq 4} |\lambda - \lambda_i| \leq .0000356$

**8.**    **(a)**    $\lambda = 15$, $x = [1, 1, 1, 1]^T$. The error decreases by a factor of $\frac{1}{3}$ with each iterate. Thus $\lambda = 5$ is likely to be the second largest eigenvalue.

**13.**    **(b)**    Using $w = [1/\sqrt{6}, -1/\sqrt{6}, -2/\sqrt{6}]^T$ to construct $P = I - 2ww^T$, we obtain $Px = e_1$. Thus

$$B = PAP^T = \begin{bmatrix} 9 & 0 & 0 \\ 0 & 18 & 0 \\ 0 & 0 & -9 \end{bmatrix}$$

In this case the matrix $B$ is diagonal, but ordinarily it wouldn't be so simple. Using $w = (1/\sqrt{30}\,)[5,1,2]^T$ will lead to a $P$ for which $Px = -e_1$. But $B = PAP^T$ will still have the same form for column and row 1. In particular,

$$B = \begin{bmatrix} 9 & 0 & 0 \\ 0 & \frac{18}{25} & -\frac{324}{25} \\ 0 & -\frac{325}{25} & \frac{207}{25} \end{bmatrix}$$

The first form of $P$ was constructed with the opposite sign to that in (9.3.9), while the second choice of $P$ was constructed according to (9.3.9). Ordinarily the form of $P$ agreeing with (9.3.9) would be the preferred form, because of the accuracy considerations pointed out following (9.3.9).

**14.**

$$Q = \frac{1}{3}\begin{bmatrix} -1 & 2 & 2 \\ -2 & 1 & -2 \\ -2 & -2 & 1 \end{bmatrix} \qquad R = 3\begin{bmatrix} -1 & 1 & -1 \\ 0 & 1 & -1 \\ 0 & 0 & 1 \end{bmatrix}$$

**15.  (a)**  $\alpha = \dfrac{b_k}{\gamma}, \ \beta = \dfrac{b_l}{\gamma}$

**(b)**

$$R^{(1,4)} = \begin{bmatrix} \dfrac{1}{\sqrt{2}} & 0 & 0 & \dfrac{1}{\sqrt{2}} \\ 0 & 1 & 0 & 0 \\ 0 & 0 & 1 & 0 \\ -\dfrac{1}{\sqrt{2}} & 0 & 0 & \dfrac{1}{\sqrt{2}} \end{bmatrix} \qquad R^{(1,3)} = \begin{bmatrix} \dfrac{\sqrt{2}}{\sqrt{3}} & 0 & \dfrac{1}{\sqrt{3}} & 0 \\ 0 & 1 & 0 & 0 \\ -\dfrac{1}{\sqrt{3}} & 0 & \dfrac{\sqrt{2}}{\sqrt{3}} & 0 \\ 0 & 0 & 0 & 1 \end{bmatrix}$$

$$R^{(1,2)} = \begin{bmatrix} \dfrac{\sqrt{3}}{2} & \dfrac{1}{2} & 0 & 0 \\ -\left(\dfrac{1}{2}\right) & \dfrac{\sqrt{3}}{2} & 0 & 0 \\ 0 & 0 & 1 & 0 \\ 0 & 0 & 0 & 1 \end{bmatrix} \qquad \hat{b} = Ub = [2,0,0,0]^T$$

$$U = R^{(1,2)}R^{(1,3)}R^{(1,4)} = \begin{bmatrix} \dfrac{1}{2} & \dfrac{1}{2} & \dfrac{1}{2} & \dfrac{1}{2} \\ \dfrac{-1}{2\sqrt{3}} & \dfrac{\sqrt{3}}{2} & \dfrac{-1}{2\sqrt{3}} & \dfrac{-1}{2\sqrt{3}} \\ \dfrac{-1}{\sqrt{6}} & 0 & \dfrac{\sqrt{2}}{\sqrt{3}} & \dfrac{-1}{\sqrt{6}} \\ \dfrac{-1}{\sqrt{2}} & 0 & 0 & \dfrac{1}{\sqrt{2}} \end{bmatrix}$$

**19.**   **(a)**   $f_0(\lambda) = 1$, $f_1(\lambda) = -\lambda$, $f_j(\lambda) = (1 - \lambda)f_{j-1}(\lambda) - f_{j-2}(\lambda)$ for $j = 2, 3, 4$, $f_5(\lambda) = (2 - \lambda)f_4(\lambda) - f_3(\lambda)$. Let $\lambda_1 < \lambda_2 < \lambda_3 < \lambda_4 < \lambda_5$. By the Gerschgorin theorem, all $\lambda_i$ lie in $[-1, 3]$. Then the roots can be separated using Theorem 9.5, as indicated in the following table. The actual roots can then be found by another method. For example, the secant method leads rapidly to $\lambda_5 \doteq 2.90211303$.

| $\lambda$ | $f_5(\lambda)$ | $s(\lambda)$ | Remark |
|---|---|---|---|
| $-1.0$ | 2.0 | 5 | $\lambda_1 > -1$ |
| 3.0 | $-2.0$ | 0 | $\lambda_5 < 3$ |
| 1.0 | 0.0 | 2 | $\lambda_3 = 1.0$, $\lambda_4 > 1$ |
| 0.0 | 1.0 | 3 | $\lambda_2 < 0.0$ |
| $-.5$ | $-1.78$ | 4 | $-1 < \lambda_1 < .5 < \lambda_2 < 0$ |
| 2.0 | $-1.0$ | 2 | $\lambda_4 > 2$ |
| 2.5 | 1.78 | 1 | $2 < \lambda_4 < 2.5 < \lambda_5 < 3$ |

**20.**   **(b)**   **(ii)**   Using Householder transformations as in (9.3.21) to (9.3.23), we obtain $T = Q^T A Q$

$$T = \begin{bmatrix} 5.0 & -4.2426406871 & 0.0 & 0.0 \\ -4.2426406871 & 6.0 & 1.4142135624 & 0.0 \\ 0.0 & 1.4142135624 & 5.0 & 0.0 \\ 0.0 & 0.0 & 0.0 & 2.0 \end{bmatrix}$$

$$Q = \begin{bmatrix} 1.0 & 0.0 & 0.0 & 0.0 \\ 0.0 & -.9428090416 & .3333333333 & 0.0 \\ 0.0 & -.2357022604 & -.6666666667 & -.7071067812 \\ 0.0 & -.2357022604 & -.6666666667 & .7071067812 \end{bmatrix}$$

Usually it would be wasteful of time to actually produce $Q$ explicitly, as done here.

**22.**   **(i)**   The eigenvalues are

$$\lambda_1 = 2 - \sqrt{3} \qquad \lambda_2 = 2 \qquad \lambda_3 = 2 + \sqrt{3}$$

**(a)**   For the $QR$ method without shift, the off-diagonal elements converge to zero linearly. The elements in the $(1, 2)$ position decrease by a factor of .536 per iterate, and those in the $(2, 3)$ position decrease by a factor of $-.134$ per iterate.

**(b)**   For the $QR$ method with shift, using the choice of

$$c_m = a_{3,3}^{(m)}$$

we have rapid convergence. After five iterates, the element in position $(2, 3)$ is less than $5 \times 10^{-11}$ in size. Denoting the original $A$ by $A_0$, we have the result

$$A_5 = \begin{bmatrix} 3.7316925974 & .0249060210 & 0.0 \\ .0249060210 & 2.0003582102 & \epsilon \\ 0.0 & \epsilon & .2679491924 \end{bmatrix}$$

with $|\epsilon| < 5 \times 10^{-11}$. When the $QR$ method with shift is applied to the reduced matrix obtained by deleting row and column 3 of $A_5$, it too converges very rapidly.

24.  For $\lambda = 7.9329$, the eigenvector is

$$x = [.7211774817, .2724739430, 1.0, .2515508351]$$

For $\lambda = 5.6689$,

$$x = [.5736191640, .5489544105, -.8148078321, 1.0]$$

26.  The values obtained are given in the following table.

| $n$ | $\lambda_1^{(n)}$ | $\lambda_2^{(n)}$ |
|---|---|---|
| 2 | .9 | .1 |
| 4 | .884566 | .103957 |
| 8 | .880971 | .105360 |
| 16 | .880085 | .105715 |
| 32 | .879865 | .105805 |

By examining ratios of successive differences, it can be seen empirically that

$$\lambda_i - \lambda_i^{(n)} = O\left(\frac{1}{n^2}\right)$$

This rate of convergence can be justified theoretically. Richardson extrapolation can then be used to produce the error estimates

$$\lambda_1 - \lambda_1^{(32)} \doteq 7.3 \times 10^{-5} \qquad \lambda_2 - \lambda_2^{(32)} \doteq 3.0 \times 10^{-5}$$

27.  **(1)**  Express $A$ and $A^+$ using their SVDs: $A = VFU^T$ and $A^+ = UF^+V^T$. Then

$$AA^+A = (VFU^T)(UF^+V^T)(VFU^T)$$

$$= VFF^+FU^T$$

$$= VFU^T \text{ because } FF^+F = F \text{ by direct computation}$$

$$= A$$

**30.**  $p(x) = -1.269091x + .392952$, root-mean-square error = .243

**33.**  **(a)**

$$QR = \begin{bmatrix} -.5179 & -.7517 & .4082 \\ .5754 & .0470 & .8165 \\ -.6330 & .6578 & .4082 \end{bmatrix} \begin{bmatrix} -1.7378 & -1.7148 \\ 0 & -.2819 \\ 0 & 0 \end{bmatrix}$$

$$SVD = \begin{bmatrix} .5774 & .7071 & .4082 \\ -.5774 & 0 & .8165 \\ .5774 & -.7071 & .4082 \end{bmatrix} \begin{bmatrix} 2.4495 & 0 \\ 0 & .2000 \\ 0 & 0 \end{bmatrix} \begin{bmatrix} .7071 & .7071 \\ -.7071 & .7071 \end{bmatrix}$$

$$A^+ = \begin{bmatrix} -\frac{7}{3} & -\frac{1}{6} & \frac{8}{3} \\ \frac{8}{3} & -\frac{1}{6} & -\frac{7}{3} \end{bmatrix} \qquad \text{cond}(A) = 12.2474$$

# INDEX

Note: (1) An asterisk (*) following a subentry name means that name is also listed separately with additional subentries of its own. (2) A page number followed by a number in parentheses, prefixed by P, refers to a problem on the given page. For example, 123(P30) refers to problem 30 on page 123.